EARTH AND ENVIRONMENTAL SCIENCES

Edited by **Imran Ahmad Dar** and
Mithas Ahmad Dar

INTECHOPEN.COM

Earth and Environmental Sciences
Imran Ahmad Dar and Mithas Ahmad Dar

CBS, Edition **2016**

Published by InTech

Janeza Trdine 9, 51000 Rijeka, Croatia

Publishing Process Manager Danijela Duric

Technical Editor Teodora Smiljanic

Cover Designer InTech Design Team

Additional hard copies can be obtained from orders@intechopen.com

Earth and Environmental Sciences, Edited by Imran Ahmad Dar and Mithas Ahmad Dar
p. cm.
978-953-307-468-9

INTECH

open science | open minds

Contents

Preface

We are increasingly faced with environmental problems and required to make important decisions. In many cases an understanding of one or more geologic processes is essential to finding the appropriate solution.

Earth and environmental sciences is by its very nature a dynamic field in which new issues continue to arise and old ones often evolve. The principal aim of this book is to present the reader with a broad overview of the subject. Since geology and earth sciences do not exist in a vacuum, this book introduces related considerations from outside geology to clarify different ramifications of the subjects discussed. Likewise, the present does not exist in isolation from the past and future; therefore, the material occasionally looks at how the earth developed into its present condition and also anticipates future development. It is hoped that this recent research will provide the reader with a useful foundation for discussing and evaluating specific environmental issues, as well as for developing ideas for problem solving.

The book has been divided into nine sections; Geology, Geochemistry, Seismology, Hydrology, Hydrogeology, Mineralogy, Soil, Remote Sensing and Environmental Sciences.

Many people have contributed to the development of this book. The editors are very much thankful to all the authors contributing to this book. Portions of the manuscript in the book "Earth and Environmental Sciences" were contributed by Dr. Imre Stephen, Dr. Meiler Miki, Dr. Smelror Morten, Prof. Oyinloye Akindele, Dr. Soua Mohamed, Dr. Somarin Alireza, Dr. Sun Weijia, Dr. Massa Marco, Prof. Longobardi Antonia, Dr. Igboekwe Magnus, Dr. Akulov Nickoloy, Dr. Rios Carlos, Dr. Tosheva Zornita, Dr. Kamona Fred, Dr. Laake Andreas, Dr. Chlachula Jiri, Prof. Zare Chahouki Mohammad Ali, Prof. Ricardi-Branco Fresia, Dr. Bernardi Massimo, Dr. Olmedo-Alvarez Gabriela, Dr. Kopačkova Veronika, Dr. Mahmoud Ahmed, Dr. Zovko Monika, Dr. Fereshte Haghighi, Dr. Huang Lin-Chong and Dr. Jimenez Oscar. The editors are very much thankful to the CEO of InTech, Aleksandar Lazinica. The editors would particularly like to extend their grateful appreciation to Ana Pantar, editorial consultant at InTech for inviting them to edit this prestigious book. Lastly, but by no means least, one must thank Ms. Danijela Duric, publishing process manager, for her timely and kind communication with the editors during the entire process. The editors wish to offer their thanks to all those involved. If any person or concern has inadvertently not been given due acknowledgment, then apologies are given.

Imran Ahmad Dar and Mithas Ahmad Dar
Department of Industries and Earth Sciences, The Tamil University,
Ocean and Atmospheric Sciences & Technology Cell,
India

Part 1

Geology

1

Geophysics and Wine in New Zealand

Stephen P. Imre[1,2] and Jeffrey L. Mauk[1]
[1]*University of Auckland*
[2]*Worley Parsons Canada*
[1]*New Zealand,*
[2]*Canada*

1. Introduction

The New Zealand wine industry is committed to producing high quality, market led wines (New Zealand Winegrowers, 2009). Consequently for the industry, it is imperative to understand what factors influence grape and wine properties in order to maximise the production of premium wines. New Zealand winegrowing regions produce different styles of wine, yet there is little data to build an understanding of what contributes to the quality of New Zealand wines both on regional and local scales. On a global scale, the New Zealand industry accounts for less than 0.03% of land under vine and less than 0.5% of global wine production (New Zealand Winegrowers, 2009), and yet it adds an estimated $4 billion to the New Zealand economy (NZIER, 2009). Global area under vine in 2006 was roughly 7,812,000 ha, and wine production was approximately 282,000,000 hl (Organisation Internationale de la Vigne et du Vin [OIV], 2006). Quantifying variables that may influence grape/wine parameters may aid in creating a better understanding of what factors influence wine characteristics. This research demonstrates that vine growth is variable in uniformly managed blocks, and geophysical tools can be used to map soil variability in vineyards, which in turn can influence grapevine vigour and presumably grape properties.

Terroir can be defined as the physical and chemical characteristics of a vineyard or region, including geology, soils, topography, and climate. Different terroirs have been delineated in Europe based largely on areas that have produced distinctive wines over a long period of time (White, 2003). These areas are typically delineated using geographical indicators such as geology, soil, topography and mesoclimate. The influence of winemaker and cultural practices can also be considered part of terroir. These environmental properties and cultural practices interact to create wine with particular characteristics, and the place of origin ultimately influences the wines produced. Terroir forms the basis for appellation d'origine controlée (AOC), the vineyard classification system used in France in areas such as Bordeaux, Burgundy, and Champagne, with wine produced from particular terroirs being associated with a certain quality and wine style (Wilson, 1998).

Several publications have emerged on different facets of terroir at various scales, indicating an increased interest in the subject. For example, Swinchatt & Howell (2004) explore the terroir of the Napa Valley examining its geology, history and environment. Haeger (2004) discusses the rise of Pinot Noir in North America and its associated terroirs. Wilson (1998) discusses the terroir of selected areas in France. Geoscience Canada has a series devoted to geology and wine from various winegrowing regions of the world (Macqueen & Meinert, 2006). Terroir studies have also been conducted on a vineyard-block scale examining local variations in properties such as microclimate and soils (e.g. Trought et al., 2008). A summary of national-regional terroir in New Zealand is described in Imre & Mauk (2009).

Grapevine trunk circumference measurements can be used as a proxy for variations in grapevine vigour on a site-specific level (e.g. Clingeleffer & Emmanuelli, 2006; Acevedo-Opazo et al., 2008; Trought et al., 2008). These localised variations in even small vineyard blocks can in turn show differences in grape quality parameters (e.g. Cortell et al., 2005; Trought et al., 2008). If blocks are uniformly managed, changes in soil properties or local topography may contribute to variations in vine trunk circumference. A rapid cost-effective method to identify areas of homogeneity in the subsurface, when coupled with GPS surveys, may provide useful input into site specific variations.

Ground penetrating radar (GPR) and electromagnetic induction (EMI) surveys are commonly used to map areas of different soil properties (e.g. Kitchen et al., 1996; Doolittle et al., 2002; James et al., 2003; Hedley et al., 2004; Carroll & Oliver, 2005; Cockx et al., 2007; Bramley, 2009). The GPR surveys can be useful in measuring depth to soil textural changes (Simeoni et al., 2009). The EMI surveys are typically single-frequency and measure soil apparent electrical conductivity (ECa) values at one or two depths (e.g. Bramley, 2001; Trought et al., 2008), and this can provide very useful data, particularly where crops have a shallow rooting depth or shallow changes in soil properties are of interest.

Grapevines have rooting depths that may exceed 6 m (Smart et al., 2006 and references therein), and therefore measuring soil ECa values at different depths may be beneficial. Electromagnetic induction surveys have been used to map soil types (e.g. James et al., 2003), and multi-frequency EMI surveys can detect vertical differences in ECa from landfill leachate or buried artefacts where differences among ECa values are large (e.g. Won et al., 1996). If soil ECa changes sufficiently with depth, multi-frequency EMI surveys may be able to provide valuable information about changes in soil with depth. This research demonstrates that geophysical tools can help map variability in soils in vineyards. Soil variability in turn influences grapevine vigour, and presumably the properties of the grapes that are harvested from different areas. Geophysical mapping can help to quantitatively evaluate differences in soils that contribute to variability in the wines that are produced from different areas.

2. Materials and methods

Three vineyard blocks with own-root pinot noir 10/5 on different soil types in Bannockburn, Central Otago were selected for detailed study (Fig. 1). The sites are characterized by a dry climate with mild temperatures and moderate solar radiation (Leathwick et al., 2002; Imre & Mauk, 2009). High vapour pressure deficits and very high annual water deficits make many plantings in the area require irrigation. The study block at Olssens in the north is located on shallow fine sandy loam from shallow loess over fine sand and gravel schist fan alluvium. The Felton Road study block is located on Waenga very deep fine sandy loam from deep alluvium derived from loess, schist and lake bed sediments. Mt Difficulty's Target Gully site is located on very shallow to shallow fine sandy loam and some deep coarse Otago Schist (Beecroft, 1988; Turnbull, 2000; Leathwick et al., 2002; Imre & Mauk, 2009).

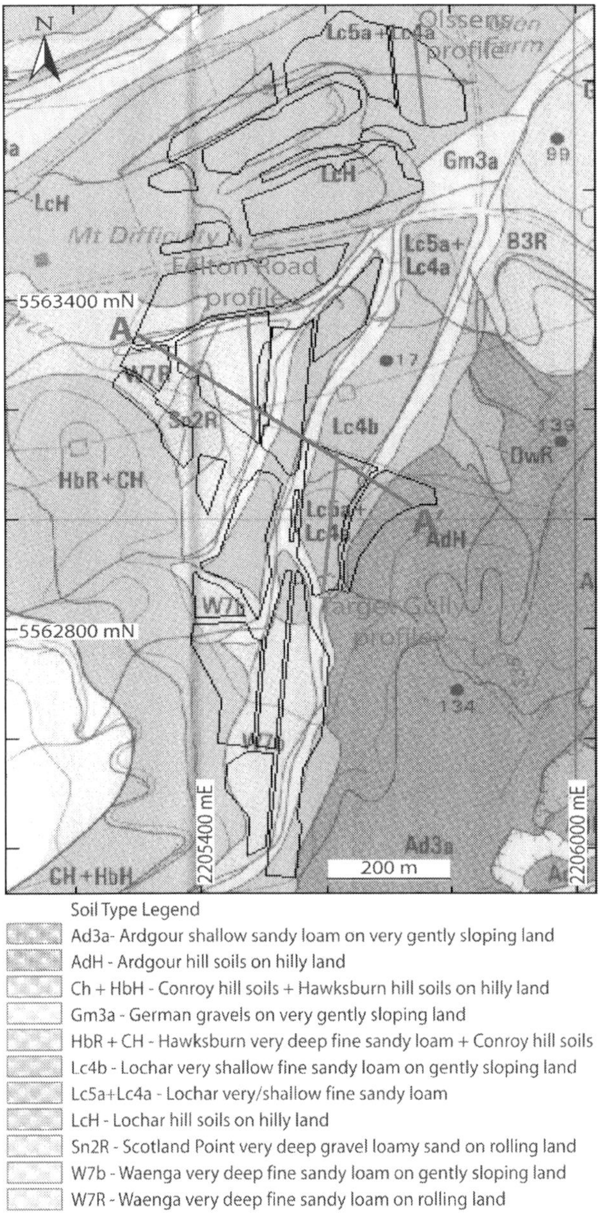

Fig. 1. Map of Bannockburn, Central Otago, showing the location of study blocks and associated transects in red, Profile A-A′ in blue. Background colour is soil type as scanned and digitized from Beecroft (1988).

2.1 Grapevine physiology

Vine trunk circumferences in a selection of rows in each study site were measured at roughly 20 cm above ground level at narrow areas of the vine, and average vine trunk circumference growth in each bay was calculated. Vine trunk circumference measurements can be used as a proxy for vigour variation (e.g. Acevedo-Opazo et al., 2008; Trought et al., 2008). The vines are all on own-root making it unnecessary to take two measurements and average the values because there is no graft union. To remove the influence of abnormally small vines, such as replantings or physically damaged vines (e.g. tractor impact), bays where the smallest vine trunk circumference was less than 70% of the bay average were omitted. The average bay trunk circumferences were then divided into five size classes (extra-small, small, medium, large and extra-large) for each study site, with approximately the same number of values in each class. All vine trunk circumference values refer to average values in each bay.

2.2 Geophysical surveys

The SIR 2000 Ground Penetrating Radar unit from Geophysical Survey Systems, Inc (GSSI; New Hampshire, USA) was used to run transects along each row in the study blocks using a 200 MHz antenna. An electromagnetic wave is emitted from a transmitting antenna, and part of the wave energy is reflected from changes in soil properties. This radar reflectance is measured by the receiving antenna and recorded. The GPR data were processed using GSSI RADAN (RAdar Data ANalyzer) version 6.5, and 3D images of the subsurface down to a depth of several meters were created. Images were then imported into a GIS (MapInfo Professional 6.5; Pitney Bowes, New York, USA) and geo-referenced using known GPS points in the study areas. This allowed for spatial comparison of survey results and vine trunk circumference data. GPR data showing vertical change are displayed using nanosecond (ns) instead of depth due mainly to the sharp change in soils and the inability of RADAN processing software to account for different dielectric properties at depth. A greyscale version of linescan mode is used to display the data because of its ability to detect gravels within a sedimentary layer. These gravels, and other isolated objects, cause a diffraction of the electromagnetic waves resulting in a ringing within the radar signal that is displayed as a downward parabola (Jol, 2009); this is discussed for each of the figures where appropriate. One representative transect for each study site is discussed further in this chapter.

For the EMI surveys, a GEM-2 (Geophex, North Carolina, USA) portable handheld broadband electromagnetic sensor connected to a handheld GPS unit was used. Each study block was surveyed using ten frequencies: 1,175 Hz, 2,525 Hz, 3,925 Hz, 7,375 Hz, 10,575 Hz, 13,575 Hz, 25,975 Hz, 35,925 Hz, 44,025 Hz and 47,025 Hz. Higher frequencies measure soil resistivity close to the surface, whereas lower frequencies penetrate to greater depths depending on soil and geology. Soil apparent electrical conductivity (ECa) is then calculated automatically by the GEM-2; the measured resistivity is the inverse of electrical conductivity. 1m x 1m grids were generated from the raw EMI data using triangulation with natural neighbour interpolation in Discover version 4.000 in MapInfo Professional 6.5. These grids allow for mapping of ECa at various depths, and to test whether soil ECa correlates with vine trunk circumference. Maps were then produced at each site from all ten frequencies corresponding to different depths. Generally patterns are clearer at the higher frequencies that represent readings close to the surface. Soil ECa at the frequencies showing the best visual results at each site are presented. Using various combinations of these frequencies as specified for each site individually (Table 1), the data were inverted based on a 1D layered earth model to create depth profiles of apparent soil ECa (Huang & Won, 2003; Huang, 2005). Frequency combinations that gave minimal fit errors when running the inversion algorithm were used. Fit errors are calculated based how the data points fit a semi-variogram of the dataset. A large difference between values will result in a larger fit error. Images were generated to 10 m depth using the inverse distance weighting method in MapInfo 6.5 Professional (Pitney Bowes, New York, USA). All results of the EMI surveys are displayed using histogram equalization stretch for colour display.

Site	Frequencies used (Hz)
Profile A-A'	2525, 3925, 7375, 10575, 13575, 25975, 35925, 44025, 47025
Felton Road Wines	1175, 3925, 7375, 25975, 35925
Target Gully	3925, 7375, 13575, 25975, 44025
Olssens	10575, 25975, 35925, 44025, 47025

Table 1. EMI frequencies used at each site for the inversion algorithm

Differential GPS surveys were conducted in all of the study blocks using a Trimble Pro XRS unit (Sunnyvale, California, USA). GPS point data was imported into MapInfo 6.5 Professional and 1m x 1m grids of elevation were created at each site. Slope and aspect were calculated using the Discover 4.000 extension. New Zealand map grid was used for all surveys in the study blocks, and all figures in this chapter are drawn with NZ map grid coordinates.

2.3 Soil sampling and analysis

We collected soil profile data from 53 trenches in the study area, and we include selected chemical data from one representative soil profile from each study site in this chapter. Samples were classified into percent gravel, sand, silt and clay using manual sieving for the >2 mm portion and using a Malvern Mastersizer (Malvern Instruments ltd., Worcestershire, UK) for the <2 mm portion. The <2 mm size fraction was analyzed for pH, P, K, Mg, Ca, CEC, Zn, Cu, Al, S, Na, Mn and Fe (Agricultural Analytical Services laboratory, Penn State University, PA, USA; USEPA, 1986; Eckert & Sims, 1995; Ross, 1995; Wolf & Beegle, 1995); the Mehlich 3 extract was used for elemental analysis. Samples were also analyzed for total N and total C at the University of Auckland using the TruSpec CN (LECO Australia Pty Ltd., Castle Hill, NSW, Australia).

2.4 Statistical analyses

Statistical analyses were conducted using SPSS 17.0 (SPSS Inc., Chicago, USA) to examine what factors significantly influence vine growth variation. A correlation matrix between all ECa values was initially created to reduce the amount of data by eliminating ECa values at different frequencies that are highly correlated at each site (adjusted-R2>0.9). Stepwise linear regression was then used with vine trunk circumference as the dependent variable and ECa, slope, elevation and aspect data as the predictor variables for each site individually as well as on the total dataset. Eight clusters of slope, elevation and aspect were used as predictor variables in a linear regression analysis both with and without the addition of ECa data. For the stepwise linear regression, the stepping method was set to use probability of F entry=0.05, exit=1.0. Due to the nature of the GPR data, it is not possible to

extract values such as those for ECa, and therefore these results were not included in the statistical analysis.
Profile A-A′ compares the EMI inversion data with the detailed soil map of Bannockburn in Central Otago (Beecroft, 1988) (Fig. 1, Fig. 9). Multinomial logistic regression analysis was done using SPSS 17.0 (SPSS Inc., Chicago, USA) in order to determine whether significant relationships exist between ECa and mapped soil type (Beecroft, 1988). Mapped soil type is the dependent variable in the analyses, and ECa values at different depths are the covariates as discussed individually for each result. The ECa survey points that are located on roads between vineyard blocks were removed from the analysis.

3. Results

3.1 Vine trunk circumference

Vine trunk circumferences vary throughout all study sites (Fig. 2), even though each block is uniformly managed. Vine trunk circumferences are largest at Felton Road with a median value of 157 mm, followed by Olssens with a median value of 138 mm, and Target Gully with the smallest median value of 126 mm. The vines at Felton Road are the largest in the study, whereas vines are smallest at Target Gully, even though the oldest vines are at Olssens. The spatial distribution of vine sizes at each site are shown in images later in this chapter.

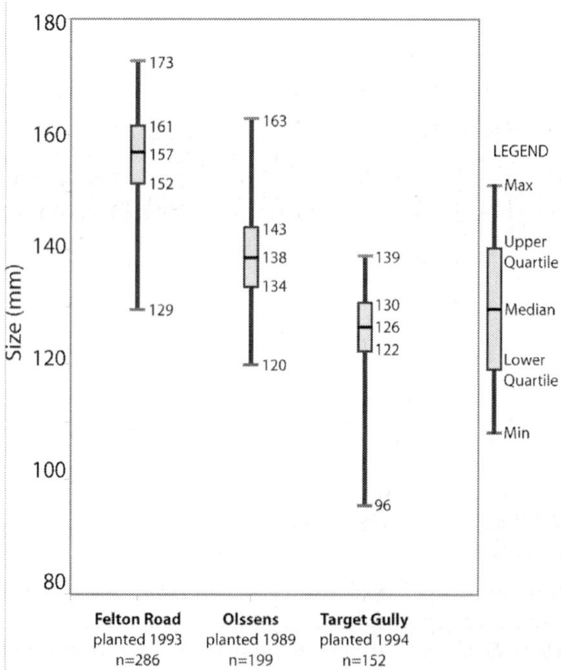

Fig. 2. Boxplots showing vine trunk circumference distribution at each block

3.2 Felton road wines

Vines with small to extra-small trunk circumferences mainly occur in the central eastern area of the block, with some in the south and northwest (Fig. 3a). Vines with medium to large trunk circumferences form a "C"-shaped pattern that surrounds all areas of the block except the central-east, south, and northwest corner. The GPR image shows distinct areas of high amplitude signals in the central-east, south and northeast areas of the study block, and low amplitude signals in a "C" shaped pattern surrounding the higher amplitude areas (Fig. 3a). Two distinct areas of high soil ECa are located in the central-east and southern areas (Fig. 3b). Smaller vines tend to occur in areas with high amplitude GPR signals and higher soil ECa values, suggesting that these soils are less favourable for vine growth. These extremely high soil ECa values generally follow the boundary of a silty loam soil type as verified by extensive soil trenching and as discussed in more detail along Profile A-A'.

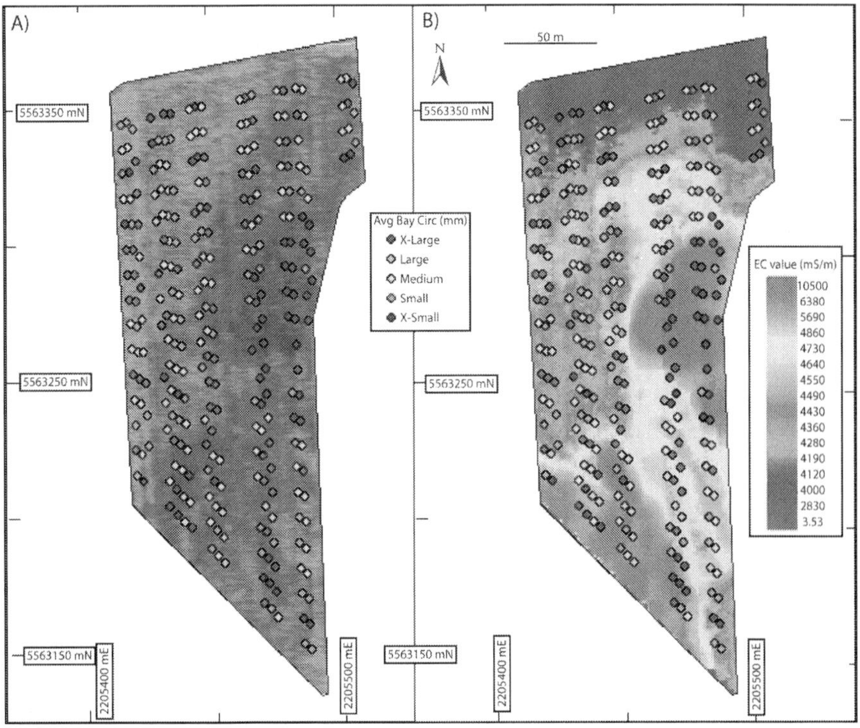

Fig. 3. Felton Road Wines showing vine trunk circumference data and a) GPR survey results, and b) EMI survey results

Soil ECa values are very high at the surface and generally decrease with depth (Fig. 4a). An area of high ECa values occurs in the middle of the row at depths greater than 2 m when compared to the rest of the row at similar depths. Soil ECa values decrease rapidly at roughly 2 m depth at the north end, decrease at a more gradual rate throughout the rest of the row, and remain relatively large in the middle of the row.

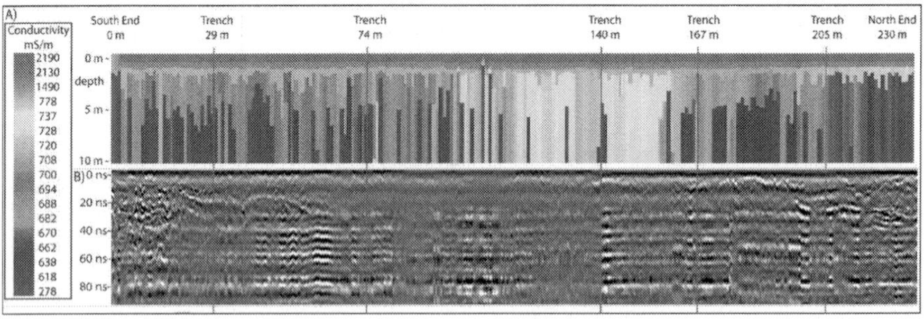

Fig. 4. Felton Road (a) inversion of EMI survey to 10m depth, (b) GPR transect

The clearest GPR signals are in the north and south ends of the row (Fig. 4b). This area is dominated by hyperbolas, whereby the size and frequency in this geophysical signature indicates the presence of gravels (Jol, 2009). The middle of the row, which contains very high ECa values, has a relatively unclear GPR record that shows alternating layers of blurry and high-amplitude signals.

3.3 Olssens garden vineyard

Extra-large vine trunk circumferences mainly occur in the central-eastern area, and extra-small vine trunk circumferences are mainly in the northern portion of the study block (Fig. 5a). The GPR image shows low amplitude signals in the west and high amplitude signals in the north, east and south (Fig. 5a). The GPR data also distinctly show a buried pipe that runs from east to west in the southern area of the study block. Two areas of high soil ECa in the west side and east side of the block coincide with areas that contain vines with large trunk circumferences (Fig. 5b). Soil ECa values at the site are all below 3.5 mS/m, which is very low compared to other sites in this study.

Soil ECa values are lowest at the Olssens study site, and are typically less than 1 mS/m. The top layer, as defined by the EMI surveys, has an average thickness of greater than 40 m, and therefore the ECa values in our 10 m deep profile do not change with depth (Fig. 6a). The great depth penetration may be due to the little variation in observed soil ECa throughout the profile; over 90% of values are less than 1 mS/m. The centre of the row generally has larger ECa values than the row ends.

The upper-most radar signature displays a uniform low-amplitude band down to roughly 20 ns, representing a homogeneous top layer (Fig. 6b). Below this depth reflections are fairly flat-lying with some variation in amplitude, dip, and direction similar to bedding plane deposits imaged using GPR in fluvial settings (Jol, 2009). There is a dipping high-amplitude reflection surface in the southern 50 meters of the image at 110 ns that may indicate a layer of more consolidated material at the base. A distinct parabolic reflection, located about 15 m from the south end at 20 ns, is the geophysical signature of a buried pipe that traverses the study block.

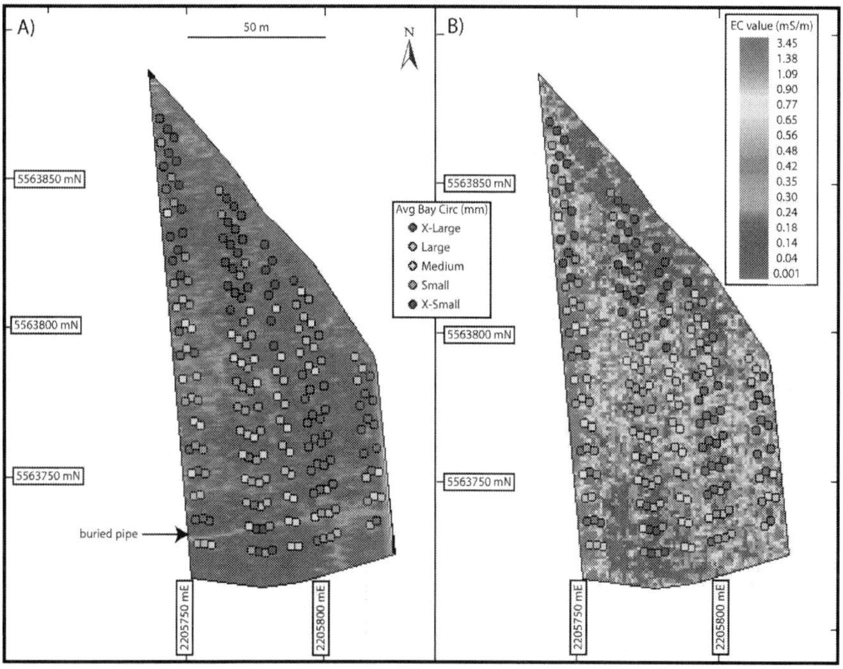

Fig. 5. Olssens Garden Vineyard showing vine trunk circumference data and a) GPR survey results, and b) EMI survey results

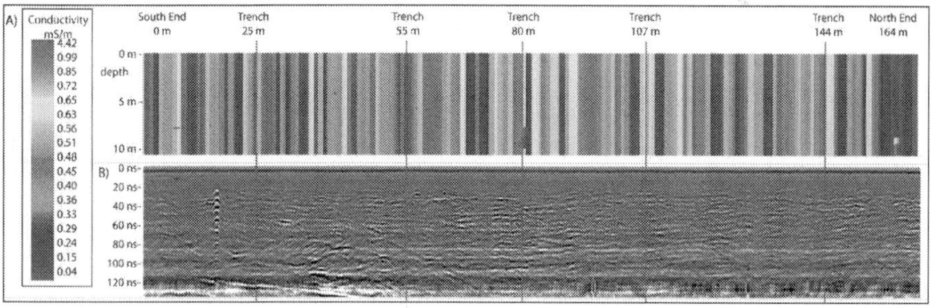

Fig. 6. Olssens (a) inversion of EMI survey to 10m depth, (b) GPR transect

3.4 Target Gully vineyard

Due in part to the long, linear shape of the Target Gully study block, vines with similar trunk circumferences are less clustered than at other sites, however some general observations can still be made. Most vines with extra-small trunk circumferences occur in the northeastern and southwestern portions of the study block, whereas most vines with extra-large trunk circumferences occur in the northern area of the block (Fig. 7a). The GPR survey shows an area of low amplitude signals from north to south in the centre of the block, and another smaller area in the west (Fig. 7a). Most high soil ECa values at 35,925 Hz occur on the eastern side of the study block (Fig. 7b).

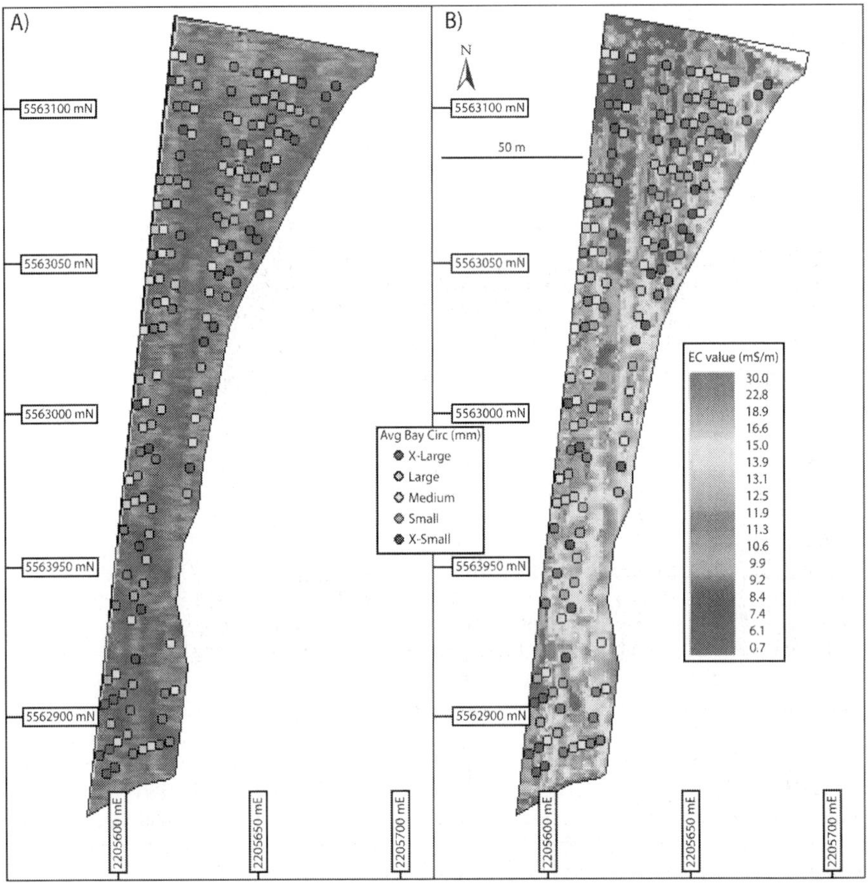

Fig. 7. Target Gully vineyard showing vine trunk circumfereces and a) GPR survey results, and b) EMI survey results

Soils are generally high in gravel content with gravels closest to the surface at the central-north end. Clay content is less than 7% close to the surface, generally decreases with depth and is largest at the north end close to the surface. Soil ECa values are similar throughout the Target Gully profile at roughly 0-5 m depth (Fig. 8a). The lowest soil ECa values are in the northern area of the block at depths of less than 5 m. There is a sharp increase in soil ECa values at roughly 5 m depth throughout the length of the row. Soil ECa values generally increase below this 5 m depth, and the largest values are in the south.

The GPR data display a uniform low-amplitude signature from the surface to about 20 ns depth indicating the presence of a homogeneous upper layer (Fig. 8b). The radar signal attenuates after encountering a flat-lying high-amplitude surface at roughly 60 ns that equates to 5 m depth using a dielectric constant of 4. This line of contrast is consistent throughout the profiles at approximately the same depth, and is also shown in the EMI surveys for Target Gully. Downward parabolas are located throughout the Target Gully profile indicating the presence of gravels, and these are located closer to the surface around the north end.

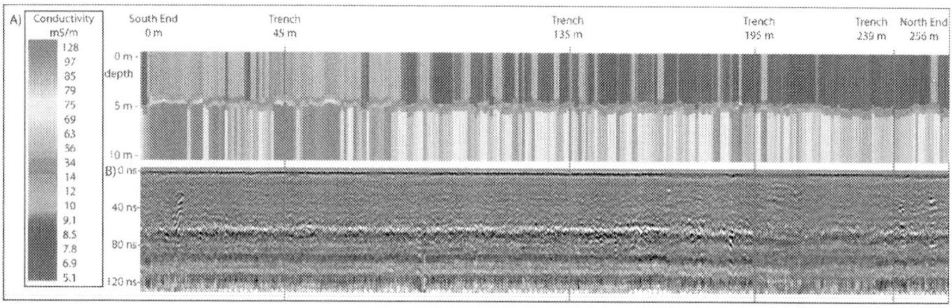

Fig. 8. Target Gully (a) inversion of EMI survey to 10m depth, (b) GPR transect

3.5 Profile A-A'

Profile A-A′ is 535 m long running from northwest to southeast from an elevation of 271 m to 298 m covering five different soil types (Fig. 1, Fig. 9). The lower elevations with very deep gravel loamy sand (Sn2R) and very deep fine sandy loam (W7R) generally have very high ECa values close to the surface that tend to decrease with depth. Shallow to very shallow fine sandy loam (Lc5a+Lc4a and Lc4b) have the lowest ECa values of generally less than 60 mS/m throughout the 10 m depth profiles. Ardgour hill soils (AdH) on the hillside generally have ECa values of 25-30 mS/m close to the surface, 30-70 mS/m in the 4-7 m depth range and lower soil ECa below that depth. The Ardgour hill soils (AdH) have increasing ECa with depth until about 7 m and then sharply decrease from 7-10 m depth; a unique trend not observed to this degree in the other soil types. Soil ECa tends to decrease with depth for deep fine sandy loam (W7R), very deep gravel loamy sand (SnR2) and German gravels (Gm3a), but values still tend to be larger than those throughout the profile for the shallow fine sandy loams (Lc5a+Lc4a and Lc4b) and Ardgour hill soils (AdH).
Significant relationships exist between soil type as mapped in Beecroft (1988) and soil ECa for the Bannockburn area along Profile A-A′. Logistic regression analysis is a technique used to predict the probability that a certain set of variables belongs to one of two classes. Multinomial logistic regression analysis is used when the dependent variable consists of more than two discrete outcomes that are nomial, such as soil type, that are not ordered in any meaningful way. Using single frequency ECa values as covariates in the multinomial regression, soil type is predicted with 37% - 46% accuracy depending on the frequency used (data not shown). Using depth-corrected ECa values as covariates, the regression analysis correctly predicts soil type 39% - 46% of the time. The models are not able to differentiate Ardgour hill soils (AdH) from German gravels (Gm3a) or shallow silty loam (Lc4b and Lc5a+Lc4a). The models are also generally unable to differentiate between Waenga very deep fine sandy loam (W7R) and Scotland Point very deep gravel loamy sand (Sn2R). Using ECa values at 1 m depth the model is able to better predict Lc4a and SnR2, albeit with low prediction accuracies of 21% and 41%,

respectively. Using ECa values below 6 m depth results in an increased prediction accuracy for German gravels. The model predicts soil type with 38% accuracy when only ECa at 1 m and 2 m depths are used in the regression analysis (data not shown). Using depth-corrected multi-frequency ECa data in 1 m intervals in the model predicts soil type with 69.2% accuracy overall, showing the added benefit of using multi-frequency ECa data.

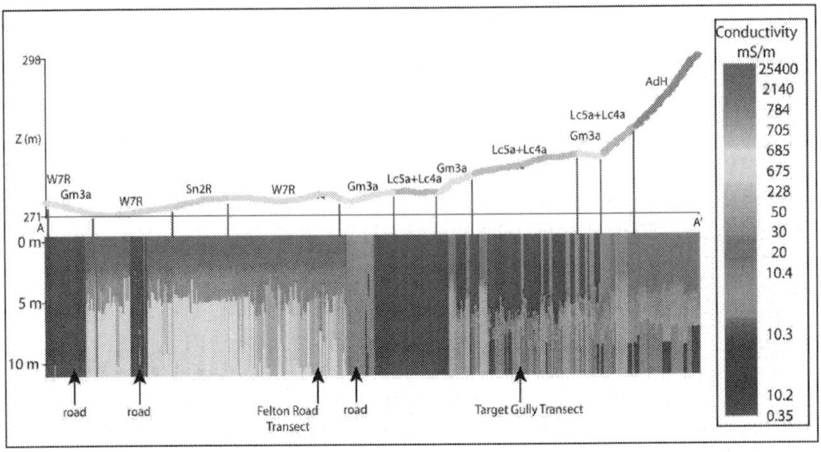

Fig. 9. Inversion of EMI survey to 10 m depth along Profile A-A'

3.6 Soil chemical analysis

Fig. 10 shows selected representative soil nutrient data for one representative trench at each site. In order to account for the relative dearth of available nutrients in gravelly soils, these values are mass-balance corrected; nutrient concentrations are multiplied by the percentage of soil that is <2 mm.

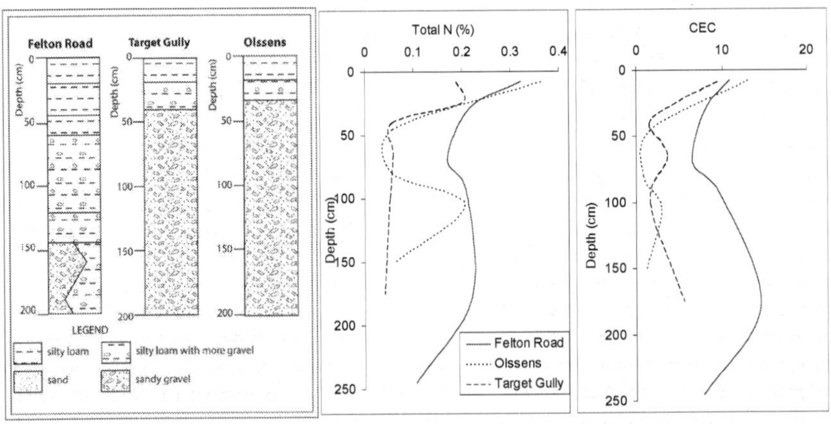

Fig. 10. Soil profiles (left) and example nutrient profiles of Total N (%) and CEC

Soil pH values tend to increase with depth at all sites and are largest at Felton Road. At 100 cm depth, the pH at Felton Road is approximately 8.5, and it is <7.0 at Target Gully. Shallow soil samples at all three sites have similar pH values. At Target Gully, Ca-rich layers occur in deeper soils. At Felton Road, Cu, Mg, K, N and Mn concentrations are typically greater throughout the soil profiles than at the other sites. Total CEC is largest at Felton Road throughout the soil profile, and is mainly comprised of 80% Ca at Felton Road, 73% Ca at Olssens and 70% Ca at Target Gully.

4. Discussion

4.1 Vine vigour

Vine trunk circumferences can be used as an indicator of vine vigour (e.g. Clingeleffer & Emmanuelli, 2006; Acevedo-Opazo et al., 2008; Trought et al., 2008). In uniformly managed blocks where vine trunk circumferences vary little, the growing conditions within the block are considered to be more uniform. Conversely, where vine sizes show more variation, growing conditions are considered to be less uniform. Average vine trunk circumferences show variation at all study sites (Fig. 2). When examining the range of average values of vine trunk circumferences in individual bays within a study block, the largest value is up to 65% larger than the smallest value, even after the data were cleaned by removing the smallest vines. Although the study sites are managed slightly differently, management within each block is uniform and therefore other variables, including soil conditions, likely control vine sizes. Variations in grapevine vigour commonly correlate with fruit/wine characteristics on a site-specific level elsewhere (e.g. Cortell et al., 2005; Trought et al., 2008), so the variation in grapevine trunk circumference observed at these study sites likely contributes to differences in grape characteristics. Yield can vary up to 10-fold in the same vineyard block, and this variation in yield has associated quality implications (Bramley, 2001). Management practices such as differential harvesting based on vigour zones may lead to more consistency of fruit composition delivered to the winery (e.g. Proffitt et al., 2006).

Felton Road had the largest vine trunk circumference values and largest vine trunk circumference growth per year. The soils at Felton Road are characterised by deep sandy to silty loams. In contrast, the Target Gully site is on soils that contain abundant schist fragments, and the Olssens site contains abundant gravels. The gravelly soils at Olssens and Target Gully produce vines with smaller trunk circumferences than the deep loams at Felton Road.

Soil fertility indicators N, P and K are generally greater at Felton Road throughout the soil profile. Deep soils are more nutrient-rich at Felton Road than at the other two sites, even though similar concentrations are in the shallow samples. This may indicate that grapevines are able to extract nutrients from deep in the soil profile, and current vineyard management practices which generally test shallow soils only, may be overlooking this important variable. These deeper soil properties may also influence grape and wine characteristics, and are an often overlooked component of soil terroir.

Some soils in this research were below recommended plant available nutrient concentrations for viticulture in New Zealand. The vines, however, did not show any obvious signs of nutrient deficiencies, presumably because they were able to extract nutrients from deep soils. More research determining plant uptake of deep nutrients is warranted, and a critical examination of recommended guidelines for viticulture in New Zealand should be undertaken.

4.2 Geophysics and vine physiology

In the GPR maps, areas of high and low amplitude signals are shown by dark and light patches, respectively. Areas of similar radar reflectance are an indication that the GPR is showing relative homogeneity in properties affecting radar reflectance at these sites. In this study, correlations vary between vine trunk circumferences and GPR data. In general, maps with distinct areas of high and low amplitude signals show visual correlations with vine trunk circumference data, but some study sites do not show much variability in GPR data.

At Felton Road (Fig. 3), areas of high amplitude signals on the GPR map correlate with areas with smaller vine trunk circumferences, suggesting that soil properties that are less conducive to vine growth produce higher amplitude signals at this site. GPR surveys over the more gravelly soils at Olssens (Fig. 5) and Target Gully (Fig. 7) do not show distinct areas of high and low amplitude signals, and therefore radar reflectance data do not correlate well with vine trunk circumference data at these sites.

Water has a high dielectric constant (k=80) when compared to common geological materials (k=5-15), and GPR field measurements are therefore largely a function of the available water in the soil (e.g. Galagdara et al. 2005). Therefore, GPR results at our study sites may be detecting changes in the available water content of different layers, rather than changes in the physical properties of the layers themselves. Nonetheless, differences in soil texture can be interpreted from water content maps (e.g. Hubbard et al., 2002; Grote et al., 2003), and future work at these study sites may improve our understanding of the causes of variability in GPR results at sites such as Felton Road.

Relative soil ECa values alone do not provide an accurate prediction of grapevine trunk circumference variation; instead these correlations must be established on a site-by-site basis. Soil ECa values can be influenced by many different properties, including soil texture, clay content, soil extractable Ca^{2+}, Mg^{2+}, Na^{+}, CEC, silt, salinity, organic matter, water content, and previous fertilizer application (e.g. Bronson et al., 2005; Corwin & Lesch, 2005). Zones of relatively high and low soil ECa are evident at Felton Road, and the areas with high soil ECa values tend to contain grapevines with relatively small vine trunk circumferences. These areas tend to have a greater silt and clay content at depth, whereas the areas with lower soil ECa values in the north of the block are derived from soils that have higher gravel contents. At Olssens, soils with relatively low ECa values in the north and south of the block tend to contain smaller vines. These areas are delineated with ECa maps showing areas of high and low soil ECa values.

Maps created from GPR and EMI surveys have the potential to measure soil properties that can influence grapevine vigour. Grapevines at Felton Road, for example, tend to form distinct clusters of similar sized vines. The maps produced by the GPR and EMI surveys also clearly delineate areas of different radar reflectance and areas of high and low soil ECa. Using such survey methods can potentially delineate areas of soil properties that may influence vine growth.

4.3 Geophysics – GPR and multi-frequency EMI surveys

Some similarities exist between aerial maps produced using the GPR and EMI data. The Felton Road surveys, for example, show similar patterns for both techniques in a large area of the study block. Areas of high amplitude signals in the GPR map are roughly cospatial with the higher soil ECa areas in the EMI map indicating soils with higher soil ECa also exhibit higher radar reflectance. Target Gully and Olssens have higher gravel content and tend to produce maps showing less distinct areas of variation making comparisons between the two techniques more difficult at these sites. Areas of high amplitude signals at Olssens are roughly cospatial with areas of both high and low soil ECa indicating different properties may influence these readings.

Depth to soil texture contrasts can be estimated using GPR (e.g. Simeoni et al., 2009), and our work suggests that EMI surveys can also yield information that can help estimate depths to horizons where soil texture changes significantly. For example, the area around the Target Gully study site was subject to sluicing activity for alluvial gold, and the distinct contrasts in the GPR and EMI surveys at roughly 5 m depth may indicate the depth of the deposited sediment during the late 1800s to early 1900s. GPR is known to detect human induced disturbances, such as this sluice deposit, because of its sensitivity to the contrast between the natural and unnatural strata in the deposits which result in a strong high-amplitude reflection (Daniels, 2004). Depth penetration is lowest at this site for the EMI survey, possibly due to relatively recent increased sediment disturbance. Conducting a single-frequency EMI survey would likely not yield data suitable to measuring thickness of the deposited sediment. High soil ECa can cause the GPR signal to attenuate quickly and produce poor quality survey results (GSSI, 2004). The Felton Road GPR survey has a poor record in the middle of the transect. High soil ECa is measured by the EMI survey at the surface, and this alone does not attenuate the GPR signal at this site. Areas of high soil ECa with depth visually correlate to locations of impeded GPR signals in the middle of the Felton Road study row indicating that a thin layer of highly conductive soil close to the surface may not necessarily result in poor quality GPR imaging. The GPR quality is lowest at this site, possibly due to very high soil ECa values. Areas of high soil ECa that affect GPR survey quality may not be detrimental to multi-frequency EMI survey quality.

When interpreting results from single-frequency EMI surveys, it is difficult to make an assessment of signal depth penetration, and consequently soil homogeneity or heterogeneity with depth is difficult to estimate. The EM38 and Veris-3100 are commonly used in soil apparent electrical conductivity mapping (e.g. Bramley, 2001; Liu et al., 2008). These methods can map soil ECa at two different depths, the EM38 in horizontal and vertical dipole mode can measure soil ECa at roughly 0-75 cm and 0-150 cm, respectively, and the Veris-3100 usually measures at 0-30 cm and 0-90 cm. These depths, however, can vary with different soil properties. Our results suggest that soil properties can vary in close proximity, and these variations can have associated effects on survey depth penetration.

Soil ECa values at Olssens are typically less than 1 mS/m. Inversion algorithms used in multi-frequency EMI surveys may have difficulty calculating depth penetration where soil ECa values are very low. With such low ECa values, little additional information is gained on soil ECa changes with depth by inverting the EMI data, and the inversion algorithm may be unable to estimate depth penetration. Changes in soil ECa values with depth must be sufficient in order for multi-frequency EMI surveys to map these changes with reasonable precision.

Soils with high clay content can decrease GPR survey quality, and increase EMI survey quality. High clay contents can produce high attenuation losses and can decrease GPR quality (Gerber et al., 2007), and our results are consistent with this observation. The north end of Target Gully contains 7.0% clay to roughly 40 cm depth, the largest in the row, and GPR quality is poor. Clay content in some Felton Road samples in the middle of the row is just less than 20%, and this location has both the largest depth-weighted percent clay and highest soil ECa with depth. The GPR signal exhibits attenuation loss to some degree where clay content is relatively large. Clay content has been positively correlated to ECa values (e.g. Hedley et al.

2004), and this may be related to greater water holding capacity and typically larger water content than surrounding soils with less clay. The highest ECa values correspond to trenches with highest percent clay content at Felton Road. The EMI survey quality is not attenuated by high clay content at any of the sites, and shows good quality where clay content is highest and GPR survey quality is poor. Low clay content at Olssens may, in part, result in overall low ECa values at the site. Multi-frequency EMI surveys may be able to detect soil ECa changes, even below clay rich layers that are more impenetrable to GPR.

Yoder et al. (2001) used EMI surveys to identify areas of high ECa, and then conducted GPR surveys to obtain more detailed information in these areas of interest. The results of this study indicate that multi-frequency EMI surveys may be able to replace the need for two surveys depending on the information required and providing that soil ECa differences with depth are sufficient. For example, the GPR survey is useful in indicating that to a depth of 5 m, Target Gully has less gravel and more uniform deposits than Olssens, which has more gravel and fluvial stratigraphy. At Target Gully, the contrast at 5 m depth is picked up by both GPR and EMI surveys. At the sites in this study, EMI survey quality was generally high providing soil ECa values are greater than roughly 1 mS/m. This demonstrates that multi-frequency EMI surveys can be used to estimate depth to layers of significant contrast within the subsurface.

Different statistical techniques have been used to predict various soil properties from a variety of data sources (e.g. James et al., 2003; McBratney et al., 2003 and references therein; Liu et al., 2008; Grunwald, 2009). For example, logistic regression can be used to make continuous soil maps of soil groups influenced by topography (Debella-Gilo & Etzelmüller, 2009) and to quantify relationships between ancillary variables and soil maps (Kempen et al., 2009). The analysis presented here uses multinomial logistic regression to predict mapped soil class using multi-frequency EMI data as covariates. No other studies were found in the literature that use depth corrected ECa values as covariates in a statistical regression analysis in an attempt to ascertain whether these data can predict soil type. Soil ECa values with depth exhibit patterns for some mapped soil types (Fig. 9). An increase in prediction accuracy from ~40% to 69% correct of the multinomial regression analysis occurs when using depth corrected ECa values instead of single frequency values. Results of the statistical analyses suggest that multi-frequency ECa data are better able to predict soil type than single frequency data alone. This may in part be due to different trends observed in ECa with depth for different soil types.

4.4 Applications in precision viticulture – Geophysics and wine

Precision agriculture incorporates widespread use of technologies that give valuable information about soil variation. Precision agriculture is beneficial in many different industries, and Bramley (2009) provides an excellent review. Mapping the variability of soil is a key component of implementing precision agriculture, and various soil characteristics can be measured. Spatial variability in soils can be reflected in EMI surveys, which in turn can be useful for delineating areas for soil sampling (e.g. Bramley, 2003) or defining water restriction zones on a vineyard scale (e.g. Acevedo-Opazo et al., 2008). Another method that can be used in vineyards is normalised difference vegetation index (NDVI; Rouse et al., 1973) imaging, although if the goal is to measure vine canopy as opposed to soil properties, spatial resolutions of 20 cm are required to differentiate vine rows from cover crop (Lamb et al., 2001). The implementation of precision viticulture can be profitable; applications in

Australia have shown a benefit of $30,000 - $40,000 / ha in some more extreme cases (Bramley 2009, and references therein).
Vine trunk circumferences vary significantly within each of our study sites. Grape quality has been linked to changes in vine vigour elsewhere (e.g. Bramley, 2003; Trought et al., 2008), and therefore grapes with different characteristics are likely produced within our sites. The EMI maps from our sites indicate that properties that affect soil conductivity may also affect vine trunk circumference growth. Electromagnetic induction surveys, which show local differences in soil apparent electrical conductivity, may provide valuable and cost-effective inputs into precision management decisions for the New Zealand viticulture industry, and these geophysical surveys should be accompanied by differential GPS surveys that can be conducted concurrently to precisely map elevation, slope and aspect.

4.5 Future research

The EMI sensor used for this research was the GEM-2 portable broadband electromagnetic sensor capable of measuring soil ECa up to roughly 48,000 Hz. The next generation GEM-2 is capable of measuring soil ECa up to roughly 96,000 Hz. The higher frequency capability measures ECa closer to the surface than the GEM-2 used in this research. By conducting multi-frequency surveys using higher frequencies, more detailed ECa profiles of the shallow soils can potentially be created. Conducting these surveys and inverting these data warrants further investigation, and may provide more precise data on soil ECa variability in soils in the rootzone. Further research in other New Zealand winegrowing regions should be conducted using the GEM-2 analysing depth penetration to test whether this survey method can be used in other soil types in New Zealand.
Our research indicates that maps produced from EMI survey data show useful correlations with vine trunk circumference data, and we recommend continued uptake of these types of surveys by the agricultural industry (e.g. Bramley, 2009 and references therein). Closer integration with yield data and data that reflect fruit characteristics and ripeness could lead to refined management of individual vineyard blocks. Such geophysical surveys can also be beneficial in the early stages of vineyard block design, where blocks are delineated according to similar environmental parameters.
The complex interactions between soil properties, grapevine physiology and must/wine properties are poorly understood. Wines from some regions, both internationally and in New Zealand, have similar characteristics leading some to assume soil imparts particular characteristics in wines. Soils are complex systems, and a wide variety of properties that may affect grape compounds can be further researched. Microvinification research should be conducted on a vineyard block scale comparing soil and wine properties in close proximity in order to determine if soil variability between adjacent grapevines influences wine properties. The lack of data that specifically links soil properties and wine characteristics indicates that far more research is required in this area. A better understanding of how soil influences specific wine properties will enhance winemakers' abilities to produce wines of desired flavour and aroma profiles.

5. Conclusions

Ground penetrating radar and electromagnetic induction surveys have the potential to be used as inputs for precision viticulture, and our results indicate that in New Zealand, both techniques can provide useful results. A wide variety of factors can affect the results of these surveys, and additional work is needed in New Zealand and elsewhere to help constrain the relative influences of different attributes on geophysical data and how these attributes influence vine growth and grape characteristics. Nonetheless, our results show that EMI surveys provide data that more clearly correlate with variations in vine trunk circumference data than GPR surveys, and including topography in such analyses may prove beneficial. As GPR surveys are also more expensive and time consuming, EMI surveys are the preferred tool for defining variation in soils in New Zealand. Multi-frequency EMI surveys can also be used to map areas of different soil type more effectively than more traditional single-frequency surveys.

Environmental properties such as slope, aspect, elevation, growing degree days and sunshine hours are similar at three adjacent vineyards in Bannockburn, Central Otago. Viticultural management practices such as trellis design, fruit exposure and bunches per shoot between these sites are also similar. However, the three sites have significantly different soils with different physical and chemical properties, and the vine trunk circumferences vary significantly among these sites. The soil component of terroir can have a significant impact on vine characteristics, and presumably also on grape and wine properties. Geophysical survey methods can provide insight into both horizontal and vertical changes in soil variability, and these data can be used to quantify and better understand physical and chemical properties of soils that affect grape and wine properties.

A more comprehensive reference list and complete information on the research presented in this chapter can be accessed at *https://researchspace.auckland.ac.nz/handle/2292/6141* (Imre 2011).

6. References

Acevedo-Opazo, C., Tisseyre, B., Guillaume, S. and Ojeda, H., 2008, The potential of high spatial resolution information to define within-vineyard zones related to vine water status: Precision Agriculture, v. 9, p. 285-302.

Beecroft, F.G., 1988, Soil taxonomic unit descriptions for Bannockburn Valley, Central Otago, South Island, New Zealand: NZ Soil Bureau, Department of Scientific and Industrial Research, Wellington, 174 p.

Bramley, R.G.V., 2001, Progress in the development of precision viticulture - Variation in yield, quality and soil properties in contrasting Australian vineyards, *in* Currie, L.D. and Loganathan, P., *eds*, Precision tools for improving land management: Occasional report No. 14, Fertilizer and Lime Research Centre, Massey University, Palmerston North. P. 25-43.

Bramley, R., 2003, Smarter thinking on soil survey: Australian and New Zealand wine industry journal, v. 18(3), p. 88-94.

Bramley, R.G.V., 2009, Lessons from nearly 20 years of precision agriculture research, development, and adoption as a guide to its appropriate application: Crop and Pasture Science, v. 60, p. 197-217.

Bravdo, B.A., 2001, Effect of Cultural Practices and Environmental Factors on Fruit and Wine quality: Agriculturae Conspectus Scientificus, v. 66 (1), p. 13-20

Bronson, K.F., Booker, J.D., Officer, S.J., Lascano, R.J., Maas, S.J., Searcy, S.W., and Booker, J., 2005, Apparent electrical conductivity, soil properties and spatial covariance in the U.S. Southern High Plains: Precision Agriculture, v. 6, p. 297-311.

Carroll, Z.L. and Oliver, M.A., 2005, Exploring the spatial relations between soil physical properties and apparent electrical conductivity: Geoderma, v. 128, p. 354-374.

Clingeleffer, P.R. and Emmanuelli, D.R., 2006, An assessment of rootstocks for Sunmuscat (Vitis vinifera L.): a new drying variety: Australian Journal of Grape and Wine Research, v. 12, p. 135-140.

Cockx, L., van Meirvenne, M. and De Vos, B., 2007, Using the EM38DD soil sensor to delineate clay lenses in a sandy forest soil: Soil Science Society of America Journal, v. 71, p. 1314-1322.

Cortell, J.M., Halbleib, M., Gallagher, A.V., Righetti, T.L. and Kennedy, J.A., 2005, Influence of vine vigor on grape (*Vitis Vinifera* L. Cv. Pinot Noir) and wine proanthocyanidins: Journal of Agricultural and Food Chemistry, v. 53, p. 5798-5808.

Cortell, J.M. and Kennedy, J.A., 2006, Effect of shading on accumulation of flavonoid compounds in (*Vitis Vinifera* L.) Pinot Noir fruit and extraction in a model system: Journal of Agricultural and Food Chemistry, v. 54, p. 8510-8520.

Corwin, D.L., and Lesch, S. M., 2005, Apparent soil electrical conductivity measurements in agriculture: Computers and Electronics in Agriculture, v. 46, p. 11-43.

Daniels, D.J., 2004, Ground Penetrating Radar 2nd Edition: IEE Radar, Sonar and Navigation series 15: Institution of Electrical Engineers, London, 726 p.

De Andres-de Prado, R., Yuste-Rojas, M., Sort, X., Andres-Lacueva, C., Torres, M. and Lamuela-Raventos, R.M., 2007, Effect of Soil Type on Wines Produced from Vitis vinifera L. Cv. Grenache in Commercial Vineyards: Journal of Agricultural and Food Chemistry 55 (3): 779-786.

Debella-Gilo, M. and Etzelmüller, B., 2009, Spatial prediction of soil classes using digital terrain analysis and multinomial logistic regression modeling integrated in GIS: Examples from Vestfold County, Norway: Catena, v. 77, p. 8-18.

Doolittle, J.A., Indorante, S.J., Potter, D.K., Hefner, S.G. and McCauley, W.M., 2002, Comparing three geophysical tools for locating sand blows in alluvial soils of southeast Missouri: Journal of Soil and Water Conservation, v. 57:3, p. 175-182.

Dry, P.R. and Loveys, B.R., 1998, Factors influencing grapevine vigour and the potential for control with partial rootzone drying: Australian Journal of Grape and Wine Research, v. 4(3), p. 140-148.

Eckert, D. and Sims, J.T., 1995, Recommended soil pH and lime requirement tests, *in* Sims, J.T., and Wolf, A. *eds*, Recommended Soil Testing Procedures for the Northeastern United States. Northeast Regional Bulletin #493, Agricultural Experiment Station, University of Delaware, Newark, DE, USA, p.11-16.

Galagdara, L.W., Parkin, G.W., Redman, J.D., von Bertoldi, P., and Endres, A.L., 2005, Field studies of the GPR ground wave method for estimating soil water content during irrigation and drainage: Journal of Hydrology, v. 301, p. 182-197.

Galet, P., 2000, General Viticulture. Translated by John Towey. Oenoplurimédia. France, 443 p.

Gerber, R., Salat, C., Junge, A. and Felix-Henningsen, P., 2007, GPR-based detection of Pleistocene periglacial slope deposits at a shallow-depth test site: Geoderma, v. 139, p. 346-356.

Gómez-Míguez, M.J., Gómez-Míguez, M., Vicario, I.M. and Heredia, F.J., 2007, Assessment of colour and aroma in white wines vinifications: Effects of grape maturity and soil type: Journal of Food Engineering, v. 79: p. 758-764.

Grote, K., Hubbard, S., and Rubin, Y., 2003, Field-scale estimation of volumetric water content using GPR groundwave techniques: Water Resources Research, v. 39-11, p. SBH5.1-5-SBH5.13.

Grunwald, S., 2009, Multi-criteria characterization of recent digital soil mapping and modelling approaches: Geoderma, v. 152, p. 195-207.

GSSI (Geophysical Survey Systems, Inc.), 2004, RADAN 6 User's Manual: Geophysical Survey Systems, Inc. United States, 135 p.

Haeger, J.W., 2004, North American Pinot Noir: University of California Press, Berkley. 445 p.

Hartsock, N.J., Mueller, T.G., Thomas, G.W., Barnhisel, R.I., Wells, K.L. and Shearer, S.A., 2000, Soil Electrical Conductivity Variability, *in* Robert P.C., et al. *eds*, Proc. 5th international conference on precision Agriculture, ASA Misc. Publ., ASA, CSSA, and SSSA, Madison, WI.

Hedley, C.B., Yule, I.J., Eastwood, C.R., Shepherd, T.G. and Arnold, G., 2004, Rapid identification of soil textural and management zones using electromagnetic induction sensing of soils: Australian Journal of Soil Research, v. 42, p. 389-400.

Huang, H., 2005, Depth of investigation for small broadband electromagnetic sensors: Geophysics, v. 70, p. G135-G142.

Huang, H., and Won, I.J., 2003, Real-time resistivity soundings using a hand-held broadband electromagnetic sensor: Geophysics, v. 68, p. 1224-1231.

Hubbard, S., Grote, K. and Rubin, Y., 2002, Mapping the volumetric soil water content of a California vineyard using high-frequency GPR ground wave data: The Leading Edge, June, p. 552-559.

Iland, P., 2004, Chemical analysis of grapes and wine: techniques and concepts. Campbelltown, SA, Australia, Patrick Iland Wine Promotions PTY LTD.

Imre, S.P., 2011, A Multi-disciplinary study to quantify terroir in Central Otago and Waipara Pinot Noir vineyards. PhD thesis, University of Auckland, New Zealand.

Imre, S.P. and Mauk, J.L. 2009, New Zealand Terroir: Geoscience Canada, v. 36(4), p. 145-159.

Imre, S.P., Mauk, J.L., Bell, S. and Dougherty, A., 2009a, Correlations among ground penetrating radar, electromagnetic induction and vine trunk circumference data: towards quantifying terroir in New Zealand Pinot Noir vineyards: Le Progrès agricole et viticole, ISSN 0369-8173, v. 126 (1), p. 8-11.

Imre, S.P., Mauk, J.L., Bell, S., and Dougherty, A., 2009b, Mapping grapevine vigour and lateral variation in soils: Journal of Wine Research, Submitted Dec 2009.

Jackson, D.I. and Lombard, P.B., 1993, Environmental and Management Practices Affecting Grape Composition and Wine Quality - A Review: American Journal of Enology and Viticulture, v. 44 (4), p. 409-430

Jackson, R., 1994, Wine science: principles and applications. Oxford, UK. Elsevier Inc., 751 p.

James, I.T., Waine, T.W., Bradley, R.I., Taylor, J.C. and Godwin, R.J., 2003, Determination of soil type boundaries using electromagnetic induction scanning techniques: Biosystems Engineering, v. 86:4, p. 421-430.

Jol, H.M., 2009, Ground Penetrating Radar: Theory and Applications. Elsevier 544 p.

Keller, M., 2005, Deficit Irrigation and Vine Mineral Nutrition: American Journal of Enology and Viticulture, v. 56 (3), p. 267-283.

Kempen, B., Brus, D.J., Heuvelink, G.B.M. and Stoorvogel, J.J., 2009, Updating the 1:50,000 Dutch soil map using legacy soil data: A multinomial logistic regression approach: Geoderma, v. 151, p. 311-326.

Kitchen, N.R., Sudduth, K.A. and Drummond, S.T., 1996, Mapping of sand deposition from 1993 midwest floods with electromagnetic induction measurements: Journal of Soil and Water Conservation, v. 51:4, p. 336-340.

Lamb, D., Hall, A. and Louis, J., 2001, Airborne remote sensing of grapevines for canopy variability and productivity. The Australian Grapegrower and Winemaker, v. 449a, 89–92.

Lambert, J.J., Dahlgren, R.A., Battany, M., McElrone, A. and Wolpert, J.A., 2008, Impact of soil properties on nutrient availability and fruit and wine characteristics in a Paso

Robles vineyard. Proceedings of the 2nd Annual National Viticulture Research Conference. July 9-11, 2008. University of California, Davis. p. 44-45.

Leathwick, J., Morgan, F., Wilson, G., Rutledge, D., McLeod, M., and Johnston, K., 2002, Land environments of New Zealand: A Technical Guide: David Bateman Ltd., Auckland, 237 p.

Liu, J., Pattey, E., Nolin, M.C., Miller, J.R. and Ka, O., 2008, Mapping within-field soil drainage using remote sensing, DEM and apparent soil electrical conductivity: Geoderma, v. 143, p. 261-262.

Mackenzie, D.E. and Christy, A.G., 2005, The role of soil chemistry in wine grape quality and sustainable soil management in vineyards: Water Science and Technology, v. 51 (1), p. 27-37.

Macqueen, R.W., and Meinert, L.D., eds., 2006, Fine wine and terroir: the geoscience perspective: Geological Association of Canada, St John's Newfoundland, Canada, 247 p.

McBratney, A.B., Mendonça Santos, M.L. and Minasny, B., 2003, On digital soil mapping: Geoderma, v. 117, p. 3-52.

New Zealand Winegrowers, 2009, Annual Report: http://nzwine.com/intro/. 20 p.

NZIER, 2009, Economic impact of the New Zealand wine industry. www.nzier.org.nz: Wellington, New Zealand, 32 p.

OIV (Organisation Internationale de la Vigne et du Vin), 2006, Situation of the world viticulture sector in 2006, http://news.reseauconcept.net/images/oiv_uk/client/Commentaire_statistiques_annexes_2006_EN.pdf

Ojeda, H., Andary, C., Kraeva, E., Carbonneau, A. and Deloire, A., 2002, Influence of Pre- and Postveraison Water Deficit on Synthesis and Concentration of Skin Phenolic Compounds during Berry Growth of Vitis vinifera cv. Shiraz: American Journal of Enology and Viticulture, v. 53 (4), p. 261-267.

Oliveira, C., Silva Ferreira, A.C., Mendes Pinto, M., Hogg, T., Alves, F. and Guedes de Pinho, P., 2003, Carotenoid Compounds in Grapes and Their Relationship to Plant Water Status: Journal of Agricultural and Food Chemistry, v. 51 (20), p. 5967-5971.

Peyrot des Gachons, C., van Leeuwen, C., Tominaga, T., Soyer, S., Gaudillegrave, J. and Dubourdieu, D., 2005, Influence of water and nitrogen deficit on fruit ripening and aroma potential of Vitis vinifera L cv Sauvignon blanc in field conditions: Journal of the Science of Food and Agriculture, v. 85, p. 73-85.

Proffitt, T., Bramley, R., Lamb, D. and Winter E., 2006, Precision viticulture - A new era in vineyard management and wine production: Winetitles: Adelaide. 92. p.

Ross, D., 1995, Recommended soil tests for determining soil cation exchange capacity, *in* Sims, J.T., and Wolf, A. *eds*., Recommended Soil Testing Procedures for the Northeastern United States. Northeast Regional Bulletin #493. Agricultural Experiment Station, University of Delaware, Newark, DE, USA., p.62-69.

Rouse, J.W., Haas, R.H., Schell, J.A. and Deering, D.W., 1973, Monitoring vegetation systems in the Great Plains with ERTS. In 'Proceedings of the 3rd ERTS Symposium' NASA SP351, 1. p, 309-317. US Government Printing Office: USA)

Sabon, I., de Revel, G., Kotseridis, Y. and Bertrand, A., 2002., Determination of Volatile Compounds in Grenache Wines in Relation with Different Terroirs in the Rhone Valley: Journal of Agricultural and Food Chemistry, v. 50 (22), p. 6341-6345.

Sarneckis, C.J., Dambergs, R.G., Jones, P., Mercurio, M., Herderich, M.J. and Smith, P.A., 2006, Quantification of condensed tannins by precipitation with methyl cellulose:

development and validation of an optimised tool for grape and wine analysis: Australian Journal of Grape and Wine Research, v. 12 (1), p. 39-49.

Simeoni, M.A., Galloway, P.D., O'Neil, A.J. and Gilkes, R.J., 2009, A procedure for mapping the depth to the texture contrast horizon of duplex soils in south-western Australia using ground penetrating radar, GPS and kriging: Australian Journal of Soil Research, v. 47, p. 613-621.

Smart D.R, Schwass, E., Lakso, A., Morano, L., 2006, Grapevine rooting patterns: a comprehensive analysis and review: American Journal of Enology and Viticulture, v. 57:1, p. 89-104.

Swinchatt, J. and Howell, D.G., 2004, The Winemaker's Dance: Exploring terroir in the Napa Valley: University of California Press, Berkley. 229 p.

Tominaga, T., Murat, M.L. and Dubourdieu, D., 1998, Development of a method for analyzing the volatile thiols involved in the characteristics aroma of wine made from Vitis vinifera L. cv. Sauvignon blanc: Journal of Agricultural and Food Chemistry, v. 46 (3), p. 1044-1048.

Trought, M. C. T., Dixon, R., Mills, T., Greven, M., Agnew, R., Mauk, J. L., and Praat, J.-P., 2008, The impact of differences in soil texture within a vineyard on vine vigour, vine earliness and juice composition: Journal International des Sciences de la Vigne et du Vin, v. 42 (2), p. 67–72.

Turnbull, I.M. (compiler), 2000, Geology of the Wakitipu area. Institute of Geological & Nuclear Sciences 1:250 000 geological map 18. 1 sheet + 72 p. Lower Hutt, New Zealand. Institute of Geological and Nuclear Sciences Limited.

USEPA, 1986, Test methods for evaluating solid waste. Volume IA: 3rd Edition, EPA/SW-846. National Information Service. Springfield, VA, USA.

Van Leeuwen, C., Friant, P., Choné, X., Trégoat, O., Koundouras, S. and Dubourdieu, D., 2004, Influence of climate, soil, and cultivar on terroir: American Journal of Enology and Viticulture, v. 55 (3), p. 207–217.

White, R. 2003. Soils for fine wines. Oxford University Press Inc. New York. 279 p.

White, R., Balachandra, L., Edis, R. and Chen, D., 2007, The soil component of terroir: Journal International des Science de la Vigne et du Vin, v. 41, p. 9-18.

Wilson, J.E., 1998, Terroir: The role of geology, climate, and culture in the making of French wines: University of California Press, Berkley. 336 p.

Winkler, A.J., Cook, J.A., Kliewer, W.M. and Lider, L.A., 1962. General Viticulture. University of California Press, USA, 710 p.

Wolf, A.M. and Beegle, D.B., 1995, Recommended soil tests for macronutrients: phosphorus, potassium, calcium, and magnesium, *in* Sims J.T., and Wolf A., *eds.*, Recommended Soil Testing Procedures for the Northeastern United States. Northeast Regional Bulletin #493, Agricultural Experiment Station, University of Delaware, Newark, DE, USA, p. 25-34.

Won, I.J., Keiswetter, D.A., Fields, G.R.A., and Sutton, L.C., 1996, GEM-2: A New Multifrequency Electromagnetic Sensor: Journal of Environmental and Engineering Geophysics, v. 1:2, p. 129-137.

Yoder, R.E. Freeland, R.S., Ammons, J.T. and Leonard, L.L., 2001, Mapping agricultural fields with GPR and EMI to identify offsite movement of agrochemicals: Journal of Applied Geophysics, v. 47, p. 251-259.

Zou, H., Kilmartin, P., Inglis, M. and Frost, A., 2002, Extraction of phenolic compounds during vinification of Pinot Noir wine examined by HPLC and cyclic voltammetry: Australian Journal of Grape and Wine Research, v. 8 (3), p. 163-174.

2

Geological Carbon Dioxide Storage in Mexico: A First Approximation

Oscar Jiménez, Moisés Dávila, Vicente Arévalo, Erik Medina and Reyna Castro
Comisión Federal de Electricidad
México

1. Introduction

Carbon dioxide (CO_2) is one of the industrial gases that contribute to the greenhouse gas (GHG) effect. During the last decades, the emissions of CO_2 due to human activity have increased significantly all over the world. There are different and important efforts to reduce or stabilize the concentrations of greenhouse gases in the atmosphere, such as improvements in the efficiency of power plants and the development of renewable energies. However, those approaches cannot deliver the level of emissions reduction needed, especially against a growing demand for energy that promotes economic growth and prosperity. Carbon capture and storage (CCS) approach encompasses the processes of capture and storage of CO_2 that would otherwise reside in the atmosphere for long periods of time. Among the different carbon capture and storage options currently in progress all over the world, the geological storage option is defined as the placement of CO_2 into an underground repository in such a way that it will remain permanently stored. Mexico is one of the countries which are signatories of different international treaties which call for stabilization of atmospheric gases emissions at a level that prevent anthropogenic interference with the world's regional climates. In Mexico CO_2 represents almost 70% of the total greenhouse gases emissions where the primary sources of CO_2 are the burning of fossil fuels for power generation. CCS is a technological approach that holds great promise in reducing atmospheric CO_2 concentrations in Mexico. This is the first coordinated assessment of carbon storage potential across the country.

1.1 Geographical location of Mexico

Mexico is a country located in the southern portion of North America, and is bordered to the north by the United States, to the southeast by Guatemala, Belize and the Caribbean Sea, to the west and south by the Pacific Ocean, and to the east by the Gulf of Mexico (Figure 1). The country's total area is about 1 972 550 square kilometers.

1.2 Previous work

With the aim of searching for places where to store carbon dioxide, Mexico was subdivided into three *exclusion zones* and four *inclusion zones* [1](Figure 2). The exclusion zones are zones A, B and G. Zone A is composed by igneous rocks with high seismic and volcanic hazard, and is not recommended for storage. Zone B encompasses also igneous rocks with less seismic and volcanic hazards than zone A, but not yet recommended for CO_2 storage. The zone G is a marine zone of exclusion comprising the ocean floor, deep marine sediments and high seismic and tectonic hazardous processes in the Pacific Ocean.

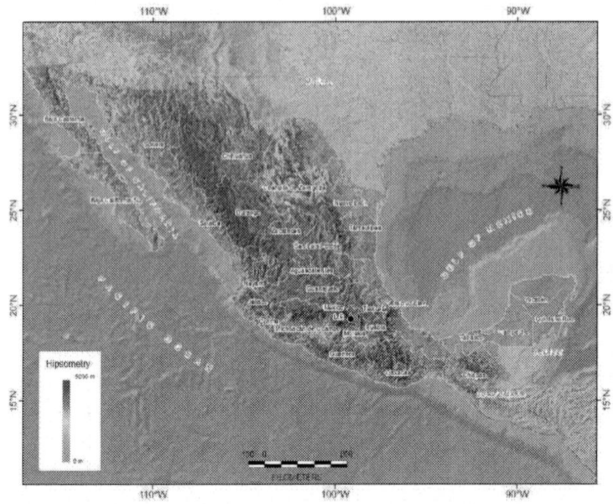

Fig. 1. Hypsographic map of Mexico displaying federal states divisions and countries' borderlines.

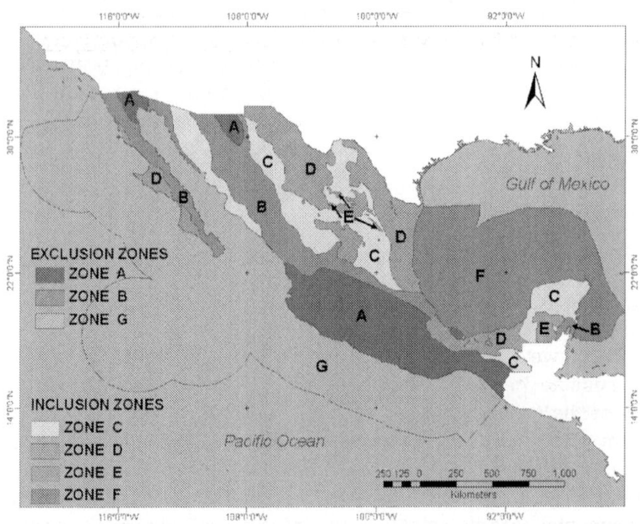

Fig. 2. Exclusion and inclusion zones for geologic CO_2 storage in Mexico. After [1].

The inclusion zones are zones C, D, E and F. Zone C represents terrigenous geological formations and mainly carbonate sedimentary rocks cropping out in the area. Zone D includes terrigenous as well as carbonate sedimentary rocks sequences. Zone E is composed of evaporitic deposits and associated sedimentary rocks. And zone F reflects sediments deposited in the marine continental shelf, slope and deep waters beneath the Gulf of Mexico. All of these zones were outlined taking into account surficial lithological features, large geological subsurface structures and recent volcanic and tectonic activity in a country scale assessment. The exclusion zones were not recommended for geologic carbon storage due to its high seismic, geothermic and active volcanic hazardous potential. On the contrary, the inclusion zones yielded the best CO_2 storage potential and were recommended for further detailed studies in order to find geological provinces with a good CCS capacity.

1.3 Purpose and scope

The purpose of this chapter is to present the analysis of different geological provinces to address the possibility of storing anthropogenic CO_2 in deep underground geologic formations, particularly in eastern continental Mexico. Up to now, the assessment has been focused on five geological provinces in order to evaluate and quantify theoretically its CO_2 storage potential and to identify prospective regions and/or sectors that should form the object of further and detailed studies.

The analysis has been considered in relation to a specific type of storage, that is, deep saline aquifers and to the location of the stationary CO_2 sources currently available for the whole nation. It must be noted though that an assessment of CO_2 storage potential is surrounded by large uncertainties, which increase in number with the lack of available data and detailed information. The proposed work in this chapter recognizes this uncertainty, and the envisaged output is an overview of possible scenarios rather than the quantification of specific areas or sites for CCS. The aim is to provide a high level summary of CO_2 geologic storage potential across Mexico where the capacity resource estimates presented are intended to be used as an initial assessment of potential geologic storage prior to a local area selection. It is expected that as new subsurface data and a more refined methodology are acquired, the CCS studies will be improved in the near future.

1.4 Methodology

The total CCS process is frequently analyzed from several viewpoints which include very wide technological, economic and environmental issues. Some of the issues are well constrained while others are poorly understood. In the particular case of CO_2 storage potential there are also various aspects involved, such as the separation and capture of CO_2 at the point of emission, the mass of CO_2 emitted by the point of emission, the infrastructure and transportation of CO_2, and the storage of CO_2 in deep underground geologic formations [2]. However, here we are only concerned with the types of CO_2 emission sources, the searching of suitable geologic reservoir rock sequences and their location, and the quantification of the theoretical capacity of storing a given volume or mass of CO_2 in selected sectors across Mexico. This pragmatic methodology was based on the public domain accessible data and present-day geological knowledge, and it does not incorporate geological constraints in the theoretical capacity estimations, nor does it incorporate risk factors, environmental hazards, solubility and mineral trapping of CO_2, or quantification of injectivity of the potential storage rock sequences.

The **first phase** included a survey of CO_2 points of emission, production information, source category, emissions factors, and annual CO_2 emissions that were obtained from the mexican Pollutant Release and Transfer Inventory (RETC by its Spanish acronym) and the Ministry of the Environment and Natural Resources (SEMARNAT, by its Spanish acronym) databases [3,4]. These databases consider the stationary sources. A compilation for the United Nations Framework Convention on Climate Change (UNFCCC) [4] includes the stationary and the non-stationary source emissions. The non-stationary source emissions such as those that come from the transportation sector, the change of land use and forestry, and some others like landfills were excluded from the analysis. The CO_2 stationary sources included power plants, oil and natural gas processing facilities, cement plants, agricultural processing facilities, iron and steel production facilities, and other industry processing facilities. The spatial location of the stationary CO_2 emission sources were calculated and compiled through different mapping tools that contain latitude and longitude information for various Mexican locations. The analysis of CO_2 stationary sources was done to provide reliable emission estimations, identify major CO_2 emission sources within each region, and to asses the applicability of the data in subsequently infrastructure analyses.

The **second phase** consisted of the identification of geological storage provinces through the careful analysis and screening of available geological data. In this regard, there are different proposed methodologies that are similar [5, 6, 7, 8, 9, 27]. Only minor differences are evident depending upon the used weights that show the relative importance of the criteria. Therefore, our selection of candidate storage provinces was according to the *basin level* of the assessment scale [10] (Figure 3). This "basin scale" exploration assessment required a little more local data categories and a better level of detail than the "country scale".

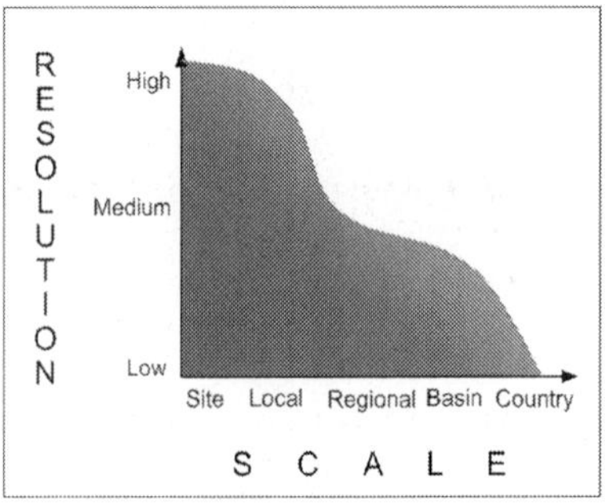

Fig. 3. Data and assessment scales for CCS geological screening studies. After [10].

In this "basin scale" assessment, both terms, *basin* and *province*, are considered synonyms. The term *basin* has different meanings depending upon geologic features of the region, such as geothermal regime, size, age, boundaries, type and thickness of sedimentary fill, geologic deformation, tectonic context, and many others parameters that can change with time [11, 12, 13, 14]. However, these variable geologic features are also possible to be applied to the meaning of the term *province*.

The assessment was focused on the previously identified inclusion zone. Within the inclusion zone, twelve provinces were defined taking into consideration the types of geomorphological developments, stratigraphic successions, major structural deformation patterns, homogeneous tectonic history, and known subsurface geological boundaries between all of them (Figure 4). Actually, their outlined boundaries are very similar with those of the petroleum basins previously named for those areas of Mexico [15, 16, 17, 18, 19]. From the twelve established provinces, at the moment, only five of them were considered to be studied in greater detail to estimate the geological resource for storing CO_2. These provinces are: Burgos, Tampico-Misantla, Veracruz, Sureste and Yucatan, all of them located in the continental and marine platform areas along Gulf of Mexico.

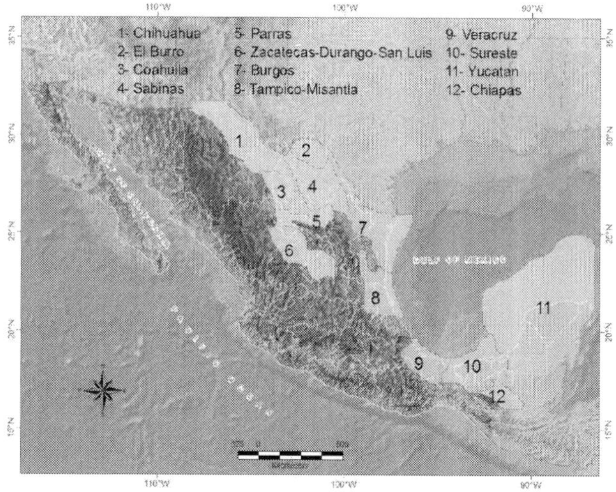

Fig. 4. Mexican geological provinces identified according to their underground potential for CO_2 storage.

The screening and selection of the provinces was based on the published geologic maps from a scale of 1:250,000 to 1:4,000,000 and reports about surface geology, stratigraphic and structural features, regional geologic cross-sections (50-200 km in length and 500m to 3 km in thickness), geophysical information and available public oil well data within each province. Three main groups of sedimentary formations for underground geologic carbon storage were observed. These groups of sedimentary formations are referred to as *carbonate, evaporite* and *terrigenous sequences* depending upon the main, respectively, carbonated, evaporitic and clastic content of the rock units. It is worth to mention that the stratigraphic uncertainty is high since the specific subsurface geologic information is quantitatively scarce and sometimes restricted and/or no detailed.

Otherwise, the disposal of CO_2 in geological formations, generally, includes unmineable coal seams, oil and gas reservoirs, and deep saline reservoirs. In Mexico unmineable coal areas are not considered as a CCS option because they are located inside the exclusion zone, that is, they are affected seismo-tectonically and located close to the surface. On the contrary, the oil and gas reservoirs are the best option, particularly the EOR (Enhanced Oil Recovery) technique in the exhausted oil fields. But, at the moment, this prospect is ruled out due to the inaccessibility to the public domain of the oil databases and information. Only PEMEX (Petróleos Mexicanos) the oil governmental industry could carry out such studies. So, based on the fact that subsurface layers of porous rocks are generally saturated with brine and that they form deep saline aquifers characterized by high concentrations of dissolved salts and unsuitable for agriculture or human consumption, they were envisaged as the favorable option for CO_2 storage in Mexico. The storing CO_2 in saline formations is achievable since there are examples from such projects [20, 21].

The **third phase** dealt with the estimation of theoretical capacity within each identified geological province. At present, various calculation methods have been proposed to know the storage capacity of a rock formation [22, 10, 23,20, 24, 25, 2]. They have been applied to different country projects within their respective areas and still there is uncertainty. The reasons for this uncertainty are diverse but they broadly comprise key aspects such as financial support, CCS technology research and development, and a real partnership between country organizations and academic teams [26, 28].

The concept of storage capacity was referred to a completely free phase of the CO_2, which means without taking into account the CO_2 reaction with the walls of the reservoirs or formations. It is considered only the volume of CO_2 that can be retained in the available porous space of the storage formation or reservoir at depths between 800 and 2500 meters. At such depths the CO_2 has some properties like a gas and some like a liquid due to the changes in temperature and pressure conditions [64]. These are known as the CO_2 *supercritical* conditions or the critical point of the CO_2. The huge advantage of storing CO_2 in the supercritical condition is that the required storage volume is much less if the CO_2 were at standard pressure conditions.

For the estimation of the theoretical capacity of storing CO_2, it was used an approach here called "parameterization". The parameterization refers to observations, deductions, and calculations derived from the physical parameters obtained from geological maps, regional stratigraphic and structural cross-sections, and well data from the public petroleum industry. Different geological variables were taken into account since the estimation was done with respect to general storage capacity resources and following the standards used in the petroleum industry, that is, stratigraphic and structural traps, as well as seal (cap) rocks that play a decisive role within any geological province.

One first step in the parameterization approach was the determination of important geological features that would fulfill the storage requirements such as structural or stratigraphic trap, seal formation, stratigraphic discontinuities, geological faults, depth conditions, appropriate porosity and thickness of the target sedimentary sequence. The critical features were: reservoir depth (more than 800 m and less than 2500 m), thickness, porosity, lithological composition (predominantly carbonates and clastic deposits) and, for effects of the volume calculation, the relationship between "net thickness" versus "total thickness". All of this, with the goal of having an expression figure of the fraction of the geological formation susceptible to become a reservoir. The previous information had to be homogeneously similar within the area with a radius between 10 and 20 kilometers around each oil well considered and the nature of trap boundaries. When the information was assumed to be minimally sufficient and it was valued as an attractive target from the point of view of the depth, thickness, porosity, and permeability, then it was selected to quantify its potential capacity to become a CO_2 storing sector. Otherwise, the portion of the regional section including the wells was discarded.

One second step of the approach was the direct application of an equation whose variables were fulfilled with the information above mentioned for deep saline aquifers. Therefore, the critical parameters obtained in the previous step were substituted in the formula proposed by Bachu *et al* in 2007 [10]:

$$VCO_2t= V\varphi(1-S_{wirr}) \ \Xi \ Ah\varphi(1-S_{wirr}) \quad (1)$$

Where A is the trap area, h is the average thickness, VCO_2t is the theoretical volume available, φ is the effective porosity, V is the volume and S_{wirr} is the irreducible water saturation. The solving of the equation yielded the theoretical storage capacity volume of the sector under consideration.

2. Estimated CO_2 emissions from stationary sources

The most recent update on the mexican national inventory (SEMARNAT) was compiled in 2006 (UNFCCC)[4]. This document shows that the total annual GHG in Mexico are above 709 million metric tons (Mt) of CO_2 equivalent. The carbon dioxide represents 69.5% out of a total of 492 Mt of emissions from stationary and non-stationary sources. There were estimated 285 Mt of CO_2 emissions from stationary sources (Figure 5).

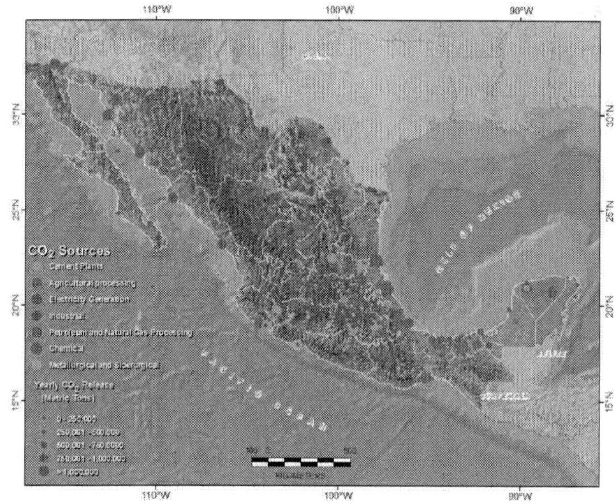

Fig. 5. Main CO_2 stationary source emissions in Mexico. Each colored dot represents a different type of stationary source by category. Dot size represents the relative magnitude of CO_2 emissions released per year.

In addition, RETC data shows approximately 216 Mt of CO_2 emitted from 1,860 stationary sources, according to the different industrial and economic activities in Mexico (Table 1).
From the above data it is evident that the electricity supplier sector is the most important contributor to CO_2 emissions from stationary sources. It releases to the atmosphere 107 Mt of CO_2, roughly 50% of the total. It includes emissions from the Federal Commission for Electricity (CFE, by its Spanish acronym) which is the national public service agency, as well as from private small electricity suppliers companies. The oil & petrochemicals facilities add another 22% and, therefore, the whole energy sector is responsible for 72% (154 Mt) of CO_2 emissions in the country. The cement, metallurgical, iron & steel industries are also major contributors to the overall CO_2 country emissions, though they are smaller in comparison to the energy industry. In fact, the electricity production industry is the largest contributor,

and it does from a small number of stationary sources (Figure 6). The industrial and chemical sectors show a much larger number of identified sources, but the relative share of their CO_2 emissions, compared to those of the energy sector, is lower.

SECTOR	CO_2 EMISSIONS (metric tons)	No. OF SOURCES
Electricity Generation	107 351 754	113
Oil & Petrochemical	47 556 986	273
Cement	26 016 726	60
Metallurgical, Iron & Steel	21 367 965	261
Industrial	8 764 815	709
Chemical	4 027 475	438
Agriculture Processing	735 319	6
TOTAL	215 821 040	1 860

Table 1. Estimations of CO_2 emissions from stationary sources by sectors. The point sources only include facilities that were reported via the *Annual Certificate of Operation* (COA, by its Spanish acronym) to RETC, managed by SEMARNAT [3].

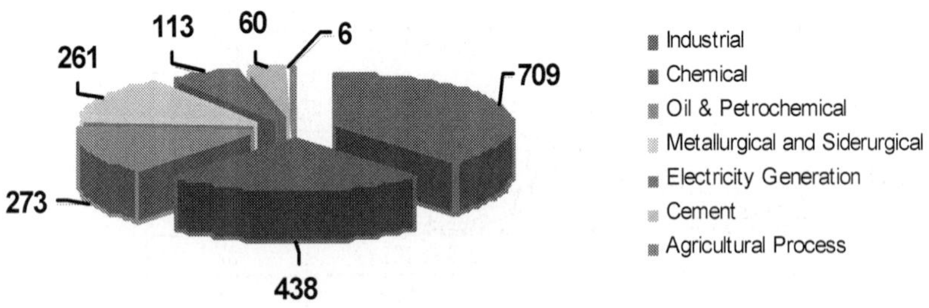

Fig. 6. Number of reported emissions from stationary sources by sector.

From the geographical point of view, the areas with higher CO_2 emissions are located in the northeastern portion of Mexico and in the ferderal states around the Gulf of Mexico. The state of Coahuila tops the list with more than 23 Mt of CO_2 released per year (Table 2). This is mostly due to the deployed coal-fired power plants and metallurgical, iron and steel facilities. The states of Nuevo León and Tamaulipas release approximately 25 Mt n of CO_2 that come from a scattered high number of source points. In the southeastern part, the states of Veracruz and Campeche together attain almost 40 Mt of CO_2.

In this context, it is advisable to apply CCS technologies in such industries, since on the one hand, the fewer number of stationary sources with a high level of CO_2 emissions, the better the opportunity to deploy CO_2 capture, injection and storage facilities. On the other, the scenario leads to an economic feasibility projects particularly at the Gulf Costal region where power generation plants, oil & petrochemical, industrial and chemical facilities share the large CO_2 emissions.

STATE	CO_2 EMISSIONS (metric tons/year)	SOURCES
Coahuila	23 219 675	66
Campeche	21 946 705	25
Veracruz	17 962 809	80
Hidalgo	16 362 111	46
San Luis Potosí	13 580 498	42
Nuevo León	12 725 855	145
Tamaulipas	12 554 901	123
Sonora	9 596 070	46
Michoacán	9 568 763	35
México	9 286 971	284
Chihuahua	8 016 227	265
Guerrero	7 286 999	4
Colima	7 040 064	11
Guanajuato	5 751 629	62
Tabasco	5 676 613	67
Baja California	4 672 787	34
Yucatán	4 214 110	13
Oaxaca	4 108 894	9
Puebla	3 982 865	53
Querétaro	3 466 122	67
Jalisco	3 301 123	87
Sinaloa	3 079 872	11
Durango	2 961 072	18
Morelos	1 805 748	18
Baja California Sur	959 132	9
Aguascalientes	799 295	32
Distrito Federal	746 588	123
Chiapas	732 172	26
Tlaxcala	203 851	43
Quintana Roo	136 962	8
Zacatecas	74 555	7
Nayarit	2	1
TOTAL	215 821 040	1 860

Table 2. Estimated CO_2 emissions by mexican state and number of point sources.

3. Geologic CO_2 storage potential

In order to estimate the CO_2 storage potential and to identify different sectors that should be the object of detailed assessment five geological provinces were analyzed. From north to south the geological provinces are: Burgos, Tampico-Misantla, Veracruz, Sureste and Yucatan (Figure 7).

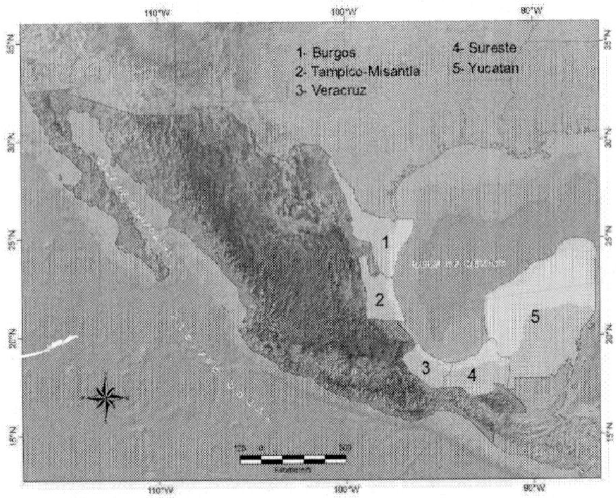

Fig. 7. Mexican geological provinces assessed for underground CO_2 storage.

3.1 Burgos province

The Burgos province is located at the most northeastern portion of Mexico. This province is bordered to the north by the United States (sharing the Rio Bravo along the borderline), to the east by the Gulf of Mexico, to the south by Tampico-Misantla province, and to the west by the first exposures that form the contact between Cretaceous and Tertiary rocks [29].

The basement of the geologic province consists of metamorphic and intrusive igneous rocks [30, 31]. However, the basement geometry and its age distribution have not been well established. On top of the basement, a sedimentary evaporitic and carbonated sequence was accumulated in Mesozoic times [50, 62]. After a period of regional subsidence a thick sequence of mainly coarse to fine grained sediments was deposited starting in the Tertiary and continuing into the Quaternary.

According to the geological analysis it is documented the existence of a thick terrigenous sequence composed by interbedded conglomerates, sandstones and shales of Cenozoic age [32]. These sequences have frequent lateral facies changes and abundant lenticular sand bodies which were deposited mainly in deltaic, shelf and deep marine environments. Exposures of these rock units extend from the Eocene to Quaternary (Figure 8).

Regional geological sections B1, B2, B3 and B4 were studied to estimate the CO_2 storage capacity on the continental portion on the Burgos province. All of them document similar stratigraphic units and characteristic sets of faults as a result of both extensional tectonic and sedimentological events [36]. Section B4 has no public subsurface geological information available, consequently, it was not considered during the assessment process.

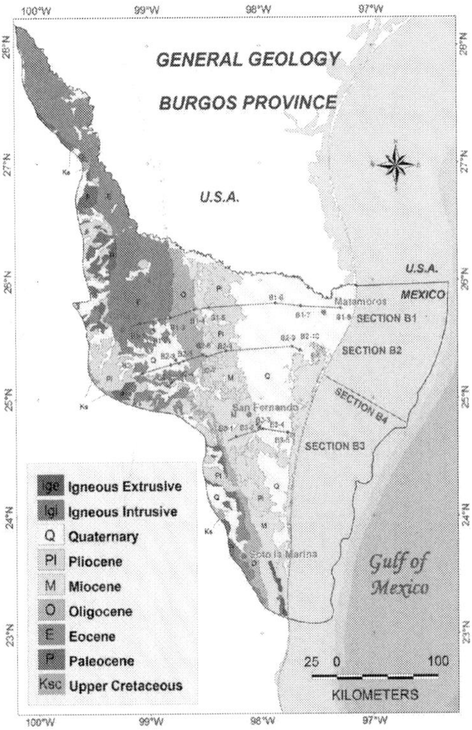

Fig. 8. Simplified geology map of Burgos province depicting geological sections and wells. After [29, 33, 34, 35, 46].

As all the sections depict similar stratigraphic and structural features, only Section B2 is presented (Figure 9). The section B2 has approximately 150 km in length and show a basement covered by slightly deformed Jurassic and Cretaceous rocks sequences. On top of it, there is a thick tertiary sedimentary and faulted sequence of rocks. The sedimentary sequence and the fault system reveal a chronological pattern from older formations and faults on the west to younger ones on the east. Across the entire section are evident the Eocene and Oligocene rocks on the west, and Miocene formations on the east.

According to the type of stratigraphical or structural trap and the lithological and petrophysical features obtained from the oil wells several extrapolations were performed along the regional geological sections in order to select the best potential sectors where saline formations could become CO_2 reservoirs.

An example of detailed description of sector B2-4 of section B2 is presented (Figure 10). The sector B2-4 displays an Eocene terrigenous sequence that is located at approximately 1500 meters depth and consists of thick bedded homogeneous sandstone layers with cross-stratification and minor amounts of intercalated, laterally discontinuous, thin bedded shale. The thickness of the unit is 880 meters but the important fraction is 0.6, therefore the considered net thickness is about 528 meters. The unit is part of a structural trap in a "rollover" anticline with a seal composed of shale from the upper limit of same sequence. The Oligocene sedimentary sequence overlies the Eocene sequence and consists of a siltstone and shale that are interpreted as a seal cap-rock.

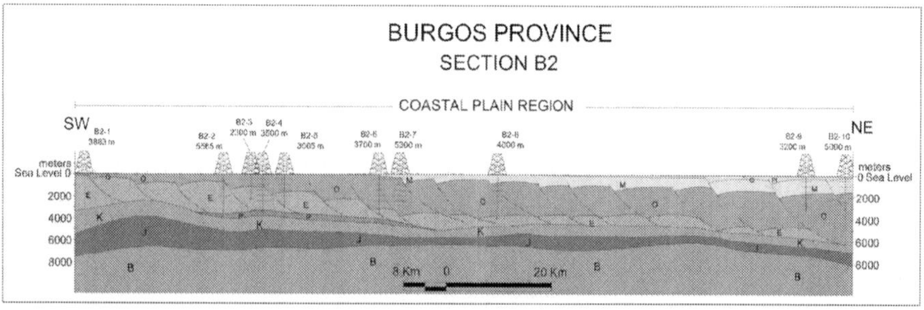

Fig. 9. Regional cross section B2. Across the section both the age of the rock units and the structural deformation are evident from west to east. B: Basement, J: Jurassic, K: Cretaceous, P: Paleocene, E: Eocene, O: Oligocene, M: Miocene, Q: Quaternary. After [31, 33, 34 y 35].

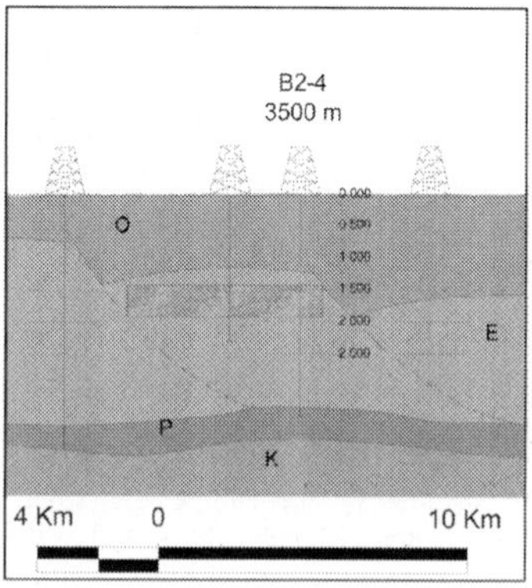

Fig. 10. Sector B2-4 from cross regional section B2. Vertical scale is in meters. K: Cretaceous, P: Paleocene, E: Eocene, O: Oligocene.

The computed petrophysical parameters are porosity 0.1, irreducible water 0.6, permeability less than 10 milidarcies (mD), density of CO2 about 675 kg/m3. The respected volume of influence is assumed based on the lithological and petrophysical homogeneities of the rock unit supported by the extrapolation of features between oil wells, and the distances imposed by stratigraphical and structural elements. The use of these parameters in the theoretical calculation of the capacity results in 1.36 giga metric tons (Gt) of CO_2 for sector B2-4 (Table 3and 4).

The same approach was used in all sections of Burgos province giving 31 potential sectors on terrigenous sequences. Sometimes several sectors are located at the same well area of influence but at different depths. The marine zone was not computerized although several projects at the shallow marine platform in the United States point out the great potential of that zone (Figure 11).

CO_2 THEORETICAL STORAGE CAPACITY IN SECTOR B2-4			
Total thickness		880	m
Net fraction		0.6	m
Net thickness		528	m
Cross section length		9 541	m
Length influence		10 000	m
Area	A	95 410 000	m^2
Volume	V	50 376 480 000	m^3
Porosity	Φ	0.1	
Irreducible water saturation	S_{wirr}	0.6	
CO_2 Density	ρCO_2	675	kg/m^3
Storage capacity in volume unit	$V_{CO2}t$	2 015 059 200.00	m^3CO_2
Storage capacity in terms of mass	MCO_2t	1.36	Gt CO_2

Table 3. Theoretical storage capacity at Sector B2-4 in the Burgos province.

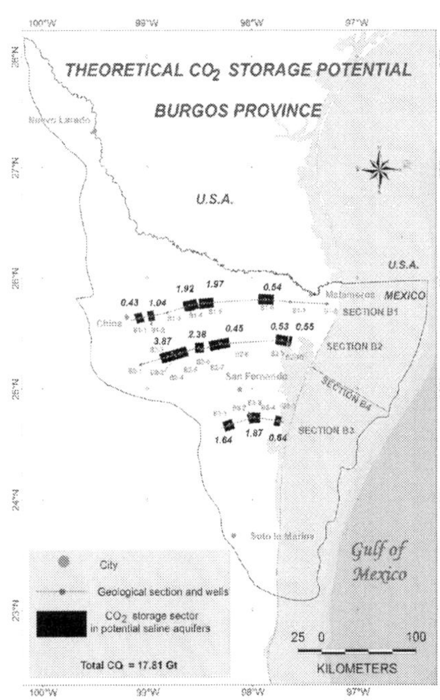

Fig. 11. Burgos province displaying the sectors (in black) of saline aquifers capable of storing CO_2. The marine zone was not quantified.

In summary, according to the geological sections, geological traps, sedimentary sequences and petrophysical parameters obtained from the Burgos province the theoretical capacity corresponds to 17.81 Gt in 31 assessed sectors (Table 4).

BURGOS PROVINCE										
CROSS SECTION	SECTOR	TRAP (*)	TARGET SEQUENCE	SIZE		GENERAL PETROPHYSICAL PARAMETERS				Partial capacity in terms of mass (Gt)
			Terrigenous	Area ($10^6 m^2$)	Thickness (m)	Effective porosity (Φ_e)	Irreducible water saturation (S_{wirr})	CO_2 Density (Kg/m^3)	Permeability (milidarcies)	
B1	B1-1	Struct	E1	76.5	402	0.05	0.6	700	<10	0.43
	B1-2	Struct	P	60.5	350	0.1	0.3	700	<30	1.04
	B1-4	Struct	E7	108.64	369.2	0.1	0.5	700	<10	1.40
	B1-4	Both	O1	60.5	93.84	0.1	0.5	650	<30	0.35
	B1-4	Both	O2	115.22	59	0.1	0.5	500	<30	0.17
	B1-5	Both	O1	117.81	376.5	0.1	0.4	700	<30	1.86
	B1-5	Both	O3	140.92	13.75	0.15	0.4	650	<60	0.11
	B1-6	Both	O3	150.57	26.5	0.08	0.3	700	<60	0.16
	B1-6	Struct	O4	82.96	110	0.1	0.4	700	<10	0.38
B2	B2-2	Both	E1	95.88	30	0.05	0.6	700	<10	0.04
	B2-2	Struct	E7	77.63	97.5	0.1	0.5	600	<10	0.23
	B2-4	Struct	E1	95.41	528	0.1	0.6	675	<10	1.36
	B2-4	Both	O1	69.7	276	0.1	0.5	600	<30	0.58
	B2-5	Both	E1	85.06	94.5	0.15	0.6	700	<10	0.34
	B2-5	Both	E7	67.38	16.25	0.1	0.5	700	<10	0.04
	B2-5	Both	O1	82.52	458	0.1	0.5	675	<30	1.28
	B2-6	Both	O1	40.68	688	0.1	0.4	700	<10	1.18
	B2-7	Both	O1	46.32	741.2	0.1	0.5	700	<30	1.20
	B2-8	Both	O2	108.2	71.5	0.1	0.5	675	<10	0.26
	B2-8	Both	O3	86.33	57.75	0.08	0.3	600	<60	0.17
	B2-8	Struct	O4	67.45	10	0.1	0.4	550	<10	0.02
	B2-9	Both	O2	111.12	77	0.1	0.5	700	<30	0.30
	B2-9	Struct	O4	57.83	97.5	0.1	0.4	690	<30	0.23
	B2-10	Struct	O4	28.42	460	0.1	0.4	700	<10	0.55
B3	B3-1	Both	O1	78.1	312	0.1	0.4	700	<30	1.02
	B3-1	Both	O2	80.91	64.4	0.1	0.5	675	<10	0.18
	B3-1	Struct	O4	44.7	250	0.1	0.4	650	<10	0.44
	B3-2	Struct	O4	36	637.5	0.1	0.4	650	<10	0.90
	B3-4	Both	O2	64.85	47	0.1	0.5	700	<10	0.11
	B3-4	Struct	O4	34.56	612.5	0.1	0.4	675	<10	0.86
	B3-5	Struct	O4	59.17	257.5	0.1	0.4	700	<10	0.64
									TOTAL	17.81

(*) Struct = Structural

Table 4. Theoretical storage capacity of the Burgos province.

3.2 Tampico-Misantla province

The Tampico-Misantla province lies in the central-east portion of Mexico. It is bordered to the north by the Burgos province and the Sierra de Tamaulipas mountain range, to the south by the mountainous fronts of the Sierra Madre Oriental folded-thrust belt and the Trans-Mexican volcanic belt, and to the east by the Gulf of Mexico [29, 37].

The deep basement of the Tampico Misantla province consists of Precambrian and Paleozoic metamorphic and granitic rocks, and faults zones caused by extensional tectonic events some of which dating back to the origin of the Gulf of Mexico [38, 39]. Also, the basement pattern shows tectonic uplifts and through structures of different shapes and sizes. Overlying the basement a thick succession of sedimentary materials have been deposited ranging from Jurassic red beds and evaporites to Cretaceous carbonate sequences originated in shelf, platform and abyssal marine facies. On top of this succession a number of terrigenous sedimentary sequences were deposited concurrently with contractional tectonic events of the Laramide orogeny, since the beginning of the Cenozoic [40]. During Cenozoic times a thick terrigenous package with minor carbonates were accumulated to fulfill the coastal plain and marine regions of the west Gulf of Mexico.

The surficial geology of the province exposes sedimentary rocks in parallel strips that run from the foothills of the Sierra Madre Oriental folded-thrust belt on the west to the existing coastal plain and marine platform regions of the Gulf of Mexico to the east. The older sedimentary rocks can be found on the west while the younger rocks are in the east. Some extrusive igneous rocks crop out on the northern and southern areas of the province (figure 12).

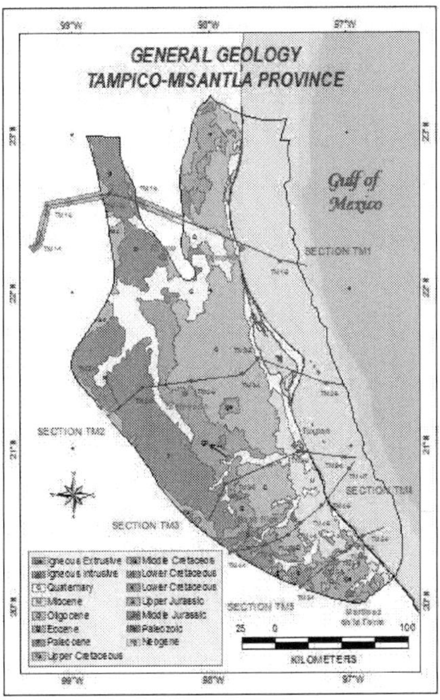

Fig. 12. Simplified geologic map of Tampico-Misantla province displaying regional cross sections and wells. After [33, 34, 35, 37, 46].

Five regional geologic cross sections were analyzed to understand the Tampico-Misantla province. Due to the similar geologic patterns showed along all regional sections, only section TM4 is presented. Section TM4 represents approximately 130 km in length of the subsurface regional geological profile, where basement faults and, horst and graben structures of different sizes are clearly revealed (Figure 13). On the western portion of section TM4 are evident the folded and thrust faulted carbonate sequences of Cretaceous age, and on the eastern side is clear the minor tectonic deformation of the Cretaceous platform carbonates as well as the Cenozoic terrigenous sequences.

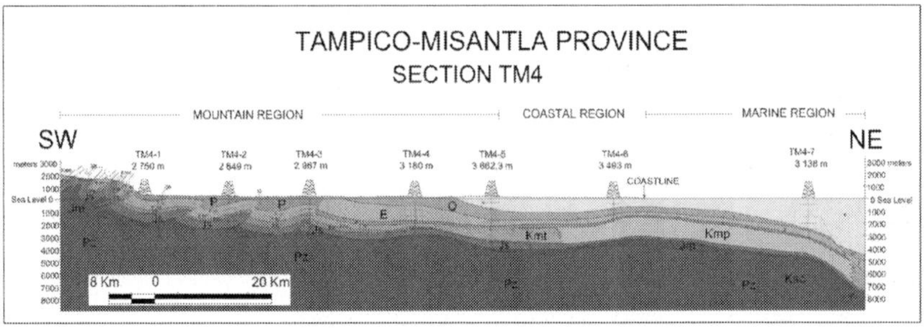

Fig. 13. Regional geologic section TM4. Mesozoic carbonate sequences are strongly deformed on the west side while Mesozoic and Cenozoic sedimentary successions are almost undeformed on the eastern side of the regional section. B: Basement, Jm: Middle Jurassic, Js. Upper Jurassic, Kic: Lower Cretaceous, Kmc: Middle Cretaceous, Ksc: Upper Cretaceous, P: Paleocene, E: Eocene, O: Oligocene, M: Miocene. After [29, 33, 34, 35, 40, 41].

In order to search sectors where saline aquifers could become potential CO_2 reservoirs the east sides of the regional sections were preferentially assessed because of their minor tectonic deformation. An example of the performed analysis is presented in sector TM4-6. Sector STM4-6 is located approximately at 2000 meters depth, and is part of carbonate reef platform sequence of Cretaceous age. The rock unit is a 635 meters package of medium to thick bedded light yellow gray fossiliferous limestone slightly deformed as an open anticline. This limestone is overlain by a sequence of thin bedded shale formed in deep basin conditions (Figure 14). The shales is interpreted as a good seal cap rock.

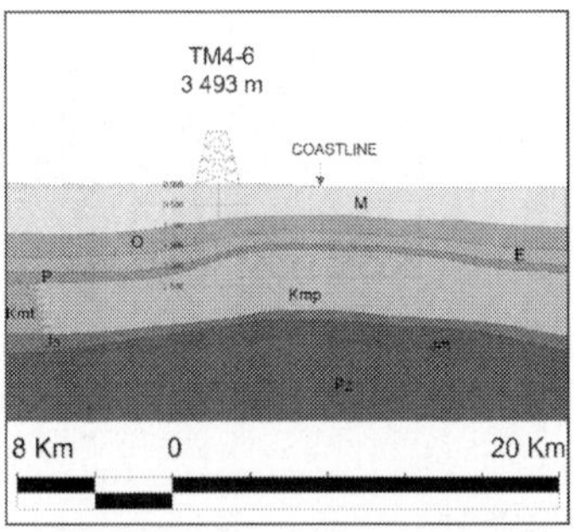

Fig. 14. Sector STM4-6 is overlaying a high basement element. Vertical scale is in meters. Pz: Paleozoic, Jm: Middle Jurassic, Js: Upper Jurassic, Kmt: Middle Cretaceous, Kmp: Middle Cretaceous, P: Paleocene, E: Eocene, O: Oligocene, M: Miocene.

The petrophysical parameters from sector STM4-6 are porosity 9%, irreducible water less than 30%, net thickness 508 meters, and CO_2 density around 693.6 kg/m^3. The use of these parameters in the theoretical calculation has resulted in 1.08 Gt (Table 5).

CO_2 THEORETICAL STORAGE CAPACITY IN SECTOR TM4-6.			
Total thickness		635	m
Net fraction		0.8	m
Net thickness		508.00	m
Cross section length		4 861.70	m
Length influence		10 000	m
Area	A	2 469 743.60	m^2
Volume	V	24 697 436 000	m^3
Porosity	Φ	0.09	
Irreducible water saturation	S_{wirr}	0.3	
CO_2 Density	ρCO_2	693.6	kg/m^3
Storage capacity in volume unit	$V_{CO2}t$	1 555 938 468.00	m^3CO_2
Storage capacity in terms of mass	MCO_2t	1.08	Gt CO_2

Table 5. Theoretical storage capacity at Sector STM4-6.

After the analysis of the entire number of regional geological sections the Tampico-Misantla province yield 12 sectors. Four of them correspond to carbonate sequences and eight to terrigenous sequences. The total CO_2 capacity estimation corresponds to 9.75 Gt (Figure 15 and Table 6).

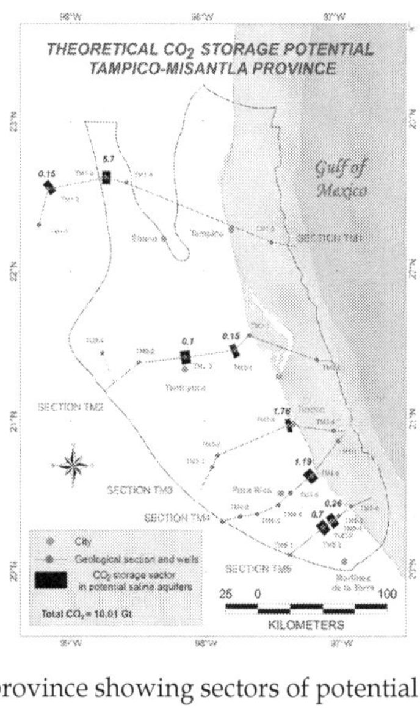

Fig. 15. Tampico-Misantla province showing sectors of potential saline aquifers capable of storing CO2 (in black).

TAMPICO-MISANTLA PROVINCE											
CROSS SECTION	SECTOR	TRAP	TARGET SEQUENCE		SIZE		GENERAL PETROPHYSICAL PARAMETERS			CO_2 Density (Kg/m^3)	Partial capacity in terms of mass (Gt)
			Terrigenous	Carbonate	Area (10^6m^2)	Thickness (m)	Effective porosity (Φ_e)	Permeability (milidarcies)	Irreducible water saturation (S_{wirr})		
TM1	TM1-3	Struct	Jm		59.2	784	0.20	300	0.20	696	5.70
	TM1-2	Struct	Ji		45.3	118.6	0.10	50	0.60	700	0.15
TM2	TM2-3	Struct		Jm2	77	26.1	0.10	60	0.30	702	0.1
	TM2-4	Strat		Jm2	0.3		0.10	60	0.30	702	0.15
TM3	TM3-3	Struct		Kmp	33.45	835.2	0.10	150	0.12	676	1.69
	TM3-3	Struct	P2 & P3		10.85	42.5	0.10	20	0.50	578	0.01
	TM3-3	Struct	E1, E2 & E3		29.2	42.7	0.12	300	0.30	426	0.06
TM4	TM4-6	Struct		Kmp	48.6	508	0.09	150	0.30	693	1.08
	TM4-6	Struct	P2 & P3		26.5	41.8	0.20	300	0.30	682	0.11
TM5	TM5-2	Strat	E1, E2 & E3		72.415	154.2	0.15	40	0.40	694	0.7
	TM5-3	Strat	E2 & E3		19.24	96.3	0.10	30	0.50	701	0.06
	TM5-3	Strat	O		41.94	95.4	0.10	30	0.30	701	0.2
										TOTAL	10.01

(*) Strat = Stratigraphic,
Struct = Structural

Table 6. Theoretical storage capacity of the Tampico-Misantla province.

3.3 Veracruz province

Veracruz province lies to the east of Mexico, sitting in the central part of the state of Veracruz. This province is bounded to the north by the Trans-Mexican volcanic belt, to the southeast by Los Tuxtlas volcanic field complex, to the west by Sierra Madre Oriental folded-thrust belt (known in this area as Sierra de Zongolica), and to the east-northeast by the Gulf of Mexico [42, 43]. The current geological context suggests a quick subsidence process along with several tectonic deformational events since Mesozoic times. The surficial geology suggests a faster subsidence process at the north of the province (Figure 16).

Six geologic sections were analyzed in order to estimate theoretical CO_2 potential capacity for this province. From the subsurface point of view, the Veracruz province can be clearly divided into two geologic subprovinces. The first subprovince is the Sierra Madre Oriental folded-thrust belt and its continuation at depth known as the "Frente Tectonico Sepultado" (Buried Tectonic Front). It is characterized by folded calcareous rocks deformed by reverse faulting. The second subprovince is known as "Cuenca Terciaria de Veracruz" (Veracruz Tertiary Basin) composed by a thick succession of interbedded shale, siltstone, sandstone and conglomerate [40, 42, 47]. This terrigenous sequence has been, in turn, affected tectonically in distinctive styles and at different depths.

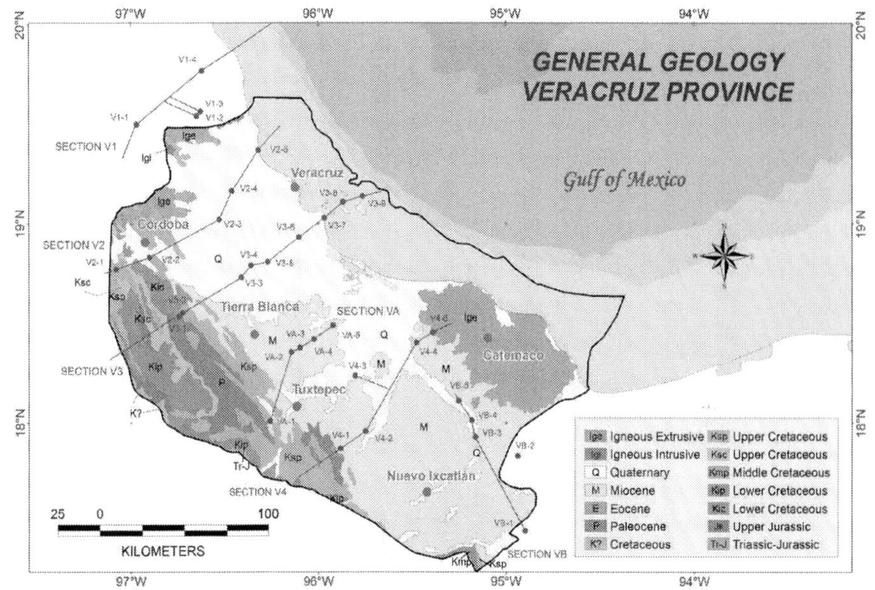

Fig. 16. Simplified geologic map of the Veracruz province, and location of regional geologic sections and wells. After [43, 33, 34, 35, 46].

For reference, figure 17 shows one of the regional sections that display structural features customarily found in the area. Section V3, about 180 km in length, lies in the middle of Veracruz province. The western half of the section displays calcareous sequences highly deformed by reverse faulting [42]. These sequences reveal Cretaceous facies from platform to basin environments. The eastern half of the section reflects terrigenous sequences wherein Paleocene and Eocene units expose reverse faulting folds.

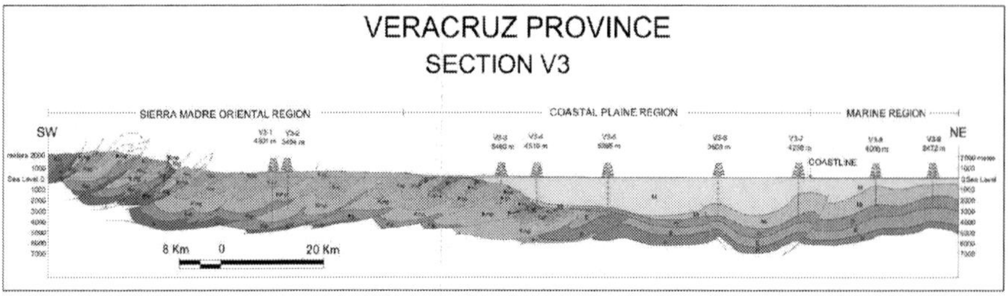

Fig. 17. Regional geological section V3. The left hand side of the regional section shows Zongolica range's Cretaceous carbonate reverse faults as well as the buried tectonic front. The opposite side reveals early Cenozoic deformed terrigenous sequences and late Cenozoic undeformed sedimentary materials. Js: Upper Jurassic, Kip: Lower Cretaceous, Kmp: Middle Cretaceous, Ksp: Upper Cretaceous, Ksc: Upper Cretaceous, P: Paleocene, E: Eocene, O: Oligocene, Mi: Lower Miocene, M: Miocene, Q: Quaternary. After [43, 33, 34, 35, 47].

Based on the regional geological sections and available oil well data, potential CO_2 storage sectors were searched in the Veracruz province. One of them is sector V2-5 in section V2. Sector V2-5 is characterized at 2450 meters depth by a lower Miocene terrigenous sequence that consists of interbedded green to gray bentonitic shale, layers of bentonite, coarse grained to conglomeratic sandstone, and conglomerate composed by fragments of gray to dark grayish brown clayey limestone and light brown bioclastic limestone [40, 43].
The conglomerate and the sandstone horizons were interpreted as potential formations to store CO_2. So, at the top of the lower Miocene sequence is a 50 meters thick horizon that is part of an anticline. It is overlain by homogeneous greenish gray shale interpreted as a good seal cap rock (Figure 18).

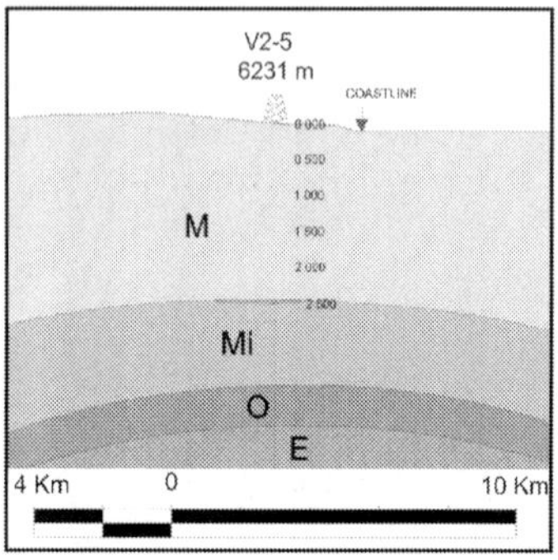

Fig. 18. Sector V2-5 showing a stratigraphic trap at the top of an anticline structure. Vertical scale is in meters. E: Eocene, O: Oligocene, Mi: Lower Miocene, M: Miocene, Q: Quaternary.

The horizon presents the following petrophysical properties, net thickness 15 meters, porosity 0.15, irreducible water 0.15, and permeability 200mD. The assumed CO_2 density for that depth of storage was 700 Kg/m^3. The use of these parameters in the theoretical calculation of the capacity resulted in 0.03 Gt (Table 7).

CO_2 THEORETICAL STORAGE CAPACITY IN SECTOR V2-5			
Total thickness		50	m
Net fraction		0.3	m
Net thickness		15	m
Cross section length		2 500	m
Length influence		10 000	m
Area	A	25 000 000	m^2
Volume	V	375 000 000	m^3
Porosity	Φ	0.15	
Irreducible water saturation	S_{wirr}	0.15	
CO_2 Density	ρCO_2	700	kg/m^3
Storage capacity in volume unit	V_{CO2t}	47 812 500.00	m^3CO_2
Storage capacity in terms of mass	MCO_2t	0.03	Gt CO_2

Table 7. Theoretical storage capacity at Sector V2-5 in the Veracruz province.

According to the theoretical calculations carried out in the Veracruz province resulted 21 sectors with CO_2 capacity potential (Figure 19). Five of the sectors correspond to carbonate sequences, and the remaining 16 are terrigenous sequences. The estimated capacity targets reach 15.23 Gt (Table 8).

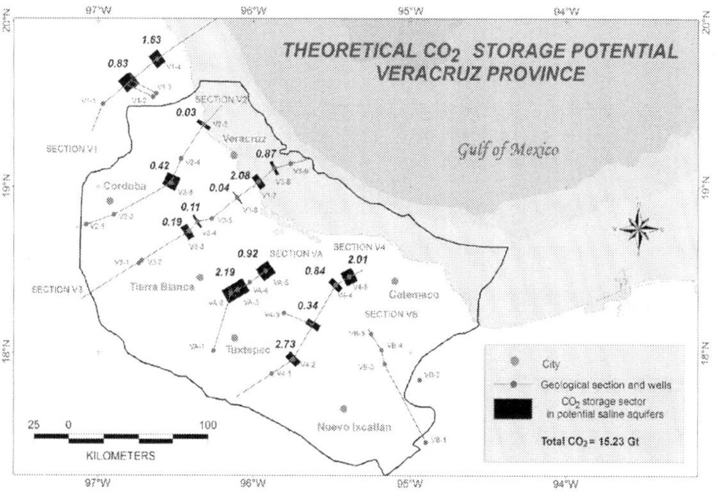

Fig. 19. Sectors with CO_2 storage potential in saline aquifers at the Veracruz province.

3.4 Sureste province

The Sureste province is situated in the southeastern region of Mexico on the southern edge of the Gulf of Mexico. This province is bordered to the south by the Sierra de Chiapas mountainous range, to the east by the Yucatan Peninsula, to the west by the Veracruz province, and to the north and northeast by the Gulf of Mexico. The Sureste province comprises both mainland and offshore areas. In mainland the extensive geological exposures show evidence of the last episode of sedimentary infilling, therefore, most of the area is covered mainly by late Cenozoic sedimentary deposits (Figure 20).

The internal subsurface configuration of the province is characterized by very deep and fragmented basement affected by different tectonic deformational events. At depth the Sureste province is divided into four subprovinces: *Salina del Istmo, Comalcalco, Reforma-Akal* and *Macuspana* [40, 44, 45]. The basement of the province consists of crystalline rocks of Precambrian and Paleozoic age [30, 49] most of which are covered by Mesozoic rock units composed of red beds, marine evaporites and carbonates of basin and platform marine facies [53]. Overlying the Mesozoic rocks are Paleogene terrigenous deposits of deep and shallow marine, deltaic, lagoonal and even alluvial facies [51, 52]. In addition, there are terrigenous sequences belonging to deltaic, lagoonal and shallow marine sedimentary facies that cover all the earlier deposits [40, 52, 54].

Six regional geologic cross sections (SE1, SE2, SE3, SE4, SE5 and SE6) were analyzed in order to estimate theoretical CO_2 potential capacity in the province. The regional cross sections show that the sedimentary sequences from Jurassic to Oligocene-Lower Miocene were folded and reversely faulted. Also, it is evident that the younger late Cenozoic terrigenous sequences were faulted, but this time, under an extensional tectonic regime. The entire province was first under contractional tectonic regimes, and then it was affected by extensional tectonic events during erosion-sedimentation stages.The position of the Sureste province could be viewed in terms of the jointly evolution of a passive continental margin associated to a strike-slip and a subduction margins both related to the plate tectonic interaction at the pacific region of Mexico. However, the complete and detailed tectonic history of the province is not yet well known. The subsurface stratigraphical and structural complexity is shown in Section SE2 which is approximately 135 kilometers long, is located in the middle of the province, and is running along a northwest-southeast line (Figure 21).

VERACRUZ PROVINCE											
CROSS SECTI-ON	SEC-TOR	TRAP (*)	TARGET SEQUENCE		SIZE		GENERAL PETROPHYSICAL PARAMETERS				Partial capacity in terms of mass (Gt)
			Terrige-nous	Carbon-ate	Area ($10^6 m^2$)	Thick-ness (m)	Effe-ctive porosi-ty (Φ_e)	Irredu-cible water sat. (S_{wirr})	CO_2 Dens-ity (Kg/m^3)	Perme-ability (mili-darcies)	
V1	V1-3	Strat		Kmp	52.7	202.5	0.1	0.04	700	<700	0.72
	V1-3	Strat	P		94.5	17.4	0.14	0.3	700	<60	0.11
	V1-4	Strat	E		78.15	285	0.15	0.25	650	<70	1.63
V2	V2-3	Struct		Kmp	17	27	0.07	0.04	700	<600	0.02
	V2-3	Strat		Ksp	56	387	0.03	0.7	700	<200	0.14
	V2-3	Strat	P		56	86.46	0.15	0.35	550	<40	0.26
	V2-5	Struct	Mi		25	15	0.15	0.15	700	<200	0.03
V3	V3-3	Struct		Kmp	26.3	10	0.07	0.2	700	<300	0.01
	V3-3	Strat		Ksp	43.6	147.2	0.08	0.4	600	<200	0.18
	V3-4	Strat	Mi		16	104	0.12	0.18	700	<300	0.11
	V3-6	Struct	Mi		10.4	54.9	0.12	0.18	700	<300	0.04
	V3-7	Struct	Mi		46	723.75	0.12	0.2	650	<300	2.08
	V3-8	Struct	Mi		21.65	698	0.12	0.2	600	<300	0.87
V4	V4-2	Strat	Mi		76.9	312	0.25	0.3	650	<80	2.73
	V4-3	Struct	Mi		43.75	115	0.12	0.2	700	<200	0.34
	V4-4	Struct	Mi		43.6	280	0.12	0.18	700	<300	0.84
	V4-5	Struct	Mi		83.7	348	0.12	0.18	700	<300	2.01
VA	VA-2	-	P		50	12	0.25	0.3	700	<20	0.07
	VA-3	-	Mi		100	75	0.12	0.1	700	<300	0.57
	VA-3	-	E		100	138	0.2	0.2	700	<50	1.55
	VA-5	-	Mi		100	133.5	0.12	0.18	700	<300	0.92
										TOTAL	15.23

(*) Strat = Stratigraphic, Struct = Structural

Table 8. Theoretical storage capacity of the Veracruz province.

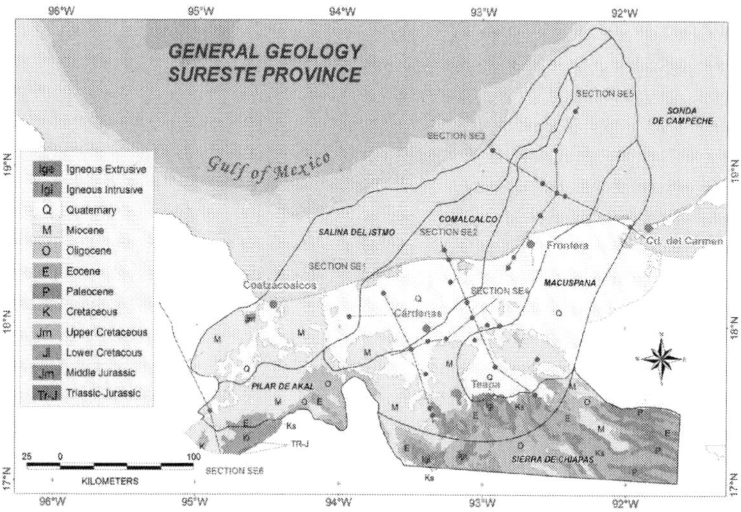

Fig. 20. Simplified geologic map of the Sureste province. It shows the location of regional geologic sections, wells, and limits of subprovinces: Salina del Istmo, Comalcalco, Macuspana and Pilar de Akal. After [33, 34, 35, 41, 44, 45, 46].

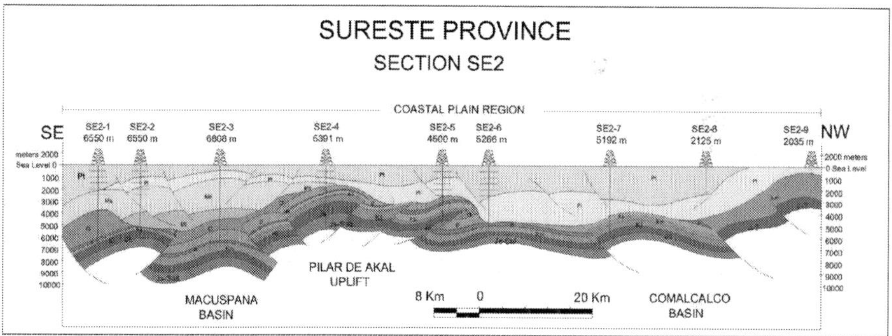

Fig. 21. Regional cross section SE3 depicting complex tectonic deformation in the Sureste province. Js: Upper Jurassic, Ki: Lower Cretaceous, Km: Middle Cretaceous, Ks: Upper Cretaceous, P: Paleocene, E: Eocene, O: Oligocene, Mi: Lower Miocene, Ms: Upper Miocene, Pl: Pliocene, Pt: Pleistocene, Q: Quaternary. After [34, 35, 40, 51].

Section SE2 traverses the Comalcalco, Macuspana and Reforma-Akal uplift subprovinces. The Comalcalco and Macuspana are sedimentary basins separated in turn by the Reforma-Akal uplift. In the three subprovinces there are from Jurassic through Oligocene folded and reverse faulted sedimentary sequences. At the Macuspana basin there are Miocene terrigenous sequences affected by both steep and gently dipping normal faults. In contrast, these terrigenous sediments are non-existent at the Comalcalco basin, therefore indicating synchronous erosion and sedimentation processes. At the Comalcalco basin the Pliocene and Plesitocene sediments can reach up to five kilometers in thickness, and the regularly spaced faults do not meet at the surface. All along the cross section is evident that the development of the basins is linked to the widespread fault systems and to subsidence mechanisms.

During the screening and selection of the sectors to estimate the CO_2 capacity, several stratigraphic and anticline traps structures were found. One of them is presented in figure 22 to illustrate the procedure. The sector SE2-4 consists of an anticline structure verging in northeast direction with an average axis orientation of N 300°. The anticline includes rock units from Jurassic to Oligocene times that are marked first by reverse faulting episode, and then by a regional unconformity. The unconformity is overlain by Miocene and Pliocene rock units.

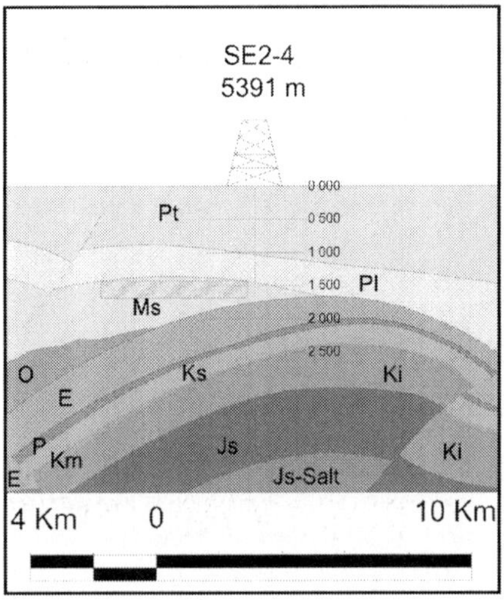

Fig. 22. Sector SE2-4 showing the location of the CO_2 storage target in cross section SE2 of the Sureste province. Vertical scale in meters. Js: Upper Jurassic, Ki: Lower Cretaceous, Km: Middle Cretaceous, Ks: Upper Cretaceous, P: Paleocene, E: Eocene, O: Oligocene, Ms: Upper Miocene, Pl: Pliocene, Pt: Pleistocene.

The CO_2 storage target is in a wedge of late Miocene well-bedded sequence about 280 meters thick and located 1550 meters deep. The storage sequence consists of a light gray, medium to coarse-grained, medium-bedded sandstone interbedded with occasional gray-greenish shale containing mollusks and lignite fragments. The sandstone is overlain by a wide package of greenish gray shale of Pliocene age and interpreted as the seal layer. The petrophysical parameters of the sandstone target sequence are net thickness about 240 meters, clay content less than 4 %, porosity (Φ_e) about 30%, irreducible water saturation (S_{wirr}) less than 20% and permeability about 60 miliDarcys (mD)(Table 9). According to the 1550 meters sandstone depth where the CO_2 density is approximately 681 Kg/m^3, the theoretical storage capacity is close to 1.84 Gt (million tons of CO_2).

CO_2 THEORETICAL STORAGE CAPACITY IN SECTOR SE2-4.			
Total thickness		283	m
Net fraction		0.85	m
Net thickness		240.55	m
Cross section length		4 573.47	m
Length influence		10 000	m
Area	A	1 100 148.21	m^2
Volume	V	11 001 482 085	m^3
Porosity	Φ	0.3	
Irreducible water saturation	Swirr	0.18	
CO_2 Density	ρCO2	681	kg/m^3
Storage capacity in volume unit	VCO2t	2 706 364 592.91	m^3CO_2
Storage capacity in terms of mass	MCO2t	1.84	Gt CO_2

Table 9. Theoretical storage capacity at Sector SE2-4, in the Sureste province, is near 1.84 million tons of CO_2.

On the basis of the estimations conducted in the Sureste province resulted 17 sectors with CO_2 capacity potential (Figure 23). Six of them are within offshore subsurface lands. The total capacity estimate is around 24.10 Gt on terrigenous sedimentary sequences (Table 10).

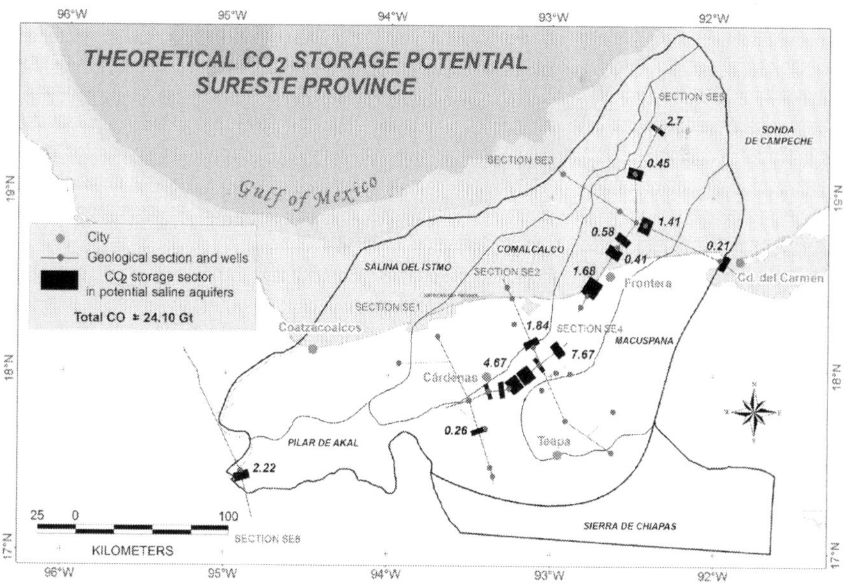

Fig. 23. Sectors shown in black with CO_2 storage potential in saline aquifers at the Sureste province.

SURESTE PROVINCE										
SECTION	SECTOR	TRAP (*)	TARGET SEQUENCE	SIZE		GENERAL PETROPHYSICAL PARAMETERS			CO_2 Density Kg/m³	Partial capacity in terms of mass (Gt)
			Terrigenous	Area ($10^6 m^2$)	Thickness (m)	Effective porosity (Φ_e)	Permeability (milidarcies)	Irreducible water saturation (S_{wirr})		
SE2	SE2-4	Struct	M	1.1	240.55	0.30	60	0.18	681	1.84
SE3	SE3-4	Struct	M	0.98		0.30	60	0.18	580	1.41
	SE3-6	Struct	O	0.3	308.70	0.05	45	0.45	591.5	0.21
SE4	SE4-1	Struct	M	0.22		0.30	60	0.18	472	0.26
	SE4-3	Struct	M	0.17		0.20	35	0.34	692.5	0.16
	SE4-3_4	Struct	M	0.25		0.20	35	0.34	682	0.23
	SE4-4	Struct	M	1.45		0.20	35	0.34	685	1.31
	SE4-4_5	Struct	M	1.72		0.30	60	0.18	688.5	2.92
	SE4-5	Struct	M	0.12		0.30	60	0.18	658.5	7.67
	SE4-6	Struct	M	4.73	811.32	0.30	60	0.18	426	0.05
SE5	SE5-2	Struct	M	0.67		0.30	60	0.18	670	1.11
	SE5-2_3	Struct	M	0.37		0.30	60	0.18	615	0.57
	SE5-3	Struct	M	0.30		0.30	60	0.18	544	0.41
	SE5-3_4	Struct	M	0.38		0.30	60	0.18	615	0.58
	SE5-5	Struct	M	0.29		0.30	60	0.18	620	0.45
	SE5-6	Struct	M	1.60	522.40	0.30	60	0.20	702	2.70
SE6	SE6-5	Struct	M	5.47	998.51	0.10	25	0.40	676.5	2.22
(*) Structural									TOTAL	24.10

Table 10. Theoretical storage capacity of the Sureste province.

3.5 Yucatan province

The Yucatan province is bounded to the northeast by the Campeche Escarpment (which is formed on the edge of the marine continental shelf), to the east by the Caribbean Sea (where the marine platform is quite narrow), to the west by the Sonda de Campeche and to the south and southeast by the Sierra de Chiapas mountain ranges, Los Chuchumatanes Dome in Guatemala, and the Maya Mountains of Belize [43, 16, 55]. The area of study comprises the onshore portion known as Yucatan Peninsula and some offshore submerged areas in the Sonda de Campeche and the Yucatan marine platform regions (Figure 24).

The geology of the province can be characterized in subsurface terms by a huge basement block composed of Paleozoic rocks [43]. This crustal tectonic element has been present since the origin of the Gulf of Mexico [56]. On top of the basement, Jurassic evaporites, Cretaceous carbonates, as well as both Tertiary carbonates and terrigenous sedimentary sequences were deposited [57, 38, 58]. The sedimentary sequences were not under intense tectonic stress since they show a nearly horizontal depositional pattern and some minor faults. However, at the surface level, the central part of the huge province presents normal faults of considerable length that could bear testimony of extensional tectonic events which affected Mesozoic and lower Tertiary rocks. Under this geological context, four long regional geologic cross sections were analyzed to estimate the CO_2 storing capacity in the Yucatan Province.

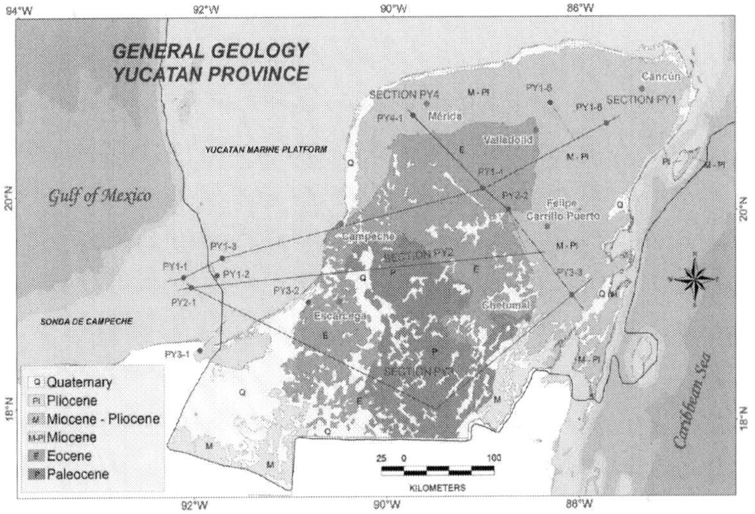

Fig. 24. Simplified geology map of Yucatan province showing regional geologic sections and wells. After [40, 43, 34, 35, 33, 55, 63].

The Yucatan province exposes a very wide and nearly horizontal sedimentary Mesozoic and Cenozoic rock sequences, where the topographic elevations rarely exceeds 200 meters above sea level. Because of this quite regular geologic homogeneity it is believed that the Yucatan peninsula remained stable throughout its geologic history. In contrast, at the edge of the basement block in the Sonda de Campeche, the offshore submerged area display Miocene contractional and extensional tectonic deformations linked to the geologic evolution of the Sureste province [59, 60]. The regional cross section Y2, approximately 400 km in length, depicts geological features frequently found in the entire province. At the offshore area within the Sonda de Campeche region gently folds structures in Mesozoic and early Cenozoic strata indicate a tectonic regime not so intense. Later, Cenozoic sequences of rocks denote normal faults systems that affected almost the complete stratigraphic column (Figure 25).

Sector PY2-1 illustrates one of the selected potential sectors where saline aquifers could eventually become CO_2 reservoirs. The Miocene terrigenous sequence is characterized by a thick succession of light colored sandstone interbedded with calcareous breccias and some layers of shale that alternate with calcareous arkoses lenses (Figure 26). Within the Miocene sequence, only the sandstone horizons were considered for the calculations of CO_2 storage. The Miocene sequence is overlain by a thick package of Pliocene sediments composed of massive carbonaceous clay interbedded with peat layers and blue color clays. This package of sediments is interpreted as the seal rock unit.

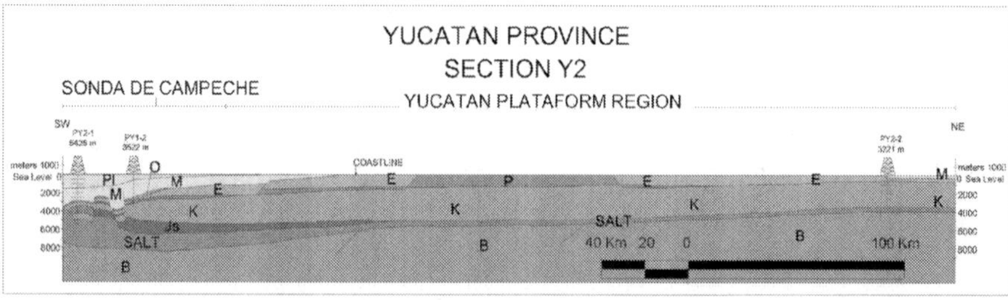

Fig. 25. Regional geological cross section Y2 showing Mesozoic sedimentary units gently deformed while the late Cenozoic sedimentary accumulations affected by extensional events within the offshore submerged region in the Sonda de Campeche. B: Basement, Js: Upper Jurassic, K: Cretaceous, P: Paleocene, E: Eocene, O: Oligocene, M: Miocene, Pl: Pliocene. After [34, 35, 33, 55].

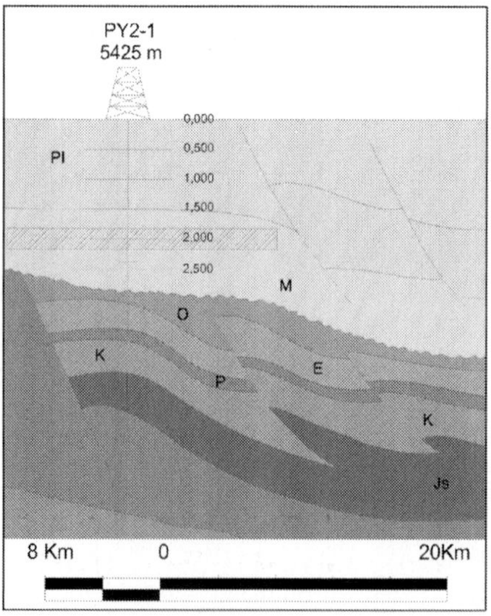

Fig. 26. Sector PY2-1 showing the location of the CO_2 storage target. Vertical scale is in meters. Js: Upper Jurassic, K: Cretaceous, P: Paleocene, E: Eocene, O: Oligocene, M: Miocene, Pl: Pliocene.

The net thickness of the target sequence is about 353 meters with porosity (Φ_e) about 10% and irreducible water saturation (S_{wirr}) 30%. Based on these parameters the theoretical capacity is 3.25 Gt of CO_2 in sector PY2-1(Table 11).

CO_2 THEORETICAL STORAGE CAPACITY IN SECTOR PY2-1			
Total thickness		884	m
Net fraction		0.40	m
Net thickness		353.60	m
Cross section length		18 793.18	m
Length influence		10 000	m
Area	A	6 645 268.45	m^2
Volume	V	66 452 684 480	m^3
Porosity	Φ	0.10	
Irreducible water saturation	Swirr	0.30	
CO_2 Density	ρCO2	699.2	kg/m^3
Storage capacity in volume unit	VCO2t	4 651 687 913.60	m^3CO_2
Storage capacity in terms of mass	MCO2t	3.25	Gt CO_2

Table 11. Theoretical storage capacity at Sector PY2-1 is near 3.25 million tons of CO_2.

The analyses of the Yucatan province yield seven sectors capable of storing CO_2 with a total theoretical capacity estimate of 14.44 Gt. Most of them are located in the offshore submerged lands of the Sonda de Campeche (Figure 27). The sectors are divided in terrigenous rock sequences with 10.46 Gt and carbonate sequences with 3.98 Gt (Table 12).

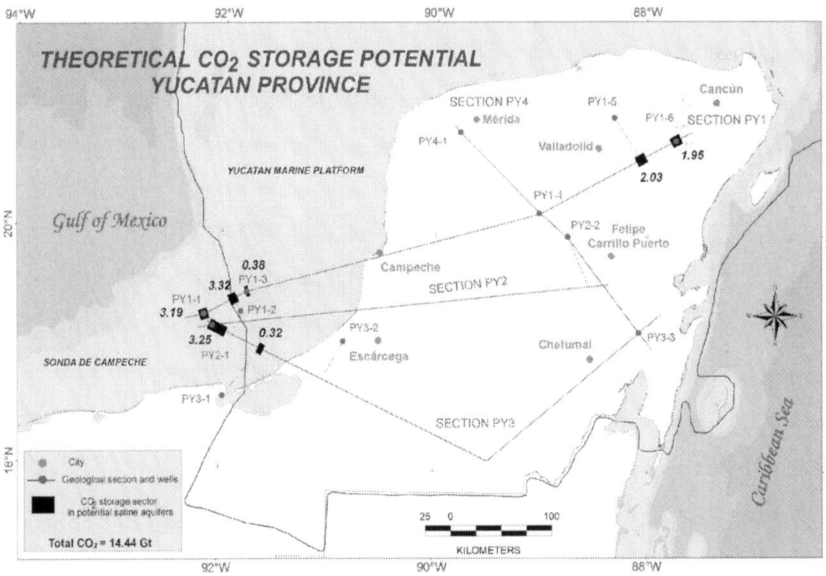

Fig. 27. Sectors (shown in black) with CO_2 storage potential in saline aquifers, Yucatan province.

YUCATAN PROVINCE											
CROSS SECTION	SECTOR	TRAP (*)	TARGET SEQUENCE		SIZE		GENERAL PETROPHYSICAL PARAMETERS			CO_2 Density (Kg/m³)	Partial capacity in terms of mass (Gt)
			Terrigenous	Carbonate	Area (10^6 m²)	Thickness (m)	Effective porosity (Φ_e)	Permeability (milidarcies)	Irreducible water saturation (S_{wirr})		
PY1	PY1-1	Strat	M		6.6	760	0.10	30	0.30	692	3.19
	PY1-2	Strat	M		7.2	837	0.10	30	0.30	653	3.32
	PY1-3	Strat	M		9.5	283.12	0.10	30	0.30	575	0.38
	PY1-5	Strat		K	3.3	320	0.10	200	0.13	702	2.03
	PY1-6	Strat		K	3.2	308	0.10	200	0.13	701.5	1.95
PY3	PY2-1	Strat	M		6.6	353.60	0.10	30	0.30	699.2	3.25
	PY3-1	Strat	M		0.65		0.10	30	0.30	691.5	0.32
(*) Stratigraphic										TOTAL	14.44

Table 12. Theoretical storage capacity in Yucatan province is 14.44 million tons of CO_2.

In summary, the theoretical CO_2 capacity estimates in Mexico stands currently at 81.59 Gt on terrigenous and calcareous sequences located within the outlined inclusion zones. The total assessed sectors are 88 with possibilities of CO_2 storage in potential saline aquifers (Table 13). The assessed sectors in terrigenous sedimentary sequences are 77 while in carbonate sequences are 11.

PROVINCE	THEORETICAL CO_2 STORAGE POTENCIAL (Gt)	SECTORS ASSESSED
Burgos	17.81	31
Tampico-Misantla	10.01	12
Veracruz	15.23	21
Sureste	24.10	17
Yucatan	14.44	7
TOTAL	81.59	88

Table 13. Summary of theoretical storage potential in saline aquifers of Mexico.

4. Conclusions

In Mexico the energy sector is responsible of more than 70% of the carbon dioxide emissions. In order to address the possibility of storing such anthropogenic CO_2 in deep underground geologic formations three lines of analysis were performed. First, the type, location and magnitude of CO_2 sources indicate approximately 216 Gt of CO_2 emissions coming from 1860 point sources. Second, five out of twelve geological provinces were analyzed. The assessed provinces are Burgos, Tampico-Misantla, Veracruz, Sureste and

Yucatan which have the best favorable conditions for underground CO_2 storage in sedimentary rock successions of Mesozoic and Tertiary age. They are geologically well defined and located within the coastal plain region around the western portion of Gulf of Mexico. Third, theoretical storage capacities in potential saline aquifers sectors were estimated for each geological province. The theoretical CO_2 storage estimates and the number of assessed sectors are: Burgos province 17.81 Gt in 31 sectors, Tampico-Misantla province 10.01 Gt in 12 sectors, Veracruz province 15.23 Gt 21 sector, Sureste 24.10 Gt in 17 sectors and Yucatan province 14.44 Gt in 7 sectors. The total theoretical CO_2 storage potential currently stands at 81.59 Gt within 88 assessed sectors for the entire nation. During the CO_2 storage capacity estimations, it became clear that some areas yielded more and better quality data than others. Therefore, it is acknowledged that these data sets are not complete. However, it is anticipated that CO_2 storage capacity estimates, geological formation maps as well as regional geological cross sections will be updated as new information, particularly oil wells data, are acquired and methodologies for CO_2 storage capacity estimates are improved in Mexico.

5. References

[1] M. Dávila, O. Jiménez, V. Arévalo, R. Castro and J. Stanley. "A preliminary selection of regions in Mexico with potential for geological carbon storage". *International Journal of Physical Science*, vol.5, num.5, pp.408-414, 2010.

[2] DOE (U.S. Departament of Energy). "2010 Carbon sequestration Atlas of the United States and Canada". Third edition, NETL (National Energy Technology Laboratory), 160p., 2011.

[3] RETC (Pollutant Release and Transfer Inventory) database. Secretaría del Medio Ambiente y Recursos Naturales, Mexico, Internal Report, 2008.

[4] SEMARNAT (Ministry of the Environment and Natural Resources). "Fourth National Communication to the United Nations Framework Convention on Climate Change (2006)". Instituto Nacional de Ecología, Mexico, 274p. Primera edición 2009. Available: http://www.ine.gob.mx.

[5] C.A. Hendricks and K. Blok. "Underground storage of carbon dioxide". *Energy Convers Manage*, vol.34, pp.949-957, 1993.

[6] S. Bachu."Sequestration of CO_2 in geological media: criteria and approach for site selection in response to climate change". *Energy Convers Manage*, vol. 41, pp. 953-970, 2000.

[7] J. Bradshaw J. and A. Rigg. "The GEODETIC Program: research into geological sequestration of CO_2 in Australia". *Environmental Geosciences*, vol. 8, pp. 166-176, 2001.

[8] S. Bachu and S. Stewart. "Geological sequestration of anthropogenic carbon dioxide in the Western Canada sedimentary basin: suitability analysis". *Canadian Journal of Petroleum Technology*, vol. 41, num.2, pp.32-40, 2002.

[9] S. Bachu. "Screening and ranking of sedimentary basins for sequestration of CO_2 in geological media in response to climate change". *Environmental Geology*, vol. 44,pp.277-289, 2003.

[10] S. Bachu, D. Bonijoly, J. Bradshaw, R. Burruss, S. Holloway, N.P. Christensen, and M. Mathiassen. "CO_2 storage capacity estimation: methodology and gaps". *International Journal Greenhouse Gas Control*, vol.1, pp. 430-443, 2007.

[11] S. Bachu. "CO_2 storage in geological media: role, means, status and barriers to deployment". *Progress Energy Combustion Science*, vol. 34, pp. 254-273, 2008.

[12] A.W. Bally."Musings over sedimentary basin evolution". *Philosophical Transactions of the Royal Society of London*, vol. 305, pp. 325-338, 1982.
[13] R. Ingersoll "Tectonics of sedimentary basins". *Geological Society of America Bulletin*, vol. 100, pp. 1704-1719, 1988.
[14] Y.L. Leonov and Y.A. Voloz. *Sedimentary basins: study methods, structure and evolution.* Nauchnyi Mir, 525p., 2004.
[15] Ch. French and Ch. Schenk (compilers). "Map showing geology, oil and gas fields, and geologic provinces of the Gulf of Mexico Region". USGS Open-File Report 97-470-L, 1997.
[16] PEMEX (Petróleos Mexicanos). "Provincias petroleras de México". Pemex Exploración y Producción, México, Versión 1.0, 11p., 2010.
[17] F. Campa and P. Coney."Tectonostratigraphic terranes and mineral resource distribution in Mexico". *Canadian Journal of Earth Sciences*, vol.20,pp.1040-1051, 1983.
[18] F. Ortega, L. Mitre, J. Roldán, J. Aranda, D. Morán, S. Alaníz, A. Nieto. "Texto Explicativo de la Quinta Edición de la Carta Geológica de la República Mexicana, Escala 1:2´000,000". Instituto de Geología, UNAM-Consejo de Recursos Minerales, SEMIP, México, 1992.
[19] R. Sedlock, F. Ortega and R. Speed. "Tectonostratigraphic terranes and tectonic evolution of Mexico". *Geological Society of America Special Paper 278*, 153 p., 1994.
[20] J. Gale, N.P. Christensen, A. Cutler and T. Torpe. "Demonstrating the potential for geological storage of CO_2: The Sleipner and GESTCO project". *Environmental Geosciences*, vol.8, num.3, pp.160-165, 2001.
[21] R. Tarkowski, B. Uliasz, and A. Wojcicki. "CO_2 storage capacity of deep aquifers and hydrocarbon fields in Poland". *Energy Procedia*, vol.1, pp.2671-2677, 2009.
[22] J. Bradshaw, C. Boreham, and F. La Pedalina (2005). "Storage retention time of CO_2 in sedimentary basins: examples from petroleum systems". Available: http:/uregina.ca/ghgt7/PDF/papers/peer/427.pdf
[23] L.G.H. van der Meer and P.J. Egberts. "A general method for calculating subsurface CO_2 storage capacity". Presented at the 2008 Offshore Technology Conference. OTC 19309, May 2008.
[24] S. Brennan and R. Burruss. "Specific Sequestration volumes: a useful tool for CO_2 storage capacity assessment". USGS Open-File Report 03-452, 2009.
[25] S. Brennan, R. Burruss, M.D. Merrill, P.A. Freeman and L.F. Ruppert. "A probabilistic assessment methodology for the evaluation of geologic carbon dioxide storage". USGS Open-File Report 2010-1127, 31p., 2010.
[26] GCCSI (Global CCS Institute). "The status of CCS projects". Interim Report 2010, 26p., 2010. www.cslforum.org
[27] DOE (U.S. Department of Energy). "Best practices for: Geologic Storage Formation Classification: Understanding Its Importance and Impacts on CCS Opportunities in the United States". NETL (National Energy Technology Laboratory), 54p., 2010.
[28] DOE (U.S. Department of Energy). "2008 Carbon sequestration Atlas of the United States and Canada". 2nd edition. NETL (National Energy Technology Laboratory), 140p., 2008.
[29] E. López Ramos. *Geologia de Mexico. Tomo II.* Edicion Escolar: Mexico, 454 p., 1979.
[30] R.T. Buffler and D.S. Sawyer. "Distribution of crust and early history, Gulf of Mexico Basin". *Gulf Coast Association Geological Societies Transactions,*vol. 35, p.333-444, 1985.
[31] PEMEX (Petróleos Mexicanos). "La Provincia Petrolera Burgos". Pemex Exploración y Producción, México, Versión 1.0, 27p., 2010.

[32] B. Ortiz."Interpretación estructural de una sección sísmica en la región Arcabuz-Culebra de la Cuenca de Burgos, NE de México". *Revista Mexicana de Ciencias Geológicas*, vol. 21, num. 2, pp. 226-235, 2007.

[33] SGM (Servicio Geológico Mexicano) Cartas geológico mineras. Escala 1:250 000. Avaible: http:// mapasims.sgm.gob.mx:8399/mapasEnLinea/

[34] CFE (Comisión Federal de Electricidad). "Integración de un Atlas de las principales cuencas sedimentarias de México". Technical Report. Convenio CFE-IPN-001/2009, enero 2010a.

[35] CFE (Comisión Federal de Electricidad). "Geología del subsuelo de las principales zonas de las cuencas sedimentarias marinas y continentales alrededor del Golfo de México". Technical Report. Convenio CFE-IPN-001/2010, diciembre 2010b.

[36] E. Lopez-Ramos and J.C. Guerrero. "Paleogeografia y tectonica del Mesozoico de Mexico". *Revista del Instituto de Geologia*, vol. 5, pp. 158-177, 1981.

[37] PEMEX (Petróleos Mexicanos). "Provincia Petrolera Tampico Misantla". Pemex Exploración y Producción, Versión 1.0. 48 p. 2010.

[38] A Salvador. "Late Triassic-Jurassic Paleogeography and Origin of the Gulf of Mexico Basin". *American Association of Petroleum Geologists Bulletin*, vol.71, p.419-451, 1987.

[39] J.L. Pindell and J. F. Dewey. "Permo-Triassic reconstruction of western Pangea and the evolution of the Gulf of Mexico/Caribbean region". *Tectonics*, vol.1, p.179-211, 1982.

[40] J. Santiago, J. Carrillo and B. Martell. "Geología Petrolera de México". In: *Evaluación de Formaciones en México*, D. Marmissolle-Daguerre, Ed. Schlumberger, 1984, p. 1-36.

[41] INEGI (Instituto Nacional de Estadística, Geografía e Informática). "Atlas de Mapas Geológicos de Mexico". Ministry of Budget and Programming, Mexico, 1981.

[42] PEMEX (Petróleos Mexicanos). "Provincia petrolera Veracruz". Pemex Exploración y Producción, México, Versión 1.0, 38 p., 2010.

[43] E. López Ramos. *Geologia de Mexico. Tomo III.* Edicion Escolar: Mexico, 453 p., 1979.

[44] W.A. Ambrose, T.F. Wawrzyniec, K. Fouad, S.C. Talukdar, R.H. Jones, D.C. Jennette, M.H. Holtz, S. Sakurai, S.P. Dutton, D.B. Dunlap, E.H. Guevara, J. Meneses, J. Lugo, L. Aguilera, J. Berlanga, L. Miranda, J. Ruiz, R. Rojas and H. Solís. "Geologic framework of upper Miocene and Pliocene gas plays of the Macuspana Basin, Southeastern Mexico". *American Association of Petroleum Geologists Bulletin*, vol.87, num.9, pp.1411-1435, 2003.

[45] PEMEX (Petróleos Mexicanos). "Provincia Petrolera Golfo de México Profundo". Pemex Exploración y Producción, México, Versión 1.0, 26p., 2010.

[46] PEMEX (Petróleos Mexicanos). "Provincias Geológicas de México". Pemex Exploración y Producción, México, Versión 1.0, 18p., 2010.

[47] PEMEX (Petróleos Mexicanos). Provincia Petrolera Cinturón Plegado de la Sierra Madre Oriental. Pemex Exploración y Producción, Versión 1.0, 14 p., 2010.

[48] SPP (Secretaría de Programación y Presupuesto). "Atlas Nacional del Medio Físico". Secretaría de Programación y Presupuesto. Gobierno de México, 224 p., 1981.

[49] J.L. Pindell. "Alleghanian reconstruction and subsequent evolution of the Gulf of Mexico, Bahamas, and Proto-Caribbean". *Tectonics*, vol. 4, pp.1-39, 1985.

[50] J. L. Pindell and L. Kennan, "Rift models and the salt-cored marginal wedge in the northern Gulf of Mexico: implications for deep water Paleogene Wilcox deposition and basinwide maturation". In: *Transactions of the 27th GCSSEPM Annual Bob F. Perkins Research Conference: The Paleogene of the Gulf of Mexico and Caribbean Basins: Processes, Events and Petroleum Systems*. L. Kennan, J. L. Pindell and N. C. Rosen (eds), pp. 146-186, 2007.

[51] F.J. Ángeles, N. Reyes, J.M. Quezada and J.R. Meneses. "Tectonic evolution, structural styles and oil habitat in the Campeche Sound, Mexico". *Transactions of the Gulf Coast Associations of Geological Societies*, vol. XLIV, pp.53-62, 1994.

[52] R. Padilla."Evolución geológica del sureste mexicano desde el Mesozoico al presente en el contexto regional del Golfo de México". *Boletín de la Sociedad Geológica Mexicana*, T. LIX, num.1, p.19-42, 2007.

[53] F.J. Angeles and A. Cantú, "Subsurface Upper Jurassic Stratigraphy in the Campeche Shelf, Gulf of Mexico". In: *The Western Gulf of Mexico Basin: Tectonics, Sedimentary Basins, and Petroleum Systems*. C. Bartolini, R.T. Buffler and A. Cantú (eds), American Association of Petroleum Geologists Memoir 75, 2001.

[54] J.Y. Narváez, J. Belenes, J. Moral, J.M. Martínez, C. Macías, O. Castillejos and M.A. Sánchez. "Bioestratigrafía de secuencias del Mioceno-Plioceno de la cuenca Macuspana, sureste del Golfo de México". *Revista Mexicana de Ciencias Geológicas*, vol.25, num.2, pp.217-224, 2008.

[55] PEMEX (Petróleos Mexicanos). "Provincia Petrolera Plataforma de Yucatán". Pemex Exploración y Producción, México, Versión 1.0, México, 17p., 2010.

[56] J. Pindell and L. Kennan. "Tectonic evolution of the Gulf of Mexico, Caribbean and northern South America in the mantle reference: an update". In *The geology and evolution of the region between North and South America*, K. James, M.A. Lorente and J. Pindell (eds), Geological Society of London Special Publication, 2009.

[57] M. Olivas. "Aspectos paleogeográficos de la región sureste de México en los estados de Veracruz, Tabasco, Chiapas, Campeche, Yucatán y el territorio de Quintana Roo". *Boletín de la Asociación Mexicana de Geólogos Petroleros*, vol. XXVI, num.10- 2, pp.323-336, 1974.

[58] S. Medina. "Tertiary zonation based on planktonic foraminifera from the marine region of Campeche, Mexico". *American Association of Petroleum Geologists*, Memoir 75, pp.397-420, 2001.

[59] R. Sánchez, "Geología petrolera de la Sierra de Chiapas". In: IX Excursión Geológica de Petróleos Mexicanos, Superintendencia General de Distritos de Exploración, Zona Sur, Libreto-Guía, 57 p. 1979.

[60] M. Guzmán and J. J. Meneses. "The North America–Caribbean plate boundary west of the Motagua–Polochic fault system: a fault jog in Southeastern Mexico". *Journal of South American Earth Sciences*, vol.13, num.4-5, pp., 2000.

[61] B. A. Méndez, "Geoquímica e isotopía de aguas de formación (salmueras petroleras) de campos mesozoicos de la Cuenca del Sureste de México: implicación en su origen, evolución e interacción agua-roca en yacimientos petroleros", Tesis Doctoral, Centro de Geociencias, UNAM, 200 p., 2007.

[62] R. K. Goldhammer and C. A. Johnson, "Middle Jurassic-Uper Cretaceous Paleogeographic evolution and sequence stratigraphic framework of the northwest Gulf of Mexico rim". In: *The western Gulf of Mexico Basin: Tectonics, sedimentary basins and petroleum systems*. C. Bartolini, T. Buffler, and A. Cantú (eds), American Association of Petroleum Geologists Memoir 75, p. 45-81, 2001.

[63] J.H. Rosenfeld. "Economic potential of the Yucatan block of Mexico, Guatemala, and Belize". In: *The Circum-Gulf of Mexico and the Caribbean-Hydrocarbon habitats, basin formation, and plate tectonics:* American Association of Petroleum Geologists Memoir 79, pp. 340–348, 2003.

[64] S. Angus, B. Armstrong and K.M. de Reuck. *International Thermodynamic Tables of the Fluid State. Volume 3. Carbon Dioxide*. Pergamon Press: IUPAC Division of Physical Chemistry, 1973, pp. 266–359.

3

Late Proterozoic – Paleozoic Geology of the Golan Heights and Its Relation to the Surrounding Arabian Platform

Miki Meiler, Moshe Reshef and Haim Shulman
Department of Geophysics and Planetary Sciences, Tel Aviv University, Tel Aviv Israel

1. Introduction

1.1 Study area

The Golan Heights (GH) is an elevated basalt-covered plateau located at the south-western tip of the Palmyrides, rising above the Sea of Galilee on the eastern side of the Jordan River (Figures 1a & 1b, Meiler, 2011; Meiler *et al.*, 2011). The Golan Plateau is covered by tens of volcanic cones and comprises the western continuation of an extensive Hauran region - a broad and flat Plio-Pleistocene volcanic province that extends eastward into Syria.

The study area is framed by prominent tectonic and geo-morphological elements (Figures 1a & 1b). To the north, the GH is bounded by the Mt. Hermon Anticline, which comprises a south-western continuation of the Palmyrides transpressive fold belt (Picard, 1943), as well as the right bend of the Dead Sea Fault System (DSFS) (Freund, 1965; 1980). To the west, the GH is delimited by the DSFS and the Jordan Rift Valley (JRV) which in this area extends along nearly 60km from the Yarmouk River in the south to the Mt. Hermon Structure in the north. The DSFS is an active plate boundary separating the Arabian plate to the east from the African plate (Sinai sub-plate) to the west (Garfunkel *et al.*, 1981). To the south, the GH is bordered by the Yarmouk River and the adjacent Jordanian Highlands, which comprises at this area an international border between Israel and Jordan. To the east, the GH is bordered by the Hauran-Jebel Druze depression, which exhibits similar morphology of an elevated plateau with prominent volcanic cones and numerous basaltic sheets. The Hauran - Jebel Druze Plateau comprises the most north-western part of the wide-scaled Harrat-Ash-Shaam volcanic field, which extends from southern Syria, across Jordan and into Saudi Arabia (Figure 1a).

The cumulative stratigraphic column and the general structure of the rocks underlying the Plio-Pleistocene basalt cover were studied by various authors. The syncline nature of the Golan Plateau was pointed out by Michelson (1979), Mor (1985), Hirsch *et al.* (2001) and Shulman *et al.* (2004). Meiler *et al.* (2011) presented first results of an extensive depth-domain seismic analysis according to which the Golan Plateau covers a large structural depression that constitutes the northern and deeper part of the extensive Irbid-Golan Syncline (Figure 2). The syncline has evolved during the Late Cretaceous - Cenozoic amid the Hermon Structure to the north and the Ajlun anticline to the south.

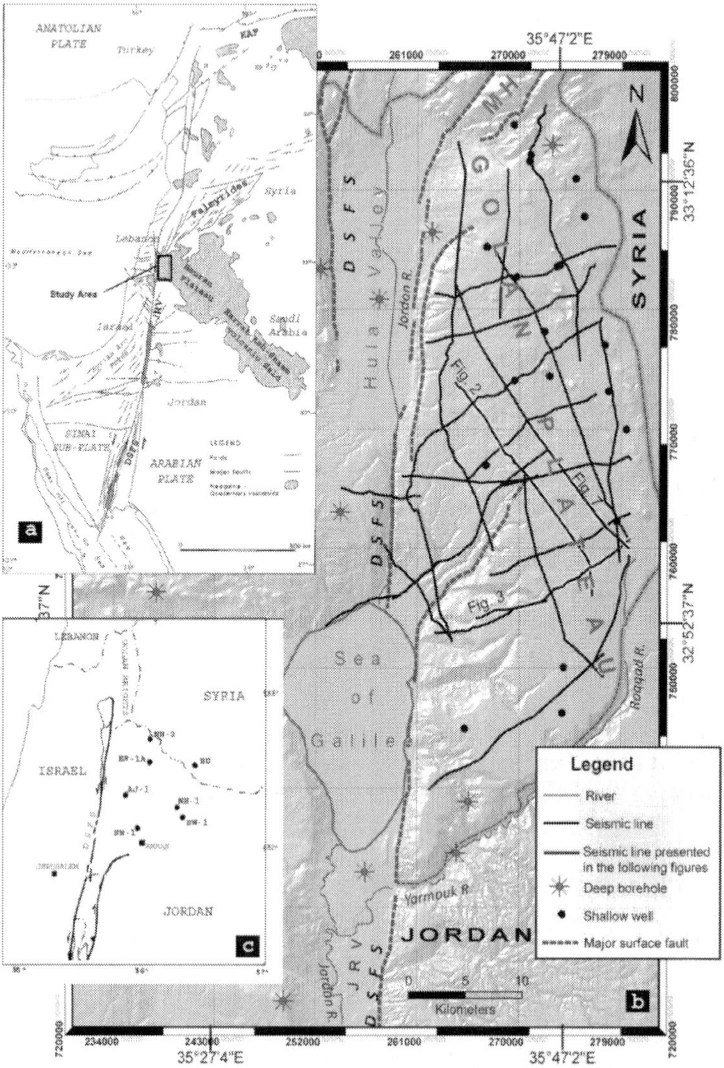

Fig. 1. **a**. Regional setting of the study area. Modified after Garfunkel *et al.*, 1981. *DSFS* - Dead Sea Fault System; *EAF* - East Anatolian Fault; *JRV* - Jordan Rift Valley; **b**. Location map of the seismic lines, deep boreholes and shallow water wells in the GH and the adjacent areas (Israeli side only), overlaying Digital Terrain Model (DTM). DTM after Hall, 1993. *MH* - Mount Hermon. The map is given in WGS-1984 and Israel-TM-grid coordinate systems. **c**. Location map of the deep boreholes drilled in the Northern Jordan and SW Syria areas. Thickness information on the principle stratigraphic units penetrated by these boreholes is presented in Table 1.

The sedimentary succession accumulated within the Golan part of the depression extends from the Late Proterozoic at the bottom to the Pleistocene basalts at the top of the section, attaining a thickness of at least 8.5km in the northern part of the plateau. The stratigraphic column beneath the basalt cover consists of up to 3,500m of Infracambrian - Paleozoic succession; up to 5,000m of Mesozoic rocks and about 1,500m of Cenozoic section (Meiler *et al.*, 2011). The thickness of the basaltic layer covering the Golan depression attains at the central Golan 1,100m (Reshef *et al.*, 2003; Meiler *et al.*, 2011).

1.2 Regional background

During Late Proterozoic - Paleozoic the areas surrounding the Golan Plateau, i.e. Levant, Arabian Platform and North Africa, constituted a part of the Gondwana continent. Following the Pan-African orogenic event and the subsequent cratonization, the region behaved typically as a stable platform during this time span. An extensive sedimentary cover of marine and continental origin has accumulated over the area in several deposition cycles. Sedimentation of mostly siliciclastic deposits has continued on the stable subsiding passive margin shelf of the Gondwanaland until Permian, when a series of rifting events related to the Neo-Tethys opening set on a new episode in the regional geological history (Garfunkel and Derin, 1984; Garfunkel, 1988; Weissbrod, 2005).

The Late Precambrian - Early Cambrian clastic cycle consists of immature, polymictic and poorly sorted conglomerates and arkose that were mostly derived from the Pan-African metamorphic and Plutonic terrain in the Arabo-Nubian Shield, to the west and south of the study area. The detritial sediments of the conglomeratic facies accumulated due to rapid and repeated subsidence episodes along major fault scarps and tectonic depressions, whereas the arcosic facies was deposited in a broad pericratonic basin, which extended from the Arabo-Nubian Shield in the south to the passive margin and the Paleo-Tethys in the north. Today, these clastics are discontinuously exposed throughout Saudi-Arabia, Egypt, Jordan and Israel, separated by erosion gaps on the elevated igneous rocks of the Arabo-Nubian Shield. The thickness of the conglomeratic facies preserved within the rift-related depressions in Northern Arabia and Eastern Desert of Egypt locally attains 5,000m, whereas the thickness of arcosic facies in Israel and Jordan attains at least 2,500m (Weissbrod, 2005).

The Paleozoic sediments are very wide spread in the north-eastern part of the Arabo-African continent, comprising one of the most voluminous bodies of sediments in the region (Garfunkel, 1988). This second sedimentary cycle continued from the Middle Cambrian to Permian, incorporating mostly siliciclastic deposits, with mixed carbonate-shale intercalations throughout the sequence. The sediments were accumulated in the fluviatile environment and shallow epicontinental shelf, attaining a thickness of almost 5,000m.

Overall, the Late Precambrian - Paleozoic sequence attains thickness of more than 10,000m. However, due at least three major uplift-and-erosion events ((1) end of Silurian; (2) end Devonian to Early Carboniferous; (3) Late Carboniferous to Early Permian) a complete time sequence is hardly found at any locality in the northern part of the Arabo-African continent (Garfunkel and Derin, 1984; Weissbrod, 2005).

1.3 The scope of the study

The purpose of the current work is to present the deepest stratigraphic section identified beneath the volcanic cover of the Golan Plateau, based on the extensive depth-domain seismic analysis, and to discuss the geological evolution of the study area during the Late Precambran - Paleozoic time span in the light of the available information from the surrounding north-western parts of the Arabian Platform.

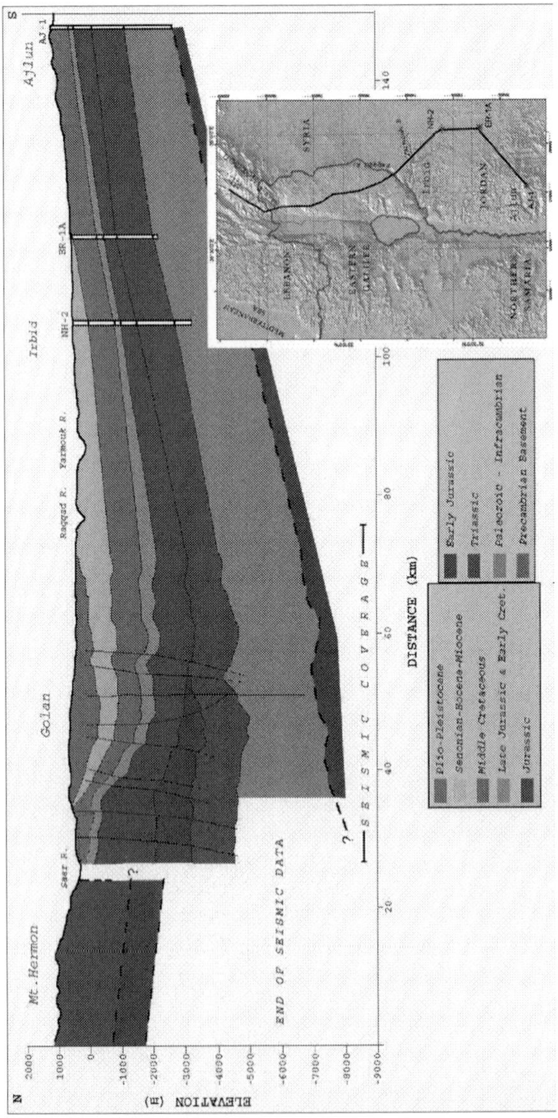

Fig. 2. Generalized cross-section showing the regional geological structure and the Late Proterozoic - Phanerozoic stratigraphic column in the area laying in between the Ajlun anticline at the south and the Mt. Hermon at the north (Modified after Meiler, 2011; Meiler *et al.*, 2011). The cross-section is based on analysis of three deep boreholes located in the Northern Jordan (AJ-1, ER-1A and NH-2) and depth-domain interpretation of seismic data that covers the GH area (Figure 1b). The cross-section outlines the syncline nature of the study area, confined by the Ajlun and Hermon anticlines. Note the similarity with respect to

the thickness of the Infracambrian - Paleozoic sections revealed in the subsurface of Northern Jordan and Golan Heights areas, suggesting analogous geological history during this time-span. On the contrary, the thickened Jurassic succession interpreted in the central and northern parts of the GH implies that significantly different geological environment prevailed in the GH with respect to that of the Jordanian Highlands during the Early - Middle Mesozoic.

2. Methods

2.1 Database (Figures 1b & 1c)

- A set of twenty five 2-D seismic reflection lines covering the GH area
- Formation tops from eighteen deep oil-exploration boreholes located in the Golan Heights area, Eastern Galilee, Northern Jordan and SW Syria. Table 1 presents the thickness information from the Jordanian and Syrian wells which penetrated the Paleozoic succession. Figure 1c indicates the location of these drillings.
- Formation tops from twenty shallow water and research wells drilled in the Golan Heights area
- Geological and topographical maps in different levels of resolution and geological cross-sections in local and regional scales

2.2 Seismic data processing

In the course of the present study, the Pre-Stack Depth Migration (PSDM) technique was utilized as the main seismic processing tool. PSDM was carried out from the surface topography, enabling an enhanced imaging of the Base-of-Basalt interface. The seismic data processing and analysis were accompanied by examination of staratigraphic information derived from the deep boreholes of Jordanian Highlands and SW Syria (Figure 1c), which penetrated the Mesozoic-Paleozoic successions, and in one case - the Precambrian basement (Ajlun-1 borehole).

Interval velocity analysis consisted of two steps:

1. 2-D velocity function construction for each of the 25 seismic lines was based on the Constant Velocity Half Space technique (Reshef, 1997).
2. 3-D interval velocity model construction, utilizing a MULTI 2-D approach. The procedure resulted in a comprehensive 3-D interval velocity model that covers the entire study area, including the subsurface parts which lay in between the seismic lines. The velocity model was then smoothed in the 3-D domain, resulting in a global interval velocity model of the study area.

The final depth sections were obtained by the Pre-Stack Explicit Finite-Difference Shot Migration and Post-stack Explicit Finite-Difference Depth Migration algorithms, employing extracted 2-D velocity functions from the global 3-D model.

2.3 Seismic data quality

Despite the thick basaltic layer entirely covering the Golan Plateau, the final depth sections show surprisingly good quality of seismic data. The final depth sections show reflections from 7 - 8 km bellow the datum (Figure 3) in the southern and central parts of the study area. There is a considerable deterioration of the seismic quality towards the Northern Golan.

2.4 Seismic interpretation

Eleven seismic markers were identified and mapped in the subsurface of the GH (Figure 3). Since the borehole information in the GH area is restricted to the upper 1,400 meters, direct correlation between the seismic data and the borehole stratigraphic information is limited to the upper two horizons only: the Base-of-Basalt (H1) and the Near Top Turonian (H2). Stratigraphic identification of the deeper seismic horizons became possible due to the fact that the seismic data was Pre-Stack depth migrated and the entire interpretation procedure took place in the geological (i.e. depth) domain. This enabled to perform an instantaneous correlation of the prominent seismic markers with the exposures of the Mesozoic section outcropping on the adjacent Mt. Hermon Anticline and to compare the intervals between the horizons with the thickness information derived from the deep boreholes of Northern Jordan. Hence, stratigraphic ascription of the LC-3 horizon (H3, electric log marker within the Lower Cretaceous) and the Near Top Jurassic horizon (H4) relies mainly on the correlation of the seismic data with the exposures of the Lower Cretaceous and the Jurassic strata outcropping on the Mt. Hermon Structure. Identification of the Near Top Triassic (H5) and the three Paleozoic - Infracambrian horizons (H6 - H8) is based on the concept that the thickness of the principle stratigraphic units in the Southern Golan should be comparable to the thickness reported in the Jordanian Highlands, across the Yarmouk River, where it is controlled by a series of deep oil-exploration boreholes. Three additional reflections with limited spatial distribution were identified in different parts of the study area; they were designated as: within the Tertiary (H1b), within the Early Jurassic (H4b) and the Near Top Precambrian basement (H9).

	AJ-1	ER-1A	KH-1	NH-1	NH-2	SW-1	BU-1
Cenozoic Basalts	-	-	-	-	-	-	507
Ajlun Group (M. Cretaceous)	546	769	427	340	801	-	418
Kurnub Group (Aptian-Albian)	159	210	235	238	217	115	228
Azab Group (Jurassic)	598	389	-	-	488	131	252
Ramtha Group (Triassic)	1043	1137	668+	542	1239	687	1119
Hudayb Group (Permian)	226	114+	-	-	439+	-	151
Amud Formation (U. Cambr.-Ordov.)	-	-	-	813	-	-	63+
Ajram Formation (M-U Cambrian)	-	-	-	253	-	213	-
Burj Formation (L-M Cambrian)	-	-	-	201	-	252	-
Salib Formation (L. Cambrian)	580	-	-	571	-	931+	-
Unassigned units + Saramuj (L. Cambrian)	634	-	-	1052	-	-	-
Total Depth	3800	2754	1333	4017	3722	2329	2938

Table 1. Thicknesses (m) of principle stratigraphic units measured within the boreholes of the Northern Jordan and Syrian Busra-1. (Summarized after Abu-Saad and Andrews, 1993 and other sources. The figures referring to Busra-1 approximately correspond to the lithostratigraphic nomenclature used for the GH and the Northern Jordan areas)

The lowest four horizons (H6 - H9) are within the scope of current study.
A detailed description of various seismic processing and interpretation aspects implemented during the study was presented by Meiler *et al.*, 2011.

3. Results

3.1 Lithostratigraphic identification

Near top basement (Horizon 9)

The deepest reflection recognizable on the depth sections was tentatively assigned as the Near Top Precambrian basement (Horizon 9). The horizon was identified on several profiles, mostly in the eastern parts of the GH. It is generally absent in the western and northern parts (Figure 3), although patches of it can be scarcely observed on some lines in these areas.
Horizon 9 is stratigraphicaly identified relying on the assumption that a smooth and gradual transition of the basement is expected between the Jordanian Highlands and the Southern Golan in the Yarmouk River area. The base of the sedimentary cover was penetrated by the AJ-1 borehole (Figures 1c & 2; Table 1), 50km south to the study area, reaching the basement at depth of nearly 3,800 meters beneath the surface. The closest boreholes to the study area drilled in the Northern Jordan and SW Syria, i.e. ER-1A, NH-2 and BU-1 (Figure 1c), did not penetrate bellow the upper Paleozoic. However, NH-1 well, located about 70km south-east to the GH, penetrated ~1,000m of the Saramuj and an unassigned clastic units, which overlay the basement. Thus, it is assumed that on the most southern profiles of the GH the basement should be found abound 1km below the Near Top Saramuj horizon (H8), the penetrated figure of the Saramuj clastics within NH-1 (Figures 3 & 4).

Sarmuj formation and the unassigned clastic unit (Horizon 8)

Horizon 8 is interpreted as the near top of the Late Precambrian - Early Cambrian sedimentary succession, known as the Saramuj Formation and the unassigned clastic unit (Figure 5). The sequence is known in the Arabian Platform region as the oldest non-metamorphosed sedimentary sequence, consisting of polymict conglomerate and poorly sorted coarse to fine grained arkose, accompanied by magmatic intrusions and extrusions (Weissbrod, 2005).

Near top burj formation (Horizon 7) and the near top paleozoic (Horizon 6)

In four, out of seven deep boreholes of Northern Jordan and Syrian Busra-1, the Paleozoic succession is topped by the Permian strata, usually limited to several hundred meters in thickness (Table 1). Therefore, it seems reasonable to assume that in the Southern GH some few tens to several hundred meters of Permian section rest at the top of the Paleozoic succession and Horizon 6 may roughly represent the Near Top Permian. The thickness of the Permian in the subsurface is expected to increase towards north, as up to 600-700m of Permian deposits were reported within the Palmyra Trough (Leonov, 2000).
Beneath the Permian section, beds of different Upper Paleozoic units are residing in the deep boreholes south to the GH (Table 1). In NH-1 and BU-1 the Permian Hudayb Group overlays the Ordovician Amud Formation (in Syrian BU-1 it is defined as "Afandy Formation"), whilst in SW-1 and AJ-1 it overlays Middle - Upper Cambrian Ajram and Lower Cambrian Salib Formations respectively. The other deep boreholes located in the vicinity of the GH did not penetrate beneath the Permian strata. However, the Middle Cambrian Burj Formation is widely distributed in the subsurface of Northern Jordan and

Syria and is known as a prominent regional seismic interface, designated as "D-reflector" (McBride *et al.*, 1990). Therefore Horizon 7 was lithostratigraphicaly assigned as the Near Top Burj Formation.

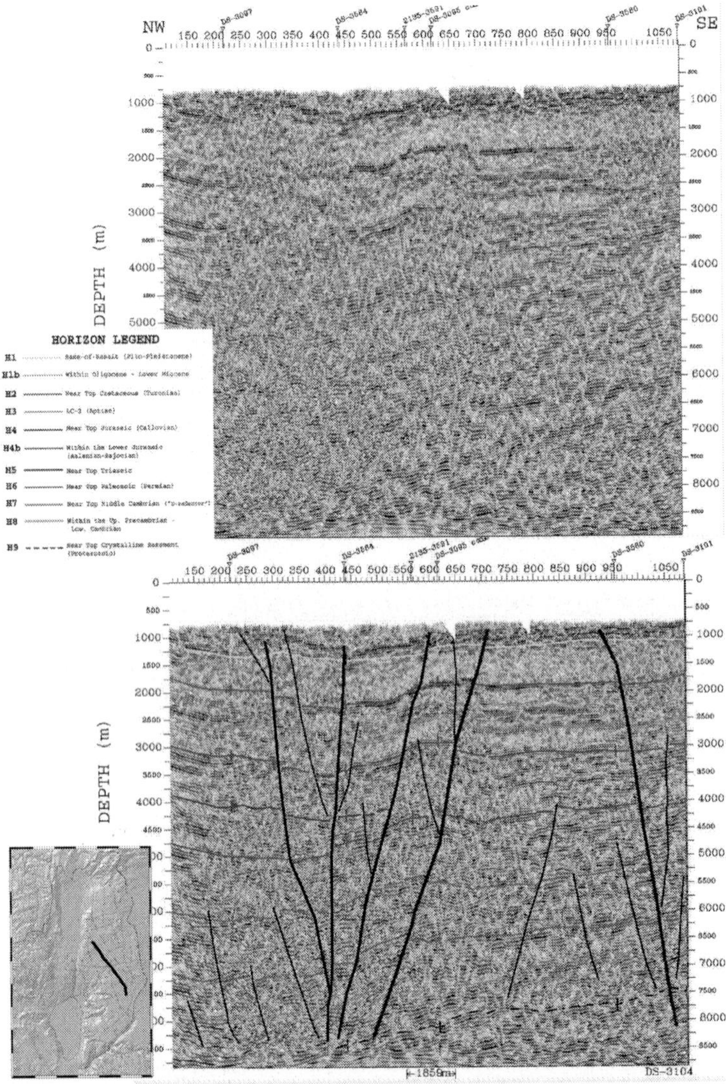

Fig. 3. DS-3104 depth section. Datum +1,150m. Note the good quality of seismic data, showing reflections from depths of 7 - 8 km bellow the datum. The gentle northward dipping of the Near Top Basement Horizon (H9), along with the overlying Paleozoic succession (H6 - H8), is observable.

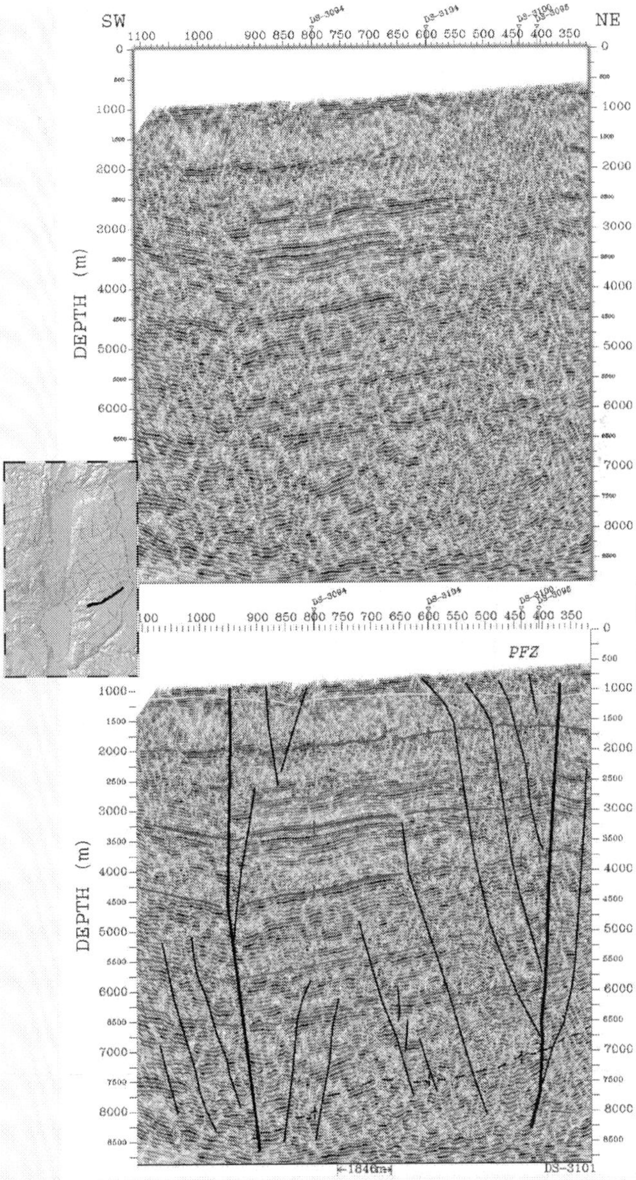

Fig. 4. DS-3101 depth section Datum +1,150m.; *PFZ* - Pezura Fault Zone. The Near Top Basement Horizon (H9) is assumed to lay ~1,000m bellow the Saramuj Fm (H8). Note the rising of the basement and the Saramuj Formation towards the PFZ on the eastern part of the profile. The western dip of the entire sedimentary column is clearly observable.

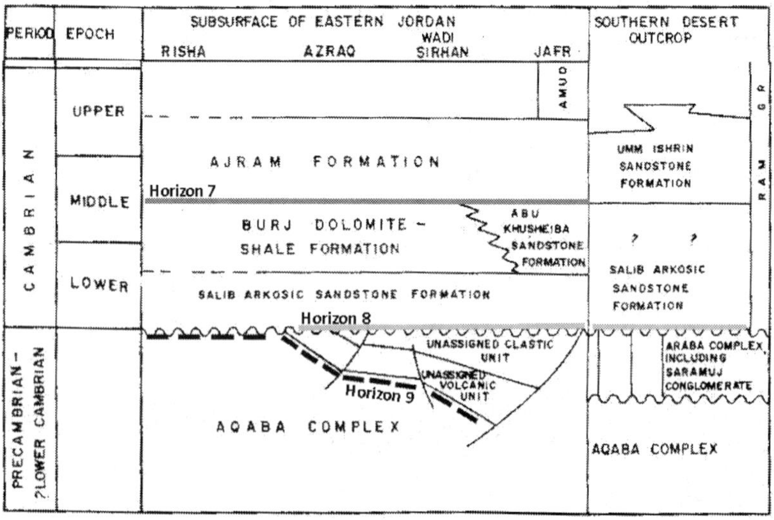

Fig. 5. Lithostratigraphy of the Lower Paleozoic (Cambrian) and the Precambrian of Jordan (Modified after Andrews, 1991). Horizons 7, 8 and 9 are tentatively assigned to the Near Top Burj Fm.; the Near Top Saramuj Fm. and/or the unassigned clastic unit; and to the Near Top Precambrian Basement, respectively (marked by the corresponding colors as presented in horizon legend on figure 3). Note the half-graben structure within the Infracambrian section.

3.2 Structural and isopach maps

Several structural and isopach maps were compiled in order to outline the geological evolution of the GH during the Late Proterozoic - Paleozoic time span. Figure 6 presents the structural map of the Near Top Basement Horizon (H9) and the isopach map compiled for the entire Infracambrian - Phanerozoic sedimentary cover.

Due to its limited appearance on the seismic sections, Horizon 9 was only partly interpreted in the subsurface of the GH area and therefore the structural and the isopach maps presented in figure 6 are restricted to the eastern and central parts of the study area. Nevertheless, the general structure and the architecture of the crystalline basement can be inferred from the maps.

The depth to the Near Top Basement Horizon, given its restricted seismic appearance and the uncertainty with respect to its stratigraphic correlation, ranges in the GH area between 5,700 to 7,700m beneath the sea level (Figure 6a) or between 6,150 - 8,500m beneath the surface topography (Figure 6b). The depth to the top of the crystalline basement in the Southern Golan is estimated to be 6 - 6.5km. The depth to the base of the sedimentary cover increases towards the Northern Golan and the Hermon Structure, where the sedimentary succession is outlined by its thickened Mesozoic sequence (Figure 2).

The thickness of the Infracambrian interval (i.e. Saramuj Formation and the unassigned clastic units of the Upper Proterozoic) in the study area varies in range from several hundred to 1,500 meters (Figure 7).

The structural map of Horizon 6 is presented in figure 8. The map displays the contemporary configuration of the Near Top Paleozoic (Permian?). The structural setting is dominated by the notable westward dipping from -3,5km in the east to -7km in the west, in the proximity of the DSFS.

Overall, based on the information derived from the deep boreholes of Northern Jordan and Syrian Busra-1, it is reasonable to assume that the seismic interval interpreted in the GH between the Near Top Paleozoic (Horizon 6) and the Middle Cambrian Burj Formation (Horizon 7) incorporates few tens to several hundred meters of Permian, overlaying additional several hundred meters of Ordovician to Middle - Upper Cambrian strata. The thickness of this interval varies between 500 - 1,100m for most of the GH area, locally attaining 1,300m (Figure 9).

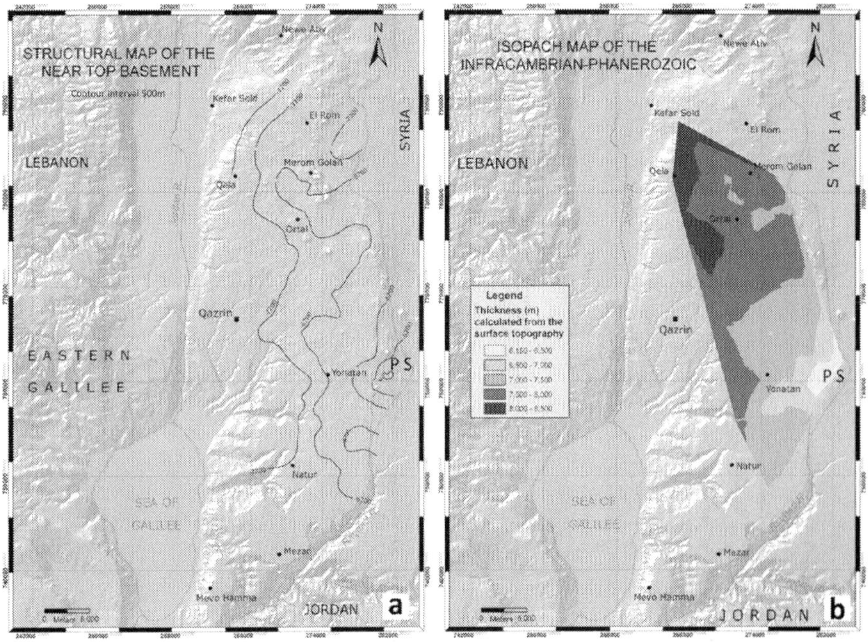

Fig. 6. **a**. Generalized structural map of the Near Top Basement Horizon (Horizon 9; faults are omitted). *PS* - Pezura Structure. Note the north-western dipping of the basement. (Modified after Meiler *et al.*, 2011). **b**. Isopach map showing the thickness of the seismic interval calculated between the Digital Terrain Model and H9. The map presents the thickness of the entire Infracambrian - Phanerozoic sedimentary cover in the central and eastern parts of the GH and indicates the depth to the crystalline basement calculated from the surface topography. (Modified after Meiler *et al.*, 2011).

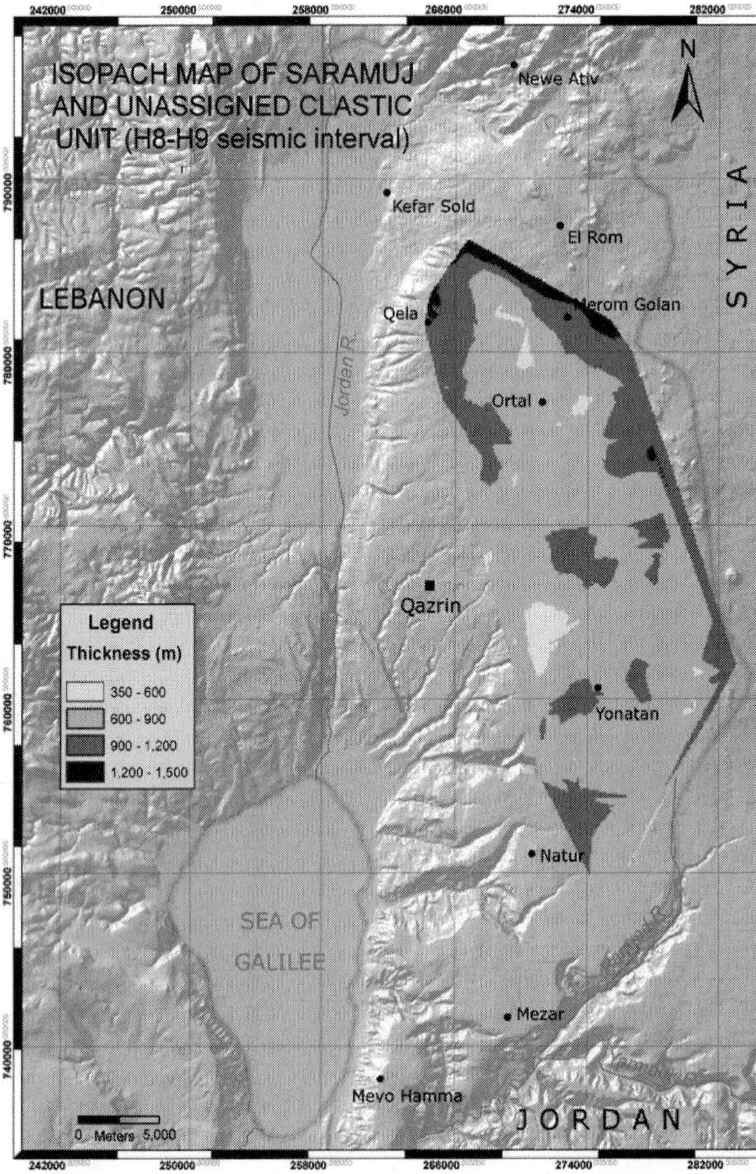

Fig. 7. Isopach map of the Saramauj and the unassigned clastic units within the Infracambrian time span. The map represents the seismic interval calculated between H8 and H9.

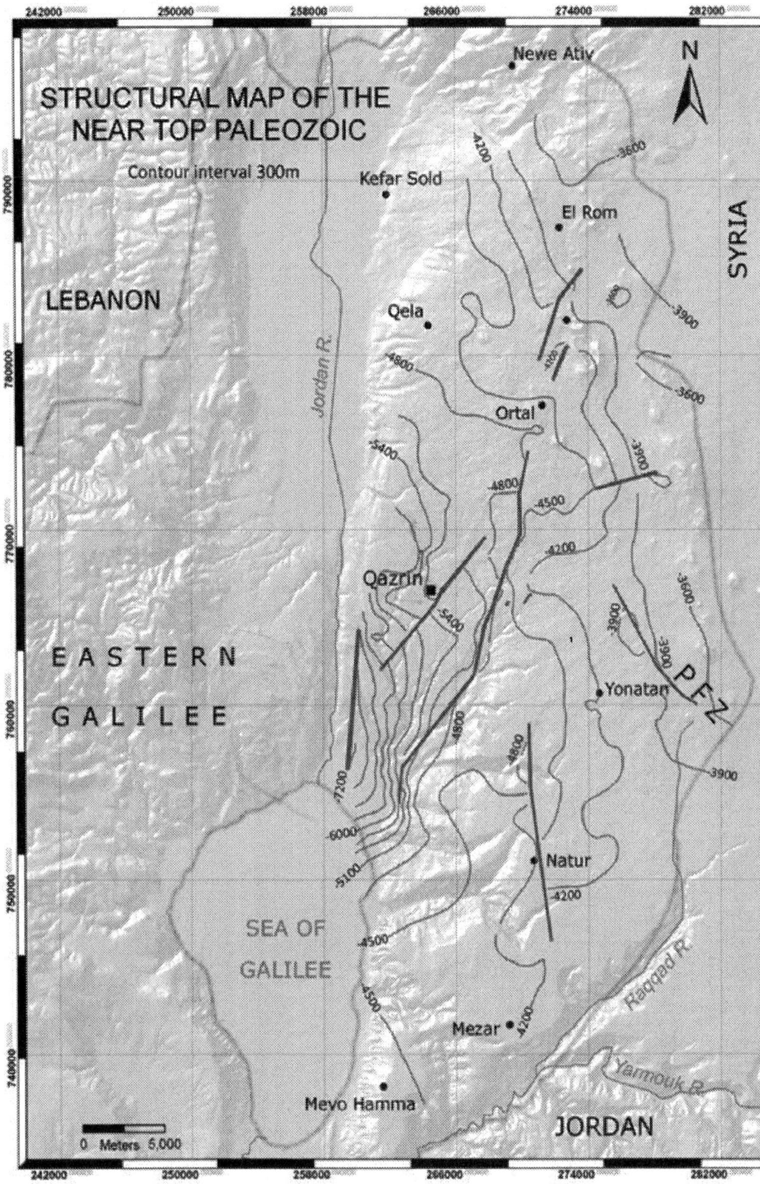

Fig. 8. Structural map of the Near Top Paleozoic Horizon (H6). After Meiler *et al.*, 2011. Note the western dip of the horizon from ~-3600m at the Syrian border in the east to about -7200m at the Sea of Galilee area in the west. Major fault zones are indicated.

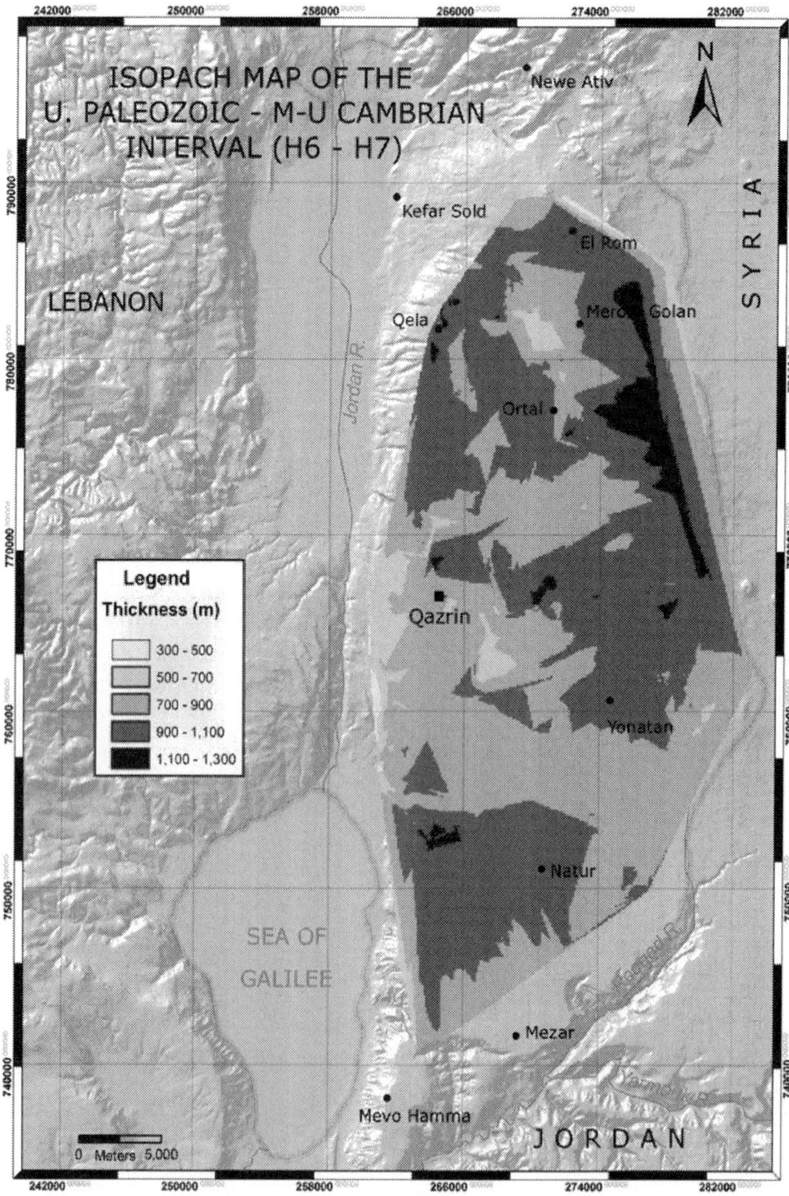

Fig. 9. Isopach map of the Upper Paleozoic (Permian?) to Middle-Upper Cambrian time span. The map represents the seismic interval calculated between H6 and H7.

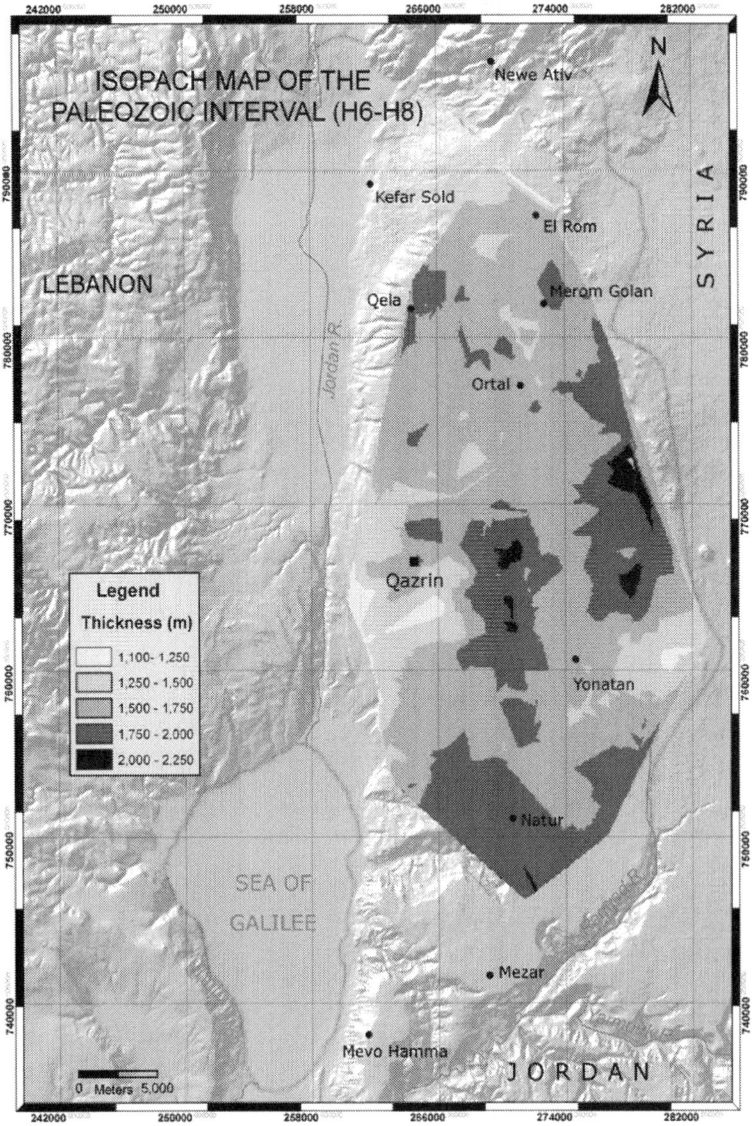

Fig. 10. Isopach map of the Paleozoic succession. The map represents the seismic interval calculated between H6 and H8.

A cumulative thickness of the Paleozoic succession is presented in isopach map calculated between the seismic intervals H6 - H8 (Figure 10). The interpreted thickness of this interval ranges from 1,100 to 2,250m.

4. Discusson

4.1 Near top basement (Horizon 9)

The depth to the base of the crystalline basement in the study area ranges between ~6km in the Southern GH to ~8.5km in the Northern Golan (beneath the surface topography). The basement dips northwards towards the Hermon Structure (Figure 2 & 3). At the foot of Mt. Hermon the thickness of the sedimentary cover is not known, but assumed to exceed the 8,500m calculated in between the Qela and the El-Rom area (Figure 6), the most northern area where the horizon is traceable and could be interpreted. In the south-western part of the Palmyride fold belt the depth to the basement was estimated to 11km within the Palmyra Trough (Seber *et al.*, 1993). East of the GH, outside of the Palmyrides, the depth to basement was estimated at 8 - 10km (Rybakov and Segev, 2004). To the west of the GH, across the DSFS in the Galilee region, the thickness of the sedimentary cover attains its regular figures of 6 - 8km (Ginzburg and Folkman, 1981). Thus, considering the northward dip of Horizon 9, it is suggested that the basement continues to deepen in the Northern Golan and the Mt. Hermon areas, whilst its depth beneath the Hermon Structure may attain 10 - 11km, as was estimated by Seber *et al.* (1993) in the south-western parts of the Syrian Palmyrides.

At the south-eastern part of the Golan the basement morphology is outlined by the significant structural uplift, referred here as the Pezura Structure (Figures 4, 6 & 8). It rises for several hundred meters above its surrounding and its structural influence can be traced upwards within the Paleozoic, Mesozoic and also Cenozoic sedimentary units. Reconstruction of seismic data to the Mid-Cambrian level (Horizon 7) indicates that this structure existed as a local high already in the Late Proterozoic - Early Cambrian (Figure 11, see).

4.2 Sarmuj formation and the unassigned clastic unit (Horizon 8)

The Late Precambrian - Early Cambrian sedimentary succession in the Arabian Platform comprises the oldest non-metamorphosed sedimentary sequence in the region, consisting of polymict conglomerate and poorly sorted coarse to fine grained arkose, accompanied by magmatic intrusions and extrusions (Weissbrod, 2005). The term "Infracambrian" describes this non-metamorphosed, mostly clastic sequence (Wolfart, 1967; Horowitz, 2001). Horizon 8 is interpreted as the near top of this sedimentary sequence in the subsurface of the Golan Heights.

In the Negev area of Southern Israel, a large Infracambrian sedimentary depression was reported overlaying the Precambrian basement (Weissbrod, 1980). It comprises a part of a broad marginal basin known as the Arabian-Mesopotamian Basin, which extends from the Arabo-Nubian Shield across Arabia, Levant and Mesopoamia to the edge of the Arabian Plate along the Bitlis Suture (Weissbrod and Sneh, 2002). The basin was filled with several kilometers of immature clastics and volcanics, defined in the Southern Israel as Zenifim and Elat Conglomerate Formations.

Saramuj Formation and the unassigned clastic unit of Northern Jordan (Figure 5) consist both of clastic sediments, mainly coarse conglomerate and arkosic sandstones as well as some volcanic components (Andrews, 1991). These units are considered both as time and lithological equivalents of the Infracambrian Elat Conglomerate and the Zenifim sandstones reported from the Southern Israel (Garfunkel, 2002; Hirsch and Flexer, 2005; Weissbrod, 2005). The overlaying Salib Formation is very similar in composition and corresponds to the Lower Cambrian Amudei Shelomo and Timna Formations (Southern Israel) composed of predominantly clastic units. Horizon 8 is hypothesized to represents the near top of this

Infracambrian sequence which is characterized by the immature clastics of Saramuj conglomerate followed by the Salib arcosic sandstones, similar to their southern contemporaneous known as Zenifim Formaion and the Lower Cambrian Amudei Shelomo and Timna Formations; all units comprising a part of the above mentioned Arabian-Mesopotamian Basin.

The thickness of the Ifracambrian interval (i.e. Saramuj Formation and the unassigned clastic units of the Upper Proterozoic) in the study area varies in range from several hundred to 1,500 meters (Figure 7). These Infracambrian units unconformably overlay the Near Top Basement Horizon (H9), filling the locally fault-bounded blocks (Figures 4 & 11a). These interpreted figures of the Infracambrian succession in the GH are comparable thickness figures to ~2,500m of Zenifim Formation estimated by Weissbrod and Sneh (2002) to overlay the basement on the regional scale.

The Infracambrian sedimentary section recognized in Jordan and Saudi Arabia is considered as a syn-rifting succession accumulated during the extensional phase of the Late Proterozoic - Early Cambrian time span (Abed, 1985; Huesseini, 1989; Best *et al.*, 1990). The period was dominated by the intra-continental rifting and wrenching (Husseini and Husseini, 1990), resulting in a series of asymmetric half-grabens with occasionally rotated basement blocks and immature syn-rift clastic deposition (Andrews, 1991; Figures 5 & 11a).

The thick Infracambrian section (Figure 7) which fills the underlying faulted blocks observable in the subsurface of the GH (Figure 11) is in agreement with the idea of possible pre-Cambrian or Early Paleozoic rifting episode that took place in the North-Western Gondwanian Arabia, as suggested by the above mentioned authors.

4.3 Pezura structure

A complex basin-and-swell configuration was proposed to prevail throughout the northern parts of the Gondwanaland during the Paleozoic (Garfunkel, 1998). Several large up-doming elements related to this Paleozoic configuration were reported in the Eastern Mediterranean region: the Hercynian Geoanticline of Helez, centred in the coastal plane of Israel (Gvirtzman and Weissbrod, 1984); Hazro structure extending across the Turkish-Syrian border (Rigo de Righi and Cortesini, 1964); Riyadh swell in central Saudi Arabia (Weissbrod, 2005).

The elevated feature interpreted in the south-eastern corner of the GH, referred here as the Pezura Structure (Figures 4 & 6), may represent one of the uplifted features which constituted a part of this basin-and-swell configuration, although in considerably smaller scale. The uplift, followed by the notable tilting and on-lapping sedimentation of younger Paleozoic strata (Figure 11c), can be related to the Hercynian Orogenic episode, which is dated in Jordan as mid-Carboniferous (Andrews, 1991) and Pre-Carboniferous or Pre-Permian (Gvirtzman and Weissbrod, 1984) event in Israel. However, figure 11b shows that the structure preceded the Middle Cambrian Burj Formation (H7) deposits, originating already in the upper Proterozoic and affecting the subsequent Paleozoic sedimentation.

This is evidenced by the on-lapping stratigraphic relations between H7 and H8. Thus, it seems that the Pezura structure was established as a tectonically active area already in the upper Proterozoic and it continued to act periodically throughout the Paleozoic, as part of the Hercynian Orogenic episode. The location of the presently elevated Pezura structure coincides with the formerly well developed fault-bounded depression (Figure 11a). This overlapping pattern in which the Upper Proterozoic rifting zones became a regional uplifts during the Early Paleozoic characterize additional regional highs, such as Rutba swell (Seber *et al.*, 1993).

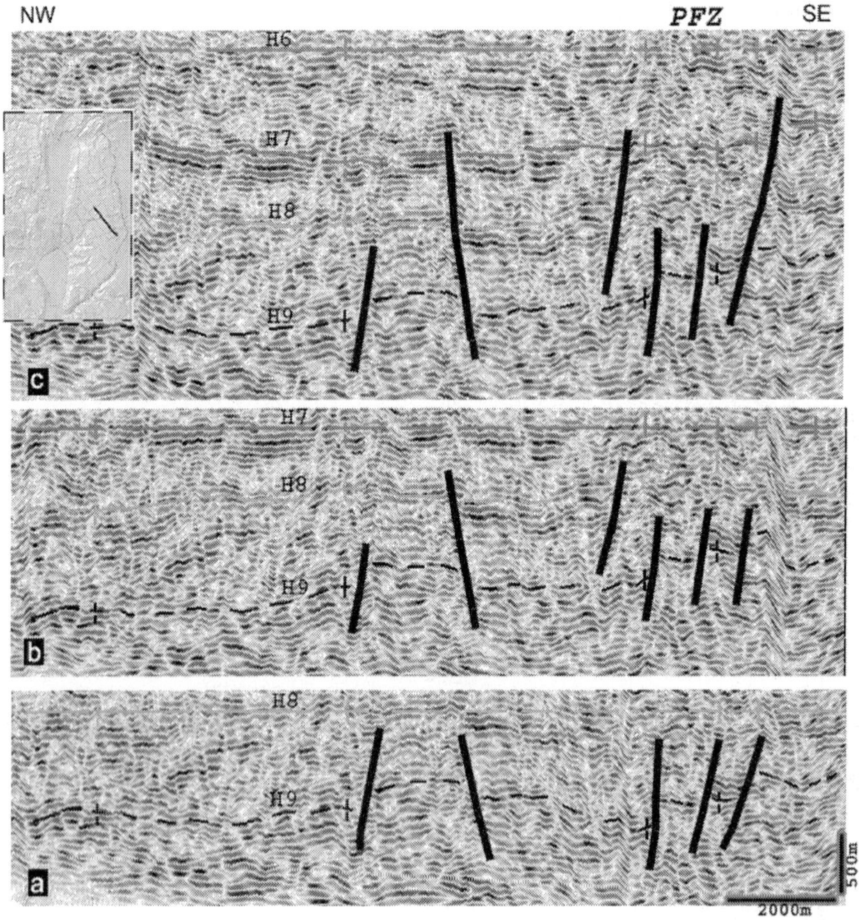

Fig. 11. Reconstruction of the Infracambrian - Paleozoic structural evolution in the Pezura Fault Zone (PFZ) area, south-eastern GH. The Early Cambian (a), Mid-Cambrian (b) and Late Paleozoic (c) stages are presented, through flattening the southern section of DS-3096 profile to H8, H7 and H6 seismic markers, respectively. (Horizon legend as in figure 3). a. The reconstruction presents the Infracambrian section overlaying the crystalline basement, as appeared at the end of the deposition of Saramuj Formation (H8). Note that in the Pezura structure area the Infracambrian Saramuj section fills the faulted blocks of the basement.b. The reconstruction presents the Late Proterozoic - Early Cambrian sections overlaying the crystalline basement, as appeared at the end of the deposition of Burj Formation (H7). Note the uplifted Pezura structure in the area formerly outlined by a series of down-faulted blocks. c. The reconstruction presents the Late Proterozoic - Late Paleozoic sections overlaying the crystalline basement, as appeared at the end of the Paleozoic (H6). Note additional faulting in the Pezura area, suggesting for an alternating tectonic activity throughout Paleozoic.

4.4 Near top burj formation (Horizon 7) and the near top paleozoic (Horizon 6)

Since in most of the deep drillings adjacent to the GH the Paleozoic succession is topped by the Permian strata, it is assumed here that in the Southern Golan some few tens to several hundred meters of Permian section rest at the top of the Paleozoic succession and Horizon 6 roughly represents the Near Top Permian.

On the regional scale the Paleozoic sediments are very widespread in the north-eastern part of the Arabo-African continent (Alsharhan and Nairn, 1997; Garfunkel, 2002; Weissbrod, 2005). Large Paleozoic basin was reported in Syria, where more than 5,000m of Cambrian - Carboniferous section was documented in the subsurface. Total thickness of the Paleozoic section in Syria locally attains 7,000m (Krasheninnikov, 2005; Leonov, 2000). In Northern Jordan, the thickness of the Paleozoic succession reaches nearly 2,000m in NH-1 borehole (Table 1). In Southern and Central Israel the Paleozoic succession is highly reduced and attains thickness of several hundred meters only (Weissbrod, 1980; Ginzburg and Folkman, 1981). Thus, the thickness of the sedimentary section interpreted in the GH within the seismic interval Horizon 6 - Horizon 8 (Figure 10) appears to be comparable to the thickness of the coeval units reported in the Northern Jordan area.

It's worth noting that the eastern regional dip of the Paleozoic strata well-documented throughout the Eastern Mediterranean (Figure 11; Gvirtzman and Weissbrod, 1984; Andrews, 1991) was not observed in the subsurface of the GH.

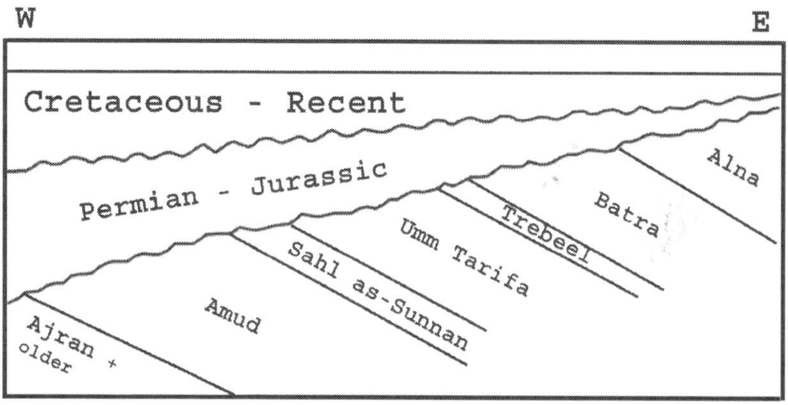

Fig. 12. Schematic stratigraphic relations between Paleozoic and Mesozoic sections in Jordan (Andrews, 1991). The sedimentary succession is outlined by the notable eastern dipping of the Paleozoic section, uncomfortably overlain by the Mesozoic units tilted towards west.

Moreover, on some profiles (Figures 3, 4 & 8) horizons attributed to the Paleozoic and Infracambtian sections (i.e. Horizons 6, 7 and 8) clearly show inclination to the opposite direction, i.e. due west, whilst a slight angular unconformity appears between the Mesozoic and the Paleozoic stratigraphic packages. A possible explanation would be an existence of an uplifted structure, like the above mentioned Pezura Structure, which locally tilted the sedimentary section to the west. However, this western inclination is clearly visible also on the northern profiles, away from the Pezura area; therefore it seems more reasonable to relate the inclination to a regional tectonic tilting which, according to the seismic data, took place during the Late Paleozoic - Early Mesozoic.

On the isopach map presenting the H6 - H7 seismic interval (Figure 9), figures of up to 1,300m are observable at the eastern edge of the GH, partly overlapping the line of the Pezura Fault Zone (The main fault plain is marked on figure 8; a number of unassigned individual fault segments related to the Pezura Fault Zone were not mapped). This increased H6 - H7 interval corresponds to the line of the volcanic cones covering the Golan Plateau and may suggest that plutonic intrusions occupy the lower parts of the Paleozoic succession. However, no definite seismic indications were observed on the depth sections to support this suggestion.

5. Summary

A series of structural and isopach maps compiled based on an extensive depth-domain seismic analysis displays the Late Proterozoic - Paleozoic evolution of the GH.
The depth to the base of the crystalline basement within the study area varies in range from 6km at the Southern Golan to 8.5km at the Northern GH (beneath the surface). As the Near Top Basement Horizon dips northward, it may attain 10 - 11km beneath the Hermon Structure, as was estimated in other parts of the Palmyrides.
The deepest sedimentary section interpreted in the subsurface of GH consists of two primary sequences:

1. Infracambrian (Late Precambrian - Early Cambrian) Saramuj Formation and unassigned clastic units which comprise the oldest non-metamorphosed sedimentary sequence in the region.
2. Paleozoic section, consisting of various units attributed to Lower Cambrian - Permian time span.

A total estimated thickness of the Infracambrian - Paleozoic succession interpreted in the subsurface of the GH varies in range of 1,800 to 3,500m (Figure 13).
About 1,000 - 1,500m of this figure corresponds to the Infracambrian deposits; its lower part (i.e. Saramuj Fm.) is interpreted as a syn-tectonic sequence, accumulated within the fault-related depressions, such as the Pezura Structure.
There is a notable contrast between the Paleozoic and the subsequent Mesozoic thickness distribution patterns within the GH. The thickness map of the Paleozoic (Figure 8) does not show the typical Mesozoic zoning and north-western thickening (Meiler, 2011), but rather characterized by a mosaic and irregular thickness distribution. This supports the findings in Syria, Jordan and Israel according to which it can be concluded that the Paleozoic structure of the northern Arabo-African Platform had very little in common with the structure that persisted during the following periods, which by the Early Mesozoic time was already greatly influenced by the establishment of the passive continental margin to the north of the Arabia shores.
Overall, it can be concluded that the stratigraphic column and the major sedimentary cycles of the Upper Proterozoic - Paleozoic interpreted in the GH closely resemble the corresponding geologic history of the adjacent Northern Jordan area. In both areas a 3 - 3.5km thick sedimentary succession of this period is preserved in the subsurface. The Paleozoic succession found in these areas attains more than 2,000m and differs significantly from the reduced Paleozoic succession exposed in the Southern Israel area, to the west and south of the GH.
This configuration has changed during the subsequent Mesozoic Era, when the deposition environment of the GH became closely affiliated to the Syrian and Israeli geologic history rather to that of the Northern Jordan.

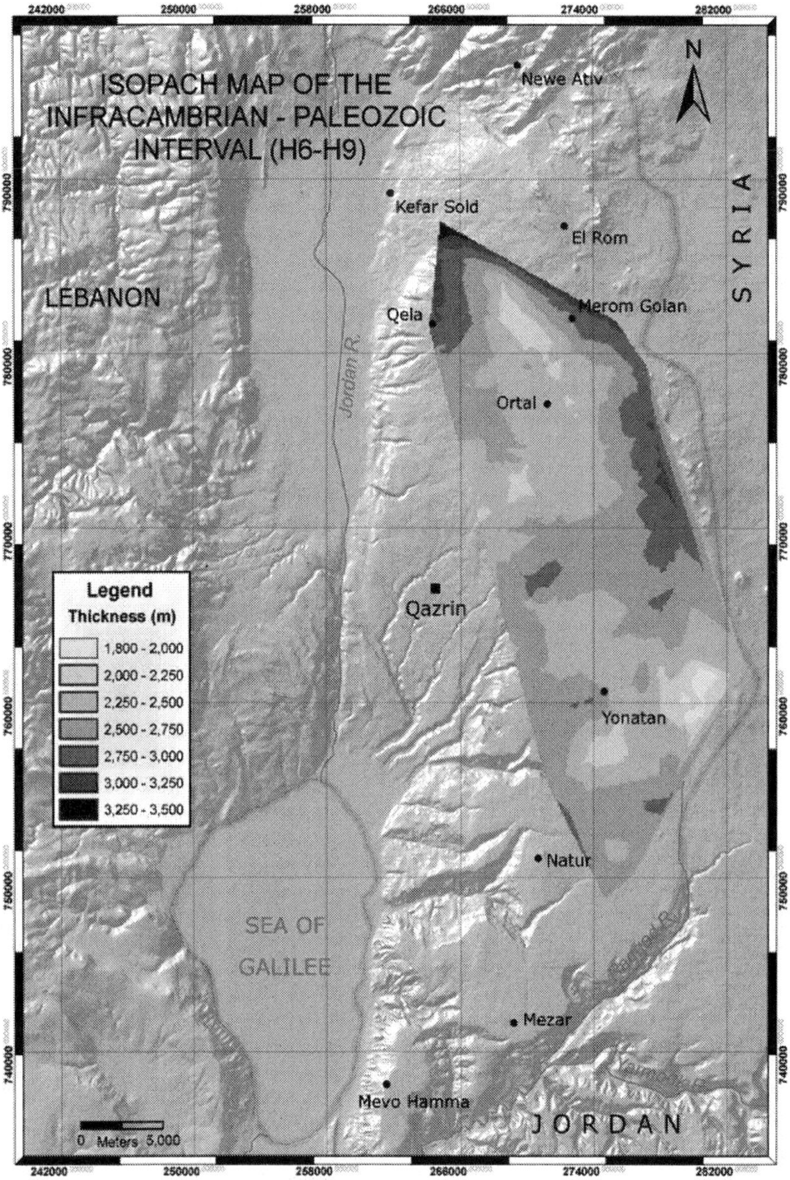

Fig. 13. Isopach map of the Infracambrian - Paleozoic succession. The map represents the seismic interval calculated between H6 - H9.

6. References

Abed, A. M., 1985. On the Supposed Precambrian Paleosuture Along the Dead-Sea Rift, Jordan. J. Geol. Soc., 142 (May): 527-531.

Abu-Saad, L. and Andrews, I. J., 1993. A Database of Stratigraphy Information from Deep Boreholes in Jordan. Amman. NRA Rep. Subsurface Geology Bulletin 6, p. 181. (in English).

Alsharhan, A. S. and Nairn, A. E. M., 1997. Sedimentary Basins and Petroleum Geology of the Middle East 843 p. Elsevier, Amsterdam, the Netherlands.

Andrews, I. J., 1991. Paleozoic Lithostratigraphy in the Subsurface of Jordan. Amman. NRA Rep. Subsurface Geology Bulletin 2, p. 75. (in English).

Best, J. A., Barazangi, M., Alsaad, D., Sawaf, T. and Gebran, A., 1990. Bouguer Gravity Trends and Crustal Structure of the Palmyride Mountain Belt and Surrounding Northern Arabian Platform in Syria. Geology, 18 (12): 1235-1239.

Freund, R., 1965. A Model of the Structural Development of Israel and Adjacent Areas Since Upper Cretaceous Times. Geol. Mag., 102: 189-205.

Freund, R., 1980. Creation processes of Hermon and Lebanon mountains *in* A. a. L. Shimda, M., ed., Mt. Hermon: Nature and Landscape: Tel-Aviv, Hakibbutz HaMeuhad (in Hebrew), p. 28-32.

Garfunkel, Z., 1988. The pre-Quarternary geology of Israel *in* Y. Yom-Tov, and E. Tchernov, eds., The zoogeography of Israel: Dordrecht, Netherlands, Dr. W. Junk, p. 7-34.

Garfunkel, Z., 1998. Constrains on the origin and history of the Eastern Mediterranean basin. Tectonophysics, 298 (1-3): 5-35.

Garfunkel, Z., 2002. Early Paleozoic sediments of NE Africa and Arabia: Products of continental-scale erosion, sediment transport, and deposition. Isr. J. Earth Sci., 51: 135-156.

Garfunkel, Z. and Derin, B., 1984. Permian-early Mesozoic tectonism and continental margin formation in Israel and its implications for the history of the Eastern Mediterranean, *in* J. E. Dixon, and A. H. F. Robertson, eds., The Geological Evolution of the Eastern Mediterranean, 17: London, Geological Society, London, Special Publications p. 187-201.

Garfunkel, Z., Zak, I. and Freund, R., 1981. Active Faulting in the Dead-Sea Rift. Tectonophysics, 80 (1-4): 1-26.

Ginzburg, A. and Folkman, Y., 1981. Geophysical investigation of crystalline basement between Dead Sea Rift and Mediterranean Sea AAPG Bull., 65 (3): 490-500.

Gvirtzman, G. and Weissbrod, T., 1984. The Hercynian Geanticline of Helez and the Late Palaeozoic history of the Levant, *in* J. E. Dixon, and A. H. F. Robertson, eds., The Geological Evolution of the Eastern Mediterranean, 17: London, Geological Society, London, Special Publications p. 177-186.

Hall, J. K., 1993. The GSI Digital Terrain Model (DTM) completed. GSI Current Researches 8. Jerusalem. p. 47-50. (in English).

Hirsch, F., Fleischer, L., Roded, R. and Rosensaft, M., 2001. The structural maps of Mount Hermon, Golan Heights and NE Jordan: stratigraphy and tectonics. Geol. Surv. Isr., Jerusalem. Rep. GSI/19/2001, p. 30. (in English).

Hirsch, F. and Flexer, A., 2005. Introduction to the Stratigraphy of Israel, *in* J. K. Hall, V. A. Krasheninnikov, F. Hirsch, H. Benjamini, and A. Flexer, eds., Geological Framework of the Levant, 2: Jerusalem, p. 269-273.

Horowitz, A., 2001. The Jordan Rift Valley, 730 p. Balkema, Lisse.

Huesseini, M. L., 1989. Tectonic and depositional model of Late Precambrian Arabian and adjoining plates. AAPG Bull., 73: 1117-1131.

Husseini, M. I. and Husseini, S. I., 1990. Origin of the Infra-cambrian salt basins of the Middle East., *in* J. Brooks, ed., Classic Petroleum Provinces, 50, Geological Society Special Publication, p. 279-292.

Krasheninnikov, V. A., 2005. Stratigraphy and Lithology of the Sedimentary Cover, *in* V. A. Krasheninnikov, J. K. Hall, F. Hirsch, H. Benjamini, and A. Flexer, eds., Geological Framework of the Levant, Vol. I: Cyprus and Syria: Jerusalem, Historical Production-Hall, p. 181-416.

Leonov, Y. G., 2000. Outline of Geology of Syria, 204 p. "Nauka", Moscow. (in Russian).

McBride, J. H., Barazangi, M., Best, J., Alsaad, D., Sawaf, T., Alotri, M. and Gebran, A., 1990. Seismic-Reflection Structure of Intracratonic Palmyride Fold-Thrust Belt and Surrounding Arabian Platform, Syria. AAPG Bull., 74 (3): 238-259.

Meiler, M., 2011. The Deep Geological Structure of the Golan Heights and the Evolution of the Adjacent Dead Sea Fault System. PhD thesis, Tel-Aviv, Tel-Aviv, 150 p.

Meiler, M., Reshef, M. and Shulman, H., 2011. Seismic Depth-Domain Stratigraphic Classification of the Golan Heights, Central Dead Sea Fault. Tectonophysics, 510: 354-369.

Michelson, H., 1979. The geology and palegeography of the Golan Heights. PhD thesis, Tel-Aviv University, Tel-Aviv, 163 p.

Mor, D., 1985. The volcanism of the Golan heights. PhD thesis, Hebrew University, Jerusalem, 125 p. Hebrew, English abstr.

Picard, L., 1943. Structure and evolution of Palestine. Geol. Dep. Hebrew Univ., Bull., p. 4: 1-134. (in English).

Reshef, M., 1997. The use of 3-D prestack depth imaging to estimate layer velocities and reflector positions. Geophysics, 62 (1): 206-210.

Reshef, M., Shulman, H. and Ben-Avraham, Z., 2003. A case study of sub-basalt imaging in land region covered with basalt flows. Geoph. Prosp., 51 (3): 247-260.

Rigo de Righi, M. and Cortesini, A., 1964. Gravity tectonics in Foothills structure belt of southwest Turkey. AAPG Bull., 48: 1911-1137.

Rybakov, M. and Segev, A., 2004. Top of the crystalline basement in the Levant. Geochem. Geophys. Geosys., 5 (9): Q09001. doi:10.1029/2004GC000690.

Seber, D., Barazangi, M., Chaimov, T. A., Alsaad, D., Sawaf, T. and Khadour, M., 1993. Upper Crustal Velocity Structure and Basement Morphology beneath the Intracontinental Palmyride Fold-Thrust Belt and North Arabian Platform in Syria. Geoph. J. Int., 113 (3): 752-766.

Shulman, H., Reshef, M. and Ben-Avraham, Z., 2004. The structure of the Golan Heights and its tectonic linkage to the Dead Sea Transform and the Palmyrides folding Isr. J. Earth Sci., 53 (3-4): 225-237.

Weissbrod, T., 1980. The Paleozoic of Israel and adjacent countries (lithostratigraphic study). PhD thesis, Hebrew University, Jerusalem, 275 p. (in Hebrew, English abstract).

Weissbrod, T., 2005. The Paleozoic in Israel and Environs, *in* J. K. Hall, V. A. Krasheninnikov, F. Hirsch, H. Benjamini, and A. Flexer, eds., Geological Framework of the Levant, 2: Jerusalem, p. 283-316.

Weissbrod, T. and Sneh, A., 2002. Sedimentology and paleogeography of the late Precambrian - early Cambrian arkosic and conglomeratic facies in the northern margins of the Arabo-Nubian Shield. Isr. Geol. Surv. Bull., 87.

Wolfart, R., 1967. Geologie von Syrien und dem Lebanon, 326 p. Gebruder Borntrager, Berlin.

4

Geology for Tomorrow's Society: Some Nordic Perspectives

Morten Smelror
Geological Survey of Norway, Trondheim
Norway

1. Introduction

To develop and sustain our societies, we need reliable access to minerals and raw materials, energy and water. With an increasing World population and emerging economies in underdeveloped parts of the World, the pressure for more natural resources will continue to rise. A good knowledge of geological resources is therefore essential for our future development. With increased urbanization and future global climatic changes, we need to better understand the human impacts on the environment, both on global and local scales. On long term the availability of geological resources and living space to house our populations are limited. Therefore, building the societies for Homo sapiens futurensis challenges us to have a holistic and international perspective of the natural resources and the environment. The present paper gives examples of how some of the key questions and challenges are addressed by the Nordic geological surveys.

2. Homo sapiens futurensis – An urban species

Over the twentieth century there has been a rapid urbanization of the world's population. In 1900 around 13% of the global population lived in urban areas, in 1950 the proportion was 49%, while today the majority (>50%) of people worldwide live in towns or cities. This number is expected to further increase to between 60-65% in the next 30-40 years. Urbanization rates vary between countries. It is estimated that more than 90% of the urban growth will occur in developing nations, and 80% will take place in Asia and Africa (UNFPA 2007, http://web.unfpa.org).

The urbanization of the World is often referred to as the "Urban Millennium" or the "tipping point". Along with this development we are also facing a global tipping point in the world's economic order. According to PricewaterhouseCoopers EU stood for 25% of the global value creation in year 2000, followed by the USA with 23%, while China was left far behind with 7%. This will change dramatically. Towards 2030 China will stand for 19% of global value-creation, USA 16% and EU 15%. India will reach a remarkably 9%. Hundreds of millions of people will be lifted from poor living conditions up to standards which we in the west take for granted. This new economic development will demand an increased global consume of natural resources.

If we also add the given the prognoses that the world population will increase from 6,9 billion today, and to 9 billion in 2050, the picture is pretty clear. There will be an enormous demand and battle for natural resources; food, water, energy and minerals. This situation will demand us to find new innovative solutions for resources management and use of materials. While we today dispose of our electronic and other modern consumables when they become tired, either for an upgrade or because it is cheaper to replace them than it is to repair them, Homo sapiens futurensis will have a different approach. In the years in front of us, the concept of urban mining will be established as a standard in the New World. Recycling of metal-containing consumables like cell-phones will not only be profitable, but most likely also be required as we only have finite resources with which to build them. According to some estimates (http://urbanmining.org) up to 30 times as much gold can be found in cell phone circuitry as can be found in the gold ore processed in gold mines (some 150 grams, or 5.3 ounces, per ton, compared to a measly 5 grams, or 0.18 ounces per ton). To add to that, the same quantity of cell phones also contains 100 kg of copper and 3 kg of silver, as well as numerous other materials. After the devices are processed and the materials separated, these valuable metals can be sold on as high quality raw materials to build new products, which in turn also can recycled.

Fig. 1. The city of Trondheim (area: 342 km^2), with a present population of 175 000 citizens. In 2050 the number of citizens is expected to be 220 000. Presently, 80% of the Norwegian population lives in urban areas (Photo: Edelpix).

As the populations are crowding up in urban areas, there will also be a shortage of space available for the physical growth. In many urban areas the only solution is to go underground. The challenge is to provide city planners, development engineers, decision makers, and the public with the geo-science information required for sound planning. This requires knowledge and competence in engineering geology, hydrology, geochemistry, as well as basic geomorphology and stratigraphy in order to build a three-dimensional model of the underground that will be taken into use. The information derived from various sources such as topographic and hydrological data, geological maps and borehole logs, have to be compiled in a digital format and stored in geo-referenced databases in the form of point, linear, and polygonal data. The data is then processed by Geographic Information Systems (GIS) to integrate the various sources of information and produce graphic 3D-models and maps describing the geological infrastructure beneath the city surface.

3. Fennoscandian mineral supplies

Today we see an increased interest in exploration and investments in mineral production all over the world. The main drivers of new investment are the high prices of raw materials, which are making new mines - and the reopening of older ones - more profitable. The use of most metals versus GDP per capita, grows almost logarithmically before it flattens at the levels of industrial countries. In simple terms; while the current per capita use of copper in China is around 2,5 kg, the similar figures in Japan and Germany are around 4-5 times as high. Consequently, the significant growth in China's economy and strong demand for minerals has lifted price of copper with more than 300% the past 10 years. Comparable situations also exist for most other metals, not to mention rare-earth-elements (REE), which are critical elements in the evolving green technology.

Since significant amounts of unexploited mineral resources are located in Nordic areas, the global urbanization and hunt for new mineral resources will be one the major driving forces leading the future development in the Nordic countries. The Fennoscandian Shield comprises a diversity of geological settings containing large resources of mineral deposits (Eilu, 2010). The resources include industry mineral for a number of applications and uses, energy minerals and not at least important metals. One could specially mention the Norrbotten and Västerbotten counties in Sweden, which are well established key mineral deposits provinces. Today, Kiruna-Malmberget is the worlds largest mining operation north of the Arctic Circle. Several new deposits are in the beeing developed, and there are planned investments for more than 30 bill. SEK in the mining sector.

Exploration and exploitation of mineral resources has for long been a priority area in Finland. As an example four gold mines have been opened in Finland since 1980. The most recent mine, Kittilä, opened in 2008. It is the largest gold deposits in western Europe with a resource of 5,7 Moz. Ore output from mines in Finland since 1950 is now at a peak, and is expected to further increase in the nearest years. Recently, Finland has launched a new national mineral strategy with the following vision towards 2050: "Finland is a global leader in the sustainable utilization of mineral resources and the minerals sector is one of the key foundations of the Finnish national economy". Sweden and Norway are now following Finland, and have started to develop their own strategies for the mineral sectors. In Norway, steps are taken to increase the coverage of relevant geological and geophysical information of the northern counties Finnmark, Troms and Nordland, where a four-year program with a total budget of 100 million NKr now is started within the frame of the government's Northern territories strategy (i.e, the MINN Program 2011-2014).

One of the primary goals for the Nordic geological surveys is to develop national and cross-border maps and databases of the bedrock geology and the mineral resources. The collective mission is to make this information and data easily accessible to all possible end users in industry, governmental agencies, public administrations and technical offices. The Fennoscandian Ore Deposit Database (FODD; http://en.gtk.fi/ExplorationFinland/fodd) is a comprehensive numeric database on metallic mines, deposits and significant occurrences in Fennoscandia. The maps and the database have been compiled in a joint project between the geological surveys of Finland, Norway, Russia and Sweden. The database contains information on 1300 mines, deposits and significant occurrences across the region. Of all deposits listed in the database, 56% have not been exploited at all. However, a number of these might well be economic in the future with additional reserves based on further exploration. FODD contains information on location, mining history, tonnages and commodity grades with a comment on data quality, geological setting, age, ore mineralogy and mineralization styles, genetic models, and the primary sources of data.

Fig. 2. Copper deposits at Repparfjord in Finmark. The Fennocandian Ore Deposit Database (FODD) contains information on 1300 mines, deposits and significant occurrences across Norway, Sweden, Finland and NW Russia (Photo: Jan Sverre Sandstad, NGU).

4. Access to clean water

Clean water is essential for any society. However, over larger parts of the globe clean water is in short supply. Many water reserves are over-exploited and polluted. Human health is endangered by the use of water which, either for natural reasons or because of pollution, contains harmful constituents. Groundwater for water supply purposes is often better and cheaper than surface water and is the most important drinking water source in many densely populated areas of the world. Groundwater reserves are renewable to the extent that the reservoirs are replenished, either directly or indirectly, by rainfall. However, groundwater is an invisible resource, which requires relatively large investments for mapping and monitoring, so the centralized management of knowledge and data concerning groundwater is of considerable economic value for society.

In Norway, NGU is responsible for mapping and monitoring of groundwater resources, and managing the national groundwater database. An important task is to develop fundamental data for groundwater management in accordance with the EU's Water Directive and associated subsidiary directives. According to the regulations, water resources shall be characterized and monitored to ensure that they have a good ecological status. In collaboration with the Norwegian Water Resources and Energy Directorate, NGU has been operating a nationwide monitoring network for untouched groundwater resources since 1977, covering groundwater levels, temperatures and water quality (Frengstad & Dagestad, 2008).

The conditions in Norway, Sweden and Finland differ considerably from most other countries within the EU as regards groundwater deposits, population density and pollution load. In general, Nordic groundwater sources provide good qualities of drinking water. Bottled mineral water is already being exported from Norway and profitable export of freshwater in bulk is in the pipeline. Europe consumes increasing amounts of bottled mineral water, and bottled water, usually derived from groundwater, is rapidly becoming the main drinking water supply.

Fig. 3. Groundwater is the largest and most reliable of all freshwater resources. In many areas most drinking water is groundwater; up to 80 % in many European countris and Russia, and even more in North Africa and the Middle East (Photo: Edelpix).

In 2010 more than 1900 "mineral water" brands were officially registered in Europe. In a recent, innovative, study conducted by EuroGeoSurveys, analysis of bottled water was used to provide indicators of the groundwater chemistry at the European scale (Reimann & Birke, 2010). The study included 1785 bottled water samples, representing 1247 locations all over Europe. The water was analyzed for more than 70 parameters (geochemical elements). The influence of geology in determining element concentrations in bottled water can be recognized for a significant number of elements. One example is the high values of chromium related to the occurrence of ophiolites, another is the high values of arsenic, fluorine, potassium, rubidium and silicon in bottled water coming from sources related to volcanic rocks. However, the natural variations are very large, usually an order of magnitude of three or four, and for some elements up to seven (Reimann & Birke, 2010). The study documented that very few analyzed samples (less than 1%) showed values exceeding maximum admission concentrations for mineral-water, as defined by the European Commission (Reimann & Birke, 2010).

The quality and hydro-chemical fingerprints of the groundwater is controlled by many factors, including rainfall chemistry, climate, vegetation and soil zone processes, the interactions between the minerals in underground and the water, groundwater residence time and mineralogy of the aquifer (Reimann & Birke, 2010). The EU Water Directive further provides scope for its practical implementation to be adapted to the natural condition in each country (Frengstad & Dagestad, 2008). Watercourses, groundwater and coastal water must be viewed in context, and the people who live upstream should resolve any problems in collaboration with those who live downstream regardless of administrative or national boundaries. Given the potential changes in watershed and the flow-regimes in our waterways due to forthcoming climatic change, access to water might be a major source to severe conflicts in the year ahead of us. Meeting these challenges, the Water Directive gives us a golden opportunity to make a common European effort to secure water resources for the future, and to provide stable and sustainable conditions for both the environment and future human generations.

5. Green energy beneath our feet

The greater use of more environmentally friendly energy is a national goal for many countries. In Norway, government has stated that it will "continue the effort to adapt national energy production and energy use, which will also have benefits in terms of climate policy, through the follow-up of the goal to introduce new environmentally friendly energy production and savings". The increased use of ground source heat—energy stored in bedrock, groundwater or sediments—will be an important contributor.

The thermal state in the shallow crust (i.e. less than 1000 m depth) is sensitive to surface effects, such as geological conditions (radiogenic heat production, terrestrial heat flow, thermal conductivity), terrain effects such as topography and slope orientation, climatic conditions (mean annual surface temperature) and human activity (land-use such as urbanization and farming).

A number of quantitative models from geothermal low activity, Nordic areas, show that at shallow depths down to a few hundred meters, mean annual surface temperature is the main factor controlling subsurface temperature (Slagstad et al., 2008). Geological variation in the underground such as heat flow, heat production and thermal conductivity first become significant at depths around 1000 m and deeper. Since ground-source heat for household heating is commonly extracted from shallow boreholes between 100 and 200 m depths, the effects of variation in heat-flow and heat-production has no impacts on the amount of heat that is extractable from the ground. This means that the key factors controlling the effect and economy of installations for extracting geothermal energy at shallow depths, are mainly linked to the overburden (cover deposits) and the hydro-geological activity in the underground (Slagstad et al., 2008). Obtaining such information is thus needed to obtain the maximum geothermal outcome from the underground.

The potential of shallow geothermal energy can be further increased by using a underground storage system. Geothermal energy, solar energy and waste-heat from large buildings and plants can be stored in the underground by a Underground Thermal Energy Storage (UTES) system, as the ground has proved to be an ideal medium for storing heat (and cold) in large quantities and over several seasons of years (Midttømme et al., 2008). In the Nordic countries UTES systems are mostly used in combination with Ground-Source Heat Pumps (GSHP). Today, more than 15 000 GSHP systems exist in Norway, extracting about 1,5 TWh heat from the underground. Two of the largest closed-loop GSHP systems in Europe, using boreholes as ground heat exchangers, are located in Norway (Akershus University Hospital and Oslo Gardermoen International Airport) (Midttømme at al., 2008).

In many countries, we are now approaching breakthroughs in utilizing geothermal sources as important energy suppliers. In addition to using the energy from the shallow-ground, many countries have areas with high thermal gradients in the underground. On average the temperature of the Earth increases with about 30°C/km (Lund et al., 2008). However, many places have significant higher gradients, for example where we have anomalous heating of rocks by decay of radioactive elements, where we have intrusions of magma from depths, or where there are have very thin crust associated with volcanic activity. One profound example is Iceland, with its sub-aerial exposures of the Mid-Atlantic Ridge, manifested by active rift zones extending from northeast to southwest on the island. The active volcanism produces a high heat flow to the surface, caused by magmas emplaced in the upper crust. This heat is extracted from both "high temperature fields" within the active volcanic zones and "low temperature fields" outside these zones. The heat is extracted as hot water and steam and is used for district heating, industrial purposes and power generation, offering a cheap and environmentally benign source of energy for the Icelandic society (Smelror et al., 2008).

Fig. 4. The Blue Lagoon on Iceland, where people can enjoy the hot water generated from the geothermal powerplant seen in the background. (Photo: Halfdan Carstens).

6. Living on polluted ground

The rapidly growing use of materials leaves larger and larger amounts of waste which have to be taken care of. Waste management and recycling is becoming increasingly important. For many years, NGU has been mapping pollution in densely populated areas. In towns and cities, there are areas where the ground is extensively polluted as a result of previous industrial discharges, fires in urban areas, road traffic and the combustion of coal and waste. The problems are linked in the first instance to heavy metals, arsenic, PAH and PCB.

Studies on the links between polluted ground and health have shown that it is not necessarily on industrial sites we have the most significant pollution problems. People come in contact with the soil pollution in the city center areas more frequent than they do with contamination from the most polluted industrial sites. Examples from the Nordic cities Oslo, Bergen, Trondheim, Tromsø and Copenhagen, as well from New Orleans have shown that it is moderately polluted urban soil in children's play areas that represents the greatest health hazard (Ottesen & Langedal, 2008; Ottesen et al., 2011). Children can come into contact with polluted ground through skin contact, by breathing in air-borne dust or gases, or by eating soil and licking their fingers. Studies have shown that around 10% of all children eat approximately 200 milligrams of soil per day, some even more (Ottesen & Langedal, 2008).

Fig. 5. Children eat soil, and in many cities moderately polluted urban soil in children's play areas represents the greatest health hazard (Photo: NGU).

The results from the studies of soil pollution at nurseries in the Norwegian cities led the Parliament in 2007 to approve a plan to map the soil pollution and to carry out clean-up operations at nurseries and school playground all over the country (Ottesen et al., 2011). Here it must be mentioned that the results of pollution-mapping in Oslo showed the greatest pollution to be in the oldest districts. All together, action was necessary at 38% of the city's 722 nurseries (Ottesen & Langedal, 2008).

The example above is just one showing the need for careful investigations of the ground of our urban areas. Another central theme is the spreading of environmental toxins from the land to the sea, where buildings in towns and cities act as an active pollution source for metals and PCB. The geological surveys and the national environmental agencies should continue to ensure that pivotal environmental problems are placed on the agenda and that the necessary measures are implemented. One such measure will be to prepare and implement the use of hazard maps for soil pollution in all the major city municipalities. In this respect, the recent EuroGeoSuveys project on mapping of the chemical environment of several major urban areas in Europe represents a major contribution (Johnson et al., 2011).

7. Facing geohazards

Human activity affects and transforms the environment around us. The land we build and live on is not always stable. The risk of natural disasters such as earthquakes, rock-falls, landslides, avalanches, floods and tsunamis must be assessed in relation to existing and planned settlements and infrastructure. One important task at the geological surveys is to evaluate and map areas with potential rock-fall and landslide hazards (Bargel et al., 2008). In Norway, this work has been intensified the recent years, and in order to assist the municipal authorities in obtaining a better overview of rock-fall and landslide processes and risk, a national landslide database (www.skrednett.no), presenting awareness- and risk-maps, has been developed.

Landslides, rock-falls and avalanches are important geological processes in the Nordic landscape. Slow displacements through time may cause instabilities and bedrock failures. If large rock-falls and landslides run into narrow fjords or alpine lakes they may trigger tsunamis, damaging near-shore settlements and infrastructure (Harbitz et al., 2006). The three natural disasters causing the largest number of deaths in Norway in the 20th century involved large rock-slides in a narrow lake in Loen (1905 and 1936) and in Tafjord (1934) (Nadim et al., 2008). Currently, several unstable mountain-sides in western and northern Norway are being monitored to follow movements. A center for monitoring the unstable mountain-side at Åkneset in Storfjorden, and other areas in the Møre-Romsdalen District, is established at Stranda (www.aknes.no). Here is also a early-warning system, which will alarm the around 3000 inhabitants in the small communities along the fjord, and the up to 30 000 tourists visiting the World Heritage Site in Geiranger per day in the summer months, when the risk of a major rock-fall and following tsunamis has reached a given threshold, and an evacuation should take place.

Fig. 6. Prekestolen at Lysefjorden, SW Norway. Rock-falls and landslides that will run into narrow fjords and create tsunamis which will damage near-shore settlements and infrastructure, represent potential large geohazards in Norway (Photo: Edelpix).

Fine-grained marine sediments cover large lowland areas of middle and eastern Norway, and large areas in southwestern and middle Sweden and Finland. Dilution of salt by groundwater flow leads to the formation of quick clay. Such processes may lead to highly unstable conditions, and fatal quick-clay slides can occur. Glacial tills on steep mountain slopes may collapse during periods of intense precipitation, and trigger debris flows. With changing climatic conditions and more extreme rainfalls in some exposed regions, the frequency of such natural hazards is expected to increase in the years to come. An important task for the geological surveys has been to produce maps showing which areas are subjected to potential quick-clay slides. Such awareness- and risk-maps have become an important tool for areal planners working on local and regional scales.

Recent incidences in Norway have demonstrated the risk for slides along the fjord-shores. Urban development and building of new infrastructure along the waterfronts creates risk for triggering slides in unstable areas. One example is the Kattmarka quick clay slide, which took place on 13 March 2009 close to the city of Namsos. The slide involved between 300 000 and 500 000 m^3 clayey soils and destroyed several homes, fortunately without serious injuries to persons. The slide was triggered by blasting taking place in connection with ongoing construction for widening the local road.

Another example is in the harbour of the Trondheim City, where a road construction triggered a sub-marine slide in 1990. In the bay of Trondheim landslides are recurrent phenomena, and recent and ongoing development of the area, including land reclamation and extension of harbor facilities, have increased concerns about the stability of the shoreline slopes (L'Heureux et al., 2010a). A recent study by L'Heureux et al. (2010b), using detailed morphological analysis of slide scars combined with limit equilibrium back-analyses, suggests that the presence of softer and more sensitive laminated clay-rich beds within the Trondheim harbour delta-deposits facilitates translational, slope failure, by acting as slip planes. Additional pre-conditioning factors promoting instability include the loading of the weaker clay-rich beds through delta progradation, and local over-steepening and artesian groundwater pressure at different underground levels. For the recent landslides in the Trondheim harbor, anthropogenic factors like embankment fillings and vibrations from construction works are considered the most likely triggering mechanisms (L'Heureux et al., 2010b). The results illustrate the importance of detailed morphological analyses, combined with a geological model including the physical/geotechnical characteristics of sediments on land and in the fjord, in order to perform a proper assessment of the shoreline slope stability.

Identification and monitoring of ground deformation can be accomplished using a number of surveying techniques. Since the early 1990's satellite-based radar interferometry has been used to identify large ground movements due to earthquakes and volcanic activity. Data stacking methods that take advantage of a growing archive of radar images, as well as increasing computing power, have led to a large increase in the precision of the technique. Both linear trends and seasonal fluctuations can be identified using the Permanent Scatterers technique. By using InSAR-technology, it has been possible to measure the degree on vertical subsidence due to compaction of the land-fills placed on the outer delta of Trondheim by an accuracy of mm per year (Dehls, 2004, 2005). By applying such novel techniques, it is possible to monitor small-scale, but critical, movements in the harbour, and other places in Trondheim City.

The cities and urban areas of the Nordic countries are small compared to megacities found in the Worlds more populated areas. But as elsewhere in the World, the major part of the population growth is within the already most populated areas, and an increasing part of the infrastructure is developed underground. The communities that invest in good knowledge-bases of geo-scientific information will have better means to secure optimal planning processes and underground operations. There are number of examples of how lack of basic geological and hydrological knowledge and data has caused serious problems. When building new road- and railway-tunnels near Oslo in Norway, cave-inn and water-leakage have been prominent problems (i.e. the Romeriksporten, Oslofjord, Hanekleiv and Hasle tunnels). The construction of the Romeriksporten tunnel led to drainage of the watercourses in Østmarka, lowering of groundwater levels and subsequent subsidence and damage to buildings (Olesen & Rønning, 2008). The final construction costs increased three-fold relative to the budget.

Fig. 7. Topographic/bathymetric view of the city of Trondheim, showing the landfills and the locations of submarine slides occurring in the years 1888, 1950 and 1990. (Illustration: NGU).

When the a central part of the Hanekleiv tunnel on the main road E18 in Vestfold south of Oslo caved-in on the 1st day of Christmas 2006 one of the national newspapers consulted a recent awareness map for tunnel-planning made by the Geological Survey of Norway (NGU) some months before. The map shows the distribution of weakness zones in the bedrock due to deep weathering, and is based on regional geophysical measurements and field observations (Olesen et al., 2007; Olesen & Rønning, 2008). The question was raised (also on the prime-time national TV-news): "Why has this geological information not been taken into account when the tunnel was built?"

Another example of critical use of geological information in an urban area comes from the UNESCO World Heritage Site Bryggen in the city of Bergen, western Norway. Since 2002, an intensive monitoring scheme has shown damaging settling rates caused by deterioration of underlying cultural deposits. Lowering of the ground-water level and increased content of oxygen in the cultural layers has caused damage of the wooden historical buildings (De Beer & Mathiessen, 2008). The monitoring has focused both on chemistry and quantity of groundwater and soil moisture content in the saturated and unsaturated zone, as well as registration of movement rates for buildings and soil surface. The documented preservation conditions within the cultural deposits as well as oxygen and moisture-content fluctuations in the unsaturated zone have a significant correlation with the different groundwater flow dynamics found throughout the site. By understanding the flow regime in the ground beneath the wooden buildings, means can be taken to stop the damaging development. The investigations demonstrated that groundwater and soil-moisture monitoring, combined with 3D transient modeling are potentially effective routines to improve the understanding of preservation conditions in complex archaeological surroundings and, therefore, protection of archaeological deposits in situ.

Fig. 8. The level and quality of groundwater are decisive for preserving valuable archaeological occupation layers in their original position, such as at UNESCO World Heritage Site Bryggen in Bergen, western Norway. "Groundwater data from NGU is helping to save Bryggen in Bergen", Jørn Holme, Head of the Norwegian Directorate for Cultural Heritage (Photo: Edelpix).

8. Managing knowledge

The Nordic geological surveys are active within almost all fields in which society has a need for geo-scientific knowledge and geological information, of which a few are described above. During the period of their existence, the surveys have generated a substantial amount of information on the Earth's crust, its natural resources, its processes, and on the geological history of Nordic areas.

The geological surveys are part of the fabric of the societies. At the core of the institution's tasks are management of knowledge that has been collected over generations and the mediation of this knowledge to the various users in society. Their collective mission is to make this geological information and data easily accessible to end users in industry, government agencies, government institutes, public administrations, technical offices, academia and research institutes, as well as for private individuals (Smelror et al., 2008). The development, operation and maintenance of national databases and maps of geological properties and processes therefore represent key tasks. From these databases, users can extract fundamental data and processed information which will help them to carry out their tasks, regardless of whether they are operating within the mineral industry, the consultancy sector, public administration or research and education.

Traditionally, geological maps and technical reports have been the main products, but today the products provided by the surveys cover a large spectrum of geological, geophysical and environmental databases, maps, models, cores and geological samples, literature and internet-based news- and service-pages. The formats and distribution protocols of geological data products have been jointly developed by the Nordic national communities and the EU (INSPIRE) spatial information community. Through the recent One-Geology project the European geological surveys have demonstrated that they can work and share data according to common standard in an interoperable way to create a common product, like on the dynamic digital (on-line) geological map of Europe (www.onegeology-europe.org).

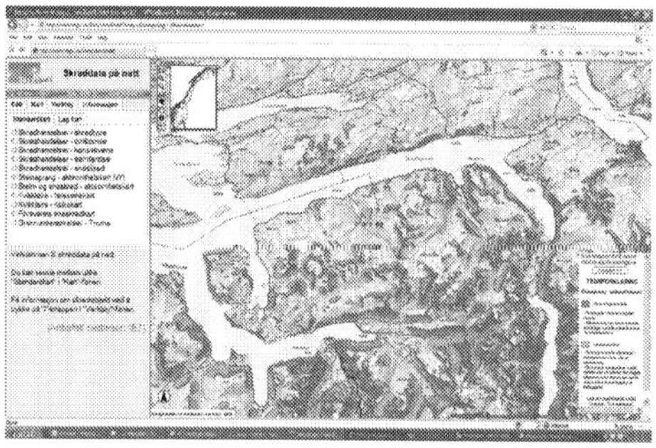

Fig. 9. The Norwegian database and map service of areas potentially at risk for landslides and avalanches provide direct access vital information to areal planners at local and national levels. (Source: www.skrednett.no).

However, in general data distribution policy varies between the countries; consequently, in some of the surveys supply of selected information or materials is chargeable by law. Others, like the Geological Survey of Norway, have made the information available on an open access basis. The internet is currently being developed as the main distribution channel as it gives easy access to key information for all users. The accessible databases are updated continuously. Over the history of the Nordic surveys, and at present, securing the growing volumes of geological and environmental information has consistently proved to be efficient and economically advantageous for society.
According to the Chinese philosopher and reformer Confucius (551 BC - 479 BC), "the essence of knowledge is to have it and to apply it". We believe is essential also to "share it".

9. References

Bargel, T., Blikra, L.H., Høst, J., Sletten, K. & Stalsberg, K. (2008). Landslide mapping in Norway. In Slagstad, T. and Dahl, R. (Eds.), *Geology for society for 150 years – the legacy after Kjerulf. Gråsteinen*, No. 12, pp. 58-75.

Cook, N.J., Karlsen, T.A. & Roberts, D. (Eds.) (2000). Industrial minerals and rocks in Norway. *Geological Survey of Norway Bulletin*, Vol. 436, pp. 1-207.

Eilu, P. (2010). Metallic mineral resources of Fennoscandia. *Geological Survey of Finland, Special Paper*, 49, 13-21.

De Beer, H. & Matthiesen, H. (2008). Groundwater monitoring and modeling from an archaeological perspective: possibilities and challenges. *Geological Survey of Norway Special Publication*, Vol. 11, pp. 67-81.

Dehls, J.F. (2004). Preliminary analysis of InSAR data over Trondheim with respect to future road development. *Geological Survey of Norway, NGU-rapport 2004.043*, pp. 1-8.

Dehls, J.F. (2005). Subsidence in Trondheim, 1992-2003: Results of PSInSAR analysis. *Geological Survey of Norway, NGU-rapport 2005.082*, pp. 1-12.

Frengstad, B. & Dagestad, A. (2008). Groundwater in Norway – A question of looking under the stream for water. In Slagstad, T. and Dahl, R. (Eds.), *Geology for society for 150 years – the legacy after Kjerulf. Gråsteinen*, No. 12, pp. 136-143.

Harbitz, C.B., Løvholt, F., Pedersen, G. & Madsson, D.G. (2006). Mechanism of tsunami generation by submarine landslides: a short review. *Norwegian Journal of Geology*, Vol. 86, pp. 355-364.

L'Heureux, J.S., Glimsdal, S., Longva, O., Hansen, L., Harbitz, C.B. (2010a). The 1888 shoreline landslide and tsunami in Trondheimsfjorden, central Norway. *Marine Geophysical Researches*, DOI 10.1007/s11001-010-9103-z.

L'Heureux, J.S., Hansen, L., Longva, O., Emdal, A. & Grande, L.O, (2010b). A multidisciplinary study of submarine landslides at the Nidelva fjord delta, Central Norway - Implications for geohazard assessment. *Norwegian Journal of Geology*, Vol. 90, pp. 1-20.

Johnson, C.C., Demetriades, A., Locutura, J. & Ottesen, R.T. (2011). *Mapping of the Chemical Environment of Urban Areas*. Wiley-Blackwell, U.K., 616 pp.

Lund, J.W., Bjelm, L., Bloomquist, G. & Mortensen, A.K. (2008). Characteristics, development and utilization of geothermal resources - a Nordic perspective. *Episodes*, Vol. 31 (1), pp. 140-154.

Midttømme, K., Banks, D., Ramstad, R.K., Sæther, O.M. & Skarphagen, H. (2008). Ground-Source Heat Pumps and underground Thermal Energy Storage - Energy for the future. In Slagstad, T. and Dahl, R. (Eds.), *Geology for society for 150 years – the legacy after Kjerulf. Gråsteinen*, No. 12, 93-98.

Nadim, F., Schank Pedersen, S.A., Schmidt-Thomé, Sigmundsson, F. & Engdahl, M. (2008). Natural hazards in Nordic Countries. *Episodes*, Vol. 31 (1), pp. 176-184.

Olesen, O., Dehls, J.F., Ebbing, J., Henriksen, H., Kihle, O. & Lundin, E. (2007). Aeromagnetic mapping of deep-weathered fracture zones in the Oslo Region - a new tool for imporved planning of tunnels. *Norwegian Journal of Geology*, Vol. 87, pp. 253-267.

Olesen, O. & Rønning, J.S. (2008). Deep weathering: Past climates cause tunnel problems. In

Slagstad, T. and Dahl, R. (Eds.), *Geology for society for 150 years – the legacy after Kjerulf. Gråsteinen*, No. 12, pp. 101-113.

Ottesen, R.T. & Langedal, M. (2008). Urban soil - A toxic history. In Slagstad, T. and Dahl, R. (Eds.), *Geology for society for 150 years – the legacy after Kjerulf. Gråsteinen*, No. 12, pp. 124-135.

Ottesen, R.T., Alexander, J., Langedal, M. & Mikarlsen, G. (2011). Clean Soil at Child-Care Centres and Public Playgrounds - An Important Part of Norway's Chemical Policy. In Johnson, C.C., Demetriades, A., Locutura, J. & Ottesen, R.T. (Eds), *Mapping of the Chemical Environment of Urban Areas*, pp. 497-520, Wiley-Blackwell, U.K.

Reimann, C. & Birke, M. (Eds.) (2010). *Geochemistry of European Bottled Water*. Borntraeger Science Publishers, Stuttgart, Germany, 268 pp (+CD-ROM).

Slagstad, T. (Ed.) (2008). Geology for Society. *Geological Survey of Norway Special Publication*, Vol. 11, pp. 1-154.

Slagstad, T., Midttømme, K., Ramstad, R.K. & Slagstad, D. (2008). Factors influencing shallow (< 1000 m depth) temperatures and their significance for extraction of ground-source heat. In Slagstad, T. (Ed.) 2008. *Geology for Society. Geological Survey of Norway Special Publication*, Vol. 11, pp. 99-1109.

Smelror, M., Ahlstrøm, A., Ekelund, L., Hansen, J.M., Nenonen, K. & Mortensen, A.K. (2008). The Nordic Geological Surveys: Geology for Society in practice. *Episodes*, Vol. 31 (1), pp. 193-200.

5

Geology and Geotectonic Setting of the Basement Complex Rocks in South Western Nigeria: Implications on Provenance and Evolution

Akindele O. Oyinloye
Department of Geology, University of Ado-Ekiti
Nigeria

1. Introduction

1.1 Regional geology of Nigeria

Nigeria lies approximately between latitudes 4°N and 15°N and Longitudes 3°E and 14°E, within the Pan African mobile belt in between the West African and Congo cratons. The Geology of Nigeria is dominated by crystalline and sedimentary rocks both occurring approximately in equal proportions (Woakes et al 1987). The crystalline rocks are made up of Precambrian basement complex and the Phanerozoic rocks which occur in the eastern region of the country and in the north central part of Nigeria. The Precambrian basement rocks in Nigeria consist of the migmatite gneissic -quartzite complex dated Archean to Early Proterozoic (2700-2000 Ma). Other units include the NE-SW trending schist belts mostly developed in the western half of the country and the granitoid plutons of the older granite suite dated Late Proterozoic to Early Phanerozoic (750-450Ma).

2. Geology of southwestern Nigeria basement complex

The area covered by the southwestern Nigeria basement complex lies between latitudes 7°N and 10°N and longtitudes 3°E and 6°E right in the equatorial rain forest region of Africa (Fig.1). The main lithologies include the amphibolites, migmatite gneisses, granites and pegmatites. Other important rock units are the schists, made up of biotite schist, quartzite schist talk-tremolite schist, and the muscovite schists. The crystalline rocks intruded into these schistose rocks. For the purpose of this chapter, discussion is limited to the crystalline basement rocks of southwestern Nigeria.

2.1 The Amphibolite and the hornblende gneiss

The amphibolite and the hornblende gneiss are the mafic and intermediate rocks in south western Nigeria. The amphibolites are made up of the massive melanocratic and foliated amphibolites. In Ilesha and Ife areas these amphibolites occur as low lying outcrops and most are seen in riverbeds. The massive melanocratic amphibolite is darkish green and fine grained. Commonly hornblende gneiss outcrops share common boundaries with the melanocratic amphibolite. This rock (hornblende gneiss) crops out at Igangan, Aiyetoro and Ifewara, along Ile-Ife road as low lying hills in southwestern Nigeria. The hornblende gneiss is highly foliated, folded and faulted in places.

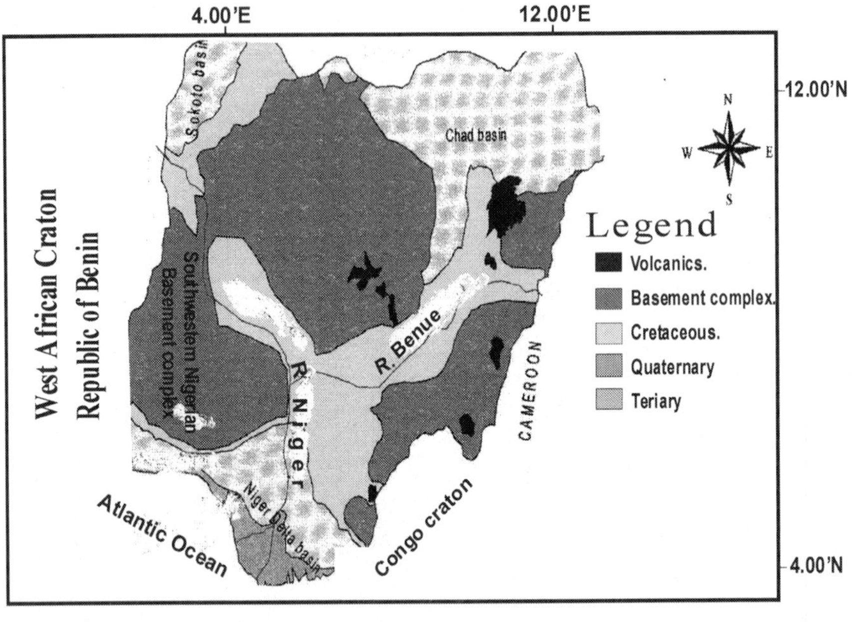

Fig. 1. a) Geological map of Nigeria; b) Geological map of Ilesha schist belt Southwestern Nigeria (modified from Elueze, 1982)

2.2 The Magmatite -gnessic complex

This geotectonic complex which constitutes over 75% of the surface area of the south-western Nigerian basement complex is said to have evolved through 3 major geotectonic events:

- Initiation of crust forming process during the Early Proterozoic (2000Ma) typified by the Ibadan (Southwestern Nigeria) grey gneisses considered by Woakes et al; (1987) as to have been derived directly from the mantle.
- Emplacement of granites in Early Proterozoic (2000Ma) and
- The Pan African events (450Ma-750Ma). Rahaman and Ocan (1978) on the basis of geological field mapping reported over ten evolutionary events within the basement complex with the emplacement of dolerite dykes as the youngest. On the basis of wide geochemical analyses and interpretation, geotectonic studies, field mapping and plumbotectonics, Oyinloye (1998 and 2011) had suggested a modified Burke et al; (1976) sequence of evolutionary events in the Southwestern Nigeria basement complex as detailed in Table1.

S/N	**Sequence of Evolutionary Events in the South Western Nigeria basement complex**
11	Shearing, Chloritic and Zeolite mineralization of uncertain age.
10	Emplacement of dolerite dykes and gold mineralization at about 550Ma **(Oyinloye, 2006b)**
9	Formation of unsheared pegmatite, unfolded granitic veins of mid-Pan African age.
8	Major remobilization and deformation in Early Pan-African
7	Minor metamorphic deformation in Kibaran
6	Emplacement of microdiorite
5	Emplacement of Ibadan-Ile-Ife-Ilesha Granite gneiss: F2 folding fabrics in Granite, Gneiss.
4	Emplacement of microdiorite dykes of uncertain age
3	Emplacement of semiconcordant aplite sheets in the banded gneiss, collision of plates subduction of ocean slab in to the mantle **(Oyinloye, 2002b)**
2	Deposition of ocean sediments covering the whole basement complex **(Oyinloye 2002a)**
1	Generation and differentiation of wet basaltic magma and formation of proto continent, **(Oyinloye, 2004a)**.

Table 1. A modified sequence of events in the basement rocks in Ibadan-Ile-Ife-Ilesha area (modified from Burk et al; 1976)

2.3 Metamorphism in the southwestern Nigeria basement rocks

On the basis of petrology a medium pressure Barrovian and Low-medium pressure types of metamorphism had been suggested for the Precambrian basement rocks in south western Nigeria (Rahaman 1988). These metamorphic types are based on the occurrence of index minerals like chlorite, biotite and sillimenite in the basement rocks of southwestern Nigeria. Rahaman (1988) therefore concluded that metamorphism in all Nigerian Precambrian complex rocks especially that of Ife-Ilesha (Southwestern Nigeria) ranges from green schist to lower amphibolite metamorphic facies. However, Oyinloye (1992) on the basis of petrology, field mapping and structural analyses reported that the prominent gneissic foliations observed on some of the gneisses suggest that metamorphism actually reached an upper amphibolite facies in the rocks of the basement complex in Southwestern Nigeria. Egbuniwe (1982) suggested 3 phases of metamorphism (M_1, M_2 and M_3) associated with 3 phases of deformation (D1, D2, D3) within the crystalline rocks of the basement complex in northern Nigeria. According to this author M1 represents a period of progressive metamorphism to lower amphibolite facies. M2 is described as retrogressive and reached only green schist grade as did M3. In the southwest Boesse and Ocan (1988) recognized 3 phases of metamorphism but only 2 phases of deformation. M1 is considered to be a syntectonic progressive phase of metamorphism to amphibolite facies with isoclinal folding, mineralogical banding and development of staurolite, sillimanite and garnet. M2 is described as syntectonic and associated with shear deformation and M3 being static retrogressing the earlier formed garnet and biotite to chlorite. Oyinloye (1992) however suggested that M1 is syntectonic and perhaps synchronous with the formation of the large scale major fault zone indicated by formation of mylonite outcrops at Iwaraja, Southwestern Nigeria. M2 is also syntectonic and contemporaneous with D2 as indicated by the development of micro faulting folding (Plate 1), fracturing, shearing, formation of phyllonite and mylonite with distorted garnet crystals surrounded by syllimanite crystals and mylonitised granite gneisses.

3. Geochronology of the basement rocks of southwestern Nigeria

It has been established that the Precambrian basement complex of Nigeria including Southwestern Nigeria is polycyclic in nature, (Ajibade and Fitches 1988). The southwestern Nigeria basement complex had undergone 4 major orogenesis in:-

i. Liberian (Archaean) 2500Ma-2750± 25Ma
ii. The Eburnean orogeny (Early Proterozoic), 2000Ma-2500Ma
iii. The Kibaran orogeny (Mid Proterozoic), 1100Ma - 2000Ma
iv. The Pan African Orogeny, 450Ma-750Ma.

Of all the above, the Eburnean and the Pan-African are major events which modified the Precambrian Geology of Nigeria including the Southwestern Nigerian basement complex. The Eburnean event is marked by the emplacement of the Ibadan granite gneiss in Southwestern Nigeria which has been dated 2500±200Ma (Rahaman 1988) and a pink granite gneiss at Ile-Ife Southwestern Nigeria dated 1875Ma using U-Pb on Zircon. Thus Archaean to Pan African ages had been suggested for the basement rocks of the Southwestern Nigeria.

Oyinloye (2006b), based on Pb-Pb model dating suggested 2750±25Ma (Archaean age) for the gneisses in Ilesha area Southwestern Nigeria. Few studies have been carried out on the basement complex due to its assumed monotonous petrology and mineralogy and the erroneous belief that it contains no mineralization. This current chapter will therefore contribute immensely to the debate on geology of the basement complex of Southwestern Nigeria.

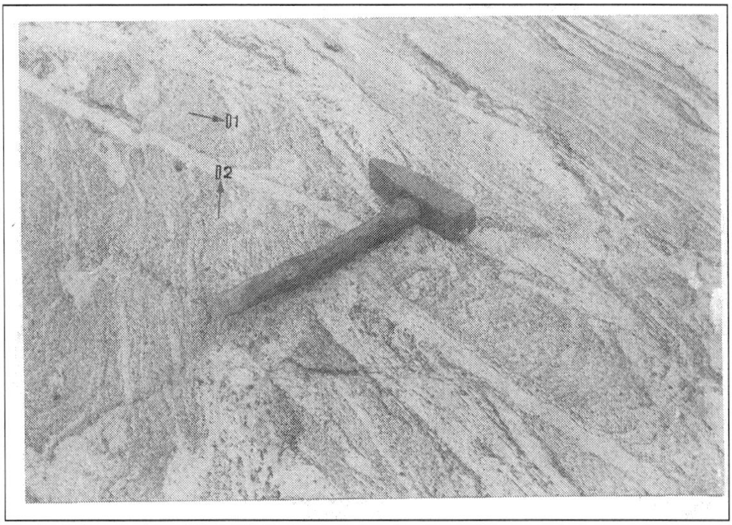

Plate 1. Deformation structures in the blotite granite gneiss, Iperindo area, Ilesha schist belt southwestern Nigeria, D2 (qtz vein) cuts across foliation planes (D1), The Hammer (2cm wide) marks a minor fault displacement

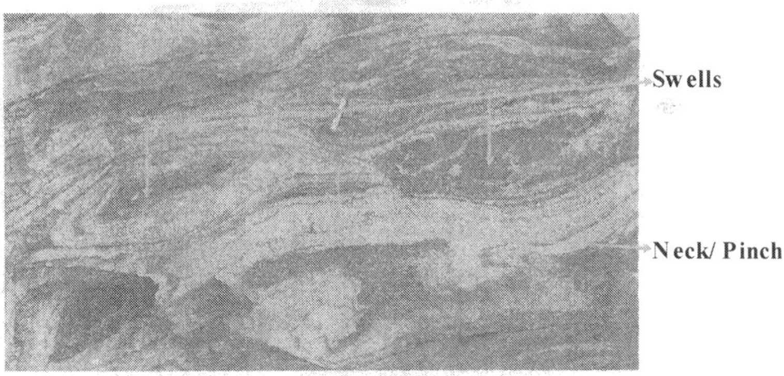

Plate 2. Showing Pinch and Swell Stuctures

4. Geological setting of southwestern Nigeria basement complex

The basement rocks which occur in southwestern Nigeria are all duplicated in the Ilesha area of southwestern Nigeria and samples of rocks here were analyzed and used as a case study of the basement rocks in Southwestern Nigeria to avoid repetition. These rocks are amphibolite, the hornblende gneiss and the granite gneisses. These rocks are described in that order.

4.1 The massive melanocratic amphibolite

Amphibolite occurs widely in southwestern Nigeria in Ile-Ife area, Ibodi, Itagunmodi in Ilesha area. Most outcrops of the massive melanocratic amphibolites are exposed in streams and river channels in these areas. The overburden soil here is strikingly red due to the presence of hematite and magmatite liberated during the weathering of the amphibolites to form the overburden soil. Two major textural varieties of amphibolites occur in this region. These are the leucocratic amphibolites and (not discussed in this study) the massive melanocractic amphibolites. The massive melanocratic amphibolite is darkish green and fine grained without any obvious folds or foliations. In places thin colourless quartz veins occur on the outcrops. This amphibolite variety is composed of hornblende, actinolite and tremolite. In thin section the mineral composition includes (apart from the above) magnetite, sphene calcite and minor monazite and zircon. The skeletal olivine contains small, opaque inclusions which are probably magnetite.

4.2 The hornblende gneiss

The hornblende gneiss shares a common boundary with the massive melanocratic amphibolite in Ilesha area, southwestern Nigeria. This rock crops out as low lying hills in Ife-Ilesha area Southwestern Nigeria. It is composed predominantly of porphy- roblastic plagioclase and hornblende phenocrysts almost in equal proportion. This rock is highly foliated folded and faulted in places and varies from medium to coarse in texture. These outcrops trend in a NE-SW direction and dip to the east at an average angle of between 50-70°. The apparent character varies from intermediate to acid. Microbands of foliation rich in plagioclase and some K-feldspars alternate with bands rich in amphiboles. In thin section this foliated hornblende gneiss consists largely of hornblende and plagioclase porphyroblats in a ground mass of ilmenite fine grained recrystallized quartz and pyroxene fragments. Brown coloured epidote (with dark cracks) apartite, sphene, zircon and monazite constitute major accessory minerals in this rock. Foliations defined by parallel arrangement of feldspars alternating with amphiboles are conspicuous in thin sections. Fine grained quartz and orthoclase feldspars are observed in the felsic microband, garnet, monazite, calcite and microcline containing well formed zircon crystals (as inclusions) occur in this rock as observed in thin sections.

4.3 The biotite granite gneiss complex

This rock group occurs widely in every part of the southwestern Nigerian basement complex. Again description is restricted to Ilesha area to avoid repetition.

Biotite granite gneiss complex occurs in the southern part of Ilesha schist belt. Outcrops of this rock group consist of high and low lying hills with myriads of flat boulders on top in places and roundish boulders on tops of hills elsewhere. This rock complex is foliated and folded with prominent synclinal and anticlinal axes. In places microfolds and microfaulting are observed (Plate1). Wide and narrow quartz veins are commonly seen on the rock and some of these are deformed to form folds and micro faults as described above. Foliations are defined by mafic (biotite rich) and felsic (quartz and feldspars) mineral bands. Drilled core samples from the biotite granite gneiss revealed that microfolds,pinkish garnet rich mylonite, greenish friable schistose phyllonite,occur in this rock.

In thin section the mylonite contains fine grains of biotite and sillimanite surrounding large crystals of garnet which show some evidence of distortion. The mylonite contains little quartz and the biotite flakes form thin foliation bands which are closely packed around garnet crystals. In some of the cores examined recovery failures are recorded indicating fracturing as described above. This biotite granite gneiss contains deformation fabrics which may be regarded as D2 and probably contemporaneous with the M2 phase of metamorphism following D1 and M1 (Plate 1). These later events may be due to movements along the major Ifewara-Zungeru fault system. The biotite granite gneiss are surrounded by muscovite-quartzite schists and in places the later are in-foled into the gneisses where they occur as reminants. At outcrop scale, the biotie granite gneiss is composed of biotite, K-feldsper, quartz and garnet. In thin section the biotite flakes are pencil-like as a result of metamorphic deformation and are aligned in parallel to sub parallel manner. The K-feldspar is mostly microcline and is porphyroblastic in texture. Well formed zircon crystals occur in association with some of the microcline grains. Apartite, monazite, magnetite, ilmenite and sphene are other accessory minerals.

In places distorted and fractured garnet grains due to metamorphic deformation are observed. Continuous well defined foliation bands of micas and felsic minerals are also common features of this gneiss. These gneissic fabrics probably indicate that metamorphism here was perhaps higher than the green schist-lower amphibolite facie regarded as the meramorphic grade for rocks in the basement complex in southwestern Nigeria. The presence of mylonite, mylonitised granite and gneissic banding are probable indications of a localized dynamic metamorphism possibly reaching an upper amphibolite facie.

4.4 The pink granite gneisses

This variety of gneiss occurs widely in the southwestern Nigeria basement complex at Ile-Ife, Ibadan, Iseyin, Eruwa and Iwaraja and in Ilesha area. The granite gneiss is pinkish with large pherocrysts of K-feldspar and porphyroblasts of hornblende. The texture of the pink granite gneiss varies from medium grained to very coarse almost becoming pegmatitic in places. Augen structures are commonly observed on the pink granite gneiss. This pink granite gneiss is fractured in places and elsewhere folded. Augen structures with clear elongate lozenges (boudins) and neck or pinch structures (pinch and swell) as a result of stressing are commonly seen on the pink granite gneiss in this region (Plate 2).

In thin section, foliation is defined by elongate hornblende and drawn out K-feldspar porphyries. Other minerals include quartz, plagioclase, some biotite flakes, garnet, apartite and zircon. Monazite forms an important accessory mineral in this rock. Commonly, phenocrysts of orthoclase occur within a matrix of recrystallised quartz and microcline. At Iwaraja, a major fault marked by a mylonite outcrop is observed within the pink granite gneiss terrain. This mylonite marks the southern extension of the Ifewara - Iwaraja-Zungeru major fault which runs in a NE-SW direction across the country. Deformation fabrics in the southwestern Nigeria basement complex are commonly aligned parallel to the direction of the Ifewara-Zungeru fault zone implying that this fault has a profound and wide influence on fabric and metamorphism in this region. According to Boesse and Ocan (1988) this major fault (marked by the mylonite outcrops) marks a break between the granite gneissic complex and the metasediments in this region.

4.5 The grey granite gneiss

The grey granite gneiss occurs prominently at Ibadan, Oyan, and in Ilesha areas of southwestern Nigeria basement complex. Usually outcrops consist of high and low hills and at Erinmo in Ilesha area occur very close to the pink granite gneiss and only separated by a narrow strip of muscovite quartzite schist. The overall colour is greyish. The texture of this variety of gneiss is fine to medium grained with well developed foliation defined by preferred orientation of biotite. This rock is mostly composed of quartz, biotite, plagioclase, K-feldspar and hornblende. In thin section recrystallised fine grained quartz covers the surface of microcline phenocrysts as overgrowths. This is a common phenomenon in all the granite gneisses investigated in this study. Mosaic textures formed by fragments of plagioclase, biotite and recrystallised quartz are also observed. Intergrowths of orthoclase and microcline forming a perthitic texture occur in places. Quartz crystals consist of fractured and recrystallised fine varieties. Well formed rod-like and fragmented zircon crystals, apatite, monazite plus minor garnet form important accessory minerals.

5. Geochemistry

The geochemical data described in this chapter are presented in the following order.

1. Massive Melanocratic Amphibolite
2. The Hornblende Gneiss
3. The Biotite Gneiss
4. The Pink Granite Gneiss

Note:- The average geochemical data discussed here are not included in this write up because of space. These are available from the author on request.

5.1 The massive melanocratic amphibolite: Major elements

In this study it is observed that element concentrations in the massive melanocratic amphibolite vary little even between samples collected from outcrops almost 1km apart. The mean SiO_2 concentration in this rock is 49% (17samples) alumina 15%, total iron 11%, MgO 10% and CaO 12%. The high iron concentration in the melanocratic amphibolite reflects the abundance of titanomagnetite and the high CaO content is an indication of the preponderance of Ca-rich pyroxene. T_iO_2 content (average 1%) reflects some sphene in addition to titanogmagnetite . The total alkaline concentration is very low reflecting the sub-alkaline nature of this rock. Na_2O is consistently higher than K_2O in this rock perhaps reflecting the dominance of albite in the massive melanocratic amphibolite.

MgO/ Fe_2O_3+MgO ratios vary between 0.45 to 0.48 with a mean of 0.46. This is considerably lower than that of a pure primitive upper mantle which has a range of 0.68-0.75 and a mean of 0.70 (Wilson, 1991).

5.2 Trace elements

In the massive melanocratic amphibolite Rb is characteristically low, 11ppm on average indicating low K-feldspar concentration as observed in the thin section studies. Sr with an average of 169ppm is relatively high in the massive melanocratic amphibolite due to substitution of Sr for Ca in the pyroxene and amphiboles as is Zr (58ppm) due to minor zircon. Zr can also substitute for Ti in accessory phase in sphene and rutile. Y concentrations are appreciable (mean 19ppm) since this element is readily accommodated in amphiboles which are the dominant minerals of the amphibolite. The low Th in this rock reflects fractionation into more felsic magmatic franctions. The average concentrations of compatible elements (Ni, Cr, and Co) in the massive melanocratic amphibolite are 102ppm, 81ppm and 54pm respectively. These values are too low for an amphibolite originating from a pure primitive upper mantle. The low concentrations of compatible elements suggest that the precursor rock of the amphibolite is from a depleted or metasomatised mantle and this has a significant implication on provenance and geotectonic setting in which the rock was formed.

5.3 Rare earth elements

Rare earth elements are significantly recorded in the massive melanocratic amphibolite. The average total REE in the massive melanocratic amphibolite is 71ppm. The dominance of light rare earth elements (59ppm) over the heavy ones average 12ppm reflects the relative abundance of monazite in the amphibotite and further suggests that this mafic rock is not from a pure primitive mantle.

5.4 The horblende gniess

Field and petrological studies revealed that this rock consists of intermediate to acid varieties. The average SiO_2 (63%) in the hornblende gneiss is much higher than that of the massive melanocratic amphibolite. TiO_2 and Fe_2O_3 averages are lower than that of the amphibolite reflecting the less mafic character of the hornblende gneiss. The relatively higher total alkalis (mean 7%) and K_2O/Na_2O ratios (0.8) are indicative of more abundant feldspars in the hornblende gneiss than the amphibolite which is consistent with field and thin section observations. The MgO/Fe_2O_3+MgO mean ratio is 0.32 and this is lower than in the amphibolite and thus further from pure upper mantle value. In this rock the average concentration of Rb (68ppm) is more than 6 times its concentration in the amphibolite paralleling the increase K-feldpar content. Sr and Ba are also strongly enriched (1266ppm and 1493ppm respectively). This is perhaps due to substitution of Ba for K in the K-feldspar and Sr for Ca in plagioclase. The higher Y (mean 37ppm) concentration in the hornblende gneiss compared with the amphibolite may be due to the presence of more hornblende (dominant mineral in the hornblende gneiss) which often concentrates this element. Low Th concent of the hornblende gneiss average (6ppm) might be due to minimal crustal contribution. In the hornblende gneiss, the mean concentraction of Ni, Cr, are less still (26ppm and 39ppm respectively) reflecting the less mafic character of this rock.

5.5 Rare earth elements (REE)

In the hornblende gneiss, there is a high concentration of REE most especially the light ones. The average total REE in the hornblende gneiss is 3232ppm. The light REE in this rock has a mean of 2174ppm. The mean concentration of the heavy REE in the hornblende gneiss (58ppm) is relatively higher than in the amphibolite reflecting more abundant REE concentrating minerals e.g. sphene and monazite.

6. The biotite granite gneiss

6.1 Major elements

The biotite granite gneiss is one of the series of granitic rocks with SiO_2 higher than 70% in southwestern Nigeria basement complex. TiO_2 average concentration in this rock (0.42%) is slightly higher than in the hornblende gneiss. Al_2O_3 mean (15%) is slightly higher in this rock than in the hornblende gneiss. Unlike the amphibolite and the hornblende gneiss, the Na_2O concentration is consistently less than K_2O in the biotite granite gneiss. K_2O/Na_2O ratios are consistently higher than 1 (one) in the biotite granite gneiss although the average total Na_2O and K_2O (8%) in the biotite gneiss is only 1% higher than that of the hornblende gneiss. The concentration of SiO_2, Al_2O_3, Na_2O and K_2O in the biotite granite gneiss indicates an abundance of felsic silicates e.g. feldspars and quartz. The consistently higher concentration of K_2O than Na_2O (thus K_2O/Na_2O ratios greater than 1) in the biotite granite gneiss reflects the abundance of K bearing rock forming silicates (i.e microcline and biotite). This trend is also characteristic of Archaean granitic rocks (Martin (1986). In the biotite granite gneiss, the concentrations of Fe_2O_3, MnO, MgO, CaO and P_2O_5 are much less than the average values in the amphibolite and hornblende gneiss. This reflects its less mafic character.

Although major elements of gneisses are sensitive to metamorphic alteration, AFM diagrams can be used to study the enrichment of these rocks in alkalis and Fe in a general way. When the AFM diagram is plotted for the hornblende gneiss and the biotite granite gneiss, both rocks plot in the calc alkali fractionation trend (Fig.2) reflecting enrichment in Al_2O_3, Na_2O and K_2O due to development of biotite and K-feldspars in the biotite granite gneiss and plagioclase in the hornblende gneiss as observed in the petrological studies.

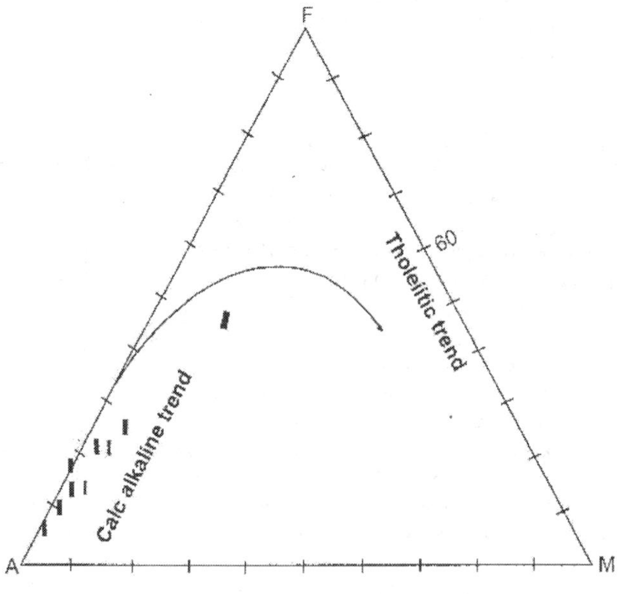

Fig. 2. A=Al_2O_3; F=FeO (total iron); M=MgO (AFM) diagram for the biotite granite gneiss, from Ilesha area

6.2 Trace elements

The average concentration of Rb in the biotite granite gneiss (182ppm) is more than double its average concentration in the hornblende gneiss. The high concentration of Rb in the biotite granite gneiss is due to substitution of Rb for K in the microcline and biotite which are abundant in this rock. The average concentration of Sr (299ppm) is less than 25% of its concentration in the hornblende gneiss. This reflects the low concentration of plagioclase,

hornblende and pyroxene in which Sr can substitute for Ca in the biotite gneiss. Zr is more concentrated in the biotite gneiss than in the hornblende gneiss. Th whose average concentration is 2ppm in the amphibiolite and 6ppm in the hornblende gneiss is relatively highly concentrated in the biotite gneiss (average 33ppm) in the biotite granite gneiss due to increased K-feldspar content which concentrates this element. The average concentration of Y (32ppm) is lower than its concentration in the hornblende gneiss reflecting less hornblende in the biotite gneiss.
The concentration of compatible heavy elements (Cr, Ni and Co) are relatively minor in the biotite gneiss reflecting its acid nature.

7. Rare earth elements (REE)

The average total REE in the biotite granite gneiss is 328ppm of which the light ones account for 320ppm, and heavy rare earth elements only 18ppm. As in the amphibolite and hornblende gneiss, higher amount of light rare earth elements are concentrated in the biotite granite gneiss than the heavy rare earth elements. However, the total REE in the biotite granite gneiss is lower than that of the hornblende gneiss due to lower abundance of sphene, plagioclase and hornblende in the biotite granite gneiss than in the hornblende gneiss.

7.1 The pink granite gneiss

The average concentrations of SiO_2 (76%) and Na_2O (2.40%) recorded in the pink granite gneiss in this region reflect higher concentration of K-feldspar in the pink granite gneiss than in the biotite granite gneiss. Al_2O_3, K_2O, CaO, P_2O_5 average concentrations in this rock are lower than in the biotite granite gneiss. The averages of K_2O/Na_2O ratios which are greater than 1 and the total K_2O+Na_2O (7%) on the average in the pink granite gneiss show the same trends as in the biotite granite gneiss. AFM plot for this rock also show a calc-alkali franctionation trend.

7.1.1 Trace elements

Ba and Sr concentrations in the pink granite gneiss are lower than in the biotite granite gneiss indicating lower plagioclase in the former than in the later. Higher concentration of Rb in the pink granite gneiss than in the biotite granite gneiss reflects more abundant K-feldspar in the pink granite gneiss. Zr has lower average in the pink granite gneiss than in the biotite granite gneiss which corresponds to lower zircon occurrence in the pink granite gneiss. Th mean concentration in the pink granite gneiss is higher than in the biotite granite gneiss which may be due to higher sedimentary contribution to the precursor of this rock. In the pink granite gneiss an increase of Y concentration is recorded compared with its average concentration in the biotite granite gneiss. This may be due to higher hornblende component of the pink granite gneiss than in the biotite granite gneiss.
The average Ni in the pink granite gneiss has diminished compared with the hornblende gneiss and biotite granite gneisses and Cr concentration in the pink granite gneiss is below the detection limit (3ppm) of the XRF used for these analyses due to implied less concentration of mafic minerals in this rock.

7.2 Rare earth elements (REE)

The total absolute REE on the average is 235ppm out of which light Rare Earth Elements account for 215 ppm and heavy rare earth elements 20ppm. The average ratio of

LREE/HREE: 11, recorded in the pink granite gneiss is lower than those of the biotite granite gneiss. However, the average HREE in the pink granite gneiss is higher than in the biotite granite gneiss. The lower concentration of REE in the pink granite gneiss than in the biotite granite gneiss reflects the less abundant REE concentrating minerals in the pink granite gneiss.

8. Geochronology

There had been some reported determination of the age of the basement complex rocks generally in Nigeria including the southwestern Nigeria basement rocks. On the basis of isotopic studies, Archaean and Proterozoic ages had been suggested as the ages of emplacement of the basement rocks in Nigeria by Dada et al; (1998) and Annor (1995). According to Ajibade et al, (1987) the southwestern Nigeria basement complex are of two age generations, one represented by migmatite gneiss complex probably of Archaean to Early Proterozoic age while the other is believed to be of Late Proterozoic age. Age determination of the southwestern Nigeria basement complex rocks has not been completed as much work needs to be done to actually date these rocks satisfactory. However Oyinloye (2006b, this author) carried out a Pb-Pb, 2-stage model age based on Stacey and Kramers (1975) on the granite gneisses in Ilesha area of southwestern Nigeria and part of the result is reproduced here.

8.1 Lead (Pb-Pb) model dates

The whole rock and feldspar samples analysed for lead isotopes in this study were from the biotite granie gneiss in Ilesha area of southwestern Nigeria. On plotting, these samples revealed limited scatter points on the Pb-Pb isochron (Fig.3) but with a well defined trend. Pb-Pb data for the six K-feldspar seperates (plotted in addition) are from the same biotite granite gneiss and are comparable to the equivalent whole rock. These results fit well to the indicated best fit line which corresponded to a two-stage isochron 2750±25Ma with an initial ratio of 12.809 and MSWD of 16 (Fig.3). On the Stacey and Kramers (1975) growth curve, the biotite granite gneiss whole rock and feldspar Pb experimental points plotted to the left of the geochron Q-P (Fig.5) crossing the growth curve at point N giving an initial ratio of 12.809 which was due to geochemical differentiation. The experimental Pb-Pb isochron yields a model age of 2750±25Ma (Fig.3). This implies that Pb was withdrawn from the unradiogenic reservoir and incorporated into the feldspars and the protolith of the biotite granite gneiss at about 2750±25Ma . This Pb-Pb age which is Archaean is therefore the age of emplacement of the precursor rock which gave rise to the biotite granite gneiss in Ilesha area of southwestern Nigeria.

On ploting the Pb-Pb data on Zartman and Doe (1981) evolutionary curve, (Fig.4) five out of the six whole rock samples and five out of the feldspar samples plot between the two curves OR and UP, (Fig.4). While only one sample of each (feldspar and whole_rock) of the feldspar and whole rock samples plot outside the curves. Samples which plot within the two curves UP and OR (Orogen) in Figure 4 indicate that their precursor rocks were derived from a tectonic environment where crustal/sedimentary and mantle materials were partially metted to generate the initial magma from which the protolith of this biotite granite gneiss was formed, (Cf. Zartman and Doe 1981). Furthermore, Pb-Pb isotope data show that the whole rock samples from the biotite granite gneiss are extremely homogenerous with only very slight deviations from the mean values. The feldspar separates show more isotopic homogeneity.(Oyinloye 2006b). This type of extreme isotopic homogeneity in rocks and feldspars is characeteristic of rocks derived from a subduction related environment like a back arc or an island arc, where mantle and upper crustal materials are thoroughly mixed to generate a magma (Billstrom 1990). Burke and Dewey (1972) had earlier described the Ilesha area in southwestern Nigeria as one that evolved in an island arc marginal basin but Oyinloye (2006b) showed that it evolved in a back arc tectonic setting.

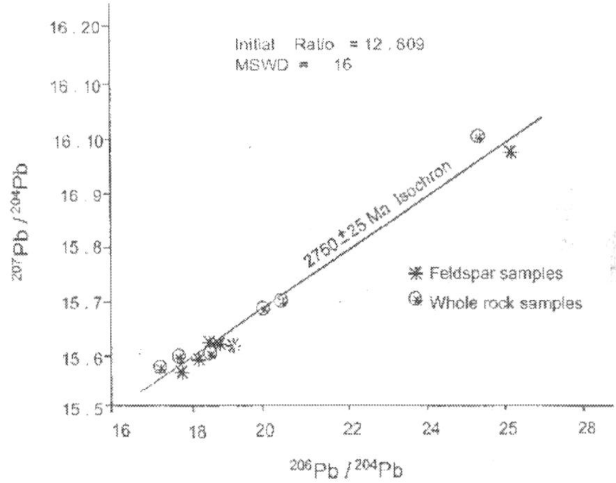

Fig. 3. Lead (Pb-Pb) whole rock and K-feldspar isochron diagram for biotite granite gneiss and K-feldspar from Ilesha schist belt , southwestern Nigeria

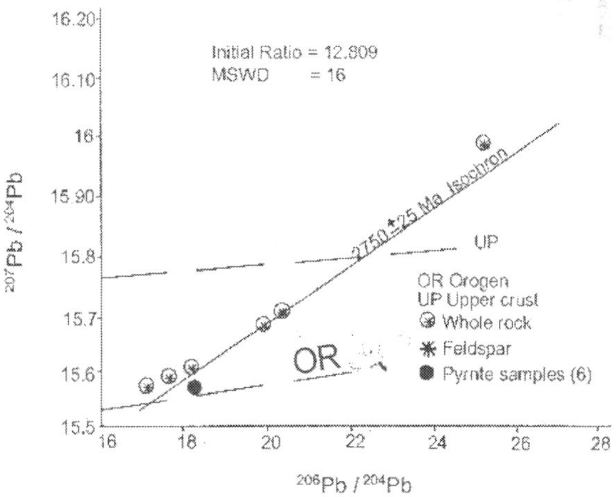

Fig. 4. Plumbotectonic plots using Pb-Pb from K-feldspar whole rock and pyrite samples (based on Zartman and Doe, 1981)

8.2 Mineralisation in Southwestern Nigeria basement complex

The crystalline rocks of the basement complex intruded the schists in the schist belts in southwestern Nigeria. The schist belts are critical to the understanding of the geology of the basement complex in Nigeria. Infact the schist belts are integral part of the southwestern Nigerian basement complex. Minerals are localized in rocks within the schist belts in southwestern Nigeria. Some of the minerals found within the southwestern Nigeria basement complex include, gold in Oyan and Ilesha schist belts, in Okolom and Gurungaji in Egbe schist belt southwestern Nigeria. Gold is the major metal found in the southwestern Nigeria basement rocks. However unproven reserves of cassiterite, columbite and tantalite are found in Ijero-Ekiti area of southwestern Nigeria. Others include gem stones - amythyst, tourmaline and quartz. Gold is the only metallic mineral of substance that has been studied in this area especially by this author. A summary of the geology and geochemistry of the gold deposit at Ilesha in southwestern Nigeria is reported here.

There are two (2) types of gold mineralization in Ilesha area. The first type is an alluvial form which occurs within the amphibolite terrain in Ilesha area southwestern Nigeria. This has not been well studied. The second one is a primary gold deposit found as auriferous quartz veins localized in a shear zone about 4km from the major Ifewara-Zungeru fault zone described within the pink grnite gneiss in this study. The biotite granite gneiss described in this report is the host rock of the Ilesha primary gold deposit, known as the Iperindo primary gold deposit. This gold deposit occurs as a system of auriferous quartz veins infillings, structurally localized at a folded boundary between biotite granite gneiss host rock and the adjacent metasedimentary complex. The granite gneiss host rock was altered by an invading hydrothermal fluid. The alteration selvages which form the hanging and footwall rocks are dominantly phyllic in nature with a minor chlorite overprinting in the foot wall rocks. These alteration selvages are relatively narrow but prominent and intensive around all the mineralized quartz veins at Iperindo. Geochemical analyses of the country and selvage rocks of this lode gold deposit show that the altered rocks are enriched in Cu, Zn, Pb and rare earth elements generally and heavy rare earth elements in particular relative to the country wall rocks reflecting presence of chalcopyrite, sphalerite, galena, and rare earth elements concentrating minerals and development of secondary alteration products such as sericite in the alteration selvages. General studies carried out using, stable carbon 13, oxygen 18 isotopes and plumbotectonics show that mineralization of gold at Iperindo near Ilesha was meteoric in origin. Ore fluid inclusion studies of selected samples from the auriferous quartz veins from Iperindo gold deposit indicate that the ore fluid was rich in carbondioxide. Microthermometric measurements show that there are two types of fluid inclusions in this gold deposit and these two homogenized at high temperatures but underwent phase separation at low temperatures. These two fluids are: 2-phases carbondioxide, (gas and liquid) and 3-phases, carbondioxide gas, carbondioxide liquid and water. Fluid inclusion studies also show that mineralization of gold took place at a temperature in excess of 286°C.

The data obtained from Pb isotope studies from pyrite which is found in Iperindo gold deposit show an extreme homogenous relationship. On Zartman and Doe (1981) evolutionary plot all the pyrite Pb plot within the two curves just like the whole rock and feldspar separates (Fig.4) indicating genetic relationship (note: only one of the pyrite Pb samples appears in Figure 4 because they all clustered at a point). The model ages calculated for each pyrite sample varies from 559Ma-573Ma with a mean of 550Ma (Oyinloye 2006b).

Also, the pyrite lead data showed that Pb isotopes in pyrites from the Iperindo lode gold are extremely homogeneous and very similar in value to those obtained from the whole rock and feldspar separates. Therefore going by the earlier interpretation of Pb homogeneity in feldspar, and hole rocks samples the Pb homogeneity observed in pyrite samples from this gold deposit might indicate derivation from a mixed crustal and mantle sources (Volcano-proto-continent precursor rocks of amphibolites and amphibolite schists) for the Pb in pyrite which forms a prominent gangue in Iperindo gold deposit.

The component of the ordinary Pb was probably withdrawn from its reservoir before 2750±25Ma as a result of magma generation and protocontinent rock formation. There was an hydrothermal invation of the volcanics leading to leaching of Au from these rocks, removal of Pb from the reservoir and incorporation into pyrite at about 550Ma, the age of gold mineralization in Ilesha area of southwestern Nigeria.

9. Provenance and evolution of the Southwestern Nigeria basement rocks

A controversial aspect of the geology of the Nigeria basement complex is its geotectonic origin. Only very few workers had applied geotectonics to interprete the origin of the basement rocks in southwestern Nigeria. In my research studies, I was able to use sophisticated equipment like scanning electron microscope Cambridge 250model in the United Kingdom to determine the spot chemical composition and empirical formulae of nearly all rock forming minerals in the rocks of the basement complex of southwestern Nigeria as represented by the amphibolite and granite gneisses in Ilesha area. A mineral known as monazite was discovered in this process. This mineral is present as a notable accessory mineral in all the crystalline rocks of the basement complex here in Ilesha area even in the amphibolite which is supposed to be purely igneous. Hither to except in the Younger Granites in the north central Nigeria and in sedimentary rocks in Lokoja and Auchi areas, monazite has not been described by any worker in the rocks of the basement rock of southwestern Nigeria in general and in Ilesha area in particular. It is in my research that monazite is being described for the first time in the rocks of the southwestern Nigeria and in Ilesha area in particular.

Monazite is a phosphate of the rare earth elements, especially the light ones e.g. (La, Ce, Nd) PO_4. Monazite is known to be a crustal or sedimentary mineral. Its presence in a supposedly igneous rocks of mantle origin therefore raises a petrogenetic question. The petrogenetic implication of the presence of monazite in the crystalline rocks of the southwestern Nigeria is that the initial magma from which the precursor rocks were formed contain some input from a crustal or sedimentary source. As described in this text. The $MgO/(Fe_2O_3+MgO)$ ratios recorded for the amphibolite are lower than that of the basalts derived from a pure primitive mantle and this ratio decays further from the hornblende gneiss to the granite gneisses.

In order to further unravel the provenance of these rocks in southwestern Nigeria, normative corundum of the hornblende gneiss and the granite gneisses were plotted against the Mol.%$Al_2O_3/(Na_2O+K_2O+CaO)$. Also, the histogram of the Mol.%$Al_2O_3/(Na_2O+K_2O+CaO)$ distribution for the gneisses were plotted (Figs.6 A and B). Most of the gneisses samples plot in S-type field while few samples plot in the I-type field (Fig. 6A). In Figure 6B, the gneisses sample show a bimodal histogram with a mode at I-type field and another at the S-Type field. Also the hornblende gneiss samples occur in both I-Type and S-Type fields. These plots imply that the magma which gave rise to the precursor rocks of these gneisses originated from a mixed source, containing igneous and sedimentary materials. Generally, all the plots show that the granitoids are very homogenous, related petrogenetically and are derived from a mixed source. The index trace elements were used to plot many discriminating diagrams (only one is shown here because of space). On plotting Ti versus Zr (Fig.8 based on Pearce et al; 1984) the massive melanocratic amphibolites data plot in the arc lavas field indicating a volcanic arc (similar to a back arc tectonic setting). Further more, it is observed that the average concentration of the compatible elements, (Ni, Cr, Co) in these rocks are extremely lower than that of the normal rocks derived from a primitive upper mantle source implying that the magmatic source had been metasomatised.

Chondrite normalized REE were plotted for all the crystalline rocks described in this study including the massive melanocratic amphibolite. The massive melanocratic amphibolite shows a slight negative Eu/Eu* anomaly and high La_N/Yb_N ratios. The gneisses show similar REE patterns and a higher negative Eu/Eu* anomally. These trends show a progressive differentiation from the basalts to the gneisses (Fig.7). The implication of these is that the precursor of these rocks originated from a basalt that differentiated progressively to the granite precursor rocks of the gneisses as shown in Figure 7.

The extreme Pb isotopic homogeneity as observed in the biotite granite gneiss samples and its feldspar separates indicates derivation from a subduction related environment like a back arc or an island arc where mantle and upper crust materials are thoroughly mixed to generate a magma (Billstron 1990). The southwestern Nigeria basement complex as typified by the Ilesha schist belt had earlier been described as one that evolved in a subduction related environment of island arc and marginal basin, Burke and Dewey (1972). But Oyinloye (2002a), Oyinloye and Steed (1996), and Oyinloye and Odeyemi (2001) on the bases of petrological, geochemical, plumbotectonics and structural analyses showed that the environment of the emplacement of the protocontinent which were the precursors of the rocks of the southwestern Nigeria basement rocks was a back arc basin. The linear trends displayed by the whole rock Pb data on the growth curve (Fig.5) reflects a mixing process between varying amounts of upper crust and mantle materials as in an island arc or a back arc environment, Billstrom (1989). The various discrimination diagrams based on the trace elements considered immobile during metamorphic alteration show that the rocks of the basement complex of southwestern Nigeria may possibly be derived from a low-K-Tholeritic magma, (Oyinloye and Odeyemi 2001). These plots also indicate a possible volcanic arc for these rocks. (Fig.8) both back arc and island arc are grouped in volcanic arch in Fig.8. A volcanic arc characteristic of the massive melanocratic amphibolite suggests that subduction tectonics was important in the formation of its parent magma. The flat shape of the REE curve (Fig.7A) with slight Eu anomaly is typical of a back arc basic rock (Wilson, 1989). Also the spider diagrams for the massive-melanocratic amphibolites (not shown) are similar to those of a spreading tectonic settings (e.g. Mid Ocean Redge basalt, or a back arc setting). But none of the samples in the discriminating diagrams plotted in the Mid Ocean Ridge Besalts Field (Fig.8). Furthermore the development of a negative Eu anomaly shown by these rocks especially the massive melanocratic amphibolite (Fig.7A) is alien to a mid-ocean ridge basalt or an island arc. A back arc tectonic setting will adequately account for the characteristics displayed by these rocks in the discrimination and REE fractionation trends as described above.

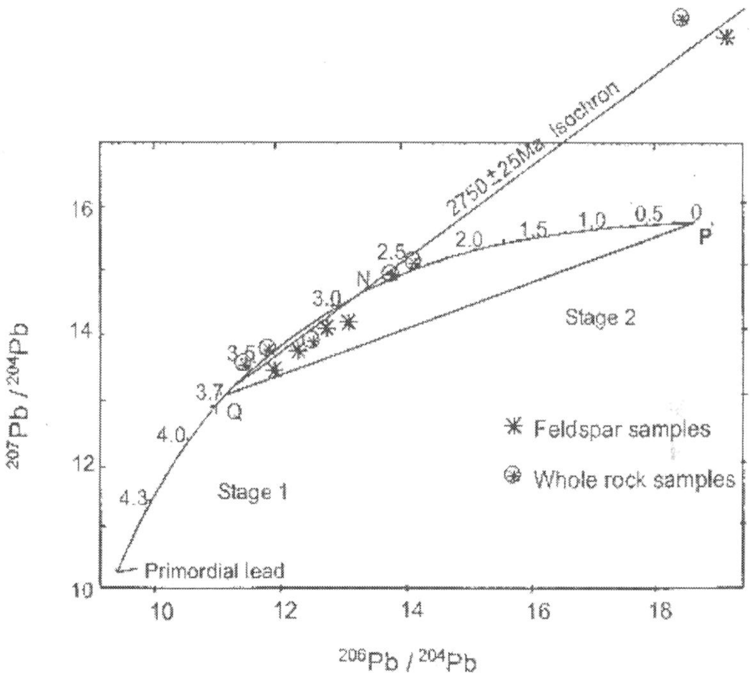

Fig. 5. Whole rock and K-feldspar Pb experimental data points on the two stage growth curve of Stacey and Kramer (1975)

The basement complex rocks of the southwestern Nigeria are believed to have developed in a back arc basin (Oyinloye and Odeyemi 2001). Rahaman et al; (1988), had suggested that an ocean was closing and opening at the West African margin. Holt et al (1978) based on geotectonic studies explained that the southwestern Nigeria basement complex resulted from the opening and closing of an ensialic basin with consequent extensive subduction during the Pan African events. Burke and Dewey (1972) from structural point of view believed that components of the schist belts containing the crystalline complex rocks in southwestern Nigeria had been formed in a back arc basin caused by the collision between the continental margin of the Tuareg shield (Hoggar belt in northwest Africa). But in this study and in the previous ones the petrological, geochemical and plumbotectonic studies revealed that these rocks originated from a mixed magma containing both mantle and sedimentary materials. Reviewing all the known tectonic environments especially island arcs and back arcs (which had been suggested as the geotectonic setting in which the rocks of the southerwestern Nigeria basement complex originated), the petrology, geochemistry and plumbotectonic studies of the rocks understudy implicate a back arc tectonic setting in which an ocean slab was subducted into the mantle. This subduction was due to a collision between an ocean slab and a continental shelf. In such an environment, the ocean slab would be subducted into the mantle with sedimentary materials and water which makes a wet mixed magma formation possible.

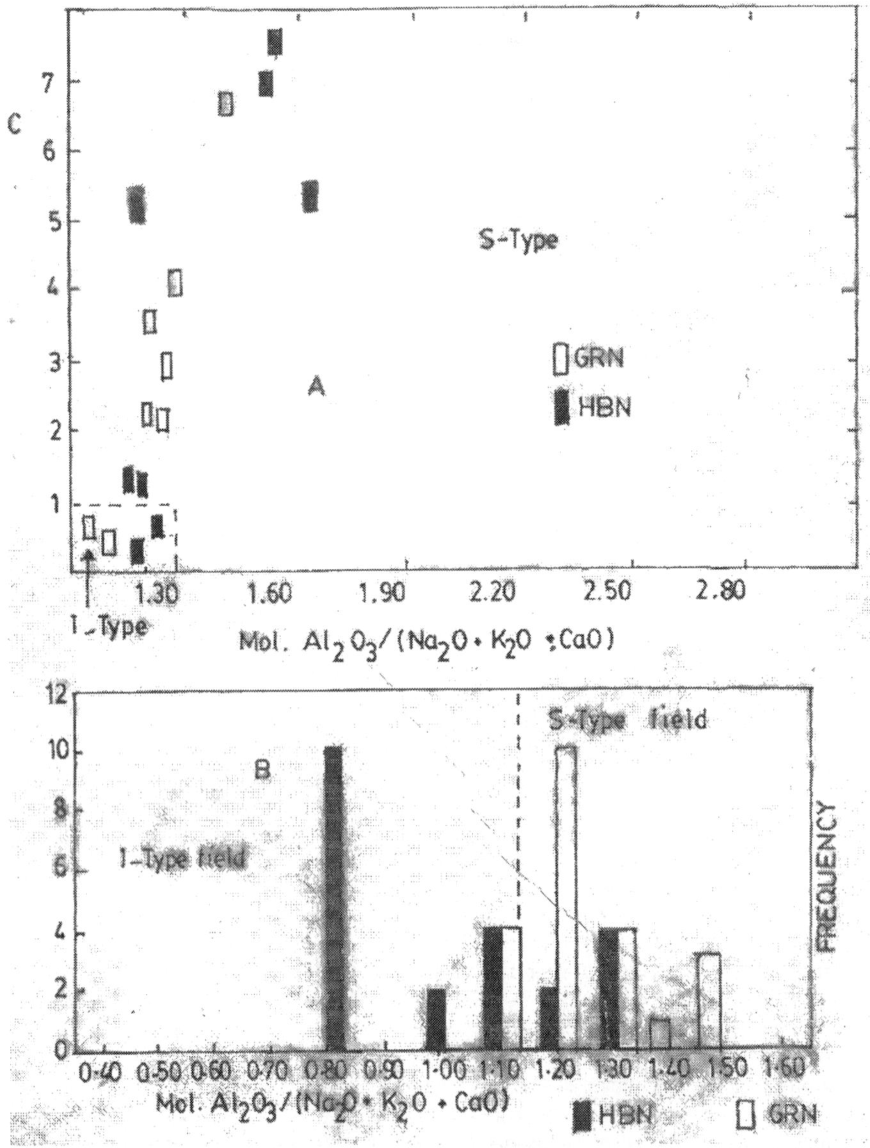

Fig. 6. a) Normative Corundum versus mol. Al_2O_3/(Na_2O+K^2O+CaO) for classification of I-Type and S-type igneous rocks, b) Histogram of mol. Al_2O_3/(Na_2O+K^2O+CaO) distribution for the Hornblende gneiss (HBN) and Granite gneiss (GRN). (Method based ov Vivaldo Waldo and David Rickard (1990)).

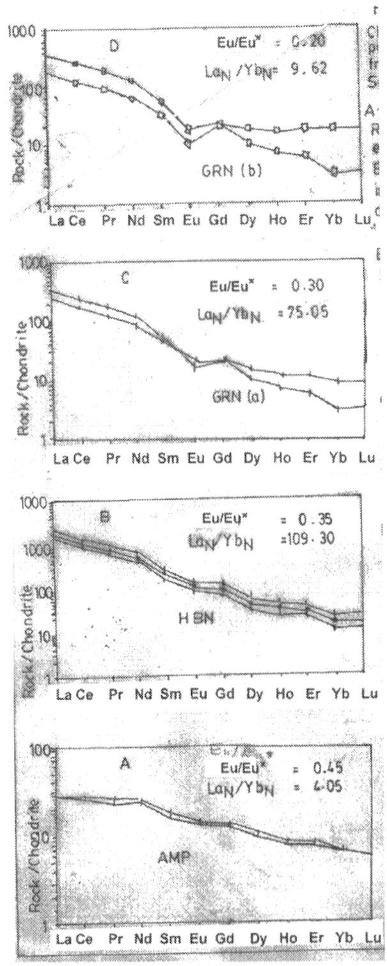

Fig. 7. Chondrite normalized REE patterns for AMP, HBN, GRN from the Ilesha schist belt Southwestern Nigeria, a) AMP chondrite normalized REE patterns showing essentially flat patterns, slight EU depletion and low La_N/Lb_N implying little or no differntation, b) HBN chondrite normalized REE patterns showing high LREE, low HREE, stepped patterns and moderate Eu depletion and very high La_N/Lb_N implying little or no differentation, c) GRN (a) chondrite normalized REE patterns showing high LREE, low HREE, pronaunced Eu depletion and high La_N/Lb_N showing very high differentiation of the source, d) GRN (b) GRN (a) chondrite normalized REE patterns showing high LREE, low HREE, high Eu depletion and moderate La_N/Lb_N showing little differentiation (last magmatic phase) These rocks shows an increase from AMP to HBN and decrease from HBN to GRN. These trends probably suggest a possible differentiation trends implicating differentiation of a basalitic magma.

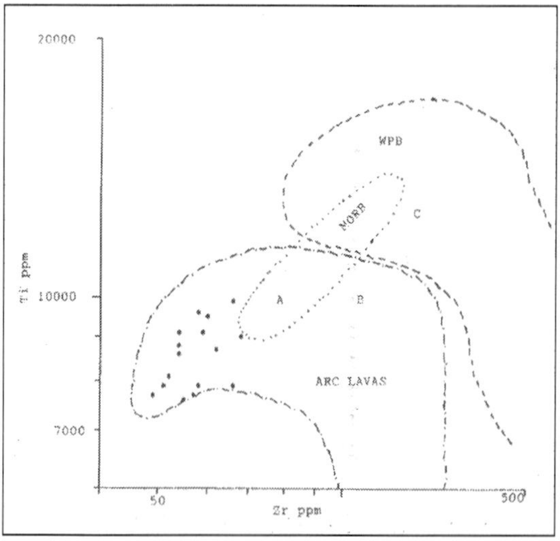

Fig. 8. Plot of Ti against Zr for the massive ampibolites in Ilesha schist belt Southwestern Nigeria. A) Mid-Ocean Ridge Basalt (MORB) field, B) Arc Lava field, C) Within Plate Basalt (WPB), (after Pearce et al. (1984)).

During collision between the continental shelf and the ocean plate, materials are scraped from the descending ocean slab and spread all over the area in southwestern Nigeria. Meanwhile, the descending ocean slab would carry sedimentary materials including water into the mantle. This is responsible for metasomatism of the mantle materials. There would be an exchange of materials in which the mantle portion of the wet magma formed would be enriched in sedimentary materials e.g. monazite and inproverIched in compatible elements e.g. Ni, Cr, Co. The basaltic wet magma thus formed intruded into the earlier laid down sedimentary cover. The magma differentiated to give rise to the amphibole rich rocks and granites which are protoliths of the later formed metamorphic rocks. After the arc magmatism, transpsressive forces operating in the magmatic chamber travelled along the magmatic channel, heat up the earlier laid down rocks and turned them into metamorphic rocks which are described in this chapter. The earlier laid down sediments were metamorphosed to give rise to the schists and metasediments found within the basement complex of southwestern Nigeria.

10. Summary and conclusions

Amphibolites, hornblende gneiss and granite gneisses are the main crystalline rocks in the basement complex of southwestern Nigeria. These rocks had undergone a polycyclic metamorphism which is mostly pervasive in Eburnean and Pan-African tectonothermal events. As a result of these, a series of deformation fabrics and evolutionary episodes had been recorded in these rocks. However, M_1 and M_2 phase of metamorphic deformation corresponding to two D_1 and D_2 phases of deformation are mostly discernible as recorded on these crystalline rocks. A two stage lead model age determination for the gneisses revealed that the protoliths of the basement rocks in southwestern Nigeria were emplaced in the Archaean (2750 ±25Ma). In this region gold mineralization was effected by the invation of a meteoric ore fluid at a temperature above 286°C. Au (gold) was probably leached from the metavolcanics in the belt and deposited as a system of auriferous quartz veins in a shear zone at about 550Ma in this region.

Geochemical (major, trace elements and REE), geological and petrological studies revealed that all these crystalline rocks are genetically related (comagmatic) and had evolved by progressive differentiation of a parent basaltic magma to give rise to the protoliths of the amphibolites probably represent the parent basaltic magma. Chemical studies also revealed that the magma of the protoliths of these rocks were from a metasomatised mantle.

Plumbotectonics, petrological, geological and geochemical analyses and interpretations carried out in this study implicate a back arc tectonic setting as the environment of emplacement of these rocks. In this type of tectonic setting an ocean slab was subducted into the mantle after colliding with a continental shelf. The subduction of an ocean slab into the mantle would enhance the formation of a mixed wet basaltic magma, consisting of both mantle and ocean sediments thoroughly mixed to form a basaltic magma. This magma extruded and intruded the earlier laid down sediments in the region, differentiated and gave rise to the protoliths of these crystalline rocks in southwestern Nigeria.

Post-magmatic transpressive forces operating in this region were responsible for the metamorphism of the protoliths of the amphibolites, hornblende gneiss and the granite gneiss (Oyinloye 2007) . Further deformation of these rocks led to faulting, fracturing, shearing, folding gneissic banding and foliation fabrics observed on some of the rocks especially, the leucocratic amphibolites, hornblende gneiss and granite gneisses in the basement complex of southwestern Nigeria.

11. References

Ajibade, A.C and Fitches W.R. (1988): The Nigerian Precambrian and the Pan -African Orogeny,Precambrian Geology of Nigeria, pp. 45-53.

Ajibade, A.C.,Woaks, M., and Rahaman, M.A. (1987): Proterozoic crustal development in Pan-African regime of Nigeria: In A. Croner (ed) Proterozoic Lithospheric Evolution Geodynamics Vol. 17, pp 259-231.

Annor , A.E. and Freeth, S.J., (1985): Thermotectonic evolution of the basement complex around Okene, Nigeria with special refrence to deformation mechanism, Precambrian Research, 28, pp. 269-281.

Billstrom, K.A. (1989): A model for the Lead isotope evolution of Early Proterozoic Svecofennian sulphide ores in Sweden and Finland. Isotopic Geology 79, pp 307-316.

Billstrom, K.A. (1990): A lead isotope study of two sulphide deposits and adjacent igneous rocks in south-central Sweden. Mineralium Deposita 25: pp 152-159.

Boesse, T.N. and Ocan, O.O. (1988): Geology and evolution of the Ife-Ilesha Schist belt southwestern Nigeria. Symposium on Benin-Nigeria geo-traverse of Proterozoic geology and tectonics of high grade terains pp. 87-107.

Burke, K.C. Dewey, J.F (1972): Orogeny in Africa . In African Geology A.J. Dessauragie, T.F.J. Whiteman (eds), pp 583-608, University of Ibadan Press, Nigeria.

Burke, K.C., Freeth S.J. and Grant, N.K. (1976): The structure and sequence of geological. events in the basement complex of Ibadan area Western Nigeria Precamb. Res.3, pp 537-545

Dada, S.S, Briqueu, K.L., Birck. J.L. (1998): Primodial crustal growth in northern Nigeria Preliminary Rb-Sr and Sm-Nd constraints from Kaduna migmatite gneiss complex J. Min. Geol. 34, pp1-6.

Egbuniwe, I.G. (1982): Geotectonic evolution of Maru Belt, northwestern Nigeria, unpublished Ph.D thesis of the University of Wales, U.K.

Holt, R.W, Egbuniwe , I.G., Fitches, W.R. and Wright J.B. (1978): The relationship between low grade metasedimentary calc-alkaline volvanics and Pan-African Orogeny, Geol. Rundsh, 67 (2), pp 631-646.

Martins, H. (1986): Progressive alteration associated with auriferous massive sulphide bodies at The Dumagami Mine, Abitibi Greenstone Belt, Quebec. Econ. Geol. Vol. 85, pp 746-764.

Oyinloye, A.O. (1992): Genesis of the Iperindo gold deposit, Ilesha schist belt, Southwestern Nigeria. Unpublished thesis of the University of Wales, Cardiff, U.K. pp. 1-267.

Oyinloye, A.O. and Odeyemi, S.B. (2001): The geochemistry, tectonic setting and origin of theMassive melanocratic amphibolites in Ilesha schist belt Southwestern Nigeria, Global Journal, Pure and Appl. Sci. (7) (1), pp.55-66.

Oyinloye,A.O. (1998): Geology, Geochemistry and origin of the banded granite gneisses in the basement complex of the Ilesha Area Southwestern Nigerian. J. Africa Earth Science, London 264, pp633-641.

Oyinloye, A.O. and Steed, G.M. (1996): Geology and Geochemistry of the Iperindo Primary gold deposits Ilesha schist belt Southwestern Nigeria. Inferences from stable carbon isotope studies. Africa J. Sc. Tech. 8 (1) pp 16-19.

Oyinloye,A.O. (2002a): Geochemical Studies of granite gneisses: the implication on source determination. Jour. Chem. Soc. Nigeria (26) (1) 131-134

Oyinloye,A.O. (2002b): Geochemical characteristics of some granite gneisses in Ilesha area southwestern Nigeria: Implication on evolution of Ilesha schist belt, southwestern Nigeria. Trends in Geochemistry India vol.2, 59-71

Oyinloye,A.O. (2004a): Petrochemistry, pb isotope systematic and geotectonic setting of granite gneisses in Ilesha schist belt southwestern Nigeria Global Jour. Geol. Sci. 2(1) 1-13.

Oyinloye, A.O. (2006b): Metallogenesis of the lode gold deposits in Ilesha Area of Southwestern Nigeria: Inferences from lead isotope systematic, Pak. J. Sci. Ind. Res. 49 (11) pp 1-11.

Oyinloye, A.O. (2007): Geology and Geochemistry of some Crystalline Basement Rocks in Ilesha area sourthwestern Nigeria: Implications on Provenance and Evolution Pak. Jour. Sci. Ind. Res. Vol. 50, No.4, 223-231.

Oyinloye, A.O. (2011): Beyond Petroleum Resources: Solid Minerals to the rescue: 31st Inaugural Lecture of the University of Ado-Ekiti, Nigeria Press, 1-36.

Pearce J.A., Harris N.W. and Tindle A.G. (1984): Trace elements discrimination diagrams for tectonic interpretation of granite rocks. Journal Petrology. Vol. 25 Par 4 956-983.

Rahaman, M.A and Ocan, O.O. (1978): On relationship in the Precambrian migmatitic gneisses of Nigeria J. Min. and Geol. Vol. 15, No.1 (abs).

Rahaman, M.A (1988):Recent advances in the study of the basement complex of Nigeria. Symposium on the Geology of Nigeria, Obafemi Awolowo University, Nigeria.

Stacey, J.S., Kramers, J.D. (1975): Approximation of terrestrial lead isotope evolution by a two-stage model. Earth Planet. Leit. 26, pp 206-221.

Vivalo, W. and Rickard D. (1990): Genesis of an early Proterozoic zinc deposit in high grade Metamorphic terrare, saxberget, central Sweden Sco. Geo Vol.85, 714-736.

Wilson, M. (1991): Igneous Petrogenesis Global Tectonic Approach, Harpar Collins Academy, London Second impression pp. 227-241.

Woakes, M. Ajibade C.A., Rahaman, M.A., (1987): Some metallogenic features of the Nigerian Basement, Jour. of Africa Science Vol. 5 pp. 655-664.

Zartman, R.E. Doe, B.R. (1981): Plumbotectonics Tectonophysics, 75, 135-162.

Part 2

Geochemistry

6

The Organic-Rich and Siliceous Bahloul Formation: Environmental Evolution Using Facies Analysis and Sr/Ca and Mn Chemostratigraphy, Bargou Area, Tunisia

Mohamed Soua[1], Hela Fakhfakh-Ben Jemia[1], Dalila Zaghbib-Turki[2], Jalel Smaoui[1], Mohsen Layeb[3], Moncef Saidi[1] and Mohamed Moncef Turki[2]
[1]*Entreprise tunisienne d'activités pétrolières, ETAP–CRDP, 4, La Charguia II, Tunisie*
[2]*Département des Sciences de la Terre, Faculté des Sciences de Tunis, Université Tunis El Manar,*
[3]*Institut supérieur arts et métiers, Siliana*
Tunisia

1. Introduction

The Cenomanian-Turonian transition deposits in Tunisia (**Figure 1; Figure 2**) were initially described by Pervinquière (1903) who distinguished black shales "*marnes schisteuses*" and laminated limestone (*calcaire en plaquettes*) in Central Tunisia representing the Turonian onset. They were also described by Berthe (1949) and by Schijfsma (1955). Then, Burollet et al. (1952) and Burollet (1956) have attributed these deposits to the Bahloul Formation which is well developed in its type locality, Oued Bahloul, located southward Kessera village in central Tunisia.

These different lithostratigraphical units are geographically relayed and are controlled by the global relative sea level rise causing the major latest Cenomanian transgression which was interpreted as a combined phenomenon between a long-term sea level rise and basin subsidence (Haq et al., 1987; Hardenbol et al., 1993; Luning et al., 2004) generated by enhanced plate tectonic activity especially in the Atlantic area and by the change in the Milankovitch frequency band (Soua, 2010). The major rise in the Cenomanian sea level was interrupted by five third order relative sea-level falls (Haq et al., 1987). In the Bargou area (**Figure 1**), the Cenomanian-Turonian deposits are mainly composed of organic-rich black shales include special siliceous radiolarian-bearing laminated beds. No previous sequence stratigraphic framework has been taken yet on such siliceous deposition. In this work, we attempt to detail sequence stratigraphic framework using facies association and Sr/Ca and Mn variation.

2. Geological setting

The Bargou area (Figure 1), connected palaeogeographically to central Tunisia, is characterized by (1) emerged palaeohighs displaying gaps and discontinuities (Turki, 1985)

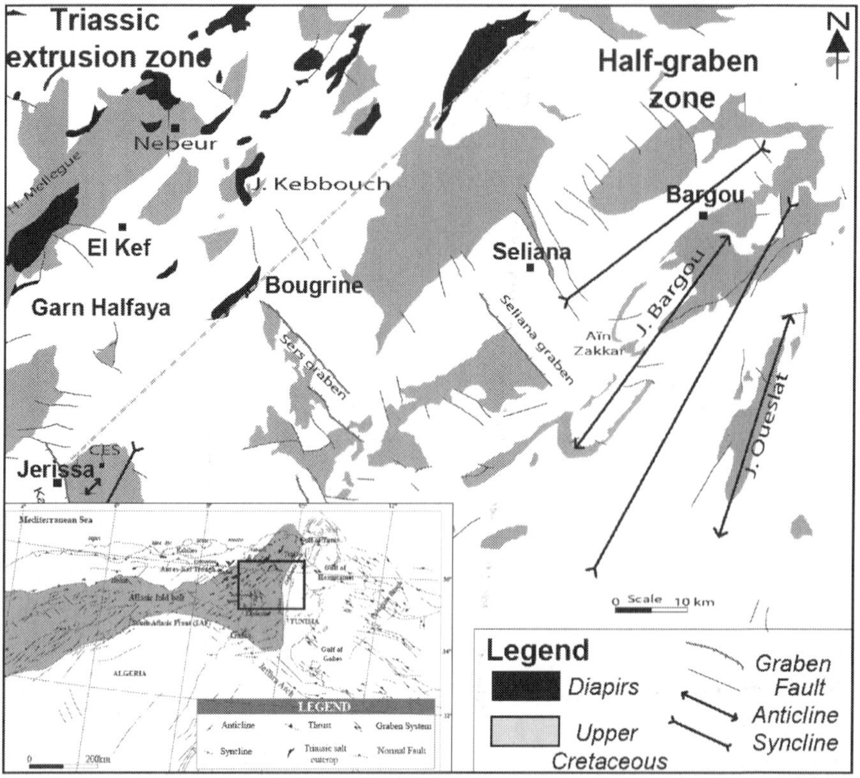

Fig. 1. Tectonic map of the central Tunisia Domain showing main upper cretaceous deposits and structures.

and (2) subsiding zones affected by deep water sedimentation. This area is dominated by N140° and N70° trend faults limiting several blocks. Cretaceous sedimentation varies on both sides. The Bargou mount is an anticlinarium structure NE-SW trending, with complex anticline folds separated by narrow synclines. This wide anticline structure is formed mostly by reefal upper Aptian to pelagic lower Albian strata. Its structural evolution may be summarized as follow: (1) during the late Jurassic to early Cretaceous, the area was subjected to a major extensional phase that delimited horst and graben systems (Martinez and Truillet, 1987) (2) In the uppermost Aptian, a regional compressional pulsation affecting the north-African platform had resulted from a transpressional scheme (Ben Ayed and Viguier, 1981) (3) New NNE-SSW trend anticline structures appeared attested by the Albian Fahdene Formation onlap features on the reefal aptian Serj deposits in subsurface (Messaoudi and Hammouda, 1994) or Albian-upper Aptian unconformity in outcrops (Ouahchi et al., 1998). (4) During the Albian, the geodynamic evolution is marked by the sealing of lower Cretaceous structures during an extensional phase that persisted to form graben systems promoting organic-rich and siliceous strata deposition throughout upper Cenomanian to Lower Turonian times. The major faults in this area are represented by

N140° and N70° trend features. The first trend borders the graben systems that extend across the Central Atlas, whereas the second one parallels the principal axe of the Bargou anticlinarium. Around the Bargou anticlinarium, the Bahloul thickness is significantly variable. It may vary from 10m to 40m in thickness. Uniquely, in this area, the top of the Bahloul represents many cenomanian olistolith levels (Soua et al., 2006) marking syndepositional tectonic activities (Turki, 1985). Elsewhere, these syn-sedimentary features are represented by local slumping. Generally speaking, in north-central Tunisia as well as on the Pelagian Province, Cretaceous diapiric movements of Triassic salt played locally an important role in controlling C/T deposition, (Patriat et al., 2003; Soua and Tribovillard, 2007; Soua et al., 2009). They are characterized by a marked thickness reduction and partly by development of detrital deposition (sandstones, conglomerates). The diapiric rise, starting from Aptian (and even before) to approximately middle Eocene was probably continuous, but it increased during periods of tectonic instability. Thus earlier diapiric movements and rise-up are super-imposed to the extensional features favouring depocenters individualisations in the central parts of rim-synclines (Soua et al., 2009).

3. Material and methods

In total 132 samples (Bahloul formation) are collected from Bargou area, Aïn Zakkar (AZ) and Dir ouled yahia section (OKS), throughout the C-T transition deposits including the Bahloul Formation (Figure 2). The high resolution sedimentological interest consists of sequence stratigraphic interpretation of the facies association inferred from classical sequence stratigraphy.

Major elements concentrations (Figure 3) were determined using Mass spectrometry (ICP-AES) process and trace elements were determined using Induced Coupled Plasma Mass Spectroscopy (ICP-MS) at ETAP. In this study Sr/Ca and Mn variation curves were determined to check relationships with sea level flucruation as described by Mabrouk et al. (2006). Thin sections were made in the hard limestone samples in order to analyze microfacies in the laboratory of sedimentology of the "Office National des Mines".

4. Facies association

Bargou area represents special organic-rich deposits showing mixed facies composed of both calcareous and siliceous material (Figure 4). In this case, at the Dir ouled Yahia (OKS) and Ain Zakkar (AZ) sections the C-T transition deposition became not conform to the typical Bahloul Formation as defined by its author (Burollet, 1956). Closer inspection of microfacies composing this mixed facies allows distinguishing diverse types (A-G) and differs from the typical Bahloul by the Facies E (see later, Figure 5).

4.1 Facies A: Calcisphaeres-rich massive limestone

It forms massive light grey limestones with nodular base. These calcisphaeres-rich are mainly packstone in texture and show bioturbations. They characterize the lower part of the studied C-T transition series equivalent to the typical Bahloul Formation. Among calcisphaeres, *Bonetocardiella sp., Calcisphaerula sp., and Pethonella sp.,* are associated with (1) diversified keeled planktonic foraminifera (*Rotalipora cushmani, Praeglobotruncana stephani, Dicarinella algeriana*) and globular forms like whiteinellids and hedbergellids (2) benthonic foraminifera (*Textularia, Lenticulina*) (3) echinoderms debris and (4) rounded phosphate and glauconite grains that may be evidence of reworking.

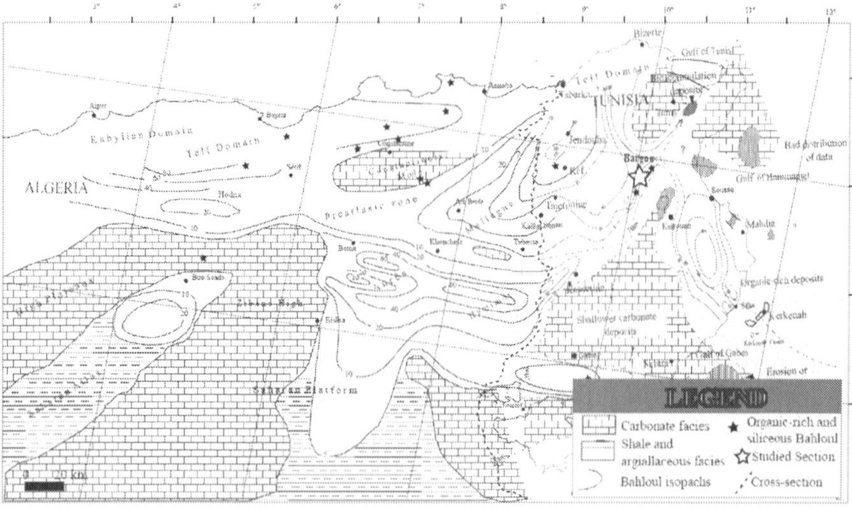

Fig. 2. Facies and thickness of the Organic-rich Late Cenomanian – Early Turonian and age equivalent units in Tunisia and Eastern Algeria superposed on the Structural setting of the eastern Maghrebian Domain during the C-T transition (inferred from the palaeogeographic map). Note the E-W trending structures are related to the Tethys rifting and the NW-SE trending structures are related to the Sirte rifting (Soua et al., 2009). Note that the isopach trends illustrated are simplified and short-distance thickness, and facies changes occur, related to the complex tectono-halokinetic palaeorelief at the time (see **Figure** 1). Organic-rich and carbonate or age equivalent units are not present on present palaeohigh.

4.2 Facies B: Pseudo-laminated limestone

They are of wackestone or mudstone texture (Figure 4). Those of wackestone texture are especially dark. In contrast those of mudstone texture are light-coloured and show frequent bioturbation marks and less varied microfauna. Similar facies are described by Layeb (1990). Within these pseudo-laminated limestones, the exclusive presence of globular planktonic foraminifera in organic-rich micritic matrix. This situation is sometimes interrupted by bioclastic material discharges (calcispheres, echinoderms debris) coming probably either from the platform or by the oxygenated episodes (supported also by ichnofossils presence).

4.3 Facies C: Laminated black shales

The Bahloul Formation is mainly composed of laminated black shales alternated with light marly levels. Microfacies analysis reveals that these black shales are constituted by tightened alternation composed of (1) light inframellimetric packstone and (2) dark mellimetric mudstone laminae.

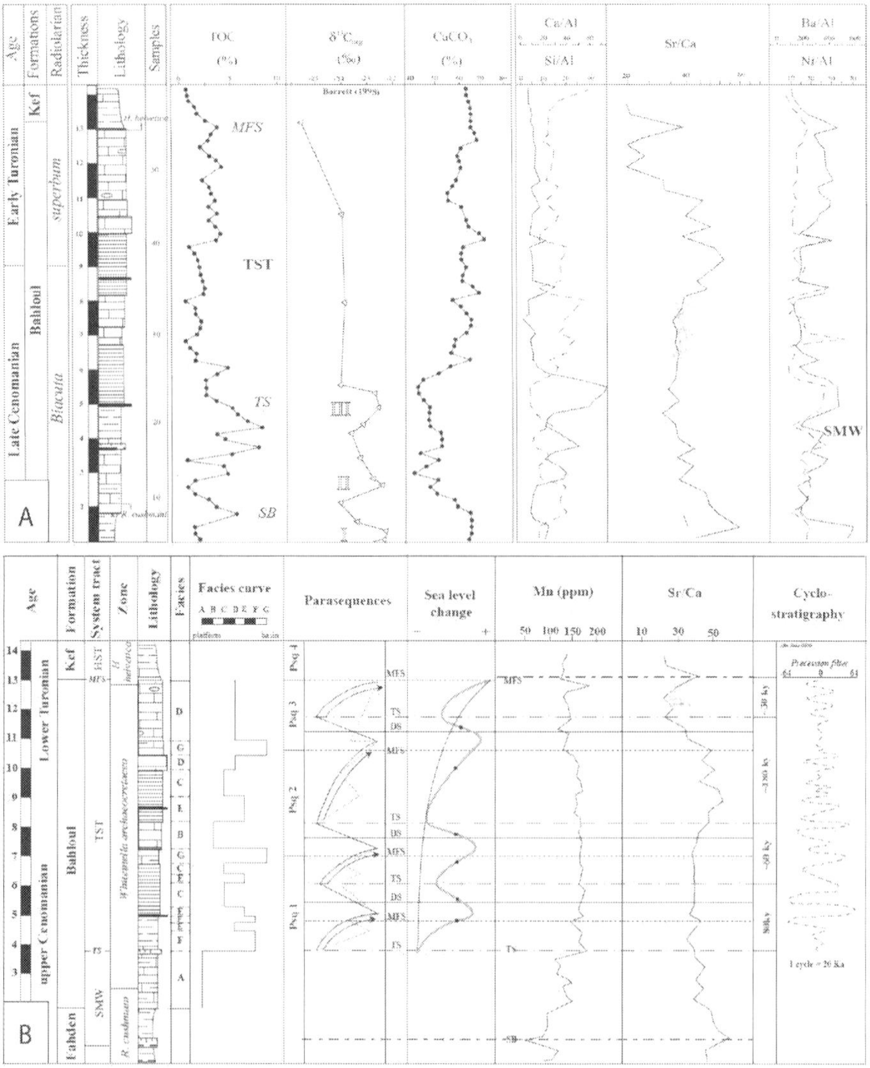

Fig. 3. **A** Lithology, Total Organic Carbon (TOC), calcimetry and chemostratigraphy data of the Ain Zakkar section combined with biostratigraphic framework. **B** Facies curve, parasequences sets, global trend and age of parasequences in Ain Zakkar section (Bargou area, central Tunisia).

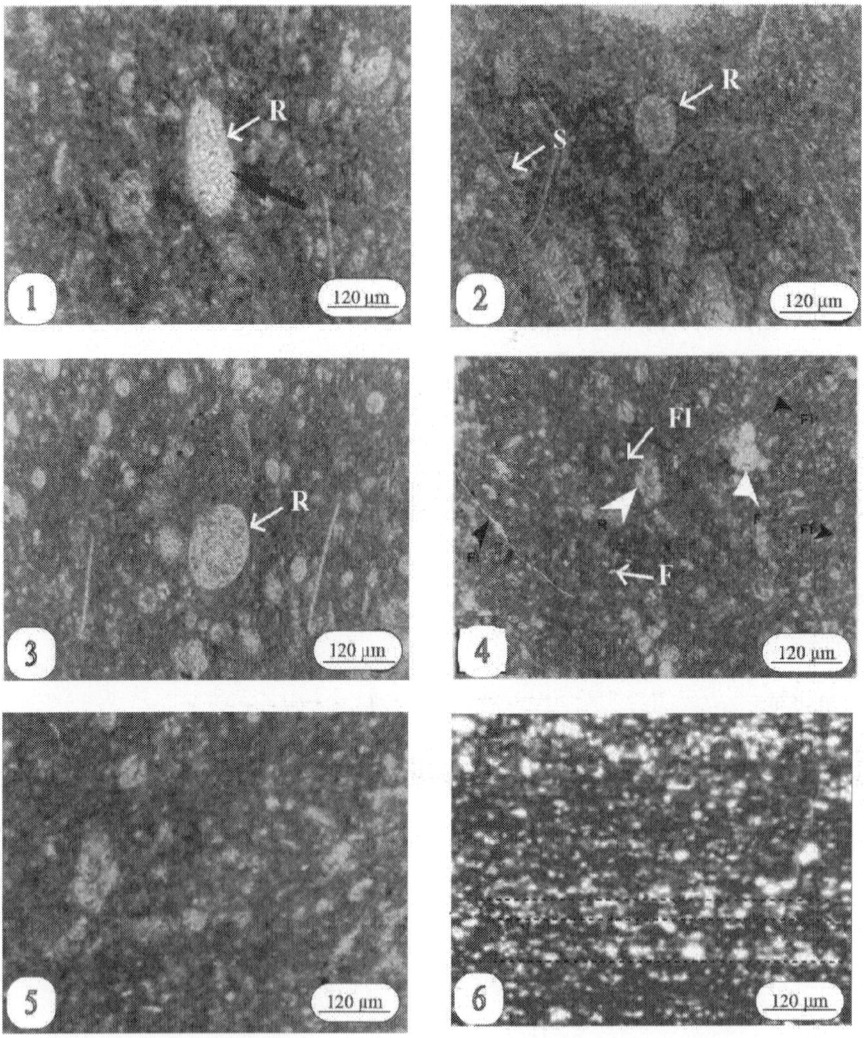

Fig. 4. (1-3) Micrographs of a radiolarian-rich layers from the Bahloul Formation at the AZ and OKS sections showing nassellarians and spumellarians, (4) micrograph across planktonic foraminiferal-rich and filaments beds; (5) microconglomeratic level at the base of the section showing broken skeltons of diverse fauna; (6) micrograph across laminated limestone beds. Note R: radiolarian, F: planktonic forminifer, Fl: filament, S: sponge spicule.

4.3.1 Light laminae

The light laminae contain abundant calcisphaeres and globular scarce biserial/triserial and trochospiral planktonic foraminiferal which may agglomerated in aggregates. The dominant biserial and triserial, *Heterohelix moremani, H. reussi* and *Guembelitria cenomana* are associated to scarce trochospiral forms belonging to whiteinellids and hedbergellids. These laminae may also contain either exclusively calcisphaeres like as within the lower and upper part of the studied section or associated whiteinellids heterohelicids or hedbergellids. Some laminae display thus, monospecific association.

4.3.2 Dark laminae

These wackestone texture laminae display significant pellets concentration and scarce globular planktonic foraminifera. The ovoid pelletoids represent no internal structure and they correspond probably to small organic-rich pellets (Purser, 1980). Their origin is discussed but probably they come from gelatinous planktonic forms analogue to the present salps. Currently, in Mediterranean Sea, salps and notably *Salpa fusiformis* can be absent during several years from the water mass, then it forms suddenly large scale population that may influence the whole pelagic community. Their diatoms-rich pellets become abundant and generate organic-rich flux involving a considerable organic matter input through the lower water mass (Morand and Dallot, 1985).

4.4 Facies D: Filament-rich laminated limestone

These filaments-rich laminated limestones occur in the upper part of the studied series at the OKS section. They display wackestone to packstone texture (figure 4). The filaments are present as elongated and thin with imbricated and tangled arrangement enveloping sparitic elements. We don't believe that these filaments are larval-planktonic stage bivalves as reported by Robaszynski et al., (1994) because in our material they display several features such as tangling and overlapping. Moreover, they display heterogeneous sizes. Sometimes these light-coloured, thin filaments are present as elongated or arched. Their abundance is related to that of planktonic foraminifera and defines pelagic facies (Soua, 2011). All these figured elements are slightly oriented with the stratification trend. Analogous facies was documented also in Juassic series. Within these series, similar filaments are attributed to debris of pelagic bivalves like as Picnodonta.

4.5 Facies E: Radiolarian-rich laminated black shales

In Bargou area, radiolarian-rich laminated black shales were reported from the Oued OKS section (Soua, 2005; Soua et al. 2006). They occur as decimetric laminated silicified limestone. They were labelled "Silexites" by Layeb and Belayouni (1989). The same facies occurs in Guern Halfaya, from Tajerouine area (Soua et al., 2008) and it is associated with diapir. The radiolarian-bearing laminated black shales in Bargou area are formed by millimetric light wackestone and dark mudstone laminae. The light laminae display a significant siliceous radiolarian concentration, however, the dark one display a dense organic rich small dark coloured pellets concentration.

4.6 Facies F: Dark marls with tiny planktonic globular foraminifera

Dark marls which don't exceed 1 m in thickness are present in AZ section in the lower part of the Bahloul Formation, just above the surface of condensation. They are dark and contain dissociated tiny globular chambers of whiteinellids, hedbergellids and *Heterohelicacea* (*Heterohelix* and *Guembelitria*) species within mud matrix. This facies suggests deep marine non agitated depositional pattern. In these marls, the eutrophic surface dweller *Guembelitria* scarcity, the benthonic foraminiferal absence and the surface dwellers whiteinellids and heterohelicids proliferation may indicate concurrent eutrophic condition at the surface sea water and anoxic condition at the sea floor (Soua, 2005). The same facies (dark marls with tiny globular chambers of planktonic foraminifera) is reported from the Hammem Mellegue section, Kef area (Nederbragt and Fiorentino, 1999; Soua et al., 2008; Soua, 2010).

4.7 Facies G: Light clayey limestones

They are of mudstone texture with scarce entire planktonic foraminifera having globular and keeled chambers. They are associated with rare pyritous ostracods and scarce benthic foraminifera (*Lenticulina* sp., *Bulimina* sp.).

5. Sr/Ca & Mn chemostratigraphy

5.1 Relationship between Sr/Ca and sea level change

In Bargou area, the Sr/Ca ratio curve (Figure 3) displays more than one short-term cycle that generally match with the depositional sequences. Generally speaking, higher Sr/Ca values broadly span the upper parts of LST's and HST's (respectively ~58 and ~56, Figure 3), with peak values (~58) very close to sequence boundaries (SB). The Sr/Ca ratio values generally fall through TST (mean value ~40) and remain relatively constant through the lower part of the Bahloul Formation, then show a significant decrease to much lower Sr/Ca values for reaching minimum values in the upper part of it (~20) close to the MFS at the top of the formation (Figure 3; Figure 5). This feature is conforming to that described by Mabrouk et al. (2006) in Bahloul Formation from Tunisian onshore and elsewhere.

5.2 Relationships between manganese (Mn) and sealevel changes

In Bargou area the Mn fluctuations (Figure 3) across each depositional sequence systems tract do not correlate well with either silicate or carbonate contents (Figure 5). This is may be due to the Mn supply that was coupled with the biogenic flux (e.g. organic carbon), which must have decreased during the SMW (Murphy, 1998; Stoll and Schrag, 2001; Mabrouk et al. 2006; Soua, 2010b).

Mn profile increased with rising sea-level, reaching a maximum around each MFS (Figure 3b), before decreasing again through the overlying HST, representing a period of relative constant carbonate supply.

6. Third order sequences

6.1 Vertical evolution of systems tract

6.1.1 Shelf margin wedge (SMW)

In the Aïn Zakkar location, the SMW is represented by the uppermost part of the late Cenomanian Fahdene Formation and the overlying calcisphaeres-rich argillaceous wackestone-packstone bioturbated limestones arranged by the authors in the Bahloul Formation or pre-Bahloul deposits (Accarie et al., 1999; Caron et al., 2006). The erosive base and the lenticular shape of this deposition interval suggest a channel sedimentary body. It is characterized by its highest abundance in calcisphaeres compared to planktonic, benthic foraminifera, bioclasts as well as fauna mixture.

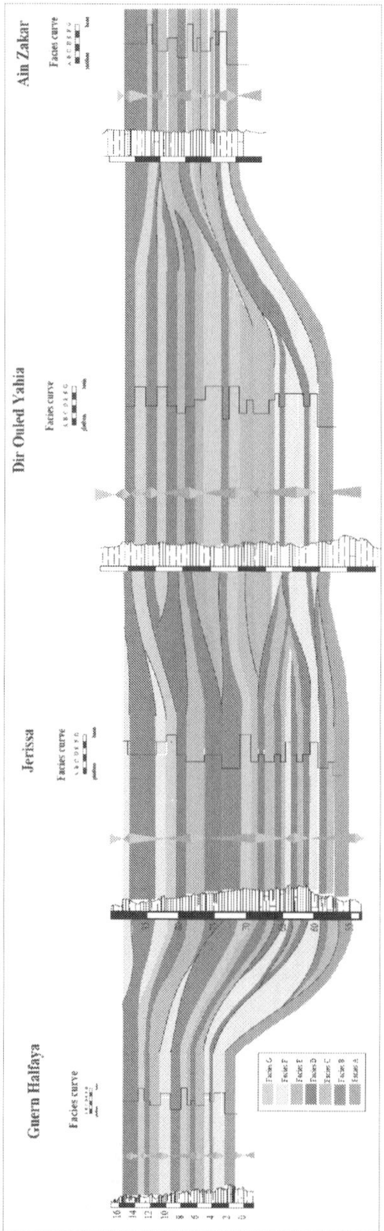

Fig. 5. Sequence stratigraphic correlation between Bargou area (OKS and AZ sections) and central Tunisia (Jerissa and Guern Halfaya, after soua, 2011)

This systems tract includes a microconglomeratic surface of condensation, rich in phosphate and glauconite grains. Internally, the microconglomeratic bed has discontinuous, wavy, parallel seams of glauconite-rich shale. Close inspection may also reveal clasts concentrations, ichnofossills (*Zoophycos* and *Thalassinoides*) and bioclasts. This surface is sharp and overlain by a (transgressive) lag of reworked bioclasts (mainly echinoderms) from the neighbor platform as well as glauconite and phosphatic grains. These data were not described by Zagrarni et al. (2008) although they are expressed in many Bahloul sections. It marks the limit between the SMW and the TST (Figure 5 Figure 6) and represents a fast shore line migration into the continent.

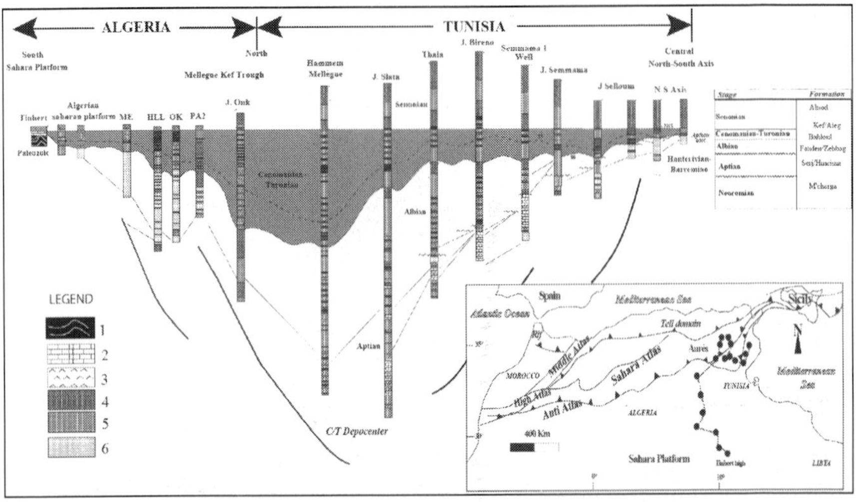

Fig. 6. Tentative correlation between several Bahloul sections in Tunisia and Algeria. Datum used is MFS (Top of Bahloul formation). 1.magmatic rocks, 2. Carbonate, 3.evaporites, 4.black shales, 5.marls/clays, 6. Sandstone.

6.1.2 Transgressive systems tract (TST)

The remainder Bahloul Formation part is considered as belonging to a transgressive systems tract (TST). Its lower boundary is represented by a transgressive surface (TS) overlying by calcareous and marly retrogradation parasequences succession (a parasequence is shallowing-upward and is bounded by a flooding surface). Indeed, following the eustatic rise, this retrogradational pattern indicates that, accommodation space creation exceeded sediment flux and the basin was underfilled. Within these parasequences, limestones are laminated plated organic-rich black shales encompassing, sometimes, pyrite nodules or iron oxides. These black shales contain tiny globular chambers of planktonic foraminifera (whiteinellids, herbedgellids, *Heterohelicacea*), and radiolarian (nassallarian), and ammonites indicating a depositional environment deepening when compared to the underlying systems tract and also an enhanced primary productivity, proved by the nassellarian prolifiration. The top of this calcareous unit, characterized by phosphate and glauconite grains abundance, suggests a maximum flooding surface (MFS). In contrast, the fissile marly levels are bioturbated. They contain a rather poor fauna of small agglutinated benthic foraminifera (e.g., neobulliminids) and weakly keeled planktonic foraminifera (e.g., dicarinellids), testifying however that oxygen-rich bottom waters influx may induced a well-oxygenated water column when compared to dark laminated limestones. The MFS is situated at the top of filaments-rich limestones. It is represented by a surface of condensation, rich in grains of glauconite and phosphorite.

6.1.3 "Early" highstand systems tract (lowermost Kef formation)

The early highstand systems tract always corresponds to retrogradational then progradational facies. In the Bargou area like in the entire megabasin spanning the central Tunisia (Figure 6), this systems tract is represented by the lowermost Kef formation marls/limestones alternations.

7. Discussion

7.1 Lateral evolution of systems tract

7.1.1 Shelf margin wedge (SMW)

This systems tract including a microconglomeratic surface of condensation, rich in phosphate and glauconite grains. Internally, the microconglomeratic bed has discontinuous, wavy, parallel seams of glauconite-rich shale. This surface is sharp and overlain by a (transgressive) lag of reworked bioclasts (mainly echinoderms) from the neighbor platform as well as glauconite and phosphatic grains. It marks the limit between the SMW and the TST (Figure 5) and represents a fast shore line migration into the continent. This systems tract is represented generally by the Facies A. Elsewhere, in the platform sequence (Razgallah et al., 1994; Abdallah et al., 2000), the SMW does not have an equivalent (Figure 6). Towards the south, the emerged Zebbag platform is subjected to sharp erosion following the fast fall relative sea level. This erosion is represented by the unconformity which overlies the limestones containing *Neolobites vibrayeanus* and *Calycoceras sp.* (Abdallah et al., 2000). In the studied area, the upper part of the SMW is related to the calcareous Bahloul Formation genesis because the Fahdene Formation was originally defined as an argillaceous unit (Burollet, 1956). Therefore, the Bahloul Formation starts by a lithoclastic calcareous unit. Elswhere, in Kalâat Senan area (Accarie et al., 1999) and in the type locality of the Bahloul Formation (Caron et al., 2006), the same facies was assigned to the "pré-Bahloul" deposits. It is generally composed by the alternation of Facies F and D. in the neighboring section these latter facies are much developed (Figure 5) while in NW Tunisia, we note the alternation of Facies B and C (in Jerissa section).

7.1.2 Transgressive systems tract (TST)

The remainder Bahloul Formation part is considered as belonging to a transgressive systems tract (TST). The top of this calcareous unit, characterized by phosphate and glauconite grains abundance, suggests a maximum flooding surface (MFS). It is represented by a surface of condensation, rich in grains of glauconite and phosphorite. Towards the south, in the adjacent platform (Gafsa basin), as well as in the Oued Bahloul location (Maamouri et al., 1994), this surface is recorded by abundant lowermost Turonian ammonites (e.g. *Fagesia sp.* and *Watinoceras sp.*). Nevertheless, this surface occurs when the accommodation space creation is faster than before. In this case there would be no progradation as mentioned by

Zagrarni et al. (2008). On the platform setting, the transgressive interval corresponds to the pelagic fauna limestones unit (Razgallah et al., 1994). Indeed, following eustatic rise, the accomodation space creation rate is much stronger than the sedimentary input, making consequently a gradual inundation of the Zebbag platform and the pelagic facies deposition. In the studied sections, we note the development of a special Facies E (siliceous and radiolarian-rich). It is worthy to mention the presence of such organic rich facies in the Algerian Tinhert belonging to the saharian platform (Figure 6).

7.1.3 Highstand systems tract (lowermost Kef Formation)

In the Bargou area like in the entire megabasin spanning the central Tunisia (Figure 5; Figure 6), this systems tract is represented by the lowermost Kef formation marls/limestones alternations. Towards the south (i.e., Gafsa basin) within the platform, this systems tract is represented by the subreefal carbonates of the Gattar member that recorded the filling of the space made available and making a progradational pattern. In this case the Gattar member is the exclusive equivalent of the Kef Formation.

7.2 Sr/Ca & Mn variation

Considering the Cenomanian-Turonian changes record related to sea-level seeing that, the Sr/Ca values change recorded in marine carbonates of pelagic environments have been considered recently as reflecting past fluctuations of the oceanic Sr and Ca budgets (Ando et al., 2006). Large and rapid increases in seawater Sr/Ca ratios and Mn profiles matches with many large Cretaceous sea level rises/falls (Stoll and Shrag, 1996). It has been suggested that sources of Sr variations in Cretaceous carbonates can be related to changes in seawater Sr/Ca or Mn.

It was reported that the Sr/Ca ratios rose progressively through the mid- to late Cretaceous, a period of generally rising eustatic sea-level (Renard, 1985; Stoll and Schrag, 2001; Hancock, 1989; Haq et al., 1988; Hancock, 1993).

In our material (Figure 3), the observed relationships between the Sr/Ca profile and the sequence stratigraphy systems tract are consistent with sea-level change forcing the short-term Sr/Ca record. Falling sea-levels during late high-stands and low stands led to exposure of carbonate shelves and pulses of aragonite-derived Sr to the oceans. Rising sea-levels during transgression promoted renewed aragonite deposition and falling seawater Sr/Ca (Stoll and Schrag, 2001). This was reversed by the development of mature carbonate platform systems with lower aragonite accumulation rates during the high stand.

Our results confirm the relationship occurrence between stratigraphic sequence sytems tract and Sr/Ca evolution as described at Culver (Murphy, 1998; Mabrouk et al., 2006) and elsewhere in Gubbio (Stoll and Schrag, 2001). The Culver Sr/Ca profile was may be compared with that of Gubbio (Italy) and Bargou (Tunisia) inspite that these three sections cannot be easily correlated biostratigraphically in detail since the biomarker among macrofossils and foraminifera are rare even absent at both Italian (Gubbio) and Tunisian (Bargou) sections seeing their siliceous facies frequency. In fact using Sr/Ca profiles, ratios maxima identified within this interval transition at Culver (England) and at Gubbio (Italy) are also identifiable at Bargou (Tunisia). Mabrouk et al. (2006) mentioned that breaks within the Sr/Ca curves may indicate sedimentological or diagenetic effect. At the Bargou section these breaks may correspond to intervals of silica-rich black shales comparable to the Livello Bonarelli.

Increasing Mn in the TST (Bahloul Formation, Figure 3) might be related to increased productivity during sea-level rise promoting an increased organic matter-associated particulate Mn flux to the seafloor. Therefore, high Mn contents might be caused by lower rates of sedimentation, with increased efficiency of Mn redox cycling leading to elevated Mn contents in the sediment (Renard, 1985).

8. Conclusion

The Late Cenomanian - Early Turonian organic-rich and Siliceous Bahloul Formation exposed in Bargou area containing planktonic foraminifera and radiolarians occurs. Seven different facies association were recognized with a special Facies E, a siliceous and radiolarian-bearing one that differs from the other Bahloul sections. Indeed, these facies and lithologic units are genetically linked and integrated in a part of third order global sequence. Therefore, important relationships exist also between Sr/Ca ratios profile and eustatic sea-level change. Within the Cenomanian-Turonian transition, Sr/Ca maxima span the upper parts of high-stand (HST) and the overlying shelf margin wedge (SMW) of the Fahdene Formation and the lower part of the siliceous Bahloul Formation, with maximum values around sequence boundaries. Sr/Ca values fall through the transgressive systems tract (TST) and attain minimum values in the upper part of it. Furthermore, manganese exhibits important relationships to sequences but differently from Sr/Ca, with minima around sequence boundaries and through SMW, rising values from the TS through TST, maxima around maximum flooding surfaces, and normally decreasing through HST.

From this high-resolution sequence sstratigraphic analysis, using facies association and Sr/Ca & Mn variations we note (1) the development of some laminated organic-rich facies in the basal part of the TST, (2) the coincidence of sudden negative shift of the Mn profile with the SB and sudden positive shifts with the TS and near to MFS respectively, (3) development of a special siliceous and radiolarian-rich facies (Facies E) in Bargou area, (4) a good correlation of Sr/Ca & Mn with eustatic sea level variarion.

9. References

Abdallah H. and C. Meister, 1997. The Cenomanian–Turonian boundary in the Gafsa–Chott area (southern part of central Tunisia): biostratigraphy, palaeoenvironments. *Cretaceous Research* 18, 197– 236.

Abdallah H., Sassi S., Meister C. and Souissi R. 2000. Stratigraphie séquentielle et paléogéographie à la limite Cénomanien–Turonien dans la région de Gafsa–Chotts (Tunisie centrale). *Cretaceous Research* 21, 35–106

Accarie H., Emmanuel L., Robaszynski F., Baudin F., Amédro F., Caron M. and Deconinck J.-F. (1996) La géochimie isotopique du carbone (δ13C) comme outil stratigraphique. Application à la limite Cénomanien/Turonien en Tunisie centrale. *Compte Rendu de l' Académie des Sciences, Paris IIa 322, 579–586.*

Accarie H., Robaszynski F., Amédro F., Caron M et Zagrarni M. F. (1999) Stratigraphie événementielle au passage Cénomanien - Turonien dans le secteur occidental de la plateforme de Tunisie centrale (Formation Bahloul, région Kalaat Senen), *Annales des mines et de la géologie N°40 - les septièmes journées de la géologie tunisienne - Tunis, 63-80*

Ando A., Kawahata H., Kakegawa T., 2006, Sr /Ca ratios as indicators of varying modes of pelagic carbonate diagenesis in the ooze, chalk and limestone realms. *Sedimentary Geology*, 191, 37–53

Ben Ayed N., and Viguier C., 1981. interprétation structurale de la Tunisie atlasique. *CRAS Paris, t 292 série II pp. 1445-1448*

Berthe 1949. Stratigraphie du Crétacé moyen et supérieur de la Tunisie, *rapport inédit SEREPT*

Burollet P.F. Dumestre A., Keppel D. et Salvador A. 1952. Unités stratigraphiques en Tunisie centrale. *pp. 1 ; 19ème congrès géol. Alger.*

Burollet P.F. 1956. Contribution a l'étude stratigraphique de la Tunisie centrale. *Ann. Mines Geol., Tunis, n° 18, 350p. IVpl.*

Caron M., Dall'Agnolo S., Accarie H., Barrera E., Kauffman E.G., Amedro F., Robaszynski F. 2006. High-resolution stratigraphy of the Cenomanian-Turonian boundary interval at Pueblo (USA) and wadi Bahloul (Tunisia): stable isotope and bio-events correlation Geobios, 39 (2), 171-200.

Hancock, J.M. 1989. Sea-level changes in the British region during the Late Cretaceous. Proc. Geol. Assoc., 100, 565-594.

Hancock, J.M. 1993. Transatlantic correlations in the Campanian-Maastrichtian stages by eustatic changes of sea-level. In: High Resolution Stratigraphy (eds. E.A. Hailwood and R.B. Kidd). Spec. Publ. Geol. Soc. London, 70, 241-256.

Haq B. U., Hardenbol J. and Vail P. R., (1987) Chronology of fluctuating sea levels, since the Triassic. *Science, Washington, pp. 1156-1165.*

Hardenbol J., Caron, M., Amedro, F., Dupuis, C. and Robaszynski, F., 1993 The Cenomanian- Turonian boundary in central Tunisia in the context of a sequence-stratigraphic interpretation. *Cretaceous Research, 14, 449-454.*

Hardenbol J., Thierry J., Farley M.B., T. Jacquin, P.-C. De Graciansky et P.R. Vail (1998) Cretaceous sequence chronostratigraphy. *In: P.-C. De Graciansky, J. Hardenbol, T. Jacquin and P.R. Vail, Editors, Mesozoic and Cenozoic Sequence Stratigraphy of European Basins Soc. Econ. Paleontol. Mineral. Spec. Publ. 60 Chart 4; Tulsa .*

Layeb M, 1990. Étude géologique, géochimique et minéralogique, régionale, des faciès riches en matière organique de la formation Bahloul d'âge Cénomano-Turonien dans le domaine de la Tunisie Nord-Centrale, *Thèse, Doct, Tunis.*

Layeb M. and Belayouni H. 1989. La formation Bahloul au Centre et au Nord de la Tunisie un exemple de bonne Roche mère de pétrole à fort intérêt pétrolier. Mémoires de l'ETAP, n°3, *Actes des II ème journées de géologie Tunisienne appliquée à la recherche des hydrocarbures (Tunis, Nov, 1989), pp, 489-503*

Luning S.,, S., Kolonic, E., M., Belhaj, Z., Belhaj, L., Cota, G., Baric and T., Wagner, 2004. An integrated depositional model for the Cenomanian-Turonian organic-rich strata in North Africa Earth Science reviews, 64, Issues 1-2. Pp 51-117.

Maamouri A. L., Zaghbib-Turki D., MatmatiI M. F., Chikhaoui M. et Salaj J., 1994 La formation Bahloul en Tunisie centro-septentrionale : variation latérales nouvelle datation et nouvelle interprétation en terme de stratigraphie séquentielle. *Journal of African Earth Sciences Vol. 18, N°1 , pp. 37-50, 1994.*

Mabrouk, A., Jarvis, I., Belayouni, Murphy A., Moody R. T. J. and Sandman R. 2006. Regional to global correlations of Cenomanian to Eocene sediments: New insights to chemostratigraphic interpretations. Proceeding of the tenth Exploration and Production Conference, Memoir N°. 26- pp. 26-45

Murphy, A.M. 1998. Sediment and Groundwater Geochemistry of the Chalk in Southern England. PhD Thesis, Kingston University, Kingston upon Thames, 288 p.

Martinez C. and Truillet R. 1987. Évolution structurale et paléogéographie de la Tunisie. Ment. *Soc. Geol. It., 38 (1987), 35-45, 4 ff.*

Messaoudi F. and Hammouda F. 1994. Evènement structuraux et types de pièges dans l'offshore Nord-Est de la Tunisie. *Proceedings of the 4th tunisian petroleum exploration conference (tunis, may 1994). pp. 55-64*

Morand, P. and Dallot, S. 1985. Variations annuelle et pluriannuelles de quelques espèces du macroplancton cotier de la Mer Ligure (1898–1914). *Rapp. Comm. Int. Mer Méd. 29: 295–297.*

Nederbragt A. J. and Fiorentino A. 1999. Stratigraphy and paleoceanography of the Cenomanian- Turonian Boundary Event in Oued Mellegue, north-western Tunisia. *Cretaceous Research, vol. 20, pp. 47–62.*

Ouahchi A., M'Rabet A., Lazreg J. Mesaoudi F. and Ouazaa S. 1998. Early structuring, paleoemersion and porosity development: a key for exploration of the aptian serdj carbonate reservoir in Tunisia. *Proceedings of the 6th Tunisian petroleum exploration and production conference (tunis May 5th – 9th (1998). pp.267-284*

Pervinquière L. 1903 Étude géologique de la Tunisie centrale. *Thèse de doc Rud paris. pp 1-360.*

Razgallah S,, Philip J,, Thomel G,, Zaghbib-Turki D, Chaabani F, Ben Haj Ali N, et M'Rabet A, 1994 La limite Cénomanien-Turonien en Tunisie centrale et méridionale : biostratigraphie et paléoenvironnements, *Cretaceous research (1994) 15, 507-533,*

Renard, M. 1985. Géochimie des Carbonates Pélagiques. Mis en Évidence des Fluctuations de la Composition des Eaux Océaniques depuis 140 Ma. Essai de Chimiostratigraphie. Doc. BRGM, 85, 650 p.

Robaszynski F., Caron M., Dupuis C., Amedro F., Gonzalez-Donso J.M., Linares D., Hardenbol J., Gartner J., Calandra F. and Deloffre R., 1990 A tentative integrated stratigraphy in the Turonian of Central Tunisia : Formations, zones and sequential stratigraphy in the Kalaat Senan area. - *Bull. Centres Rech. Explor. Prod. Elf-Aquitaine, 14 / 1, 213-384.*

Robaszynski F., Amedro, F. and Caron, M., 1993 La limite Cénomanien-Turonien et la Formation Bahloul dans quelques localités de Tunisie Centrale. *Cretaceous Research, n°14, pp, 477-486.*

Soua M., 2005. Biostratigraphie de haute résolution des foraminifères planctoniques du passage Cénomanien Turonien et impact de l'événement anoxique EAO-2 sur ce groupe dans la marge sud de la Téthys, exemple régions de Jerissa et Bargou. *Mémoir de Mastère, Univ., de Tunis El Manar, 169p. 10pl.*

Soua M, Zaghbib-Turki D, O'Dogherty L., 2006. Radiolarian biotic responses to the Latest Cenomanian global event across the southern Tethyan margin (Tunisia). Proceeding of the tenth Exploration and Production Conference, Memoir N°. 26- pp. 195-216

Soua M., and Tribovillard N. 2007. Modèle de sédimentation au passage Cénomanien /Turonien pour la formation Bahloul en Tunisie. *Compte Rendu Geoscience 339, 10, 692-701*

Soua M., Zaghbib-Turki D., Tribovillard N. 2008. Riverine influxes, warm and humid climatic conditions during the latest Cenomanian-early Turonian Bahloul deposition. Proceeding of the tenth Exploration and Production Conference, Memoir N°. 27- 201-212.

Soua, M., Echihi, O. Herkat, M., Zaghbib-Turki, D., Smaoui, J., Fakhfakh-Ben Jemia, H., Belghaji, H., 2009. Structural context of the paleogeography of the Cenomanian - Turonian anoxic event in the eastern Atlas basins of the Maghreb. C. R. Geoscience, 341, 1029–1037

Soua M., 2010 Time series (orbital cycles) analysis of the latest Cenomanian - Early Turonian sequence on the southern Tethyan margin using foraminifera. Geologica Carpathica, 61, 2/2010

Soua, M., 2010b Productivity and bottom water redox conditions at the Cenomanian-Turonian Oceanic Anoxic Event in the southern Tethyan margin, Tunisia.*Revue Méditerranéenne de l'Environnement 4 (2010) 653-664*

Soua M., Zaghbib Turki D., Smaoui J., 2010 Application of Time series analysis to the latest Cenomanian - Early Turonian sequence on the southern Tethyan margin using foraminifera. Society of Petroleum Engeneers, 2/2010? Cairo Conference.

Soua M. 2011 Le Passage Cénomanien-Turonien en Tunisie: biostratigraphie, chimiostratigraphie, cyclostratigraphie et stratigraphie séquentielle. *pp 420, 73fig, 16pl. PhD thesis, Université de Tunis El Manar, Tunisia.*

Stoll, H.M. and Schrag, D.P. 2000. High-resolution stable isotope records from the Upper Cretaceous rocks of Italy and Spain: Glacial episodes in a greenhouse planet? Geol. Soc. Am. Bull., 112, 308-319.

Stoll, H.M. and Schrag, D.P. 2001. Sr/Ca variations in Cretaceous carbonates: relation to productivity and sea level changes. Palaeogeogr. Palaeoclimatol. Palaeoecol., 168, 311-336.

Wignall, P.B. 1991 Model for transgressive black shales? *Geology* 19, 167–170.

Zagrarni, M. F., Negra, M. H. & Hanini, A. 2008. Cenomanian-Turonian facies and sequence stratigraphy, Bahloul Formation, Tunisia, *Sedimentary Geoogy* 204, 18-35.

7

Petrography, Geochemistry and Petrogenesis of Late-Stage Granites: An Example from the Glen Eden Area, New South Wales, Australia

A. K. Somarin
Department of Geology, Brandon University, Brandon, Manitoba, Canada

1. Introduction

The Glen Eden area is located within the New England Orogen (also known as New England Fold Belt). This orogen is one of the major structural elements within the extensive Tasman Orogenic Province which comprises the eastern part of the Australian continent (Hensel, 1982). The present length of this orogen is about 1500 km from Townsville to Newcastle. It is separated from the Thomson and Lachlan fold belts to the west by the Permian and Triassic strata of the Bowen-Gunnedah-Sydney Basin. The Mesozoic Clarence-Moreton and Great Artesian basins separate the northern and southern parts of this orogen.
The New England Orogen was the site of the extensive episodic calc-alkaline magmatism related to west-dipping subduction from middle Paleozoic to Early Cretaceous time. The oldest rocks might have formed at least partly in a volcanic island arc, but from the Late Devonian, the orogen developed as a convergent Pacific-type continental margin. During Late Devonian-Carboniferous time, parallel belts representing continental margin, volcanic arc, forearc basin and subduction complex assemblages can be recognized (Murray, 1988).
More than one hundred plutons were emplaced from the Late Carboniferous to the Triassic in the southern NEO. These intrusions have been attributed to two major periods of plutonism, the first during Late Carboniferous time and the second during the Late Permian and Triassic. The resulting plutons comprise the New England Batholith. Although volcanogenic massive sulfides and volcanic-hosted epithermal gold-silver ore deposits occur in older rock sequences (Murray, 1988), almost all of the other ore deposits of this region, including the Glen Eden Mo-W-Sn deposit, have a genetic or paragenetic relationship with plutons of the New England Batholith which is one of the largest Paleozoic-Mesozoic batholiths in eastern Australia. It underlies an area of almost 20000 km^2 and is composed of more than one hundred N-S-trending plutons which include all of the granitoids in the southern part of the NEO. These granitoids intruded into the tectono-stratigraphic terranes (Flood and Aitchison, 1993a, b) and deformed trench-complex metasedimentary rocks (Shaw and Flood, 1981). The composition of this batholith is 80% monzogranite, 18% granodiorite, 1% diorite and tonalite, 1% quartz-bearing monzonite and <0.2% gabbro (see Shaw and Flood, 1981).
On the basis of petrography, geochemistry and isotopic characteristics, Shaw and Flood (1981) subdivided the granitoids of the New England Batholith into five intrusive suites and

a group of leucoadamellites. They pointed out that the differences between these six groups reflect differences in their source rock types.

The Glen Eden Granite (GEG) occurs as dykes at depths of more than 80 m and is not exposed at the surface (Fig. 1). Mineralogical studies and field evidence indicate that the observed dykes have intruded after initiation of the hydrothermal activity. Based on petrographic studies, three types of GEG can be recognized: microgranite porphyry, micrographic granite, and aplite. Petrographic features of these granites are discussed below.

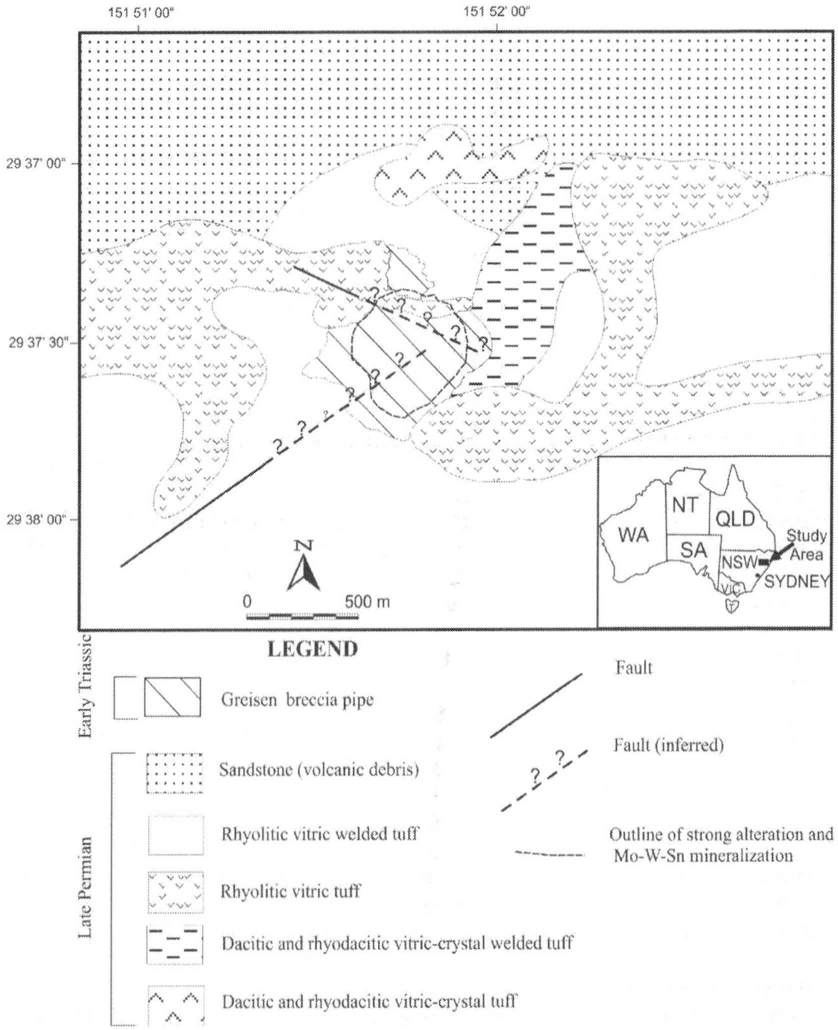

Fig. 1. Geological map of the Glen Eden area (after Somarin and Ashley, 2004).

2. Petrography of the Glen Eden granite

2.1 Microgranite porphyry

Microgranite porphyry of GEG is composed of quartz, K-feldspar and plagioclase as major minerals and biotite, zircon, xenotime, monazite and fluorite as accessory phases. Its texture is granular with quartz, K-feldspar, plagioclase and biotite as phenocrysts up to 8 mm in size. The groundmass is composed of quartz, K-feldspar and plagioclase, typically 50 to 300 μm, average 200 μm, in size. Most of the phenocrysts have irregular margins due to resorption and replacement by the groundmass. The cracks and embayments in these phenocrysts have been filled by the groundmass.

Quartz occurs as anhedral to euhedral grains, commonly rounded in shape and forming mosaics within feldspathic matrix. Some quartz grains form well-developed euhedral crystals, possibly due to secondary overgrowth. The presence of quartz as inclusions within biotite, plagioclase and fluorite and replacement of quartz phenocrysts by groundmass suggest that quartz crystallized relatively early.

K-feldspar is mostly orthoclase ($Or_{86}Ab_{14}$ to $Or_{98}Ab_2$) and mainly occurs as cloudy or perthitic anhedral crystals up to 5 mm in size. Rare microcline occurs as anhedral to subhedral grains 200-300 μm across. Perthitic hydrothermal K-feldspar in veins is common.

Plagioclase is mostly albitic ($Ab_{85}Or_{13}An_2$ to $Ab_{99}Or_1$) and occurs as subhedral to euhedral crystals and varies in size from 80-200 μm in groundmass up to 1.5-2.2 mm as phenocrysts. There is no zoning. In altered samples, the presence of K-feldspar as replacement rims around plagioclase implies sub-solidus alteration of plagioclase.

Biotite is dark brown to brown in color, strongly pleochroic and is mainly siderophyllite in composition. This mineral occurs as euhedral flakes, 50 μm up to a few millimeters in size. Commonly, biotite flakes have inclusions of magmatic quartz, zircon, xenotime, monazite and, in some samples, fluorite, rutile and secondary goethite accompany these flakes. These features are similar to those of Climax-type intrusives (e.g. White et al., 1981). Biotite is inferred to have been the most unstable mineral during hydrothermal alteration and commonly is replaced by sericite and goethite. Mostly, due to this replacement, only relicts of biotite can be seen and its color changes from brown to cream. Based on textural criteria, the position of biotite in the crystallization sequence cannot be determined unequivocally. However, the interstitial nature of biotite in GEG and occurrence of other minerals as inclusions within it are indicative of its late crystallization which is consistent with a high activity of F during crystallization (see below; Munoz and Ludington, 1974; Tischendorf, 1977; Collins et al., 1982).

Locally, muscovite occurs as flakes in samples adjacent to hydrothermal veins; they are interpreted to be of hydrothermal origin. Fluorite occurs as anhedral, interstitial grains with a purple tint in plane-polarized light and 70 μm up to 1 mm across. Locally, it occurs as inclusions within biotite flakes where biotite is unaltered. This indicates that fluorite in granite porphyry of GEG has a magmatic origin and reflects high activity of fluorine in the GEG magma. Micrographic intergrowth of quartz and K-feldspar in granite porphyry is common. Commonly the contact between a granitic dyke and surrounding rhyolitic volcanic rocks is marked by quartz veins. It seems that these contacts had the role of conduits for later hydrothermal fluids from the dykes or a deeper source.

2.2 Micrographic granite

Mineralogy and appearance of micrographic granite is similar to that of microgranite porphyry, however, the former can be distinguished by lower biotite contents and finer grain size. Its K-feldspar ($Or_{87}Ab_{13}$ to $Or_{95}Ab_5$) and plagioclase ($Ab_{90}Or_2An_8$ to $Ab_{98}An_2$) composition is similar to those in the microgranite porphyry. The intensity of micrographic growth varies. In some samples, there are discrete crystals of quartz and K-feldspar in addition to micrographic intergrowths, whereas in other samples almost all of the rock is composed of micrographic intergrowth of quartz and K-feldspar, and biotite and plagioclase are less abundant.

2.3 Aplite

Aplite at Glen Eden occurs as dykes up to 10 cm wide at a depth of ~85 m. It has granular texture and is composed of quartz, plagioclase and K-feldspar with grain size ranging from 50-400 μm, average 150 μm. No biotite or other accessory phases occur in aplite samples. Plagioclase is albitic ($Ab_{97}Or_2An_1$ to Ab_{100}) and K-feldspar (orthoclase, $Or_{86}Ab_{14}$ to $Or_{94}Ab_6$) grains are cloudy. These dykes have experienced potassic alteration and contain quartz-K-feldspar veins. The contact of aplite dykes with volcanic wall rock is sharp. Along these contacts, rhyolite groundmass has recrystallized, suggesting interaction of hot aplitic magma with cooler wall rock. Aplitic materials, in addition to aplite dykes, occur also in crenulate quartz layers and parting veins.

2.4 Crenulate quartz layers (comb layering)

Comb layering was defined by Moore and Lockwood (1973) as 'relatively unusual type of layering in granitoid rocks in which constituent crystals (plagioclase and hornblende in their study) are oriented nearly perpendicular to the planes of layering'. The types with ductile deformation are called 'crenulate quartz layers' (White et al., 1981; Kirkham and Sinclair, 1988). Comb layers are also referred to as ribbon rock, ribbon banded structures, rhythmically banded textures, brain rock, ptygmatic veins, wormy veins, vein dykes, unidirectional solidification textures and Willow Lake-type layering. Because of its deformed character, comb layering is called crenulate quartz layers, herein.

The crenulate quartz layers mainly occur within 5-10 m from the GEG dykes at depths >300 m. They are composed of quartz layers ranging in thickness from 2 mm to 3 cm. Quartz crystals in these layers are anhedral and they do not show perpendicular growth against the layer walls, possibly due to deformation and recrystallization. The quartz layers typically alternate with layers of aplitic material 1 mm to 2 cm thick. Some aplitic layers are discontinuous and terminate sharply within quartz layers. This suggests that the relative content of melt in the comb layer-forming system was low. Ptygmatic folding does not occur. Some quartz crystals in quartz layers are bent and elongate due to deformation. Similar deformation has been reported from Climax, Colorado (e.g. White et al., 1981), Hall, Nevada (Shaver, 1984a) and Anticlimax, British Colombia (Kirkham and Sinclair, 1988). Aplitic layers are composed of fine-grained quartz and feldspars, including orthoclase ($Or_{91}Ab_9$ to $Or_{97}Ab_3$) and albite ($Ab_{98}Or_1An_1$ to $Ab_{99}An_1$). Locally, quartz phenocrysts, up to 2 mm across, occur in aplitic layers. Based on microscopic and macroscopic studies, these conclusions can be made.

- The broken and bent quartz and aplite layers imply formation in a dynamic environment. Although subsequent deformation could produce partly similar features in GEG and wall rock, such features are not seen in these rocks. Also, if subsequent deformation was the main cause of bending, all layers should show this bending, whereas some of them are undeformed.

- Ductile deformation of these layers indicates that they were not completely solidified at the time of deformation. Also, deformation of some layers while the others are undeformed, suggests successive precipitation and deformation.
- The absence of a sharp boundary between quartz and aplite layers, and replacement of aplitic material by quartz suggest disequilibrium conditions during formation of quartz layering.
- The magma or fluid from which aplitic material precipitated was saturated with the components of sodic plagioclase and K-feldspar. The presence of some quartz phenocrysts in aplitic layers indicates that the magma crystallized in at least two stages, in which formation of groundmass followed crystallization of phenocrysts. The fine-grain size of aplitic material shows that the temperature difference between magma and the surrounding environment was large and magma crystallized rapidly.
- Delicate aplitic layers and close spatial relationship between crenulate quartz layers and parting veins indicate that the parent magma had very low viscosity. A similar conclusion was reached by Kirkham and Sinclair (1988).
- The low volume of aplitic materials and their mineralogical composition, which is similar to GEG, may imply that they represent a small portion of highly fractionated melt, possibly carried by escaping hydrothermal fluids. The association of aplitic material of crenulate quartz layers with quartz pods, parting veins, breccia zone and resorbed crystals suggests overlapping of magmatic processes by hydrothermal activity. Association of the crenulate quartz layers with Mo mineralization and silicification has been reported by other investigators and these layers have been considered as a prospecting guide (e.g., Povilaitis, 1978).
- The presence of primary two-phase fluid inclusions within quartz layers and quartz phenocrysts in the aplitic layers indicates the presence of hydrothermal fluid at the time of formation of these layers. Also, the similarity of formation temperature and salinity of these layers to those of other hydrothermal assemblages (Somarin and Ashley, 2004) indicates that at least the quartz layers and quartz phenocrysts in the aplitic layers have precipitated from fluid, not melt.
- Common occurrence of crenulate quartz layers in the apical parts (close to contact) of felsic intrusions related to porphyry deposits (White et al., 1981; Carten et al., 1988; Kirkham and Sinclair, 1988) may indicate that the main body of GEG is in the vicinity of these layers.

Generally there are two ideas regarding the genesis of crenulate quartz layers.

1. They have crystallized from the melt (White et al., 1981)
2. They have precipitated from the aqueous phase (Moore and Lockwood, 1973; Stewart, 1983; Shaver, 1984 a, b).

White et al. (1981) proposed that P_{H_2O} and P_{HF} increased during crystallization of the magma due to lack of hydrous minerals. The increased P_{H_2O} and P_{HF} would expand the quartz field in the ternary Q-Ab-Or system and lower the thermal minimum. They suggested that the combined effect of increasing P_{H_2O} and P_{HF} caused the precipitation of quartz without feldspar. Release of volatiles due to fracturing of wall rocks shrinks the quartz field and allows the crystallization of feldspar with quartz. This cycle occurs repeatedly to produce crenulate quartz layers. Based on this model, crenulate quartz layers have a magmatic source. It appears that even under high P_{H_2O} and P_{HF}, precipitation of pure quartz cannot be expected and some feldspar will crystallize as well. However, no feldspar occurs in quartz layers. Furthermore, 7 to 19 wt% F is needed in the system to destabilize feldspars (Glyuk and Aufiligov, 1973). This amount of F should cause movement of the eutectic point toward the Ab apex in the Q-Ab-Or system, which is not evident in the Glen Eden Granite. Therefore, it is unlikely that the formation of crenulate quartz layers of the Glen Eden Mo-W-Sn deposit can be explained by this model.

Based on observations mentioned above, it seems that for aplitic and quartz layers, two different sources should be considered. Aplitic layers indicate evidence of crystallization from a very low-viscosity melt, whereas quartz layers have crystallized from an aqueous fluid. It is more likely that aplitic material represents the relicts or parts of the highly fractionated low-viscosity melt in a dynamic moving, mainly upward, fluid which has separated from the melt. Kirkham and Sinclair (1988) suggested that the rapid drop in fluid pressure due to brecciation and fracturing of surrounding rocks quenches the adjacent silicate melt along the roof and walls of the magma chamber. This results in the formation of aplitic or porphyritic aplitic layers between the comb quartz layers. Occurrence of this process, successively, explains the rhythmic repetition of layers. The successive brecciation at Glen Eden was able to release pressure alternately and cause upward quenching of the melt. High fluid pressure and continued movement of magma, probably, resulted in the ductile deformation of the layers (Kirkham and Sinclair, 1988). The absence of thick comb quartz layers and pegmatitic lenses may indicate trapping of a large volume of volatiles (testified by pervasive hydrothermal brecciation) and relatively rapid build-up of fluid pressure. This could prevent the growth of layer crystals before fluid escape. However, occasionally coarse-grained K-feldspar and ore minerals can be seen in the breccia pipe, indicating less rapid build-up of fluid pressure, permitting the growth of these minerals.

3. Genetic implications of micrographic texture

In the Glen Eden Granite, micrographic texture occurs as the main texture of the micrographic granite and as a texture of some phenocrysts in microgranite porphyry. Generally there are two ideas about the genesis of graphic texture.

1. Infiltration and replacement of one mineral (host) by another mineral (guest) (e.g. Augustithis, 1973).
2. Eutectic crystallization of intergrowth-forming minerals (e.g. Fenn, 1979; Kirkham and Sinclair, 1988).

Graphic textures most commonly develop in water-rich magmas, generally in the presence of a separate aqueous phase (Nabelek and Russ-Nabelek, 1990), even though studies by Fenn (1979) have shown that a separate aqueous phase is not always required. In experiments using crushed glass from bulk samples of Spruce Pine pegmatite, Burnham (1967) found that in the presence of H_2O alone, the melts crystallized to an assemblage of alkali feldspar, quartz and muscovite. However, with a solution containing 6.2 wt% total dissolved alkali feldspar, muscovite did not appear and the melt crystallized to a graphically intergrown assemblage of alkali feldspar and quartz. Based on these studies, White et al. (1981) concluded that graphic textures represent zones of accumulation of a separate, Cl-rich aqueous phase. However, the presence of F in magma, which increases the amount of water in the separate phase by decreasing its solubility in the melt, may also help the formation of graphic texture. Also, pressure-quenched crystallization is able to produce micrographic texture (Kirkham and Sinclair, 1988).

Petrographic studies show that, genetically, there are two kinds of graphic texture at Glen Eden.

1. Graphic texture in the GEG. The following observations imply that this texture is the result of eutectic crystallization rather than replacement.
   a. Absence, in fresh rocks, of replacement of other minerals, such as plagioclase, by either quartz or K-feldspar.
   b. The occurrence of micrographic granite in which the entire rock is composed of micrographic intergrowth of quartz and K-feldspar.
   c. Absence of evidence of infiltration of quartz-forming solutions and replacement of K-feldspar. Although there are some low-temperature fluid inclusions in quartz phenocrysts of GEG, there is no clear evidence of replacement of other minerals by quartz.
   d. Absence of reaction margins in host K-feldspar or other minerals.
   e. The presence of graphic grains in which K-feldspar patches occur as inclusions within quartz. In the replacement model, in which quartz has been introduced by a solution, euhedral quartz grains should have formed by progressive replacement, rather than a groundmass for K-feldspar patches (Augustithis, 1973). Furthermore there is no evidence of infiltration of K-feldspar-forming solutions into quartz grains and replacement of quartz by K-feldspar.
2. Graphic texture in potassic alteration zone. Microscopic studies show that infiltration of quartz-forming solutions into fractures, intergranular spaces and cleavages of K-feldspar resulted in the replacement of K-feldspar by quartz. This replacement looks like a graphic intergrowth and clearly is the result of post-magmatic hydrothermal activity.

It seems that the presence of crenulate quartz layers, micrographic texture and hydrothermal breccia at Glen Eden indicates saturation of magma from water and the presence of a fluid-rich environment. The presence of free vapor and aqueous phases during graphic crystallization of quartz and K-feldspar is proved by the presence of fine (2-5 μm) primary two-phase fluid inclusions within quartz of the graphic texture.

4. Emplacement of the Glen Eden Granite

The presence of topaz, fluorine-rich biotite and widespread occurrence of fluorite in all alteration assemblages indicate that the Glen Eden Granite magma was uncommonly fluorine-rich. Since fluorine has significant effects on the physico-chemical properties of granitic magma, these effects are discussed below.

4.1 Effects of fluorine on the magma

High F content of GEG and presence of magmatic fluorite provide links between this granite and other F-rich rocks, such as topaz granite, ongonites and topaz rhyolites (Kovalenko et al., 1971; Pichavant and Manning, 1984; Taylor, 1992; Kontak, 1994). The effects of fluorine in magma have been studied by many investigators. These effects can be summarized as follows.

- Fluorine decreases the solubility of water in the melt (Dingwell, 1985, 1988), so water exsolution may occur earlier during crystallization of F-rich melts (Strong, 1988). The presence of breccia pipes testifies that magma had become saturated in water and volatiles.

- Both fluorine and water lower the crystallization temperature of granitic magmas (Bailey, 1977). Manning (1981) has documented the persistence of melt at 550°C in a granite with 4% F. This effect of fluorine would allow melt to fractionate more. The occurrence of mineral deposits similar to that at Glen Eden with highly fractionated granitic rocks may suggest that this factor (more fractionation) is important for the evolution of ore-bearing vapor phase from melt, since incompatible elements, including metals, would concentrate in residual melt.
- Fluorine changes the order of crystallization by promoting quartz, topaz and feldspars above biotite (Bailey, 1977; Hannah and Stein, 1990). This could be the result of increasing the thermal stability of hydrous phases by fluorine (Hannah and Stein, 1990).
- Fluorine lowers the viscosity of melt (Dingwell et al., 1985; Hannah and Stein, 1990). At 1000°C, addition of 1 wt% F to a melt of albitic composition results in an order of magnitude decrease in melt viscosity (Dingwell, 1988). Because of the smaller temperature dependence of viscosity in F-bearing melts versus F-free melt, the effects of F on melt viscosity is greatest at 600° to 800°C (Dingwell, 1988). The lower viscosity could cause higher migration of melt and replacement into shallow levels. Approach of the melt to shallow levels in the crust and hence decreasing pressure and the escape of water and volatiles may lead to increasing viscosity. High-level emplacement of the Glen Eden Granite, along with the presence of crenulate quartz layers and parting veins indicate low viscosity of the melt.
- By decreasing the solidus temperature of the magma, the assimilation ability of the magma may be increased (Keith and Shanks, 1988).
- Due to the decreased solidification temperature of the melt, F-bearing magmas may show extreme differentiation. The solidification temperature could be as low as 550-600°C in the presence of various volatiles (Strong, 1988). The solidus of an acid melt will decrease by 60°C in the presence of a vapor phase containing 5% HF (Schroecke, 1973). The association of Sn, Mo and W ore deposits with highly fractionated granites implies that extreme differentiation is essential for the concentration of these elements in evolved aqueous phase. This explains why intrusions at high levels have more potential to associate with rare-element mineralization in comparison to those intruded at low levels, since the high-level intrusions have been differentiated more than deep-level ones (Tischendorf, 1977). Also, water saturation develops through extreme differentiation. Intrusions without high concentration of magmatic water are typically barren (Strong, 1988). So it seems that the presence of volatiles, which affect the physical and chemical properties of the magma, is crucial for the formation of rare-metal ore deposits. A strongly depolymerized F-rich melt is more capable of hosting incompatible elements than a polymerized volatile-poor melt (Webster and Holloway, 1990).
- Fluorine increases cation diffusion in silicate melts (Dingwell, 1985) which is important for the transportation of the constituents necessary for ore deposition.
- The various effects of F could cause changes in commencement of the late-magmatic metasomatic processes (Tischendorf, 1977).
- Fluorine could change the solid/melt partition coefficients of elements because the stability of each element's site within the melt is altered (Hannah and Stein, 1990).
- Fluorine increases the solubility of silicate melt in the fluid phase (Hannah and Stein, 1990).
- Fluorine increases Ab content of the near-minimum melts (Manning and Pichavant, 1988).

4.2 Emplacement of GEG

Field evidence, including presence of the breccia pipe, crenulate quartz layers and parting veins, which commonly occur in the roof of the intrusion, indicate that the Glen Eden Granite, like other leucogranites of the New England Batholith, is a high-level intrusive body. The high-level emplacement of GEG indicates that the magma was water-poor, since the main control on depth of crystallization of a rising body of granitic magma is its H_2O content (Burnham, 1979; Wyllie, 1979). Burnham (1979) pointed out that for felsic magma to attain a volcanic or sub-volcanic environment, the initial water content cannot be greater than about 3 wt%. Magma with higher initial water content would become completely crystallized after boiling of its volatiles at a depth of several kilometres (Sheppard, 1977). In addition to water, fluorine also affects the emplacement of granitic magmas. Fluorine depolymerizes the structure of the melt and decreases its viscosity which would allow higher migration and shallow-level emplacement of the magma. Also, fluorine decreases the solubility of water in the melt. This water can escape from melt as a result of pressure drop, but fluorine does not, because it enters the OH sites of biotite and possibly exists in the melt as alkali-LILE-fluoride complexes (Collins et al., 1982) or alkali-aluminium-fluoride complexes (Velde and Kushiro, 1978). Therefore, the viscosity of magma will decrease progressively while water is released, and this magma can reach epizonal environments (Plimer, 1987). The formation of massive greisen (Somarin and Ashley, 2004) before intrusion of the Glen Eden granitic dykes might be due to this released water.

The path of movement, initially, is mainly dependent on the direction of weak zones, such as faults. A velocity of 1-2 cm/year, as proposed by Bankwitz (1978), may be enough to cause upward and outward movement of melt without complete crystallization. The prolonged period of tectonic activity in the New England area during Permo-Triassic compression and extension (Collins et al., 1993) could produce suitable structures, such as faults, for the rise of plutons. Also, fracturing of roof rocks by heat flow from the melt, which increases the amount of elastic energy, helps the movement of the melt (Bankwitz, 1978). As mentioned above, high content of F would retard crystallization of melt and allow it to move away from the magma chamber. The intense veining of parts of GEG, while the other parts show less or no veining, may reflect that the outer vein-bearing parts became colder than inner parts due to encountering cold wall rock.

The intrusive body utilizes structures which are later utilized by metal-bearing hydrothermal fluids (Plimer and Kleeman, 1985). The presence of quartz veins at boundaries between granitic dykes and wall rock at Glen Eden supports this idea and indicates that these boundaries were relatively weak zones, along which hydrothermal fluids could easily move.

On the whole, the high-level emplacement of the GEG and its highly differentiated character reflect high content of fluorine in the magma. Phosphorus, like F, also decreases the liquidus and solidus temperatures of the melt by modifying the silica network with the formation of phosphate-oxygen-metal complexes (London, 1987; Hannah and Stein, 1990). However, the low concentration of P in the GEG and absence of apatite in the hydrothermal assemblages indicate low P content of magma.

5. Geochemistry

5.1 Major element geochemistry

The Glen Eden Granite is highly felsic, as indicated by SiO_2 contents between 76 and 78 percent (Table 1). Aplite samples show potassic alteration. The chemical compositions of

	Granite porphyry										Micrographic granite			
	R75283	R75284	R75285	R75286	R75287	R75288	R75289	R75290	R75291	**Average**	R75292	R75293	R75294	**Average**
SiO_2	76.27	76.24	76.97	76.33	76.77	77.60	76.70	76.12	77.23	**76.69**	76.66	77.78	77.72	**77.39**
TiO_2	0.06	0.07	0.08	0.06	0.05	0.09	0.05	0.06	0.05	**0.06**	0.10	0.06	0.04	**0.07**
Al_2O_3	12.82	12.34	12.15	13.04	13.15	12.63	12.66	12.43	12.39	**12.62**	12.53	12.29	12.34	**12.39**
Fe_2O_3	0.11	0.18	0.30	0.13	0.08	0.10	0.17	0.18	0.16	**0.16**	0.25	0.08	0.06	**0.13**
FeO	0.43	0.82	0.70	0.38	0.21	0.22	0.60	0.63	0.55	**0.50**	0.71	0.48	0.63	**0.61**
MnO	0.02	0.03	0.03	0.02	0.01	0.01	0.03	0.03	0.02	**0.02**	0.03	0.02	0.03	**0.03**
MgO	0.12	0.04	0.06	0.08	0.07	0.08	0.05	0.05	0.04	**0.06**	0.06	0.07	0.10	**0.08**
CaO	0.27	0.33	0.40	0.26	0.20	0.23	0.31	0.35	0.39	**0.30**	0.43	0.38	0.36	**0.39**
Na_2O	2.78	3.17	3.67	3.05	3.22	3.04	3.58	3.46	3.71	**3.30**	3.86	3.45	2.94	**3.42**
K_2O	5.25	5.29	4.56	5.94	5.43	4.83	4.54	4.96	4.70	**5.05**	4.32	4.17	4.20	**4.23**
P_2O_5	0.01	0.01	0.01	0.01	0.01	0.01	0.01	0.01	0.01	**0.01**	0.01	0.00	0.02	**0.01**
S	0.03	0.01	0.01	0.02	0.02	0.02	0.05	0.01	0.01	**0.02**	0.01	0.02	0.02	**0.02**
LOI	1.01	1.40	0.54	0.70	0.54	0.84	0.92	0.78	0.53	**0.81**	0.54	0.87	1.33	**0.91**
Total	99.15	99.92	99.47	100.00	99.74	99.68	99.62	99.06	99.78	**99.58**	99.50	99.65	99.77	**99.66**
K_2O/Na_2O	1.89	1.67	1.24	1.95	1.69	1.59	1.27	1.43	1.27	**1.53**	1.12	1.21	1.43	**1.24**
Q	39.02	36.25	36.83	34.97	36.66	40.81	37.47	35.82	36.45	**37.14**	36.24	40.52	43.20	**39.99**
C	2.10	0.82	0.47	1.14	1.63	2.01	1.33	0.77	0.52	**1.20**	0.73	1.42	2.34	**1.50**
Or	31.03	31.27	26.95	35.11	32.09	28.55	26.80	29.32	27.75	**29.87**	25.53	24.65	24.82	**25.00**
Ab	23.52	26.78	31.05	25.81	27.25	25.68	30.25	29.24	31.39	**27.89**	32.66	29.19	24.88	**28.91**
An	1.27	1.60	1.92	1.22	0.93	1.08	1.45	1.65	1.88	**1.44**	2.10	1.86	1.66	**1.87**
Di	0.00	0.00	0.00	0.00	0.00	0.00	0.00	0.00	0.00	**0.00**	0.00	0.00	0.00	**0.00**
Hy	0.86	1.38	1.06	0.70	0.38	0.35	0.95	1.07	0.89	**0.85**	1.11	0.89	1.30	**1.10**
Mt	0.16	0.26	0.43	0.19	0.12	0.14	0.25	0.23	0.26	**0.23**	0.36	0.12	0.09	**0.19**
Ilm	0.10	0.12	0.14	0.10	0.09	0.16	0.09	0.10	0.09	**0.11**	0.19	0.11	0.08	**0.13**
Ap	0.02	0.01	0.02	0.02	0.02	0.02	0.02	0.02	0.01	**0.02**	0.01	0.00	0.05	**0.02**
Py	0.06	0.02	0.02	0.03	0.04	0.04	0.09	0.02	0.02	**0.04**	0.02	0.04	0.04	**0.03**
Total	98.15	98.51	98.92	99.30	99.21	98.84	98.71	98.24	99.26	**98.79**	98.96	98.79	98.45	**98.73**
Ab/An	18.52	16.74	16.17	21.16	29.30	23.78	20.86	17.72	16.70	**19.37**	15.55	15.69	14.99	**15.46**
100Mg/Mg+Fe	40	9	16	34	49	61	16	13	14	**23**	17	24	24	**22**
A.S.	1.19	1.07	1.04	1.09	1.14	1.19	1.12	1.06	1.04	**1.10**	1.06	1.13	1.23	**1.14**
DI	94	94	95	96	96	95	95	94	96	**95**	94	94	93	**94**

Table 1. Major element analyses and CIPW norms of the Glen Eden Granite, compared with average calc-alkaline and alkaline granites of Nockolds (1954), and average granite of Le Maitre (1976) and average I-, S-, and A-type granites of Whalen et al. (1987) and biotite porphyry of Climax (White et al., 1981).

	1	2	3	4	5	6	7
SiO_2	75.70	72.08	73.86	72.04	73.39	73.39	73.81
TiO_2	0.56	0.37	0.20	0.30	0.26	0.28	0.26
Al_2O_3	12.70	13.86	13.75	14.42	13.43	13.45	12.40
Fe_2O_3	0.47	0.86	0.78	1.22	0.60	0.36	1.24
FeO	0.57	1.67	1.13	1.68	1.32	1.73	1.58
MnO	NA	0.06	0.05	NA	0.05	0.04	0.06
MgO	0.37	0.52	0.26	0.71	0.55	0.58	0.20
CaO	1.07	1.33	0.72	1.82	1.71	1.28	0.75
Na_2O	3.10	3.08	3.51	3.69	3.33	2.81	4.07
K_2O	5.60	5.46	5.13	4.12	4.13	4.56	4.65
P_2O_5	ND	ND	ND	ND	0.07	0.14	0.04
S	ND	ND	ND	ND	ND	ND	ND
LOI	ND	ND	ND	ND	ND	ND	ND
Total	100.14	99.29	99.39	100.00	98.84	98.62	99.06
K_2O/Na_2O	1.81	1.77	1.46	1.12	1.24	1.62	1.14
Q	33.63	28.80	31.34	29.13	33.20	35.24	30.19
C	0.00	0.46	1.11	0.58	0.54	1.90	0.00
Or	33.10	32.27	30.32	24.35	24.41	26.95	27.48
Ab	26.23	26.06	29.69	31.22	28.18	23.78	34.44
An	4.20	6.60	3.57	9.03	8.03	5.44	1.83
Di	0.87	0.00	0.00	0.00	0.00	0.00	1.40
Hy	0.52	3.15	1.84	3.35	2.96	3.94	1.34
Mt	0.21	1.25	1.13	1.77	0.87	0.52	1.80
Ilm	1.06	0.70	0.38	0.57	0.49	0.53	0.49
Ap	0.00	0.00	0.00	0.00	0.17	0.33	0.09
Py	0.00	0.00	0.00	0.00	0.00	0.00	0.00
Total	100.14	99.29	99.39	100.00	98.84	98.62	99.06
Ab/An	6.25	3.95	8.32	3.46	3.51	4.37	18.82
100Mg/Mg+Fe	100	48	42	60	53	43	30
A.S.	0.97	1.03	1.09	1.04	1.03	1.13	0.95
DI	93	87	91	85	86	86	92

NA = Not analyzed; DI = Differentiation index
ND = No data; LOI = Loss on ignition
A.S. = Degree of aluminum saturation (molecular proportion of $Al_2O_3/CaO+Na_2O+K_2O$)

1) Biotite porphyry-Climax (White et al., 1981)
2) Average of calk-alkaline granite (Nockolds, 1954)
3) Average of alkali granite (Nockolds, 1954)
4) Average of granite recalculated to 100% (Le Maitre, 1976)
5) Average of felsic I-type granites (Whalen et al., 1987)
6) Average of felsic S-type granites (Whalen et al., 1987)
7) Average of A-type granites (Whalen et al., 1987)

Cont. Table 1.

	Granite Porphyry										Micrographic Granite						
	R752 83	R752 84	R752 85	R752 86	R752 87	R752 88	R752 89	R752 90	R752 91	**Average**	R752 92	R752 93	R752 94	**Average**	1	2	3
Nb	27	40	27	26	38	26	42	41	36	**34**	28	23	13	**21**	12	13	37
Zr	90	103	96	89	113	115	126	110	96	**104**	107	86	81	**91**	144	136	528
Y	54	83	76	58	58	66	114	106	109	**80**	79	83	87	**83**	34	33	75
Sr	11	3	4	17	3	4	3	6	5	**6**	5	5	17	**9**	143	81	48
Rb	488	510	365	552	434	408	309	458	427	**439**	379	324	321	**341**	194	277	169
Th	50	53	48	44	50	52	52	50	48	**50**	50	46	44	**47**	22	18	23
Pb	37	38	33	43	43	38	38	36	39	**38**	35	36	22	**31**	23	28	24
As	2	4	5	3	4	8	3	2	8	**4**	5	5	5	**5**	ND	ND	ND
U	9	14	15	12	11	12	12	12	17	**13**	17	11	7	**12**	5	6	5
Ga	24	25	21	23	25	25	24	27	26	**24**	21	25	20	**22**	16	17	25
Zn	5	17	47	6	3	7	26	8	8	**14**	25	13	19	**19**	35	44	120
Cu	5	8	3	<2	<2	<2	<2	<2	<2	**3**	<2	2	3	**2**	4	4	2
Ni	20	18	6	3	3	25	10	19	11	**13**	4	2	7	**4**	2	4	<1
Ce	38	29	51	39	40	41	43	34	37	**39**	58	50	62	**57**	68	53	137
Nd	21	18	22	22	26	21	24	24	23	**22**	28	34	38	**33**	ND	ND	ND
Ba	38	19	10	105	20	16	15	16	10	**28**	31	19	42	**31**	510	388	352
V	<2	2	4	5	<2	<2	3	5	<2	**3**	<2	<2	4	**3**	22	23	6
La	16	12	18	19	15	13	13	18	10	**15**	25	16	22	**21**	ND	ND	ND
Sc	<2	4	6	6	<2	6	<2	7	3	**4**	8	4	5	**6**	8	8	4
Sn	<3	4	3	<3	<3	<3	<3	<3	<3	**<3**	3	5	3	**4**	ND	ND	ND
Mo	1	2	10	5	2	2	1	2	1	**3**	2	1	2	**2**	ND	ND	ND
W	10	65	18	286	17	16	21	32	24	**54**	12	14	8	**11**	ND	ND	ND
F	NA	1580	1530	NA	1180	480	NA	1430	NA	**1240**	3200	3400	<2500		ND	ND	ND
K/Rb	89	86	104	89	104	98	122	90	91	**96**	95	107	109	**103**	177	137	229
Rb/Sr	44	170	91	32	145	102	103	76	85	**71**	76	65	19	**38**	1.36	3.42	3.52
Rb/Ba	13	27	37	5	22	26	21	29	43	**16**	12	17	8	**11**	0.38	0.71	0.48
Ga/Al	3.54	3.83	3.27	3.34	3.59	3.74	3.59	4.11	3.97	**3.66**	3.17	3.85	3.07	**3.36**	2.25	2.39	3.75

NA = Not analyzed ND = No data

1) Average Felsic I-type (Whalen et al. 1987) 2) Average Felsic S-type (Whalen et al. 1987) 3) Average A-type (Whalen et al. 1987)

Table 2. Trace element abundances in the Glen Eden Granite and average I-, S-, and A-type granites of Whalen et al. (1987). F values of less than 2500 ppm are from sodium peroxide fusion/SIE method whereas values shown by <2500 ppm (less than detection limit) are from XRF.

microgranite porphyry and micrographic granite are similar, however micrographic granite has lower K_2O and higher F, Nd and, in some samples, Ce (Table 2). The characteristics of GEG are similar to granites associated with Climax-type molybdenum ore deposits (White et al., 1981). The GEG, like Climax-type rocks, shows enrichment in silica and depletion in Ca, Al, total Fe, and Mg with respect to both the average calc-alkaline and alkaline granites of Nockolds (1954). Average K_2O/Na_2O in microgranite porphyry is close to that of average alkaline granite (Nockolds, 1954) and average normal granite (Le Maitre, 1976), but this ratio in micrographic granite is less than that of alkaline granite and is more than the average of normal granite. Normative Ab/An ratio is high and reflects the low Ca content of the GEG.
The samples of microgranite porphyry and micrographic granite show little chemical variation and no clear trends on Harker-type diagrams using SiO_2 or MgO as an index of possible fractionation (Fig. 2). Although the least-altered samples were chosen for analysis,

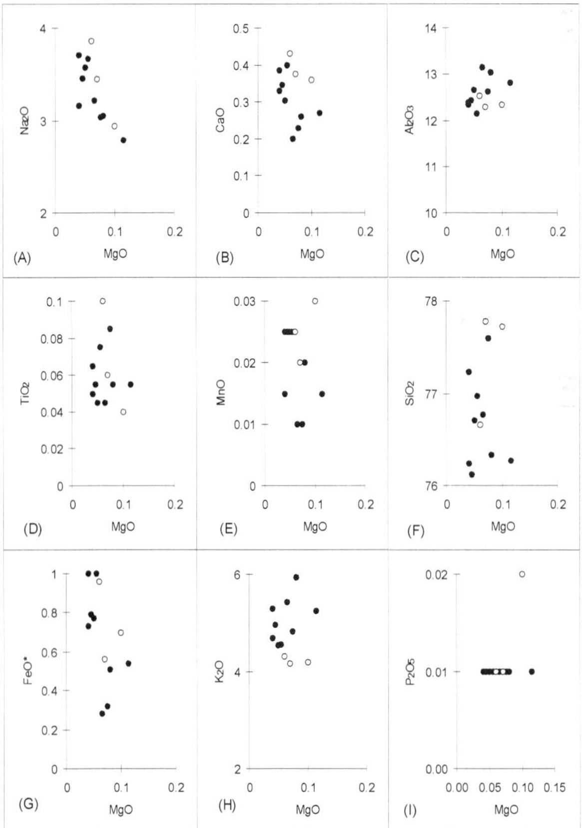

Fig. 2. Chemical variation diagrams for major elements of the GEG. None of these elements defines a well-developed trend. All oxides are in percent. Open circle = micrographic granite, closed circle = granite porphyry.

some scattering in Harker diagrams may be due to slight hydrothermal alteration. However, one of the features of Climax-type rocks is that almost none of them has completely escaped interaction with hydrothermal fluids (White et al., 1981). The GEG contains very low concentrations of P_2O_5 and CaO. Although low concentrations of CaO, Sr and Ba (Table 2) may be due to post-magmatic hydrothermal alteration, it seems that they reflect strong fractionation of the GEG magma. The geochemical changes accompanying progressive fractionation include enrichment of melt in alkali elements (Fig. 3A).

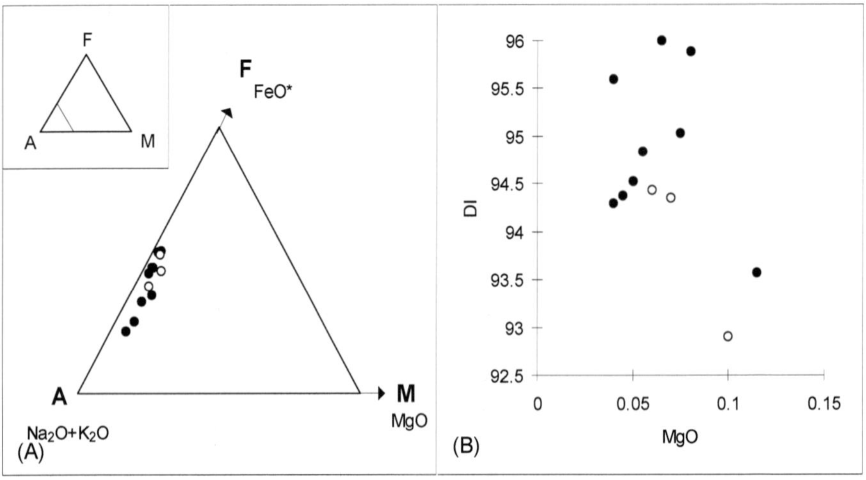

Fig. 3. A - FeO* - Na_2O+K_2O - MgO diagram for GEG samples showing evolution of the GEG toward the alkali apex. B - Plot of DI versus MgO for GEG samples showing no clear trend. All oxides are in percent. Open circle = micrographic granite, closed circle = granite porphyry.

Differentiation indices (DI= normative quartz + albite + orthoclase) for the GEG range from 93 to 97 (Table 1) which is like that of Climax granite (91-94, White et al., 1981). Inasmuch as the differentiation index represents the degree of magmatic evolution, and the normative constituents considered represent minerals with low entropies of melting (Carmichael et al., 1974; White et al., 1981), the high differentiation indices of the GEG suggest crystallization from highly differentiated, low-temperature melts. Like major elements, DI does not show any definite trend when plotted against MgO (Fig. 3B). Although the GEG, like Climax-type granites (White et al., 1981), is alkali rich, its molecular proportion of Al_2O_3 is a little more than its molecular proportion of $CaO+Na_2O+K_2O$, and so it is corundum normative (Table 1). Thus GEG is Al-saturated rather than peralkaline, like other Climax-type rocks (White et al., 1981). The GEG has peraluminous nature, however the samples show a trend toward the peralkaline field (Fig. 4A). This along with trends in Fig. 4B-C suggests that with increasing fractionation (i.e. decreasing MgO) peralkalinity increases and peraluminousity decreases. This trend (enrichment in alkali elements with fractionation), also can be seen in Fig. 3A.

A low initial H_2O content for the GEG magma can be inferred from the high-level emplacement of this granite. Furthermore, chemical composition of the GEG shows low

CaO, FeO and MgO contents of the magma. Under these conditions, high fluorine contents would not crystallize as fluorite nor substitute in the structure of ferromagnesian minerals, such as biotite. This would indicate that extreme enrichment in F (>4%) and Cl (>5000 ppm) could occur in the magma and in associated hydrothermal fluids during the late stages of the crystallization of the magma (Hannah and Stein, 1990; Webster and Holloway, 1990).

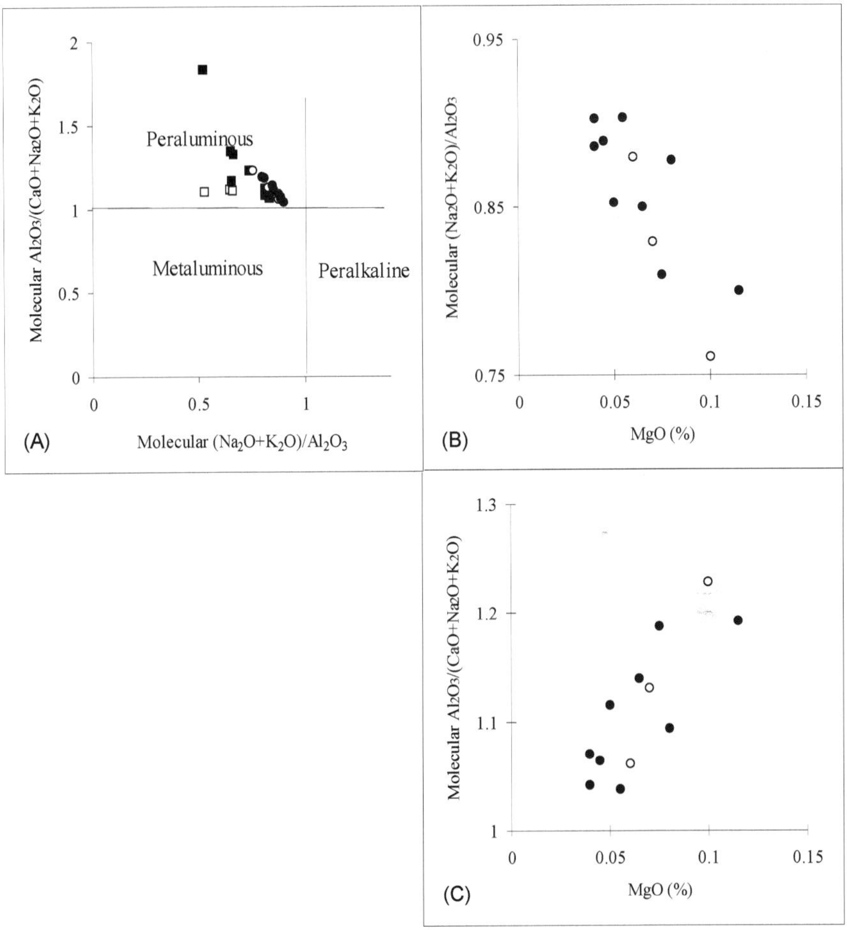

Fig. 4. A- Plot of peraluminousity index (molecular Al_2O_3/(CaO+Na_2O+K_2O)) versus peralkalinity index (molecular (Na_2O+K_2O)/Al_2O_3) for the GEG and volcanics samples showing peraluminous nature of these rocks. B- Plot of peralkalinity index versus MgO showing increasing peralkalinity with fractionation. C- Plot of peraluminousity index versus MgO showing decreasing peraluminousity with fractionation. Closed circle = granite porphyry, open circle = micrographic granite, closed square = rhyolite, open square = dacite and rhyodacite (date for volcanic rocks from Somarin, 1999).

5.2 Trace element geochemistry

Trace element abundances in GEG are presented in Table 2. Some trace elements such as Nb, Y, Ga, Zr, U, Nd, La and Ce show poorly developed trends in Harker-type diagrams (Fig. 5).

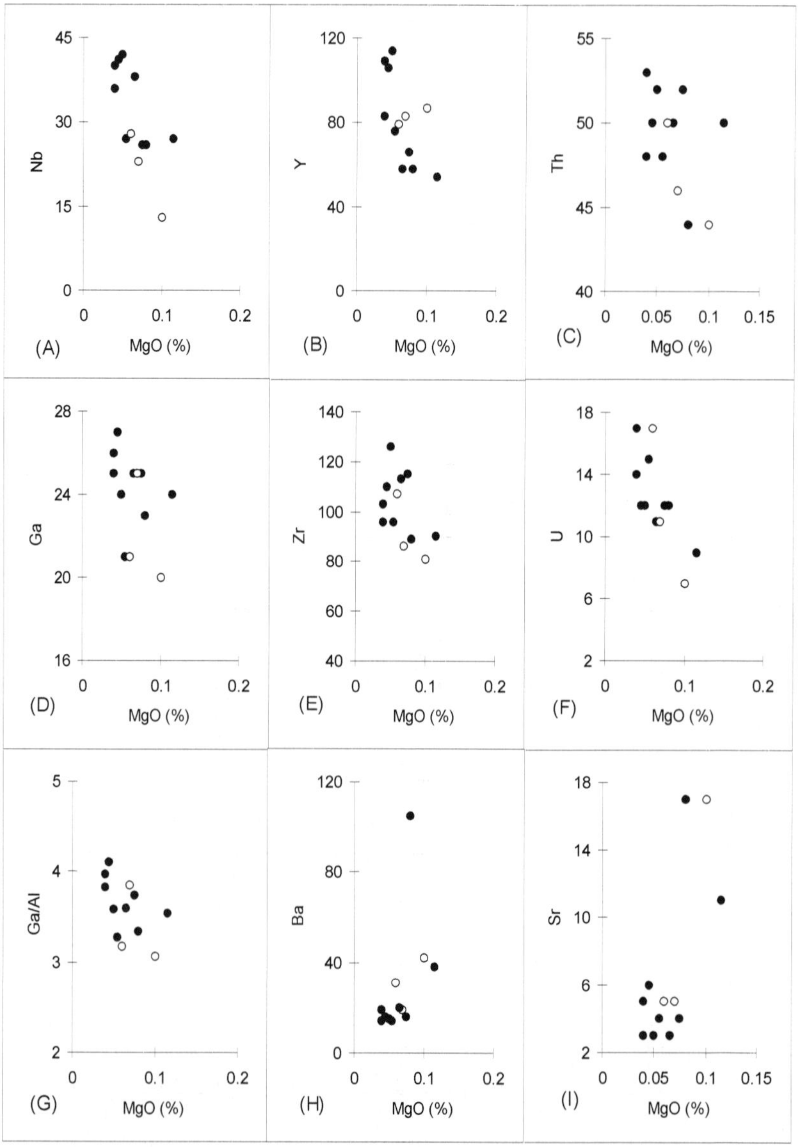

Fig. 5. Chemical variation diagrams for trace elements (in ppm) of the GEG. Open circle = micrographic granite, closed circle = granite porphyry.

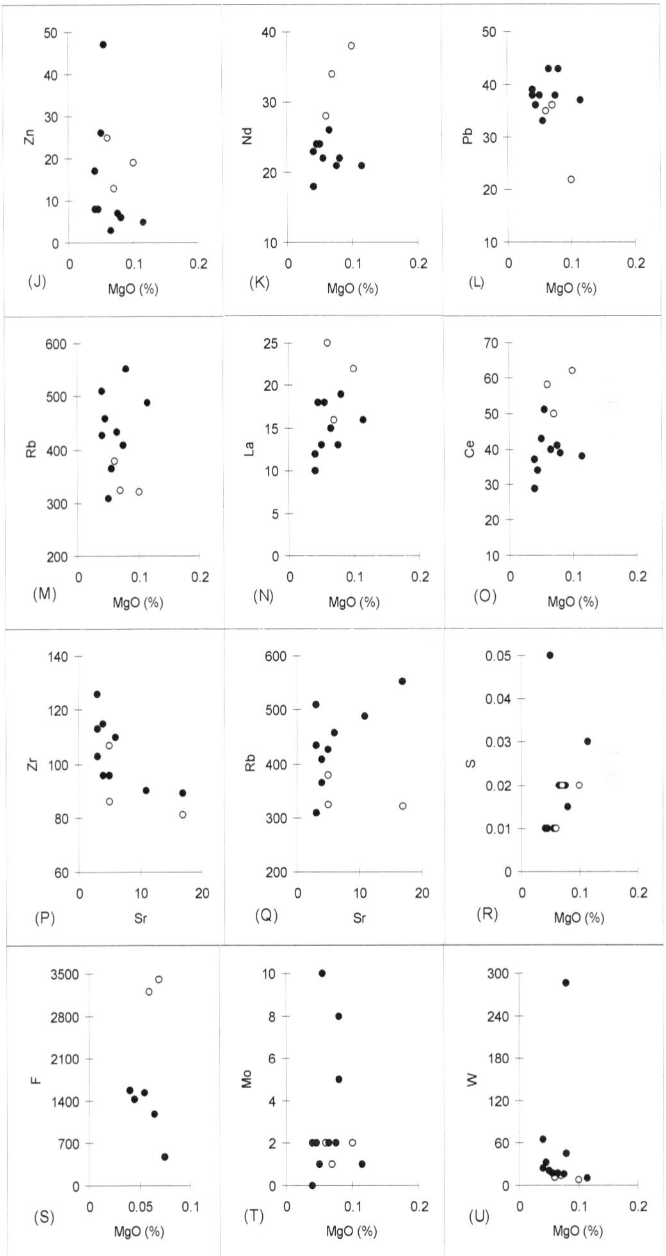
(J)
Zn
MgO (%)
(K)
Nd
MgO (%)
(L)
Pb
MgO (%)
(M)
Rb
MgO (%)
(N)
La
MgO (%)
(O)
Ce
MgO (%)
(P)
Zr
Sr
(Q)
Rb
Sr
(R)
S
MgO (%)
(S)
F
MgO (%)
(T)
Mo
MgO (%)
(U)
W
MgO (%)

Cont. Fig. 5.

However, due to narrow range and low concentrations (near detection limit) of MgO, these trends may not be significant. The important trace element features of the GEG are low concentrations of Sr, Ba, Zr and Zn and high concentrations of Y, Th, U, and Ga relative to average A-type, felsic I- and S-type granites. Also GEG contains high concentrations of F and W, similar to other ore-associated granites (Tischendorf, 1977). High values and wide ranges of Rb/Ba (5 to 43) and Rb/Sr (18 to 170) in microgranite porphyry and micrographic granite indicate crystal fractionation in the magma (Chappell and White, 1992). High Rb/Sr (commonly over 25), very low CaO (<0.7%) and pronounced Eu anomalies (<0.3) are characteristics of granitoids associated with Sn and W ore deposits (Stemprok, 1990). Fluorine has the highest enrichment in micrographic granite (Fig. 5S). The high F concentrations in the GEG and micas (Somarin and Ashley, 2004) resemble those of topaz granites, topaz rhyolites, ongonites and other volatile-enriched granitic rocks (see Manning and Pichavant, 1988). It is suggested that strong fractionation in the melt caused extraction of metal-bearing fluid from the parent magma. GEG contains high W concentrations and it is possible that sub-solidus leaching of this granite could provide additional W for mineralization. However, the volcanic wall rocks have low concentrations of Mo, W and Sn (Somarin and Ashley, 2004) and could not be a major source of these metals.

5.3 Glen Eden Granite in the Q-Or-Ab system

Average normative Q:Ab:Or in microgranite porphyry is Q_{39}, Ab_{29}, Or_{32} and in micrographic granite is Q_{42}, Ab_{31}, Or_{27}. These approximate the eutectic composition Q_{39}, Ab_{30}, Or_{31} for the calcium-poor granite system at P_{H_2O}=0.5 kbar (Winkler, 1974). Holtz et al. (1992) showed that decreasing H_2O content of the melt causes a rise in liquidus temperatures and a progressive shift of minimum and eutectic compositions toward the Q-Or join at approximately constant normative quartz content. The GEG does not show such shift.

Microgranite porphyry and micrographic granite samples plot around the minimum melt composition on the Q-Or-Ab ternary diagram for F-poor Q-Or-Ab-H_2O systems, whereas aplite samples plot toward the Or apex, reflecting the potassic alteration of these samples (Fig. 6). The minimum melt composition of these samples explains the absence of well-defined trends in Harker diagrams. The samples do not have Ab-enriched compositions expected of near-minimum melts in F-rich Q-Or-Ab-H_2O systems (Fig. 6A) (Manning and Pichavant, 1988), but this does not necessarily prove that the GEG melt was F-poor. There is a possibility that the GEG was a minimum melt at P less than 1 kbar but with higher concentrations of fluorine. However, in a calcium-poor granite system, like GEG, with Ab/An >15, as little as 0.5% fluorine in the melt could significantly affect the crystallization processes, since crystallization of fluorite will not occur until the late stages.

The pattern of data in the Q-Or-Ab system in the GEG is very similar to that of East Kemptville, Nova Scotia, in which data points plot around the minimum in the F-poor system, suggesting F concentration of less than 1% in the melt, while the effects of F in the various geochemical trends and greisen formation are clear (Richardson et al., 1990). The large amount of F as topaz and fluorite which accompany all the alteration assemblages in the Glen Eden Mo-W-Sn deposit and the presence of primary magmatic fluorite and F-rich biotite in this granite indicate the presence of F in the magma and its effects on the magmatic differentiation of GEG. F-rich granitic rocks typically contain between 0.5 and 2.5 wt% F (Keppler, 1993). F contents measured in granitic rocks should be considered as lower limits for the original F contents of the respective melts as large amounts of F could have been lost by the evolution of a fluid phase (Keppler, 1993). In the Glen Eden prospect, the widespread occurrence of fluorite in all assemblages indicates that a large amount of F has been concentrated in the late-stage fluids.

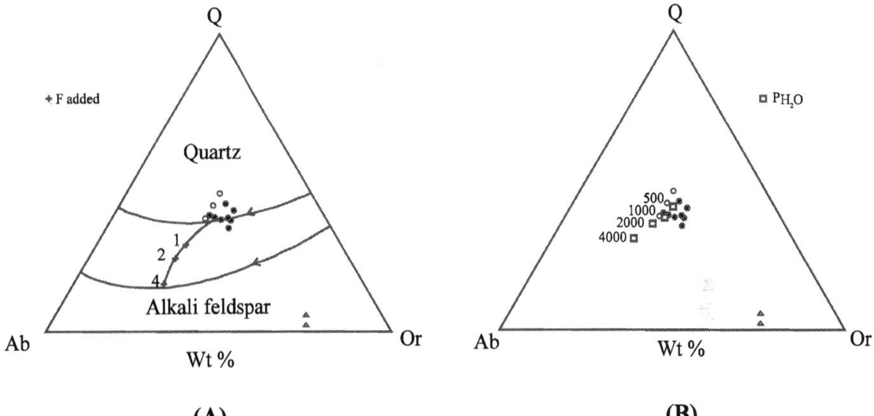

Fig. 6. A- The GEG compositions (triangle = aplite, open circle = micrographic granite, closed circle = granite porphyry) compared with liquidus phase relationships in the system Q-Ab-Or at 1 kbar total pressure with excess water (Tuttle and Bowen, 1958) and with F added (in %) under water-saturated conditions (after Manning and Pichavant, 1988). B- Position of minima and eutectics in the system Q-Ab-Or at various P_{H_2O} (Tuttle and Bowen, 1958; Luth, 1976). H_2O pressures are given in bars.

Crystallization of GEG has probably occurred between 500 and 1000 bars (Fig. 6B) which suggests high-level emplacement of the GEG. Also, compositional uniformity of analyzed granite samples and their similarity to minimum melt composition at low P_{H_2O} may imply that at least this part of the GEG has crystallized under low P_{H_2O}, since with increasing P_{H_2O} there is more potential for differentiation. This may suggest that these dykes have crystallized after separation and escape of the first episode of aqueous phase. The absence from the GEG of older hydrothermal alterations, which are recognized in the volcanic wall rock, supports this idea and indicates that there was an activation of the still-unsolidified magma chamber which yielded the emplacement of granitic and aplitic dykes within the alteration products. These observations do not imply that water content of the melt was low, since the presence of breccia pipe and the great quantity of hydrothermal veins and assemblages in the central zone reflects a high content of water.

6. Classification of the Glen Eden granite

It seems that three factors have influenced the concentrations of major and trace elements in the GEG.

1. The composition of the protolith: High concentrations of elements such as Th and U, which are high not only in the GEG but also in other I-type leucogranites of the New England area [e.g. Gilgai (Walsh, 1991; Stroud, 1995; Vickery et al., 1997), Kingsgate and

Mount Jonblee (Plimer, 1973), Stanthorpe (Bampton, 1988), Mole (Brodie, 1983; Stegman, 1983; Kleeman, 1985; Vickery et al., 1997), Dumboy-Gragin (Vickery et al., 1997) and Oban River (Le Messurier, 1983)] most probably reflect the composition of the protolith.

2. Post-magmatic hydrothermal alteration: The concentrations of CaO, Sr and possibly Ba are less than those that can be attained by fractional crystallization alone and it seems that destruction of feldspars by hydrothermal solutions can account for their low concentrations. Since P_2O_5 is more immobile than those components mentioned above, it seems that the low concentrations of P_2O_5 are unlikely to be the result of hydrothermal leaching.
3. Fractionation in the melt: It seems that fractionation was the most important factor controlling the composition of the GEG. Almost all of the geochemical characteristics of the GEG, such as high SiO_2, Rb, U, Th, Nb, Y, Ga and W and low concentrations of CaO, P_2O_5, Ba, Sr, Zr and Zn and high values and wide ranges of Rb/Ba and Rb/Sr can be explained by various degrees of fractionation of the magma.

Comparison of the GEG with well-known I-, S- and A-type granites is complicated since various investigators have reported different average values for some elements and other features of these granites (e.g. Whalen et al, 1987; Chappell and White, 1992). For example DI of the GEG resembles that of average A-type granite of Whalen et al. (1987) (Table 1), but is more similar to that of average fractionated I-type granite of Chappell and White (1992) (Table 3). The problem of determining I-, S- or A-type affinities of highly felsic granites (such as GEG) has been addressed by several authors (e.g., Whalen et al., 1987; Eby, 1990; Chappell and White, 1992). Aluminium saturation index (ASI; Zen, 1986), molecular $Al_2O_3/(Na_2O+K_2O+CaO)$, in the GEG varies between 1 and 1.2. Avila-Salinas (1990) used ASI=1.1 as a boundary between I- and S-type granites. Chappell and White (1992) showed that ASI in S-type granites of Lachlan Fold Belt (LFB), Australia, are always greater than 1 whereas I-type granites generally show ASI<1.1, although 2.8% of them have ASI>1.1. They explained higher ASI in S-types to be a reflection of sedimentary source rocks which contain more clay and so more Al. In contrast, lower ASI in I-types results from lower Al contents in igneous sources. However, they suggested that compositionally very similar felsic granites can be produced from these two quite different source rocks. In such circumstances, the only clue to the nature of a granite protolith might well be isotopic compositions (Chappell and White, 1992). Non-diagnostic values of ASI in the GEG and the very felsic composition of this granite may suggest that it cannot be classified on this criterion alone, and also no conclusion can be made about the source rocks, without isotopic data. Low concentrations of CaO and Sr and resultant higher values of ASI in the GEG may partly reflect slight leaching of these elements by hydrothermal solutions.

The average compositions of microgranite porphyry and micrographic granite are compared, in Table 3, with the average compositions of unfractionated and fractionated felsic I- and S-type granites and A-type granites (data from Chappell and White, 1992). As can be seen, GEG in both major and trace elements is mainly similar to fractionated I- and A-type granites. However, in some elements, GEG shows similarity to other types as well. GEG has very low concentrations of Fe_2O_3 in comparison with others, which indicates the reduced character of this granite. K_2O concentrations of the GEG overlap all types of granites. CaO, Sr and Ba concentrations are more similar to fractionated I-type, but actually

	Average GEG (porphyry)	Average GEG (micro-graphic)	Unfracti-onated I-type	Fracti-onated I-type	Unfracti-onated S-type	Fracti-onated S-type	A-type	Similar type
SiO_2	76.69	77.39	72.90	76.17	71.58	74.40	73.47	*FI*
TiO_2	0.06	0.07	0.30	0.10	0.42	0.16	0.30	*FI*
Al_2O_3	12.62	12.39	13.48	12.51	13.83	13.50	12.88	*FI, A*
Fe_2O_3	0.16	0.13	0.54	0.32	0.45	0.28	0.90	*?*
FeO	0.50	0.61	1.47	0.71	2.38	1.14	1.63	*FI*
MnO	0.02	0.03	0.05	0.04	0.05	0.04	0.06	*FI, FS*
MgO	0.06	0.08	0.66	0.12	1.02	0.27	0.30	*FI*
CaO	0.30	0.39	1.63	0.61	1.74	0.67	1.06	*FI*
Na_2O	3.30	3.42	3.27	3.37	2.57	3.06	3.50	*UI, FI, A*
K_2O	5.05	4.23	4.42	4.92	4.33	4.84	4.62	*?*
P_2O_5	0.01	0.01	0.09	0.02	0.14	0.18	0.07	*FI*
FeO*	0.64	0.73	1.96	1.00	2.79	1.39	2.44	*FI*
Nb	34	21	14	21	12	19	26	*FI, A*
Zr	104	91	151	116	168	92	322	*FS*
Y	80	83	38	75	34	28	71	*FI, A*
Sr	6	9	147	31	114	43	96	*FI*
Rb	439	341	219	424	221	475	188	*FI, FS*
Th	50	47	25	47	19	17	24	*FI*
Pb	38	31	29	35	28	25	27	*?*
U	13	12	6	16	4	11	5	*FS*
Ga	24	22	16	19	17	21	22	*A*
Zn	14	19	38	29	53	46	95	*FI*
Cu	3	2	6	2	7	3	5	*FI, A*
Ni	13	4	5	<1	10	2	2	*UI, US, FS,A*
Ce	39	57	74	79	63	37	130	US, FS
Ba	28	31	488	99	512	150	547	FI
V	3	3	25	3	41	7	9	FI
La	15	21	35	35	28	16	55	FS
Sc	4	6	8	6	10	5	11	FI, FS
Sn	<3	4	7	13	8	23	8	?
Mo	3	2						
W	54	11						
F	1240							
As	4	5						
Nd	22	33						
S (%)	0.02	0.02						
Q	37.00	40.00	31.89	35.87	33.65	35.98	32.06	FI, FS
Or	30.00	25.00	26.12	29.08	25.59	28.61	27.31	?
Ab	28.00	29.00	27.67	28.52	21.75	25.89	29.62	UI, FI, A
An	1.44	1.87	7.50	2.90	7.72	2.15	4.80	FS
Hy	0.85	1.10	3.49	1.25	5.94	2.34	2.61	FI
Mt	0.23	0.19	0.78	0.46	0.65	0.41	1.30	?
Ilm	0.11	0.13	0.57	0.19	0.80	0.30	0.57	FI

Ap	0.02	0.02	0.21	0.05	0.33	0.42	0.17	FI
C	1.20	1.50	0.57	0.58	2.09	2.44	0.36	?
Py	0.04	0.03						
DI	95	94	86	93	81	90	89	FI
ASI	1.10	1.14	1.03	1.04	1.15	1.17	1.02	?
100Mg/ Mg+Fe	23	22	54	29	50	35	34	FI
K_2O/ Na_2O	1.53	1.24	1.35	1.46	1.68	1.58	1.32	?
Rb/Sr	71	38	1.50	14	2	11	2	FI
Rb/Ba	16	11	0.50	4.30	0.40	3	0.30	FI
10000* Ga/Al	3.66	3.36	2.24	2.87	2.32	2.93	3.22	A

*Total Fe as FeO, FI= Fractionated I-type, FS= Fractionated S-type, A= A-type, UI= Unfractionated I-type, US= Unfractionated S-type.

Table 3. Comparison of the GEG with average compositions of various types of granites (data from Chappell and White, 1992).

they are very low in GEG due to hydrothermal leaching. A-type granites have higher Nb, Y, La, Ce, Sc, Zn, Zr and Ga in comparison to all I- and S-types. However, for Nb and Ga, the fractionated I- and S-type averages move towards the A-type values, relative to unfractionated values, as also does Y for the I-types (Chappell and White, 1992). These changes due to fractionation also increase Rb, U and Sn and decrease Ba and Sr concentrations in fractionated I- and S-types relative to A-types. It seems that increasing Ga and decreasing Al in fractionated granites, especially in fractionated I-types which contain less Al_2O_3, would cause highly fractionated granites to plot in the A-type field in discrimination diagrams of Whalen et al. (1987). Whalen et al. (1987) showed that A-type granites have a high ratio of 10000Ga/Al (>2.6) and they used this ratio for the construction of discrimination diagrams. On these diagrams, high concentrations of Ga and resultant high Ga/Al ratios cause the GEG to plot within the A-type granite field (Fig. 7). However,Whalen et al. (1987) stated that highly fractionated felsic I- and S-type granites can have high Ga/Al ratios and overlap with A-types. They suggested that these fractionated rocks can be distinguished from A-types using Zr+Nb+Ce+Y as a discriminator. Use of this discriminator is based on the principle that at any given degree of fractionation, the A-type granites would contain higher abundances of these elements. On the FeO_{total}/MgO versus Zr+Nb+Ce+Y diagram, GEG plots in all fields. This may be due to post-magmatic alteration. On the K_2O+Na_2O/CaO versus Zr+Nb+Ce+Y diagram of Whalen et al. (1987), FeO_{total}/MgO versus SiO_2 and 10000Ga/Al versus Zr+Nb+Ce+Y diagrams of Eby (1990) (Fig. 8), who used a higher minimum Ga/Al ratio for A-type granites, GEG samples plot in both 'Fractionated Granite' and A-type granite fields. In the multicationic diagram of Batchelor and Bowden (1985) (Fig. 9), GEG plots in 'Anorogenic' and 'Post-orogenic' granitoids fields which mainly include A-type granites. Based on these observations, a few conclusions can be made.

- The GEG is highly fractionated. High Ga concentrations have been considered as a diagnostic feature of A-type granites by many investigators (e.g., Collins et al., 1982; Whalen et al., 1987; Eby, 1990; Haapala and Ramo, 1990; Whalen and Currie, 1990). As stated by Whalen et al. (1987), Sawka et al. (1990) and Chappell and White (1992), high fractionation of I- and S-type granites could enrich Ga in the magma. So high Ga concentration is not exclusively a feature of A-type granites.

- The GEG most probably is not S-type, as S-type granites have high concentrations of P_2O_5 which increase with fractionation in this type of granite (Sawka et al., 1990; Chappell and White, 1992). Very low concentrations of P_2O_5 in the GEG and rarity of apatite in hydrothermal assemblages of the Glen Eden prospect indicate a low content of this component in the GEG melt.
- The discrimination diagrams (Figs 8 and 9) cannot unequivocally classify the GEG. There are two main possibilities.

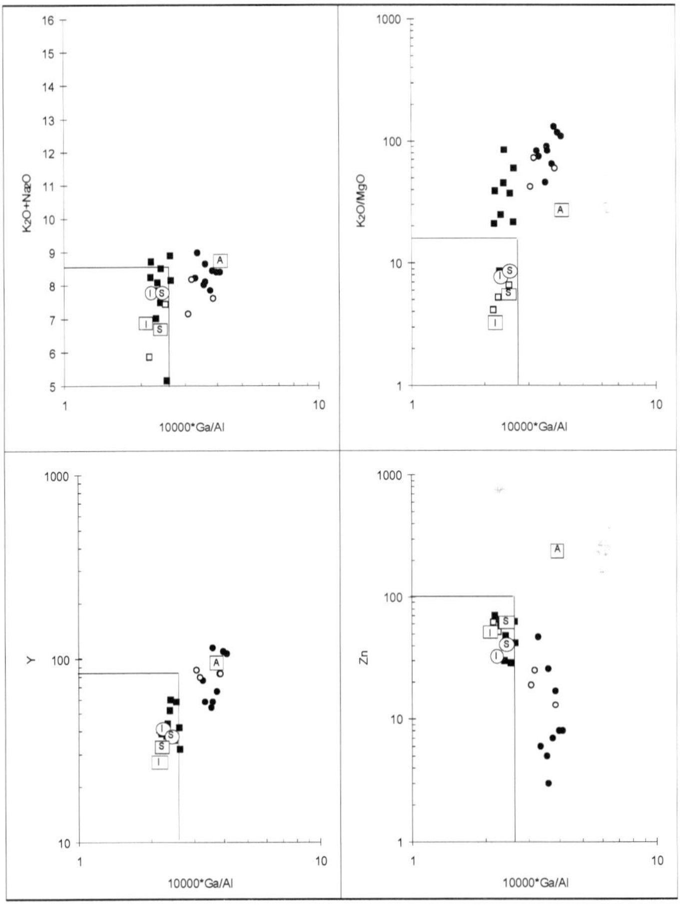

Fig. 7. Data from the GEG and Emmaville Volcanics in discrimination diagrams of Whalen et al. (1987), showing the possible A-type characteristics of this granite. [I] I-type granite average, [S] S-type granite average, (I) felsic I-type granite average, (S) felsic S-type granite average, [A] A-type granite average. . Oxides and trace elements are in percent and ppm, respectively. Closed circle = granite porphyry, open circle = micrographic granite, closed square = rhyolite, open square = dacite and rhyodacite.

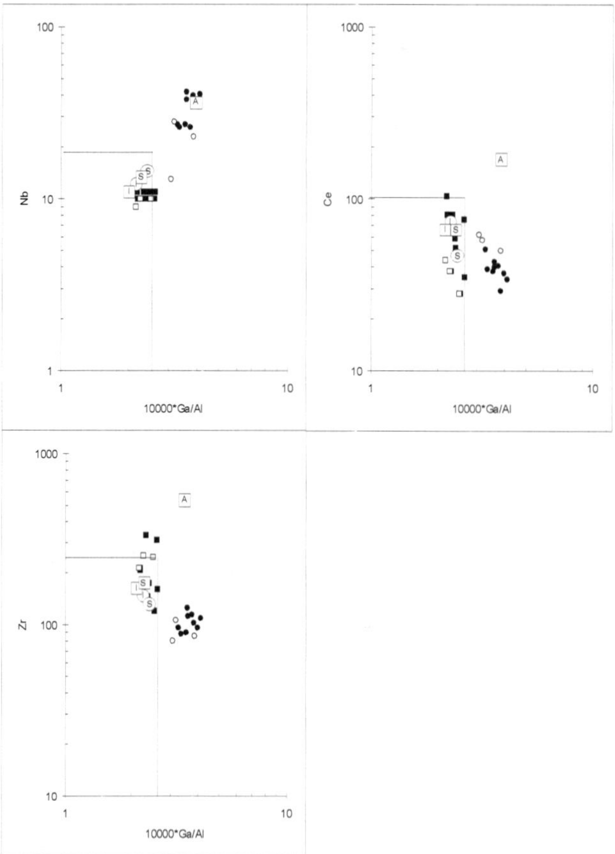

Cont. Fig. 7.

a. The GEG is A-type. In this case, very low concentrations of Zr and Zn (Fig. 10) could be due to low concentrations of these elements in the source rocks or a result of peraluminousity of the GEG, as non-peralkaline granites contain much lower Zr and Zn than peralkaline types. Watson and Harrison (1983) stated that peralkaline melts in comparison with peraluminous ones can maintain extremely high zircon solubility by complexing Zr^{4+} with free alkalies that are not associated with Al. High mobility of Zn during hydrothermal alterations may account for its low concentrations in the GEG.
b. The GEG is fractionated I-type: It has been found that a large degree of fractional crystallization of I- and even S-type granite magmas can produce a minor amount of evolved magma with high Ga/Al, very low concentrations of Ba and Sr and large variation in Rb/Sr and Rb/Ba ratios (Whalen and Currie, 1990). As can be seen in Table 3, fractionation of I-type granite magmas in the Lachlan Fold Belt increased SiO_2, Rb, Pb, Th, U, Nb, Y, Ce, Ga, Sn, DI, Rb/Sr, Rb/Ba and partially Na_2O and K_2O and decreased TiO_2, Al_2O_3, Fe_2O_3, FeO, MnO, MgO, CaO, P_2O_5, Ba, Sr, Zr, Sc, V, Ni, Cr, Co, Cu, and Zn. Also, with fractionation, the Fe_2O_3/FeO ratio increases (Fig. 11). So it seems

that fractionation of I-type granitic magma could produce the GEG. This is consistent with the field occurrences of the GEG as dykes. These dykes may represent the last stage of fractional crystallization of a major pluton at greater depth.

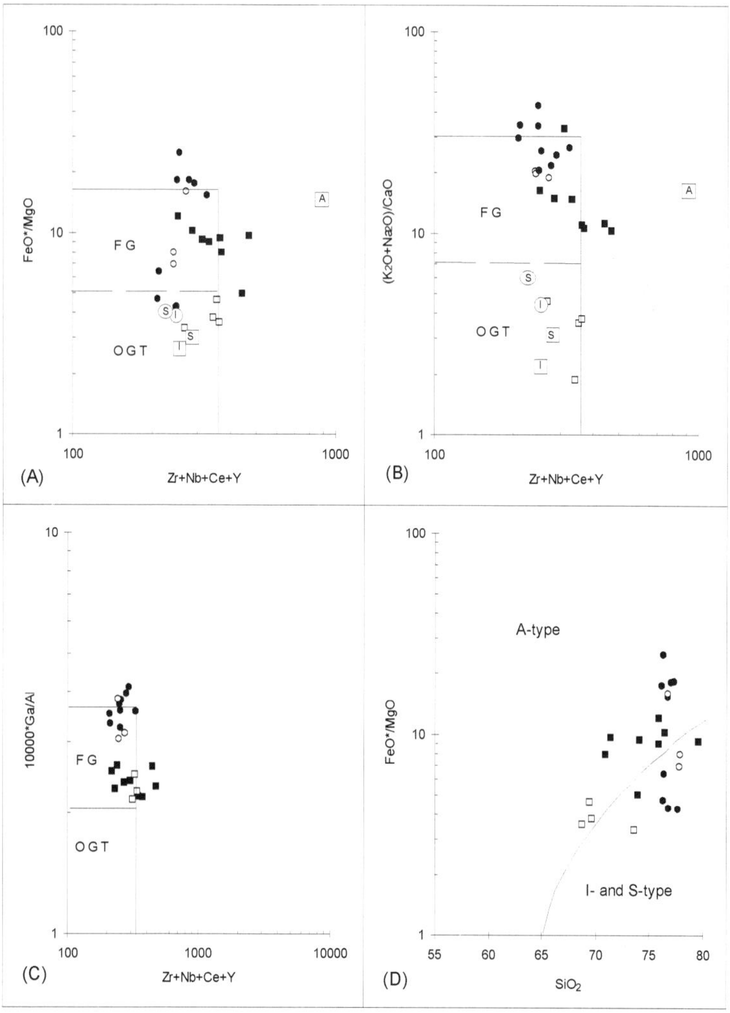

Fig. 8. The GEG and Emmaville Volcanics in discrimination diagrams of Whalen et al. (1987) (A and B) and Eby (1990) (C and D) plot in 'Fractionated Granites' and 'A-type Granites' fields. FeO*= total FeO, F G = Fractionated felsic Granite, O G T = Unfractionated Granite. Oxides and trace elements are in percent and ppm, respectively. Closed circle = granite porphyry, open circle = micrographic granite, closed square = rhyolite, open square = dacite and rhyodacite..

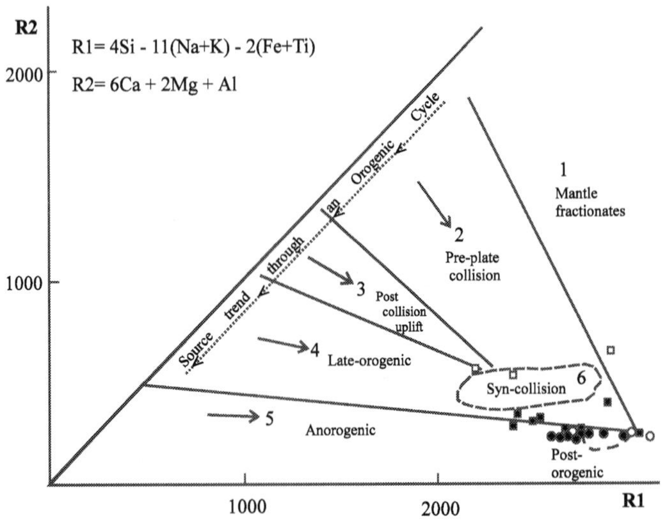

Fig. 9. Data from the GEG and Emmaville Volcanics in the multicationic discrimination diagram for the major granitoids (after Batchelor and Bowden, 1985). Analyzed samples plot mainly in the 'Post Orogenic' and 'Anorogenic' fields. Closed circle = granite porphyry, open circle = micrographic granite, closed square = rhyolite, open square = dacite and rhyodacite.

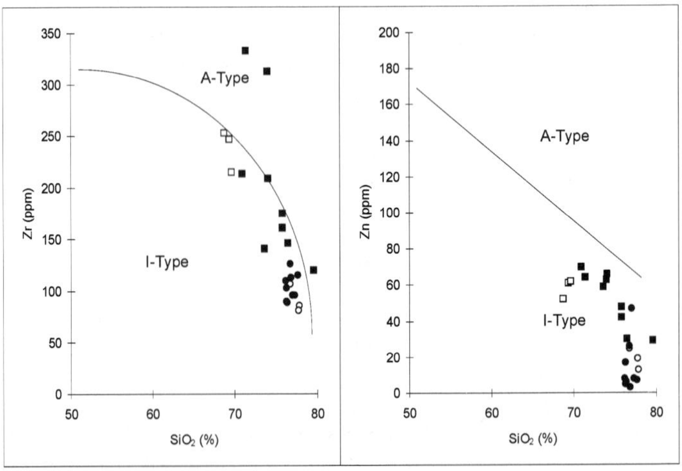

Fig. 10. Data from GEG and Emmaville Volcanics in Zr-SiO_2 and Zn-SiO_2 diagrams (after Newberry et al., 1990). Low concentrations of Zr and Zn in the GEG cause GEG to plot in the I-type field. Closed circle = granite porphyry, open circle = micrographic granite, closed square = rhyolite, open square = dacite and rhyodacite.

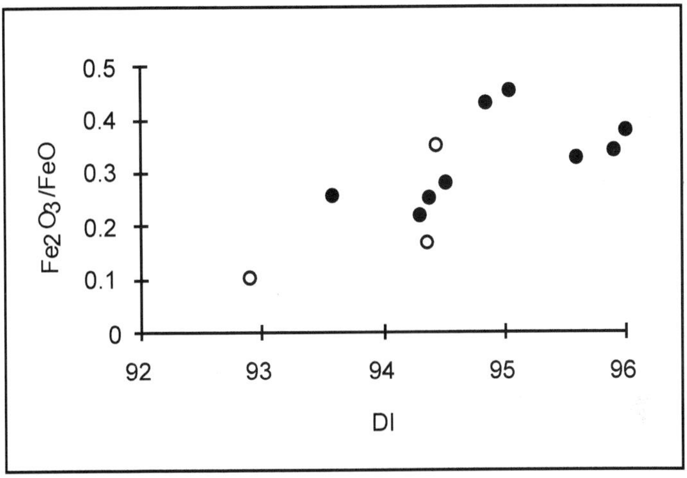

Fig. 11. Plot of DI versus Fe_2O_3/FeO for the GEG, showing increasing Fe_2O_3/FeO with fractionation. Open circle = micrographic granite, closed circle = granite porphyry.

Due to removal of some feldspars and accessory minerals, these dykes show low concentrations of CaO, Sr, Ba, Zr, Zn and high concentrations of U, Th, Nb, Rb, Y, Ga and W. This is consistent with field evidence of other granites associated with Mo-Sn-W ore deposits, wherein fine-grained granites mostly occur as 'carapace' facies formed near the upper contacts of the pluton (e.g., at Krusne hory, Erzgebirge; Stemprok, 1985), or as dyke-like products at the apical part of the larger plutons (Tischendorf, 1977; Blevin and Chappell, 1996a). Either way, fine-grained granites as marginal carapace (Plimer, 1987) or as late-stage phases (Stemprok, 1990; Blevin and Chappell, 1995, 1996a) are one of the common features of plutons associated with Sn, W and Mo ore deposits. Most of these plutons show vertical zoning as enrichment in some elements such as W, Be, Sn, F, Cs, Rb and Li at the top with an impoverishment of Ba, Sr, Ni, Cr and V (Tischendorf, 1977). It is noteworthy that the GEG contains high concentrations of Ni, relative to fractionated I- and A-type granite compositions, which are inconsistent with normal fractionation, as Ni concentration decreases with fractionation. Since Ni is relatively immobile, it is unlikely that this element has been added during slight hydrothermal alteration. It is more likely that the high concentrations of Ni and low concentrations of Ce, La, and Sn in the GEG, relative to other fractionated I-type granites, reflect the compositional features of the source.

Regarding the classification of Ishihara (1977), the GEG has formed under reduced conditions (Fig. 12) and belongs to the ilmenite series, since its Fe_2O_3/FeO ratio is <0.5 (Ishihara, 1981). Generally, I-type granites show higher f_{O_2} than S-types and Chappell and White (1992) consider this feature to be inherited from the source. The GEG shows a wide range of Fe_2O_3/FeO, overlapping with unfractionated and fractionated I- and S-type granites. Fe_2O_3/FeO ratio increases with fractionation in the GEG samples (Table 3, Fig. 11).

In summary, the highly fractionated character of the GEG makes classification difficult. The geochemical features and field observations show that GEG could be A-type or fractionated I-type. As stated by Chappell and White (1992), fractionated I-type granites are similar to fractionated A-types and they can be mistaken.

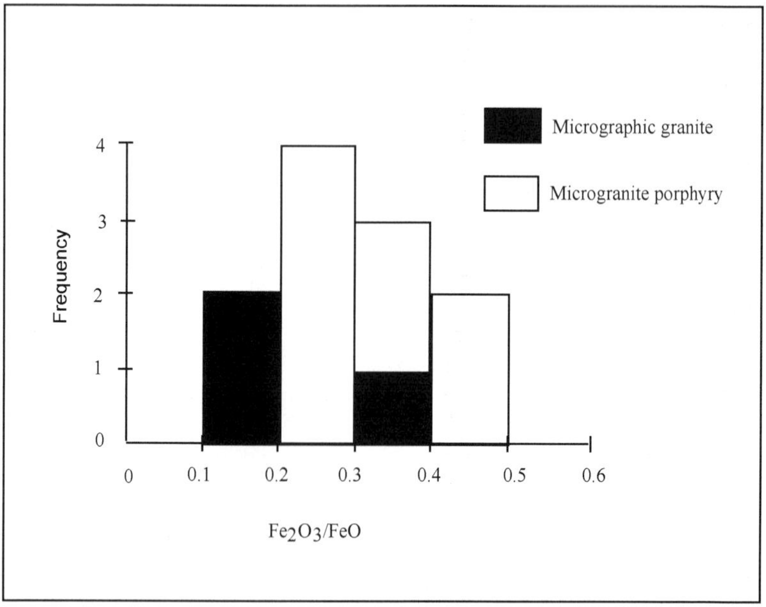

Fig. 12. Histogram of Fe_2O_3/FeO in the GEG, showing the reduced character of the Glen Eden Granite.

6. Tectonic setting

On tectonic discrimination diagrams of Pearce et al. (1984), data from the GEG, but not the associated volcanics, mostly plot in the 'Within Plate' field (Fig. 13). This is typical for A-type granites (Pearce et al., 1984; Whalen et al., 1987), but does not mean that the GEG is necessarily A-type (Whalen and Currie, 1990). Although I- and S-type granites mostly plot in the 'Volcanic Arc' field (Whalen et al., 1987), they may also plot in the 'Within Plate' field (Whalen, 1988). It seems that the high fractionation of I-type granites would increase the concentrations of Nb, Y and Rb and would cause these rocks to plot in the 'Within Plate' field. For example, average fractionated I-type granites of the Lachlan Fold Belt (Table 3) plot in the 'Within Plate' field, as does the GEG.

There seems to be general agreement that A-type granites were emplaced into tensional (or non-compressive) environments either at the end of an orogenic cycle in continental rift zones or in oceanic basins (Eby, 1990). The Glen Eden Granite plots in 'Post-orogenic' and 'Anorogenic' fields on the multicationic diagram of Batchelor and Bowden (1985) (Fig. 9) and it seems that this granite was emplaced into an unstable active margin. This tectonic setting is similar to that proposed for the A-type Topsails Granite, western Newfoundland (Whalen and Currie, 1990).

7. Summary

The various features of the GEG can be summarized as follows.

- The compositional features of the GEG are enrichment in SiO_2, Rb, U, Th, Nb, Y, Ga and W and impoverishment in CaO, P_2O_5, Ba, Sr, Zr and Zn and high values and wide range of Rb/Ba and Rb/Sr.
- The GEG, like other mineralization-associated plutons of the New England Batholith (Kleeman, 1978; Blevin and Chappell, 1996a, b; Vickery et al., 1997), is a high-level leucogranite of near minimum melt composition.
- The presence of crenulate quartz layers, micrographic texture and hydrothermal breccia at Glen Eden suggests saturation of magma from water and the presence of fluid-rich environment.
- The highly fractionated character of the GEG does not allow unequivocal classification but it has strong similarities to fractionated I-type and A-type granites.
- The tectonic setting of the GEG, based on geochemical criteria, is 'Within Plate' and possibly it has been emplaced into an unstable active margin.
- Strong fractionation of the granitic magma increased concentration of incompatible elements, including metals such as Sn, W and Mo, in the final melt and magmatic solution. Increasing the pressure of this fluid eventually caused brecciation of the cap rocks and formed a breccia pipe wherein Mo-W-Sn mineralization occurred.

8. References

Augustithis, S.S. 1973. Atlas of the textural pattern of granites, gneisses and associated rock types. Elsevier, Amsterdam, London, New York, 378pp.

Avila-Salinas, W.A. 1990. Tin-bearing granites from the Cordillera Real, Bolivia: a petrological and geochemical review. Geol. Soc. Am., Special Paper, 145-159.

Bailey, J.C. 1977. Fluorine in granitic rocks and melts: a review. Chem. Geol., 19, 1-42.

Bampton, M.D. 1988. Alteration and mineralisation of the southern part of the Stanthorpe Adamellite, near Tenterfield, New South Wales. Unpublished BSc (Hons) thesis, University of Sydney, 163pp.

Bankwitz, P. 1978. Remarks concerning the development of the Erzgebirge pluton. In M. Stemprok, L. Burnol and G. Tischendorf (eds), Metallization associated with acid magmatism, 3, 156-167.

Batchelor, R.A. and Bowden, P. 1985. Petrogenetic interpretation of granitoid rock series using multicationic parameters. Chem. Geol., 48, 43-55.

Blevin, P.L. and Chappell, B.W. 1995. Chemistry, origin, and evolution of mineralized granites in the Lachlan Fold Belt, Australia: the metallogeny of I- and S-type granites. Econ. Geol., 90, 1604-1619.

Blevin, P.L. and Chappell, B.W. 1996a. Internal evolution and metallogeny of Permo-Triassic high-K granites in the Tenterfield-Stanthorpe region, southern New England Orogen, Australia. In Mesozoic geology of the eastern Australia plate conference, Geol. Soc. Aust., Abs., 43, 94-100.

Blevin, P.L. and Chappell, B.W. 1996b. Permo-Triassic granite metallogeny of the New England Orogen. In Mesozoic geology of the eastern Australia plate conference, Geol. Soc. Aust., Abs., 43, 101-103.

Brodie, R.S. 1983. Geology and mineralization of the Mole River-Silent Grove area, near Tenterfield, northern New South Wales. Unpublished BSc (Hons) thesis, University of New England, 149pp.

Burnham, C.W. 1967. Hydrothermal fluids in the magmatic stage. In H.L. Barnes (ed), Geochemistry of hydrothermal ore deposits, New York, John Wiley, 34-76.

Burnham, C.W. 1979. Magmas and hydrothermal fluids. In H.L. Barnes (ed), Geochemistry of hydrothermal ore deposits, 2nd edition, New York, John Wiley, 71-136.

Carmichael, I.S.E., Turner, F.J. and Verhoogen, J. 1974. Igneous Petrology. New York. McGraw-Hill Book Co., 739pp.

Carten, R.B., Walker, B.M., Geraghty, E.P. and Gunow, A.J. 1988. Comparison of field-based studies of the Henderson porphyry molybdenum deposit, Colorado, with experimental and theoretical models of porphyry systems. In R.P. Taylor and D.F. Strong (eds), Recent advances in the geology of granite-related mineral deposits, Can. Ins. Min. Metall., 39, 1-12.

Chappell, B.W. and White, A.J.R. 1992. I- and S-type granites in the Lachlan Fold Belt. Trans. Roy. Soc. Edin.: Earth Sci., 83, 1-26.

Collins, W.J., Beams, S.D., White, A.J.R. and Chappell, B.W. 1982. Nature and origin of A-type granites with particular reference to southeastern Australia. Contrib. Mineral. Petrol., 80, 189-200.

Collins, W.J., Offler, R., Farrell, T.R. and Landenberger, B. 1993. A revised Late Palaeozoic-Early Mesozoic tectonic history for the southern New England Fold Belt. In P.G. Flood and J.C. Aitchison (eds), New England Orogen, eastern Australia, University of New England, 69-84.

Dingwell, D.B. 1985. The structure and properties of fluorine-rich silicate melts: implications for granite petrogenesis. In R.P. Taylor and D.F. Strong (eds), Granite-related mineral deposits geology, petrogenesis and tectonic setting, CIM Conference on Granite-related Mineral Deposits, 72-81.

Dingwell, D.B. 1988. The structure and properties of fluorine-rich magmas: a review of experimental studies. In R.P. Taylor and D.F. Strong (eds), Recent advances in the geology of granite-related mineral deposits. Can. Inst. Min. Metall., Special Volume 39, 1-12.

Dingwell, D.B., Scarfe, C.M. and Cronin, D.J. 1985. The effect of fluorine on viscosities in the system Na_2O-Al_2O_3-SiO_2 - implications for phonolites, trachytes and rhyolites. Am. Mineral., 70, 80-87.

Eby, G.N. 1990. The A-type granitoids: a review of their occurrence and chemical characteristics and speculations on their petrogenesis. In A.R. Woolley and M. Ross (eds), Alkaline igneous rocks and carbonatites. Lithos, 26, 115-134.

Fenn, P.M. 1979. On the origin of graphic intergrowth [abs.]. Geol. Soc. Am., Abstr. Programs, 11, 424.

Flood, P.G. and Aitchison, J.C. 1993a. Understanding New England geology: the comparative approach. In P.G. Flood and J.C. Aitchison (eds), New England Orogen, eastern Australia, University of New England, 1-10.

Flood, P.G. and Aitchison, J.C. 1993b. Recent advances in understanding the geological development of the New England Province of the New England Orogen. In P.G. Flood and J.C. Aitchison (eds), New England Orogen, eastern Australia, University of New England, 61-67.

Glyuk, D.S. and Anfiligov, V.N. 1973. Phase equilibria in the system granite-H_2O-HF at a pressure of 1000 kg/cm^2. Geochem. Internat., 10, 313-317.

Haapala, I. and Ramo, O.T. 1990. Petrogenesis of the Proterozoic rapakivi granite of Finland. In H.J. Stein and J.L. Hannah (eds), Ore-bearing granite systems ; petrogenesis and mineralizing processes. Geol. Soc. Am., Special Paper, 246, 275-286.

Hannah, J.L. and Stein, H.J. 1990. Magmatic and hydrothermal processes in ore bearing systems. In H.J. Stein and J.L. Hannah (eds), Ore-bearing granite systems; petrogenesis and mineralizing processes. Geol. Soc. Am., Special Paper, 246, 1-10.

Hensel, H.D. 1982. The mineralogy, petrology and geochronology of granitoids and associated intrusives from the southern portion of the New England Batholith. Unpublished Ph.D. thesis, University of New England, 273pp.

Holtz, F., Pichavant, M., Barbey, P. and Johannes, W. 1992. Effects of H_2O on liquidus phase relations in the haplogranite system at 2 and 5 kbar. Am. Mineral., 77, 1223-1241.

Ishihara, S. 1977. The magnetite-series and ilmenite-series granitic rocks. Min. Geol., 27, 293-305.

Ishihara, S. 1981. The granitoid series and mineralization. Econ. Geol., 75th Anniversary Volume, 458-484.

Keith, J.D. and Shanks, W.C. 1988. Chemical evolution and volatile fugacities of the Pine Grove porphyry molybdenum and ash-flow tuff system, southwestern Utah. In R.P. Taylor and D.F. Strong (eds), Recent advances in the geology of granite-related mineral deposits. Can. Inst. Min. Metall., Special Volume, 39, 402-423.

Keppler, H. 1993. Influence of fluorine on the enrichment of high field strength trace elements in granitic rocks. Contrib. Mineral. Petrol., 114, 479-488.

Kirkham, R.V. and Sinclair, W.D. 1988. Comb quartz layers in felsic intrusions and their relationship to porphyry deposits. In R.P. Taylor and D.F. Strong (eds), Recent advances in the geology of granite-related mineral deposits. Can. Inst. Min. Metall., Special Volume, 39, 50-71.

Kleeman, J.D. 1978. Tin mineralizing granites in New England [abs]. Australian Geology Convention, 3rd, Townsville, August 1978, Abstracts and Programs, 37.

Kleeman, J.D. 1982. The anatomy of a tin-mineralizing A-type granite. In P.G. Flood and B. Runnegar (eds), New England Geology., University of New England and AHV Club, 327-334.

Kontak, D.J. 1994. Geological and geochemical studies of alteration processes in a fluorine-rich environment: the east Kemptville Sn-(Zn-Cu-Ag) deposit, Yarmouth Country, Nova Scotia, Canada. In D.R. Lentz (ed), Alteration and alteration processes associated with ore-forming systems, Geological Association of Canada, Short Course Notes, 11, 261-314.

Kovalenko, V.I., Kuz'min, M.I., Antipin, V.S. and Petrov, L.L. 1971. Topaz bearing keratophyre (ongonite), a new variety of subvolcanic igneous vein rock. Doklady Academy Science, U. S. S. R., Earth Science Section, 199, 132-135.

Le Maitre, R.W. 1976. The chemical variability of some common igneous rocks, Jour. Petrology, 17, 589-637.

Le Messurier, L.A. 1983. The genetic relationships between two alkali granites and associated enclosing I-type plutons within the New England region. Unpublished BSc (Hons) thesis, University of New England, 131pp.

London, D. 1987. Internal differentiation of rare-element pegmatites: effects of boron, phosphorus, and fluorine. Geochim. et Cosmochim. Acta, 51, 403-420.

Luth, W.C. 1976. Granitic rocks. In D.K. Bailey and R. Macdonald (eds), The evolution of the crystalline rocks, Academic Press, London, 335-417.

Manning, D.A.C. 1981. The effect of fluorine in liquidus phase relationships in the system Qz-Ab-Or with excess water at 1 kb. Contrib. Mineral. Petrol., 76, 206-215.

Manning, D.A.C. and Pichavant, M. 1988. Volatiles and their bearing on the behavior of metals in granitic systems. In R.P. Taylor and D.F. Strong (eds), Recent Advances in the Geology of Granite-Related Mineral Deposits, Can. Ins. Min. Metall., Special Volume, 39, 13-24.

Moore, J.G. and Lockwood, J.P. 1973. Origin of comb layering and orbicular structure, Sierra Nevada Batholith, California. Geol. Soc. Am. Bull., 48, 1-20.

Munoz, J.L. and Ludington, S.D. 1974. Fluorine-hydroxyl exchange in biotite. Am. Jour. Sci., 274, 396-413.

Murray, C.G. 1988. Tectonic evolution and metallogenesis of the New England Orogen. In J.D. Kleeman (ed), New England Orogen - tectonics and metallogenesis. University of New England, 204-210.

Nabelek, P.I. and Russ-Nabelek, C. 1990. The role of fluorine in the petrogenesis of magmatic segregations in the St. Francois volcano-plutonic terrane, southeastern Missouri. In H.J. Stein and J.L. Hannah (eds), Ore-bearing granite systems; petrogenesis and mineralizing processes. Geol. Soc. Am., Special Paper, 246, 71-87.

Newberry, R.J., Burns, L.E., Swanson, S.E. and Smith, T.E. 1990. Comparative petrologic evolution of the Sn and W granites of the Fairbanks-Circle area, interior Alaska. In H.J. Stein and J.L. Hannah (eds), Ore-bearing granite systems; petrogenesis and mineralizing processes. Geol. Soc. Am., Special Paper, 246, 121-142.

Nockolds, S.R. 1954. Average chemical compositions of some igneous rocks. Geol. Soc. Am. Bull., 65, 1007-1032.

Pearce, J.A., Harris, N.B.W. and Tindle, A.J. 1984. Trace elements discrimination diagrams for the tectonic interpretation of granitic rocks. Jour. Petrology, 25, 956-983.

Pichavant, M. and Manning, D. A. C. 1984. Petrogenesis of tourmaline granites and topaz granites; the contribution of experimental data. Physics of the Earth and Planetary Interiors, 35, 31-50.

Plimer, I.R. and Kleeman, J.D. 1985. Mineralization associated with the Mole Granite, Australia. In: High heat production (HHP) granites, hydrothermal circulation and ore genesis. St. Austell, England, Inst. Mining Metallurgy, 563-570.

Plimer, I.R. 1973. The pipe deposits of tungsten-molybdenum-bismuth in eastern Australia. Unpublished PhD. thesis, Macquarie University, 288pp.

Plimer, I.R. 1987. Fundamental parameters for the formation of granite-related tin deposits. Geologische Rundschau, 76, 23-40.

Povilaitis, M.M. 1978. Effect of the conditions of magmatic emplacement on high-temperature postmagmatic ore mineralization. In M. Stemprok, L. Burnol and F. G. Tischendorf, Metallization associated with acid magmatism, 3, 375-384.

Richardson, J.M., Bell, K., Watkinson, D.H. and Blenkinsop, J. 1990. Genesis and fluid evolution of the East Kemptville greisen-hosted tin mine, southwestern Nova Scotia, Canada. In H.J. Stein and J.L. Hannah (eds), Ore-bearing granite systems; petrogenesis and mineralising processes. Geol. Soc. Am., Special Paper, 246, 181-203.

Sawka, W.N., Heizler, M.T., Kistler, R.W. and Chappell, B.W. 1990. Geochemistry of highly fractionated I- and S-type granites from the tin-tungsten provinces of western Tasmania. In H.J. Stein and J.L. Hannah (eds), Ore-bearing granite systems; petrogenesis and mineralizing processes. Geol. Soc. Am., Special Paper, 246, 161-179.

Schroecke, H. 1973. Grundlagen der magmatogenen lagerstättenbildung. Enke Verlag. Stuttgart, 287pp.

Shaver, S.A. 1984a. Origin of crenulate quartz layers - evidence from the Hall (Nevada Moly) molybdenum deposit, Nevada [abs]. Geol. Soc. Am., Abstr. Programs, 16, 254-255.

Shaver, S.A. 1984b. The Hall (Nevada Moly) molybdenum deposit, Nye Country, Nevada: geology, alteration, mineralization and geochemical dispersion. Unpublished Ph.D. thesis, Stanford Univ., 261pp.

Shaw, S.E. and Flood, R.H. 1981. The New England Batholith, eastern Australia: geochemical variations in space and time. Jour. Geoph. Res., 86, 10530-10544.

Sheppard, S.M.F. 1977. Identification of the origin of ore-forming solutions by the use of stable isotopes. In: Volcanic processes in ore genesis, Geol. Soc. London, Special. Publication, 7, 25-41.

Somarin, A.K. 1999. Mineralogy, geochemistry and genesis of the Glen Eden Mo-W-Sn deposit, New England Batholith, Australia. Unpublished PhD thesis, University of New England, Armidale, Australia, 340pp.

Somarin, A.K., and Ashley, P., 2004. Hydrothermal alteration and mineralization of the Glen Eden Mo-W-Sn deposit: a leucogranite-related hydrothermal system, Southern New England Orogen, NSW, Australia. Mineralium Deposita, 39, 282-300.

Stegman, C.L. 1983. The Mole Granite and its Sn-W-Mo-base metal mineralization - a study of its southern-central margin. Unpublished BSc (Hons) thesis, University of New England, 177pp.

Stemprok, M. 1985. Vertical extent of greisen mineralization in the Krusne hory/Erzgebirge granite pluton of central Europe. In: High heat production (HHP) granites, hydrothermal circulation and ore genesis. St. Austell, England, Inst. Mining Metallurgy, 41-54.

Stemprok, M. 1990. Intrusion sequences within ore-bearing granitoid plutons. Geological Jour., 25, 413-417.

Stewart, J.P. 1983. Petrology and geochemistry of the intrusives spatially associated with the Logtung W-Mo prospect, south-central Yukon Territory. Unpublished M.Sc. thesis, University of Toronto, 243pp.

Strong, D.F. 1988. A review and model for granite-related mineral deposits. In R.P. Taylor and D.F. Strong (eds), Recent advances in the geology of granite-related mineral deposits. Can. Inst. Min. Metall., Special Volume, 39, 424-445.

Stroud, W.J. 1995. Inverell 1: 250000 metallogenic map.

Taylor, R.P. 1992. Petrological and geochemical characteristics of the Pleasant Ridge zinnwaldite-topaz granite, southern New Brunswick, and comparison with other topaz-bearing felsic rocks. Can. Mineral., 30, 895-921.

Tischendorf, G. 1977. Geochemical and petrographic characteristics of silicic magmatic rocks associated with rare-element mineralization. In M. Stemprok, L. Burnol and G. Tischendorf (eds), Metallization associated with acid magmatism, 2, 41-96.

Tuttle, O.F. and Bowen, N.L. 1958. Origin of granite in the light of experimental studies in the system $NaAlSi_3O_8$-$KAlSi_3O_8$-SiO_2-H_2O. Geol. Soc. Am. Mem., 74.,153pp.

Velde, B. and Kushiro, I. 1978. Structure of sodium aluminosilicate melts quenched at high pressure; infrared and aluminum K radiation data. Earth Planet. Sci. Lett., 40, 137-140.

Vickery, N.M., Ashley, P.M. and Fanning, C.M. 1997. Dumboy-Gragin Granite, northeastern New South Wales: age and compositional affinities. In P.M. Ashley and P.G. Flood (eds), Tectonics and Metallogenesis of the New England Orogen, Geological Society of Australia Special Publication, 19, 266-271.

Walsh, J. 1991. Two distinctive granitoids from the Copeton region: a mineralogical, geochemical and mineralization study. Unpublished BSc (Hons) thesis, University of New England, 200pp.

Watson, E.B. and Harrison, T.M. 1983. Temperature and compositional effects in a variety of crustal magma type. Earth Planet. Sci. Lett., 64, 295-304.

Webster, J.D. and Holloway, J.R. 1990. Partitioning of F and Cl between magmatic hydrothermal fluid and highly evolved granitic magmas. In H.J. Stein and J.L. Hannah (eds), Ore-bearing granite systems; petrogenesis and mineralizing processes. Geol. Soc. Am., Special Paper, 246, 21-34.

Whalen, J.B. and Currie, K.L. 1990. The Topsails igneous suite, western Newfoundland, fractionation and magma mixing in an "orogenic" A-type granite suite. In H.J. Stein and J.L. Hannah (eds), Ore-bearing granite systems; petrogenesis and mineralizing processes. Geol. Soc. Am., Special Paper, 246, 287-299.

Whalen, J.B., Currie, K.L. and Chappell, B.W. 1987. A-type granites; geochemical characteristics, discrimination, and petrogenesis. Contrib. Mineral. Petrol., 95, 407-419.

Whalen, J.B. 1988. Granitic rocks of New Brunswick and Gaspe, Quebec: a transect across the southern Canadian Appalachians. Geol. Assoc. Can. Prog. Abst., 13, A133.

White, W.H., Bookstrom, A.A., Kamilli, R.J., Ganster, M.W., Smith, R.P., Ranta, D.E. and Steininger, R.C. 1981. Character and origin of Climax-type molybdenum deposits. Econ. Geol., 75th Anniversary Volume, 270-316.

Winkler, H.G.F. 1974. Petrogenesis of Metamorphic Rocks. Berlin, Springer Verlag.

Wyllie, P.J. 1979. Magmas and volatile components. Am. Mineral., 64, 469-500.

Zen, E. 1986. Aluminum enrichment in silicate melts by fractional crystallization, some mineralogic and petrographic constraints. Jour. Petrology, 27, 1095-1117.

Part 3

Seismology

8

Seismic Illumination Analysis with One-Way Wave Propagators Coupled with Reflection/Transmission Coefficients in 3D Complex Media

Weijia Sun[1], Li-Yun Fu[1], Wei Wei[1] and Binzhong Zhou[2]
[1]*Key Laboratory of the Earth's Deep Interior, Institute of Geology and Geophysics, Chinese Academy of Sciences, Beijing,*
[2]*CSIRO Earth Science and Resource Engineering, Kenmore,*
[1]*China*
[2]*Australia*

1. Introduction

Seismic illumination analysis can help us to better understand how various seismic acquisition parameters and configuration affect seismic image qualities. It allows us to design effective seismic survey systems to provide reliable high-resolution seismic images for complex structures in both theoretical and exploration seismology. It is also a useful tool for estimation of the potential detecting power of a specific acquisition system for a given velocity structure of the medium.

Most existing techniques for calculating illumination intensity distributions can be divided into two categories: one is based on ray tracing methods (Berkhout, 1997; Schneider, 1999; Bear et al., 2000; Muerdter et al., 2001a, 2001b, 20001c); and the other based on wave equation methods (Wu & Chen, 2002; Xie et al., 2006; Wu & Chen, 2006; Sun et al., 2007). The ray tracing is quite efficient and inexpensive. However, it bears large errors in complex areas because of the multi-path problem, the high frequency approximation and the singularity problem of the ray theory (Hoffmann, 2001). These factors limit its accuracy severely in complex areas.

The wave-equation-based methods can provide reliable and accurate illumination intensity distributions in complex media. Full-wave finite-difference (FD) method is commonly employed for seismic forward modelling, i.e., wave propagation simulation for a known interval velocity model. But it is too expensive for illumination analysis in industrial application, especially for 3D case. Recently, One-way wave propagators have been widely used to seismic modelling and migration for its huge memory-saving and high computational efficiency. Localized illumination methods based on the Gabor-Daubechies frame decomposition or the local slant stack have been presented, which can give local information both in space and direction (e.g., Xie and Wu, 2002; Wu et al., 2000; Xie et al., 2003; Wu & Chen, 2006; Mao & Wu, 2007; Cao & Wu, 2008). However, its computational burden is still unacceptable due to the wavefield decomposition in every spatial point.

In this chapter, we developed a 3D one-way wave-equation-based illumination method for complex media. It combines the one-way propagators coupled with reflection/transmission (R/T) coefficients and the phase encoding techniques. The one-way propagators coupled with R/T coefficients were firstly developed by Sun et al. (2009) for the 2D case. The R/T operators not only account for amplitude variations with incident angles across interfaces, but also accommodate to complex media with steep dip angle and large lateral velocity contrast. Firstly, we extended the method to the 3D case in complex structures starting from generalized Lippmann-Schwinger equations and applied the method to calculate illumination intensity distributions. Then, the phase encoding technique in frequency domain (Romero et al., 2000) was employed to reduce the computational cost further by calculating a number of shots together, which can apply to any frequency domain illumination methods.

In the following sections, the accuracy of the R/T propagators, defined by the corresponding dispersion relationship, was analyzed by comparison with the exact solution and other popular one-way propagators known as the split-step Fourier (SSF) method (Stoffa et al., 1990) and the generalized screen propagator (GSP) (de Hoop et al., 2000). Then, special attention was given to the implementation of numerical procedures to improve efficient computation efficiency and huge memory savings. Several numerical examples were given to show its capabilities to handle complex media.

As the actual interval velocity model or geological model cannot be known exactly, the effects of velocity error on illumination intensity distributions were evaluated by numerical examples. Finally, we showed some practical applications of target-oriented illumination analysis in seismic acquisition and survey design, which is fundamental to exploration geophysics.

2. Methodology

In this section, we first derive 3D version of one-way propagators coupled with reflection/transmission (R/T) coefficients. Then, we briefly introduce the definition of seismic illumination and theories of phase encoding. The accuracy of R/T propagators is analyzed by comparison with the exact solution and other popular one-way propagators. Finally, we present some applications of seismic illumination.

2.1 One-way Lippmann-Schwinger wave equation

Generally, most media can be sliced into heterogeneous slabs perpendicular to the major propagation direction. Figure 1 depicts the geometry of such a heterogeneous slab denoted by Ω with the top interface Γ_0, the bottom interface Γ_1, and the thickness Δz. The velocity distribution in the slab is denoted by $v(\mathbf{r})$ where $\mathbf{r}$ is the position vector, and its reference velocity is v_0. We start with the scalar Helmholtz equation for a time-harmonic wavefield $u(\mathbf{r})$

$$\nabla^2 u(\mathbf{r}) + k^2 u(\mathbf{r}) = 0, \tag{1}$$

where the wavenumber $k = \omega / v(\mathbf{r})$. The total wavefield $u(\mathbf{r})$ at location $\mathbf{r} \in \Omega$ is composed for scattering problems of

$$u(\mathbf{r}) = u_1^s(\mathbf{r}) + u_2^s(\mathbf{r}). \tag{2}$$

$u_1^s(\mathbf{r})$ is the scattered field by the boundary structure $\Gamma = \Gamma_0 + \Gamma_1$ and satisfies the following boundary integral equation

$$u_1^s(\mathbf{r}) = \int_\Gamma \left[G(\mathbf{r},\mathbf{r}') \frac{\partial u(\mathbf{r}')}{\partial n} - u(\mathbf{r}') \frac{\partial G(\mathbf{r},\mathbf{r}')}{\partial n} \right] d\mathbf{r}' , \tag{3}$$

where $\partial / \partial n$ denotes differentiation with respect to the outward normal of the boundary Γ.

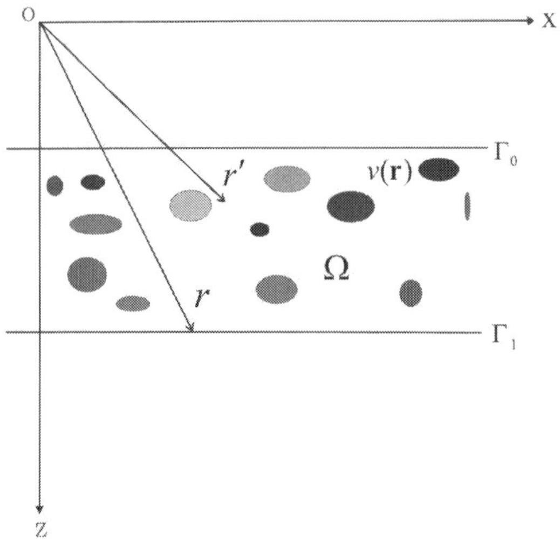

Fig. 1. Geometry of a heterogeneous slab

$u_2^s(\mathbf{r})$ is the scattered field by the volume heterogeneities within the slab and satisfies the following Lipmann-Schwinger integral equation

$$u_2^s(\mathbf{r}) = k_0^2 \int_\Omega O(\mathbf{r}') u(\mathbf{r}') G(\mathbf{r},\mathbf{r}') d\mathbf{r}' , \tag{4}$$

where the reference wavenumber $k_0 = \omega / v_0$, and $O(\mathbf{r})$ is the relative slowness perturbation defined as $O(\mathbf{r}) = n^2(\mathbf{r}) - 1$ with the acoustic refractive index $n(\mathbf{r}) = v_0 / v(\mathbf{r})$.
These Helmholtz integral representation formulae are derived using the Green's function $G(\mathbf{r},\mathbf{r}')$ in the background medium, that is, $G(\mathbf{r},\mathbf{r}') = \exp(ik_0|\mathbf{r}-\mathbf{r}'|)/(4\pi|\mathbf{r}-\mathbf{r}'|)$ for 3-D problems. Substituting Equation (3) and (4) into (2) and considering the "boundary naturalization" of the integral equations, that is, a limit analysis when the "observation point" $\mathbf{r}$ approaches the boundary Γ and tends to coincide with the "scattering point" $\mathbf{r}' \in \Gamma$, we obtain the following generalized Lipmann-Schwinger integral equation

$$\int_\Gamma \left[G(\mathbf{r},\mathbf{r}') \frac{\partial u(\mathbf{r}')}{\partial n} - u(\mathbf{r}') \frac{\partial G(\mathbf{r},\mathbf{r}')}{\partial n} \right] d\mathbf{r}' + k_0^2 \int_\Omega O(\mathbf{r}') u(\mathbf{r}') G(\mathbf{r},\mathbf{r}') d\mathbf{r}' = \begin{cases} u(\mathbf{r}) & \mathbf{r} \in \Omega \\ C(\mathbf{r}) u(\mathbf{r}) & \mathbf{r} \in \Gamma \\ 0 & \mathbf{r} \notin \bar{\Omega} \end{cases} , \tag{5}$$

for all $\mathbf{r}' \in \bar{\Omega} = \Omega + \Gamma$, where the coefficient $C(\mathbf{r}) = 1/2$ for a flat Γ. This is a wave integral equation that is equivalent to the Helmholtz equation (1) and describes two-way wave propagation in the heterogeneous slab. It is a Fredholm integral equation of the second kind with the existence and uniqueness of its solution for both interior and exterior Helmholtz problems assured by the classical Fredholm's theorems of integral equations. Numerical studies of this equation based on the discrete-wavenumber boundary methods have been conducted for wave propagation simulation (Fu and Bouchon, 2004) in piecewise heterogeneous media that the earth presents. Because numerous matrix operations are involved and the matrix for each frequency -component computation must be inverted, the numerical methods are computationally intensive at high frequencies.

Equation (5) describes wave propagation in the space-frequency domain. Formulating it in the frequency-wavenumber domain using a plane-wave expansion will lead to quite different numerical schemes. In Figure 1, we assume wave propagation along the z-axis, crossing the slab from the slab entrance Γ_0 to the exit Γ_1. Let $q(\mathbf{r}) = \partial u(\mathbf{r})/\partial n$ indicate the acoustic pressure gradient, $\mathbf{r} = (x, z)$ represent the observation point, and $\mathbf{r}' = (x', z')$ denote the scattering point. Substituting the 3D Green function into Equation (5), we obtain

$$\int_{\Gamma_0} \left[G(\mathbf{r},\mathbf{r}')q(\mathbf{r}') - u(\mathbf{r}')\frac{\partial G(\mathbf{r},\mathbf{r}')}{\partial n} \right] d\mathbf{r}' = \frac{1}{4\pi}\int_{-\infty}^{\infty} \left[(ik_z^{-1}q(\mathbf{k}_\mathrm{T},z) + u(\mathbf{k}_\mathrm{T},z))\exp(ik_z\Delta z) \right] \exp(i\mathbf{k}_\mathrm{T}\cdot\mathbf{x})d\mathbf{k}_\mathrm{T}\,, \quad (6)$$

and

$$\int_{\Gamma_1} \left[G(\mathbf{r},\mathbf{r}')q(\mathbf{r}') - u(\mathbf{r}')\frac{\partial G(\mathbf{r},\mathbf{r}')}{\partial n} \right] d\mathbf{r}' = \frac{i}{4\pi}\int_{-\infty}^{\infty} \left[k_z^{-1}q(\mathbf{k}_\mathrm{T},z+\Delta z) \right] \exp(i\mathbf{k}_\mathrm{T}\cdot\mathbf{x})d\mathbf{k}_\mathrm{T}\,, \quad (7)$$

where $\mathbf{k}_\mathrm{T} = (k_x, k_y)$ and $\mathbf{x} = (x, y)$. Using the rectangular rule to evaluate the volume integration over the slab in Equation (5) yields

$$k_0^2\int_{\Omega} O(\mathbf{r}')u(\mathbf{r}')G(\mathbf{r},\mathbf{r}')d\mathbf{r}' = \frac{k_0}{8\pi}\int_{-\infty}^{\infty} k_z^{-1}[F(\mathbf{k}_\mathrm{T},z)\exp(ik_z\Delta z)]\exp(i\mathbf{k}_\mathrm{T}\cdot\mathbf{x})d\mathbf{k}_\mathrm{T}\,. \quad (8)$$

This is actually the Born approximation applied to the slab. It implies that the heterogeneity of the slab is represented by its top/bottom interfaces and consequently requires the slab is thin enough with respect to the wavelength of incident waves. Substituting equations (6), (7), and (8) into (5) and noting that each inner integral is a Fourier transforms, we obtain

$$k_z u(\mathbf{k}_\mathrm{T},z+\Delta z) - iq(\mathbf{k}_\mathrm{T},z+\Delta z) = [k_z u(\mathbf{k}_\mathrm{T},z) + iq(\mathbf{k}_\mathrm{T},z) + k_0F(\mathbf{k}_\mathrm{T},z)]\exp(ik_z\Delta z)\,. \quad (9)$$

Equation (9) is a wavenumber-domain wave equation that describes two-way wave propagation in the heterogeneous slab, including multiple forward and back scatterings between Γ_0 and Γ_1.

For one-way wave propagation using the matching solution techniques, further simplification should be made to Equation (5) by reducing it to one-way version. From Equation (9), we see that two-way wave propagation involves two terms: the displacement $u(\mathbf{r})$ and acoustic pressure gradient $q(\mathbf{r})$. In practical, we do not often measure both $u(\mathbf{r})$ and $q(\mathbf{r})$ at a given level. The pressure gradient $q(\mathbf{r})$ at the slab entrance Γ_0 can be dropped by choosing Γ_0 as an acoustically soft boundary (Dirichlet boundary condition) which leading to one-return approximation shown in Figure 2. This Rayleigh-type integral representation is valid if we neglect back scatterings. With this choice, no energy returns from the upper boundary Γ_0 and multiple reflections between Γ_0 and Γ_1 can be avoided. This choice updates Equation (9) to

$$k_z u(\mathbf{k}_\mathrm{T}, z+\Delta z) - iq(\mathbf{k}_\mathrm{T}, z+\Delta z) = [2k_z u(\mathbf{k}_\mathrm{T}, z) + k_0 F(\mathbf{k}_\mathrm{T}, z)]\exp(ik_z \Delta z)\,. \tag{10}$$

To account for the effect of transmission and refraction at Γ_1 in a natural manner, we need to build a boundary integral equation in the medium immediately below the slab with the radiation conditions applied to the bottom boundary of the medium,

$$u(\mathbf{r}) + \int_{\Gamma_1}\left[G(\mathbf{r},\mathbf{r}')q(\mathbf{r}') + u(\mathbf{r}')\frac{\partial G(\mathbf{r},\mathbf{r}')}{\partial n}\right]d\mathbf{r}' - k_0^2\int_\Omega O(\mathbf{r}')u(\mathbf{r}')G(\mathbf{r},\mathbf{r}')d\mathbf{r}' = 0\,. \tag{11}$$

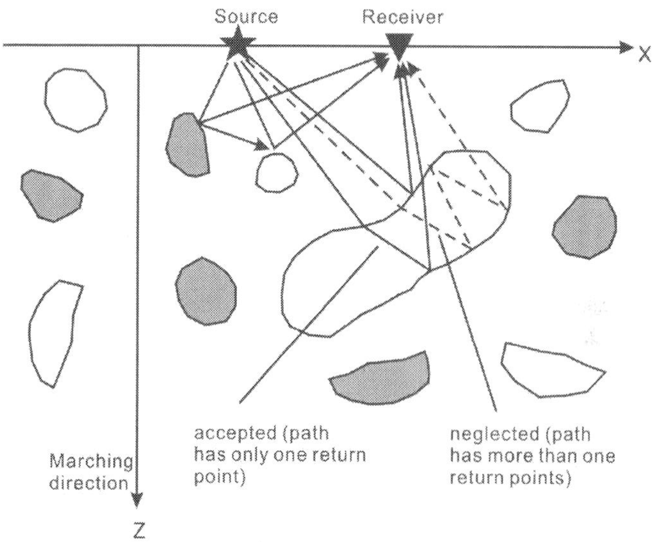

Fig. 2. Sketch showing the meaning of the one-return approximation (modified from Wu et al. (2006))

Applying the plane wave representation of the Hankel function to Equation (11) results in

$$iq(\mathbf{k}_\mathrm{T}, z+\Delta z) = -k_z' u(\mathbf{k}_\mathrm{T}, z+\Delta z) + k_0 F(\mathbf{k}_\mathrm{T}, z+\Delta z)\,,$$

where k_z' is the wavenumber related to the medium immediately below Γ_1, and $F(K_T, z)$ is the form in wavenumber domain of the velocity-weighted wavefield $F(\mathbf{r}) = ik_0 \Delta z O(\mathbf{r})u(\mathbf{r})$ defined in the space domain. Substituting in Equation (10) gives

$$(k_z + k_z')u(\mathbf{k}_\mathrm{T}, z+\Delta z) - k_0 F(\mathbf{k}_\mathrm{T}, z+\Delta z) = [2k_z u(\mathbf{k}_\mathrm{T}, z) + k_0 F(\mathbf{k}_\mathrm{T}, z)]\exp(ik_z \Delta z)\,. \tag{12}$$

Using the approximate calculation $F(\mathbf{k}_T, z+\Delta z) \approx F(\mathbf{k}_T, z)\exp(ik_z\Delta z)$, we have:

$$u(\mathbf{k}_T, z+\Delta z) = \frac{2k_z}{k_z + k_z'}[u(\mathbf{k}_T, z) + \frac{k_0}{k_z}F(\mathbf{k}_T, z)]\exp(ik_z\Delta z). \tag{13}$$

Equation (13) is one-way downward wave equation with one-return approximation that accounts for both the accumulated effect of forward scattering by volume heterogeneities inside the slab and the transmission between adjoining slabs on wave amplitude and phase. Next, we derive the formula of one-way upward waves.

For upward wave propagation, it has $q_{\text{up}}(\mathbf{k}_T, z) = -q_{\text{down}}(\mathbf{k}_T, z)$. We rewrite the volume integration over the slab in Equation (5) using the rectangular rule

$$k_0^2\int_\Omega O(\mathbf{r}')u(\mathbf{r}')G(\mathbf{r},\mathbf{r}')d\mathbf{r}' = \frac{k_0}{4\pi}\int_{-\infty}^{\infty} k_z'^{-1}[F(\mathbf{k}_T, z+\Delta z)]\exp(i\mathbf{k}_T\cdot\mathbf{x})d\mathbf{k}_T. \tag{14}$$

Substituting equations (6), (7) and (14) into (5), and noting that each inner integral is a Fourier transform, we have

$$k_z'u(\mathbf{k}_T, z) - iq(\mathbf{k}_T, z) = [k_z u(\mathbf{k}_T, z+\Delta z) + iq(\mathbf{k}_T, z+\Delta z) - k_0F(\mathbf{k}_T, z+\Delta z)]\exp(-ik_z'\Delta z). \tag{15}$$

To account for the effects of the reflection and transmission at Γ_0 and Γ_1 on backward wave propagation, we build two boundary integral equations:

$$u(\mathbf{r}) + \int_{\Gamma_0}\left[G(\mathbf{r},\mathbf{r}')q(\mathbf{r}') + u(\mathbf{r}')\frac{\partial G(\mathbf{r},\mathbf{r}')}{\partial n}\right]d\mathbf{r}' - k_0^2\int_\Omega O(\mathbf{r}')u(\mathbf{r}')G(\mathbf{r},\mathbf{r}')d\mathbf{r}' = 0, \tag{16}$$

and

$$\frac{1}{2}u(\mathbf{r}) + \int_{\Gamma_1}\left[G(\mathbf{r},\mathbf{r}')q(\mathbf{r}') + u(\mathbf{r}')\frac{\partial G(\mathbf{r},\mathbf{r}')}{\partial n}\right]d\mathbf{r}' = 0. \tag{17}$$

Substituting the Green function to equations (16) and (17), and using the approximation $F(\mathbf{k}_T, z) \approx F(\mathbf{k}_T, z+\Delta z)\exp(-ik_z'\Delta z)$, substituting into equation (15), we obtain

$$u(\mathbf{k}_T, z) = \frac{k_z' - k_z}{k_z' + k_z}u(\mathbf{k}_T, z+\Delta z)\exp(-ik_z'\Delta z). \tag{18}$$

This is one-way upward wave equation with one-return approximation. Equations (13) and (18) account for both the effects of forward and backward scattering by volume heterogeneities inside a slab and the R/T between adjoining slabs on wave amplitude and phase.

2.2 One-way propagators

2.2.1 First-order separation-of-variables screen propagators

We consider the following splitting operator decomposition

$$\bar{k}_z(\bar{k}_x, n) \approx \sum_{j=1}^{m} f_j(\bar{k}_x)g_j(n). \tag{19}$$

For this we need to deal with the following two problems: (1) construction of the splitting operators $f_j(\bar{k}_x)$ and $g_j(n)$, and (2) implementation of the Fourier transform algorithms to equation (19). We will introduce some rational approximations for the construction of the splitting operators.

For convenience, we normalize the wavenumbers $\bar{\mathbf{k}}_\mathrm{T} = \mathbf{k}_\mathrm{T} / k_0$ and $\bar{k}_z = k_z / k_0$. Then equation (13) becomes a standard equation for one-way propagation in heterogeneous media

$$u(\mathbf{k}_\mathrm{T}, z+\Delta z) = \frac{2k_z}{k_z + k_z'}[u(\mathbf{k}_\mathrm{T}, z) + \frac{1}{\bar{k}_z}F(\mathbf{k}_\mathrm{T}, z)]\exp(ik_z\Delta z). \tag{20}$$

with $F(\mathbf{k}_\mathrm{T}, z) = FT_\mathbf{x}[ik_0\Delta z(n(\mathbf{r})-1)u(\mathbf{r})]$ where $FT_\mathbf{x}$ denotes the Fourier transform from $\mathbf{x}$ to $\mathbf{k}_\mathrm{T}$. Because of the second term inside the bracket, this equation takes account of the accumulated effect of forward scatterings by volume heterogeneities in the slab. The corresponding dispersion equation can be written as

$$\bar{k}_z = \sqrt{1-\bar{\mathbf{k}}_\mathrm{T}^2} + (n-1)\left(\sqrt{1-\bar{\mathbf{k}}_\mathrm{T}^2}\right)^{-1}. \tag{21}$$

Since $\left|\bar{\mathbf{k}}_\mathrm{T}\right| \le 1$ for one-way propagation, the term $(1-\bar{\mathbf{k}}_\mathrm{T}^2)^{-1/2}$ in equation (21) can be approximated by the following rational expansion

$$(1-\bar{\mathbf{k}}_\mathrm{T}^2)^{-1/2} = 1 - \sum_{j=1}^{m}\frac{a_j\bar{\mathbf{k}}_\mathrm{T}^2}{1+b_j\bar{\mathbf{k}}_\mathrm{T}^2}, \tag{22}$$

where the coefficients a_j and b_j are independent of n. Submitting this equation into Equation (21), we have

$$\bar{k}_z = \sqrt{1-\bar{\mathbf{k}}_\mathrm{T}^2} + n - 1 - (n-1)\sum_{j=1}^{m}\frac{a_j\bar{\mathbf{k}}_\mathrm{T}^2}{1+b_j\bar{\mathbf{k}}_\mathrm{T}^2}. \tag{23}$$

Coefficients in Equation (23) can be determined numerically by an optimization procedure using the least-squares method. Because of the mathematical properties and approximation behavior of rational functions (Trefethen and Halpern, 1986), equation (23) should be well-posed especially for the lower-order terms. In practice, its first-order equation or at most the second-order equation is adequate for common one-way propagation in large-contrast media with wide propagation angles in seismology.

In what follows, we formulate these separation-of-variables screen propagators (SVSP) by a Fourier-transform-based representation for numerical implementation. Substituting Equations (23) into (20) and using the approximate calculation $e^{i\zeta} \approx 1 + i\zeta$, we obtain

$$u(\mathbf{k}_\mathrm{T}, z+\Delta z) = \frac{2k_z}{k_z + k_z'}\left[\sum_{j=1}^{m}\frac{a_j\bar{\mathbf{k}}_\mathrm{T}^2}{1+b_j\bar{\mathbf{k}}_\mathrm{T}^2} + (1 - \sum_{j=1}^{m}\frac{a_j\bar{\mathbf{k}}_\mathrm{T}^2}{1+b_j\bar{\mathbf{k}}_\mathrm{T}^2})\exp[ik_0\Delta z(n-1)]\right]u(\mathbf{k}_\mathrm{T}, z)\exp[ik_z\Delta z]. \tag{24}$$

We see that advancing wavefields through a slab becomes a linear interpolation in the *f-k* domain between the reference phase-shift solution and the split-step solution. Setting

$C_j = a_j \bar{\mathbf{k}}_T^2 / (1 + b_j \bar{\mathbf{k}}_T^2)$ and taking its first-order term, then the separation-of-variables Fourier solution to Equation (24) can be expressed as:

$$u(\mathbf{k}_T, z + \Delta z) = \frac{2k_z}{k_z + k_z'} \{C_1 u(\mathbf{k}_T, z) + (1 - C_1) FT_{\mathbf{x}}[u(\mathbf{x}, z)\exp(ik_0 \Delta z(n-1))]\} \exp[ik_z \Delta z]. \quad (25)$$

The computational time with Equation (25) is almost the same as traditional split-step Fourier solutions, but with high accuracy close to the generalized screen propagator (GSP). Note that there is a singularity in equation (25) when k_z and k_z' approach zero simultaneously which leads to an instability of the algorithm. This can be avoided using the following relation (Huang *et al.*, 1999):

$$k_z = \sqrt{\frac{\omega^2}{v_0^2} - (1 + i\eta)^2 \mathbf{k}_T^2}\,, \quad (26)$$

where η is a small real number.

The R/T progators can be also applied to other known one-way propagators, e.g., SSF, FFD, due to the same or similar operator structure. Here, we give the common form in the frequency-wavenumber domain as

$$u(\mathbf{k}_T, z + \Delta z; \omega) = T(k_z, k_z') P\left[u(\mathbf{k}_T, z; \omega)\right], \quad (27)$$

and

$$u(\mathbf{k}_T, z; \omega) = R(k_z, k_z') P\left[u(\mathbf{k}_T, z + \Delta z; \omega)\right], \quad (27)$$

where $P[\cdot]$ is one-way propagators. The transmission coefficients $T(k_z, k_z')$ and the reflection coefficients $R(k_z, k_z')$ are given by

$$T(k_z, k_z') = \frac{2k_z}{k_z + k_z'}, \quad (28)$$

and

$$R(k_z, k_z') = \frac{k_z' - k_z}{k_z + k_z'}. \quad (29)$$

The present method is a separation-of-variables Fourier marching algorithm. The whole medium is sliced into a stack of thin slabs perpendicular to the main propagation direction. The implementation procedures may be summarized as follows:

1. Slice the medium into a stack of slabs perpendicular to the propagation direction (i.e., along the z-axis direction).
2. Choose a reference velocity in each slab to make it a perturbation slab represented by the acoustic refractive index.
3. Interact with the split-step terms and Fourier transform the wave fields at the entrance of each slab into the frequency-wavenumber domain.

4. Conduct the linear interpolation and propagate the wave fields through the slab in the frequency-wavenumber domain by multiplying a phase shift using the reference velocity.
5. Interact with the R/T coefficients to get the transmitted fields and reflected fields and inverse Fourier transform the wave fields at the exit of the slab into the frequency-space domain.
6. Repeat steps 3 to 5 slab by slab until the last slab.

2.2.2 Analysis of relative phase errors

The corresponding dispersion relation is shown in Equation (23). The relative phase error is defined by $E = |\varepsilon| / \bar{k}_z$, where

$$\varepsilon = \bar{k}_z - \sqrt{1-\bar{k}_x^2} - (n-1) + (n-1)\sum_{j=1}^{m} \frac{a_j \bar{k}_x^2}{1+b_j \bar{k}_x^2} . \tag{30}$$

Figure 3a compares the angular spectra of the first-order separation-of-variables screen propagator (SVSP1) with the GSP and SSF propagators under a 5% relative phase error. We see that the SVSP1 works perfectly for all the n values and all propagation angles, almost approaching that of the GSP. To investigate the global properties of equation (30), we evaluate the first- and second-order separation-of-variables approximations across three different n values by comparing their dispersion circles with the exact, GSP, and SSF dispersion relations (Figure 3b). As expected, the approximation can be an exact expression in the small-angle pie slice $(\bar{k}_x \approx 0)$ for all the n values. This accuracy comparison of the first- and second-order screen propagators with other Fourier propagators demonstrates a quick converge of the regional separation-of-variables approximation in the low-order terms.

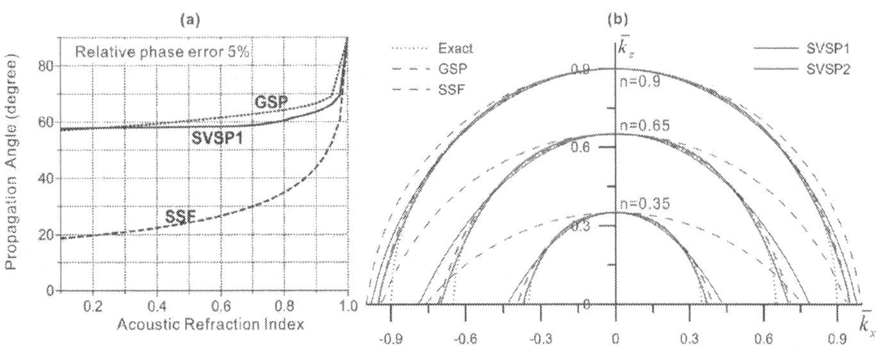

Fig. 3. Accuracy analysis of the SVSP with different orders. (a) Angular spectra of the first order (solid line) compared with the GSP (dotted line) and the SSF (dashed line). (b) Dispersion circles of the first order (thin solid line) and second order (thick solid line) compared with the exact (dotted line), the GSP (thick dashed line), and the SSF (thin dashed line) for $n = 0.35$, $n = 0.65$ and $n = 0.9$.

2.2.3 Definition of seismic illumination

Seismic illumination stands for the distribution of seismic wave energy underground. Since seismic wave energy is located at a frequency band cantered at peak frequency, the illumination map can be obtained by summing up seismic wavefields over a small frequency band. We define multi-frequency point source illumination $I_s(\mathbf{x},z)$ as

$$I_s(\mathbf{x},z)=\left[\sum_{\omega=\omega_0-\Delta\omega}^{\omega_0+\Delta\omega}\left|u_s(\mathbf{x},z;\omega)\right|^2\right]^{1/2}, \tag{31}$$

where ω is angular frequency, ω_0 is peak frequency, and $\Delta\omega$ is half of the frequency band. For multi-frequency point source-geophone illumination $I_{sg}(\mathbf{x},z)$, it can be expressed as

$$I_{sg}(\mathbf{x},z)=\left[\sum_{\omega=\omega_0-\Delta\omega}^{\omega_0+\Delta\omega}\left|u_s(\mathbf{x},z;\omega)u_g(\mathbf{x},z;\omega)\right|^2\right]^{1/2}, \tag{32}$$

where $u_s(\mathbf{x},z;\omega)$ and $u_g(\mathbf{x},z;\omega)$ are source and geophone wavefields at location $(\mathbf{x},z)$ for a frequency, respectively.

For plane-wave source case, the definition of multi-frequency source illumination $I_s(\mathbf{x},z)$ and multi-frequency source-geophone illumination $I_{sg}(\mathbf{x},z)$ can be given in the same way:

$$I_s(\mathbf{x},z)=\left[\sum_{\omega=\omega_0-\Delta\omega}^{\omega_0+\Delta\omega}\left|S(\mathbf{x},z;\omega)\right|^2\right]^{1/2}, \tag{33}$$

and

$$I_{sg}(\mathbf{x},z)=\left[\sum_{\omega=\omega_0-\Delta\omega}^{\omega_0+\Delta\omega}\left|S(\mathbf{x},z;\omega)G(\mathbf{x},z;\omega)\right|^2\right]^{1/2}, \tag{34}$$

where $S(\mathbf{x},z;\omega)$ and $G(\mathbf{x},z;\omega)$ are synthetic plane-wave wavefields of source and geophone wavefields at location $(\mathbf{x},z)$ for a frequency, respectively.

2.2.4 Computational efficiency

The one-way propagators are shuttled between the space domain and the wavenumber domain via fast Fourier transform (FFT). Thus, the computational efficiency of our method is dependent on the performance of FFT. It is necessary to pick an efficient FFT for fast implementation of our method. As known, lots of FFT codes are contributed to us, which have different computation cost. Here, we test two of the fastest FFT packages, i.e., FFT in west (FFTW) (Frigo and Johnson, 2005) and Intel Math Kernel Library (MKL). Both are free packages under the terms of the GNU General Public License.

Although computer technologies have been advanced greatly, people do not satisfy with computational cost yet. Thus, it is still necessary to choose FFT as fast as possible. The benchmark is performed on a personal computer. The configuration and compiler environment is shown in Table 1.

To test the performance of the two FFT packages, we take a 2D Gaussian function as input data, defined as

$$f(x,y) = e^{-\left[\frac{(x-x_0)^2}{100}+\frac{(y-y_0)^2}{100}\right]}, \tag{35}$$

where $x = i\Delta x$, $y = j\Delta y$, $i = 1,2,\cdots,Nx$, $j = 1,2,\cdots,Ny$, $x_0 = Nx \times \Delta x / 2$ and $y_0 = Ny \times \Delta y / 2$. We compute the floating point operations per second (FLOPS). The formula is given by

$$mflops = 5N\log_2(N)/t, \tag{36}$$

where $N = Nx \times Ny$ is the length of FFT, t is the CPU time of performing forward and inverse FFT, its unit is second. Noting that the FLOPS is not true one, which is a normalized value by $5N\log_2(N)$, which is the redundancy of the radix-2 Cooley-Tukey algorithm.

Type	Lenovo Thinkpad X61 7673AN6
CPU	Intel(R) Core(TM)2 Duo CPU T7300@2.00GHz
Memory	3GB DDR2 667MHz
Operating System	Red Hat Enterprise Linux
Compiler	Intel C++ Compiler 11.0.072

Table 1. Computer configuration and compiler environment

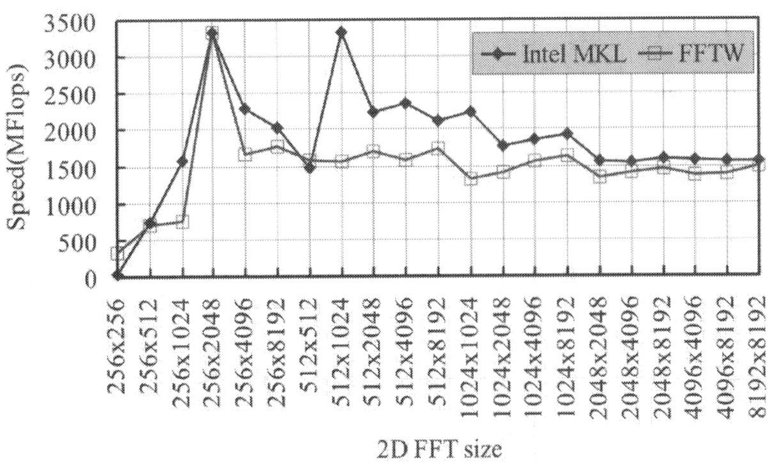

Fig. 4. Comparison of computational efficiency of Intel MKL and FFTW

Figure 4 shows the computational speed of the two FFT packages. The FLOPS of Intel MKL is much higher than the one of FFTW, especially at location of commonly used 2D FFT size (512x1024-512x8192). Thus, we choose Intel MKL to perform FFT in our method.

2.3 Applications

In this section, we will demonstrate the numerical and practical applications of seismic illumination in exploration geophysics. Firstly, we verify its accuracy by comparing illumination map with prestack depth migration for a 2D salt model. Secondly, we take 2D SEG/EAGE salt model as an example to obtain a criterion for seismic illumination and migration. Finally, we take a 3D fault block model in eastern China to show its powerful applications in industry.

2.3.1 Numerical verification

Figure 5 displays a theorectical model with a salt body at the center. The horizontal distance is 12000m and the depth is 1440m. The space sampling is 25m and 6m in the horizontal and depth directions, respectively. There are 120 receivers per shot. The shot interval is 50m and the receiver interval is 25m. The plane-wave source illumination map and prestack depth migration (PSDM) profile are shown in Figure 6. The result of illumination map and migration profile are quite similar. While it shows larger energy in the illumination map, it shows good phases and amplitudes in PSDM profile, and vice versa. In Figure 6a, there is an obvious illumination shade due to the exsitance of salt body. Accordingly, it has poorly imaged at the same position in figure 6b.

Geologic structures have great influence on waves propatation udnerground. This leads to uneven distribution of seismic energy. To demonstrate the influence of geologic structures on illumination energy, we give two point source illumination maps at different location (4000m and 8000m) in Figure 7. In Figure 7a, the shot is located at leftside of salt body, where the energy distributes uniformly. However, the distribution of seimic wave energy is seriously uneven while the shot locates at the rightside of salt body (see figure 7b) and will lead to poor images.

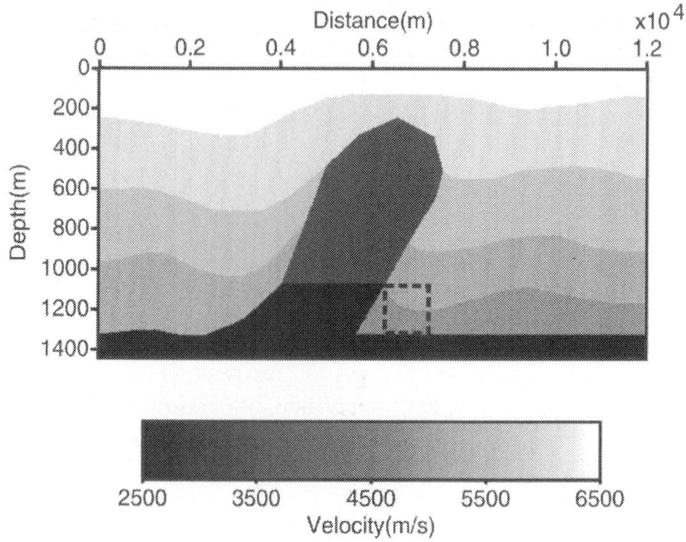

Fig. 5. A theorectical model with a salt body

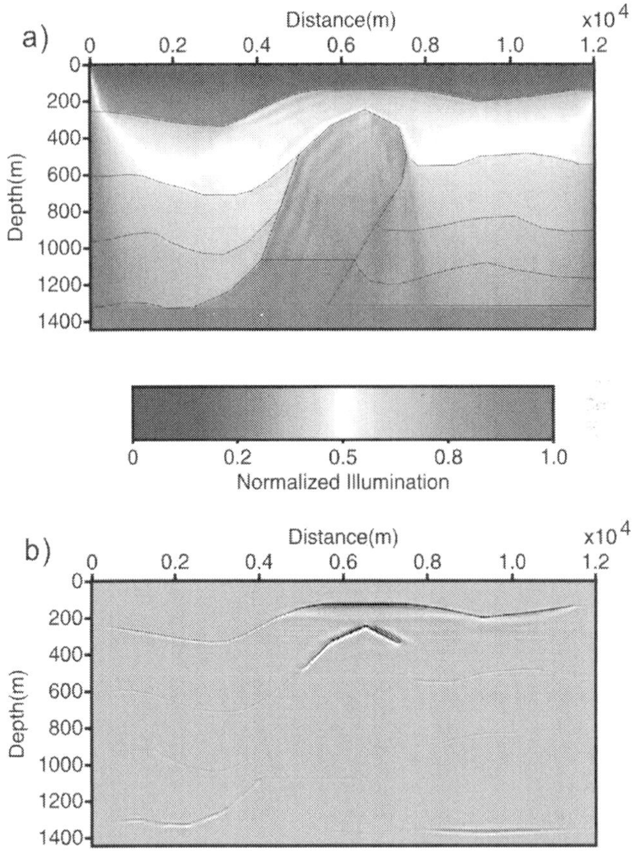

Fig. 6. (a) illumination intensity for the model shown in Figure 5, and (b) corresponding prestack depth migration

2.3.2 Seismic illumination and PSDM

In this section, we clarify two criterions of seimic illumination by comparing illumination map with PSDM profile. With the help of the two criterions, people can predict the results of seismic acquisition and migration via illumination distribution. The two criterions are 1) strong illumination energy and 2) uniform distribution of illumination. Here, we take 2D SEG/EAGE salt model as an example for simplicity. The most important reason to pick 2D SEG/EAGE salt model is that the model is a standard model used to test various migration algorigthms.

Figure 8a shows the 2D SEG/EAGE salt model. The horizontal and vertical sampling numbers are 1290 and 300, respectively. The spatial sampling interval is 12.5m. We use Fourier finite-difference method (FFD) (Ristow and Rühl, 1994) to obtain the post-stack migraion result shown in Figure 8b. Figure 8c shows the illumination map calculated by plane wave source. The source function is Ricker wavelet with dominant frequency 15Hz.

The frequency band is 2Hz, from 14Hz to 16Hz. The incident angle of plane wave ranges from -50° to 50°. In Figure 8c, two shade areas subsalt can be observed obviously. This is because seismic waves cannot penetrate the salt and are reflected. Comparing the migration results and the illumination map, we find that both have good consistence with each other. That means, the poorer the illumination is, the worse the migration is, and vice versa.

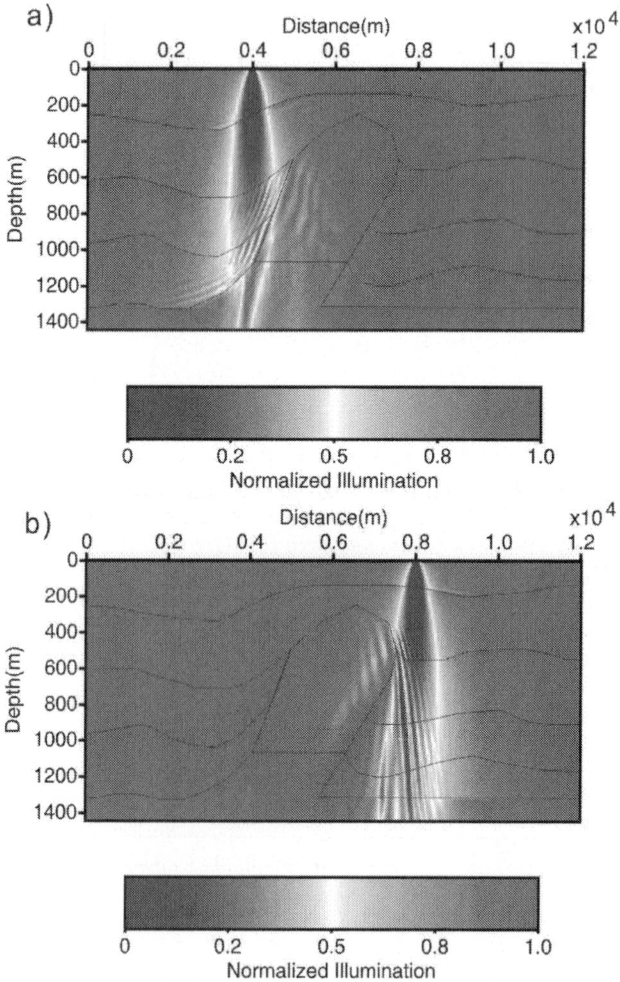

Fig. 7. Normalized illumination of point sources at different positions: (a) 4000m and (b) 8000m

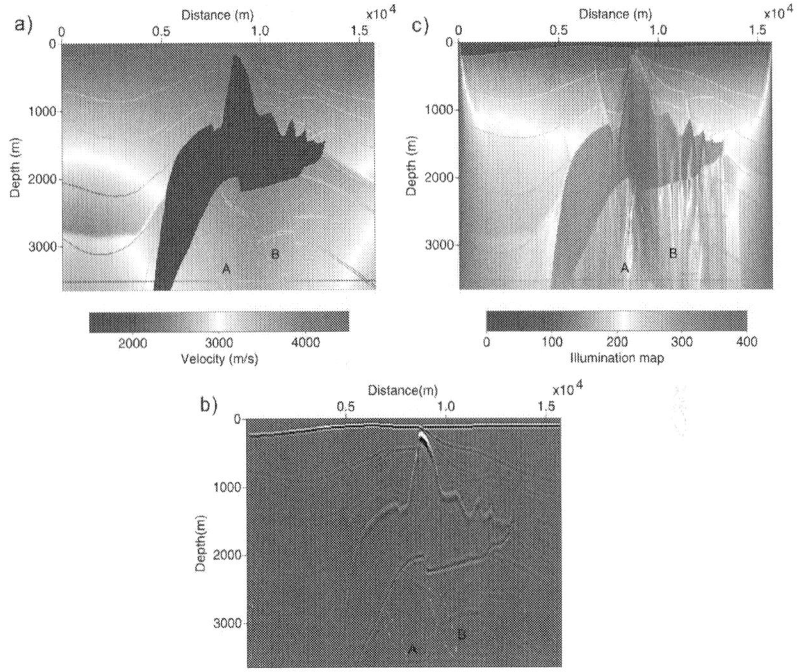

Fig. 8. (a) SEG/EAGE 2D salt velocity model, (b) post-stack depth migration for the model shown in Figure 8a and (c) the corresponding illumination intensity map

2.3.3 Effects of velocity uncertainties

Same as the prestack depth migration (PSDM), the illumination maps are also calculated based on interval velocity model in depth domain. In general, the actual interval velocity model or geological model cannot be known exactly. The quality of PSDM is dependent on the accuracy of interval velocity model. Thus, it is necessary to evaluate the effects of velocity errors on illumination intensity distributions.

In this section, we take a 3D real geologic model (Figure 9) in Eastern China as an example to investigate the influence of velocity error on the illumination intensity. The geologic structure in this region is very complex. Small faults/blocks are quite common. The migration images from this area are relatively poor. The vertical slices of the model are shown in figure 9a and 9b. The horizontal sampling numbers are 501 and 201, respectively. The depth sampling number is 209. The spatial sampling interval is 25m in all directions. The depth of target zone is 3000-4000m. The source function is Ricker wavelet with dominant frequency 15Hz. The frequency band is 2Hz, from 14Hz to 16Hz. The strike of the geologic structure is along the x-axis. The incident angle of plane wave ranges from -50° to 50° in the x direction, and from -20° to 20° in the y direction. Here, we first give an illumination map with a point source in Figure 10. In this example, we use correct interval velocity. The energies in target zone are distributed unuiformly, which illustrates complex structures have serious effects on wave propagations.

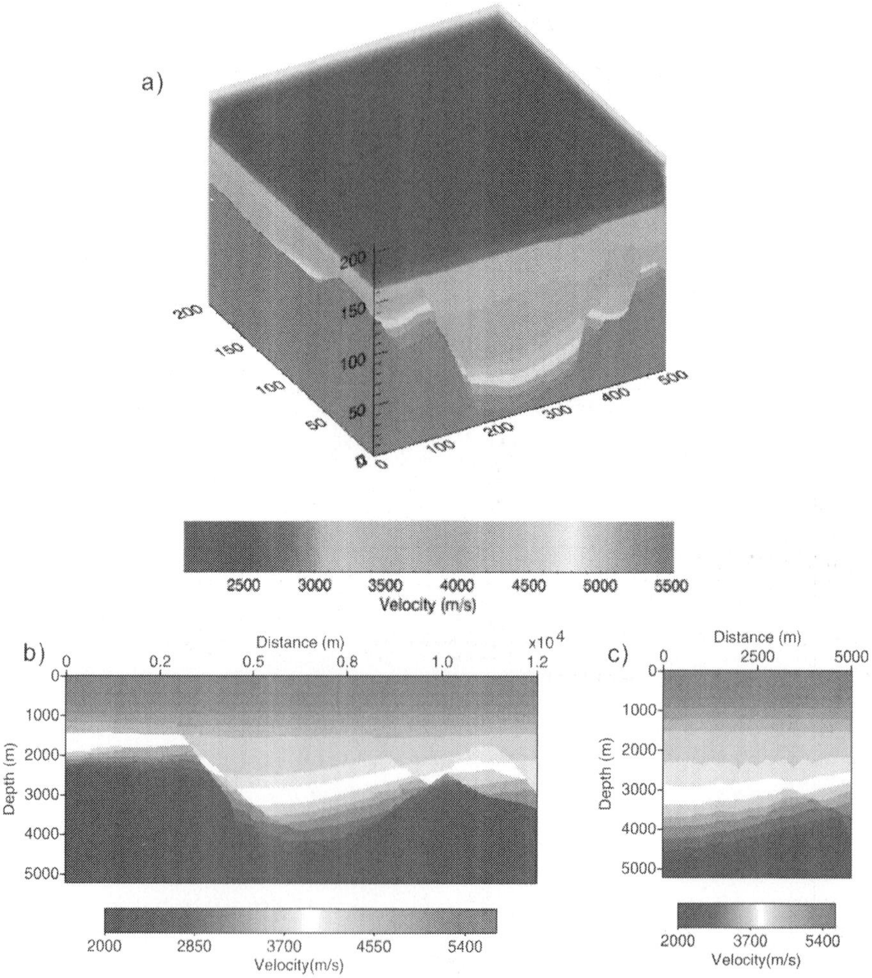

Fig. 9. (a) A 3D real geologic model in Eastern China, and its vertical velocity slices at (b) y=2500m and (c) x=6250m

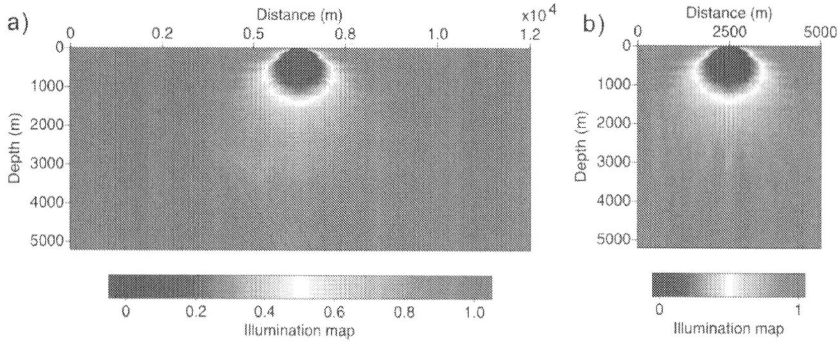

Fig. 10. Vertical slices of illumination intensity for a point source at (a) y=2500m and (b) x=6250m. The illumination is calculated with 100% correct velocity

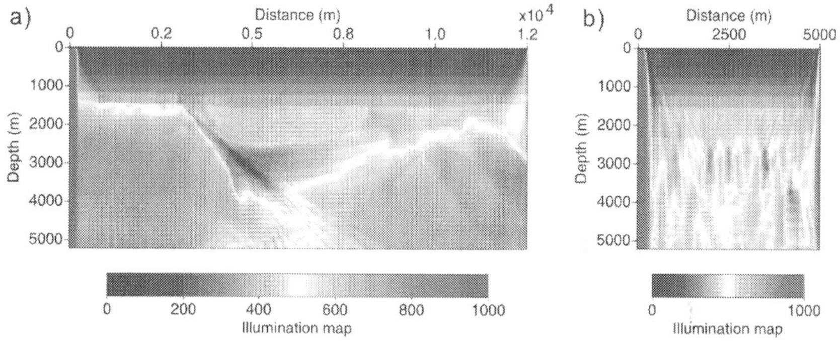

Fig. 11. Vertical slices of plane-wave illumination intensities at (a) y=2500m and (b) x=6250m. The illumination is calculated with 90% correct velocity

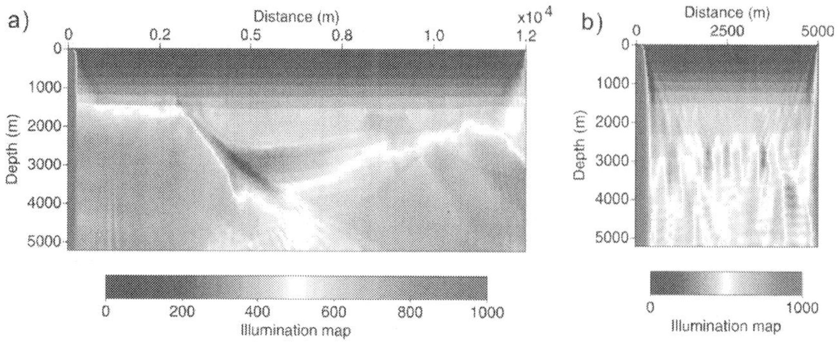

Fig. 12. Vertical slices of plane-wave illumination intensities at (a) y=2500m and (b) x=6250m. The illumination is calculated with 100% correct velocity

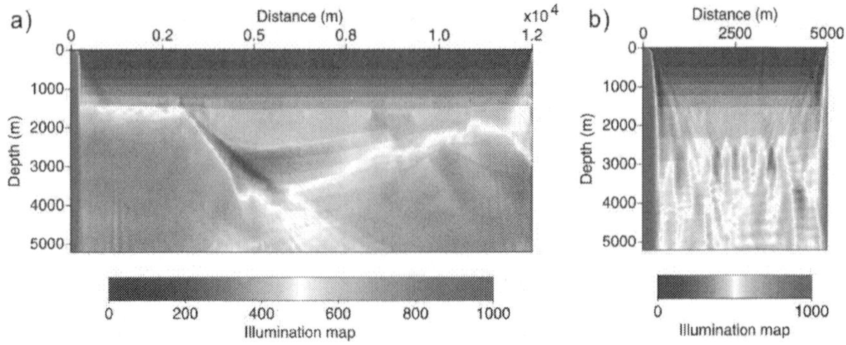

Fig. 13. Vertical slices of plane-wave illumination intensities at (a) y=2500m and (b) x=6250m. The illumination is calculated with 110% correct velocity

Now, the effects of velocity uncertainties on illumination are investigated in complex media. We compare the illumination maps calculated with 90%, 100% and 110% correct velocity, shown in Figure 11-13, respectively. Although the illumination maps are obtained by using incorrect velocities, their characteristics are almost same with each other. Thus, we can conclude that the small velocity error has little influences on illumination intensity.

2.3.4 Target-oriented illumination analysis

As the criterions stated above, we can obtain a good quality image of subsurfaces, if the illumination intensity is strong and distribute uniformly. Or else a poor quality image will be resulted. From Figure 12, the illumination in target zone is weak and does not distribute uniformly, which will lead to poor image underground.

To obtain an image with good quality, people are always trying to satisfy the two criterions. An direct way is to enhance the illumination intensity by adding shot excitations in the field. The key of the method is where to place the shots and receivers in order to enhance the illumination in target zones.

Now, we try to locate effective shot postions according to the reciprocity theory. The illumination at the surface can be considered as the contribution to the target zone while shots are excited at the surface. We firstly place the seismic sources in the target zone. Then, we propagate the wavefields to the surface and calculate the illumination intensity at the surface. Figure 14 shows us the illumination at the surface where the source are placed in the target zone. It can be found that the illumination does not distribute uniformly, since the wavepaths are distorted while waves run in the complex media. The results of Figure 14 can help us to determinate where to add shots.

After the effective locations of shots excitation are obtained, we next ensure locations of corresponding receivers. Firstly, we put the source at where has largest contributions to improve the illumination intensities of target zones. Then, the reflectivities in the complex media are set to zero except in the target zones. According to equations (13) and (18), the wavefields are propagated in the complex media. The wavefields recorded at the surface are reflected only from the target zone. Thus, we can use the target-reflected illumination intensities to locate where to receive the wavefields. Figure 14 show the optimal receivers position for a given source at (6200,1500)m. In Figure 13 and 14, both optimal sources and receivers positions are irregular, since the wavefields are distorted while running in the compelx media. Thus, target-oriented illumination analysis would totally change the conventional way of designing seismic acquisition system.

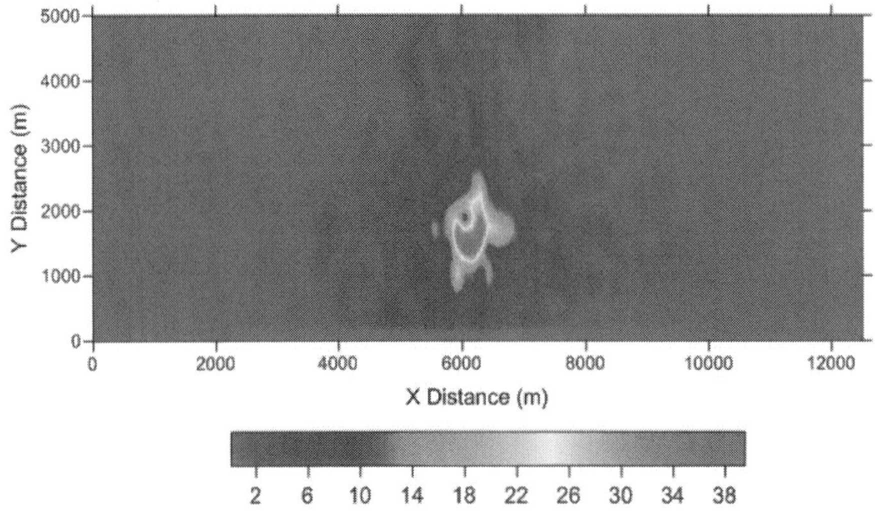

Fig. 14. The optimal positions of shots to illuminate the target zones

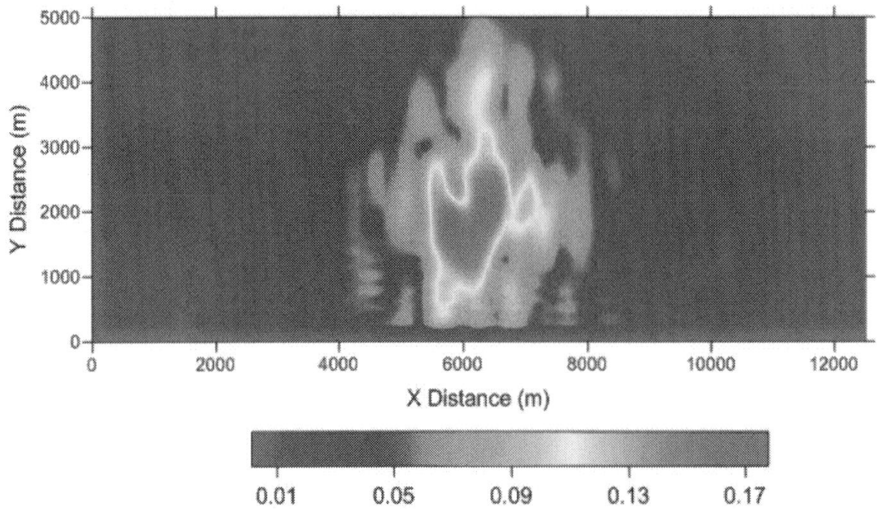

Fig. 15. The optimal positions of receivers to record signals reflected from the target zones

3. Conclusion

Starting from generalized Lippmann-Schwinger equations, we developed a 3D one-way wave-equation-based illumination method in complex media. It combines the one-way propagators coupled with reflection/transmission (R/T) coefficients and the phase encoding techniques. The R/T operators not only account for amplitude variations with incident angles across interfaces, but also accommodate to complex media with steep dip angle and large lateral velocity contrast.

Several numerical examples are given to illustrate its resolving capabilities for complex media. In this chapter, we have demonstrated the numerical and practical applications of seismic illumination in exploration geophysics. Numerical examples shows that the illumination maps and results of post-and pre-stack depth migration are consistent with each other. Two criterions i.e. strong illumination energy and uniform distribution of illumination are obtained by comparing the illumination intensity map with the prestack depth migration for 2D SEG/EAGE salt model.

The velocity errors to the illumination intensities are investigated, which shows the velocity error has little influences on illuminations. The target-oriented illumination analysis has been applied to design seismic acquisition, which is fundamental to the exploration geophysics.

4. Acknowledgment

The helpful discussions with T. Jiang and X. J. Wan are greatly appreciated. This work was supported by China Postdoctoral Science Foundation (Project No. 20100480447).

5. References

Bear, G. ; Lu, C. ; Lu, R. & Willen, D. (2000). The construction of subsurface illumination and amplitude maps via ray tracing, *The Leading Edge*, Vol.19, No.7, pp. 726-728

Berkhout, A. J. (1997). Pushing the limits of seismic imaging, Part I: Prestack migration in terms of double dynamic focusing, *Geophysics*, Vol.62, No.3, pp. 937-954

Cao, J. & Wu, R. S. (2008). Local-angle domain illumination for full-wave propagators, *78th Annual International Meeting, SEG, Expanded Abstracts*, pp. 2246-2251

de Hoop, M. V.; Rousseau, J. H. & Wu, R. S. (2001) Generalization of the phase-screen approximation for the scattering of acoustic waves, *Wave Motion*, Vol.31, No.1, pp. 285-296

Frigo, M. & Johnson, S. G. (2005). The design and implementation of FFTW 3, *Proceedings of the IEEE*, Vol.93, No.2, pp. 216-231

Fu, L. Y. & Bouchon, M. (2004). Discrete wavenumber solutions to numerical wave propagation in piecewise heterogeneous media-I. Theory of two-dimensional SH case, *Geophys J Int*, Vol.157, pp. 481-491

Hoffmann, J. (2001). Illumination, resolution and image quality of PP- and PS-waves for survey planning, The leading edge, Vol.20, No.9, pp. 1008-1014

Huang, L. J.; Fehler, M. & Wu, R. S. (1999). Extended Local Born Fourier Migration Method, *Geophysics*, Vol.64, pp.1524-1534

Mao, J. & Wu, R. S. (2007). Illumination analysis using local exponential beamlets, *77th Annual International Meeting, SEG, Expanded Abstracts*, pp. 2235-2239

Muerdter, D. & Ratcliff, D. (2001). Understanding subsalt illumination through ray-trace modeling, Part 1: Simple 2-D salt models. The Leading Edge, Vol.20, No.6, pp. 578-594

Muerdter, D. ; Kelly, M. & Ratcliff, D. (2001). Understanding subsalt illumination through ray-trace modeling, Part 2: Dipping salt bodies, salt peaks, and nonreciprocity of subsalt amplitude response, *The Leading Edge*, Vol.20, No.7, pp. 688-697

Muerdter, D. & Ratcliff, D. (2001). Understanding subsalt illumination through ray-trace modeling, Part 3: Salt ridges and furrows, and the impact of acquisition orientation, *The Leading Edge*, Vol.20, No.8, pp. 803-816

Ristow, D. & Rühl, T. (1994). Fourier finite-difference migration, *Geophysics*, Vol.59, pp. 1882-1893

Romero, L. A.; Ghiglia, D. C. & Ober, C. C. (2000) Morton S A. Phase encoding of shot records in prestack migration, *Geophysics*, Vol.65, No.2, pp.426-436

Schneider, W.A. & Winbow, G.A. (1999). Efficient and accurate modeling of 3-D seismic illumination , *69th Ann. Internat. Mtg. Soc. Expl. Geophys., Expanded Abstracts*, pp. 432-435

Stoffa, P. L.; Fokkema, J. T.; Freire, R. M.; de Luna, F. & Kessinger, W. P. (1990). Split-step Fourier migration. *Geophysics*, Vol.54, No.4, pp. 410-421

Sun, W. ; Fu, L. Y. & Wan, X. (2007). Phase encoding-based seismic illumination analysis, *Oil Geophysical Prospecting*, Vol.42, No.5, pp. 539-543

Sun, W.; Fu, L. Y. & Yao, Z. X. (2009). One-way propagators coupled with reflection/transmission coefficients for seismogram synthesis in complex media. *Chinese J Geophys. (in Chinese)*, Vol.52, No.5, pp. 1044-1052

Trefethen, L. N. & Halpern, L. (1986). Well-Posedness of One-Way Equations and Absorbing Boundary Conditions, *Math. Comput*, Vol.147, pp.421-435

Wu, R. S. & Chen, L. (2002). Mapping directional illumination and acuqsition-aperture efficacy by beamlet propagators, *72nd Annual International Meeting, SEG, Expanded Abstracts*, pp. 1352–1355

Wu, R. S. & Chen, L. (2006). Directional illumination analysis using beamlet decomposition and propagation, *Geophysics*, Vol.71, No.4, pp. S147–S159

Wu, R. S.; Wang, Y. & Gao, J. H. (2000). Beamlet migration based on local perturbation theory, *70th Annual Meeting, SEG, Expanded Abstracts*, pp. 1008–1011

Wu, R. S.; Xie, X. B. & Wu, X. Y. (2006). One-way and one-return approximations for fast elastic wave modeling in complex media, *Advances in wave propagation in heterogeneous earth: Elsevier*, pp. 266~323

Xie, X. B.; Jin, S. W. & Wu, R. S. (2006). Wave-equation-based seismic illumination analysis, *Geophysics*, Vol.71, No.5, pp. S169~S177

Xie, X. B. & Wu,R. S. (2002). Extracting angle domain information from migrated wavefield, *72nd Annual International Meeting, SEG, Expanded Abstracts*, pp. 1360–1363.

Xie, X. B.; Jin, S. W. & Wu, R. S. (2003). Three-dimensional illumination analysis using wave-equation based propagator, *73rd Annual International Meeting, SEG, Expanded Abstracts*, pp. 989–992

9

Seismic Ground Motion Amplifications Estimated by Means of Spectral Ratio Techniques: Examples for Different Geological and Morphological Settings

M. Massa[1], S. Lovati[1], S. Marzorati[2] and P. Augliera[1]
[1]*Istituto Nazionale di Geofisica e Vulcanologia, Sezione Milano-Pavia, Milano*
[2]*Istituto Nazionale di Geofisica e Vulcanologia, Centro Nazionale Terremoti, Passo Varano (Ancona), Italy*

1. Introduction

One of the most important issue in seismic hazard and microzonation studies is the evaluation of local site response (i.e. the tendency of a site to experience during an earthquake greater or lower levels of ground shacking with respect to another). In general site effects reflect all modifications (in amplitude, frequency content and duration) of a wave-field produced by a seismic source during the propagation near the surface, due to particular geologic (stratigraphy and morphology), geotechnical (mechanical properties of deposits) and physical (e.g. coupling of incident, diffracted and reflected seismic waves) conditions of a particular site.

Actually local seismic amplification represents one of the main factors responsible for building damage during earthquakes: this statement is supported by well documented evidences of structural damages during past moderate to high energy events occurred both in Italy (e.g. 23th November 1980, Mw 6.9, Irpinia earthquake, Faccioli, 1986; 26th September 1997, Mw 6.0, Umbria-Marche earthquake, Caserta et al., 2000; the 31th October 2002, Mw 5.7, Molise earthquake, Strollo et al., 2007; 6th April 2009, Mw 6.3, L'Aquila earthquake, Cultrera et al., 2009) and in other worldwide countries (e.g. 3rd March 1985, Mw 7.8, Chile earthquake, Celebi, 1987; 17th August 1999, Mw 7.6, Izmit earthquake, Sadik Bakir et al., 2002). For this reason the site effects evaluation, performed by experimental methods but also through numerical simulations, has attracted the attention of engineering seismology and earthquake engineering communities. Of consequence, in the last decade many experiments were performed in correspondence of different setting such as alluvial basins (Parolai et al., 2001 and 2004; Ferretti et al., 2007; Massa et al., 2009; Bindi et al., 2009) or topographies (e.g. Pischiutta et al., 2010; Massa et al., 2010; Buech et al., 2010; Marzorati et al., 2011; Lovati et al., 2011).

The present work has the aim to evaluate the capabilities of the most common passive methods at present used in seismology to evaluate the site response: HVSR (Horizontal to Vertical Spectral Ratio technique on seismic noise, Nakamura, 1989, or earthquakes, Lermo & Chavez Garcia, 1993) and SSR (Standard Spectral Ratio, Borcherdt, 1970). The reliability of the considered techniques was evaluated by comparing the results obtained analysing different seismic signals (noise, local earthquakes and teleseisms) recorded in different geological and morphological setting and by using different instrumentation (weak motion and strong motion sensors).

In order to obtain the aforementioned scope, 5 Italian test sites housing, at present, a seismic station (permanent and temporary networks) managed by Italian National Institute for Geophysics and Volcanology [INGV], were selected. The site selection was performed in order to evaluate possible local site effects in different conditions: stations numbered as 1 and 2 (figure 1) are located in the centre of the Po Plain and in correspondence of its edge respectively, station 3 represents a station that, being installed in the central Alps, could in general represent a reference site, while two stations (4 and 5, figure 1) are installed at the top of topographies, the first located in North-Est Italy and the second, more characterized by a clear 2D configuration, in the central Apennines respectively.

It is worth noting that in correspondence of areas characterized by a low rate of seismicity, but potentially able to suffer energetic seismic events (Gruppo di Lavoro CPTI 2004, http://emidius.mi.ingv.it/CPTI/), such Northern Italy, the capability of different techniques to estimate local response, also evaluated using different type of instruments, might represent a fundamental step in order to avoid some practical problems such as long in time field experiments due to the lack of recordings related to local events (e.g. noise measurements or analyses on teleseisms might be able to provide good results in particular frequency ranges). In any case, the results coming from spectral analysis, in particular if they are obtained without a reference site, have to be always read combined to other detailed geological, geothecnical and geophysical information related to the investigated site (stratigraphy, shear wave velocity etc.).

2. Geological and geomorphological settings

In order to highlight the capabilities of the considered spectral technique to estimate the site response, sites characterized by geological and/or geomorphological setting were selected.

The sites were selected both in Northern (station 1, 2, 3 and 4, see figure 1 and 2) and central Italy (station 5, see figure 2).

The site where station 1 is installed (figure 1, top) is located in the foreland of the Central-Alps, in correspondence of morainic deposits with depth of dozen of meters (Regione Lombardia, 2003). The site where station 2 (figure 1, middle) is installed is located in the central area of the Po Plain, one of the more extended alluvial basin at global scale (surface of about 46.000 km^2) characterized by thickness of deposits up to some kilometers (Regione Lombardia, 2003). Stations 3 (figure 1, bottom) and 4 (figure 2, top) are installed in correspondence of stiff formations: site 3, located in the Central Alps (about 800 m of quota) is characterized by the presence of massive limestone (and/or dolomite, Regione Lombardia, 2002), while site 4, even if from a stratigraphical point of view is characterized by compact sandstone and clay (sheet 037 of the 1:100.000 Geological Map of Italy), represents an interesting case study being the station installed at the top of a hill. Station 5 (figure 2, bottom) is installed at the top of a very steep ridge chracterized by a pronunced 2D morphology. From a geological point of view the site is characterized by massive limestone formation (sheet 138 of the 1:100.000 Geological Map of Italy; Amanti et al., 2002). The areas surrounding the ridge at NE of the ridge is characterized by alluvial, lacustrine and fluvial deposits that overlap locally the limestone massif.

In Bordoni et al. (2003), on the base of the geological information of the 1:500.000 Italian Geological Map, the authors classified the Italian territory following the provision reported in the Eurocode8 (CEN, 2004). For the considered site, the errors associated to the 1:500.000 scale have been checked by comparing this map with very detailed geological maps (scale 1:10.000 and 1:5.000, figure 1, right panels). From such a comparison no significant differences has been observed. On the basis of this classification (after adopted also in the new Italian code for buildings NTC, 2008) site 3, 4 and 5 are included in A soil-category (Vs30 > 800 m/s), station 1 is include in B soil-category (360 < Vs30 < 800 m/s) and station 2 is included in C soil-category (Vs30 < 360 m/s). Moreover, following NTC, 2008, site 4 and 5 are included in T2 (average slope > 15°) and T3 (15° < average slope < 30°) topographic-category respectively.

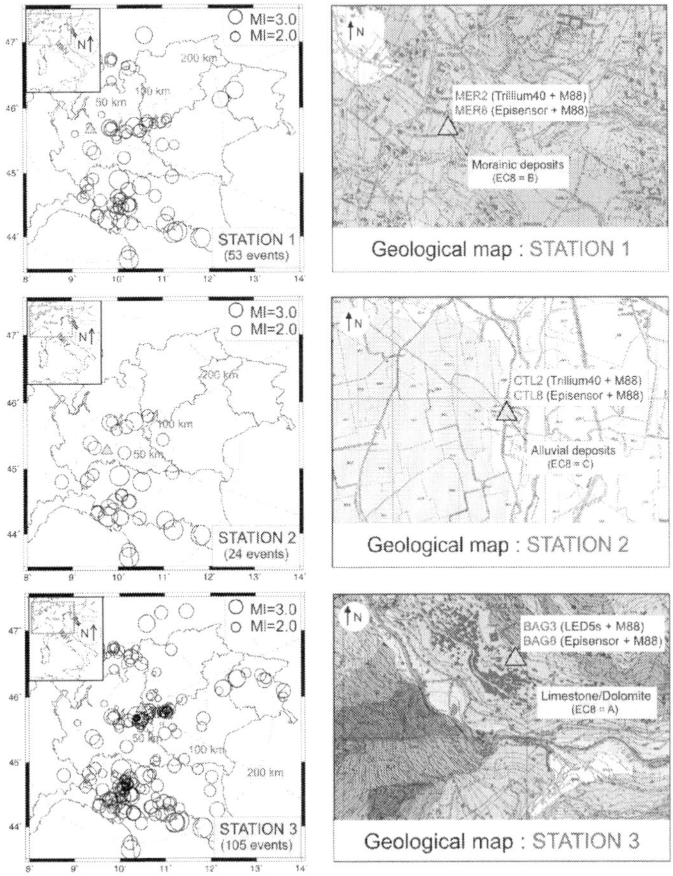

Fig. 1. Sites selected for the analyses (yellow triangles): on the left panels circles represent weak-motion (M_L between 2.0 and 3.0) recorded at each site (53 for station 1, 24 for station 2 and 105 for station 3). In the right panels the relative geological maps are reported.

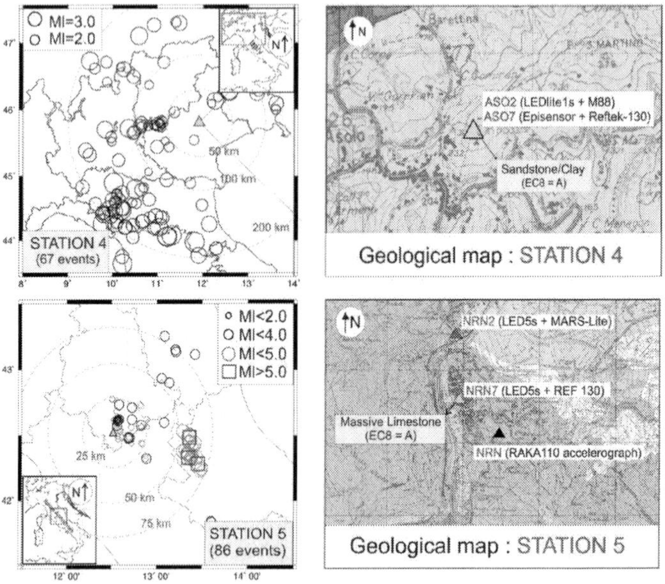

Fig. 2. Sites selected for the analyses (yellow triangles). Top panels: as explained in figure 1 but for station 4 (67 events), located at the top of a topography. Left bottom panel: data set available for station 5 (circles are 29 events with $M_L < 4$ and squares are 12 events with $M_{L} \geq 4$). Right bottom panel: geological map and available stations (in red the reference site, used for SSR analyses, in black the permanent RAN strong-motion station).

3. Data set and data processing

For the analyses a relevant data set collected in the last 5 years was taken into account (Figure 1 and 2 left panels). It is composed by microtremor recordings, local events occurred in Northern and central Italy (M_L up to 5.3 and epicentral distance up to 200 km) and teleseisms.

Stations 1, 2, 3 and 4 (Figure 1 and 2, right panels) belong to the permanent strong motion network of Northern Italy (RAIS, http://rais.mi.ingv.it). They are equipped both with strong-motion (Kinemetrics Episensor FBA ES-T) and velocimetric sensors (broad band Trillium 40s for station 1 and 2, semi broad-band Lennartz LE3D-5s for station 3 and short period Lennartz LE3D-1s for station 4). Station 5 (NRN7, see figure 2), located at the top of Narni ridge (central Italian Apennines) belongs to a temporary velocimetric array, composed by 10 stations, installed in correspondence of the ridge and surroundings in the period March-September 2009 (Massa et al., 2010; Lovati et al., 2011). In this study, in order to make considerations about SSR technique, the station NRN2, located at the base of the ridge (red triangle in figure 2) was considered as reference site. In figure 2 also the location of the permanent RAN (Italian Accelerometric Network, www.protezionecivile.gov.it) strong-motion station installed at the top of the ridge (NRN, black triangle in figure 2) is indicated. All selected stations are equipped by 20 or 24 bits recording systems (Lennartz Mars88 and Reftek 130 respectively). The sampling rate range from 100 sps to 125 sps.

The selection of stations characterized by different type of sensors allows us to record in a wide dynamic range of amplitude: indeed, while on one hand, a weak motion sensor assures high quality records related to seismic background noise, local earthquakes of low magnitude and teleseismic events, on the other hand the strong motion sensor avoids the loss of recordings in the case of high magnitude events with epicenter close to the stations.
Data processing was computed following standard procedures that include: mean removal (on the whole signal), baseline correction (least square regression), removal of instrument response by deconvolution with the instrument response curve (for strong-motion sensors computed just for NRN, being the RAN strong-motion sensor analogue, Massa et al., 2010) also in the case of analogue instrument), time domain cosine tapering (5%), selection of high and low pass filter (band-pass Butterworth 4 poles). Considering the recorded signal the low-pass frequency were set at 30 Hz while the high pass cut off was choosen by a visual inspection of data considering both the magnitude of the sensor of the station. Noise recordings were in general windowed in time series of 60 s (in case of semi broad band stations) or 120 s (in case of broad band stations) length, while for local events different portion of S-phase (5 s, 10 s, 15 s starting from the S onset) and 20 s of coda were considered. Finally, teleseismic events were analysed considering 80 s windows length selected on both Pn and Sn phases. For all considered signals, for each time window the FFT was calculated and then smoothed using the Konno & Ohmachi (1998) window (b=20). In order to detect possible polarization effects, in particular for 2D configurations, the NS and EW horizontal components of ground motion were clockwise rotated of 180°, by step of 5°.

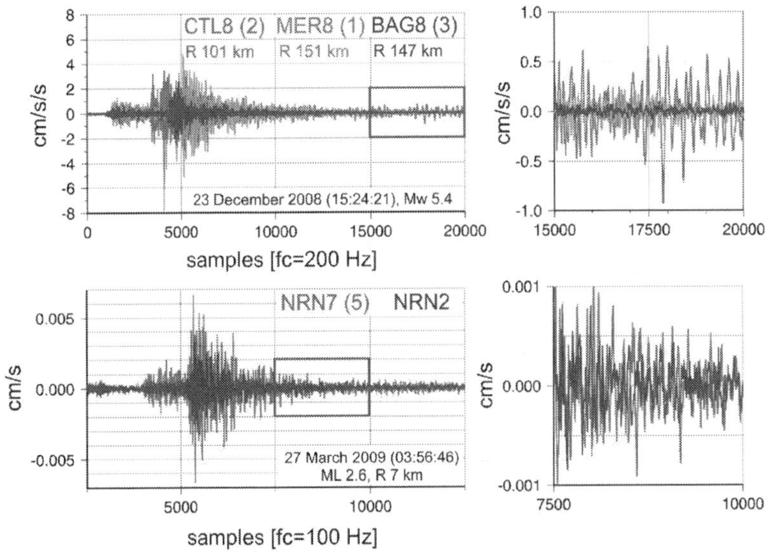

Fig. 3. Top left panel: waveforms of 23rd December 2008, Mw 5.4, Parma earthquake recorded by station 1 (green), station 2 (red) and station 3 (blue). On the right a more detailed image of coda waves. Bottom left panel: waveforms of a weak motion of M_L 2.6 (epicentral distance 7 km) as recorded at Narni ridge by station 5 (red) and by the reference one (blue). Also in this case on the right is reported a detailed of coda amplitudes.

The processed signals were analyzed applying the single station Horizontal to Vertical Spectral Ratio technique (HVSR), both on noise (Nakamura technique, Nakamura, 1989) and earthquakes (Lermo & Chavez Garcia, 1993) and, also for station 5 (availability of a reference site), the Standard Spectral Ratio technique (SSR, Borcherdt, 1970).
In the top panels of Figure 3 an example of waveforms related to the 23th December 2008, Mw 5.4, Parma earthquake as recorded by stations 1, 2 and 3 is illustrated: it is possible to appreciate the difference both in peak ground acceleration (PGA) between station 3 (blue), located on hard rock, and the others installed on alluvial (station 2, red) and morainic (station 1, green) deposits respectively and, in particular the amplitude of surface waves affecting the station (2) installed in the middle of the Po Plain (red in the top right panel). In the bottom panels of figure 3, weak motion waveforms (M_L 2.6) as recorded at Narni ridge by the station 5 (at the top, NRN7 in red) and by the reference station (at the base, NRN2 in blue, see figure 2) are reported. Even in this case, the right panel shows, as for coda waves, the amplitudes recorded at the station installed at the top appear to be higher than those recorded at the base (even if less evidence with respect to the S-phase amplification).

4. Experimental techniques for seismic site response evaluation

The evaluation of local seismic site response is usually estimated through different spectral techniques applied both on background noise and earthquakes data. In an optimal condition the operator has the possibility to integrate results coming from different approaches in order to assure reliable responses in a broad range of frequencies. Unfortunately this condition is usually an exception due to the lack of earthquake recordings (especially in areas characterized by low seismicity rate), unavailability of reliable reference site or very high level of background noise (low signal to noise ratio). The main scope of the studies concerning site effects is to identify the fundamental frequencies of a site and, if possible, to provide the related amplification factor. A careful knowledge of the resonance frequency of a soil, coupled to the information about the predominant period of a structure can give a reliable idea of potential damages that we can expect for a site in case of an earthquake (in particular if the predominant frequency of the source reflects that detected for the site of interest). Nowadays in seismology the more commonly used techniques are:

1. single station Horizontal to Vertical spectral ratio on noise (NHVSR, Nakamura, 1989);
2. single station Horizontal to Vertical spectral ratio on earthquakes (HVSR, Lermo & Chavez Garcia, 1993);
3. Standard Spectral Ratio (SSR) using a reference site (Borcherdt, 1970).

4.1 Nakamura technique

The Nakamura technique consists in performing the spectral ratio between the horizontal and the vertical component of a selected window of background noise recorded at a particular site. Being the seismic noise characterized by a low frequency (< 1 Hz) natural component (ocean storm or meteorological perturbations) and a high frequency (> 1 Hz) anthropic component (Gutenberg, 1958; Asten, 1978), the related analyses allow to obtain information in a broad range of frequency. From a theoretical point of view, considering that the seismic noise is a continuous and stationary phenomena, the spectral analyses have to be computed considering the ratio between the Power Spectral Density (PSD) calculated on each single component for the considered window. In spite of this consideration, also for noise analyses the Fourier spectrum is usually adopted, being easier the direct comparison to earthquake spectra.

In general noise measurements, being fast and cheap, are usually used for local seismic site response in particular in areas characterized by a low rate of seismicity. The results coming from Nakamura technique have, if possible, to be always supported by further analyses, in particular if the considered site does not show 1D configuration (i.e. non negligible influence of surface waves) or it is characterized by low impedance contrast between bedrock and overlapping soft-layer (usually 4 is considered as a lower bound) or it is characterized by an increase of velocity with depth described by a gradient. Detailed information concerning the data processing that the operator have to follow during noise measurements are provided by the SESAME Project guide line (SESAME, 2003).

4.2 HVSR technique

Non reference site technique or single station Horizontal to Vertical Spectral Ratio (Lermo & Chavez-Garcia, 1993), follows the same idea that is at the base of Nakamura technique: in the case of a soft-layer that overlaps a generic stiff bedrock the incident vertical wave field does not undergo significant modification along the whole source to site path with respect to the horizontal one. In this wave supposing a 1D configuration of the considered site, the simple ratio between the Fourier spectrum of the horizontal component and Fourier spectrum of the vertical component (both selected on S-phase) allows us to detect the real response of the site (due to the body wave only). The local site response computed by HVSR might be affected by the window selection: in this case longer in time is the S-phase selected window and more probable is the contamination of other phases, in particular if the site does not well reflect a real 1D configuration. The consequence is a contamination of the vertical component that can lead of consequence to an underestimation of the amplification factor. As well as for Nakamura technique, HVSR results, both in terms of frequency and amplification factor have to be verified (if possible) by those coming from techniques based on the reference site.

4.3 SSR technique

The Reference site technique or Standard Spectral Ratio (Borcherdt, 1970) is one of the most widely approach used to estimate site effects using earthquake recordings. The site response is evaluated by the ratio between the Fourier spectrum calculated on the horizontal (or vertical) component recorded at a generic site of interest (supposed to be a generic soft soil) and the Fourier spectrum of the same component recorded at the reference site (the outcropping rock is assumed as a generic bedrock). The main difficult concerning this method is, at first, the availability of a good quality reference site (i.e. avoiding for example fractured or weathered rock formations) and, at second, the difficult to have a relevant number of good quality signals recorded at the same time by the stations of interest and by the reference site: usually this condition arose in correspondence of very urbanized areas characterized by a high level of background noise and at the same time by a low rate of seismicity. Of consequence this is the case where it is faster and easier to apply the non reference site technique previous described.

5. Analyses from seismic noise: Results

Figure 4 and 5 show examples of results deduced from seismic noise recorded at the considered stations. Data were analyzed by Nakamura technique as explained in the previous paragraph. In this case, with the aim to also investigate possible polarization phenomena affecting the wave-field propagation, NHVSRs were calculated rotating the NS component from 0° to 180°, by step of 5°. In this way for each of the 36 obtained directions the ratio between the Power Spectral Density calculated for the horizontal and for the vertical components were calculated. In the left panels of Figure 4 the directional amplification functions obtained for stations 1, set on morainic deposits, station 2, set on hundred meters of alluvial deposits, and station 3, installed on rock, are shown. In this case blue and green lines represent the results considering as input seismic noise recorded during winter and summer time respectively. In each panel the single amplification functions represent the average NHVSR calculated considering 120s or 60s signal (for broad band and semi broad band sensors respectively) recorded in different time during the 24 h (night and day) and in different days of the year (seasonal variations). Each window of signal was processed as explained in the previous paragraph.

Station 1 (Figure 4, top panels), installed in correspondence of a site that approximates enough a 1D configuration, exhibits a clear amplification peak at frequency between 2 and 3 Hz that (also considering the results showed in the next paragraphs) reflects the site response of the site. On the contrary the very narrow peak at 1 Hz reflects probably amplification of cultural noise typical of Northern Italy region (Marzorati & Bindi, 2006). The site does not show preferential direction of amplification (see polar plot in the top right panel of figure 4, but also the low dispersion of amplification functions) and no differences between results obtained considering signals recorded during winter and summer.

On the contrary station 2 (Figure 4, middle panels) shows a low frequency amplification peak (around 0.2 Hz) clearly dependent of seasonal cycles, being the peak detected in winter more amplified with respect to the same related to summer: this phenomena is explained with microseisms amplification due to meteorological perturbations during winter (Marzorati & Bindi, 2006). The right middle panel of figure 4 shows a slight polarization effects at low frequency direction 60°-240°.

The station 3 (figure 4, bottom panels), being installed on rock, does not show particular evidence of amplification.

Station 4 (figure 5, top panels), even if from a geological point of view has characteristic similar to station 3, shows a response that undergoes the influence of topography. One of the main marker for the presence of topographic effects is a strong polarization of the amplification function, more evident if the morphology has a clear 2D configuration. The figure shows two peaks: the first around 5 Hz clearly polarized in 5°-185° (see the polar plot in the top right panel of figure 5) and the second between 7-8 Hz without particular preferential directions of amplification. Being the station 4 installed in a tower of an ancient stronghold, as already demonstrated in Massa et al. (2010) , the peak around 8 Hz describes a typical case of interference between ground motion and the vibration of the structure. Bottom panels of Figure 5 depict the results for NRN7 station, located at the top of Narni ridge (central Italian Apennines). In this case distinction between noise recorded during winter and in summer was not possible because of the field experiment was exploited from the end of March to September 2009. However, it is worth mentioning that the long period fluctuations of seismic noise, as demonstrated also from the analyses on station 2, are only detectable in particular configurations (e.g. wide alluvial basin) able to show low frequency responses. On the contrary, topographies are structures that usually are characterized by short predominant period of vibration (< 1 s).

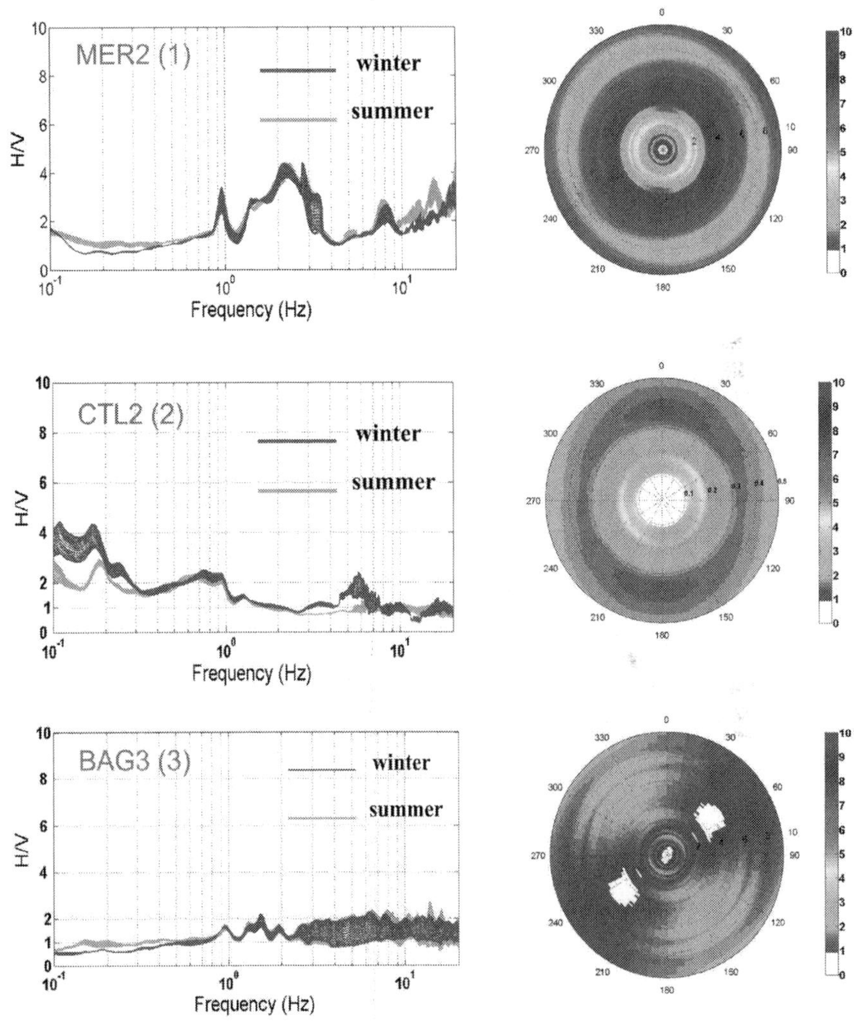

Fig. 4. Left panels: directional NHVSRs for station 1 (morainic deposits), 2 (deep alluvial deposits) and 3 (rock). Blue and green lines represent the average NHVSRs calculated for winter and summer noise recordings. Right panels: corresponding polar plots.

The main evidence for station 5 is a typical response of a 2D elongated ridge showing an amplification peak around 4 Hz clearly polarized in EW direction, that represents the azimuth perpendicular to the main elongation of the ridge (NS).

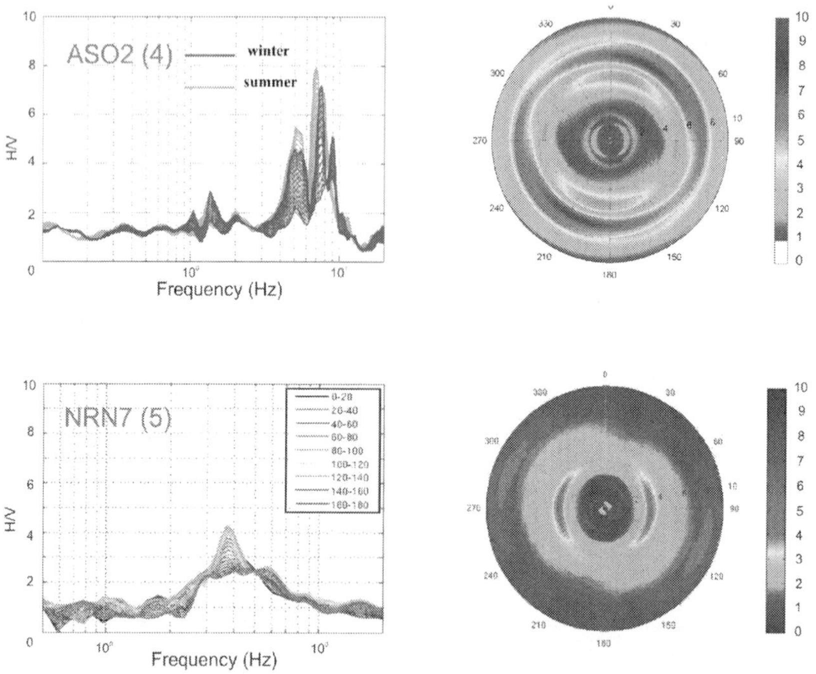

Fig. 5. Left panels: directional NHVSRs for stations 4 (limestone and dolomite 3D hill) and 5 (limestone 2D ridge). In the top panel blue and green lines represent the average NHVSRs calculated for winter and summer noise recordings, while in the bottom each different colour refer to groups of different directions (step of 20°). Right panels: corresponding polar plots.

6. Analyses from local earthquakes: Results

In this paragraph HVSR and SSR (only for station 5) results, obtained considering local earthquakes recorded by the four RAIS stations (stations 1, 2, 3 and 4) and by station 5 (NRN7), are presented and discussed in comparison to those obtained from seismic noise analyses.

Figure 6 and 7 show the results, in terms of directional HVSRs, obtained for all analysed stations. In the left panels of figure 6 and in the left top panel of figure 7 each single blue, green and red line represents the average amplification functions (HVSRs) obtained considering windows of 5 s and 15 s on S-phase and 20 s on coda. For each station the data set of events showed in figure 1 and 2 was considered. In particular, the averaged amplification functions were calculated for different azimuths, by rotating the NS horizontal component between 0° and 180° (by step of 5°).

The bottom left panel of figure 7 is dedicated to the results obtained for station 5, also in this case showed in term of averaged directional HVSRs calculated on 10 s of S-phase. All right panels of figure 6 and 7 show the related polar plot calculated for 15 s of S-phase.

Station 1 (Figure 6, top panels), in agreement to the results obtained from noise analyses, confirms an amplification peak at frequency around 2 Hz without differences considering the three selected windows of signal, even if HVSRs on coda seem to slight underestimate the amplification factor (probable contamination of seismic noise involving the coda). The corresponding polar plot does not show particular preferential directions of amplification.

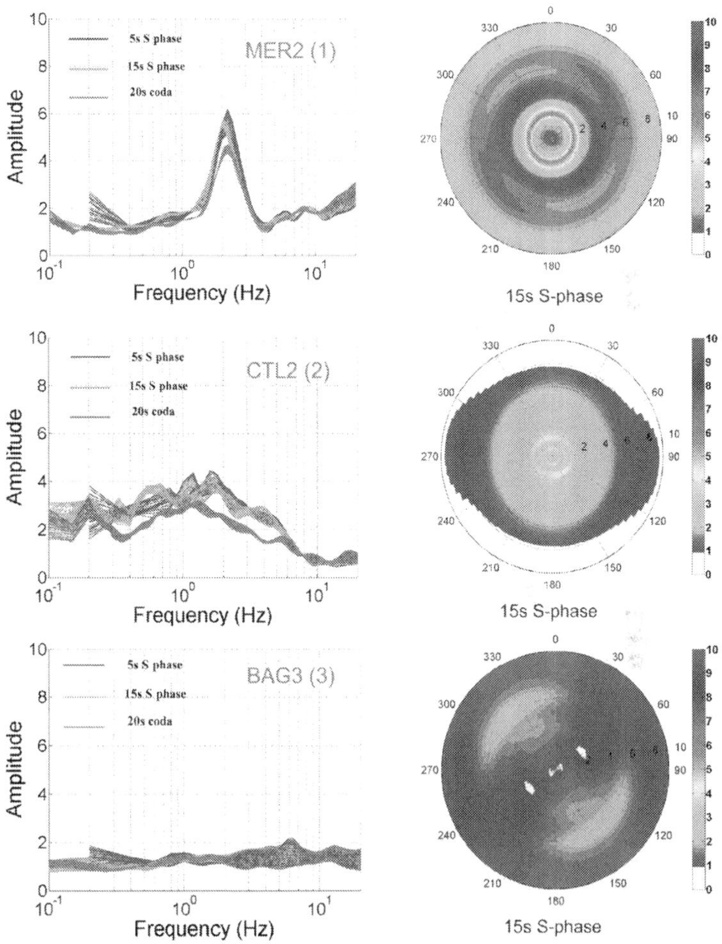

Fig. 6. Left panels: directional HVSRs for station 1 (top), 2 (middle) and 3 (bottom) calculated considering 5 s (blue) and 15 s (green) of signal selected on S-phase and 20 s (red) of signal selected on coda. Right panels: polar plots calculated for 15s of S-phase.

More complicated is the interpretation of the results for station 2 (Figure 6, middle panels), installed in correspondence of thick alluvial deposits (the stratigraphy is characterized by an absence of relevant impedance contrasts). In general, in particular considering S-phase selections, the results disagree with those obtained from noise, highlighting the presence of slight broad peak of amplification at frequencies between 1 and 2 Hz. Also in this case the analysis on coda shows the lower amplification factors but at the same time is able to reproduce the low frequency peak around 0.2 Hz; considering both the low magnitude of events and the very urbanized area where the station is installed, it is obvious that, being the station characterized by a high level of background noise, the coda of the events is characterized by a predominant percentage of noise itself. In this case the polar plot does not show preferential directions of amplification.

Bottom panels of figure 6 show the results for station 3, that being installed on rock, shows, in agreement to the results obtained from noise analyses, flat HVSRs in the whole frequency range.

Top panels of figure 7 show the results for station 4, installed at the top of a 3D hill. Excluding the peak around 8 Hz, due to the soil-structure interaction (discussed in the previous paragraph), the HVSRs results highlight the presence of two main amplification peaks, the first around 2.5 Hz and the second, already showed in the noise analyses, at 5 Hz. Concerning the second peak, polarized, as already pointed out from noise, in NS direction (5°-185°), the phase selection appears to be not influent on final results. Regarding the peak around 2 Hz (less polarized), the analysis on coda underestimates, in agreement to the results obtained from noise, the amplification factor and at the same time slightly moves the resonance frequency towards lower frequencies.

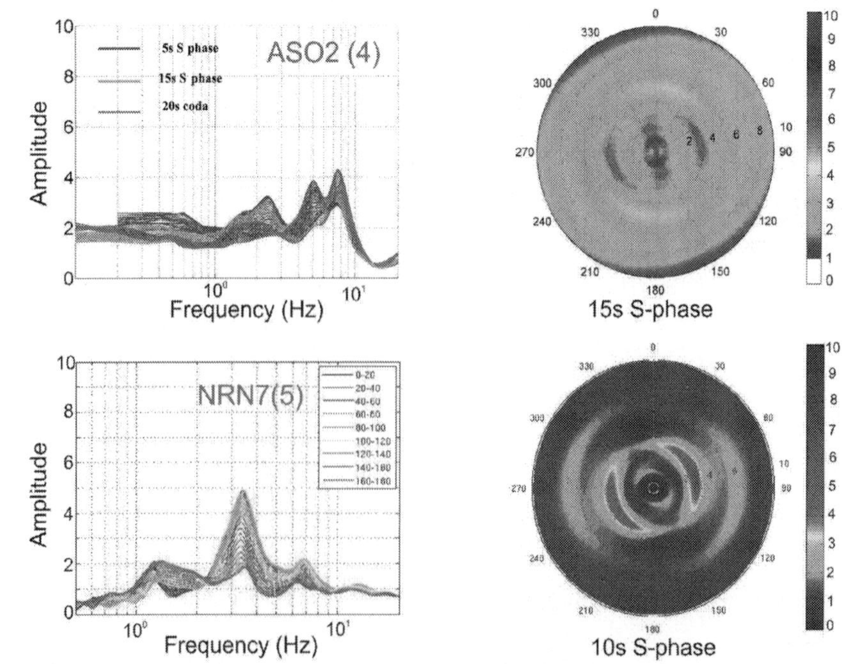

Fig. 7. Top left panels: directional HVSRs for station 4 calculated considering 5 s (blue) and 15 s (green) of signal selected on S-phase and 20 s (red) of signal selected on coda. Top right panels: polar plot calculated at station 4 for 15s of S-phase.

The bottom panels of figure 7 show the results for station 5 (NRN7), installed at the top of a pronounced 2D topography. In this case the main cause for ground amplification is clear and of consequence is not surprising the agreement between the results obtained from noise or earthquakes. Also in this case the figure highlights a clear amplification peak between 3 and 4 Hz strongly polarized (difference in amplification factors up to 3) in the EW direction that is perpendicular to the main elongation of the morphology.
Being Narni ridge monitored by ten temporary velocimetric stations during 2009 (Massa et al., 2010), we have the possibility to compare, at station 5, the HVSR results to those obtained by SSRs technique (in the bottom panel of figure 2 the location of the reference station, NRN2, is shown). Figure 8 shows the SSR results considering for NRN7 (top) and NRN2 (bottom) 29 events with M_L between 1.5 and 3.6 and epicentral distance up to 30 km.
In this case each SSR curve represents the averaged directional spectral ratio between the Fourier spectra of the considered station and that calculated for the reference (the meaning of the different colours showed in the figure 8 is the same as explained for HVSRs).
As it is possible to note, SSRs highlight the presence of a peak around 2 Hz, not showed by single station HVSR techniques and also a slight shift in frequency, towards higher values, concerning the second peak (between 4 and 5 Hz). Also the polarization analyses, even if the higher values are detected for azimuth about perpendicular to the elongation of the ridge, are slightly different with respect to those obtained from HVSRs. In this case, with respect to the other described techniques, being a reference site available and of consequence the amplification functions more approximable to theoretical transfer functions, it is possible to make reliable consideration also about the amplification factor of the site: for the investigated station the amplification value was found to be around 4.
Bottom left panel: directional HVSRs for stations 5 calculated on 10 s of S phase selected considering only events with $M_L < 3$ and distances lower than 50 km (see bottom panel of figure 2): different colours refer to groups of different directions (step of 20°). Bottom right panels: corresponding polar plots.

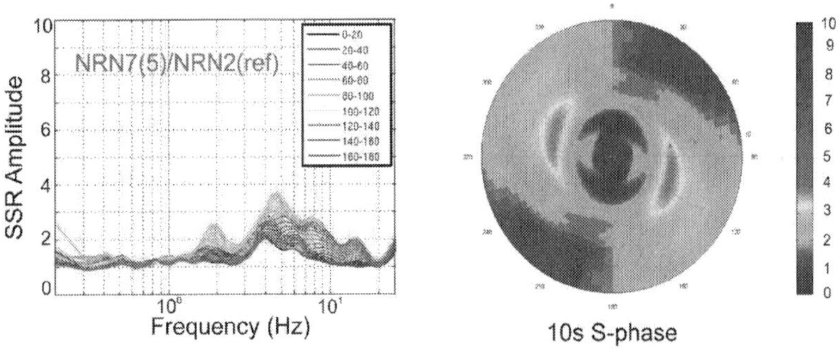

Fig. 8. Left: directional SSRs performed for station 5 (NRN7) considering 10s of S-phase selected on weak motion (29 events with M_L between 1.5 and 3.6 and epicentral distance up to 30 km). NRN2 (see bottom panel of figure 2) is the reference station. Right: related polar plot.

Finally, being the four RAIS stations (1, 2, 3 and 4) characterized both by accelerometric and velocimetric sensors, installed one close to the other, the reliability of a strong-motion sensor

to evaluate the site response was investigated in term of rotational HVSRs comparing, for station 1, the results obtained using, as input, the recordings coming from different type of instruments. The analyses were computed considering the same data set (figure 1) and the same selection of windows.
Figure 9 shows the HVSR results for station 1 considering the strong motion sensor. In this case, the coupling strong-motion sensor vs. digitizer assuring, in the frequency of interest, a good signal to noise ratio, the HVSR results well reflect those obtained considering the broad band velocimetric sensor (see top panels of figure 6), both in terms of amplified frequency and amplification factors. Also in term of difference among considered phases and polarization effects, in this case the performance of the considered sensor appears to be very similar. In general problem might be arose, in particular at low frequencies, when the same sensor is coupled to digitizer characterized by different dynamic range, as discussed in the following ad-hoc paragraph.

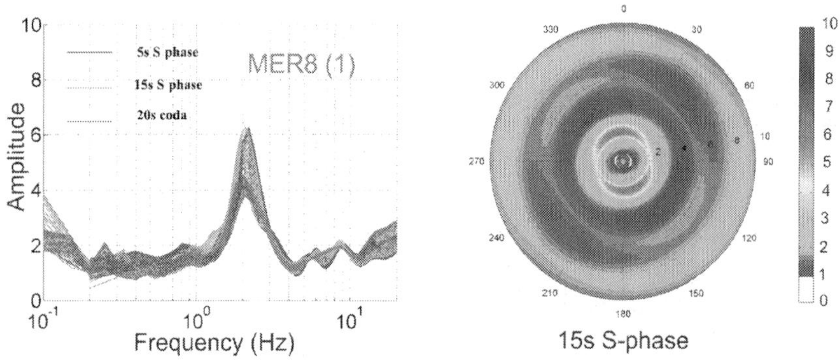

Fig. 9. Left: directional HVSRs calculated for station 1 considering the data set showed in the top right panel of figure 1 but recorded by the strong-motion sensor (Kinemetrics Episensor), for 5 s (blue) and 15 s (green) of signal selected on S-phase and 20 s (red) of signal selected on coda. Right: polar plots calculated for 15s of S-phase.

6.1 Comparison between HVSR and SSR techniques in seismic site evaluation

Data collected during the Narni experiment (Massa et al., 2010) give us the opportunity to verify the reliability of HVSR and SSR results for site response evaluation. For this purpose we collected a data set composed by the strongest (M_L > 4.0, see right bottom panel of figure 2) aftershocks of the 6th April 2009, Mw 6.3, L'Aquila mainshock (Ameri et al., 2009). Figure 10 shows a directional HVSRs calculated for the reference station (NRN2, top panel) and for the station 5 (NRN7, bottom panel) and the related directional SSRs (bottom panel). Even in this case amplification functions with different colours indicate all investigated azimuths, as indicated in the legend. This example allows us to point a warning in the use of only techniques without reference site in the estimation of site response. HVSRs reported in the top and middle panel of figure 10 show in particular the presence, for both stations, of a low frequency peak around 0.6 Hz (see grey area in the figure). The SSR obtained between NRN7 and NRN2 reported in the bottom panel gives completely different results and/or interpretation: the low frequency peak, suffered also by the reference station (that on the contrary shows flat response considering noise and local data, non showed here), disappears in the SSR analysis, that, on the contrary well highlights, at station 5, the peak around 4 Hz due to the topography response, in good agreement to HVSRs computed both on noise and local events (see figures 5 and 7, bottom panels). Even if the interpretation about the origin of the low frequency peak is difficult to explain (probably non correlated to the site response), this example points out a warning about the use of just HVSR (both in frequency and amplitude) in the estimation of site response: the results of a single station analyses can be biased by other phenomena that might mask the real seismic local amplification due to the site.

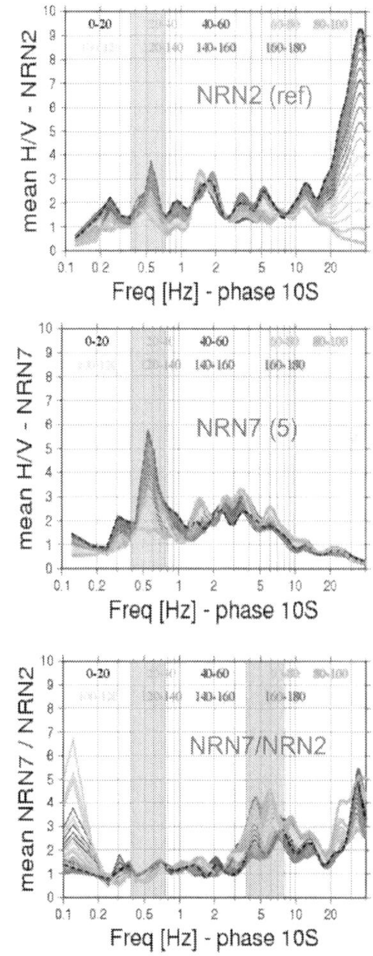

Fig. 10. Directional HVSRs obtained for Narni reference site (NRN2, top panel) and for the investigated station 5 (NRN7, middle panel) and related SSRs (bottom panel). Results were obtained analysing 10 s of S-phase selected on the strongest aftershock of L'Aquila sequence.

7. Analyses from teleseisms: results

In order to provide further data to use for site response analyses in areas characterized by a low rate of seismicity, HVSRs were calculated also considering the recordings related to four teleseisms: 13th January 2007, Mw 8.2, Kurili earthquake (figure 11, left panel), 25th March 2007, Mw 5.8, Greece earthquake, 8th June 2007, Mw 6.2, Greece earthquake (figure 11, right panel), 12th September 2007, Mw 7.9, Sumatra earthquake. In the last years some papers (Riepl et al., 1998, Dolenc & Dreger, 2005, Ferretti et al., 2007) demonstrated the capability of teleseisms to well predict the frequency response of a site in the range where the recordings are characterized by a good signal to noise ratio (usually up to about 2 Hz). In this way directional horizontal to vertical spectral ratio of teleseismic recordings (calculated by rotating the NS component from 0° to 180°, by step of 5°) are computed selecting 80 s P and S phases and the results are compared with the same non-reference site technique applied to local earthquakes (Lermo & Chavez-Garcia, 1993) and seismic noise (Nakamura, 1989).

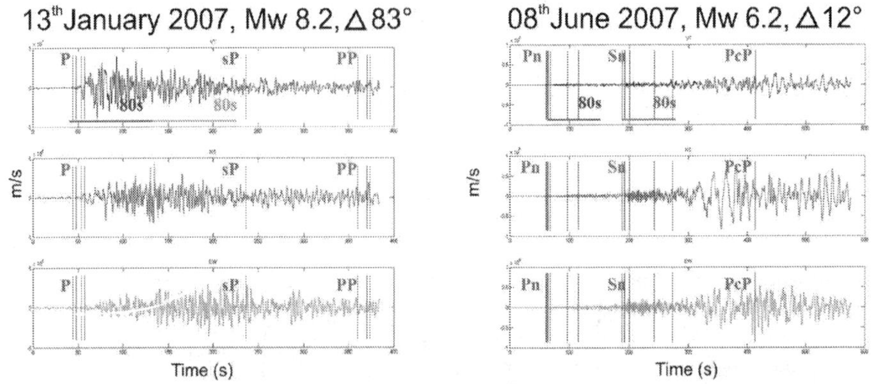

Fig. 11. Teleseismic waveforms (top, middle and bottom are vertical, NS and EW components) related to the 13th January 2007, Mw 8.2, Kurili earthquake (left) and 8th June 2007, Mw 6.2, Greece earthquake (right) recorded at station 2 by the broad-band sensor (Nanometrics Trillium 40 s). The selected portions of windows for HVSR analyses are also indicated.

The HVSR were calculated for station 1 (figure 12, right panels), were a clear target peak was detected both by noise and local events and for station 2 (figure 12, left panels) characterized, in general, by low frequency responses.

In particular one scope was to verify possible improvement concerning the resolution at low frequency (< 1 Hz) for station 2, installed in the central part of a wide alluvial basin. The processing was computed as well as for noise and local earthquake and also in this case the influence of the azimuth on the amplification was investigated.

Regarding station 2 it is clear as HVSRs calculated on the considered teleseisms well agree to the response obtained by noise and local events: the analyses highlight a clear peak around 2 Hz. In this case considering each single event, it is possible to note that non particular differences are detected considering P or S-phases. The level of amplification in this case is probably more influenced by each single source to site path. For this site ratio higher than 3 Hz simply reflects the behaviour of background noise. Even in this case the results for station 2 are more complicated, where, on the base of signal to noise ratio, only considerations up to 1.5 Hz are possible. In this case while Kurili and Sumatra events do not show particular evidences, the two Greece earthquakes better highlight the presence of amplification around 0.2 Hz, detected both on P and S-phase. In particular, even if the main peak was detected considering the P-phase of the "Greece 1" event, in general considering a broader band of frequency (up to 1 Hz), the S-phase for both earthquakes appears more amplified with respect to P-phase (bottom left panel of figure 12).

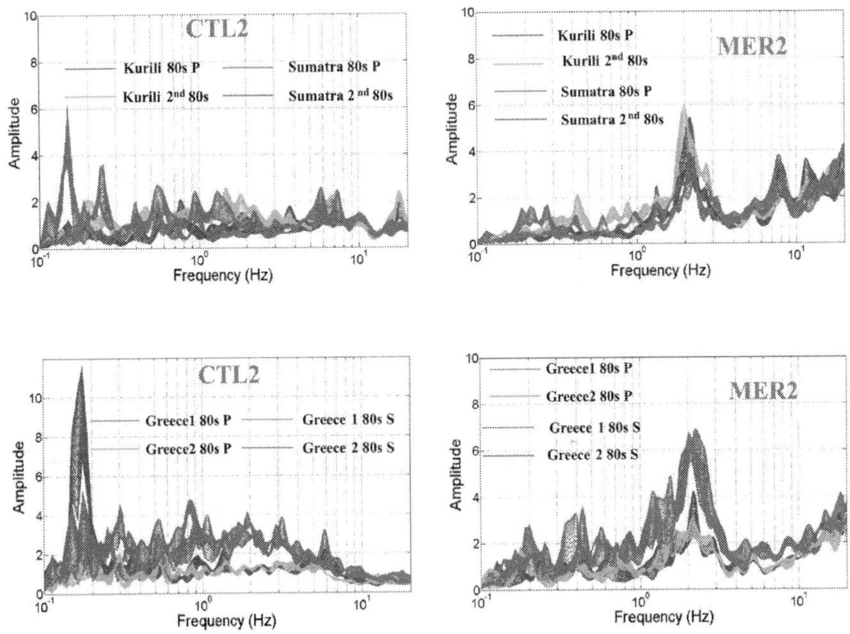

Fig. 12. Directional HVSRs calculated for 13th January 2007, Mw 8.2, Kurili earthquake, for 12th September 2007, Mw 7.9, Sumatra earthquake, for 25th March and 8th June 2007, Mw 5.8 and Mw 6.2, Greece earthquakes (indicated in the bottom panels as "Greece 2" and "Greece 1" respectively) recorded at stations 2 (left) and 1 (right).

8. Open issues about site effects evaluation

In this paragraph the attention is addressed to two common situations that can be encountered during the execution of experimental measurements for site response and consequently can affect the results: the soil structure interaction and the minimum resolution assured by the used instrumentation.

8.1 Soil-structure interaction

Soil structure interaction is a phenomenon that arises if the predominant period of vibration of a structure where the seismic sensor is installed covers the frequency response of the related site. As demonstrated in Massa et al. (2010) this phenomenon is able to bias the recordings if the sensor is directly connected to the foundation of a building; on the contrary, even if the sensor is installed inside a structure but directly connected to the ground the soil-structure interaction decreases a lot. Concerning the investigated site, station 4, installed at the base of the ancient stronghold gives recordings that show example of soil structure interaction.

The fortress, characterized by a rectangular-shaped with side-length of about 50 and 30m respectively is elongated in the NE–SW direction, very similar to the orientation of the hill (azimuth of about 45°N). Close to the NE corner of the structure is present a tower, where the station 4 is installed.

Figure 13 shows the directional NHVSR obtained from two contemporary seismic noise measurements (at least 30 minutes) performed at the base (in proximity of station 4, Aso-1, green lines) and at the top (Aso-2, blue lines) of the structure. The data processing was performed as explained in the previous paragraphs. On the basis of the results showed for Aso-2 measurement, it is possible to suppose that the ground shaking recorded at the base is probably influenced by the free oscillation of the housing structure, in particular at frequency where no particular polarization phenomena are detected (between 7 and 9 Hz). Bottom panels of figure 13 show the polar plots corresponding to Aso-1 (on the right) and Aso-2 (on the left). Aso-2 polar plot highlights directions of preferential amplification polarized in different ways: the first peak, around 5 Hz, in NS direction (it can be also noticed considering the polar plot for Aso-1, on the right), the second, around 7 Hz , about 150°-330° while the third, around 9 Hz, about 90°-270°.

8.2 Comparison between velocimeter and accelerometer

The second issue regards the minimum resolution that the coupling between a sensor and a digitizer is able to assure. This fact it is important in particular when the site response can be evaluated only by using Nakamura technique. In same cases the instrument resolution is not able to correctly resolve the real background noise (in particular for site characterized by a low level of noise). In order to verify how much the signal recorded by a velocimeter and by an accelerometer performs during noise measurements, station 1, where two 20 bits Lennartz Mars88 are coupled to an accelerometer (Kinemetrics Episensor) and to a broad band sensor (Nanometric Trillium 40 s) respectively are installed, was investigated. The Probability Density Function (PDF) of the Power Spectral Density (PSD) of noise window (1 minute long in time) were calculated by Mc Namara & Buland method (2004) and compared to the Peterson noise curve (figure 14) and the spectra of 5 local weak events recorded in the first 50 km. Figure 14 shows as, only the noise recordings from broad band sensor are able to well reproduce the trend of real background noise at the site while the coupling of a Lennartz Mars88 to a Kinemetrics Episensor is not able to assure at this site reliable noise recordings for frequencies up to 10 Hz. In this range we have clearly recorded the instrumental noise (flat PDF), while for frequencies higher than 10 Hz the trend is similar to those shown for the coupling digitizer and velocimeter, assuring real noise recordings. Concerning the local events behaviour the figure shows that using accelerometric data we obtain unreliable results for frequencies lower than 1 Hz: this is the reason for what the HVSR peak at 2 Hz obtained for station 1 (top panel of figures 6 and 9) is coherent, even if it is calculated using data from weak or strong motion sensor.

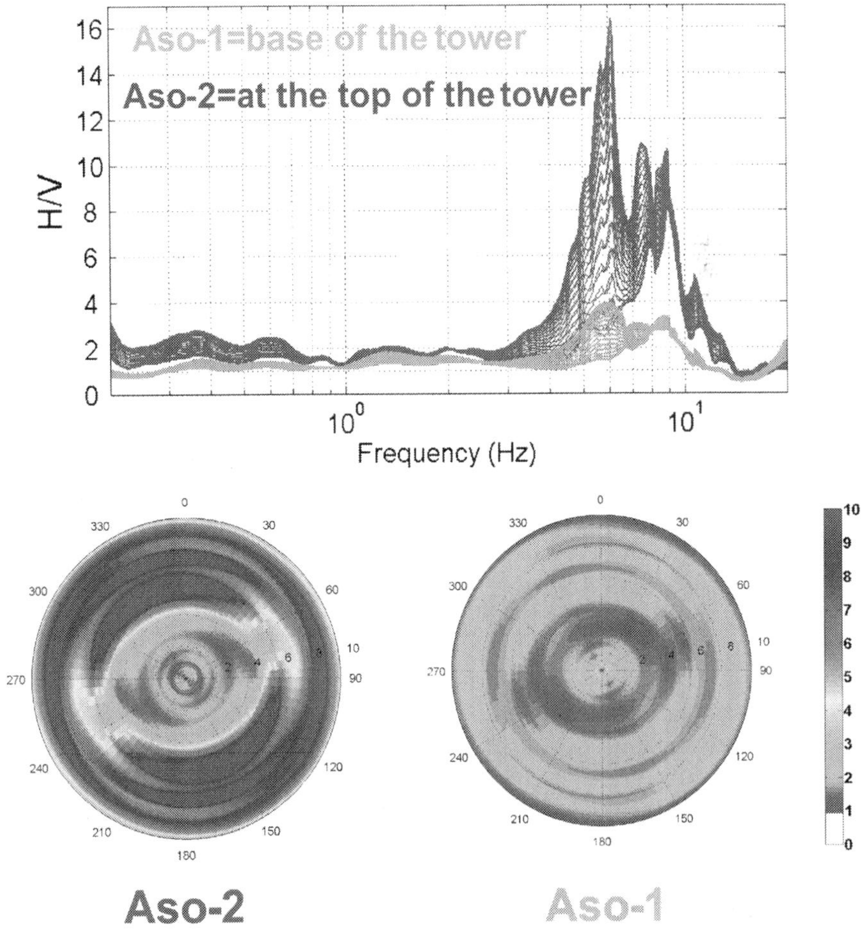

Fig. 13. Top panel: directional NHVSR results obtained from synchronized noise measurements performed, the first (green lines), where the sensors of station 4 are installed (Aso-1, at the base of the tower) and the second (blue lines) at the top of the structure (Aso-2, at the top of the tower). Bottom panels: corresponding polar plots.

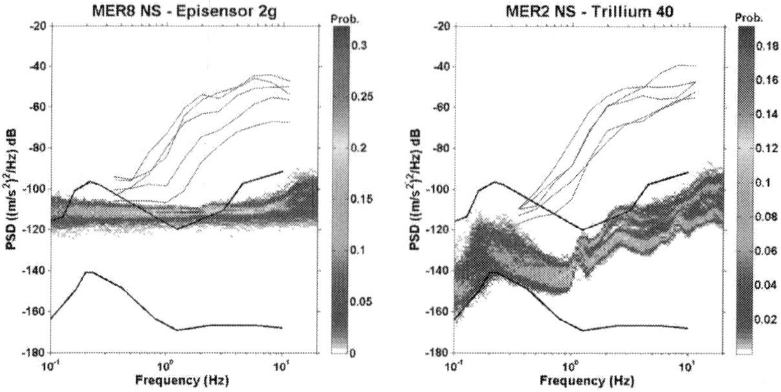

Fig. 14. Probability Density Function calculated averaging Power Spectral Density calculated from many windows of seismic noise (1 minute) recorded at station 1 by a Lennartz Mars88 digitizer coupled to a Kinemetric Episensor (left) and a Nanometrics Trillium 40s (right). In figure also the New High Noise Level and New Low Noise Level Peterson curves are reported (thick black lines). The thin black lines are spectra calculated for weak events selected as example.

9. Conclusions

The present chapter focuses the attention on the evaluation of the capabilities of the most common experimental methods used in seismology to estimate the ground motion amplification due to different geological or geomorphological features of a site.

Different techniques were evaluated: single station Horizontal to Vertical Spectral Ratio on noise (Nakamura technique, Nakamura 1989), single station Horizontal to Vertical Spectral Ratio on earthquakes (Lermo & Chavez Garcia, 1993) and Standard Spectral Ratio technique (Borcherdt ,1970).

While the first two approaches do not take into account a reference site, the last is based on signals contemporary recorded both in the site of interest and in correspondence of a station installed on hard rock (approximably to a bedrock).

In order to investigate a broad frequency responses due to different stratigraphy and morphological setting, 5 Italian sites where seismic stations managed by the Italian National Institute for Geophysics and Volcanology (INGV, department of Milano-Pavia) are installed, were investigated using seismic noise, local (weak motions) and teleseismic events occurred in the last years.

The general considerations, deduced from comparisons made in terms of HVSR and SSR might be summarized as follow:

noise measurements are a cheap and quick tool for seismic site response under particular conditions such as a simple configuration of the site of interest (similar to 1D model) characterized by a high impedance contrast between the soft soil layer and the bedrock (in theory at least > 4).

In general noise measurements give information about the first resonance frequency of the site and tend to underestimate the amplification function obtained from earthquakes (on S-phase) at the same site.

Amplification functions obtained from noise are more similar to those obtained analyzing the coda of events: in any case no particular consideration about amplification factor is possible in absence of a reference site.

On the basis of the results reported in figure 14, accelerometric sensor is not a good choice for noise analyses, being the results strongly dependent on the available instrumentation.

In a simple configuration and for frequency higher than 1 Hz there is a good agreement between HVSR results coming from accelerometric and velocimetric data recorded at the same site using the same digitizer. More complicated are the interpretations for stations where possible influence of other phases (i.e. surface waves) are present, such as stations installed in correspondence of alluvial basin.

The good agreement in terms of HVSR obtained comparing local events and teleseisms indicates that for regions characterized by low rate of seismicity, but potentially able to suffer energetic events (such as Northern Italy), the use of teleseisms can give a further improvement to the analyses at low frequency (usually lower than 2 Hz, but however depending on the noise level of each analysed site).

As demonstrated in the example of figure 10, the only use of HVSR, in particular for complicated settings such a topography, can lead to completely biased interpretations: in general, if possible, the use of a reference site technique is strongly encouraged.

If a field experiments is performed in correspondence of urban areas the results can be biased by possible soil-structure interactions, in particular if the sensors are installed inside buildings and directly connected to their foundations.

10. References

Amanti, M.; Bontempo, R., Cara, P., Conte, G., Di Bucci, D., Lembo, P., Pantaleone, N.A. & Ventura, R. (2002). Carta Geologica d'Italia Interattiva 1:100,000 (Interactive geological map of Italy, 1:100,000), SGN, SSN, ANAS, 3cd-rom.

Ameri, G.; Massa, M., Bindi, D., D'Alema, E., Gorini, A., Luzi, L., Marzorati, M., Pacor, F., Paolucci, R., Puglia, R. & Smerzini, C. (2009). The 6 April 2009, Mw 6.3, L'Aquila (Central Italy) earthquake: strong-motion observations, *Seismological Research Letters*, Vol. 80, No. 6, pp. 951-966.

Asten, M.W. (1978). Geological control of the three-component spectra of rayleigh-wave microseisms. *Bull. Seism. Soc. Am.*, Vol. 68, No. 6, pp. 1623-1636.

Athanasopoulos, G. A.; Pelikis, P. C. & Leonidou, E. A. (1999). Effects of surface topography on seismic ground response in the Egion (Greece) 15-6-1995 earthquake. *Soil Dynamics and Earthquake Engineering*, Vol. 18, No. 2, pp. 135-149.

Bard, P. Y. (1982). Diffracted waves and displacement field over two-dimensional elevated topographies. *Geophysical Journal Int.* , Vol. 71, No. 3, pp. 731-760.

Bard, P. Y. (1998). Microtremor measurement: a tool for site effect estimation? In: Second International Symposium on the Effects of the Surface Geology on the Seismic Motion, EGS98, Japan.

Bindi, D.; Parolai, S., Spallarossa, D. & Cattaneo, M. (2000). Site effects by H/V ratio: comparison of two different procedures, *J. Earthquake Eng.*, Vol. 4, No. 1, pp. 97–113.

Borcherdt, R.D. (1970). Effects of local geology on ground motion near San Francisco Bay. *Bull. Seism. Soc. Am.*, Vol. 60, pp. 29-61.

Bordoni, P.; De Rubeis, V., Doumaz, F., Luzi, L., Margheriti, L., Marra, F., Moro, M., Sorrentino, D. & Tosi, P. (2003). "Geological class map", In: *Terremoti probabili in Italia tra l'anno 2000 e 2030: elementi per la definizione di priorità degli interventi di riduzione del rischio sismico*, Annex 1, Task 3.2, pp. 3-4, GNDT Proj., Rome.

Caserta, A.; Bellucci, F., Cultrera, G., Donati, S., Marra, F., Mele, G., Palombo, B. & Rovelli, A. (2000),Study of site effects in the area of Nocera Umbra (Central Italy) during the 1997 Umbria-Marche seismic sequence, *J. of Seismology*, Vol. 4, No. 4, pp. 555-565.

CEN (Comité Européen de Normalisation) (2004). Eurocode 8: Design of structures for earthquake resistanc - Part 5: Foundations, retaining structures and geotechnical aspects. Brussels, Belgium.

Dolenc, D. & Dreger, D. (2005). Microseismic observations in the Santa Clara Valley, California, *Bull. Seism. Soc. Am.*, Vol. 95, No. 3, pp. 1137–1149.

Donati, S.; Marra, F. & Rovelli, A. (2001). Damage and ground shaking in the town of Nocera Umbra during Umbria-Marche, central Italy, earthquakes: the special effect of a fault zone. *Bull. Seism. Soc. Am.*, Vol. 91, No. 3, pp. 511-519.

Faccioli, E.; Vanini, M. & Frassine, L. (2002). "Complex" Site Effects in Earthquake Ground Motion, including Topography. *12th European Conference on Earthquake Engineering*, Barbican Centre, London, UK.

Ferretti, G.; Massa, M., Isella, L. & Eva, C. (2007). Site amplification effects based on teleseismic wave analysis: the case of Pellice Valley (Piedmont, Italy), *Bull. Seism. Soc. Am.*, Vol. 97, No. 2, pp. 605-613.

Géli, L.; Bard, P. Y. & Jullien, B. (1988). The effect of topography on earthquake ground motion : a review and new results, *Bull. Seism. Soc. Am.*, Vol. 78, No. 1, pp. 42-63.

Gutenberg, B. (1958). Microseisms. *Advan. Geophys.*, Vol. 5, pp. 53-92.

Kallou, P.V.; Gazetas, G. & Psarropoulos, P.N. (2001). A case history on soil and topographic effects in the 7th September 1999 Athens earthquake, *Proceedings of 4th Int. Conf. on Recent Advances in Geotechnical Earthquake Engineering and Soil Dynamics*, San Diego, California.

Lachet, C. & Bard, P. Y. (1994). Numerical and theoretical investigations on the possibilities and limitations of Nakamura's technique, *J. Phys. Earth*, Vol. 42, pp. 377–397.

LeBrun, B.; Hatzfeld, D., Bard, P.Y. & Bouchon, M. (1999). Experimental study of the ground motion on a large scale topographic hill al Kitherion (Greece), *J. of Seismology*, Vol. 3, pp. 1-15.

Lermo, J. & Chavez-Garcia, F. J. (1993). Site effect evaluation using spectral ratio with only one station, *Bull. Seism. Soc. Am.*, Vol. 83, No. 5, pp. 1574-1594.

Lovati, S.; Bakavoli, M.K.H., Massa, M., Ferretti, G., Pacor, F., Paolucci, R., Haghshenas, E. & Kamalian, M. (2011). Estimation of topographical effects at Narni ridge (Central Italy): comparisons between experimental results and numerical modelling, submitted to *Bull. of Earthquake Engineering*.

Marzorati, S. & Bindi, D. (2006), Ambient noise levels in North-central Italy, *Geochem. Geophys. Geosyst.*, Vol. 7, Q09010, 14 pp., ISSN 1525-2027.

Marzorati, S.; Ladina, C., Falcucci, E., Gori, S., Saroli, M., Ameri, G. & Galadini, F. (2011). Site effects "on the rock": the case of Castelvecchio Subequo (L'Aquila, Central Italy), *Bull. of Earthquake Engineering*, Vol. 9, No. 3, pp. 841-868.

Massa, M.; Ferretti, G., Cevasco, A., Isella, L. & Eva, C. (2004). Analysis of site amplification phenomena: an application in Ripabottoni for the 2002 Molise, Italy, earthquake, *Earthquake Spectra*, Vol. 20, Issue S1, pp. S107-S118.

Massa, M.; Marzorati, S., Ladina, C. & Lovati, S. (2010). Urban seismic stations: soil-structure interaction assessment by spectral ration analyses, *Bulletin of Earthquake Engineering*, DOI 10.1007/s10518-009-9138-1, Vol. 8, No. 3, pp. 723-738.

Massa, M.; Lovati, S., D'Alema, E., Ferretti, G. & Bakavoli, M. (2010). An experimental approach for estimating seismic amplification effects at the top of a ridge, and the implication for ground-motion predictions: the case of Narni (central Italy), *Bull. Seism. Soc. Am.*, Vol. 100, No. 6, pp. 3020-3034.

McNamara, D.E. & Buland, R.P. (2004). Ambient noise levels in the Continental Unites States, *Bull. Seism. Soc. Am.*, Vol. 94, No. 4, pp. 1517-1527.

Nakamura, Y. (1989). A method for dynamic characteristics estimations of subsurface using microtremors on the ground surface, *Quarterly Rept. RTRI Japan*, Vol. 30, pp. 25-33.

NTC (Nuove Norme Tecniche per le Costruzioni) (2008). Part 3: Categorie di sottosuolo e condizioni topografiche, *Gazzetta Ufficiale della Repubblica Italiana*, No. 29 del 4 febbraio 2008.

Parolai, S.; Bormann, P. & Milkereit, C. (2001). Assessment of the natural frequency of the sedimentary cover in the Cologne area (Germany) using noise measurements, *J. Earthquake Eng.*, Vol. 5, No. 4, pp. 541-564.

Parolai, S.; Richwalski, S.M., Milkereit C. & Bormann, P. (2004). Assessment of the stability of H/V spectral ratios from ambient noise and comparison with earthquake data in the Cologne area (Germany), *Tectonophysics*, Vol. 390, No. 1-4, pp. 57-73.

Peterson, J. (1993). Observation and modeling of background seismic noise. Open File Report 93-322, USGS, Albuquerque, New Mexico.

Regione Lombardia (2002). Progetto cartografia geoambientale, 1:25000 vers. 1.0.

Regione Lombardia (2003). Progetto cartografia geoambientale, 1:25000 vers. 1.0.

Riepl, L.; Bard, P.Y., Hatzfeld, D., Papaioannou, C. & Nechtschein, S. (1998). Detailed evaluation of site response estimation methods across and along the sedimentary Valley of Volvi (EURO-SEISTEST), *Bull. Seism. Soc. Am.*, Vol. 88, No. 2, pp. 488-502.

Rovelli, A.; Caserta, A., Marra, F. & Ruggiero, V. (2002). Can seismic waves be trapped inside an inactive fault zone? The case study of Nocera Umbra, central Italy, *Bull. Seism. Soc. Am.*, Vol. 92, No. 6, pp. 2217-2232.

SESAME, Site Effects Assessment Using Ambient Excitations (2003). European Commission Research General Directorate. Project No. EVG1-CT-2000-00026 SESAME. Final report WP08 - Nature of noise wavefield.

Strollo, A.; Richwalski, S. M., Parolai, S., Gallipoli, M. R., Mucciarelli, M. & Caputo, R. (2007). Site effects of the 2002 Molise earthquake, Italy: analysis of strong motion, ambient noise, and synthetic data from 2D modelling in San Giuliano di Puglia, *Bull. Earth. Eng.*, Vol. 5, No. 3, pp. 347-362.

10

Slope Dependent Morphometric Analysis as a Tool Contributing to Reconstruction of Volcano Evolution

Veronika Kopačková, Vladislav Rapprich,
Jiří Šebesta and Kateřina Zelenková
Czech Geological Surve
Charles University in Prague,
Faculty of Science, Department of Applied Geoinformatics and Cartography
Czech Republic

1. Introduction

People have been fascinated by volcanoes since time immemorial. This is mainly due to the serious consequences that volcanic eruptions represent for human society. Volcanic activity develops in various ways. Therefore, evolutionary trends and the history of the volcanic system should be well understood when future hazards must be predicted and their impact on human society reduced. A volcano's history can be reconstructed from its deposits, their superposition and spatial relationships. Geological mapping is the crucial method for acquiring this information. Unfortunately, large areas in volcanic zones are inaccessible for research directly in the field. In these areas, the geological setting must be investigated by a combination of remote sensing methods and field observations from accessible outcrops.
Surface methods such as remote sensing and morphological analysis provide fast and relatively cheap information, complementary to classical field geology for studying the subsurface geology. These methods can be beneficial, especially for areas with poor accessibility and/or dense vegetation cover. Volcanoes or volcanic complexes quite often represent such areas. Land forms are a result of geologic and geomorphologic processes that occur on the earth's surface thus land forms are not chaotic, but have been structured by geologic and geomorphologic processes over time. The geomorphology of volcanic formations as a whole seems to be a reflection of the underlying geology with steep-sided land forms occurring at each of the "strong" rock units and long, with gentle slopes and topographic breaks found on „soft" rocks. To support this theory, we employed and tested new methodology combining information arising from field surveys together with visual interpretation and statistical spatial analysis of morphometric slope-depending classes to define the spatial extent of various volcanic formations and to identify major tectonic phenomena from features derived from the geomorphology in more accurate way.

2. Study areas

Morphometric analysis was applied to two case study areas, two volcanic complexes of distinct geotectonic setting, age and volcanic evolution. Selected volcanic areas encompass a number of features and rock types associated with volcanic activity. The first case study was carried out in the Conchagua Volcanic Complex, El Salvador (Central America), while the second one was performed in the Doupovské hory Volcanic Complex, Czech Republic (Central Europe).

2.1 Conchagua Volcano

The Pacific coast of Central America is bordered by a chain of active subduction-related volcanoes. This chain is called the Central American Volcanic Arc (CAVA) and extends from Guatemala via El Salvador, southern Honduras, Nicaragua and Costa Rica to western Panama (e.g., Carr et al., 2003). The volcanic arc is associated with the subduction of the Cocos plate beneath the Caribbean plate and it is divided into several segments by traverse faults. Conchagua Volcano is located near one of these segment boundaries (Carr, 1984).

Conchagua Volcano (Fig. 1), on which our research has been focused, is the easternmost volcano of the Salvadorian mainland. Conchagua volcano is located on the Conchagua Peninsula surrounded by the Pacific Ocean and the Gulf of Fonseca. The area of the Gulf of Fonseca including the Conchagua Peninsula is characterized by the presence and intersection of three important tectonic structures. The Median Trough (syn. Salvadorian Depression called the Nicaraguan Depression further to the SE) is parallel to the Middle America Trench. The Trough originated in response to extension related to the subduction roll-back of the Cocos Plate (Phipps-Morgan et al., 2008; Funk et al., 2009). The tension on oblique subduction is accommodated by dextral strike-slip movements on the El Salvador Fault Zone (the northern edge of the Salvadorian Depression - Corti et al., 2005). Extension related to eastward escape of the Chortis Block is thought to be the main reason for formation of the Comayagua Graben (Burkart & Self, 1985). The Guayape Fault running from the Gulf of Fonseca to the northeast (Finch & Ritchie, 1991) is interpreted as a Mesozoic terrane boundary, originally being part of the Guayape–Papalutla Fault Zone (Silva-Romo, 2008). Early studies assumed sinistral movement on the Guayape Fault (Burkart & Self, 1985), but sinistral displacement exceeding 50 km was documented by Finch & Ritchie (1991). The latter authors have also observed several dextral strike-slip basins providing evidence for a later dextral movement phase. Dextral movements on this fault may result from anticlockwise rotation of the Chortis Block (Gordon & Muehlberger, 1994).

The eruptive history of the Conchagua Peninsula has been recently reconstructed by Rapprich et al. (2010). The oldest rocks cropping out in this area are Playitas welded rhyolitic ignimbrites of Miocene age. The next stage is represented by non-welded pyroclastic deposits of La Unión unit (mean K–Ar age: 13.3 ± 3.7 Ma). The presence of banded pumice, deposits containing both mafic scoria and felsic pumice fragments is interpreted as being a result of mingling between basaltic and dacitic magmas. Eruptions of this unit were most likely triggered by injection of basaltic magma into a dacitic magma chamber. Rocks of the subsequent Pozo unit are poorly exposed and strongly altered. Andesite lavas alternate with mafic pyroclastic flow deposits. As the non-welded pyroclastic and strongly altered effusive and pyroclastic rocks have similar surface features, the products of these two phases were combined in this study. Subsequent activity became much calmer and was predominated by effusions of basaltic andesite to andesite lavas. The lava sequences were subdivided into two formations in relation to their geochemical constraints (Rapprich et al., 2010). The earlier of the two formations, Pilón Lavas, were dated at 8.4 ± 1.2 Ma (Quezada & García, 2008), whereas the younger lavas of Pre-Conchagua – Juana-Pancha were dated at 1.6 ± 0.6 to 1.3 ± 0.4 Ma (Quezada & García, 2008; Rapprich et al., 2010). Identical physical properties make these two formations indistinguishable on the basis of their morphology. Hence, both lava formations were combined for the purpose of this study. The volcanic evolution terminated with the formation of two subsequent composite scoria cones in the Pleistocene (0.15 ± 0.02 and 0.41 ± 0.1 after Quezada & García, 2008). Similar physical properties led us again to combine the two cones in a single unit. Since the Pleistocene, the complex has been quiet in terms of volcanic eruptions, but the volcanic forms have been modified by erosion and post-volcanic tectonics. Tectonic depressions were filled with sediments and distal ash-fall during the Holocene. The most prominent ash layer in these depressions is white in colour and has rhyolitic composition. It is interpreted as distal fallout of the Tierra Blanca Joven eruption of the Ilopango Caldera (Rapprich et al., 2010).

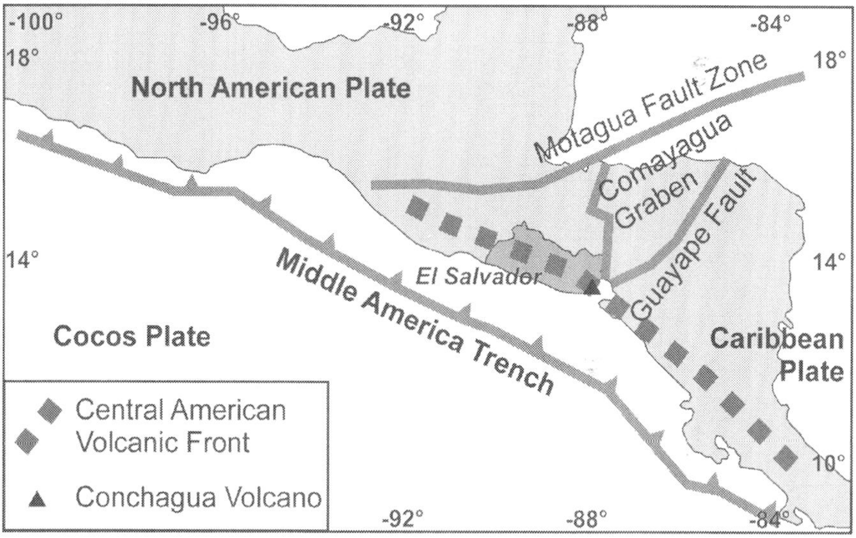

Fig. 1. Location of the Conchagua Volcano - black triangle

2.2 Doupovské hory Volcanic Complex

The Dopovské hory Volcanic Complex (DHVC) belongs to a system of Cenozoic intra-plate volcanic/magmatic complexes in Central-Western Europe (e.g., Lustrino & Wilson, 2007). Similarly to other Cenozoic volcanic complexes in Central-Western Europe, the magmas of the DHVC can be classified as within-plate alkaline in character and were derived from the sublithospheric mantle.

The DHVC is located in the western part of the Eger Graben, which is the easternmost branch of the European Cenozoic Rift System (ECRIS, Dèzes et al., 2004). The Eger Graben runs across the north-western part of the Bohemian Massif in the ENE-WSW direction (Fig. 2), roughly following the Variscan suture between the Saxothuringian and Teplá-Barrandian domains of the Bohemian Massif (Babuška et al., 2010; Mlčoch & Konopásek, 2010). It is interpreted as an incipient rift structure formed during two distinct extensional phases (Rajchl et al., 2009). The Late Eocene to Early Miocene phase was characterised by

NNE-SSW to N-S oriented horizontal extension, oblique to the rift axis. The palaeostress field of this phase as well as OIB-like magmatism within the Eger Graben most probably reflected lithospheric doming due to thermal perturbation of the astenosphere (Dèzes et al., 2004). Later, lithospheric folding in the Alpine-Carpathian foreland and stretching along the crest of a growing regional-scale anticlinal feature resulted in an orthogonal extensional phase (Dèzes et al. 2004; Bourgeois et al. 2007; Rajchl et al. 2009).

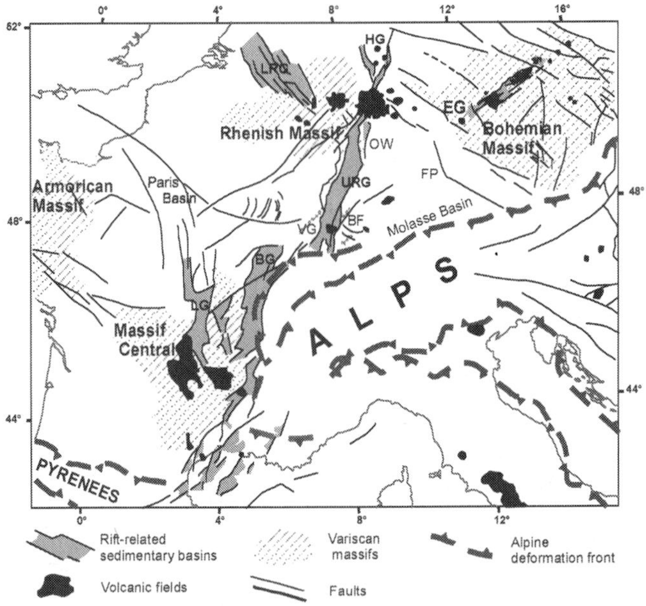

Fig. 2. Location of the Doupovské hory Volcanic Complex within the ECRIS (after Dézes et al., 2004)

Several geological units meet in the basement of the DHVC in the form of a heterogeneous crustal mosaic (Mlčoch & Konopásek, 2010; Valenta et al., 2011). Paragneisses, felsic granulites, orthogneisses, mica schists, and phyllites of the Saxothuringian Unit extend from the north in the basement of the DHVC. Variscan Nejdek-Eibenstock granitic pluton plunges beneath the DHVC from the west. Gneisses, micaschists and phyllites of the Teplá-Barrandian Unit form the basement to the south. Amphibolites, eclogites and peridotites (equivalent to the Mariánské Lázně Complex) occur beneath the central part of the DHVC. The south-eastern sector of the DHVC (delimited by the Střezov and Liboc faults) is underlain by an up-to-800 m-thick sequence of Late-Palaeozoic sedimentary rocks covering the Teplá-Barrandian Unit basement.

The Doupovské hory Volcanic Complex was formerly interpreted as a huge stratovolcano (e.g., Zartner, 1938; Kopecký, 1988). However, voluminous pyroclastic material was produced only during the early stages of the volcanic evolution (Hradecký, 1997). Volcaniclastic material of subsequent phases was not produced by explosive activity. The volcanic evolution of the DHVC started in the very Early Oligocene with eruptions of

Strombolian and phreatomagmatic types. These eruptions buried the fauna of mammal zone MP-21 (Fejfar & Kaiser, 2005), which facilitated dating of this initial phase. Up to 80 m of pyroclastic deposits were produced during the Early Oligocene eruptions. Effusive activity soon predominated over volcanic explosions. The lavas emitted from a set of subsequent shield volcanoes separated from one another by phases of volcano edifice decay. The complex consists of alkaline volcanic rocks, namely foidites, basanites and tephrites. Weak erosion preserved superficial products dominated by sequences of gently dipping mafic lavas with subordinate concomitant volcaniclastics (Rapprich & Holub 2008). The subvolcanic rocks are exposed on a small area in the central part of the DHVC (Holub et al., 2010; Haloda et al., 2010). The activity terminated in the Early Miocene when several small monogenic volcanoes were formed on the northern periphery of the complex (Sakala et al., 2010).

3. Data and methods

3.1 Introduction to morphometric analysis

Automatic methods of analyzing DEM have been increasingly used in geomorphological (Dikau et al., 1989; Kaab et al., 2005; Hancock et al., 2006) and morphotectonic research (Jordan et al., 2005). A quantitative technique for analysis of land surface parameters is known as morphometry / geomorphometry. In simple terms, morphometry aims at extracting (land) surface parameters (e.g., morphological, hydrological) and objects (watershed, stream networks, landforms) using a set of numerical characteristics such as slope, profile curvature, plan convexity, cross-sectional curvature minimum and maximum curvature derived from DEM (Wood, 1996; Pike, 2000; Fisher et al., 2004). Landform and lithological units differ in their geotechnical properties (e.g., rock strength) and in the degree of weathering and rock disorganisation resulting from diverse erosion processes; therefore, they display statistically significant compositional differences with respect to their proportions of morphometric classes. Morphometric analysis can provide unique information that can be linked to land erosion conditions, landform characteristics, morphologic and tectonic evolution. Various approaches have been employed to link morphometry with the geomorphological and volcanological conditions (Ganas et al., 2005; Bolongaro-Crevenna at al., 2005; Liffton et al., 2009; Passaro et al., 2010; Altin & Altin, 2011).

3.2 DEM inputs

For both case study areas, comparable elevation data were utilized and later processed in the same way. Vector topographic base maps on a scale 1:25,000 (Servicio Nacional de Estudios Territoriales, SNET and the military topographic maps called DMÚ25) were used to reconstruct the Conchagua and Doupovské hory volcanic complexes. A Digital Elevation Model (DEM) has been prepared from vector contours at 10 m intervals from the 1:25 000 topographic map, interpolated and resampled to 5x5 m^2 pixel size.

3.3 DEM data processing

In morphometric approaches, the first and second order derivatives of DEM's are the key components which can be related to geomorphological features and processes. While Evans (1972) separated curvatures into two orthogonal components (e.g., profile and plane curvature), Wood (1996) proposed an algorithm for measures of the surface convexity/concavity. As a result, one component, the cross sectional curvature, is calculated instead of the profile and plane curvature components. This parameter is computed in a more simple way and can be directly linked to geomorphological phenomena. Like Evan's profile and plane curvatures, this parameter can be calculates as long as the slope differs from zero (slope=0, the cross-sectional curvature (crosc) and longitudinal curvature (longc) remain undefined). In these cases, two alternative measures of the convexity (e.g., minimum and maximum curvatures) are determined. To calculate the morphometric features, a local window passes over the DEM and the changes in the gradient of a central point in relation to its neighbors are extracted using the approximations given in Table 1.

Morphometric parameter	Approximation
Slope	$\arctan(\mathrm{sqrt}(d^2 + e^2))$
Cross-sectional curvature	$n * g * (a * x^2 + by^2 + cxy + dx + ey + f)$
Maximum curvature	$n * g * (-a - b + \mathrm{sqrt}((a - b)^2 + c^2))$
Minimum curvature	$n * g * (-a - b - \mathrm{sqrt}((a - b)^2 + c^2))$

Table 1. DEM pixel size; n: local window size; x,y: local coordinates; a-f: quadratic coefficients.

Based on the DEM derivatives specified above, Wood defines a set of criteria (e.g., slope, cross-sectional curvature, maximum and minimum curvature) to identify morphometric classes (Tab. 2). For features with positive values (+) of the slope, the cross sectional curvature should be considered and, for features with zero slope value (0), the cross section curvature is undefined (x) and the maximum and minimum curvatures become to be the main classification criteria.

We constructed the morphometric maps utilizing Wood's algorithm. First, the algorithm was used pixel by pixel to calculate the topographic slope and the maximum and minimum convexity values. Then, for each pixel, the variation in these parameters was quantified with respect to neighboring pixels (in orthogonal directions), and then, based on a set of tolerance rules (Tab. 2), each pixel was assigned to one of six possible elemental forms or morphometric classes: ridge, channel, plane, peak, pit and pass.

Wood's algorithm offers the option of parametrizing the relief on the basis of changes in the tolerance of the topographic slope and convexity for assigning to morphometric classes. Slope change tolerance values are used to decide if a pixel qualifies as a peak or a pit, whereas convexity tolerance values are used to determine if a pixel has enough curvature to qualify as a channel or a ridge.

The constructed model was calibrated by running the algorithm with slope tolerance values varying between 0.3 and 3.5 and convexity tolerance values set from 0.001 to 1.000. The resultant morphometric classes were color-coded and visualized; the best result was achieved by draping the color-coded morphometric classes over a three-dimensional (3D) map formed by the fusion of an altitudinal map using ArcGIS 3D Analyst SW. The best fit occurred with slope tolerance values of 3.0 and convexity tolerance values of 0.02.

The relationship was assumed to exist between a geotechnical property of the studied rock formations and the slope angles. Additionally, a systematic break in the slope angles matching the elevation change across tectonic features (e.g., faulting, fracture jointing) could be observed (Gamas et al., 2005). To test the feasibility of linking the geomorphological and tectonic features with the morphometric features classified into defined classes based on their steepness (slope degree), a new product, a slope-dependent morphometric map, was constructed.

Morphometric Feature	Description	Slope	Cross-sectional curvature	Maximum curvature	Minimum curvature
Peak	Point that lies on a local convexity in all directions (all neighbours lower).	0	x	+va	+va
Ridge	Point that lies on a local convexity that is orthogonal to a line with no convexity/concavity.	0	x	+va	0
		+va	+va	*	*
Pass	Point that lies on a local convexity that is orthogonal to a local concavity.	0	x	+va	-va
Plane	Points that do not lie on any surface concavity or convexity.	0	x	0	0
		+va	0	*	*
Channel	Point that lies in a local concavity that is orthogonal to a line with no concavity/convexity.	0	x	0	-va
		+va	-va	*	*
Pit	Point that lies in a local concavity in all directions (all neighbours higher).	0	x	-va	-va

Table 2. Classification criteria for morphometric features (modified from Wood): va: derivative values, x: undefined value, *: not a part of the selection criteria.

For the Conchagua volcano, a thematic raster was created from the DEM by grouping the slope values together into six classes: class 1: flat terrain (inclination < 5°), class 2: very low-steep slopes (inclination 5 - 10°), class 3: low-steep slopes (inclination 10 - 15°), class 4: moderate-steep slopes (inclination 15 - 20°), class 5: steep slopes (inclination 20 - 25°) and class 6: very steep slopes (inclination > 25°). The topography of the Doupovské hory volcanic complex does not exhibit such high altitudes and flat to moderate slope terrain is characteristic for this area rather than steep slopes; therefore the slope values were classified into four classes as follows: class 1: flat terrain (inclination < 5°), class 2: very low-steepness slopes (inclination 5 - 10°), class 3: low-steepness slopes (inclination 10 - 15°), class 4: moderate- high steepness slopes (inclination >15°).

In order to classify the areal morphometric classes (ridge, plane, peak, pit and pass) with respect to the slope gradient, a matrix analysis was applied. Matrix analysis produced a new thematic layer (matrix of 6x6 classes for the Conchagua and 6x4 classes for the DHVC, respectively) that contained a separate class for every coincidence of selected classes in the morphometric map (e.g., peak, ridge, pass, plane, channel, pit) and also a thematic slope map (slope classes 1-6 and 1-4, respectively). As result, maps classifying the six morphometic features according to the slope gradient of the relief from flat to very steep peaks, ridges, passes, channels, planes and pits were constructed.

3.4 Interpretation and further geostatistical analysis

Slope-dependent morphometric maps calculated for the both test sites were correlated with the available geological maps. Visual analysis of the spatial occurrence of the newly derived morphometric parameters within the diverse litho-stratigraphic formations of the Conchagua Volcanic Complex, El Salvador (e.g., La Union pyroclastic deposits versus mafic lavas) and the Doupovské hory Volcanic Complex, Czech Rep. (e.g., lahar deposits versus lavas) clearly showed that a spatial distribution (pattern) of these morphometric features reflects variations in the rock strength, resistance, tectonics, and volcanic topography. As result, the morphometric map became a basis for delineating major geomorphological entities (Figs. 4, 10).

Zonal statistics analysis was employed to study the morphometric pattern and its statistical differences within the geomorphologic units. Zonal functions were used to compute an output dataset, where for each zone (in our case each morphologic unit) the following statistical variables were computed based on the morphometric feature values of the cells, on their location and the association that the location has within a geomorphological zone: i) MAJORITY – Determined the value that occurred most often of all the cells in the input dataset (morphometric map) that belonged to the same zone (morphological unit); ii) MINORITY – Determined the value that occurs least often of all the cells in the input dataset (morphometric map) that belonged to the same zone (morphological unit); iii) MEDIAN – Determined the median value of all the cells in the input dataset (morphometric map) that belonged to the same zone (morphological unit).; iv) VARIETY – Calculated the number of unique values for all the cells in the input dataset (morphometric map) that belonged to the same zone (morphological unit).

4. Results

The morphometric spatial pattern of each geomorphological entity was assessed; frequency graphs showing the abundance of the morphometric matrix classes within each geomorphological unit are given in Figs. 8 and 13. The results from the zonal statistics are depicted in Figs. 9 and 14. In both study areas, the peaks and passes showed none or very sparse (minor) abundance thus cannot be distinguished either in the morphometric maps or in the frequency charts.

4.1 Conchagua Volcano

The morphometric analysis produced an image enhancing different morphologies in the area of Conchagua Volcano (Fig. 4). The loose, non-welded pyroclastic deposits display high variability of morphometric features resulting from the intense erosion (grooves) of ephemeral streams. The hard rocks have significantly more equable morphology with short steep slopes defining the fronts of lava flows or even lava lobes.

Six distinct morphologies were identified in the area of the Conchagua volcano (Fig. 4). Flat surfaces (slope < 5°) predominate in welded rhyolitic ignimbrites (I), non-welded pyroclastic and altered volcanic rocks (II), and monogenetic cone lithologies (V). However, these lithologies still contain such morphometric features as ridges and channels. Holocene post-volcanic sediments (VI) have overall aligned smooth relief with slope of < 5°; very steep slopes (ridges) are encountered least frequently.

In rhyolitic ignimbrits (I), the ridges and channels characterize margins of welded ignimbrite exposures, whereas the surface of these resistant rocks creates flat plains. The channels result partly from tectonic disturbances of the oldest rock sequence and also from prolonged erosion. On the other hand, the sturdiness of these rocks prevents the edges of channels and ridges from being smoothed down. Consequently, ridges and channels with low slopes occur only rarely.

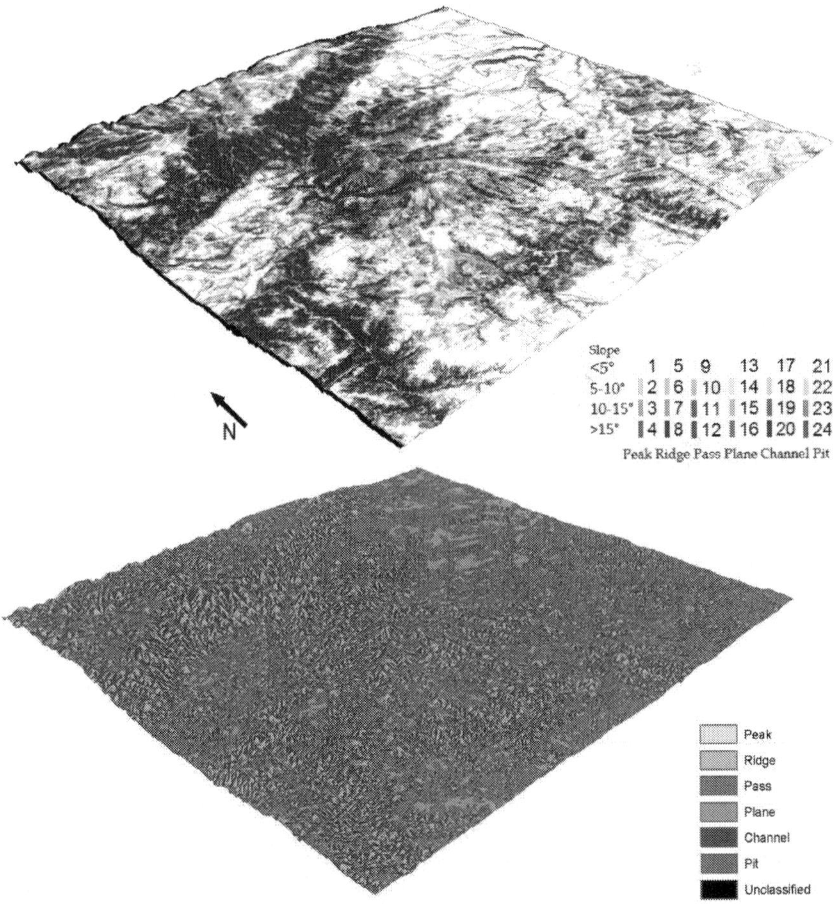

Fig. 3. DHVC: The morphometric map (lower layer) and its derived product: slope-dependent morphometric map (upper layer).

Different compositions of morphological features characterize the Miocene non-welded pyroclastic and altered sequences of lavas (II) alternating with pyroclastics. All landforms were easily smoothed down by erosion. Low-angle dipping planes, ridges, passes and channels strongly dominate over steep-slope forms (Figs. 4, 8).

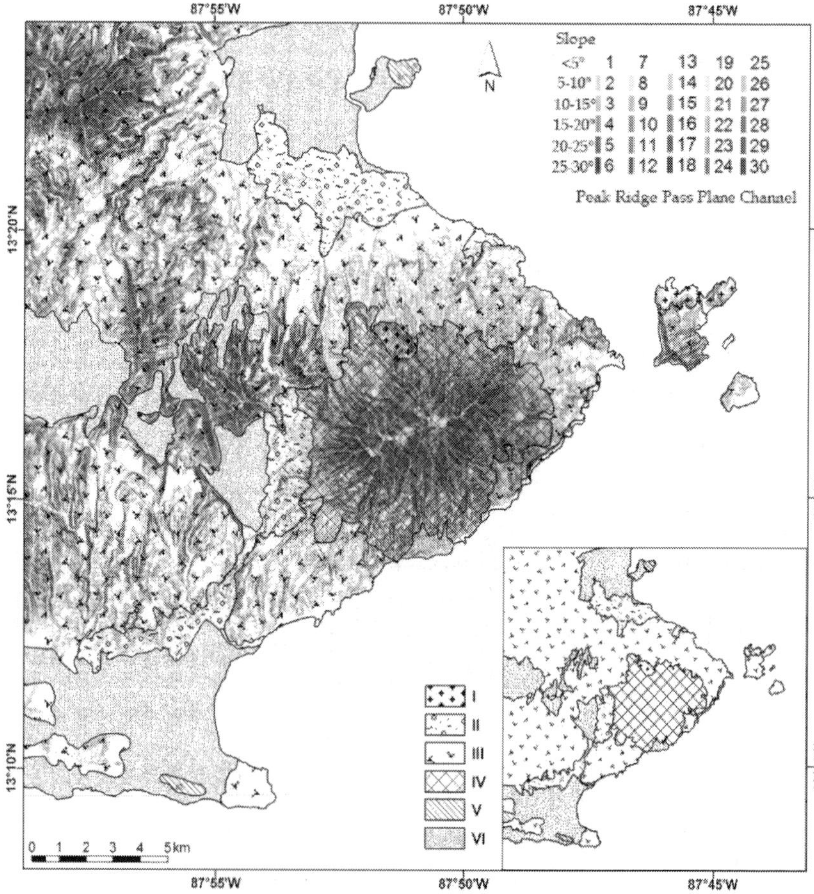

Fig. 4. Morphometric map and morphological units: I - welded rhyolitic ignimbrites; II - non-welded pyroclastic and altered volcanic rocks (Miocene); III - lava sequences (Miocene to Pliocene); IV - composite scoria cones (Pleistocene); V - monogenetic cones (Pleistocene); VI - post-volcanic sediments (Holocene).VI - post-volcanic sediments (Holocene).

The lava sequences of Miocene to Pliocene age (III) dip gently from the source vents; therefore, in contrast to the lithologies described above, the lava sequences have a characteristic flat terrain while short, low to moderate steep and steep ridges and channels characterize the lava flow fronts and sides (Fig. 4). The traverse section of a typical lava flow is concave and the steep sides of the lava continuously pass into a plateau at the top of the lava. The morphology of the lavas combines planes with ridges and channels of variable steepness. The resistance of the (basaltic) andesite lavas against common erosion preserves the original morphology long after the lava emplacement. Hence the lava-front can still be identified in the morphometric map (Figs. 4, 5).

The Conchagua Volcano (IV), i.e. both its cones Ocotal and Banderas, are characterized by very steep (> 25°) to moderate-steep (15°-25°) ridges with a dense network of erosion very steep (> 25°) to moderate-steep (15°-25°) grooves represented in the morphometric map as channels (Figs. 4, 8). The majority and median (Fig. 9) point of the same morphometric feature - ridges with very steep slopes, flat (slope < 5°) passes are the least frequent. This distinctive morphology, where very steep to steep slopes are predominant is clearly visible in the field, topographic maps, DEM and this observation is sufficiently confirmed by numerical evaluation of morphometric analysis. Steeply inclined channels and ridges distinctly predominate over low-angle landforms. Bimodal distribution of morphometric classes (Fig. 8.) is characteristic for this unit.

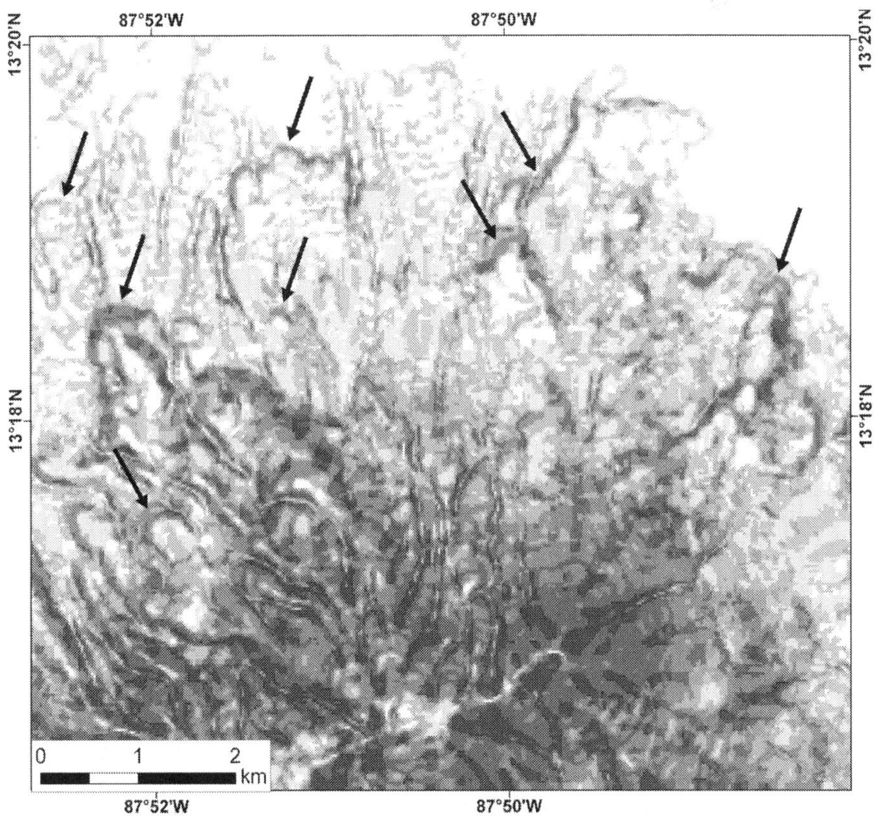

Fig. 5. Detail of the lava flow fronts (black arrows) from the slope-dependent morphometric map.

The monogenetic volcanic cones (V) did not produce a prominent morphology. Consisting of non-welded pyroclastic deposits, the volcanic forms are characterized by morphological composition comparable with Miocene non-welded pyroclastics and altered volcanic rocks. Planes are combined with low-angle, inclined ridges and channels.

Several tectonic depressions are associated with N-S to NNE-SSW trending faults filled with post-volcanic sediments and distal ash fall-out deposits (VI). Sedimentation strongly outweighs erosion in these areas and sediments level the surface. The morphology is therefore dominated by horizontal planes and low-angle structures.

Fig. 6. Dense network of erosion grooves on the Ocotal (centre and right) and Banderas (to the left) cones of the Conchagua Volcano.

Fig. 7. Post-volcanic sediments and distal ash fall-out deposits from 130 km distant Ilopango Caldera filling up the tectonic depressions.

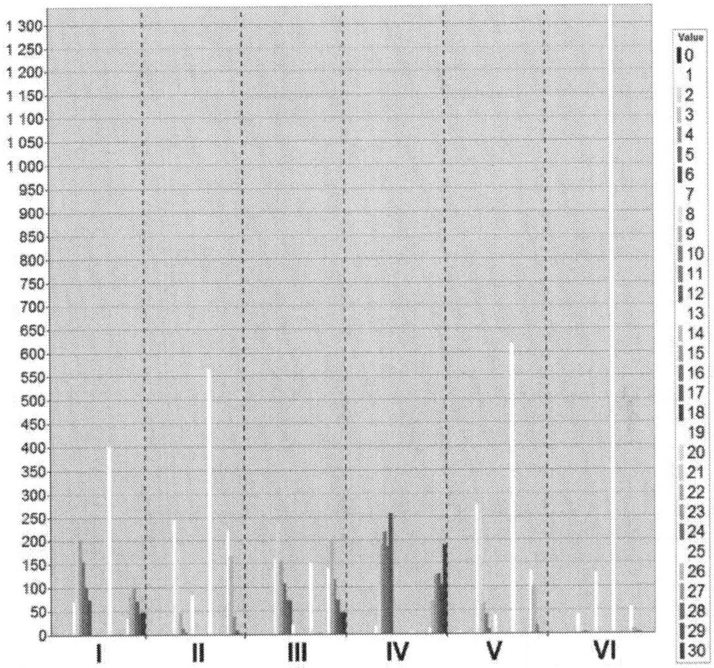

Fig. 8. Frequency of morphometric classes within the defined geomorphological features normalized by the areal extent of the geomorphological features. Pits were not identified.

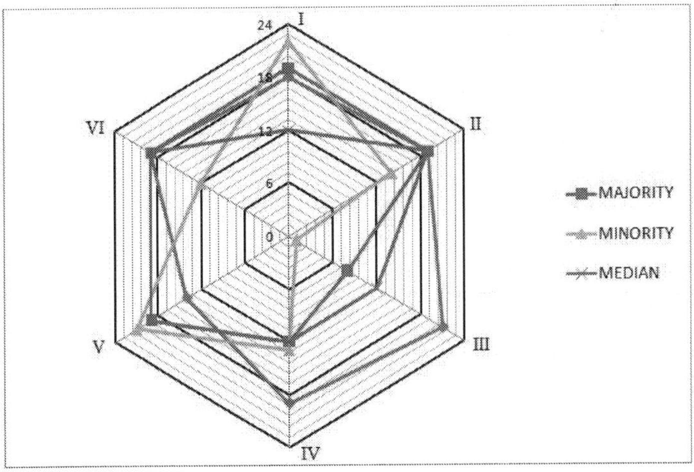

Fig. 9. Conchagua volcano: Graph showing the zonal statistics parameters for the six principal geomorphological units.

4.2 Doupovské hory Volcanic Complex

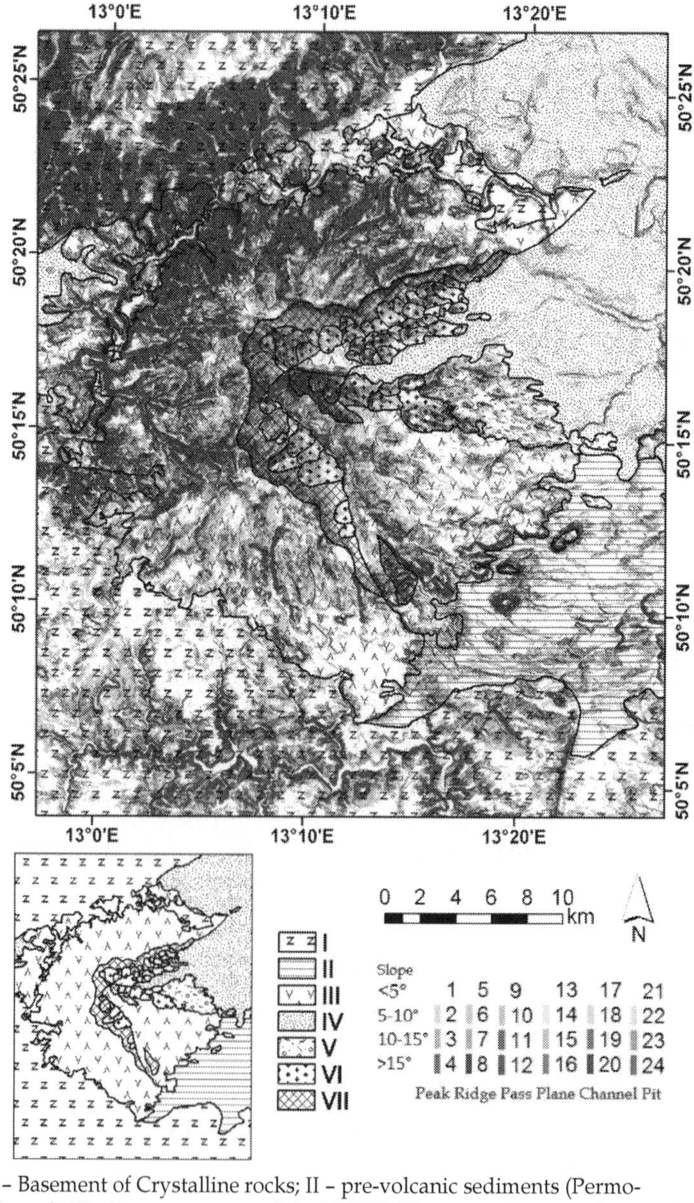

Fig. 10. I - Basement of Crystalline rocks; II - pre-volcanic sediments (Permo-Carboniferous); III - Lava sequences of the DHVC; IV - post-volcanic sediments (Miocene); V - lahar sequences; VI - diastrophic blocks (4, diastr_block); VII - fault scarp.

The area of the Doupovské hory Volcanic Complex exhibits seven principal geomorphological units comparing the geomorphological results with the geological data. (Fig. 10). The basement of the DHVC (I) is built of crystalline rocks, namely granites, gneisses, shists and metabasic rocks. These rocks are resistant to erosion and this is reflected in the morphology, which therefore contains extensive planes, as a peneplenized pre-volcanic landscape, combined with steep-sided gorges cut into these plateaus. These canyons are characterized by unsmoothed edges and therefore low-angle ridges and channels are in a minority compared to steeply inclined ones. The flat relief is locally disturbed by isolated remnants of scattered monogenetic volcanoes penetrating through the metamorphic rocks (Fig. 10).

Fig. 11. Small isolated volcanic bodies rising above the flat relief on crystalline rocks south of the DHVC.

Fig. 12. Lava sequences of the DHVC form tabular rocks with flat apical plateaus.

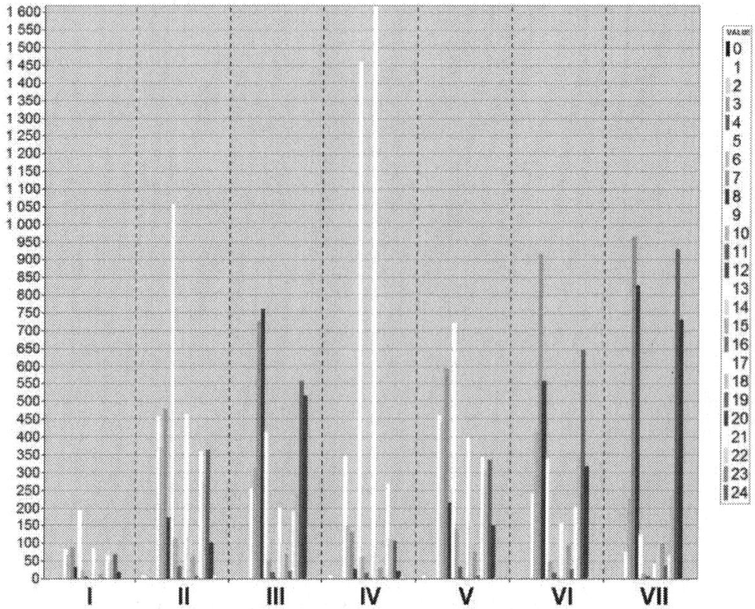

Fig. 13. DHVC: Frequency of morphometric classes within the defined geomorphological features normalized to the areal extent of the geomorphological feature. Pits were identified but had a minor occupancy.

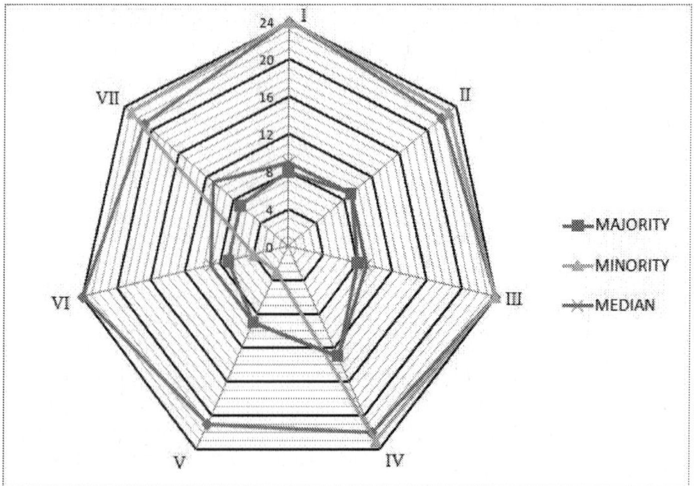

Fig. 14. DHVC: Graph showing the zonal statistics parameters for the seven DHVC principal geomorphological units.

Pre-volcanic sediments (II) represented by Permo-Carboniferous siltstones, sandstones and conglomerates created part of the flat, pre-volcanic morphology. The most frequent morphometric class is a flat slope (< 5°) ridge. In contrast to crystalline rocks, the sediments are much more prone to erosion. Valleys cut into these rocks do not have such steep walls and sharp edges, as these are effectively smoothed down by erosion of the soft rocks. Consequently, deep valleys and pronounced ridges with steep walls occur rarely in areas formed by these rocks. On the other hand, gently dipping channels and ridges are frequent as are planes.

The proper Doupovské hory Volcanic Complex consists of lava sequences (III) displaying flat, weakly periclinally inclined relief hold by structural surfaces of lava flows. Steep (most freuqeunt) to moderate ridges and channels create a characteristic morphometric pattern. The morphology can be readily compared with the results observed on lava sequences in the area of the Conchagua Volcano. The margins of the lava plateaus combine gentle and steep channels and ridges, whereas the surface of the lavas is characterized by horizontal to gently dipping planes.

The areas of post volcanic (Miocene) sediments (IV) are characterized by sedimentation prevailing over erosion. Therefore, these areas are dominated by flat low-angle (<5°) forms, namely planes whereas steep-sided ridges and channels are only occasionally represented by a few river valleys.

Fig. 15. Hill resembling "hummock" built of lahar deposits on the eastern foothills of the DHVC.

The landscape on the eastern foothills exhibits hummocky-like relief of debris avalanche deposits. However, each "hummock" is stratified and consists of several lahar units (V) and this morphology can be explained as being a result of erosion of a large lahar-based alluvial fan, where the erosion is controlled by the geometry of the lahar-lobes. In contrast to the other lithologies, flat passes are the most frequent morphometric features of the lahars (the majority is also the median in this particular case). These also have the lowest variety of morphometric features of all the studied formations.

Sharp and steep (> 15°) slopes (channels, ridges) correspond to fault scarps (VII). The original steeply inclined planes were affected by erosion and transformed to set of parallel channels and ridges running down the slopes. The fault-scarps observed in the morphology correspond well to important regional faults described for the DHVC basement (e.g., Valenta et al. 2011). The fault scarps are locally associated with areas dominated with large boulders - diastrophic blocks (VI), resulting from local gravity instabilities along the fault scarps. The diastrophic blocks exhibit morphology dominated by steep channels and ridges. We assume that peak classes will dominate in this type of morphology if we use a lower resolution grid.

5. Conclusion

At the present time, DEM's are type of information that is available for much of the globe (SRTM DEM: 90-m spatial resolution, ASTER DEM: 30-m spatial resolution). Slope-dependent morphometry proved to be an efficient and low-cost method that provided unique and valuable information that could not be gathered by any of the field-based methods, especially in terrains with dense vegetation and lacking outcrops. Use of this method permits classification of large formations and algorithmic parameterization facilitates the creation of maps at a different level of generalization. The four morphometric parameters (e.g., slope, cross-sectional curvature, maximum and minimum curvature) led to description of the landforms as ridge, plane, channel, peak, and pass. Additional matrix transformation combining the depicted morphometric parameters with defined slope classes (very low to very high slope) enhanced differences in the rocks based on their geotechnical properties (e.g., rock strength, degree of weathering) and enabled finer sub-level classification which, in combination with the available geological information, allowed us to delineate major geomorphological units for the studied areas, the Conchagua volcano and Doupovské hory Volcanic Complex (DHVC). Due to the exogenic and endogenic processes volcanoes can be quite frequently characterized with high dynamics and slope instability (Kopačková & Šebesta, 2007). As our results combine the relief parameters together with the physical properties of the rocks, they can be further utilized for delineating hazard zones prone to landslides.

Slope-dependent morphometric analysis clearly separates areas of distinct lithologies, as these produce different morphologies. The histograms of morphometric classes may identify, whether the studied area is dominated by erosion or by sedimentation. Additionally, areas of rocks with similar properties can be compared in terms of relative age, as the older ones are affected by erosion more intensively.

The morphometric analysis contributed in geological research of the Conchagua Volcano. Field geological mapping was insufficient for precise definition of lithological boundaries. The boundary lines were improved using morphometric analysis as individual lithologies displayed significantly distinct morphologies. Specific morphologies were observed in the

Doupovské hory Volcanic Complex for lahar sequences and diastrophic blocs bordering fault scarps.

6. Acknowledgment

The described studies were carried within the framework of the Research Plan of the Czech Geological Survey (MZP0002579801). Additionally, research was supported by research projects 205/06/1811 and 205/09/1989, both covered by grants from the Czech Science Foundation (GAČR). The authors would also like to thank to their colleagues from Servicio Nacional de Estudios Terriotoriales, namely Walter Hermandez Geology Unit, and Giovanni Molina, GIS Unit, for their cooperation and data sharing.

7. References

Altın, T.B. & Altın, B.N. (2011). Development and morphometry of drainage network in volcanic terrain, Central Anatolia, Turkey. *Geomorphology*, Vol. 125, No. 4, pp. 485–503, ISSN 0169-555X

Babuška, V., Fiala, J. & Plomerová, J. (2010). Bottom to top lithosphere structure and evolution of western Eger Rift (central Europe). *International Journal of Earth Sciences*, Vol. 99, No. 4, pp. 891–907, ISSN 1437-3254

Bolongaro-Crevenna, A., Torres-Rodríguez, V., Sorani, V., Frame, D., Ortiz, M. & A. (2005). Geomorphometric analysis for characterizing landforms in Morelos State, Mexico. *Geomorphology*, Vol. 67, No. 3-4, pp. 407-422, ISSN 0169-555X

Burkart, B. & Self, S. (1985). Extension and rotation of crustal blocks in northern Central America and effect on the volcanic arc. *Geology*, Vol. 13, No. 1, pp. 22–26, ISSN 0091-7613

Carr, M.J. (1984). Symmetrical and segmented variation of physical and geochemical characteristics of the Central American volcanic front. *Journal of Volcanology and Geothermal Research*, Vol. 20, No. 3-4, pp. 231–252, ISSN 0377-0273

Carr, M.J., Feigenson, M.D., Patino, L.C. & Walker, J.A. (2003). Volcanism and geochemistry in Central America: progress and problems. In: *Inside the Subduction Factory*, Eiler, J. (Ed.), pp. 153–174, Geophysical Monograph, Vol. 138. American Geophysical Union, Washington, DC

Corti, G., Carminati, E., Mazzarini, F & Garcia, M.O. (2005). Active strike-slip faulting in El Salvador, Central America. *Geology*, Vol. 33, No. 12, pp. 989–992, ISSN 0091-7613

Dèzes, P., Schmid, S.M. & Ziegler, P.A. (2004). Evolution of the European Cenozoic Rift System: interaction of the Alpine and Pyrenean orogens with their foreland lithosphere. *Tectonophysics*, Vol. 389, No. 1-2, pp. 1-33, ISSN 0040-1951

Dikau, R. (1989). The application of a digital relief model to landform analysis in geomorphology. In: Raper, J. (Ed.), *Three Dimensional Applications in Geographical Information Systems*, Taylor & Francis, London, pp. 51–77

Evans, I.S. (1972). General geomorphology, derivatives of altitude and descriptive statistics. In: Chorley, R.J. (Ed.), *Spatial Analysis in Geomorphology*. Harper and Row, NY, pp. 17–90

Fejfar, O. & Kaiser, T.M. (2005). Insect bone-modification and paleoecology of Oligocene mammal-bearing sites in the Doupov Mountains, northwestern Bohemia. Palaeon Electron 8.1.8A:
1–11, http://palaeo-electronica. org/2005_1/fejfar8/fejfar8.pdf

Fisher, P., Wood, J. & Cheng, T. (2004). Where is Helvellyn? Fuzziness of Multiscale Landscape Morphometry. *Transactions of the Institute of British Geographers*, Vol. 29, No. 1, pp. 106-128, ISSN 0020-2754

Finch, R.C. & Ritchie, A.W. (1991). The Guayape fault system, Honduras, Central America. *Journal of South American Earth Sciences*, Vol. 4, No. 1-2, pp. 43–60, ISSN 0895-9811

Funk, J., Mann, P., McIntosh, K. & Stephens, J. (2009). Cenozoic tectonics of the Nicaraguan depression, Nicaragua, and Median Trough, El Salvador, based on seismic-reflection profiling and remote-sensing data. *Bulletin of the Geological Society of America*, Vol. 121, No. 11-12, pp. 1491–1521, ISSN 0016-7606

Ganas, A., Pavlidesb, S. & Karastathisa, V. (2005). DEM-based morphometry of range-front escarpments in Attica, central Greece, and its relation to fault slip rates. *Geomorphology*, Vol. 65, No. 3-4, pp. 301–319, ISSN 0169-555X

Gordon, M.B. & Muehlberger, W.R. (1994). Rotation of the Chortis Block causes dextral slip on the Guayape Fault. *Tectonics*, Vol. 13, No. 4, pp. 858–872, ISSN 0278-7407

Haloda, J., Rapprich, V., Holub, F.V., Halodová, P. & Vaculovič, T. (2010). Crystallization history of ijolitic rocks from the Doupovské hory Volcanic Complex (Oligocene, Czech Republic). *Journal of Geosciences*, Vol. 55, No. 3, pp. 279-297, ISSN 1802-6222

Hancock, G.R., Martinez, C., Evans, K.G. & Moliere, D.R. (2006). A comparison of SRTM and high-resolution digital elevation models and their use in catchment geomorphology and hydrology: Australian examples. *Earth Surface Processes and Landforms*, Vol. 31, No. 11, pp. 1394-1412, ISSN 0197-9337

Holub, F.V., Rapprich, V., Erban, V., Pécskay, Z. & Mlčoch, B. (2010). Petrology and geochemistry of alkaline intrusive rocks at Doupov, Doupovské hory Mts. *Journal of Geosciences*, Vol. 55, No. 3, pp. 251-278, ISSN 1802-6222

Hradecký, P. (1997). The Doupov Mountains. In: Vrána S, Štědrá V (eds) Geological Model of Western Bohemia Related to the KTB Borehole in Germany. *Sborník geologických věd, Geologie*, Vol. 47, pp. 125–127, ISSN 0581-9172

Kaab, A. (2005). Combination of SRTM3 and repeat ASTER data for deriving alpine glacier flow velocities in the Bhutan Himalaya. *Remote Sensing of Environment*, Vol. 94, No. 4, pp. 463-474, ISSN 0034-4257

Kopačková, V., Šebesta, J., (2007). An approach for GIS-based statistical landslide susceptibility zonation - With a case study in the northern part of El Salvador. In Proceedings of SPIE - Remote Sensing 2007, Vol. 6749 - Remote Sensing for Environmental Monitoring, GIS Applications and Geology. EHLERS, M., MICHEL, U.,(Eds), SOCIETY OF PHOTO-OPTICAL INSTRUMENTATION ENGINEERS - SPIE, (Publ.), Bellingham, USA. Paper No. 6749-105. ISSN 9780819469076.

Kopecký, L. (1988). Young volcanism of the Bohemian Massif, Part 3. *Geologie a Hydrometalurgie Uranu*, Vol. 12, No. 3, pp. 3–40, UVTEI 76065

Lifton, Z.M., Thackray, G.D., Van Kirk, R. & Glenn, N.F. (2009). Influence of rock strength on the valley morphometry of Big Creek, central Idaho, USA. *Geomorphology*, Vol. 111, No. 3-4, pp. 173–181, ISSN 0169-555X

Lustrino, M. & Wilson, M. (2007) The circum-Mediterranean anorogenic Cenozoic igneous province. *Earth-Science Reviews*, Vol. 81, No. 1-2, pp. 1–65, ISSN 0012-8252

Mlčoch, B. & Konopásek, J. (2010). Pre-Late Carboniferous geology along the contact of the Saxothuringian and Teplá–Barrandian zones in the area covered by younger sediments and volcanics (western Bohemian Massif, Czech Republic). *Journal of Geosciences*, Vol. 55, No. 2, pp. 81–94, ISSN 1802-6222

Passaro, S., Milano, G., D'Isanto, C., Ruggieri, S., Tonielli, R., Bruno, P.P., Sprovieri, M. & Marsella, E. (2010). DTM-based morphometry of the Palinuro seamount (Eastern Tyrrhenian Sea): Geomorphological and volcanological implications. *Geomorphology*, Vol. 115, No. 1-2, pp. 129–140, ISSN 0169-555X

Phipps-Morgan, J., Ranero, C.R. & Vannucchi, P. (2008). Intra-arc extension in Central America: links between plate motions, tectonics, volcanism, and geochemistry. *Earth and Planettary Science Letters*, Vol. 272, No. 1-2, pp. 365–371, ISSN 0012-821X

Pike, R.J. (2000). Geomorphology - Diversity in quantitative surface analysis. *Progress in Physical Geography*, Vol. 24, No. 1, pp. 1-20, ISSN 0309-1333

Quezada, A.M. & García, O.M. (2008). Edades 40Ar/39Ar de rocas de las áreas de Conchagua y Chilanguera (El Salvador). *Unpublished Report*, pp. 1-19, LaGeo, Santa Tecla, El Salvador.

Rajchl, M., Uličný, D., Grygar, R. & Mach, K. (2009). Evolution of basin architecture in an incipient continental rift: the Cenozoic Most Basin, Eger Graben (Central Europe). *Basin Research*, Vol. 21, No. 3, pp. 269-294, ISSN 0950-091X

Rapprich, V. & Holub, F.V. (2008). Geochemical variations within the Upper Oligocene-Lower Miocene lava succession of Úhošť Hill (NE margin of Doupovské hory Mts., Czech Republic). *Geological Quarterly*, Vol. 52, No. 3, pp. 253-268, ISSN 1641-7291

Rapprich, V., Erban, V., Fárová, K., Kopačková, V., Bellon, H. & Hernández, W. (2010). Volcanic history of the Conchagua Peninsula (eastern El Salvador). *Journal of Geosciences*, Vol. 55, No. 2, pp. 95-112, ISSN 1802-6222

Sakala, J., Rapprich, V. & Pécskay, Z. (2010). Fossil angiosperm wood and its host deposits from the periphery of a dominantly effusive ancient volcano (Doupovské hory Volcanic Complex, Oligocene-Lower Miocene, Czech Republic): systematics, volcanology, geochronology and taphonomy. *Bulletin of Geosciences*, Vol. 85, No. 4, pp. 617–629, ISSN 1214-1119

Silva-Romo, G. (2008). Guayape–Papalutla fault system: a continuous Cretaceous structure from southern Mexico to the Chortis Block? Tectonic implications. Geology, Vol. 36, No. 1, pp. 75–78, ISSN 00917613

Valenta, J., Brož, M., Málek, J., Mlčoch, B., Rapprich, V., Skácelová, Z. & Doupov Working Group (2011). Seismic model and geological interpretation of the basement beneath the Doupovské Hory Volcanic Complex (NW Czech Republic). Acta Geophysica, Vol. 59, No. 3, pp. 597-617, ISSN 1895-6572

Wood, J.D. (1996). The geomorphologic characterization of digital elevation models. *PhD Dissertation*, University of Leicester, UK

Zartner, W.R. (1938). Geologie des Duppauer Gebirges. I. Nördliche Hälfte. *Deutsche Gesellschaft der Wissenschaften und Künste,* Vol. 2, pp. 1-132

11

Beneficiation of Talc Ore

Mahmoud M. Ahmed[1], Galal A. Ibrahim[2] and Mohamed M.A. Hassan[3]
[1]*Mining and Metallurgical Engineering Department,*
Faculty of Engineering, Assiut University, Assiut,
[2]*Mining and Petroleum Engineering Department*
Faculty of Engineering, Al-Azhar University, Qena,
Egypt

1. Introduction

Talc is an industrial mineral, which is composed of hydrated magnesium sheet-silicates with theoretical formula of $Mg_3Si_4O_{10}(OH)_2$ that belongs to the phyllosilicate family (Fuerstenau and Huang, 2003; Ozkan, 2003; Yehia and AL-Wakeel, 2000; Boghdady et al, 2005). Talc may have white, apple green, dark green or brown colors, depending on its composition. Talc is the softest one in all minerals, which has Mohs hardness ranges from (1–1.5) and a greasy feel (Boghdady et al, 2005). The specific gravity of talc is about 2.75; it is relatively inert, and water repellent (Engel and Wright, 1960). Talc is formed by the alteration of serpentine. The resulting talc contains magnesia and water but relatively more silica than serpentine (Andrews, 1985).

Talc surface is comprised of two types of surface area, the basal cleavage faces and the edges. The faces surface has no charged group, therefore, it is believed that the talc faces are non-polar and hydrophobic, whereas the edges are hydrophilic due to the presence of charged ions (Mg^{2+} and OH^-) (Kusaka, et al, 1985; Sarquis and Gonzalez, 1998). The major gangue minerals of talc are carbonates, magnesite, dolomite, serpentine, chlorite and calcite, which contribute to the production of undesirable characteristics. The trace minerals in talc include magnetite, pyrite, quartz and tremolite (Andrews, 1985; Sarquis and Gonzalez, 1998; Simandle and Paradis, 1999; Al-Wakeel, 1996; Schober, 1997).

1.1 Petrographical and geochemical characterization of talc

Talc is an extremely versatile mineral which is composed of hydrated magnesium sheet-silicates with a theoretical chemical formula of $Mg_3(Si_2O_5)_2(OH)_2$ that belongs to the phyllosilicate family (Ozkan, 2003; Yehia and AL-Wakeel, 2000; Shortridge, et al 2000). It is formed by the alteration of serpentine. The resulting talc contains magnesia and water with a relatively more silica than serpentine (Andrews, 1985). The talc particles are composed of hydrophobic and hydrophilic surfaces, faces and edges; the former are created by cleavage whereas the latter ones are created by a spontaneous hydrolysis to form oxides sites (Boghdady et al, 2005; Kusaka, et al, 1985). Talc edges consist of charged ions (Mg^{2+} and OH) and therefore, number of bonding possibilities exist between water molecules and talc edges. Hence the edges of talc are likely to be hydrophilic and the talc surface is hydrophobic (Khraisheh, et al 2005). Commercial talc may contain related sheet silicates such as chlorite and serpentine, as well as, carbonates, such as magnesia, dolomite and calcite (Ozkan, 2003; Sarquis and Gonzalez, 1998).

Okunlda et. al. (2003) worked at Baba talc occurrence (Nigeria) with preliminary quantities estimation of 3 million tons. By the aid a thin section examination, they showed that talc, tremolite and chlorite are the main minerals. Talc content ranged from 14% to 72% and occurred as fibrous aggregates and sometimes as a platy.

Gondim and Loyola (2002) mentioned that talc deposits of the Parana district (Brazil) occurred as layers, lenses and veins. They attributed the formation of two types of talc mineralization processes in the deposit of Parana district to the regional dynamo-thermal metamorphism (organic metamorphism) and hydrothermalism.

Simandle and Paradisl (1999) stated that the age of talc mineralization (Ontario, Canada and New York State, USA) is mainly Precambrian. Most carbonate hosted talc deposits are believed to be formed of dolomite with silica and water.

The origin of talc deposits and their associated minor sulfide occurrences in Eastern Desert of Egypt has been a controversy topic from the time of their discovery (Schandl, et al 2002, 1999a, 1999b; Helmy and Kaindl, 1997; El Bahariya and Arai, 2003). El Sharkawy (2000) has reported that, the majority of talc occurrences in Egypt may be derived and hosted by ultramafic rocks, mainly serpentinite. Serpentinite bodies characteristically occur in belts of low- grade metamorphic sedimentary and volcanic rocks. Talc deposits are widely variable in shape and are mostly pod-shaped, lenticular, thin shells and irregular masses.

The majority of talc exploitation in Egypt is wadi El-Allaqi, Derhib, wadi El-Atshan, Gabal El-Angoria and wadi Eggat (Kamel, et al 2001).

Nasr and Masoud (1999) investigated the area which lies between latitude 22° 30′ 00″ - 22° 37′ 30″ N and longitudes 33° 22′ 30″ - 33° 32′ 30″ E, covering about 190km^2. They found that talc deposit lies between latitude 22° 30′ 00″ and longitude 33° 29′ 50″, and talc occurred as lenses along the shear zone which were affected by hydrothermal solution rich in magnesium or as pockets enclosed within the shear zones. X-ray diffraction of twelve representative samples of talc lenses in wadi El-Allaqi area showed that the talc mineral represented 99% of the samples with rare carbonate and illite but other samples showed that the talc mineral represented 95% of the samples with traces of carbonate, illite, magnesium, iron, aluminum clinchlore and manganese.

Attia (1960) have studied the talc deposits in Aswan district, which occurred in a metamorphic schist area at the head of wadi Um Guruf, a tributary of wadi Abu Agag. The working area (Latitude 24° 02′ N and longitude 33° 05′ E) is East of El-Shallal railway station at a distance of 17 km. The schist in this area is dominantly plagioclase-quartz-biotite schist. Talc exists beneath the ground surface and is found in the form of bands or lenses in the schist. It seemed evident that the talc of this locality is an alteration product of the schist.

Yousef (2003) studied the characterization of the talc varieties in El-Allaqi and Abu-Dahr areas. Firstly, El-Allaqi samples were characterized with white to white-grey talc flakes and spherulitic structure intersected with the carbonate crystals. The X-ray diffraction revealed the presence of dolomite, kaolinite, chlorite and quartz in these samples. The chemical analysis indicated that carbonates are the main gangue minerals in the samples (the percent of CaO is equal to 4.64% and the percent of loss on ignition equal to 12.22%). Abu-Dahr samples showed a formation of vein of chlorites talc and highly pyritized talc; the X-ray diffraction of samples showed the presence of chrome and chlorite. The chemical analysis of samples showed that, iron oxide was the main gangue mineral in samples (the percent of Fe_2O_3 = 4.96%).

Boulos et. al. (2004) have performed X-ray diffraction (XRD) of Shalatin samples, which showed a higher percentage of loss on ignition and CaO but for El-Allaqi samples showed higher quality with minor amount of carbonates and traces of quartz. The presence of chlorite in both samples was also confirmed by X-ray diffraction. The total amount of exposed talc in wadi El-Allaqi is more than 165000 tons of very high grade talc quality, but no accurate estimation of talc reserves has been calculated for shalatin locality.

1.2 Processing of talc

Boulos et. al. (2004) have applied the wet attritioning technique as a substitution of the conventional ball or rod milling in talc beneficiation plants because of the friable nature of talc. The objective of this process was to achieve preconcentration of talc by differential grinding from harder carbonate impurities. Optimization of this process included verification of some parameters such as; attritioning time, attritioning speed, and pulp density. Attritioning was carried out on two samples obtained from Shalatin and El-Allaqi regions.

For Shalatin sample, an attritioning scrubbing of -11 mm crushed talc ore was carried out at 60% solids, 1500 rpm motor speed and an attritioning time of 60 minutes. The product has 8.4% loss on ignition with 74.7% mass recovery.

For El-Allaqi sample, an attritioning scrubbing was executed at 60% solids, 2100 rpm motor speed and an attritioning time of 60 minutes. The product has 5% loss on ignition and about 87.5% mass recovery (Boulos et. al. 2004).

Piga and Maruzzo (1992) tried the attritioning of talc-carbonates. Because the carbonates are harder than talc, they are used as a grinding medium for the slurry formed of the ore to be treated. So the fine fraction should be enriched in talc and the coarser fraction enriched in carbonates. A selective attritioning carried out with 76% pulp density, at 20 minutes attritioning time, and a 1 kg/t sodium hexametaphosphate as a dispersant gave a concentrate grading of 82% talc-chlorite and 18% carbonates. The recovery of talc-chlorite was around 74% from a crude ore containing 67% talc-chlorite and 33% carbonates. This product may be sent to flotation for further removal of carbonates and separation of talc from chlorite.

Yousef (2003) used attritioning scrubber as a preconcentrator for talc of El- Allaqi locality. The selective attrition executed with a 50% solid/liquid ratio and an attritioning time of 20 minutes to obtain a concentrate with 9.1% loss on ignition, the percent of CaO was 1.07% and the whiteness increased from 74% in the original ore to 80% in the concentrate.

Flotation is the preferred concentration technique to remove impurities from talc Kho and Sohn, 1989). Various factors that control the flotation process of talc include particle size, pH values, collector dosage, depressant dosage, pulp density and frother dosage which were studied by many authors (Fuerstenau and Huang, 2003; Boulos et. al. 2004; Kho and Sohn, 1989; Andrews, 1989; Feng and Aldrich, 2004).

When adjusting the pH values of the system this can enhance or prevent the flotation of a mineral. Thus, the point of zero charge (ZPC) of the mineral is an important mineral property in such systems (Wills, 1992).

The critical pH is a value below which any given mineral can float, and above which it will not float. This critical pH value depends on the nature of the mineral, the particular collector, its concentration, and temperature (Wills, 1992).

Chang (2002) beneficiated talc at the Gouverneur district in New York. Talc was crushed firstly by jaw crushers, and then by gyratory crushers, conveying to the plant, storage of wet ore, the ore was ground to minus 0.95 cm, tertiary crushing, grinding of coarse product by using a pebble mill in a closed circuit with Raymond mill, and finally grinding of the fine product with fluid energy mill. Concentrating tables were installed to remove high-gravity product containing Ni, Co and iron minerals. He used a flotation plant for the production of high-grade talc. Combination of froth flotation and high intensity magnetic separation has been studied for the removal of iron-bearing minerals.

Wills (1992) reported that increasing concentration of collector tends to float other minerals and reduce selectivity. It is always difficult to eliminate a collector already adsorbed. An excessive concentration of a collector has also an adverse effect on the recovery of valuable minerals; this fact may be due to the development of collector multi-layers on the particles or by reducing the proportion of hydrocarbon radicals oriented into the bulk solution. The hydrophobicity of the particles is thus reduced, and hence their floatability. The flotation limit can be extended without loss of selectivity by using a collector with a longer hydrocarbon chain, which produces greater water-repulsion, rather than by increasing the concentration of a shorter chain collector.

It is common to add more than one collector to the flotation system. A selective collector may be used at the head of the circuit, to float the highly hydrophobic minerals, then after a more powerful, but less selective one is added to promote recovery of the slower floating minerals (Wills, 1992).

Fuerstanau and Pradip (2005) have revealed that; adsorption of collectors in the flotation of silicate minerals is controlled by the electrical double layer at the mineral-water interface. In the systems where the collector is physically adsorbed, the flotation process with anionic or cationic collectors depends on the mineral surface which is being charged oppositely.

In beneficiation of Egyptian talc-carbonate ore, Yehia and Al-Wakeel (2000) applied flotation process at 25% solids, airflow rate = 1000 L/min., pH value = 7, using 0.1 kg/ton of polypropylene glycol as a frother. They obtained concentrate with 90% recovery and 60% grade of talc. The final product was treated by using diluted hydrochloric acid of 10% and 300 ppm of Tin chloride ($SnCL_2$). This product may be suitable one for cosmetic, paint and paper industries.

Al-Wakeel (1996) treated talc ore from wadi El-Baramiya having a size fraction of (-50+45 µm). A selective flotation of talc was applied at a pH value = 6, a frother dosage of polypropylene glycol (AF65) = 0.1 kg/t and an impeller speed =1100 rpm. The pH value of the pulp was adjusted before the addition of frother which in turn is followed by aeration. The grade of talc was about 72.5%. By applying another cleaning stage using 0.075 kg/t a frother dosage at a pH = 7, the grade of talc was increased to 93.5% with a recovery of 70%.

Yousef (2003) has applied more than one technique for the flotation of -75 µm scrubbed talc obtained from attritioning process. The obtained results showed that:

1. Using natural floatability of talc, the final product obtained has loss on ignition about 6.85% and a recovery of 50%.
2. With the addition of 0.05 kg/t of frother (Aerofroth 73), increased the recovery to 65.8% with a slight decreasing in grade is observed.
3. The best result was achieved by using 1 kg/t oleic acid as a collector in the presence of 0.4 kg/t hexametaphosphate as a depressant and at pH value = 10. The final concentrate having 6.6% loss on ignition and a recovery of 61.8%.

Andreola et. al., (2006) have reported that more increasing of sodium hexametaphosphate (used as a depressant) may lead to an increase in the final percent of CaO. This trend can be interpreted to the ability of sodium hexametaphosphate (SHMP) to sequester the calcium cations (Ca^{+2}) forming with the calcium a strong hydrophilic complex compound. But the effect of sodium hexametaphosphate towards aluminum is weak to sequester the aluminum sites. This may be interpreted to the interaction of sodium hexametaphosphate (SHMP) anions with the exposed atoms of Al giving complexes anions.
Khraisheh et. al. (2005) revealed that the adsorption of carboxymethyl cellulose (CMC) depressant onto talc can be increased by increasing the molecular weight of CMC depressant and by the addition of magnesium, potassium and calcium to carboxymethyl cellulose.
Derco and Nemeth (2002) treated three types of talcose rocks (Slovakia): talc-magnesite, talc-dolomite, and talc-magnesite-dolomite. The flotation process was applied for talc-magnesite rock at pH value = 6, pine oil was used as a frother by an amount of 0.5 g/L, and Na_2CO_3 (0.2 g/L) was used as a depressant. The product fulfilling the requirements of pharmaceutical usage. Then the product was treated with polygradient electromagnetic mud separator to decrease the percent of Fe_2O_3 from 1.38% to 1.00% which is suitable for electro ceramic technology. The flotation of talc-dolomite rock was carried out using sodium hexametaphosphate to depress dolomite. This gave a product is used for pharmaceutical purpose. Dressing of talc-magnesite-dolomite rocks produced talc suitable for electro ceramic technology.
Leaching process with diluted acid solutions has some advantages over the other techniques. It is cheap and the acid can be easily recovered from the beneficiated solid ores by filtration (Rizk, et al 2001).
Sarquis and Gonzalez (1998) have reported that the chemical treatments using acids may be applied for further increasing of the grade of concentrate. The basis of the proposed technique lies in the fact that talc is inert with most chemical reagents. The final concentrate of flotation process was leached with diluted hydrochloric acid having a concentration of 10%. The residues of leaching process were washed first with acidulated hot water and then with pure water. The obtained results showed that the whiteness increased from 65.5% to 70.2% at 60 °C., while the loss on ignition decreased from 18% to 6.3% at the same temperature and the assays of CaO, MgO and Fe_2O_3 also decreased.
Al-Wakeel (1996) and Roe (1983) have treated the final cleaned product with a diluted hydrochloric acid having a concentration of 10% and $SnCl_2$ (300 ppm) to produce talc free from carbonates. The iron content was nearly removed and the whiteness increased to 93%. Their last product was suitable for different purposes for paper, cosmetic, paint, roofing, ceramic and rubber filling industries.

Aim of the work

1. Evaluation of the petrographical properties and geochemical characteristics of the talc samples. This evaluation was carried out on four representative samples of different types of talc carbonates to determine the possibility of improving the talc quality.
2. The possibility to improve talc quality by using flotation and determination of the optimum values of operating variables of flotation process such as pH value, depressant dosage, collector dosage and pulp density.
3. The possibility of using leaching process to improve the quality of talc to increase its suitability for different industrial purposes.

2. Experimental

2.1 Materials

2.1.1 Talc ore sample

The head sample used in the present work is a mixing of four samples obtained from different sites in the Eastern Desert of Egypt (Shalatin area). Petrographically, the talc samples are classified into four different types. From the geochemical point of view, the samples are correlated together with high percentages of aluminum oxide and iron oxide. The details and characteristics of these samples were discussed in a previous published paper (Boghdady, et al. 2005). The petrographical types and chemical analysis of the samples, showing the major constituents of talc (MgO and SiO_2) and some associated minerals which constitute the gangue (Fe_2O_3, Al_2O_3 and CaO) are tabulated in Table 1. Mixing of the four samples was carried out to achieve a minimum percent of main wastes in talc (Fe_2O_3, Al_2O_3 and CaO), and a maximum percent of talc constituents (SiO_2 and MgO). Equal mixing of samples did not result any improvement of constituents in the head sample. The percentage of constituents was improved by mixing 5% of sample 1, 5% of sample 2, 45% of sample 3 and 45% of sample 4. The percentage of magnesium oxide and silicon dioxide increased compared with the equal mixing; the percentage of calcium oxide decreased from 2.2% to 0.8%; the percentage of aluminum oxide decreased from 7% to 5.5% and the percentage of iron oxide decreased from 8.6% to 8.2%. The final head sample was crushed to minus 35 mm in a semi-industrial jaw crusher, then to minus 4.75 mm in a laboratory jaw crusher. A wet attrition scrubbing was used owing to the friable nature of talc. The attrition scrubbing conditions were as follows: pulp density = 50%, motor speed = 1720 rpm and attrition time = 45 minutes. A particle size of minus 75 μm (the desired size for flotation process) was obtained (Boulos, 2004). The flowsheet of crushing and attritioning processes is shown in Fig 1. The final product was collected, filtered and dried. The chemical analysis of the studied head sample is given in Table 2. The particle size distribution of the flotation feed is shown in Table 3.

Sample No.	Petrographically	*Constituent, %*				
		MgO	*SiO_2*	*Fe_2O_3*	*Al_2O_3*	*CaO*
1	Tremolite-talc-chlorite-schist	29.0	50.8	9.67	6.95	2.05
2	Antigorite-serpentinite	28.5	45.2	8.68	10.6	5.81
3	Talc-schist	30.4	58.2	7.04	3.02	0.18
4	talc-chlorite-schist	29.6	51.3	9.12	7.35	0.81

Table 1. The petrographical types and chemical analysis of different samples

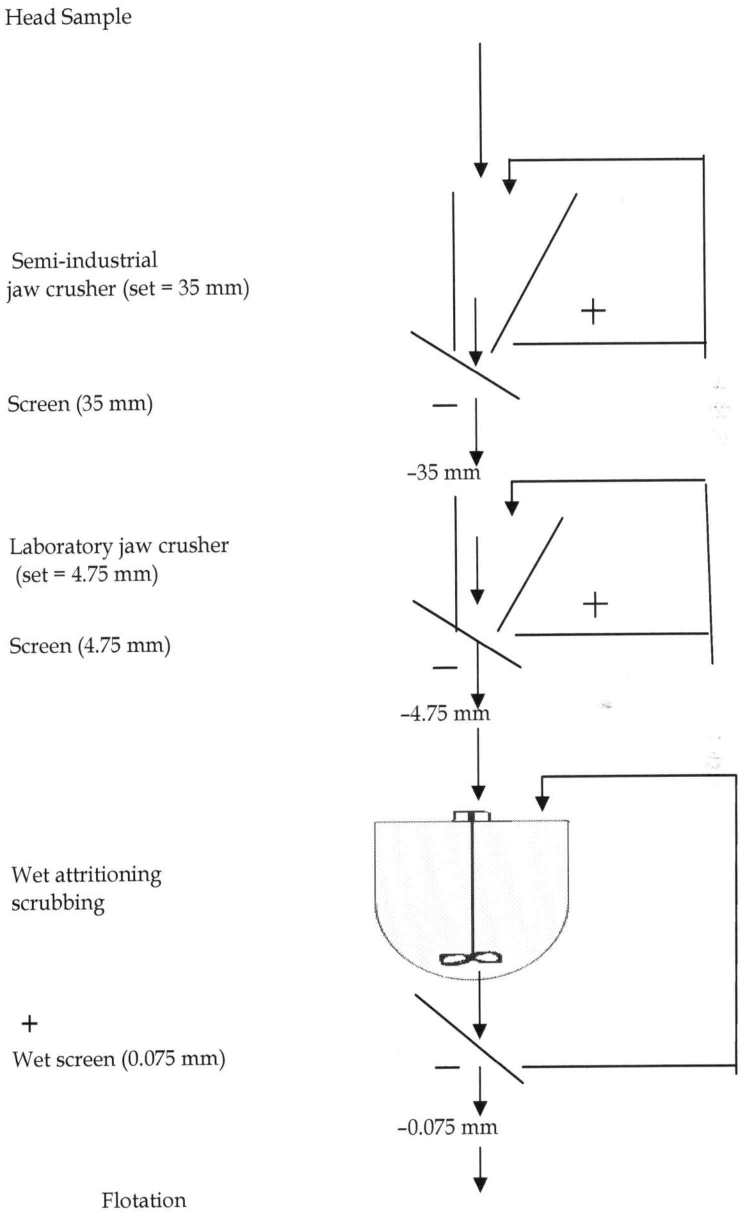

Fig. 1. Flowsheet of crushing and attritioning processes

Assay (A_f), %									
SiO_2	MgO	CaO	Al_2O_3	Fe_2O_3	P_2O_5	Na_2O	K_2O	MnO	SO_3
54.10	29.90	0.80	5.50	8.20	0.52	0.12	0.15	0.16	0.43

Table 2. Chemical analysis of the studied head sample

Size fraction, µm	wt. ret., %	cum. wt. ret., %
-75+53	10.0	10.0
-5.3+45	42.0	52.0
-45+38	8.0	60.0
-38	40.0	100.0
Σ	100.0	

Table 3. The particle size distribution of the flotation feed

2.1.2 Reagents

The flotation tests were carried out using oleic acid (in an equal mixture with kerosene) as a pure collector (iodine value 85-95, acid value 196-204 and molecular weight 282.47). The dosage was varied from 0.6 to 1.4 kg/t (Boulos, 2004). Sodium hexametaphosphate (SHMP) were employed as a selective depressant of carbonates. It was changed from 0.4 to 1.2 kg/t (Boulos, 2004; Andreola et al. 2006a, 2006b). The frother agent used for all tests was pine oil (a dosage of 0.1 kg/t). Sodium hydroxide and hydrochloric acid (30%-34%) were used to adjust the pH of the medium. Tap water was used to maintain the flotation pulp level. All other conditions were kept constant.

2.2 Apparatus

Laboratory flotation tests were carried out in a 2800 cm3 Wemco Fagergren cell. The impeller speed was fixed at 1100 rpm. An aeration rate of 6 L/min was used. A water Perspex tank was used to maintain the pulp level at a constant value. Hand skimming was used to collect the froth overflow.

Leaching test was executed in a glass reactor of 1000 cm3 capacity situated on a heater. A thermometer was used for adjusting the required temperature. The pulp was stirred with a mechanical stirrer fitted with a stainless steal impeller.

The chemical analysis of the samples was done by XRF analysis carried out at the central laboratories sector of the Egyptian Mineral Resources Authorities,

Giza, Egypt. Whiteness of talc was determined by the apparatus of Dr Lang Micro Color V2.0 (Kho and Sohn, 1989; Boulos, 2004; Al-Wakeel, 1996).

2.3 Procedure

All flotation tests were carried out at room temperature. The total conditioning time was 10 min. The talc sample was added slowly and conditioned with water for five min. Further water was then added to bring the liquid level to 10 mm below the overflow lip. The pH modifiers were added to adjust the required pH. The depressant dosage (sodium hexametaphosphate) was added at the end of the initial conditioning period and was allowed to condition for 2 min. with the pulp. The collector dosage (oleic acid and kerosene mixture) was added at the end of the previous period and was allowed to condition for 2 min. with the pulp. The frother dosage was then added and a further 1 min. of conditioning was allowed prior to aeration. The air supply was gradually opened. The required pulp level was maintained constant.

In each experiment, after allowing 15 seconds for froth to form, a hand skimmer was used to collect the froth over until the froth is stopped. After the process being finished, the products (concentrate and tailings) were dried, weighed and chemically analyzed.

The final concentrate of flotation was leached with a diluted hydrochloric acid having a concentration of 10% using a solid liquid ratio of (1:2) and at a temperature of 60 0C for a period of 30 min. The residues of leaching process were washed with acidulated hot water and then with pure water. After the process being finished, the concentrate was dried, weighed, and chemically analyzed.

3. Results and discussions

3.1 Calculations of experimental mass and component recoveries of flotation products

Using the mass percent and assays of different constituents in the feed, concentrate, and tailings, the experimental values of mass and component recoveries of the flotation products can be calculated as follows:

$$\text{Mass recovery of concentrate} = R_m(c) = 100 . \frac{C}{F} \tag{1}$$

$$\text{Mass recovery of tailings} = R_m(t) = 100 . \frac{T}{F} \tag{2}$$

$$\text{Component recovery in concentrate} = R_c(c) = 100 . \frac{C.c}{F.f} \tag{3}$$

$$\text{Component recovery in tailings} = R_c(t) = 100 . \frac{T.t}{F.f} \tag{4}$$

3.2 Effect of pH

Table 4a includes the chemical analysis of constituents in flotation products of experiments carried out at different pH values. These tests were carried out at collector dosage = 1.0 kg/t, depressant dosage = 0.8 kg/t, and pulp density = 200 g/L. This table showed that the percentages of SiO_2 and MgO (the major constituents of talc) decreased in the concentrate product with increasing pH value from 4 to 12. Conversely, the percentages of CaO, Al_2O_3 and Fe_2O_3 (the main wastes in talc) increased.

These results are illustrated also in Figures 2a through 2e. From Figures 2a and 2b, it can be seen that, the SiO_2 assay decreased from pH 4 to pH 12 (57.02% to 54.30%), as well as, MgO assay decreased from pH 4 to pH 12 (34.0% to 31.8%). This may be interpreted due to the tendency of magnesium to precipitate or hydrolysis with increasing pH, which leads to decrease its content, especially at the alkaline values (Rath et al. 1995; Bremmell and Addai-Mensah, 2005).

Figures 2c through 2e showed that with increasing the value of pH from 4 to 12, CaO increased (from 0.35% to 0.72%), Al_2O_3 increased (from 2.17% to 4.30%), and Fe_2O_3 increased (from 6.22% to 8.10%). This may be attributed to that, the carbonates content increase with increasing pH. These results are in agreement with the work of Al-Wakeel, 1996.

Exp. No.	pH	Concentrate Assay (A_c), %					Tailings Assay (A_t), %				
		SiO_2	MgO	CaO	Al_2O_3	Fe_2O_3	SiO_2	MgO	CaO	Al_2O_3	Fe_2O_3
1	4	57.02	34.00	0.35	2.17	6.22	50.81	25.28	1.31	9.26	10.43
2	7	56.15	33.39	0.42	2.62	6.53	50.41	23.61	1.48	10.69	11.21
3	9	55.9	33.03	0.44	2.78	6.85	50.45	23.55	1.53	11.02	10.94
4	11	55.71	32.92	0.50	3.03	7.14	45.32	13.44	2.44	18.97	13.98
5	12	54.3	31.81	0.72	4.3	8.10	53.40	23.24	1.08	9.68	8.55

Table 4a. The chemical analysis of constituents in flotation products of experiments carried out at different pH values

The percentage of iron oxide is still high due to the association of iron oxide with all mineral components of the sample, which sometimes appearing as yellowish brown threads along its schistose structure and in other cases, associating with the banded talc- chlorite structure (Boulos, 2004).

The effect of pH values on the mass recovery of concentrate is shown in Table 4b and Fig. 2f. From this figure, it can be revealed that, the mass recovery increased from 53% to 84.5%, as the value of pH increased from pH 4 to pH 11. More increase of the pH value, above 11, decreased the mass recovery.

pH	Concentrate Component recovery $R_c(c)$, %					Concentrate $R_m(c)$, %	Tailings Component recovery R_c (t), %					Tailings $R_m(t)$, %
	SiO_2	MgO	CaO	Al_2O_3	Fe_2O_3		SiO_2	MgO	CaO	Al_2O_3	Fe_2O_3	
4	55.9	60.3	23.2	20.9	40.2	53.0	44.1	39.7	76.8	79.1	59.8	47.0
7	66.7	71.8	33.8	30.6	51.2	64.3	33.3	28.2	66.2	69.4	48.8	35.7
9	69.2	74.0	36.9	33.9	56.0	67.0	30.8	26.0	63.1	66.1	44.0	33.0
11	87.0	93.0	52.8	46.6	73.6	84.5	13.0	7.0	47.2	53.4	26.4	15.5
12	78.0	82.7	69.9	60.7	76.8	77.7	22.0	17.3	30.1	39.1	23.2	22.3

Table 4b. The mass recoveries of flotation products and the component recoveries of constituents of experiments carried out at different pH values

The effect of pH values on the component recoveries of constituents in concentrate is illustrated also in Table 4b and Fig. 2f. From this figure, it can be shown that, the component recovery of SiO_2 increased from 56% to 87%, as well as, the component recovery of MgO increased from 60% to 93% as the pH value increased from pH 4 to pH 11. Any increase of pH value (above pH 11) decreased the component recoveries of these two constituents. On the other hand, the component recoveries of CaO, Al_2O_3 and Fe_2O_3 increased with increasing the pH value.

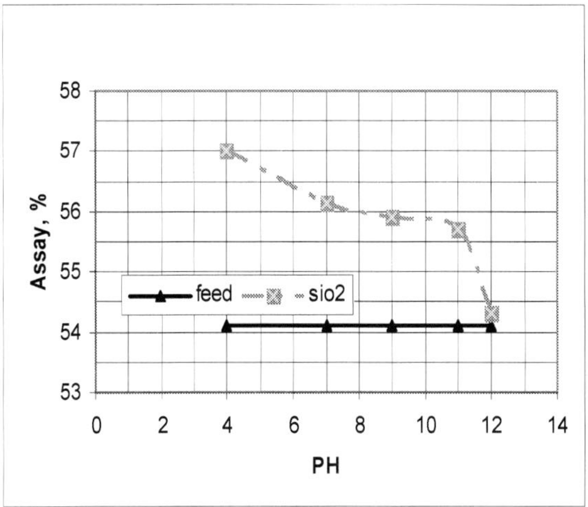

a

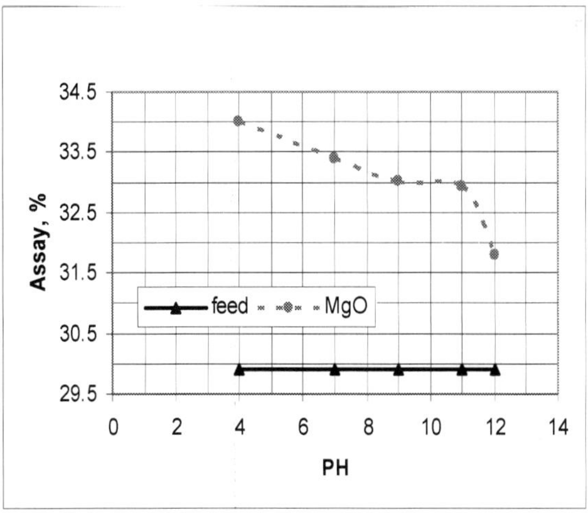

b

Fig. 1. a. SiO_2 assay with pH b. MgO assay with pH

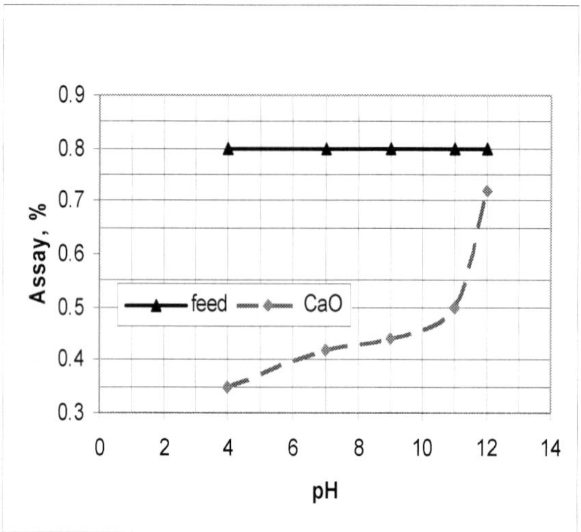

c

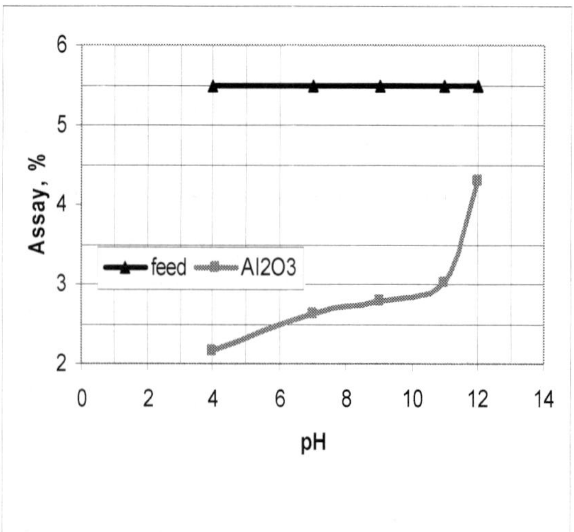

d

Fig. 1. c. CaO assay with pH , d. Al_2O_3 assay with pH

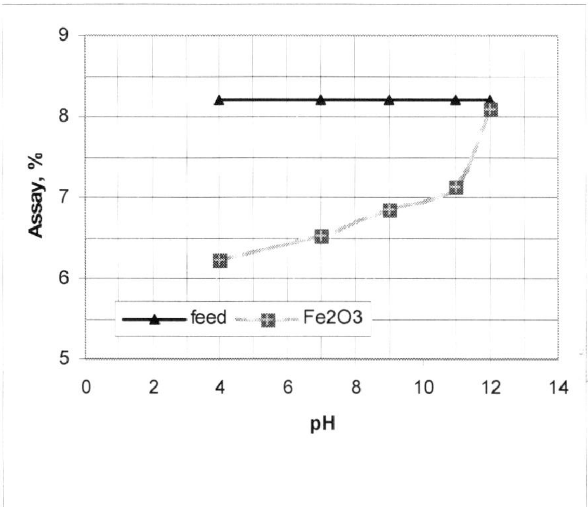

e

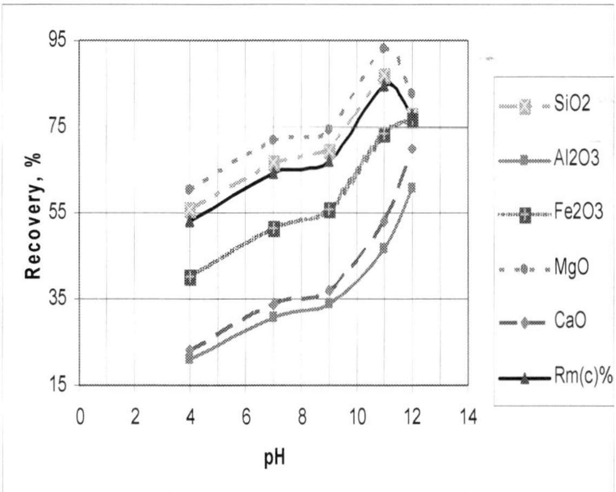

f

Fig. 1. e. Fe_2O_3 assay with pH, f. Mass and component recoveries with pH

Fig. 1. Effect of pH values on the mass recovery, as well as, on the assays and component recoveries of different constituents in concentrate

The above results can be interpreted such that, by increasing the pH, the effective coverage by the collector is enhanced and the resulting talc surface therefore becomes more hydrophobic leading to greater talc recovery. The native talc surface does not become more hydrophobic (Fuersrenau and Huang, 2003) but in fact above pH 9 becomes more hydrophilic and the recovery drops as pH goes up. The recovery of the oleic acid treated talc was shown to increase as the pH increased from 2 to 11 however the rate of increase from pH 7 to 11 was much greater than the rate of increase from 2 to 7 (see Fig 2f) (Al-Wakeeel, 1996). The anisotropy nature of talc revealed that their crystals consist of faces, which are being hydrophobic and not charged, and edges being hydrophilic and charged. The bubbles will attach to the hydrophobic faces where they are not affected by the edge charge. At a higher pH value, over 11, the overall charge on the particle may give a rise to a high repulsion of air bubbles. This may be due to that some dissolution of particles takes place at high pH and adsorption of hydrolyzed species may contribute to increase of hydrophilicity (Fuerstenau and Huang, 2003). Although the talc has a point of zero charge (PZC) at pH 1.8 (Al-Wakeeel, 1996), the using of oleic acid, as an anionic collector, will change zeta potential of talc from a negative to a positive value at pH 11 (Xu et al. 2004).
From Table 4a, it can be seen that the assays of SiO_2 and MgO were at their lowest values in tailings, as well as, the assays of CaO, Al_2O_3 and Fe_2O_3 were at their highest ones at the same value of pH 11. This assures that the optimum conditions of these experiments can be obtained at pH 11.

3.3 Effect of depressant dosage

Table 5a contains the chemical analysis of constituents in flotation products of experiments executed at different depression dosages of sodium hexametaphosphate (SHMP). These experiments were executed at pH 11, collector dosage = 1.0 kg/t, and pulp density = 200 g/L. The table showed that the percentages of SiO_2 and MgO (the major constituents of talc) decreased in the concentrate product with increasing the depression dosage from 0.4 to 1.2 kg/t. Conversely, the percentages of CaO, Al_2O_3 and Fe_2O_3 (the main wastes in talc) increased. These results are also revealed in Figures 3a through 3e. From Figures 3a and 3b, it can be shown that, SiO_2 decreased (from 57.23% to 54.46%) with increasing the depression dosage from 0.4 to 1.2 kg/t, as well as MgO decreased (from 34.13% to 31.29%) with the same dosage.
Figure 3c shows that CaO decreased (from 0.72% to 0.42%) if the depression dosage was increased from 0.4 Kg/t to 1.0 Kg/t. More increasing of the dosage will lead to increase the final percent of CaO. This trend can be explained due to the ability of sodium hexametaphosphate (SHMP) to sequester the calcium cations (Ca2+) and forming with the calcium a strong hydrophilic complex compound (Andreola et al. 2006a, 2006b). Figures 3d and 3e showed that with increasing the depression dosage from 0.4 to 1.2 kg/t, Al_2O_3 increased (from 1.98% to 4.21%), as well as, Fe_2O_3 increased (from 5.88% to 8.07%). This could possibly be due to the interaction of SHMP anions with the exposed Al atoms to give complexed anions but, as Andreola et. al. (2006a) has shown, SHMP is too weak to sequester the aluminum sites.
The effect of depression dosages on the mass recovery of concentrate is shown in Table 5b and Fig. 3f. From this figure, it can be seen that, the mass recovery increased from 71.2% to 88.0%, as the value of depression dosage increased from 0.4 to 1.0 kg/t. More increase of the depression dosage, above 1.0 kg/t, decreased the mass recovery. The effect of depression dosages on the component recoveries of constituents in concentrate is illustrated also in Table 5b and Fig. 3f. From this figure, it can be revealed that, the component recovery of

SiO_2 increased (from 75.3% to 89.4%), as well as, the component recovery of MgO increased (from 81.3% to 93.9%) as the depression dosage increased from 0.4 to 1.0 kg/t. Any increase of depression dosage above this value decreased the component recoveries of these two constituents. From Fig. 3f, it can be also shown that the component recovery of CaO decreased (from 64.0% to 46.2%) as the depression dosage increased from 0.4 to 1.0 kg/t and then increased by increasing of depression dosage above 1.0 kg/t. On the other hand, the component recoveries of Al_2O_3 and Fe_2O_3 increased with increasing the depression dosage value.

Exp. No.	Depression dosage, kg/t	Concentrate					Tailings				
		Assay (A_c), %					Assay (A_t), %				
		SiO_2	MgO	CaO	Al_2O_3	Fe_2O_3	SiO_2	MgO	CaO	Al_2O_3	Fe_2O_3
1	0.4	57.23	34.13	0.72	1.98	5.88	46.36	19.44	1.00	14.20	13.94
2	0.6	56.57	33.67	0.63	2.14	6.35	45.59	16.91	1.39	17.07	14.57
3	0.8	55.71	32.92	0.50	3.03	7.14	45.32	13.44	2.44	18.97	13.98
4	1.0	54.97	31.89	0.42	3.08	7.21	41.36	15.31	3.59	23.25	15.46
5	1.2	54.46	31.29	0.68	4.21	8.07	52.96	25.5	1.18	9.59	8.61

Table 5a. The chemical analysis of constituents in flotation products of experiments carried out at different depression dosages

Depression dosage, kg/t	Concentrate						Tailings					
	Component recovery $R_c(c)$, %					$R_m(c)$, %	Component recovery R_c (t), %					$R_m(t)$, %
	SiO_2	MgO	CaO	Al_2O_3	Fe_2O_3		SiO_2	MgO	CaO	Al_2O_3	Fe_2O_3	
0.4	75.3	81.3	64.0	25.6	51.1	71.2	24.7	18.7	36.0	74.4	48.9	28.8
0.6	81.0	87.3	61.0	30.2	60.0	77.5	19.0	12.7	39.0	69.8	40.0	22.5
0.8	87.0	93.0	52.8	46.6	73.6	84.5	13.0	7.0	47.2	53.4	26.4	15.5
1.0	89.4	93.9	46.2	49.3	77.4	88.0	10.6	6.1	53.8	50.7	22.6	12.0
1.2	76.5	79.5	64.6	58.2	74.8	76.0	23.5	20.5	35.4	41.8	25.2	24.0

Table 5b. The mass recoveries of flotation products and the component recoveries of constituents of experiments executed at different depression dosages

From Table 5a, it can be seen that the assays of SiO_2 and MgO were at their lowest values in tailings, as well as, the assays of CaO, Al_2O_3 and Fe_2O_3 were at their highest ones at the same value of depression dosage (1.0 kg/t). This assures that the optimum conditions of these experiments may be obtained at a depression dosage of 1.0 kg/t.

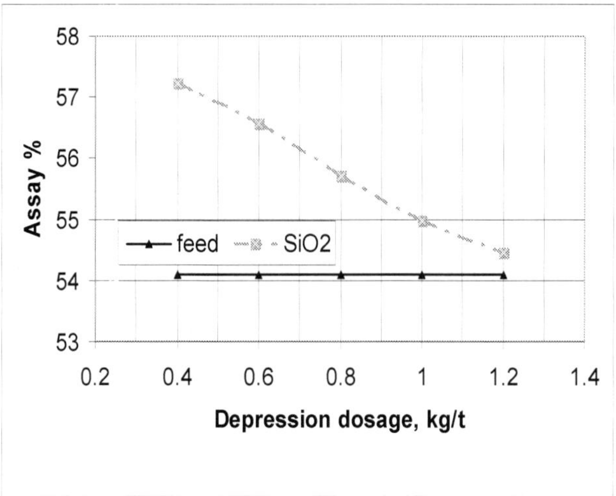

a

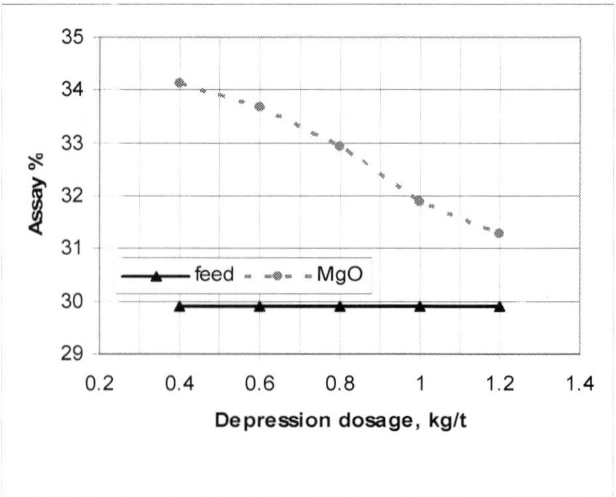

b

Fig. 2. a. SiO_2 assay with depression dosage , b: MgO assay with depression dosage

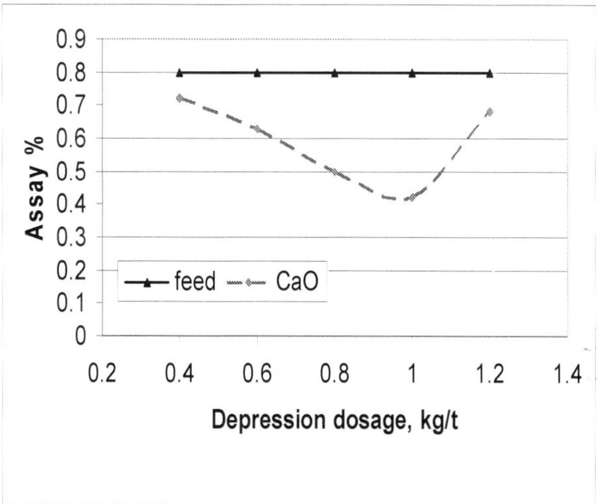

c

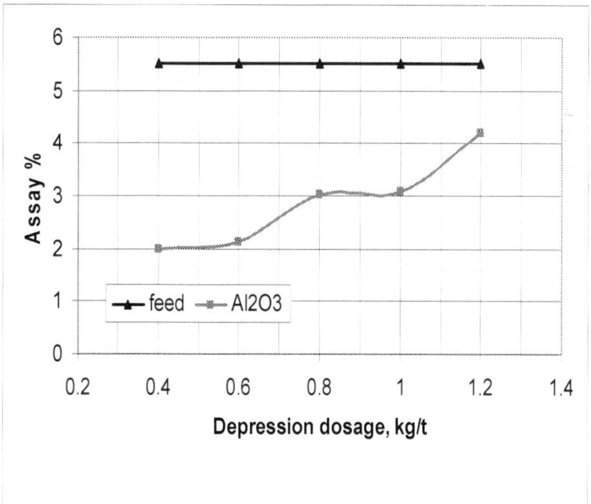

d

Fig. 2. C. CaO assay with depression dosage , d. Al_2O_3 assay with depression dosage

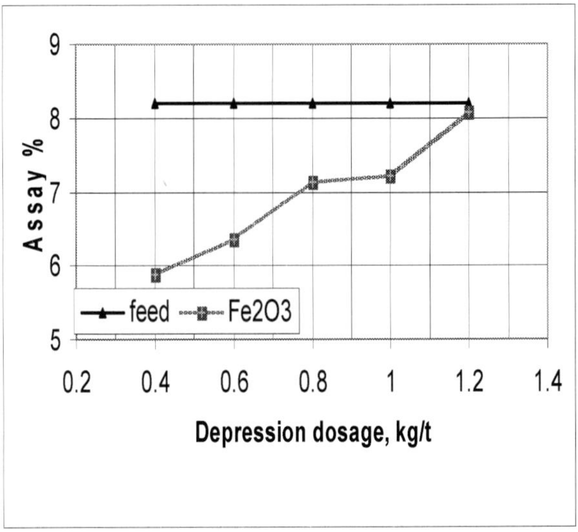

e

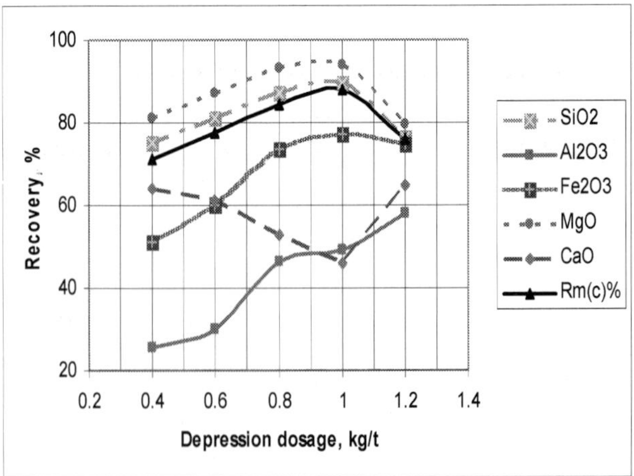

f

Fig. 2. E. Fe_2O_3 assay with depression dosage, f. Mass and component recoveries with depression dosage

Fig. 2. Effect of depression dosage on the mass recovery, as well as, on the assays and component recoveries of different constituents in concentrate

3.4 Effect of collector dosage

The experiments of collector dosage parameter were carried out at pH 11, depressant dosage of 1.0 kg/t, and pulp density of 200 g/L. Table 6a included the chemical analysis of constituents in flotation products of experiments carried out at different collector dosages of oleic acid (in an equal mixture with kerosene). The table showed that the percentages of SiO_2 and MgO (the major constituents of talc) decreased in the concentrate product with increasing the collector dosage from 0.6 to 1.4 kg/t. Conversely, the percentages of CaO, Al_2O_3 and Fe_2O_3 (the main wastes in talc) increased. These results are illustrated also in Figures 4a through 4e. From Figures 4a and 4b, it can be shown that, SiO_2 decreased (from 57.46% to 54.15%) at increasing of the collector dosage from 0.6 to 1.4 kg/t, as well as MgO decreased (from 34.22% to 30.69%) at the same collector dosage. This may be attributed to the power of oleic acid (anionic collector) to produce a water-repulsion and monomolecular layer on particle surfaces (starvation level), thereby imparting hydrophobicity to the particles (Wills and Napier-Munn, 2006).

Exp. No.	Collector dosage, kg/t	Concentrate Assay (A_c), %					Tailings Assay (A_t), %				
		SiO_2	MgO	CaO	Al_2O_3	Fe_2O_3	SiO_2	MgO	CaO	Al_2O_3	Fe_2O_3
1	0.6	57.46	34.22	0.30	2.09	5.90	45.42	18.74	2.09	14.31	14.14
2	0.8	55.78	32.83	0.36	2.75	6.78	47.16	17.80	2.62	16.85	14.06
3	1.0	54.97	31.89	0.42	3.08	7.21	41.36	15.31	3.59	23.25	15.46
4	1.2	54.94	31.76	0.42	3.11	7.24	41.35	7.29	5.42	24.20	19.87
5	1.4	54.15	30.69	0.54	3.95	7.97	49.13	24.98	2.14	13.52	8.05

Table 6a. The chemical analysis of constituents in flotation products of experiments carried out at different collector dosages

Collector dosage, kg/t	Concentrate Component recovery R_c (c), %					Concentrate R_m (c),%	Tailings Component recovery R_c (t), %					Tailings R_m(t), %
	SiO_2	MgO	CaO	Al_2O_3	Fe_2O_3		SiO_2	MgO	CaO	Al_2O_3	Fe_2O_3	
0.6	76.6	82.5	27.0	27.4	51.9	72.1	23.4	17.5	73.0	72.6	48.1	27.9
0.8	83.0	88.4	36.2	40.3	66.6	80.5	17.0	11.6	63.8	59.7	33.4	19.5
1.0	89.4	93.9	46.2	49.3	77.4	88.0	10.6	6.1	53.8	50.7	22.6	12.0
1.2	93.8	98.2	48.5	52.3	81.6	92.4	6.2	1.8	51.5	47.7	18.4	7.6
1.4	83.9	86.0	56.6	60.2	81.5	83.8	16.1	14.0	43.4	39.8	18.5	16.2

Table 6b. The mass recoveries of flotation products and the component recoveries of constituents of experiments carried out at different collector dosages

Figures 4c through 4e showed that as the collector dosage increased from 0.6 to 1.4 kg/t, CaO increased (from 0.30% to 0.54%), Al_2O_3 increased (from 2.09% to 3.95%), and Fe_2O_3 increased (from 5.90% to 7.97).

The effect of collector dosages on the mass recovery of concentrate is shown in Table 6b and Fig. 4f. From this figure, it can be revealed that, the mass recovery increased from 72.1% to 92.4%, as the collector dosage increased from 0.6 to 1.2 kg/t. More increase of the collector dosage above 1.2 kg/t decreased the mass recovery.

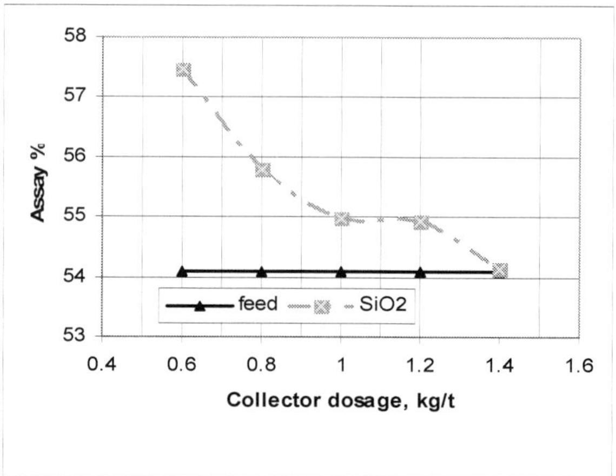

a

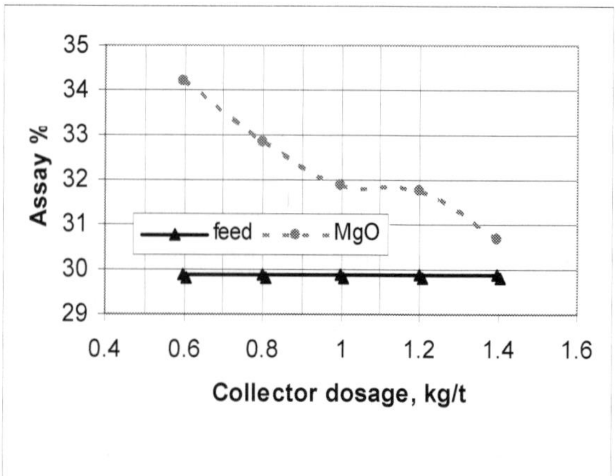

b

Fig. 3. a. SiO_2 assay with collector dosage, b. MgO assay with collector dosage

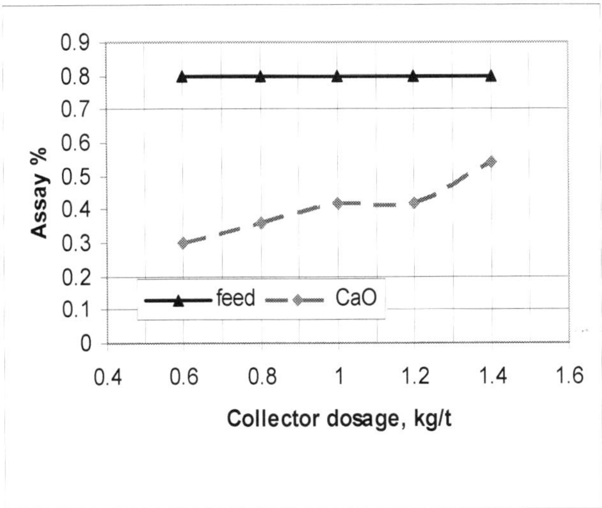

c

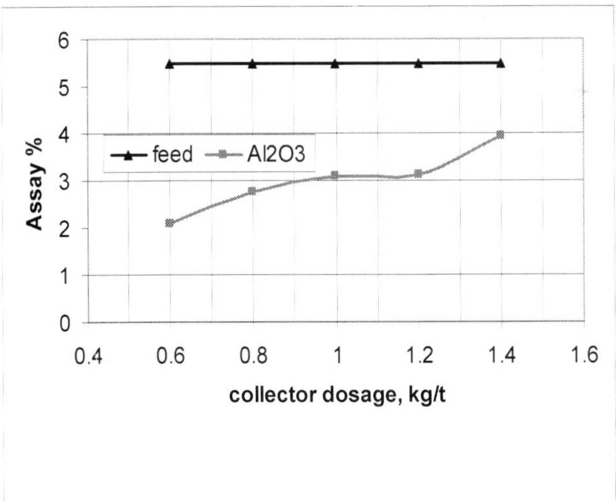

d

Fig. 3. C. CaO assay with collector dosage , d. Al_2O_3 assay with collector dosage

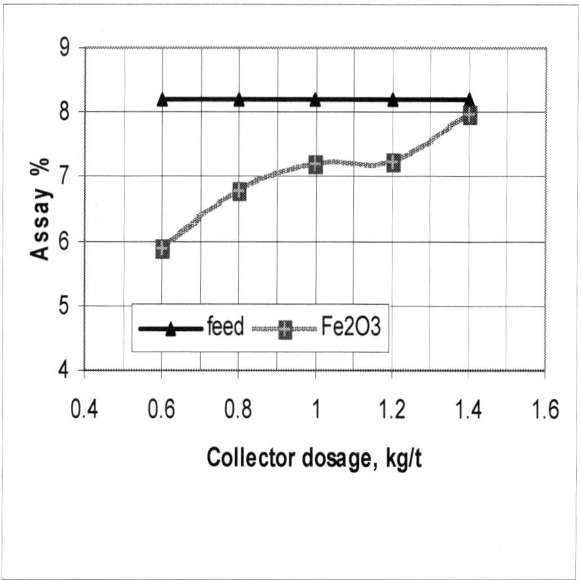

e

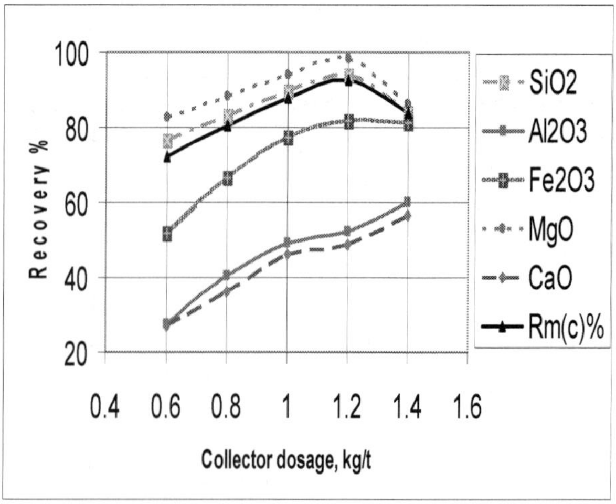

f

Fig. 3. e. Fe_2O_3 assay with collector dosage, f. Mass and component recoveries with collector dosage

Fig. 3. Effect of collector dosage on the mass recovery, as well as, on the assays and component recoveries of different constituents in concentrate

The increase of the flotation recovery with the collector dosage may be explained due to the more rapid reaction at higher concentration or due to more rapid approach of the exchanger adsorption equilibrium at higher concentration. This trend can be explained also due to the displacement of the exchange-adsorption equilibrium more and more toward complete collector adsorption as concentration of the collector is increased (Ahmed, 1995). The effect of collector dosages on the component recoveries of constituents in concentrate are illustrated also in Table 6b and Fig. 4f. From this figure, it can be shown that, the component recovery of SiO_2 increased (from 76.6% to 93.8%), as well as, the component recovery of MgO increased (from 82.5% to 98.2%) as the collector dosage increased from 0.6 to 1.2 kg/t. This may be interpreted due to that, the faces of the talc are mildly hydrophobic and thus contribute largely to the floatability of the talc. At low collector concentration, only the faces are covered making them more hydrophobic leading to an increase in recovery (Ahmed, 1995). At higher dosages, the hydrophilic edges start to be covered and recovery is further enhanced (Fuerstenau and Huang, 2003). At dosages higher than 1.2 kg/t recovery begins to decrease possibly due to a build up of multilayers of collector on the surface. At these high dosages collector starts to build up on the other minerals leading to a reduction in selectivity and a lowering of the grade (Wills and Napier-Munn, 2006).

On the other hand, the component recoveries of CaO, Al_2O_3 and Fe_2O_3 increased with increasing the collector dosages.

From Table 6a, it can be seen that the assays of SiO_2 and MgO were at their lowest values in tailings, as well as, the assays of CaO, Al_2O_3 and Fe_2O_3 were at their highest ones at the same value of collector dosage (1.2 kg/t). This assures that the optimum conditions of these experiments are obtained at a collector dosage of 1.2 kg/t.

3.5 Effect of pulp density

Table 7a contains the chemical analysis of constituents in flotation products of experiments executed at different pulp densities. These experiments were carried out at pH 11, 1.0 kg/t depressant dosage and a collector dosage of 1.2 kg/t. The table showed that the percentages of SiO_2 and MgO (the major constituents of talc) decreased in the concentrate product with increasing the pulp density from 100 to 300 g/L. Conversely, the percentages of CaO, Al_2O_3 and Fe_2O_3 (the main wastes in talc) increased. These results are illustrated also in Figures 5a through 5e.

Figures 5c through 5e revealed that CaO increased (from 0.29% to 0.68%), Al_2O_3 increased (from 2.02% to 4.86%), and Fe_2O_3 increased (from 5.59% to 8.04%) as the pulp density increased from 100 to 300 g/L.

The effect of pulp density on the mass recovery of concentrate is shown in Table 7b and Fig. 5f. From this figure, it can be shown that, the mass recovery increased from 70.0% to 92.4%, as the value of pulp density increased from 100 to 200 g/L. More increase of the pulp density, above 200 g/L, decreased the mass recovery.

The effect of pulp density on the component recoveries of constituents in concentrate is illustrated also in Table 7b and Fig. 5f. From this figure, it can be revealed that the component recovery of SiO_2 increased from 74.5% to 93.8%, as well as, the component recovery of MgO increased from 80.3% to 98.2% as the pulp density increased from 100 to 200 g/L. Any increase of pulp density above 200 g/L decreased the component recoveries of these constituents. Feng and Aldrich, (2004) stated that an 8% solids concentration was optimal with respect to the lowest recovery of talc. Higher pulp concentrations had a detrimental effect on flotation. Wills, (2006) has reported that the denser the pulp, the less cell volume is required in the commercial plant and fewer reagents are required, since the effectiveness of most reagents is a function of their concentration in solution. The optimum pulp density is of great important, as in general the more dilute the pulp, the cleaner the separation.

Exp. No.	Pulp density, g/L	Concentrate					Tailings				
		Assay (A_c), %					Assay (A_t), %				
		SiO_2	MgO	CaO	Al_2O_3	Fe_2O_3	SiO_2	MgO	CaO	Al_2O_3	Fe_2O_3
1	100	57.60	34.30	0.29	2.02	5.59	45.93	19.63	1.99	13.62	14.29
2	150	56.24	33.07	0.34	2.63	6.49	43.43	14.09	3.09	19.82	16.73
3	200	54.94	31.76	0.42	3.11	7.24	41.35	7.29	5.42	24.20	19.87
4	250	54.75	31.65	0.52	3.69	7.83	50.56	20.36	2.33	15.37	10.22
5	300	54.23	30.88	0.68	4.86	8.04	53.62	26.26	1.25	7.88	8.79

Table 7a. The chemical analysis of flotation products of different experiments carried out at different pulp densitiesFrom Figures 5a and 5b, it can be revealed that, as the pulp density increased from 100 to 300 g/L, SiO_2 decreased (from 57.60% to 54.23%), as well as, MgO decreased (from 34.30% to 30.88%).

Pulp density, g/L	Concentrate						Tailings					
	Component recovery $R_c(c)$, %					$R_m(c)$, %	Component recovery R_c (t), %					$R_m(t)$, %
	SiO_2	MgO	CaO	Al_2O_3	Fe_2O_3		SiO_2	MgO	CaO	Al_2O_3	Fe_2O_3	
100	74.5	80.3	25.4	25.7	47.7	70.0	25.5	19.7	74.6	74.3	52.3	30.0
150	86.6	92.1	35.4	39.8	65.9	83.3	13.4	7.8	64.6	60.2	34.1	16.7
200	93.8	98.2	48.5	52.3	81.6	92.4	6.2	1.8	51.5	47.7	18.4	7.6
250	85.5	89.5	54.9	56.7	80.7	84.5	14.5	10.5	45.1	43.3	19.3	15.5
300	78.0	81.4	67.0	69.6	77.3	78.8	22.0	18.6	33.0	30.4	22.7	21.2

Table 7b.The mass recoveries of flotation products and the component recoveries of constituents of experiments executed at different pulp densities

From Figure 5f, it can be seen that the component recoveries of CaO, Al_2O_3 and Fe_2O_3 increased with increasing pulp densities.

From Table 7a, it can be shown that the assays of SiO_2 and MgO were at their lowest values in tailings, as well as, the assays of CaO, Al_2O_3 and Fe_2O_3 were at their highest ones at the same value of pulp density (200 g/L). The whiteness improved from 75.4% to 83.7% and the loss on ignition decreased from 6.64% to 5.21% in the final product. This assures that the optimum condition of these experiments is obtained at a pulp density of 200 g/L.

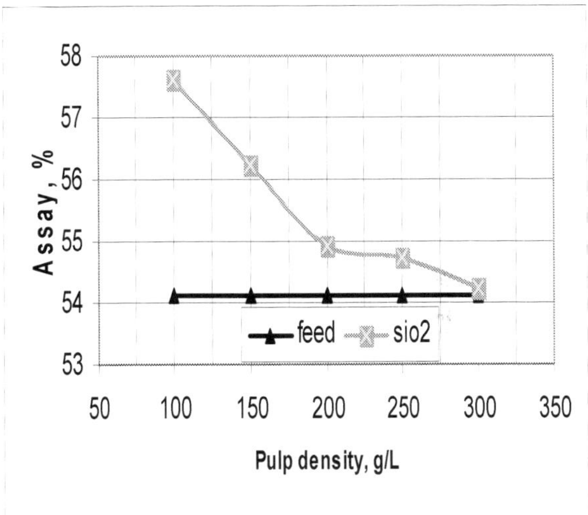

a

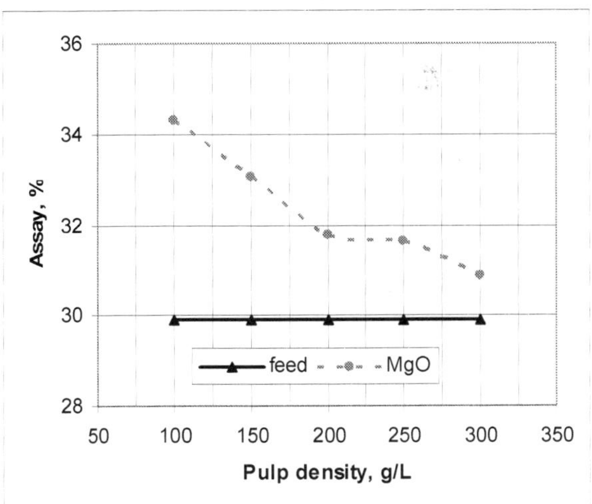

b

Fig. 4. a. SiO_2 assay with pulp density , b. MgO assay with pulp density

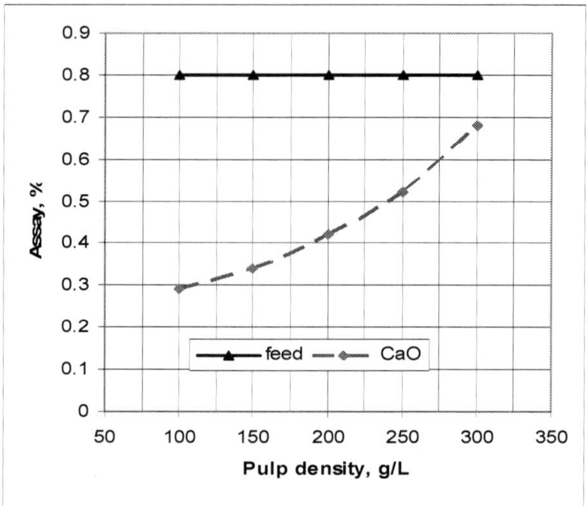

c

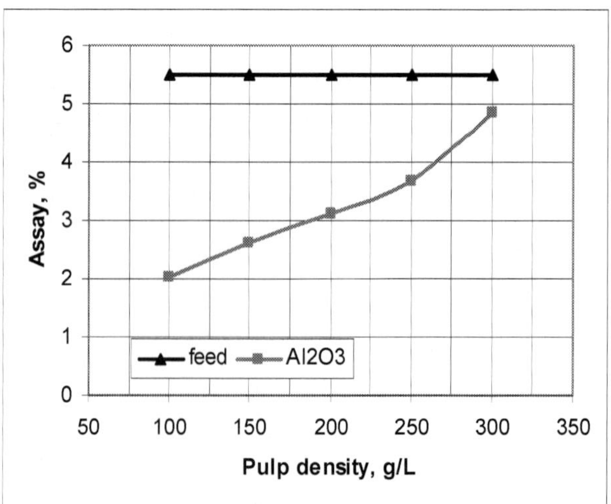

d

Fig. 4. c. CaO assay with pulp density , d. Al_2O_3 assay with pulp density

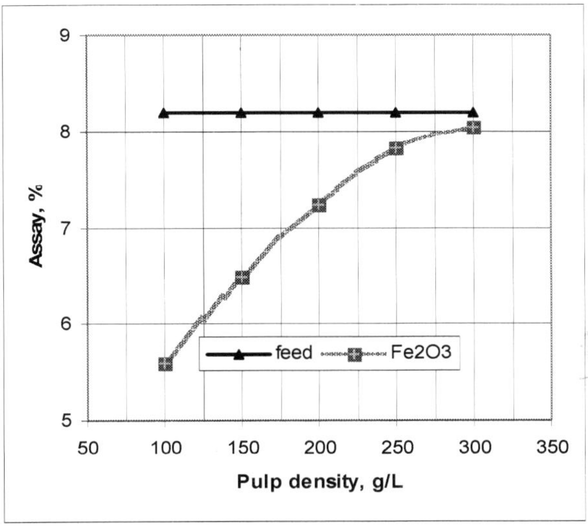

e

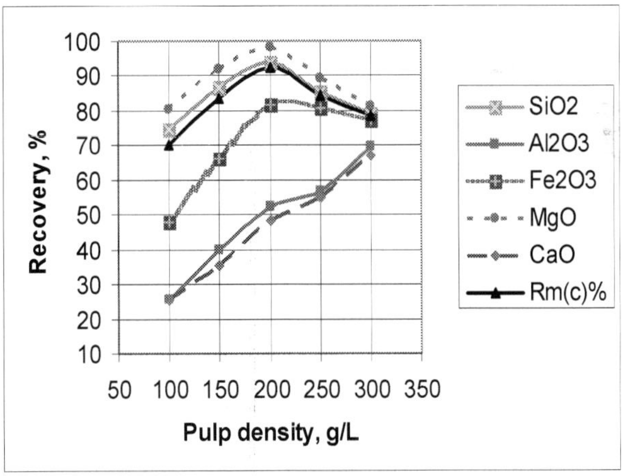

f

Fig. 4. e. Fe_2O_3 assay with pulp density, f. Mass and component recoveries with pulp density

Fig. 4. Effect of pulp density on the mass recovery, as well as, on the assays and component recoveries of different constituents in concentrate

From Table 7 and Fig. 5, it can be revealed that there are another operating conditions, which give better result of the characteristics of talc in concentrate but with lower mass recovery. This conclusion was obtained at a pulp density of 100 g/L. At this condition, the assays of SiO_2 (57.60%) and MgO (34.30%) were at their highest values in concentrate, as well as, the assays of CaO (0.29%), Al_2O_3 (2.02%) and Fe_2O_3 (5.59%) were at their lowest ones. The mass recovery was the lowest one (70%), as well as, the component recovery of talc constituents were also the lowest ones (74.5% and 80.3% respectively).

3.6 Leaching of final concentrate of flotation

More improvement of talc quality was obtained by leaching of final concentrate of flotation with a dilute hydrochloric acid having a concentration of 10%, solid-liquid ratio (1:2), at a temperature of 60° C for a period of 30 minutes using a mechanical stirrer. The residues of leaching process were washed with acidulated hot water and then with pure water. After the process being finished, the concentrate was dried, weighed and chemically analyzed. The final results of chemical analysis of talc constituents, mass recovery, whiteness, and loss on ignition in feed, final concentrate of flotation, and leaching product are shown in Table 8. The mass recovery of final product was about 85% of initial feed. In this final product, Fe_2O_3 decreased to 3.12%, CaO decreased to 0.38%, and Al_2O_3 decreased to 3.01%. The assay of SiO_2 increased to 58.96%, MgO percent was nearly the same, the whiteness increased to 88.5%, and the loss on ignition decreased to 4.03%. The improvement of whiteness may be attributed to the reduction of iron content in the final product (Hassan, 2007 and Ahmed, et al 2007). The experimental errors of all experiments were within the permissible limits, i.e. lesser than 5%.

		Feed	Final concentrate of flotation	Leaching product
Assay, %	**SiO_2**	54.10	54.94	58.96
	MgO	29.90	31.76	31.73
	CaO	0.80	0.42	0.38
	Al_2O_3	5.50	3.11	3.01
	Fe_2O_3	8.20	7.24	3.12
Mass recovery, %		100	92.40	85.00
Whiteness, %		75.4	83.7	88.5
Loss on ignition, %		6.64	5.21	4.03

Table 8. The final results of chemical analysis of talc constituents, mass recovery, whiteness, and loss on ignition in feed, final concentrate of flotation, and leaching product

4. Conclusions

From the results of this investigation, the following conclusions can be drawn:

1. The floatability of talc increased as the pH increased up to a value of 11. Above this value, the floatability of talc decreased.
2. The CaO and Al_2O_3 can be partially sequestered using sodium hexametaphosphate as a selective depressant at a certain limit. The optimum value was found at 1.0 kg/t.
3. Using oleic acid in conjunction with kerosene resulted in an increase in talc recovery up to a maximum of 1.2 kg/t after which recovery decreased.

4. The recovery of talc increased as the pulp density increased up to 200 g/L, above which the recovery of talc decreased.
5. The final concentrate of flotation was leached with a dilute hydrochloric acid (10%). In the final product, SiO_2 assay increased to 58.96%, and Fe_2O_3 decreased to 3.12%. The mass recovery was about 85% of the initial feed, and whiteness improved to 88.5%.
6. The final talc product of this research is suitable for many industrial uses such as low-loss electronics (a type of ceramics), paints, rubber, plastics, roofing, textiles, refractories, insecticides and coating of welding rods.

5. Nomenclature

C	=	mass of concentrate, g
T	=	mass of tailings, g
F	=	mass of feed, g
c	=	assay of constituent in concentrate, %
t	=	assay of constituent in tailings, %
f	=	assay of constituent in feed, %
Rm(c)	=	mass recovery of concentrate, %
Rm(t)	=	mass recovery of tailings, %
Rc(c)	=	component recovery in concentrate, %
Rc(t)	=	component recovery in tailings, %

6. References

[1] Ahmed, M.M., Ibrahim, G.A. and Hassan, M.M.A., 2007, "Improvement of Egyptian talc quality for industrial uses by flotation process and leaching" International Journal of Mineral Processing, Vol. 83, pp 132-145.

[2] Ahmed, M.M., 1995, "Kinetics of Maghara coal flotation", M.Sc. Thesis, Assiut University, Egypt, pp. 73-78.

[3] Al-Wakeel, M.I., 1996, "Geology and beneficiation of some Egyptian talc- carbonate rocks", Ph.D. Thesis, Ain Shams University, Egypt, pp. 313-365.

[4] Attia, M.I., 1960, "Topography, Geology and iron-ore deposits of the district East of Aswan", Ministry of Commerce and Industry, Mineral Resources Department, Geological Survey, p. 232.

[5] Andreola, F., Castellili, E., Manfredini, T. and Romagnoli, M., 2006a, "The role of sodium hexametaphosphate in the dissolution process of kaolinite and kaolin", Journal of the European Ceramic Society, Vol. 24, pp. 2113-2124.

[6] Andreola, F., Castellili, E., Ferreira, J., Olhero, S. and Romagnoli, M., 2006b, "Effect of sodium hexametaphosphate and ageing on the rheological behavior of kaolin dispersions", Applied Clay Science, Vol. 31, pp. 56-64.

[7] Andrews, P.R.A., 1985, "Laboratory study of the flotation circuit at Baker talc Inc., high water, Quebec", CIM Bulletin, Vol. 78, No. 884, pp. 75-78.

[8] Andrews, P.R.A., 1989, "Pilot-plant treatment of Quebec talc ore", CIM Bulletin, Vol. 82, No. 932, pp. 76-81.

[9] Boghdady, G.Y., Ahmed, M.M., Ibrahim, G.A. and Hassan, M.M.A., May 2005, "Petrographical and geochemical characterisation of some Egyptian talc samples

for possible industrial applications", Journal of Engineering Science, Assiut University, Vol. 33, No. 3, pp. 1001-1011.

[10] Boulos, T.R., 2004, "Transforming upgrading of talc for different industrial application", Final Report supmeted to the Egyptian Academy for Scientific Research and Technology, pp. 1-54.

[11] Bremmell, K.E. and Mensah, J., 2005, "Interfacial-chemistry mediated behavior of colloidal talc dispersions", Journal of Colloid and Interface Science, Vol. 283, Issue 2, pp. 385-391.

[12] Chang, L.L.Y., 2002, "Industrial mineralogy", Prentice Hall, New Jersey, pp. 398-407.

[13] Derco, J. and Nemeth, Z., 2002, "Obtaining of high quality talc from talcose rocks: a case study from the Sinec Kokava deposits (Slovakia)", Boletin Paranaense de Geociencias, No. 50, pp. 119-130.

[14] El Bahariya, G.A. and Arai, S., 2003, "Petrology and origin of Pan-African serpentinites with particular reference to chromian spinel composition, Eastern Desert, Egypt: Implications for supra-subduction zone ophiolite", The Third International Conference on the Geology of Africa, Faculty of Science, Assiut University, Egypt, Vol. 1 (A), Dec. 7-9, pp. 371-388.

[15] El-Sharkawy, M.F., 2000, "Talc mineralization of ultramafic affinity in the Eastern desert of Egypt", Mineralium Deposita, Vol. 35, pp. 346-363.

[16] Engel, A.E.J. and Wright, L.A., 1960, "Talc and Soapstone", In Gillson, J.L. (Ed.), Industrial Minerals and Rocks, The American Inst. of Mining, Metallurgical and Pet. Engineering (AIME), New York, 3rd ed., pp. 835-850.

[17] Feng, D. and Aldrich, C. 2004, "Effect of ultrasonication on the flotation of talc", Ind. Eng. Chem., Vol. 43, pp. 4422-4427.

[18] Fuerstenau, D.W. and Huang, P., 2003, "Interfacial phenomena involved in talc flotation and depression", XXII International Mineral Processing Congress, South Africa, pp. 1034-1043.

[19] Fuerstanau, D.W. and Pradip, 2005, "Zeta potential in the flotation of oxide and silicate minerals", Advanced in Colloid and Interface Science, Vol. 114, pp. 9-26.

[20] Hassan, M.M.A., 2007, "Beneficiation of Egyptian Talc Ore", M.Sc. Thesis, Assiut University, Egypt, pp. 80-85.

[21] Helmy, H.M. and Kaindl, R., 1997, "Contribution to the mineralogy and petrogenesis of the talc-base metal sulfide", The Geol. Soc. Egypt, Abstract, 35th Annual Meeting, Cairo.

[22] Kamel, A., Abuzeid, A.M., Moharram, M.R. and Mahmoud, D.M., 2001, "Ceramics raw materials workshop", Proceedings of the 7th International Conference on Mining, Petroleum and Metallurgical Engineering (MPM'7), Assiut University, Assiut, Egypt, Vol. IV, pp. 82-96, February 10-12.

[23] Kho, C.J. and Sohn, H.J., 1989, "Column flotation of talc", International Journal of Mineral Processing, Vol. 27, pp. 157-167.

[24] Khraisheh, M., Holland, C., Creany, C., Harries, P. and Parolis, L., 2005, "Effect of molecular weight and concentration of the adsorption of CMC onto talc at different ionic strengths", International Journal of Mineral Processing, Vol. 75, pp. 197-206.

[25] Kusaka, E., Amano, N. and Nakahiro, Y., 1997, "Effect of hydrolysed aluminum (III) and chromium (III) cations on the lipophilicity of talc", Int. J. Miner. Process, Vol. 50, pp. 243-253.

[26] Nasr, B.B. and Masoud, M.S., 1999, "Geology and genesis of wadi Allaqi talc deposit, South Eastern desert, Egypt", Annals Geol. Survey. Egypt, Vol. XXII, pp. 309-317.

[27] Okunlola, O.A., Ogedengbe, O. and Ojutalyo, A., 2003, "Composition features and industrial appraisal of the Babe Ode talc occurrence, South Western Nigeria", Global Journal of Geological Science, Vol. 1, No. 1, pp. 63-72.

[28] Ozkan, A., 2003, "Coagulation and flocculation characteristics of talc by using different flocculants in presence of cations", Minerals Engineering, Vol. 16, pp. 59-61.

[29] Piga, L. and Marruzz, G., 1992, "Preconcentration of an Italian talc by magnetic separation and attrition", International Journal of Mineral Processing, Vol. 35, pp. 291-297.

[30] Rath, R.K., Subramnian, S. and Laskowski, J.S., 1995, "Adsorption of guar gum onto talc", Processing of Hydrophobic Minerals and fine Coal, Proceeding of the 34th Annual Conference of Metallurgies, CIM, pp. 105- 119.

[31] Rizk, A.M.E., Ahmed, M.M. and Ahmed, A.A., , May 2001, "Application of a factorial method on leaching process of calcareous phosphate ore", Bulletin of the Faculty of Engineering, Assiut University, Egypt, Vol. 29, No. 2, pp. 185- 197.

[32] Roe, L.A., 1983, "Talc", In: Lefond, S.J. (Ed.), Industrial Minerals and Rocks, The American Inst. of Mining, Metallurgical and Pet. Engineering (AIME), New York, N.Y.,5th ed., pp. 1275-1301.

[33] Sarquis, P.E. and Gonzalez, M., 1998, "Limits of the use of industrial talc-the carbonate effect", Minerals Engineering, Vol. 11, No. 7, pp. 657-660.

[34] Schandl, E.S., Gorton, M.P. and Bleeker, W., 1999, "A systematic study of rare earth and trace element geochemistry of host rocks to the Kidd Creek volcanogenic massive sulfide deposit", In: Hannington, M.D., and Barrie, C.T. (Eds.), Economic Geology Monograph, Vol. 10, pp. 309-334.

[35] Schandl, E.S., Sharara, N.A. and Gorton, M.P., 1999, "The origin of the Atshan talc deposit in the Hamata area, Eastern Desert, Egypt: A geochemical and mineralogical study", Canadian Mineralogist, Vol. 37, pp. 1211-1227, (1999).

[36] Schandl, E.S., Gorton, M.P. and Sharara, N.A., 2002, "The origin of major talc deposits in the Eastern Desert of Egypt: relict fragments of a metamorphosed carbonate horizon", Journal of African Earth Sciences, Vol. 34, pp. 259-273, (2002).

[37] Schober, W., 1997, "Quality compounds require premium talc grades and sophisticated formulations", Eurofillers97-filler. Doc, Manchester (UK), pp. 1-12, September 8-11.

[38] Shortridge, P.G., Harris, P.J., Bradshaw, D.J. and Koopal, L.K., 2000, "The effect of chemical composition and molecular weight of polysaccharide depressants on the flotation of talc", International Journal of Mineral Processing, Vol. 259, pp. 215-224.

[39] Simandle, G.J. and Paradis, S.P., 1999, "Carbonate-hosted talc", Industrial Minerals, British Columbia Mineral Deposits Profiles Doc., Ministry of Energy and Mines (Canada), Vol. 3, pp. 1-6.

[40] Wills, B.A., 1992, "Mineral processing technology", Pergamon Press., Great Britain, 5th ed., pp. 491-644.

[41] Xu, Z., Plitt, V. and Liu, Q., 2004, Recent advances in reverse flotation of diasporic ores, Minerals Engineering, 17, 1007-1015.

[42] Yehia, A. and AL-Wakeel, M.I., 2000, "Talc separation from talc-carbonate ore to be suitable for different industrial applications", Minerals Engineering, Vol. 13, No. 1, pp. 111-116.

[43] Yousif, A.A., 2003, "The national project for upgrading the Egyptian ores required by the local industry", Final Report supmeted to the Egyptian Academy for Scientific Research and Technology, pp. 49-100.

Part 4

Hydrology

12

Water Resources Assessment for Karst Aquifer Conditioned River Basins: Conceptual Balance Model Results and Comparison with Experimental Environmental Tracers Evidences

Antonia Longobardi, Albina Cuomo, Domenico Guida and Paolo Villani
University of Salerno, Department of Civil Engineering, Fisciano (SA) Italy

1. Introduction

Water resources management, more and more limited and poor in quality, represents a present key issue in hydrology. The development of a community is highly related to the management of the water resources available for the community itself and there is a need, for this reason, to rationalize the existing resources, to plan water resources use, to preserve water quality and, on the other hand, to prevent flood risk. The importance of decision support systems tools, such as hydrological models, generating streamflow time series which are statistically equivalent to the observed streamflow time series, is even more important considering the combination of multiple and complex issues concurring in the definition and optimization of water resources management practices.

When river basins with particular features have to be modelled, both traditionally conceptually based models and more recent sophisticated distributed models appear to give not very reliable results. In those cases it is possible to take advantage of a semi-distributed formulation, where every sub-catchment is modelled to account for its features and informations coming from all the sub-catchments are related to each other in order to improve the system description.

In this study, starting from the application of a catchment scale modelling tool, we propose a semi-distributed conceptually based framework, able to describe the sub-catchment scale systems hydrological response. The modelling approach is supported by field measurements collected within several seasonal campaigns, that has been set up for the Bussento river basin, located in Southern Italy, well known to hydrogeology and geomorphology scientists for its karst features, characterized by soils and rocks with highly different hydraulic permeability and above all an highly hydrogeological conditioning. The groundwater circulation is very complex, as it will be later discussed, and groundwater inflows from the outside of the hydrological watershed and groundwater outflows toward surrounding drainage systems frequently occur. With the aim to enhance the knowledge of the interaction between the groundwater and surface water and acknowledged the substantial help given by natural isotope tracers experiments to solve hydrological complex systems circulations problems, radon-in water concentrations have also been collected, in a limited number of cross sections, along the upper Bussento river reach.

Even though the proposed approach has some similarity with a few well known conceptually schemes, based on the existence of linear reservoirs and liner channel to describe the different components the streamflow can be decomposed in, it is valuable because of the possibility, which is in this case the necessity, to join all together hydraulic, hydrological and geological data to achieve reliable results.

2. Hydro-geomorphological setting of the Bussento river basin

The Bussento river basin, located in the Cilento and Vallo di Diano National Park (figure 1), is one of the major and more complex fluvial systems of the southern Campania region (Southern Italy).

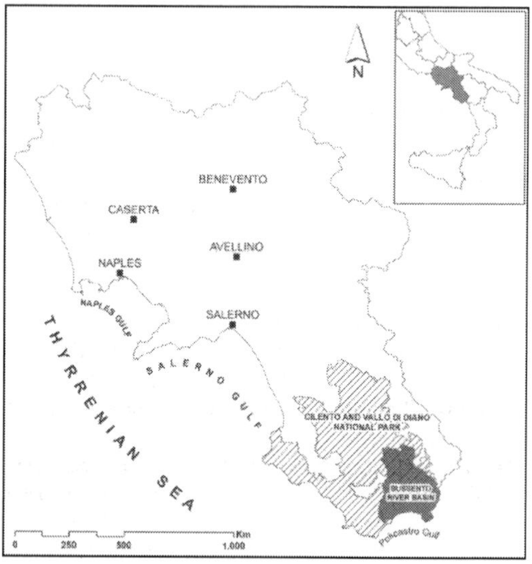

Fig. 1. Location of the Bussento river basin.

The main stream originates from the upland springs of Mt. Cervati (1.888 m), one of the highest mountain ridges in Southern Apennines. Downstream, the river flows partly in wide alluvial valleys (i.e., Sanza valley) and, partly, carving steep gorges and rapids, where a number of springs, delivering fresh water from karst aquifers into the streambed and banks, increase progressively downstream the river discharge.

The hydro-geomorphological setting of the river basin is strongly conditioned by a complex litho-structural arrangement derived from geological, tectonic and morphogenetic events occurred from Oligocene to Pleistocene along the Tyrrhenian Borderland of the southern Apennine chain (Bonardi et al., 1988). The chain is a NE-verging fold-thrust belt derived from an orogenic wedge, accreted by deformation and overthrust shortening of the sedimentary covers of several paleogeographic domains: Internal Sedimentary Domain in the Ligurian oceanic crust on the External Sedimentary Domain of Carbonate Platform-Continental Basin along the passive margin of the African plate (D'Argenio et al., 1973; Ippolito, F. et al., 1975).

The sedimentary basin successions related to the Internal Domain can be grouped in the following tectonic units (Cammarosano et al., 2000, 2004) (figure 2):

Castelnuovo Cilento Unit (Mid Eocene-Lower Miocene), constituted, from the top, by the Pianelli Formation (PNL), micaceous fine sandstone, siltites and shales; Trenico Formation (TNC), marls and calcarenites and Genesio Formation (GSO), argillites and calcilutites. Widely outcroping in the western sector of the basin (Sciaratopamo torrent sub-basin);

North Calabrian Unit (Upper Eocene-Lower Miocene), from the top, the Saraceno Formation (SCE), cherty calcarenites, marls and argillites and Crete Nere Formation (CRN), calciluties, black marls and argillites. The Unit is widely outcroping in the southern sector of the basin and, partially, in the upper and eastern Bussento river basin;

Cilento Group, a torbidite sequence, represented, from the top, by Monte Sacro Formation (SRO), conglomerates and sandstones; San Mauro Formation (MAU), sandstones, marls and conglomerates; Pollica Formation (PLL), sandstones and silty clay.

The tectonic units related to deformation of the above cited passive African continental margin are represented in the study area only by the:

Alburno-Cervati - Pollino Unit (Upper Trias-Middle Miocene), constituted from the top by: Bifurto formation (BIF), marls, quarz-sandstones and fine limestone breccias; Cerchiara formation (FCE), glauconite calcarenites; Trentinara formation (TRN), calcirudites and *Spiruline* calcilutite; *Radiolitides* Limestones (RDT); *Requienie* Limestones (CRQ); *Cladocoropsis* and *Clypeina* Limestones (CCM). At the top of the sequence, in disconformity, follows the Piaggine formation, calcirudites, sandstones and wildflysch succession, as low-standing olistostrome, announcing the arrival of Internal Units.

In the Plio-Pleistocene times, the above cited fold-thrust belt is affected by polyphase uplift transtensive and trans-pressive movements, with general lowering toward the Tyrhenian sea and juxtaposition of the clayey-marly successions in the erosional grabens, to the carbonate sequence, as karst summit horsts (Brancaccio et al., 1991; Cinque et al., 1993; Ascione A.&Cinque A., 1999).

In the study area, the main structural features are the overthrust of the Internal Units on the Bifurto/Piaggine formations at the NE piedmont of the M.nt Centaurino and the Sanza trans-tensive line, along the southern piedmont of M.nt Cervati massif.

The complexity of the geological setting gives to river basin an analogous complexity in hydrological response, due to the space-time variability of the river-aquifer interactions, conditioned by the hydro-structural setting and the karst landforms and processes highly affecting hydro-geomorphological behaviour (figure 3).

In general, in the Bussento river basin main and secondary aquifers can be recognized..

The **M.nt Cervati karst carbonate aquifer**, located in the northern side of the river basin, is one of the main aquifer of the southern Apennine; it is delimited at the North and N-E, by regional hydro-tectonic lines and at the SW and South by clayey aquicludes and highly fratured carbonate aquitards; minor hydro-structural lines induces multilayered and compartmented aquifers (sub-aquifers), with centrifugal directions of the groundwater flows (table 1).

M.nt Forcella karst carbonate aquifer, located in the eastern sector of the Bussento river basin, having 75% in area outside of the Upper Bussento, feeds only the 13 Fistole Spring Group, emerging a few hundred meters upstream the end of the river segment, with a M.A.D. 3 m^3/s.

M.nt Alta karst carbonate aquifer, located on the N-E sector of the basin feeds only the Farnetani Spring Group, with a M.A.D. 1.5 m^3/s, is interconnected with the Sanza Endorheic

Basin and related sinkhole-cave system, feeding the Bonomo Watermill seasonal springs and resurgences.

M.nt Centaurino multilayered terrigenous aquifer feeds several spring with a total M.A.D. 0.1 m^3/s.

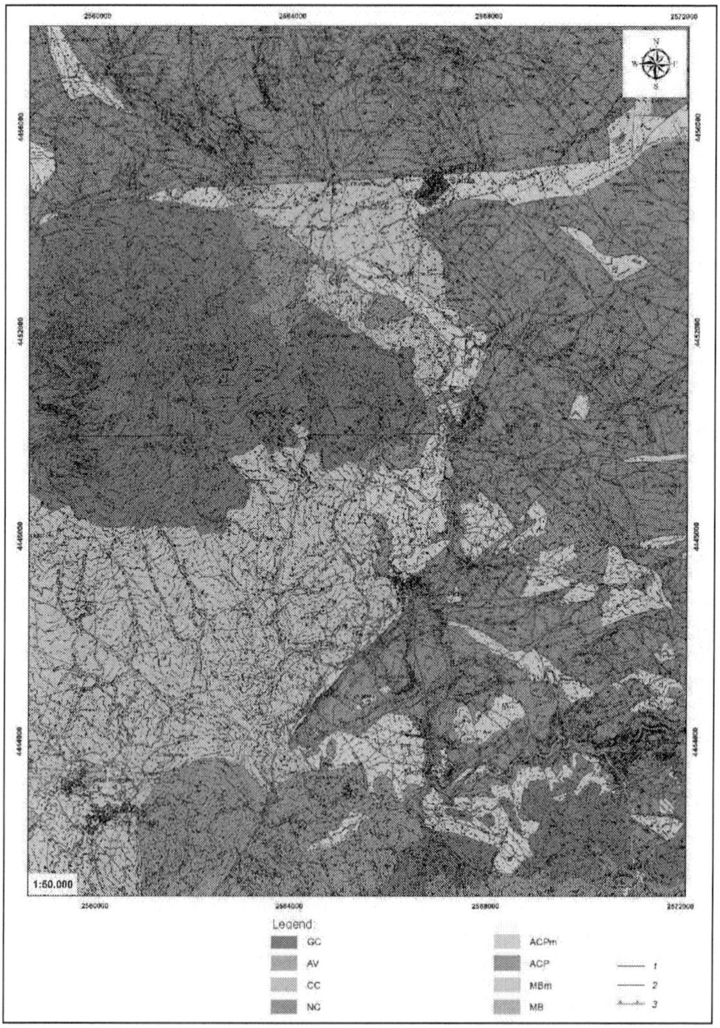

Fig. 2. Geological map of the Upper and Middle Bussento river (1:50.000 scale).

Legend: GC. Cilento Group; AV. Sicilide Unit, included in CC. Castelnuovo unit; NC. Nord Calabrian Unit; ACPm. Piaggine formation; ACP. Alburni-Cervati -Pollino Unit; MBm. Bulgheria-Roccagloriosa Unit - clayey marls; MB. Bulgheria-Roccagloriosa Unit- Limestone; 1. Fault; 2. Stratigraphic boundary 3. Overthrust

Spring name	Sub-aquifer name	Elevation (m a.s.l.)	M.A.D. (l/s)	GWFD	Receiving River Basin
Rio Freddo	M.nt Arsano	470	750	East	Tanagro
Fontanelle Soprane	M.nt Arsano	470	800	N-E	Tanagro
Fontanelle Sottane	M.nt Arsano	460	400	N-E	Tanagro
Varco la Peta	Vallivona	1200	40	Southern	Bussento
Montemezzano	Inferno creek	900	100	Southern	Bussento
Sanza Fistole Group	Basal Southern Cervati	550-470	300	Southern	Bussento
Faraone Fistole Group	Pedale Raia	450	400	S-W	Mingardo
Calore Group	Neviera	1150	100	North	Calore
Sant'Elena Group	Rotondo	420	400	N-W	Calore
Laurino Group	Scanno Tesoro	330-400	600	N-W	Calore
Capodifiume Group	Chianiello-Vesole	30-35	2900	N-W	Capodifiume
Paestum-Cafasso Group	Chianiello-Vesole	1-10	750	N-W	Capodifiume
Acqua Solfurea Group	Chianiello-Vesole	5	250	N-W	Capodifiume

Table 1. Hydrogeological characteristics of the springs from Cervati aquifer. M.A.D.: Mean Annual Discharge; GWFD: GroundWater Flow Direction.;

The mainstream originates from south-western summit mountain slope of the Mount Cervati, where many, low discharge springs from shallow aquifer in debris cover laying on marly-clayey bedrock originate ephemeral creek inflowing into the Vallivona Affunnaturo sinkhole. From the Varco la Peta spring-resurgence, the Inferno creek flows southward, carving steep gorges in form of a typical bedrock stream, with cascade and rapids, where further springs (Montemezzano spring), along the streambed, increase progressively the river discharge (table 1), as well as along the piedmont (Sanza Fistole spring groups).
The true Bussento river begins downstream the junction of the above cited Inferno creek and the Persico creek. This last flows at the bottom of an asymmetric valley, characterized at the left side by the above cited southern steep mountain front of M.nt Cervati and at the right by the gentle northern mountain slope of the M.nt Centaurino (1551 m asl). The middle right side of the basin is characterized by marly-arenaceous rocks outcrops (M.nt Marchese hilly ridge), while the left middle side is characterized by karst limestone sequences (M.nt Rotondo and Serra Forcella).

Fig. 3. Hydro-geomorphological map of the Bussento river and related hydro-geomorphological features.

Legend: Hydrogeological complexes: s. Sandy conglomerate; gsl. Gravelly sandy silty; dt. Debris; Ol. Blocky clayey olistostrome; Ar. Sandstone; MAr. Marly sandstone; CMAg. Marly conglomerate sandstone; Am. Silty Sandstone; M. Marly; Cm. Marly limestone; C. Limestone; D. Dolomite. Symbology: 1. Permeability limit; 2. Buried permeability limit; 3. Overtrusth hydro-geological limit; 4. Syncline hydro-geological limit; 5. Overturned strata; 6. Horizontal strata; 7. Sincline; 8. Karst summit; 9. Groundwater flow direction; 10. Sinkhole; 11. Main spring; 12. Submarine spring.

3. The streamflow and geo-chemical database

3.1 Historical streamflow data

Historical streamflow data consist of two short streamflow time series, recorded at the Caselle in Pittari and Sicilì gauging stations, providing daily data, respectively, from 1952-1968 and from 1952-1957.

The lack of historical adequate streamflow time series, both on a temporal and on a spatial point of view, makes even more difficult a realistic calibration of a modelling approach. For this reason, an intensive monitoring campaign, illustrated in the following paragraphs, was planned to temporally and spatially extend the streamflow database.

3.2 The catchment and sub-catchment monitoring campaign

On January 2003, the Sinistra Sele River Watershed Regional Agency, started a monitoring campaign with the aim to measure, in many different cross sections and on a monthly time base, the Bussento river discharge. Based on the above described geomorphological and hydrogeological settings, 25 gauge stations were indicated as significant to define the river and springs hydrological regime (figure 4).

Fig. 4. The Bussento river basin monitoring network. BS is the symbol for streamflow stations whereas PG is the symbol for marine stations, for discharge and radon measurements.

Since December 2009 a further monitoring campaign, focused on the upper Bussento river basin (figure 5 and Table 2), was started and managed by CUGRI (Centro Universitario per la Previsione e Prevenzione dei Grandi Rischi) on behalf of the Regional Agency for the environmental protection of Campania region (ARPAC), within a more comprehensive study on the radon-222 activity concentration in stream and spring waters. Besides radon concentration, more chemical and physical variables have been measured, such as pH, water temperature, dissolved oxygen, atmospheric pressure, water conductivity and water resistivity.

Monitoring stations locations have been carefully identified to investigate the complex interaction between groundwater and streamflow, caused by the complex karst hydrogeological structure and system hydrodynamics. The monitoring timing of the river discharge was oriented to measure the delayed sub-surface flow and the baseflow component of the hydrograph. For this reason, several recession curves of historical data were analysed, deriving the more appropriate time from the flood peak discharge at which the delayed sub-surface and baseflow occur. Consequently, the monitoring campaigns were planned to measure the stream discharge at least seven days after the end of the rainfall event, while in dry periods the measures were conducted two times a months.

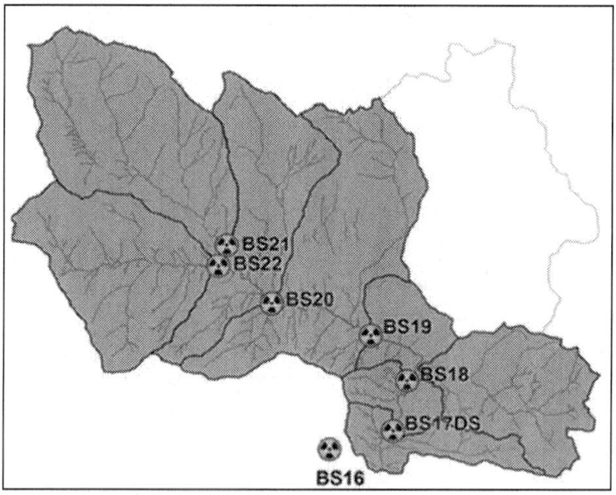

Fig. 5. Monitoring stations in the upper Bussento river basin. In blank Sanza Endorheic Basin.

Station Code	Drainage area (Km^2)	Elevation (m.a.s.l.)	Pervious drainage area (Km^2)	Impervious drainage area (Km^2)
BSU17	85.15	912	64.08	21.07
BSU18	82.13	927	62.43	19.70
BSU19	66.84	927	49.49	17.35
BSU20	47.20	1079	38.74	8.46
BSU22	14.73	926	11.12	3.61

Table 2. Bussento river sub-catchments main characteristics.

4. Conceptual hydro-geological modelling

The hydro-geomorphological settings, above briefly illustrated, induce a very complex surface-groundwater interaction and exchanges, with groundwater inflows, from outside of the hydrological watershed, and groundwater outflows, towards surrounding drainage systems. The hydro-geomorphological domain includes karst and fluvial landforms and processes conditioning groundwater recharge ("karst input control", sensu Ford and Williams, 2007), by means of the infiltration and runoff processes, including: a) allogenic recharge from surrounding impervious drainage basins into deep and shallow sinking stream infiltration points, and fractured bedrock stream infiltration; b) autogenic recharge, including sub-soil and bare diffuse epikarst infiltration, endorheic runoff infiltration in dolines and poljes; c) groundwater discharge ("karst ouput control", sensu Ford and William, 2007), differentiated in the groundwater-river interactions within the aquifer-river domain. The last includes the complex interactions between the streambed-springs system, which generally results in a downstream river discharge increasing, occurring generally in typical bedrock streams, flowing in gorge and canyons carved in enlarged fractured limestone sequences.

Each of the mentioned components corresponds, in the modelling conceptualization of the scheme, to a linear storage, which releases streamflow as a function of the water storage and of a characteristic delay time. The characteristic time indicates that there is a delay between the recharge to the system and the output from the system itself, and this delay is greater for deeper aquifers. The number of storages, each representing, thus, a different process, contributes to the total streamflow through a recharge coefficient, that is a measure of the magnitude of the single storage capacity.

The application of a conceptual model, such as the one briefly described, requires the calibration of the model parameters, and in particular of the characteristic delay time and of the recharge coefficient of each single storage. In complex catchments, such as the Bussento River System, characterized by a large impact of karstic phenomena, raw streamflow data are not sufficient to the quantification of the contribute and magnitude of the single storage, and, therefore, are not sufficient to calibrate the model. To this aim, the use of Radon activity concentration measurements could represent a valuable future perspective.

In this river basin, the results of previous hydrogeological and hydrological studies (Iaccarino G., 1987; Iaccarino G. et al., 1988; Guida D. et al. 2006) indicate a weak correspondence between recorded data and model simulations, due to the strongly conditioning of deep karst circulation on the hydrological response, with an alternation between gaining river reaches from groundwater, and losing river reaches towards the karst aquifers, and also towards external watershed.

Due to these karst-induced features, the surface and groundwater recharge, circulation and discharge turns out to be very complex, and, therefore, a conceptual hydro-geomorphological model has been developed as a physical context in assessing basin and sub-basin water budget by a semi-distributed hydrological model (Todini E, 1996; Franchini M., et al, 1996). Following White (2002), the basic components of the generic karst aquifer flow system can be sketched as in figure 6.

Clearly, not all of these components are present in all aquifers, and their presence and relative importance is a fundamental point of distinguishing one aquifer from another. With reference to Iaccarino et al. (1988) and White (2002), this general conceptual model has been applied to the Bussento Hydro-geological System (BHS), recognizing the following recharge-discharge components (figure 6).

Starting from the catchment scale characterization and modelling of the Bussento river system, and taking advantage of a consistent long term monitoring campaign, mainly operated over the Upper Bussento river system, it is possible to calibrate an hydro-geological framework to assess the hydrological response at the sub-catchment scale. With the aim to enhance the knowledge and approach a quantification of the interaction between the groundwater and surface water, natural isotope environmental tracers technique have also been used. Radon-in water concentrations have been collected, in a limited number of cross sections, along the upper Bussento river reach, represented, along with the monitoring sections, in the following figure 7. Sub-catchment scale modelling for the upper Bussento river reach will be later compared to experimental environmental tracers evidences, with the intent to set up a modelling framework that, starting from a limited (in space and time) number of observation, concerning hydraulic, hydrological, chemical and geological data, is able to assess water resources systems for karst aquifer highly conditioned river basins.

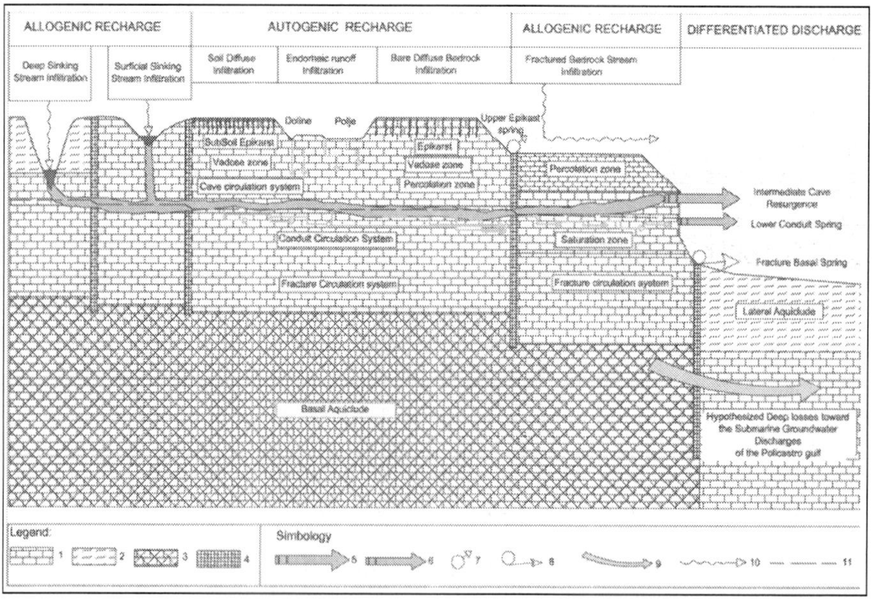

Fig. 6. Specific conceptual model of the karst aquifers in the Bussento Hydrological System (BHS). Legend: 1. Limestone fractured Aquifer; 2. Marly clayey aquitard; 3. Cataclastic Basal Lime-Dolostone Aquitard; 4. Lateral Limestone very fractured aquitard; 5. Intermittent or seasonal growndwater flow from Cave System; 6. Perennial groundwater flow from Conduit System; 7. Secondary springs ; 8. Spring Group; Losses toward

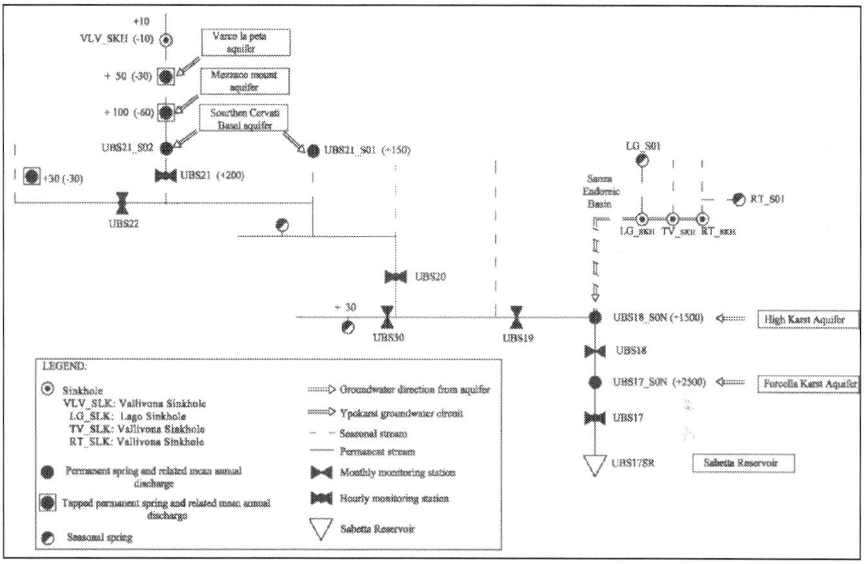

Fig. 7. Upper Bussento river reach network hydro-geological schematization and significant monitoring cross sections.

4.1 Conceptual modelling calibration based on streamflow database

Given the Bussento catchment geomorphological and hydro-geological features described in the previous paragraphs, a lumped model cannot guarantee reliable results. For this reason and taking advantage of the dense monitoring campaign, a semi-distributed formulation, accounting for each sub-basin particular characteristics, seems to be more appropriate.

When dealing with the monthly time scale, each sub-basin can be described (figure 8) as two linear reservoirs in parallel, representing the groundwater flow and the deep subsurface flow, whereas the rainfall contributes, which are characterized by delay times smaller than a month, are supposed to reach the outlet through a linear channel (Claps et al., 1993).

The scheme is also supported by the conceptual hydro-geological model described in the previous paragraph. In this case coupling the linear reservoirs balance equations with the whole system balance equation, total streamflow D at each time step is related to the net input by means of an ARMA (2,2) model, which stochastic formulations corresponds to:

$$D(t) - \Phi_1 D(t-1) - \Phi_2 D(t-2) = \varepsilon(t) - \Theta_1 \varepsilon(t-1) - \Theta_2 \varepsilon(t-2) \tag{1}$$

where ε is the model residual, related to the net input I, that is then a periodic independent random process, and Θ_1, Θ_2, Φ_1, and Φ_2 are the model stochastic parameters, related to the model conceptual parameters K_1, K_2 (reservoirs response times, respectively of the ground water system and of the subsurface plus surface water system), a and b (recharge coefficients, respectively to the ground water system and to the subsurface plus surface water system) according to the following equations:

In its original formulation the model algorithm, starting from an observed streamflow time series, apply a maximum likelihood procedures to estimates the model parameters and, because of the univariate approach, with an inverse procedure, the net rainfall input is also estimated. As an example, model performance are illustrated for the Bussento at Caselle historical time series, at the monthly scale, in figure 9. The linearity of the quantile - quantile plot entails the good performances of the linear applied model, when historical time series are available.

$$\theta_1 = \frac{e^{-1/k_1} + e^{-1/k_2} - ar_{k1}(1+e^{-1/k_2}) - br_{k2}(1+e^{-1/k_1})}{(1-ar_{k1}-br_{k2})}$$
$$\theta_2 = \frac{ar_{k1}e^{-1/k_2} + br_{k2}e^{-1/k_1} - e^{-1/k1}e^{-1/k_2}}{(1-ar_{k1}-br_{k2})}$$
$$\Phi_1 = -e^{-1/k_1} + e^{-1/k_2} \quad \Phi_2 = -e^{-1/k_1}e^{-1/k_2} \tag{2}$$
$$r_{k1} = k_1(1-e^{-1/k_1}) \quad r_{k2} = k_2(1-e^{-1/k_2})$$

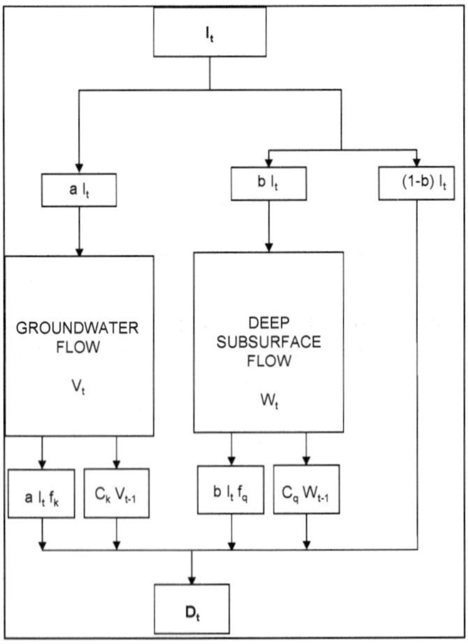

Fig. 8. Linear system for monthly time series.

But to calibrate a semi-distributed model, a number of streamflow recorded time series are needed for a number of nested catchments. Even if the gauging stations planned in the monitoring campaign, resemble a nested catchments scheme, collected streamflow data consist of discharge instantaneous data measured within a month time window, at each section, which does not represent a monthly discharge recorded time series, thus the data are not available to calibrate the ARMA(2,2) model with its inverse procedure at the sub-basin scale.

With the aim to set up a modelling semi-distributed approach able to reproduce observed discharge values along the river network, both short historical streamflow time series and streamflow data collected during the monitoring campaign have been used to a priori estimate the model parameters and net rainfall input.

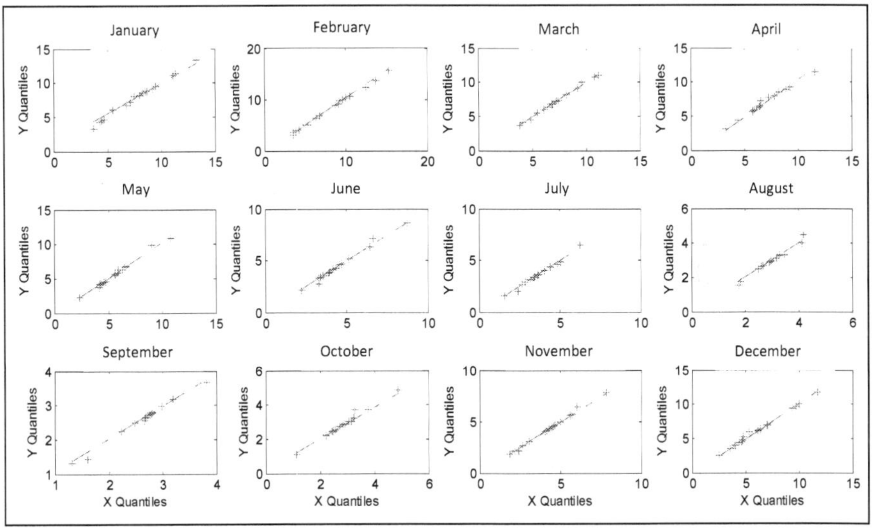

Fig. 9. Modelled and observed discharge data at Bussento at Caselle: quantile - quantile plot.

The recharge coefficients a and b have been estimated on the base of previous studies (Celico, 1978; Iaccarino et al., 1988) focused on the assessment of potential infiltration coefficients based on surface soils properties. The recharge coefficients, initially computed on the base of surface soil properties, are then susceptible to a successive calibration aimed at the optimization of the mean annual water balance. More insights aimed at their calibration would also come from the analysis of radon in water concentration, monitored on a particular river reach as later commented.

The K_1 delay response time has been estimated, during non-raining periods, applying the base flow recession equation:

$$Q(t) = Q_0 e^{-t/K} \tag{3}$$

which describe the event descending hydrograph limb, to two successive events, in order to remove the unknown streamflow data Q_0 at time t_0:

$$\begin{gathered} Q(t_1) = Q_0 e^{-t_1/K_1} \qquad Q(t_2) = Q_0 e^{-t_2/K_1} \\ \frac{Q(t_1)}{Q(t_2)} = \frac{Q_0 e^{-t_1/K_1}}{Q_0 e^{-t_2/K_1}} \Rightarrow K_1 = \frac{(t_2 - t_1)}{\log[Q(t_1)/Q(t_2)]} \end{gathered} \tag{4}$$

The K_2 delay response time has been instead estimated during raining periods, when both deep and surface components are detectable in streamflow data. Considering t1 and t2 as two successive time instants for which both streamflow, Q(t1) and Q(t2), and rain values, p1 and p2, are known, and initially retaining that:

$$\frac{Q_0(t_1)}{Q_0(t_2)} = \frac{p1}{p2} \tag{5}$$

then the K_2 response time can be evaluated from the following equation:

$$\ln\frac{Q(t_1)}{Q(t_2)} = \ln\frac{p1}{p2} - \frac{t_1}{K_2} + \frac{t_2}{K_2} \tag{6}$$

The described a-priori estimation procedure set up to compute the fast delay response time, make K_2 the parameter which is likely to be affected by the largest uncertainty, that would in the end also compromise the coupled hydro-geological and hydrological model performances.

Estimated conceptual parameters K_1, K_2, a and b (b = 1-a) are indicated, as an example, for the Upper Bussento river reach gauging stations, in table 32.

Table 3. A priori estimates of delay times and recharge coefficients model parameters, for the Upper Bussento river basin cross sections.

Station Code	K_1 (days)	K_2 (days)	a	1-a
BSU17	288	40	0.70	0.30
BSU18	204	50	0.65	0.35
BSU19	120	39	0.60	0.40
BSU20	123	27	0. 60	0.40
BSU22	115	50	0.60	0.40

4.2 Conceptual modelling simulation results

With regard to the model rainfall net input, the pursued procedure has been to generate it from its probability density distribution, with given parameters. The I(t) probabilistic representation is the Bessel distribution, which is the sum of a Poissonian number of events with exponentially distributed intensity:

$$\begin{aligned} &P[I=0] = e^{-\upsilon} && I=0 \\ &f_1(I) = e^{-\lambda I-\upsilon}(\upsilon\lambda / I)\Im_1[2(\upsilon\lambda I)^{1/2}] && I>0 \end{aligned} \tag{7}$$

where λ=1/β is the exponential parameter, v is the Poisson parameter and $\Im_1$ is the modified Bessel function of order 1. The rationale for such probabilistic representation is given by the positive values and finite probability at zero that I(t) has to present.

Parameters β and v are estimated from the existing two streamflow time series. The temporal patterns found for the two series are rather similar, thus we assumed β and v spatially invariant over the catchment (table 4).

	Jan	Feb	Mar	Apr	May	Jun	Jul	Aug	Sep	Oct	Nov	Dec
ν	1.05	1.19	1.21	1.39	1.48	1.82	2.07	1.66	1.28	1.09	1.04	1.03
β mm/day	5.29	2.65	1.52	1.43	6.33	0.73	0.30	0.86	2.85	2.69	1.77	3.63

Table 4. Net rainfall input distribution β and ν parameters.

Equation (1) as been then used to generate 1000 years monthly streamflow time series at each section, comparing thus the discharge probability distributions, at each section and for each month, with the occurred observed values.
Figure 10 and figure 11 show, as an example, the generated streamflow probability distribution, at cross section BSU19 respectively for the summer (month of august) and the winter season (month of February). If we assume as acceptable the region of data included between the 25° ed il 75° percentile, that is the 50% of data located in the middle of a sorted sample data, the proposed conceptual bivariate hydro-geological modelling approach shows reasonable performances both during the summer season (observed discharge corresponds to the 50° percentile) and the winter season (observed discharge corresponds to 56° percentile).

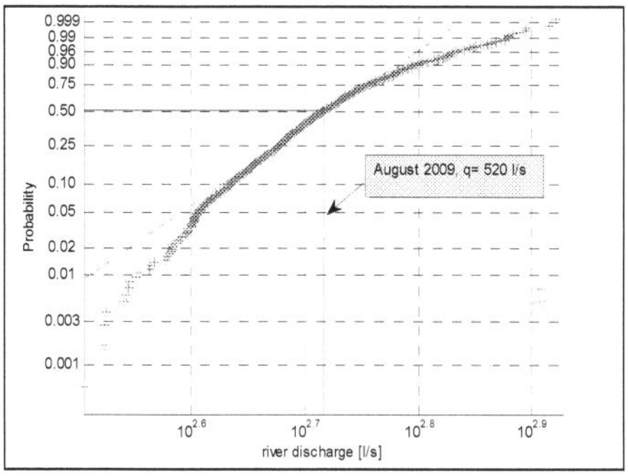

Fig. 10. Generated streamflow probability distribution compared with occurred value, cross section BSU19, for the summer season (August).

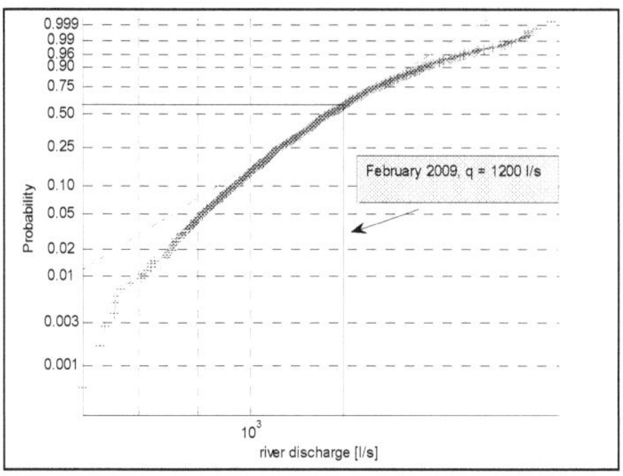

Fig. 11. Generated streamflow probability distribution compared with occurred value, cross section BSU19, for the winter season (February).

5. Environmental tracers experimental evidences

Isotopic tracers studies was introduced into catchment hydrology research in the 1960s as a complementary tools to conventional hydrological methods, to address questions about the pathways taken from precipitation infiltrating water to the stream network and about the water residence times within the catchment boundary (McDonnell, 2003). Especially in the Mediterranean environments, where karst aquifer groundwater represents more than 98% of the available fresh-water supply, the study of the interaction between groundwater and surface water is particularly important and difficult, at the same time, because of the complex hydraulic interconnections between fractured carbonate rocks and watershed network (Brahana and Hollyday, 1988). Deep water resources system discharge is of particular importance during the summer period, when, because of the rainfall deficiency, it contributes to the total streamflow in a measure of 70% (Dassonville and Fé d'Ostiani, 2003; Tulipano et al., 2005; Longobardi and Villani, 2008).

In the last decades, the use of isotopic tracers has been of substantial need in many problems concerning the decomposition of total streamflow into its main components, such as the surface, the sub-surface and deep flows, both in experimental and laboratory experiments (Levêque et al., 1971; McDonnell, 2003; Solomon et al. 1993, 1995, 1997; Goldscheider and Drew, 2007). Besides the traditional and long time use of natural isotopes in hydrology and hydrogeology (Flora and Longinelli, A., 1989; Emblanch et al., 2003), one of the most interesting, promising and innovative approach to quantitatively assess the groundwater contributions to streamflow and seawaters in natural environments, consists in measuring radon-in-water activity concentrations (Andrews and Wood, 1972; Shapiro, 1985). The principle at the base of this technique is the larger concentration of Radon in groundwater compared to surface waters (Rogers, 1958).

Radon-222 (simply 'radon' in the following) is a volatile gas with a half-life of 3.8 days, moderately soluble in water and atmosphere. It is released to groundwater from Radium-226 alpha decay, by means of permanent alpha recoil in micro-pore or fracture walls (Rama and Moore, 1984) and progressive dissolution of the aquifer-forming-material that supplies more and more soluble Ra-226, subsequently decaying to radon (Ellins et al., 1990). Due to its volatility, radon gas quickly degasses into the atmosphere producing a significant disequilibrium between concentrations in groundwater and surface water.

From the seminal work of Rogers (1958), the assessment of spatial-temporal variations in radon concentrations between surface and groundwater (Ellins, 1990; Lee and Hollyday, 1987, 1991) have provided insights in: i) testing infiltration-filtration models (Genereaux and Hemond, 1990; Genereaux et al., 1993; Gudzenko, 1992; Kraemer and Genereux, 1998), ii) performing hydrograph separation (Hooper and Shoemaker, 1986), iii) calculating water residence times (Sultankhodzhaev et al., 1971), iv) interpreting the role of "old water" in non-linear catchments hydrological response, v) estimating shallow and deep water mixing (Hoehn and von Gunten, 1989; Hamada, 2000; Hakl et al., 1997; Semprini, 1987; Gainon et al., 2007), and vi) calculating flow velocities in homogeneous aquifers (Kafri, 2001).

In addition, the use of radon enables the researchers to trace groundwater migration pathways (Hoehener and Surbeck, 2004), and to assess the time dependence of groundwater migration processes (Schubert et al., 2008). Infiltration of surface waters from a river to groundwater (Hoehn and von Gunten, 1989) as well as flow dynamics in a karst system (Eisenlohr and Surbeck, 1995) are just a few examples of applications where radon-based methodology has been used successfully to gain additional information on environmental processes.

In the current project, the general objectives of the radon in-water monitoring program are (i) to localize and quantify the contributions of groundwater along the main stream riverbed and banks, (ii) to set up an adaptive methodology, based on monthly radon activity concentration measurements in streamflow and springs, for the baseflow separation from other streamflow components; (iii) to verify the hydro-dynamical behaviour of the karst circuits and their influence on streamflow and iv) to calculate the downstream groundwater influence on streamflow. The project has been also planned in order to implement and improve this approach in the conventional regional public practice, to compliance the suggestions given from the European Water Framework Directive (EWFD, 2000) and to apply the methodology to other similar karst-conditioned river basin in Southern Italy.

5.1 Illustration of radon-in-water concentration collected data

The monitoring campaign of Rn-222 concentration is oriented to investigate the variability of radon gas in stream water and stream inflowing springs water and to separate the total streamflow in the subsurface and baseflow components.

Rn-222 concentration in stream water in a particular cross section along the river network is tightly related to the residence times of water collected at that particular section: the longer is the journey made by each drop of water through the rock formations, the larger is the Radon-in water concentration. According to this criteria, waters flowing from different source systems are characterized by different isotopic labels, that allow then distinguishing waters from different origins. As an example, in figure 12, the temporal variability of radon concentration is given for a number of gauging stations along the upper Bussento river network. It is interesting to highlight how radon concentration temporal variability can be dramatically different in different cross sections. For the particular river reach under consideration, it is possible to observe that there exist two significantly different temporal patterns: a first behavioural pattern, that records a temporal large fluctuation of radon concentration around a mean value and a second behavioural pattern, that records a mainly constant radon concentration during the time. The existence of these two patterns is strictly related to the presence or not of significant inflowing springs water contributions in particular gauging stations. Opposite to stations BSU17 and BSU18, where substantial stream inflowing springs water occur, stations BSU19, BSU20 and BSU22 are indeed featured by the absence of springs feeding the streams along the correspondent river reach. This physical characterization would explain the larger and constant concentration of radon for the first two stations and the lower and fluctuating radon concentration for the remaining stations. Radon concentration fluctuation detected for the stations BSU19, BSU20 and BSU22 can be explained by the fact that, in these particular sections, surface flow component, which is the results of the fastest transformation of rainfall and is the poorest radon concentration stream water fraction, is significantly contributing to the total discharge.

The effect of the increase in the surface component respect to total discharge is also detectable for stations significantly affected by springs feeding, as a function of the proportion of groundwater versus surface water: stations BSU18, compared to station BSU17, receives a large fraction of groundwater contributing to total streamflow and is not thus affected from rainfall events, only increasing the surface component of total discharge.

As a proof, in figure 13, the temporal pattern of radon concentration measured in sections BSU17 and BSU18, is compared to the temporal pattern of radon concentration in spring water feeding the stream, in cross sections BS17S0N and BS18S0N, immediately upstream sections BSU17 and BSU18. It is evident, in particular for the station BSU18 for the same reasons previously indicated, that the temporal pattern of radon concentration in stream water (BSU18) strongly resembles the temporal pattern of radon in spring water (BS18S0N).

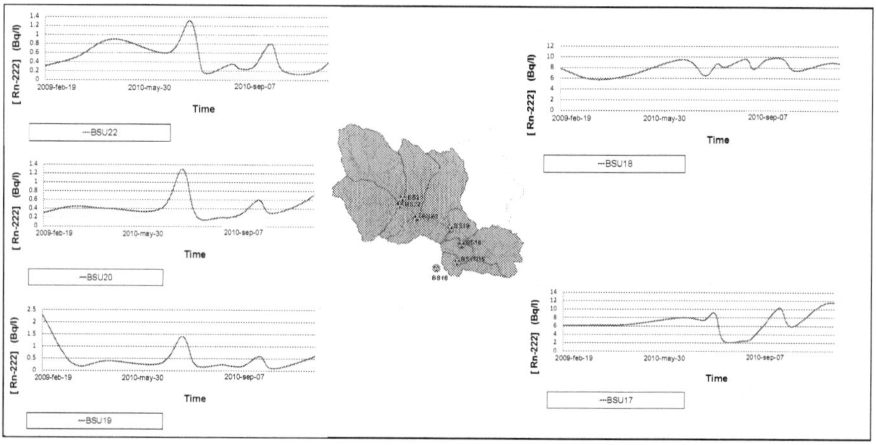

Fig. 12. Temporal variation of radon concentration measured in the gauging stations along the river network.

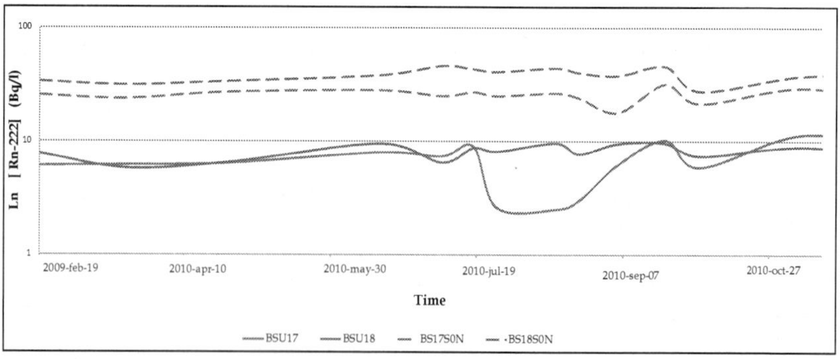

Fig. 13. Temporal pattern of radon concentration in stream water (BSU17 and BSU18) and in spring water feeding the stream (BS17S0N and BS18S0N).

More insights about the radon concentration dynamics could be achieved by comparing, at the annual scale, the temporal pattern of precipitation, discharge and radon concentration (figure 14).

During the winter season, the abundant precipitations recharge the deep water resources system and at the same time produce a significant surface component of total streamflow (flood conditions), with radon concentration approaching a rather constant and average value during the whole period. During the summer dry season, instead, when low flow conditions occur, the river discharge is mainly sustained by the baseflow, that is the outflow of deep water resources systems, characterized by the larger radon concentration because of the long residence times. Measured data confirm indeed that, in the period from May to September, the river discharge decrease and a consequent increment in radon concentration is instead detected.

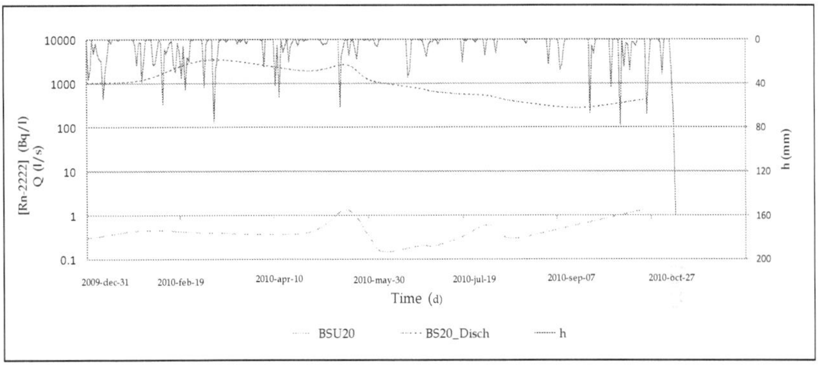

Fig. 14. Precipitation, discharge and radon concentration temporal patterns vs. daily rainfall.

5.2 Preliminary combined analysis of radon data and hydro-geological modelling

Starting from the well-known assumption that water is composed of a set of well mixed end members, the collected data of radon concentration are used to illustrate an example of hydrograph separation into different flow components. To this aim, mass balance and mixing equations can be written, as described in Kendall and McDonnell (1998):

$$\begin{aligned} Q_T &= Q_{SSF} + Q_{GW} \\ C_T Q_T &= C_{SSF} Q_{SSF} + C_{GW} Q_{GW} \end{aligned} \tag{8}$$

where:

Q_T is the total streamflow, Q_{SSF} is the sub-surface delayed flow, Q_{GW} is the groundwater flow, C_T is the Rn-222 value in total streamflow, C_{SSF} is the Rn-222 value in sub-surface delayed flow, C_{GW} is the Rn-222 value in groundwater flow.

As an example, the mixing equations (8) are applied at cross section BSU18, which is one of the gauging sections where groundwater contributions are extremely large, to derive the Q_{SSF} and Q_{GW} components of total discharge. If Q_{SSF} and Q_{GW} are the unknown variables, application of equations (8) requires observation and measures of all other variable. Q_T and C_T are indeed the only measured variables, whereas values for C_{SSF} and C_{GW} are inferred from measurements referred to different cross sections.

The Rn-222 content of river water is strongly affected by volatilization to the atmosphere, and this must be accounted for in using radon data to estimate a possible groundwater influx from subsurface water sources (Kies A., 2005). If C_{DS} and C_{US} are the radon concentration measured in a downstream and upstream cross sections, and L is the length of the river segment between the mentioned cross sections, the relationship between radon concentrations is described by the following equation (9), from (Wu Y. et al., 2004):

$$C_{DS} = C_{US} \times e^{-\alpha L} \tag{9}$$

Model equation (9) is applied between sections BS18_S0N and BSU18, assuming C_{US} as the radon concentration in section BS18_S0N, which is the river inflow spring water section, and determining the C_{DS} as the radon concentration in section BS18, at a distance of about 1 Km, also representing the C_{GW} concentration in section BSU18. Application of the volatilization model requires a value for the parameter α, previously calibrated on a specific river reach of the Bussento network (Guadagnuolo D., 2009), whose hydro-geomorphological settings are similar to the one of the river reach investigated in this report and resulting in an α ccoefficient equal to 0.9 (l/km). CSSF radon concentration at cross section BSU18, that is the concentration of sub-surface flow, is computed as the mean value of radon concentration measure in sections BSU22, BSU20 and BSU19, where deep water resources contribution are negligible and representative of the sub-surface flow. Results are illustrated in figure 15.

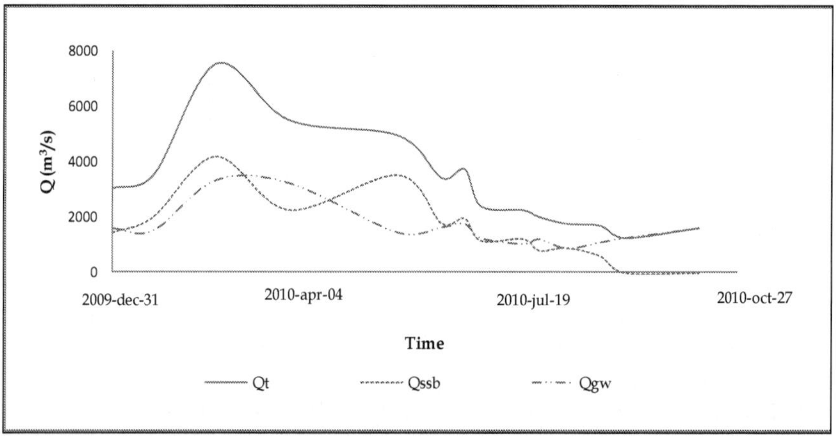

Fig. 15. Hydrograph separation based on mixing equation solution and environmental tracers measurements.

Total streamflow components hydrograph patterns appear realistic: groundwater component has a smoother pattern compared to subsurface flow, the most responsive component to rainfall event between the two, and essentially represents total streamflow during the summer season, when rainfall contribution are negligible or absent. With regard to the quantification of the deep water resources, the mixing equations separation performance, based on the number of hypothesis at the base of the application, lead to quite comparable volume of total groundwater and subsurface flow. For the same cross section, instead, the hydro-geological conceptual approach would indicate a larger contribution of groundwater to total streamflow, of about 60-70%, as indicated from the a-priori estimation of the a model parameter, representing the groundwater coefficient of recharge and, approximately, also the proportion of groundwater flow versus total streamflow. Similar quantifications are moreover confirmed by the analysis of observed streamflow time series for the Bussento river catchment. As already stressed, the qualitative and quantitative results of hydrograph separation based on the use of environmental tracers rely on a number of assumptions (e.g. the impossibility to measure radon concentration relative to the single hydrograph component and calibration of the volatilization model) obviously impacting the assessment uncertainty. Besides its potential capability then, attempts to achieve an improvement in the calibration of the technique are to be found. With reference to the illustrated modeling conceptual approach, major concernments has to be focused on a more realistic hydrograph separation, preserving a weight of the different streamflow components, in the water balance, which would tightly fits the system hydro-geological characteristics. Further data analysis and collection could also be planned to estimate the residence times of total streamflow components, which would give a quantitative assessment of delay times K_1 and K_2 model parameters.

6. Conclusion

In this paper modelling difficulties that have to be faced when water basins with particular hydro-geological features are under investigation have been highlighted and suggestions of alternative, and in somehow integrated, approaches have been proposed in order to assess the potentiality of water resources systems. The presented case study is the Bussento river basin, located in Southern Italy, which is well known to hydrogeology and geomorphology scientists for its karst features, characterized by soils and rocks with highly different hydraulic permeability and, above all, an highly hydrogeological conditioning.

Traditional hydrological modelling relies on the existence of, at least, recorded streamflow time series, but when dealing with complex system watershed, lumped modelling, at the catchment scale, would not be satisfactory and a semi-distributed methodology, at the sub-basin scale, would be more appropriate. But such an approach would require even more data, that is even more streamflow time series in many different cross sections along the river network, enabling model calibration for homogeneous sub-catchment areas.

The methodology we have presented, has undoubtedly some similarity with a few well known conceptually hydrological and hydrogeological schemes, based on the existence of linear reservoirs and liner channel to describe the different components the streamflow can be decomposed in. However, it benefits from the coupling of hydraulic, hydrological and geological data, to set up a parsimonious model which, based on an a-priori calibration procedure requiring poorly dense time sampled data, is able to reasonable simulate total discharge.

The hydro-geological conceptual modelling approach has been compared to a water resources assessment methodology based on the use of environmental tracers, in particular of radon in water concentration. The preliminary hydrograph separation, into main components, performance based on radon concentration sampled data is affected by a number of assumptions (e.g. the impossibility to measure radon concentration relative to the single hydrograph component and calibration of the volatilization model) at the base of the application of the mixing equation model. At this stage, water resources assessment results from geo-chemical method, even though reasonable, is not indeed really comparable to the results of the conceptually modeling technique, which results, at the sub-basin scale, are instead confirmed by time series analysis at the catchment scale. Geo-chemical methods have however a great potential to be a fast and poorly dense time sampled data requirement method for water resources assessment and major attempts to achieve an improvement in the calibration of the technique are then to be found.

7. Acknowledgment

The authors wish to thank the Regional Water Basin Authority, Sinistra Sele, and the CUGRI, Centro Universitario per la Previsione e Prevenzione dei Grandi Rischi, for their support. The research was partially supported by Italian Ministry for the University and the Research grant. Pluviometric data have been provided by the CERIUS (Centro di Eccellenza per i Rischi Idrogeologici dell'Università di Salerno).

8. References

Anderson, M.P. & Woessner, W.W. (1992). *Applied Ground Water Modelling*, Academic Press, San Diego.

Andrews, J. N. & Woods, D. F. (1972). Mechanism of radon release in rock matrices and into groundwaters, *Transactions of the Institution of Mining and Metallurgy*, B81, pp. 198-209.

Ascione, A. & Cinque, A. (1999). Tectonics and erosion in the long term relief history of the Southern Apennines (Italy). *Zeitschrift für Geomorphologie*, Suppl.-Bd., Vol.118, pp. 1-16;

Bonardi, G., Ciampo, G. & Perrone, V. (1985). La formazione di Albidona nell'Appennino calabro-lucano: ulteriori dati stratigrafici e relazioni con le unità esterne appenniniche, *Bollettino della Società Geologica Italiana*, vol. 104, Roma.

Brahana, J.V., & Hollyday, E.F. (1988). Dry stream reaches in carbonate terranes — Surface indicators of ground-water reservoirs, *American Water Resources Association Bulletin*, Vol.24, No.3, pp. 577 - 580.

Brancaccio, L., Cinque, A., Romano, P., Rosskopf, C., Russo, F., Santangelo, N. & Santo, A. (1991). Geomorphology and neotectonic evolution of a sector of the Tyrrhenian flank of the southern Apennines (Region of Naples, Italy). *Zeitschrift für Geomorphologie N.F.*, Suppl.-Bd., Vol. 82, pp. 47-58.

Cammarosano, A., Danna, M., De Rienzo, Martelli, L., Miele, F. & Nardi, G. (2000). Il substrato del gruppo del Cilento tra M. Vesalo e il monte Sacro (Cilento, Appennino Meridionale). *Bollettino Società Geologica Italiana*, Vol.119, No.2, pp. 395-405, fig.4.

Cammarosano, A., Cavuoto, G., Danna, M., De Capoa, P., De Rienzo, F., Di Staso, A., Giardino, S., Martelli, L., Nardi, G., Sgrosso, A., Toccacelli, R. M. & Valente, A. (2004). Nuovi dati sui Flysch del Cilento (Appennino Meridionale, Italia). *Bollettino Società Geologica Italiana*, Vol.123, No.2, pp. 253-273, fig.6.

Claps, P., Rossi, F. & Vitale, C. (1993). Conceptual- stochastic modelling of seasonal runoff using autoregressive moving average models at different scales of aggregation, *Water Resources Research*, Vol.29, No.8, pp. 2545–2559.

Celico, P. B., (1978). Schema idrogeologico dell'Appennino Meridionale, In: *Memorie e Note dell' Istituto di Geologia Applicata*, Vol.19, Napoli, Italia.

Cinque, A., Patacca, E., Scandone, P. & Tozzi, M. (1993). Quaternary kinematic evolution of the Southern Apennines. Relationship between surface geological features and lithospheric structures. *Annali di Geofisica*, Vol.36, pp. 249-260.

D'Argenio, B., Pescatore, T.S. & Scandone P. (1973). Schema geologico dell'Appennino meridionale (Campania e Lucania), *Proceedings of* Moderne vedute sulla geologia dell'Appennin, Vol. 183, pp. 49-72.

Dassonville, L. & Fé d'Ostiani, L. (2003). Mediterranean watershed management: overcoming water crisis in the Mediterranean, *Proceedings of Watershed Management: Water Resources for the Future*, Chapter 6 "Watershed Management & Sustainable Mountain Development", Porto Cervo, Sassari, Sardinia, Italy, October 22-24, 2003.

Eisenlohr, L. & Surbeck, H. (1995). Radon as a natural tracer to study transport processes in a karst system. An example in the Swiss Jura, *Comptes Rendus de l'Académie des Sciences*, Paris, Vol.321, série IIa, pp. 761-767.

Ellins, K. K., Roman-Mas, A. & Lee, R. (1990). Using 222Rn to examine Groundwater/surface discharge interaction in the Rio Grande de Manati, Puerto Rico, *Journal of Hydrology*, Vol.115, pp. 319-341.

Emblanch, C., Zuppi, G.M., Mudry, J., Blavoux, B. & Batiot, C. (2003). Carbon-13 of TDIC to quantify the role of the unsaturated zone: the example of the Vaucluse karst systems Southeastern France, *Journal of Hydrology*, Vol.279, pp. 262-274.

EWFD, (2000). Directive 2000/60/EC of the European Parliament and of the Council of 23 October 2000 establishing a framework for Community action in the field of water policy, In: *Official Journal L 327*, 22.12.2000, Available from http://ec.europa.eu/environment/water/water-framework/index_en.html

Gainon, F., Goldscheider, N. & Surbeck H. 2007. Conceptual model for the origin of high radon levels in spring waters -the example of the St. Placidus spring, Grisons, Swiss AlpsSwiss, *Journal of Geosciences*, vol.100, pp. 251-262.

Genereaux, D. P., & Hemond, H. F. (1990). Naturally occurring radon-222 as a tracer for streamflow generation: steady state methodology and field example, *Water Resources and Research*, Vol.26, No.12, pp. 3065 - 3075.

Genereaux, D. P., Hemond, H. F. & Mulholland, P. J. (1993). Use of radon-222 and calcium as tracers in a three-end-member mixing model for streamflow generation on the wet fork of Walker Branch watershed, *Journal of Hydrology*, Vol.142, pp. 167- 211.

Goldscheider, N. & Drew, D. (2007), *Methods in Karst Hydrogeology*. Taylor & Francis Group, London, UK.

Guadagnuolo, D. (2010). *Investigation of the groundwater-river interaction, using Radon-222 as a natural tracer, in a karst Mediterranean environment like in the case study of the Bussento river basin*, PhDThesis. University of Salerno.

Guida, D., Iaccarino, G. & Perrone, V. (1998). Nuovi dati sulla successione del Flysch del Cilento nell'area di M.te Centaurino: relazioni fra Unità Litostratigrafiche, Unità Litotecniche e principali Sistemi Franosi", *Memorie della Società Geologica*, Vol.41.

Hakl, J., Hunyad,i I., Csige, I., Geczy, G., Lenart, L. and Varhegyi, A. (1997). Radon transport phenomena studied in karst caves: International experiences on radon levels and exposures, *International Conference on Nuclear Tracks in Solids* N. 18, vol. 28, No. 1-6 , pp. 675 - 684, Cairo , Egypte, September 01-1996.

Hamada, H. (2000). Estimation of groundwater flow rate using the decay of 222Rn in a well, *Journal of Environmental Radioactivity*, Vol.47, pp. 1-13.

Hoehener, P. & Surbeck, H. (2004). 222Rn as a tracer for nonaqueous phase liquid in the vadose zone: experiments and analytical method, *Vadose Zone Journal*, Vol.3, pp. 1276-1285.

Hoehn, E. & von Gunten, H. R. (1989). Radon in Groundwater: a tool to assess infiltration from surface waters to aquifers, *Water Resources and Research*, Vol.25, pp. 1795 - 1803.

Hooper, R.P. & Shoemaker, C.A. (1986): A comparison of chemical and isotopic streamflow separation, *Water Resources Research*, Vol.22, No.10, pp. 1444-1454.

Iaccarino, G., Guida, D. & Basso, C. (1998). Caratteristiche idrogeologiche della struttura carbonatica di Morigerati, *Memorie Società Geologica Italiana*, Vol. 41, pp. 1065-1077, Roma, 1988.

Ippolito, F., D'Argenio, B., Pescatore, T. & Scandone, P. (1975). Structural-stratigraphic units and tectonic framework of Southern Apennines. In: *Geology of Italy*, Squyres, C. (Ed.), Earth Science Society Lybian Arab Republic, pp. 317-328.

Kafri, U. (2001). Radon in Groundwater as a tracer to assess flow velocities: two test cases from Israel, *Environmental Geology*, Vol. 40, No. 3, pp. 392-398.

Kendale, C. & McDonnell, I.J. (1998). *Isotope tracers in catchment hydrology*, Elsevier, New York, pp. 40 – 41, ISBN 0-444-50155-X.

Kies, A., Hofmann, H., Tosheva, Z., Hofmann, L. & Pfister, L. (2005). Using Radon-222 for hydrograph separation in a micro basin (Luxembourg), *Annals of geophysics*, Vol.48, No.1, pp. 101-107

Kraemer, T.F. & Genereux, D.P. (1998). Applications of Uranium- and Thorium-Series Radionuclides in Catchment Hydrology Studies, In: *Isotope Tracers in Catchment* Hydrology, Kendall, C. and McDonnell, J.J. (Eds.), 679-722, Elsevier, Amsterdam.

Lee, R. & Hollyday, E. F. (1987). Radon measurement in streams to determine location and magnitude of ground-water seepage, In: *Radon, radium, and other radioactivity in groundwater*, Graves B. (Ed.), 241-249, Lewis Publishers, Chelsea, Michigan.

Lee, R. & Hollyday, E. F. (1991). Use of radon measurements in Carters Creek, Maury County, Tennessee, to determine location and magnitude of groundwater seepage, In: *Field studies of radon in rocks, soils and water*, Gundersen, L.C. and Wanty, R.B. (Eds.), 237-242. U. S. Geological Survey Bulletin.

Levêque, P. S., Maurin, C. & Severac, I. (1971). Le 222Rn traceur naturel complementaire en hydrologie souterranie, *Comptes Rendus Hebdomadaires des Seances de l Academie des Sciences*, Vol.272, No.18, p. 2290.

Longobardi, A. & Villani, P. (2008). Baseflow index regionalization analysis in a Mediterranean environment and data scarcity context: role of the catchment permeability index, *Journal of Hydrology*, Vol. 355, pp. 63-75.

Loucks, D. & Gladwell, J. (1999). *Sustainability criteria for water resource systems*, Cambridge University Press, ISBN 0-521-56044-6, United Kingdom.

McDonnell, I. J. (2003). Where does water go when it rains? Moving beyond the variable source area concept of rainfall-runoff response, *Hydrological Processes*, Vol.17, pp. 1869- 1875.

Rama & Moore, W. S. (1984). Mechanism of transport of U–Th series radioisotopes from solids into ground water, *Geochimica et Cosmochimica Acta*, Vol. 48, No. 2, pp.395-399.

Rogers, A. (1958). Physical behaviour and geologic control of radon in mountain streams, *U.S. Geological Survey Bulletin* 1052 – E.

Schubert, M. (2008). Personal Communication.

Semprini, L. (1987). Radon-222 concentration in groundwater from a test zone of a shallow alluvial aquifer in the Santa Clara Valley, California, In: *Radon in Groundwater*, Graves B., (Ed.), 205-218, Lewis Publishers, Clelsea, Michigan.

Shah,T., Molden,D., Sahthiradelvel R. & Seckler D. (2001). The global situation of groundwater: overview of opportunities and challenges, *International Water Management Institute*, ISBN 92-9090-402- X.

Shapiro, M.H., Rice, A., Mendenhall, R., M.H., Melvin, D. & Tombrello, T.A. (1984). Recognition of environmentally caused variations in radon time series, *Pure and Applied Geophysics*, Vol.122, pp. 309-326.

Shuster, E. T. & White, W. B. (1971). Seasonal fluctuations in the chemistry of limestone springs. A possible means for characterizing carbonate aquifers, *Journal of Hydrology*, Vol.14, pp. 93-128.

Simonovic, S. (1998). Water resources engineering and sustainable development, *Proceedings of the XXVI Congress of Hydraulics*, Catania, Italy.

Solomon, D. K., Schiff, S. L., Poreda, R. J. & Clarke, W. B. (1993). A validation of the 3H/3He method for determining groundwater recharge, *Water Resources and Research*, Vol.29, No.9, pp. 2851-2962.

Solomon, D. K., Poreda, R. J., Cook, P. G. & Hunt, A. (1995). Site characterization using 3H/3He groundwater ages (Cape Cod, MA.), *Ground Water*, Vol.33, pp. 988-996.

Solomon, D. K., Cook, P. G. & Sanford, W. E. (1997). Dissolved Gases in Subsurface Hydrology, In: *Isotope Tracers in Catchment Hydrology*, Kendall C. and McDonnell J.J. (Eds.), 291-318, Elsevier.

Sultankhodzhaev, A. N., Spiridonov, A. I. & Tyminsij, V. G. (1971). Underground water's radiogenic and radioactive gas ratios (He/Rn and Xe/Rn) in groundwaters and their utilization for groundwater age estimation, *Uzbekistan Journal of Geology*, Vol. 5, p. 41.

Tulipano, L., Fidelibus, D. & Panagopoulos, A. (2005). *Groundwater management of coastal karstic aquifers*, COST Action 621, Final report.

White, W. B. (1969). Conceptual models for limestone acquifers, *Groundwater*, Vol.7, No.3, pp. 15-21.

White, W. B. (1977). Conceptual models for carbonatee acquifers: revised, In: *Hydrologic Problems in Karst Terrain,* Dilamarter, R. R. and Casallany, S. C. (Eds), 176-187, Western Kentucky University, Bowling Green, KY.

White, W. B. (2003). Conceptual model for karstic acquifers. Speleogenesis and Karstic Aquifers, *The virtual Scientifical Journal*, Vol.1, pp. 1-6.

Wu, Y., Wen, X. & Zhang, Y. (2004). Analysis of the exchange of groundwater and river water by using Radon-222 in the middle Heihe Basin of northwestern China, *Environmental Geology*, Vol.45, No.5, p.p. 647-653.

Yoneda, M., Inoue, Y. & Takine, N. (1991). Location of groundwater seepage points into a river by measurement of 222Rn concentration in water using activated charcoal passive collectors, *Journal of Hydrology*, Vol.124, 307-316.

13

Fundamental Approach in Groundwater Flow and Solute Transport Modelling Using the Finite Difference Method

M.U. Igboekwe and C. Amos-Uhegbu
Department of Physics,
Michael Okpara University of Agriculture, Umudike,
Nigeria

1. Introduction

Water below the ground surface is usually referred to as groundwater. Groundwater has recently become a major source of water supply in almost every sector.

Over-dependence on it for many purposes has led to its' over-exploitation, and this has led to much concern for groundwater assessment and management.

For a proper assessment and management of groundwater resources, a thorough understanding of the complexity of its processes is quite essential. Expansion of human activities causes dispersion of pollutants in the subsurface environment. The fate and movement of dissolved substances in soils and groundwater has generated considerable interest out of concern for the quality of the subsurface environment

Groundwater flow and transport analysis have been an important research topic in the last three decades. This is as a result of many geo-environmental engineering problems having direct or indirect impact by groundwater flow and solute transport. Solute transport by flowing water (dissolved suspended particles) has broad impact in environmental protection and resource utilization via groundwater contamination. The leaching (displacement) of salts and nutrients in soils also has an impact on agricultural production.

To predict the contaminant migration in the geological formation more accurately, many analytical solutions for partial differential equations exist but because of the difficulty in obtaining analytical solutions, numerical solutions are more generally used. Numerical solutions are often more difficult to verify, so mathematical model error has to be kept as small as possible.

Many techniques for solving numerically the solute transport (advection-dispersion) equation are used such as the Finite Element Method (FEM), Finite Difference Method (FDM), Boundary Element Method (BEM), Fuzzy Sets Approach (FSA), Artificial Neural Networks method (ANN), Particle Tracking Method (PTM), Random Walk Method(RWM), Integrated Finite Difference Method (IFDM) etc.

Conventional analysis of groundwater flow is generally made by using the relevant physical principles of Darcy's law and mass balance.

1.1 Review of Darcy's law

In 1856, French engineer Henry Darcy was working on a project involving the use of sand to filter the water supply. He performed laboratory experiments to examine the factors that govern the rate of water flowing through the sand.
The results of his experiments defined the empirical principle of groundwater flow, in an equation now known as Darcy's law which states that "the saturated flow of water through a column of soil is directly proportional to the head difference and
inversely proportional to the length of the column". Darcy's apparatus consisted of a sand-filled column with an inlet and an outlet for water.

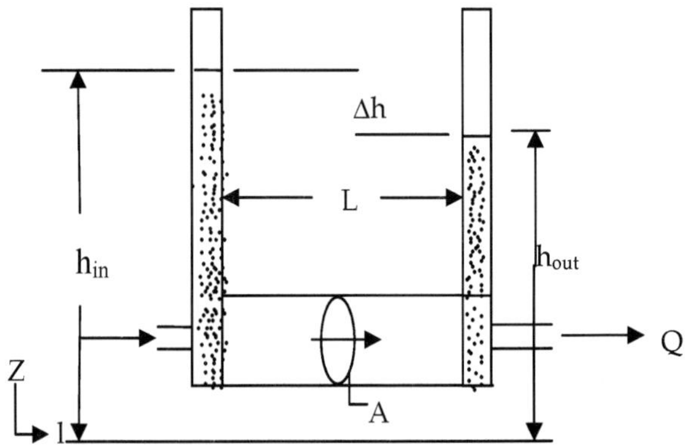

Fig. 1. Darcy's law experiment

Two manometers measure the hydraulic head at two points within the column (*h*1 and *h*2). The sand is saturated, and a steady flow of water is forced through it at a volumetric rate of Q [L3/T] (Q is sometimes called the volumetric flow rate or the discharge rate). Darcy found that Q was proportional to the head difference *dh* between the two manometers, inversely proportional to the distance between manometers *l*, and proportional to the cross sectional area of the sand column (A).
This can be mathematically written as:

$$Q \,\alpha\, (h_1 - h_2)$$

$$Q \,\alpha\, 1/\,l$$

$$Q \,\alpha\, A$$

Therefore we can say, $Q \,\alpha\, A\, dh/\,l$
Combining these observations and writing an equation in differential form gives Darcy's law for one-dimensional flow:

$$Q = K(A\, dh/l) \tag{1}$$

Where Q = volumetric flow rate or the discharge rate(m^3/s), Cross sectional flow area perpendicular to l(m^2), K=hydraulic conductivity (m/s), and d = denotes the change in h over the path l.

It can be re-written in differential form as:

$$Q = -KA(dh/dl) \tag{2}$$

The minus sign is necessary because head decreases in the direction of flow (i.e., water is always flowing from higher hydraulic head to lower hydraulic head). If there is flow in the positive l direction, Q is positive and dh/dl is negative[2]. Conversely, when flow is in the negative l direction, Q is negative and dh/dl is positive. The constant of proportionality K is the hydraulic conductivity in the l direction, a property of the porous medium and the fluid (water) filling the pores. The common units for hydraulic conductivity are meters/year for regional studies, m/day for local aquifer-scale studies, and cm/sec for laboratory studies. Therefore, in some analysis, we often deviate from the rule of using the SI unit.
Another form of the Darcy's law is written for the Darcy flux (or the Darcy Velocity, or, the Specific Discharge) (q) which is the discharge rate per unit cross-sectional area:

$$\begin{aligned} q &= Q/A \\ &= -KA(dh/dl)/A \\ &= -K(dh/dl) \end{aligned} \tag{3}$$

The Darcy flux q has unit of velocity [L/T] and assumes that flow occurs through the entire cross section of the material without regard to solids and pores.
However, Darcy flux is not the actual fluid velocity in the porous media; it is just discharge rate (Q) per unit cross-sectional area.
In a Cartesian x, y, z coordinate system, it is commonly expressed as:

$$qx = -Kx\, dh/dx \tag{4}$$

$$qy = -Ky\, dh/dy \tag{5}$$

$$qz = -Kz\, dh/dz \tag{6}$$

where Kx, Ky, and Kz are the hydraulic conductivity in each of the coordinate direction, respectively. qx, qy, and qz are 3 components of the Darcy flux $\rightarrow q^3$ .The Kx, Ky, and Kz are the directional hydraulic conductivity evaluated along each of the coordinate axis. To estimate these directional conductivities, Darcy test can be conducted along the x axis, in which case a horizontal hydraulic conductivity along the x direction can be determined: Kx. Same idea applies to estimating conductivities in the y and z directions.

1.2 Groundwater velocity / limits of Darcy's law

Application of Darcy's law has both upper and lower limits, it does not hold at very high fluid velocities. Hence, the need to calculate the groundwater velocity because the Specific Discharge (Darcy's flux) of equation 3 which is sometimes called Darcy's velocity is not actually groundwater velocity but discharge rate per unit cross-sectional area (Q/A).
Flow of water is limited in pore spaces only, Hence in calculating for the actual seepage velocity of groundwater, the actual cross-sectional area through which the flow is occurring must be accounted for as follows:

$$V = \frac{Q}{\varnothing A} \tag{7}$$

where V = Groundwater velocity, also known as the seepage velocity, and commonly called average linear velocity (m/s),
Q = Volumetric flow rate or the discharge rate(m^3/s), A= Cross sectional area perpendicular to l (m^2), $\varnothing$ = porosity.

Recall that in equation 3, $\left(\frac{Q}{A}\right) = q = -K(dh/dl)$. Substituting equation 3 into equation 7, we have $V = -K(dh/dl)/\varnothing$

This is differentially re-written as

$$V = \frac{q}{s}\frac{-K}{s}\left(\frac{dh}{dx}\right) \tag{8}$$

s is the effective porosity of the porous medium and x the dimension (m) while other symbols remain unchanged.
The nature of flow as quantified by the Reynold's Number is expressed as:

$$(Re) = \frac{\rho v d}{\mu} \tag{9}$$

This is a dimensionless ratio where v = the velocity (m/s), ρ = fluid density (kg/m^3), μ = the fluid velocity (kg/m/s), and d = diameter of the pipe (m).
Experimental evidence indicates that Darcy's law is valid as long as *Re* does not exceed a critical value between (1 and 10).This law holds for low velocity and high gradient (Shazrah et al 2008).

2. Groundwater modelling

Modeling has emerged as a major tool in all branches of science (Igboekwe et al.,2008).
Models are conceptual descriptions, tools or devices that represent or describe an approximation of a field situation, real system or natural phenomena.
Models are applied to a range of environmental problems mainly for understanding and the interpretation of issues having complex interaction of many variables in the system. They are not exact descriptions of physical systems or processes but are mathematically representing a simplified version of a system. This mathematical calculation is referred to as simulations. Groundwater models are used in calculating rates and direction of groundwater flow in an aquifer.
The simulation of groundwater flow needs a proper understanding of the complete hydrogeological characteristics of the site because the applicability, reliability or usefulness of a model depends on how closely the mathematical equations approximate the physical system being modelled. So, the following should be well investigated and understood:

- Hydrogeologic framework (surface thickness of aquifers and confining units, boundary conditions that control the rate and direction of groundwater flow).
- Hydraulic properties of aquifer(a depiction of the lateral and vertical distribution of hydraulic head i.e initial, steady-state and transient conditions; distribution and

quantity of groundwater recharge / discharge; and sources or sinks being referred to as stresses and they may be constant or transient).

Groundwater modelling entails simulation of aquifer and its response to input/output systems.

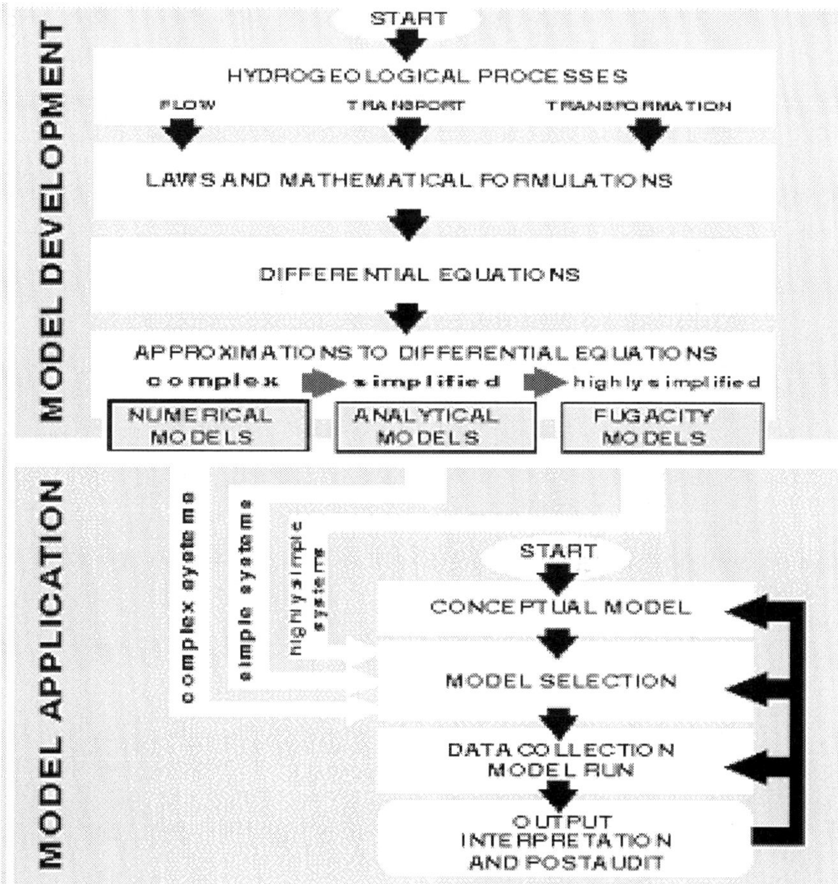

Fig. 2. Development process of a model (Adapted from Kumar 2002).

By mathematically representing a simplified version of a hydrogeological condition, realistic alternative settings can be predicted, tested, and compared.

Groundwater models are used to predict/illustrate the groundwater flow and transport processes using mathematical equations based on certain simplifying assumptions. These assumptions normally involve the direction of flow, geometry of the aquifer, the heterogeneity or anisotropy of sediments or bedrock within the aquifer, the contaminant transport mechanisms and chemical reactions. Because of the simplifying assumptions embedded in the mathematical equations and the many uncertainties in the values of data required by the model, a model must be viewed as an estimate and not an accurate replication of field settings.

Groundwater models, though they represent or approximate a real system are investigation tools useful in many applications.

Application of existing groundwater models include water balance (in terms of water quantity), gaining knowledge about the quantitative aspects of the unsaturated zone, simulating of water flow and chemical migration in the saturated zone including river-groundwater relations, assessing the impact of changes of the groundwater regime on the environment, setting up monitoring networks and groundwater protection zones (Kumar 2002).

2.2 Types of models

Models could also be classified based on their typical applications; an illustration is below in table 1.

MODEL	APPLICATION
Groundwater flow The problem of water supply is normally described by one equation mainly in terms of hydraulic head. The resultant model providing the solution to this equation is the groundwater flow model.	Water supply, Regional aquifer analysis, Near-well performance, Groundwater/surface water interactions, Dewatering operations.
Solute transport When the problem involves water quality, then an additional equation to the groundwater flow equation is needed to solve for the concentration of the chemical species. The resultant model is the solute transport model.	Sea-water intrusion, Land fills, Waste injection, Radioactive waste storage, Holding ponds, Groundwater pollution.
Heat transport Problems involving heat require a set of equations similar to solute transport equation, expressed in terms of temperature. The resultant model is the heat transport model.	Geothermal, Thermal storage, Heat pump, Thermal pollution.
Deformation Deformation model combines a set of equations that describe aquifer deformation.	Land subsidence.

Table 1. Classification of groundwater models based on their applications (modified after Shazrah et al 2008)

There are other subdivisions of models such as those describing porous media and those of fractured media namely physical scale model, analog model (Hale-Shaw model) etc.

3. Model concept

Groundwater modelling begins with a conceptual understanding of the physical problem. The next step in modelling is translating the physical system into mathematical terms. In general, the results are the familiar groundwater flow equation and transport equations.

3.1 Groundwater flow equation

It is an established hydraulic principle that groundwater moves from areas of higher potential i.e recharge areas (higher elevation or higher pressure/hydraulic head) to areas of lower pressure or elevation. This implies that direction of flow of groundwater ideally follows the topography of the land surface. Cracks, inter-connected pore spaces make a rock material permeable. Some permeable materials may allow fluid to move several metres in a day; while some may move a few centimeters in a century.
In the real subsurface, groundwater flows in complex 3D patterns. Darcy's law in three dimensions is analogous to that of one dimension. This is often derived using a representative fixed control volume element (RFCVE) of fixed dimensions usually a cube.

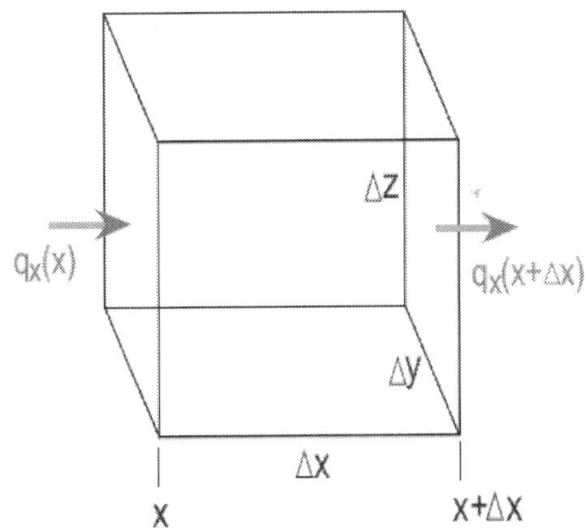

Fig. 3. A Representative Fixed Control Volume Element RFCVE (Mass balance diagram).

Thus on the basis of Darcy's law, the volume of the

$$\text{RFCVE} = \Delta X \Delta Y \Delta Z \tag{10}$$

In this typical mass balance analysis, the water flowing in and out (net flux of mass) through the boundary of the RFCVE is equated to the (time) rate of change of mass within the element:
Mass influx(M/T) - Mass outflux (M/T) = rate of change of mass(M/T).
For example, if 8litres of water flows into the RFCVE every minute(mass influx), and 5litres of water leaks out of it, then the water in the RFCVE experiences a change in storage(net flux

of mass) that is 8litres minus 5litres which is 3litres every minute. This implies that at every minute, the water in the RFCVE increases by 3litres in volume. This net flux of mass (change in volume) is known as change in storage (Ss). But on the other hand, the source of the recharge looses to the RFCVE about 8litres of water per minute, thus giving rise to the terms "source" and "sink".

Let q be the specific discharge, so that the rate of flow of water through the RFVCE can be expressed in terms of the three components $q_x, q_y, q_z.$

In accordance to the mass balance equation, influx minus outflux is equal to change in storage i.e (q_{in} - q_{out}) = (Ss). Now considering the flow along x-axis of the RFVCE, the influx through the face, ΔYΔZ is the same with the specific recharge q_x from the x-axis which is $q_x(x)$; the outflux is denoted as $q_x(x+\Delta x)$.

Therefore, $q_x(x)\,\Delta Y\Delta Z - q_x(x+\Delta X)$ = change in flow along x-axis.

Bringing like terms together

$$q_x(x) - q_x(x+\Delta X)\,\Delta Y\Delta Z$$

$$q_x(x - x + \Delta X)\,\Delta Y\Delta Z$$

$$q_x(\Delta X)\Delta Y\Delta Z$$

This is differentially written as; $\frac{dqx}{dx}\Delta X\Delta Y\Delta Z$ = change in flow rate along x axis.

Therefore, change in flow rate along x-axis =

$$\frac{dqx}{dx}\Delta X\Delta Y\Delta Z \tag{11}$$

Also, change in flow rate along the y-axis is specified as

$$\frac{dqy}{dy}\Delta X\Delta Y\Delta Z \tag{12}$$

While that of z-axis is correspondingly given as

$$\frac{dqz}{dz}\Delta X\Delta Y\Delta Z \tag{13}$$

At this point, the total change in volumetric flow rate is obtained by summing the individual changes in flow rate of the three axes i.e equations (9), (10), and (11)

$$\left(\frac{dqx}{dx}\Delta X\Delta Y\Delta Z\right)+\left(\frac{dqy}{dy}\Delta X\Delta Y\Delta Z\right)+\left(\frac{dqz}{dz}\Delta X\Delta Y\Delta Z\right)$$

Bringing like terms together,

$$\left(\frac{dqx}{dx}+\frac{dqy}{dy}+\frac{dqz}{dz}\right)\Delta X\Delta Y\Delta Z = \text{total change in flow rate(Change in Storage)} \tag{14}$$

The specific change in storage (Ss) is defined as the volume of water released from storage per unit change in head (h) per unit volume of aquifer i.e.

$$Ss = \frac{\Delta V}{\Delta h \Delta X \Delta Y \Delta Z} \tag{15}$$

Ss ΔhΔXΔYΔZ = ΔV

Then, the rate of change of storage is given by:

$$\frac{\text{Ss } \Delta h \Delta X \Delta Y \Delta Z}{\Delta t} = \frac{\Delta V}{\Delta t}$$

$$Ss \ \Delta h \Delta X \Delta Y \Delta Z \frac{\Delta h}{\Delta t} = \frac{\Delta V}{\Delta t}$$

So,

$$Ss \ \Delta h \Delta X \Delta Y \Delta Z \frac{dh}{dt} = \frac{\Delta V}{\Delta t}$$

This could similarly be re-written as

$$-\frac{\Delta V}{\Delta t} = -Ss \Delta X \Delta Y \Delta Z \frac{dh}{dt} \tag{16}$$

Where $\frac{\Delta V}{\Delta t}$ is the rate of change in storage, so by bring equation (16) into equation (14),we get

$$-\left(\frac{dqx}{dx} + \frac{dqy}{dy} + \frac{dqz}{dz} \right) \Delta X \Delta Y \Delta Z = -Ss \Delta X \Delta Y \Delta Z \frac{dh}{dt}$$

By cancelling out like terms on both sides, we get

$$-\left(\frac{dqx}{dx} + \frac{dqy}{dy} + \frac{dqz}{dz} \right) = -Ss \frac{dh}{dt} \tag{17}$$

Going back to equations(4),(5),and (6), we recall that

$$qx = -kx \frac{dh}{dx}, \ qy - ky \frac{dh}{dy}, \ qz = -kz \frac{dh}{dz}$$

Substituting for qx, qy, qz respectively in equation (17), we get

$$-\left\{ \frac{d}{dx}\left(-kx \frac{dh}{dx} \right) + \frac{d}{dy}\left(-ky \frac{dh}{dy} \right) + \frac{d}{dz}\left(-kz \frac{dh}{dz} \right) \right\} = -Ss \frac{dh}{dt}$$

Or

$$\left\{\left(kx\frac{d^2h}{dx^2}\right)+\left(ky\frac{d^2h}{dy^2}\right)+\left(kz\frac{d^2z}{dz^2}\right)\right\}=Ss\frac{dh}{dt} \tag{18}$$

Equation (16) is the governing equation of a 3D groundwater flow in a confined aquifer because herein, recharge (source) and leakage (sink) is ignored.
As earlier on stated, movement of groundwater is based on a hydraulic principle, therefore a source or sink is usually associated within an elemental volume which is expressed as Q, so the general governing equation of a 3D groundwater flow in an unconfined aquifer (which is the basic assumption) in this context is expressed as:

$$\left\{\left(kx\frac{d^2h}{dx^2}\right)+\left(ky\frac{d^2h}{dy^2}\right)+\left(kz\frac{d^2z}{dz^2}\right)\right\}-Q=Ss\frac{dh}{dt} \tag{19}$$

where,
Kx, Ky, Kz = hydraulic conductivity along the x, y ,z axes which are assumed to be parallel to the major axes of hydraulic conductivity;
h = piezometric (hydraulic) head;
Q= volumetric flux per unit volume representing source/sink terms;
Ss = specific storage coefficient defined as the volume of water released from storage per unit change in head per unit volume of porous material.
Equation (19) is the general governing equation of groundwater flow.
Groundwater flow in aquifers is often modelled as two-dimensional in the *horizontal* plane. This is due to the fact that most aquifers have large aspect ratio like a laminated plane sheet of paper, with horizontal dimensions hundreds of times greater than the vertical thickness. In such a setting, groundwater relatively flows along the horizontal plane, which implies that the z component of the velocity is comparatively small.
Therefore, a two-dimensional analysis is carried out in conjuncture with the use of transmissivity, by assuming that $h=h(x,y,t)$ only (h does not vary with z, thus (dh/dz = 0). This simplification of modeling 3D aquifer flow as horizontal two-dimensional flow is called the Dupuit-Forchheimer approximation.
Thus the groundwater flow equation for *confined* aquifer is

$$\left\{Tx\frac{d^2h}{dx^2}+Ty\frac{d^2h}{dy^2}\right\}\pm Q=Ss\frac{dh}{dt} \tag{20}$$

Where, T_x and T_y is transmissivity in the x and y direction and T = k * b = (m/day)(m) = m^2/day, while b is the saturated thickness of the aquifer.
But a situation, where the aquifer becomes *unconfined*, $b = h$, therefore $T = k * h$, so the equation (20) reduces to

$$\left\{K_{xh}\frac{d^2h}{dx^2}+K_yh\frac{d^2h}{dy^2}\right\}\pm Q=Ss\frac{dh}{dt} \tag{21}$$

Where h is the water level function of x and y.
Then, the partial differential equation of second order for groundwater flow is given as:

$$\frac{d}{dx}\left(K_x h\frac{dh}{dx}\right)+\frac{d}{dy}\left(K_y h\frac{dh}{dy}\right)+\frac{d}{dz}\left(K_z h\frac{dh}{dz}\right)\pm Q = Ss\frac{dh}{dt} \quad (22)$$

where S = storage coefficient; and h = hydraulic head.

3.2 Solute transport equation

The transport of solutes in the saturated zone is governed by the advection-dispersion equation which for a porous medium with uniform porosity distribution is formulated as follows:

$$\frac{\partial c}{\partial t}=-\frac{\partial}{\partial xt}(cvi)+\frac{\partial}{\partial xi}\left[D_{ij}\frac{\partial c}{\partial xj}\right]+Rc \qquad i,j=1,2,3 \quad (23)$$

Where,
∂ = delta function meaning change in;
x = the dimension (m);
t = the time (s)
c = concentration of the solute (kg/m^3);
Rc = reaction rate (concentration of solute)
in the source or sink (kg m^3/s);
Dij = dispersion coefficient tensor (m^2/s);
vi = velocity tensor (m/s).

An understanding of these equations and their associated boundary and initial conditions is necessary before a modelling problem can be formulated.

From equation 23 above, the first term on the right hand side represents advection transport and describes movement of solutes at the average seepage velocity of the flowing groundwater.

The second term represents the change in concentration due to hydrodynamic dispersion. Hydrodynamic dispersion is defined as the sum of mechanical dispersion and molecular diffusion.

The third term represents the effects of mixing with a source fluid of different concentration from the groundwater at the point of recharge or injection.

Chemical attenuations of inorganic chemicals can occur by sorption/desorption, precipitation/dissolution, oxidation /reduction, etc.

On the other hand, organic chemicals could be absorbed or degraded through microbial processes thus giving rise to a new set of equation such as:

$$\frac{\partial c}{\partial t}=-\frac{\partial}{\partial xt}(cvi)+\frac{\partial}{\partial_{xt}}\left[D_{ij}\frac{\partial c}{\partial xj}\right]+Rc+Y \quad (24)$$

Where Y represents all the chemical, geochemical and biological reactions responsible for the transfer of mass between the liquid and solid phase or the conversion of dissolved chemical species from one form to another.

Assuming that the reactions are limited to equilibrium-controlled sorption or exchange and first-order irreversible rate (decay) reactions in an isotropic homogenous porous medium, then equation 24 may be written as:

$$\frac{\partial c}{\partial t}+\frac{(\rho b)\partial \overline{c}}{\varepsilon(\partial t)}=-\frac{\partial}{\partial xt}(cvi)+\frac{\partial}{\partial xt}\left[D_{ij}\frac{\partial c}{\partial xj}\right]+\frac{Rc}{\varepsilon}-\lambda C\frac{\rho b}{\varepsilon}\lambda\overline{C} \tag{25}$$

Where
ρb = bulk density (M L^{-3});
λ = decay constant, or reaction rate (T^{-1});
$\overline{C}$ = is the solute concentration sorbed on the subsurface solids (M L^{-3}]);
C= is the dissolved concentration of a solute ([M L^{-3}] [M L^{-3}]);
ε = porosity (L^3 L^{-3});
The temporal change in sorbed concentration in equation 25 can be represented in terms of the solute concentration by using the chain rule of calculus as follows:

$$\frac{\partial \overline{C}}{\partial t}=\frac{\partial C}{\partial C}\frac{\partial C}{\partial t} \tag{26}$$

For equilibrium sorption and exchange reaction $\frac{\partial C}{\partial t}$ as well as C is a function of C only. Hence equilibrium reaction for C and $\frac{\partial C}{\partial t}$ can be substituted into the governing equation to develop a partial differential equation in terms of C alone. The resultant single transport equation is solved for solute concentration, sorbed concentration can be calculated using the equilibrium reaction. The linear sorption reaction considers that the concentration of solute sorbed to the porous medium is directly proportional to the reaction.

$$\overline{C}=K_dC \tag{27}$$

where K_d = distribution coefficient ($M^{-1}L^3$)
This reaction is assumed to be instantaneous and reversible. The relationship between the adsorbed and dissolved concentrations can be described using three possible isotherms: linear, Langmuir and Freundlich.
Thus in terms of the linear isotherm,

$$\frac{\partial \overline{C}}{\partial t}=\frac{\partial C}{\partial C}\frac{\partial C}{\partial t}=Kd\frac{\partial C}{\partial t} \tag{28}$$

Isotherms can be incorporated into the transport model using a retardation factor, R [M L^{-3}] (Kutílek and Nielsen 1994; Zheng and Bennett 2002):

$$R=1+\frac{\rho b}{s}K_d \tag{29}$$

By substituting equation 27 into equation 24, we get

$$\frac{\partial c}{\partial t}+\frac{(\rho b)}{\varepsilon}K_d\frac{\partial C}{\partial t}=-\frac{\partial}{\partial xt}(cvi)+\frac{\partial}{\partial xt}\left[D_{ij}\frac{\partial c}{\partial xj}\right]+\frac{Rc}{\varepsilon}-\lambda C\frac{\rho b}{s}\lambda K_dC \tag{30}$$

Now, substituting equation 28 into equation 29, the resultant equation is as follows:

$$R\frac{\partial C}{\partial t}=-\frac{\partial}{\partial xt}(cvi)+\frac{\partial}{\partial xt}\left[D_{ij}\frac{\partial c}{\partial xj}\right]+\frac{Rc}{s}-R\lambda C \tag{31}$$

This equation 31 is the solute transport (advection-dispersion) governing equation based on the assumption that the reactions are limited to equilibrium-controlled sorption.

The solution to this governing equation is identical to the solution for the governing equation without sorption effects, except the velocity, dispersive flux and source are deduced by the retardation factor "R".

For a given general equation, there is an infinite number of possible solutions. For steady-state flow, the unique and appropriate solution is one that matches the particular boundary conditions (BC) of the conceptual model. (For transient flow system and for solute transport, both initial condition and BC are required to obtain the unique solutions of head and concentrations.)

3.3 Boundary conditions

Boundary conditions indicate how an aquifer interacts with the environment outside the model domain. They include things such as heads at surface waters in contact with the aquifer, the location and discharge rate of a pumping well or a leaching irrigation field. For a distinct solution, at least one distinctive boundary condition is specified.

There are four types of boundary conditions which are derived from the most common two: *specified head* and *specified flux* conditions.

i. **Constant Head Boundary**: This is a type of specified head boundary condition, in which the head is known and the source of water has a constant water level at the model boundary. This condition is used in modelling an aquifer that is in good interaction with a lake, river or another external aquifer. These are usually where the groundwater is in direct contact with surface water such as a lake or a river and drains interact freely with the aquifer.

This condition is also known as the *first type* boundary condition and it is mathematically referred to as *Dirichlet* boundary.

It is stated mathematically as $h_{(x)}=h_{o(x)}, x\in\delta\cap$, *Dirichlet* Where h_o = the specified head along the boundary segment $\delta\cap$ of the modeled domain $\cap$.

ii. **Constant Flux Boundary**: This is a type of specified flux boundary condition also known as the second type of boundary condition, and mathematically known as Neumann's condition or recharge boundaries. Entering or leaving the aquifer is prescribed/constant flux. This boundary condition is used in simulating rainfall or distributed discharge for instance evaporation and also used in specifying known recharge to the aquifer owing to induced recharge or reticulation.

It is stated mathematically as $\frac{\delta h(x)}{\delta n}=-k\frac{\delta h(x)}{\delta n}, x\in\cap_2$ Neumann

Where $\frac{\delta h(x)}{\delta n}$ is the specified outward normal gradient to the boundary segment $\delta\cap_2$.

iii. **Mixed Boundary:** This type of boundary condition involves some combination of head and flux specification whereby, the rate of flow in and out of the aquifer is a function of the elevation of the stream bed, aquifer head, and leakage between the aquifer and the stream. For instance, leakage through a silty river bed to an underlying aquifer is represented by a flux that is proportional to the vertical conductivity of the silt layer and proportional to head difference from the river to the underlying aquifer. It is referred to as the *third type* of boundary condition and mathematically known as Cauchy Robbins condition.

This less common, mixed boundary or induced flux condition is also known as Stream or River Head Dependent Boundary. This boundary condition is used in the modelling of streams in poor connection with the aquifer, upward leakage in artesian aquifers, drains and overlying aquitards.

It is stated mathematically as $\alpha h(x) + \beta \frac{\delta h(x)}{\delta n} = C_0, \; x \in$

Where C_0 is specified function value along the boundary segment δn , α and β are specified functions.

iv. **No flow Boundary:** This is a very special type of the prescribed flux boundary and is referred to as no-flux, zero flux, impermeable, reflective or barrier boundary.

No flow boundaries are impermeable boundaries that allow zero flux. They are physical or hydrological barriers which inhibit the inflow or outflow of water in the model domain. No flow boundaries are specified either when defining the boundary of the model grid or by setting grid blocks as inactive (i.e hydraulic conductivity = 0)

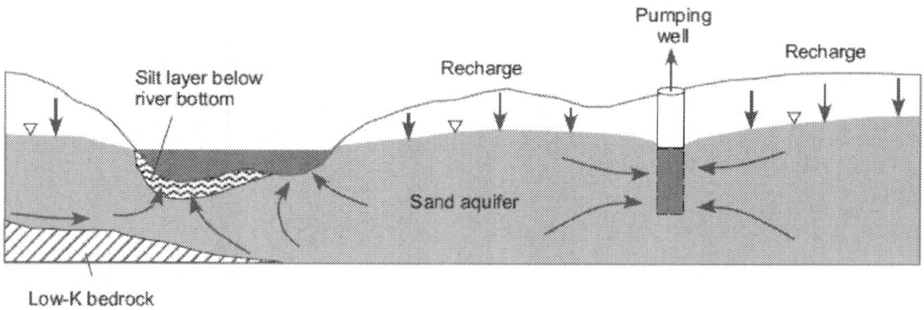

Fig. 4. Examples of flow domain boundary conditions. (Adapted from Zhang 2009).

In the figure above, the pumping well is a specified flux condition at the permeable section of the well. The recharge is a specified flux applied at the water table. The low-K bedrock is considered a special specified flux boundary (no-flow boundary). The leaky silt layer below the river is a mixed condition where the flux through the layer is proportional to the difference between the head in the river and the head in the aquifer beneath the silt layer. Where the river is in direct contact with the aquifer, there is a specified head condition.

4. Mathematical model

Mathematical models are used in simulating the components of the conceptual model and comprise an equation or a set of governing equations representing the processes that occur, for example groundwater flow, solute transport etc.

The differential equations are developed from analyzing groundwater flow (and transport) and are known to govern the physics of flow (and transport). The reliability of model predictions depends on how well the model approximates the actual natural situation in the field. Certainly, simplifying assumptions are made in order to construct a model, because the field situations are usually too complicated to be simulated exactly.

Mathematical models of groundwater flow and solute transport can be solved generally with two broad approaches:

Analytical solution of the mathematical equation gives *exact* solution to the problem, i.e., the unknown variable is solved continuously for every point in space (steady-state flow) and time (transient flow).Analytical models are exact solution to a specified, well simplified groundwater flow or transport equation. Because of the complexity of the 3D groundwater flow and transport equations, the simplicity inherent in analytical model makes it impossible to account for variations in field conditions that occur with time and space. For these problems, (variations in field conditions) such as changes in the rate/direction of groundwater flow, stresses, changes in hydraulic, chemical and complex hydrogeologic boundary conditions, the assumptions to be made to obtain an analytical solution will not be realistic. To solve mathematical models of this nature, we must resort to approximate methods using numerical solution techniques.

Numerical solution of the mathematical equation gives *approximate* solution to the problem, i.e., the unknown variable is solved at discrete points in space (steady-state flow) and time (transient flow). Numerical models are able to solve the more complex equations of multi-dimensional groundwater flow and solute transport. Many numerical solutions to the advection-dispersion equation have been reported. The most popular techniques are as the Finite Difference Method (FDM) and Finite Element Method (FEM).

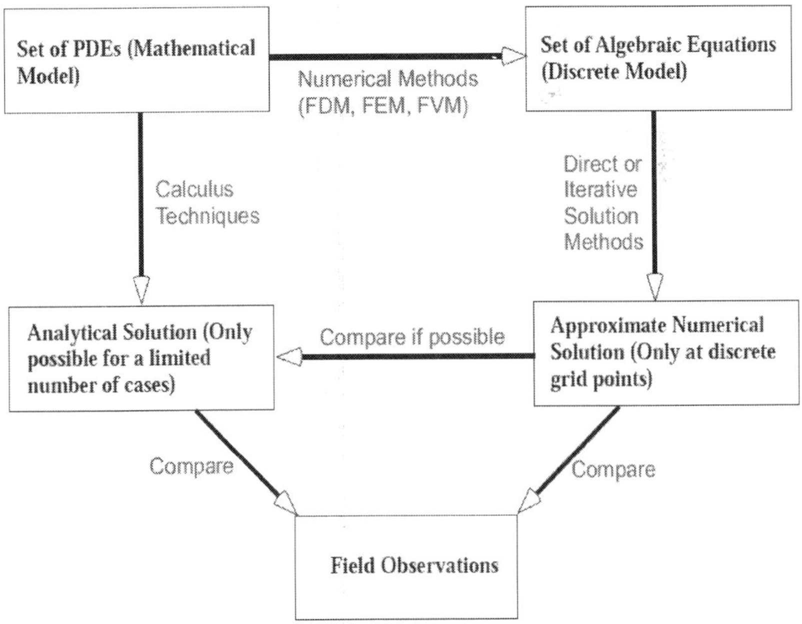

Fig. 5. Flow chart showing the approach to Mathematical model solutions (After Zhang 2009).

The characteristics of the various numerical models in comparison with analytical models are:

- The analytical solution exists at any point in the solution domain, while the numerical methods yield values for only a fixed (specified) finite number of grid points in the solution (space and time) domain.
- Using either FDM or FEM, the partial differential equation is converted to a set of algebraic equations involving unknown values at these particular points. This implies that the partial differential equations representing the balances of considered extensive quantities are replaced by a set of algebraic equations.
- Analytical solutions are not available for many problems of practical interest while the numerical methods allow us to solve the governing equation in more than one dimension, for complex boundary conditions and for heterogeneous and anisotropic aquifers, whereas most analytical solutions are restricted to consideration of homogeneous, isotropic aquifers. This implies that the solution is obtained for specified set of numerical values of the various model coefficients.

4.1 Finite difference method

A **finite difference** is a mathematical expression of the form $f(x + b) - f(x + a)$. If a finite difference is divided by $b - a$, one gets a *difference quotient*.

The primary vehicle of calculus and other higher mathematics is the function. Its "input value" is its argument usually a point ("P") expressible on a graph. The difference between two points, themselves, is known as their Delta (ΔP),as is the difference in their function result, the particular notation being determined by the direction of formation:

- Forward difference: $\Delta F(P) = F(P + \Delta P) - F(P)$;
- Central difference: $\delta F(P) = F(P + \frac{1}{2}\Delta P) - F(P - \frac{1}{2}\Delta P)$;
- Backward difference: $\nabla F(P) = F(P) - F(P - \Delta P)$.

The general preference is the forward orientation, as F(P) is the base, to which differences (i.e "ΔP"s) are added to it.

Furthermore,

- If [ΔP] is finite (meaning measurable), then ΔF(P) is known as the finite difference with specific denotations of DP and DF(P);
- If [ΔP] is infinitesimal (an infinitely small amount -*l*- usually expressed in standard analysis as a limit; $(\lim_{\Delta \to 0})$, then ΔF(P) is known as the infinitesimal difference with specific denotations of dP and dF(P) (in calculus graphing,the point is almost exclusively identified as '*x*' and *F*(*x*) as '*y*'

The function difference divided by the point difference gives the difference quotient (Newton's quotient): $\frac{\Delta F(P)}{\Delta P} = \frac{F(P+\Delta P) - F(P)}{\Delta P} = \frac{\nabla F(P+\Delta P)}{\Delta P}$

If ΔP is infinitesimal, then the difference quotient is a derivative, otherwise it is a divided difference

The approximation of derivatives by finite differences plays a central role in finite difference methods for the numerical solution of differential equations, especially boundary value problems.

An important application of finite differences is in numerical analysis, especially in numerical differential equations, which aim at the numerical solution of ordinary and partial differential equations respectively. The idea is to replace the derivatives appearing in the differential equation by finite differences that approximate them. The resulting methods are called finite difference methods.

A **forward difference** is an expression of the form

$$\Delta_h[f](x) = f(x+h) - f(x) \tag{32}$$

Depending on the application, the spacing h may be variable or constant.

A **backward difference** uses the function values at x and $x - h$, instead of the values at $x + h$ and x:

$$\nabla_h[f](x) = f(x) - f(x - h). \tag{33}$$

Finally, the **central difference** is given by

$$\delta_h[f](x) = f(x+½h) - f(x - ½h) \tag{34}$$

The derivative of a function f at a point x is defined by the limit

$$f'(x) = \lim_{k \to 0} \frac{f(x+h) - f(x)}{h} \tag{35}$$

If h has a fixed (non-zero) value, instead of approaching zero, then the right-hand side is

$$\frac{f(x+h) - f(x)}{h} = \frac{\Delta h[f](x)}{h} \tag{36}$$

Hence, the forward difference divided by h approximates the derivative when h is small. The error in this approximation can be derived from *Taylor's theorem*. Assuming that f is continuously differentiable, the error is

$$\frac{\Delta h[f](x)}{h} - f'(x) = 0(h) \ (h \to 0) \tag{37}$$

The same formula holds for the backward difference:

$$\frac{\nabla h[f](x)}{h} - f'(x) = 0(h) \tag{38}$$

However, the central difference yields a more accurate approximation. Its error is proportional to square of the spacing (if f is twice continuously differentiable):

$$\frac{\delta h[f](x)}{h} - f'(x) = 0(h^2) \tag{39}$$

The main problem with the central difference method, however, is that oscillating functions can yield zero derivative. If f(nh)=1 for n uneven, and f(nh)=2 for n even, then f'(nh)=0 if it is calculated with the central difference scheme. This is particularly troublesome if the domain of f is discrete.

In an analogous way one can obtain finite difference approximations to higher order derivatives and differential operators. For example, by using the above central difference formula for $f'(x + h / 2)$ and $f'(x - h / 2)$ and applying a central difference formula for the derivative of f' at x, we obtain the central difference approximation of the second derivative of f:

$$f''(x) \approx \frac{\delta^2 h[f](x)}{h^2} = \frac{f(x+h) - 2f(x) + f(x-h)}{h^2} \tag{40}$$

More generally, the nth-order forward, backward, and central differences are respectively given by:

$$\Delta_h^n[f](x) = \sum_{i=0}^{n} (-1)^i \binom{n}{i} f(x + (n-i)h) \tag{41}$$

$$\nabla_h^n[f](x) = \sum_{i=0}^{n} (-1)^i \binom{n}{i} f(x - ih) \tag{42}$$

$$\delta_h^n[f](x) = \sum_{i=0}^{n} (-1)^i \binom{n}{i} f\left(x + \left(\frac{n}{2} - i\right) h\right) \tag{43}$$

Note that the central difference will, for odd n, have h multiplied by non-integers. This is often a problem because it amounts to changing the interval of discretization. The problem may be remedied taking the average of $\delta^n[f](x - h / 2)$ and $\delta^n[f](x + h / 2)$.
The relationship of these higher-order differences with the respective derivatives is very straightforward:

$$\frac{d^n}{dx^n}(x) = \frac{\Delta_h^n[f](x)}{h^n} + 0(h) = \frac{\nabla_h^n[f](x)}{h^n} + 0(h) = \frac{\delta_h^n[f](x)}{h^n} + 0(h^2) \tag{44}$$

Higher-order differences can also be used to construct better approximations. As mentioned above, the first-order difference approximates the first-order derivative up to a term of order h. However, the combination

$$\frac{\Delta h[f](x) - \frac{1}{2}\Delta_h^2[f](x)}{h} = \frac{f(x+2h) - 4f(x+h) + 3f(x)}{2h} \tag{45}$$

approximates $f'(x)$ up to a term of order h^2. This can be proven by expanding the above expression in *Taylor series*, or by using the calculus of finite differences. If necessary, the finite difference can be centered about any point by mixing forward, backward, and central difference.

4.2 Basics of finite difference method

The partial differential equation describing the flow and transport processes in groundwater include terms representing derivatives of continuous variables. Finite difference method is based on the approximations of these derivatives (slopes or curves) by discrete linear changes over small discrete intervals of space and time. A situation where the intervals are adequately small, then all of the linear increment will represent a good approximation of the true curvilinear surface.

Following an illustration in figure 6a below, Bennett (1976) used observation wells in a confined aquifer to show a reasonable approximation for the derivative of head, $\partial h / \partial x$, at a point (d) midway between wells 1 and 0 is given as follows:

$$\left[\frac{\partial h}{\partial x}\right] d \approx \frac{h_0 - h_2}{\Delta x} \tag{46}$$

It is worthy to note that the observation wells are spaced at equal distance apart, likewise a reasonable approximation for second derivatives, $\partial^2 h / \partial x^2$ at point 0 (centre of the well) is:

$$\frac{\partial^2 h}{\partial x^2} = \frac{\left[\frac{\partial h}{\partial x}\right] \text{e-} \left[\frac{\partial h}{\partial x}\right] \text{d}}{\Delta x} = \frac{\frac{h_2 - h_0}{\Delta x} - \frac{h_0 - h_1}{\Delta x}}{\Delta x} = \frac{h_1 + h_2 - 2h_0}{(\Delta x)^2} \tag{47}$$

Considering also wells 3 and 4 in figure 6b located on a line parallel to the y-axis, we can also approximate $\partial^2 h / \partial y^2$ at point 0 (the same point 0 as figure 6a) as follows:

$$\left[\frac{\partial^2 h}{\partial y^2}\right] \approx \frac{h_3 - h_4 - 2h_0}{(\Delta y)^2} \tag{48}$$

A situation where the spacing of the well in figure 6b is equidistant ($\Delta x = \Delta y = a$), then the following approximation is as follows:

$$\frac{\partial^2 h}{\partial x^2} + \frac{\partial^2 h}{\partial y^2} \approx \frac{h_1 + h_2 + h_3 + h_4 - 4h_0}{a^2} \tag{49}$$

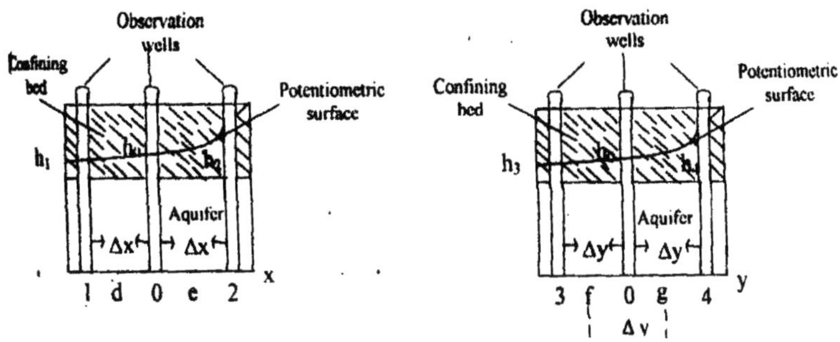

Fig. 6. Schematic cross-section through a confined aquifer to illustrate numerical approximation to derivatives of head, $\partial h / \partial x$ (a) and $\partial h / \partial y$ (b).(Modified after Bennett 1976).

These approximations can also be obtained by Taylor series expansion. A certain error is involved in the approximation of the derivatives by finite differences, but this error will generally decrease as Δx and Δy are given small values. This error is called 'truncation error' because the replacement of a derivative by a difference quotient is equivalent to a truncated Taylor series.

4.3 Taylor Series and finite difference approximations

Taylor Series Expansion: Approximation of a function h at a point $h(x + \Delta x)$ using derivatives of the function at $h(x)$.

- Forward Approximation:

$$h(x+\Delta x)=h(x)+\frac{\Delta x}{1}\frac{\partial h(x)}{\partial x}+\frac{\Delta x}{2}\frac{\partial h(x)}{\partial x^2}+\frac{\Delta x}{3}\frac{\partial^2 h(x)}{\partial x^2}+\ldots \tag{50}$$

If we truncate the higher order terms,

$$\text{i.e } \frac{\Delta x}{n}\frac{\partial = h(x)}{\partial x^x}, n \geq 2 \quad n\ 2, \text{we get:}$$

$$h(x+\Delta x)\approx h(x)+\Delta x\frac{\partial h(x)}{\partial x}\Rightarrow\frac{\partial h(x)}{\partial x}\approx\frac{h(x+\Delta x)-h(x)}{\Delta x} \tag{51}$$

- Backward Approximation:

$$h(x-\Delta x)=h(x)-\frac{\Delta x}{1}\frac{\partial h(x)}{\partial x}+\frac{\Delta x}{2}\frac{\partial^2 h(x)}{\partial x^2}-\frac{\Delta x}{3}\frac{\partial^2 h(x)}{\partial x^2}+\ldots\ldots \tag{52}$$

If we truncate the higher order terms(n=2), we get;

$$h(x-\Delta x)=h(x)-\Delta x\frac{\partial h(x)}{\partial x}\Rightarrow\frac{\partial h(x)}{\partial x}\approx\frac{h(x-\Delta x)-h(x)}{\Delta x} \tag{53}$$

The Taylor Series is key to understanding the finite difference method.
Considering also the discretization of time, which may be viewed as another dimension and hence represented by another index. Illustrating using a hydrograph in figure 8 below, the head is plotted against time for a transient flow system; n is the index or subscript used to denote the line at which a given head is with respect to time, and it can be approximated as $\partial h/\partial t \approx \Delta h/\Delta t$. In terms of the heads calculated at specific time increments (or time slope), the slope of the hydrograph at time n can be approximated as follows:

$$\left[\partial h/\partial t\right]_{n\Delta t}\approx\frac{hn_{-1}-hn}{\Delta t} \tag{54}$$

or

$$\left[\partial h/\partial t\right]_{n\Delta t}\approx\frac{hn-hn_{-1}}{\Delta t} \tag{55}$$

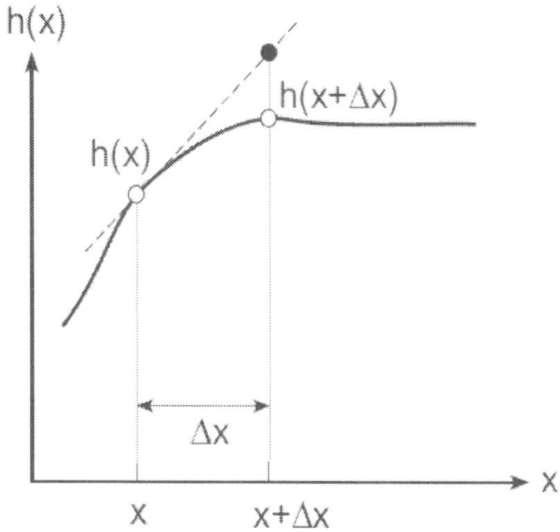

Fig. 7. Taylor series expansion along x-axis (After Zhang 2009).

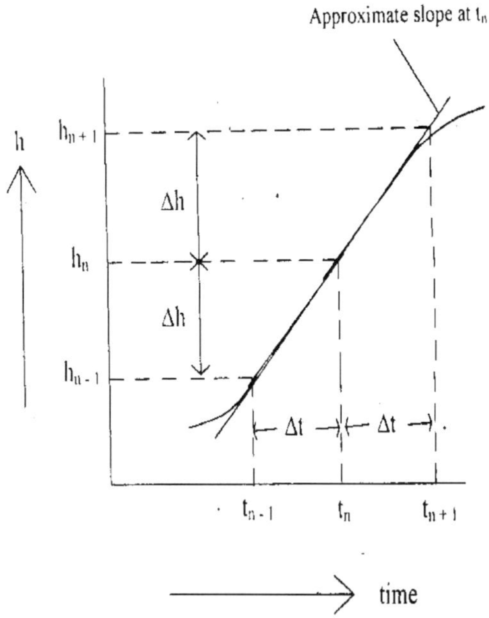

Fig. 8. Part of a hydrograph showing that the derivative (or slope $\partial h / \partial t$) at time node t_n may be approximated by $\Delta h / \Delta t$ (Konikow,1996).

4.4 The finite-difference grid

The application of a numerical model to the solution of a ground-water problem is a creative process.

There are many different techniques that can be applied to solve the same problem and each modeller has developed preferred ways of approaching a model design and the software package to use in terms of flexibility. However, no software package can be totally flexible.

Many software packages are been used by modellers to interactively design a three dimensional finite-difference ground-water flow and contaminant transport models namely GV(Groundwater Vista), MODFLOW, MT3D, and MODPATH. GV model design is generic because it can be used to create data sets for MODFLOW, MT3D, and MODPATH. While each of these specific models has its own data input format, they all have key features in common. The most important features in common are the physical layout of the grid or mesh, the specification of boundary conditions, and the definition of hydraulic properties.

A finite-difference model is constructed by dividing the model domain into square or rectangular regions called **blocks** or **cells**. Head, drawdown, and concentration are computed at discrete points within the model called **nodes**. The network of cells and nodes is called the **grid** or **mesh**. These terms are used. There are two main types of finite-difference techniques, known as block-centered and mesh-centered. The name of the technique refers to the relationship of the node to the grid lines. Head is computed at the centre of the rectangular cell in the block-centered approach. Conversely, head is computed at the intersection of grid lines (the mesh) in the mesh-centered technique. The figure below illustrates this concept graphically.

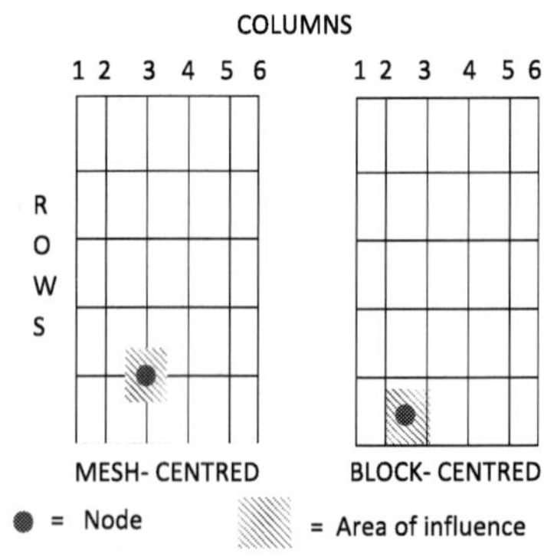

Fig. 9. The finite-difference grid.

In this figure, note that the dependent variable (head or concentration) is computed at the centre of cells in the block-centered technique but may be offset from the centre in the mesh-centered approach.

In each technique, the head and all physical properties are assumed to be constant throughout the cell region surrounding the node.

The finite-difference grid is designed by manipulating rows, columns, and layers of cells. A series of cells oriented parallel to the x-direction is called a row. A series of cells along the y-direction is called a column. A horizontal two-dimensional network of cells is called a layer. Cells are designated using the row and column co-ordinates, with the origin in the upper

left corner of the mesh. That is, the upper left cell is called (row 1, column 1). The upper layer is layer 1and layers increase in number downward.
For example, a block-centered finite-difference grid is created in GV by first specifying the number of rows, columns, and layers.
The user also provides the initial row and column widths or spacings. GV then creates a mesh with uniform row and column widths. This is called a **regular mesh**. While the regular mesh represents the most accurate form of the finite-difference solution (Anderson and Woessner 1992), it is often necessary or desirable to refine the mesh in areas of interest. In this manner, more accuracy is achieved in key areas at the expense of lower accuracy at the edges of the model grid.

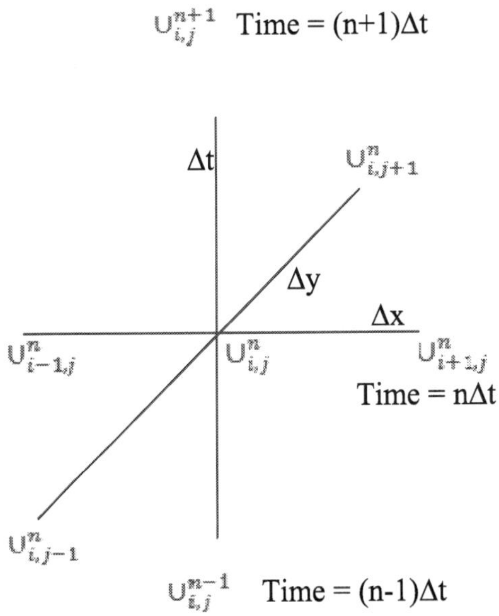

Fig. 10. Notation for forward and backward simulation of the time derivative. (Modified after Bear, 1979).

Based on the notation in the figure above, the derivatives at (i, j) for the forward, backward, central and second order of the finite difference formulas of equations (32), (33), (34) and (40) are as follows respectively:

$$\Delta h[f](x) = f(i+1,j) - f(i,j) \tag{56}$$

$$\nabla_h[f](x) = f(i,j) - f(i-1,j) \tag{57}$$

$$\delta_h[f](x) = \frac{f(i+1,j) - f(i-1,j)}{2} \tag{58}$$

$$f''(x) \approx \frac{f(i+1,j)-2f(i,j)+(i-1,j)}{h^2} \tag{59}$$

Explicit Approximation: Using a nodal point (U_i^n) anywhere on the grid where 'n' is the time interval. The forward difference approximation is:

$$\frac{U_i^{n+1}-U_i^n}{\Delta t} = \frac{U_{i+1}^n - 2U_i^n + U_{i-1}^n}{(\Delta x)^2} \tag{60}$$

Therefore ,

$$U_i^{n+1} = U_i^n + \frac{\Delta t}{(\Delta x)^2}\left[U_{i+1}^n - 2U_i^n + U_{i-1}^n\right] = U_i^n + r\left[U_{i+1}^n - 2U_i^n + U_{i-1}^n\right] \tag{61}$$

Where $r = \dfrac{\Delta t}{(\Delta x)^2}$

$$U_i^{n+1} = rU_{i+1}^n + (1-2r)U_i^n + rU_{i-1}^n \tag{62}$$

r depends on interval levels assigned to Δx and Δt. Here unknown values can be calculated from known values of previous levels. Explicit method is stable if $\dfrac{\Delta t}{(\Delta x)^2} = \frac{1}{2}, or\, 0 < r < 0.5$.

it is seen from the above conditions that Δt must be very small since $\Delta t \le \frac{1}{2}(\Delta x)^2$, and the number of computations needed to reach a time level are extremely large.

Implicit Approximation: Here backward difference approximation is used in place of forward difference, so we have

$$\left[\frac{\partial u}{\partial t}\right]_i^n = \left[\frac{\partial^2 u}{\partial x^2}\right]_i^n$$

where 'n' is the time interval and 'I' is the space interval. Therefore, the equation for backward difference approximation at t = $(n + 1)\,\Delta t$, we get for constant Δx as follows:

$$\frac{U_i^{n+1}-U_t^n}{\Delta t} = \frac{U_{i+1}^{n+1} - 2U_i^{n+1} + U_{i-1}^{n+1}}{(\Delta x)^2} \tag{63}$$

It is seen here that only one value is associated with the previous time level while three are associated with the present time level unlike the Explicit method.

Assuming that the values U_t^n are known at all nodes at time $n\Delta t$, equation (63) is a single with three unknowns U_i^{n+1}, U_{i+1}^{n+1} and U_{i-1}^{n+1}. Equations for each node in the flow domain can be written like that of equation (63). Since there is now one known value of head, for time $(n+1)\Delta t$, at each node, then we shall have a system of equations whereby the total number of the unknowns is equal to the total number of equations. Now, we are able to solve the

entire set of equations, getting the new value Ui^{n+1} at each node i, this is the Implicit method.
Implicit method requires much work than Explicit method in solving the set of simultaneous equations, and it is unconditionally stable irrespective of the time step Δt .

4.5 Crank-Nicholson method

In an Implicit method, the choice of Δt can be independent of Δx. Further refinement is possible. Also, the smaller the truncation error, the faster is the convergence of the finite difference equations to the differential equation.
In Implicit and Explicit methods for the time derivative $\partial u/\partial t$,the truncation error was of the order of $0(\Delta t)$. By replacing $\partial u/\partial t$ with the central difference, the truncation error associated with the time term would be reduced from $0(\Delta t)$ to $0\{(\Delta r)^2\}$. This is known as the Crank-Nicholson method.
By averaging the approximations $\partial^2 u/\partial x^2$ at n and n+1 time levels, the implicit method is obtained. At times, a weight $\lambda(0<\lambda<1)$ is used here. Then value for $\partial u/\partial t$ evaluated at (n + ½) is obtained.
Therefore,

$$\frac{Ui^{n+1}-Ut^{n}}{\Delta t}=\frac{\lambda\left(U_{i+1}^{n+1}-2U_{i}^{n+1}+U_{i-1}^{n+1}\right)+(1-\lambda)\left(U_{i+1}^{n}-2U_{i}^{n}+U_{i-1}^{n}\right)}{(\Delta x)^{2}} \tag{64}$$

Explicit approximation is obtained by setting $\lambda = 0$ in equation (17) while $\lambda = 1$gives the Implicit scheme.
On the other hand, Crank-Nicholson approximation is obtained by setting $\lambda= ½$ in equation (64).

$$\frac{Ui^{n+1}-Ui^{n}}{\Delta t}=\frac{\frac{1}{2}\left(U_{i+1}^{n+1}-2U_{i}^{n+1}+U_{i-1}^{n+1}\right)+\frac{1}{2}\left(U_{i+1}^{n}-2U_{i}^{n}+U_{i-1}^{n}\right)}{(\Delta x)^{2}} \tag{65}$$

Modification of equation (65) to handle problems of saturated and unsaturated flow is possible and the extension to more than one space dimension poses no difficulty.

5. Conclusion

Mathematical models are efficient tools commonly used in studying groundwater systems. Generally, mathematical models are used to simulate (or to predict) the groundwater flow and in some cases the solute and/or heat transport conditions. They may also be used in the evaluation of remediation alternatives.
Finite difference methods can be used to solve the solute transport equation, either using backward, forward or central differencing. These methods however can result in artificial oscillation (under or over shooting) or numerical dispersion due to truncation errors of the discretization. Application of backwards finite difference, results in an implicit scheme

which is always convergent but is computationally expensive, and introduces considerable numerical dispersion (Zheng and Bennett 2002).

The use of central differencing (such as Crank Nicholson schemes) in the discretization can cause numerical oscillation in the form of "wiggles" when implicit schemes are used. If an explicit formulation is used instead the solution often is non convergent (Leonard 1979). Numerical oscillation can be minimized by the use of upstream weighting, but this leads to considerable numerical dispersion owing to truncation errors (Zheng and Bennett 2002).

Another solution for artificial oscillation is the use of finer grids, with a choice based on the Peclet number: $Pe = u * \Delta x / D$

where u is the flow velocity [L T^{-1}] , Δx is the grid spacing [L] and D is the diffusivity [L^2 T^{-1}]. A Pe number < 2 can greatly reduce or eliminate numerical oscillation, but usually the associated computational cost due to excessively fine grids is impractical (Zheng and Bennett 2002).

For stability of numerical techniques, the diagonal dominance check is performed, where the coefficient of the left hand side must be equal to the coefficient on the right hand side (Osuwa, J.C 2011).

Because of the associated complexities in numerical modelling, models approximate a real system and therefore are not exact descriptions of physical systems or processes. They're mathematically representing a simplified version of a system, thus expressing a range of possible outcomes reflecting the inherent assumptions.

Predictive simulations must be viewed as estimates, dependent upon the quality and uncertainty of the input data.

While using models as predictive tools, field monitoring must be incorporated to verify model predictions.

To determine model error, the examination of numerical methods through an analysis of their ability compared with analytical methods is strongly recommended.

6. References

Anderson, M.P. and Woessner, W.W. 1992, Applied Groundwater Modeling: Simulation of Flow and Advective Transport, Academic Press, Inc., New York, 381 pp.

American Society for Testing and Materials, 1995, Standard Guide forSubsurface Flow and Transport Modeling. ASTM Standard D 5880-95, West Conshohocken, PA, 6 p.

Bear, J., 1979, Hydraulics of Groundwater. McGraw-Hill, New York

Bear, J., Beljing, M. S., and Ross, R.R., 1992; Fundamentals of GroundwaterModeling, US.EPA

Bear, J., and A. Verruijt, 1987. Modeling Groundwater Flow and Pollution. D.Reidel Publishing Company, 414 p.

Bennett, G.D 1976. Introduction to Groundwater Hydraulics: A Programmed Text for Self-Instruction. Techniques of Water Resources Investigations of the US Geological survey, Book 3, Ch, B2; p 172.

Chong-xi, C., and Guo-min, L., 1996; Theory and Model of Groundwater Flow and Solute Transport M]. Wuhan: Press of China University of Geosciences, 1996 (in Chinese).

Elango, L 2005. Numerical Simulation of Groundwater Flow and Solute Transport, Allied Publishers, Chennai.

El-Sadek, A 2007 Assessing The Impact On Groundwater Of Solute Transport In Contaminated Soils Using Numerical And Analytical Models Researcher, National Water Research Center, Fum Ismailia Egypt. *Eleventh International Water Technology Conference, IWTC11 2007 Sharm El-Sheikh, Egypt* 649.

Franke, O.L., Bennett, G.D., Reilly, T.E., Laney, R.L., Buxton, H.T., and Sun, R.J., 1991, Concepts and Modeling in Ground-Water Hydrology -- A Self-Paced Training Course. U.S. Geological Survey Open-File Report 90-707.

Grift B. V., and Griffioen J. 2008, Modelling Assessment of Regional Groundwater Contamination due to Historic Smelter Emissions of Heavy Metals [J]. Journal of Contaminant Hydrology, 2008, 96: 48-6

Igboekwe , M.U., Gurunadha Rao, V.V.S and Okwueze., E.E. 2008. Groundwater flow modeling of Kwa Ibo River watershed, southeastern Nigeria. Hydrological Processes 22, 1523-1531.

James R. C., and Alan, J. R., 2006, Finite Difference Modeling of Contaminant Transport Using Analytic Element Flow Solutions [J]. Advances in Water Resources, 2006, 29:1075-1087.

Kashyap, D, 1989, Mathematical Modellingfor Groundwater Management- Status in India. Indo-French Seminar on Management of Water Resources, 22- 24 September, 1989,Festival of France-1989, Jaipur, pp. IV-59 to IV-75.

Kinzelbach, W., 1986, Groundwater Modeling: An Introduction with Sample Programs in BASIC. Elsevier, New York, 333 p.

Konikow, L.F., 1996. Numerical models of groundwater flow and transport. In : Manual on Mathematical Models in Isotope Hydrogeology, international Atomic Energy Report. IAEA-TECDOC-910, Vienna, Austria: 59-112.

Kumar, C. P., 1992, Groundwater Modelling - In. Hydrological Developments in India Since Independence. A Contribution to Hydrological Sciences, National Institute of Hydrology, Roorkee, pp. 235-261.

Kumar, C. P., 2001, Common Ground Water Modelling Errors and Remediation. Journal of Indian Water Resources Society, Volume 21, Number 4, October 2001, pp. 149-156.

Kumar, C. P., 2002, Groundwater Flow Models. Scientist 'E1' National Institute of Hydrology Roorkee - 247667 (Uttaranchal) publication.

Kutílek, M. and Nielsen, D. R. (1994). Soil hydrology, Catena Verlag Cremlingen-Destedt, Germany.

Leonard, B. P. (1979). "A stable and accurate convective modelling procedure based on quadratic upstream interpolation." *Computer Methods in Applied Mechanics and Engineering,* 19: 59-98.

McDonald, M.G. and Harbaugh, A.W. 1988, *A Modular Three-Dimensional Finite-Difference Ground-Water Flow Model,* USGS TWRI Chapter 6-A1, 586 p.

Neumann, L.N., Cook, F., Western, A.W. and Verburg, K. (2009) A one dimensional solute transport model for hydrological response units. 18th World IMACS / MODSIM Congress, Cairns, Australia 13-17 July 2009.

Osuwa, J.C 2011, Unpublished lecture notes on Mathematical and Numerical Methods (PHY 712). Depatrment of Physics. Michael Okpara University of Agriculture , Umudike.

Pinder, G.F., and Bredehoeft, J.D. 1968, *Application of the Digital Computer for Aquifer Evaluation*, Water Resources Research, Vol. 4, pp. 1069-1093.

Rai, S. N., 2004, Role of Mathematical modeling in Groundwater Resources management, Sri Vinayaka Enterprises Hyderabad

Shazrah, O., Atai, S., and Sreevdi, P.D., 2008, Governing Equations of Groundwater Flow and Aquifer Modelling Using Finite Difference Method. In: GROUNDWATER DYNAMICS IN HARD ROCK AQUIFERS.-Sustainable Management and Optimal Monitoring Network Design. P186-224

Wang, H.F. and Anderson, M.P., 1982, *Introduction to Groundwater Modeling*. W.H. Freeman and Company, San Francisco, CA, 237 p.

Zhang, Y., 2009 Groundwater Flow and Solute Transport Modeling. Draft lecture note on GEOL 5030, Dept. of Geology & Geophysics University of Wyoming

Zheng, C. and Bennett, G. D. (2002). Applied Contaminant Transport Modeling. New York, John Wiley & Sons.

14

Deposits from the Glacial Age at Lake Baikal

N.I. Akulov and M.N. Rubtsova
Institute of the Earth's Crust, Siberian Branch of the RAS
Russia

1. Introduction

Lake Baikal is one of the unique sites on the globe. Its huge freshwater reserves were accumulated over the course of several geological epochs, with Pleistocene glaciations having had an important role in this respect. Clean melt waters from the numerous glaciers that were descending down the surrounding mountain ranges serves as the main supplier of freshwater. The Quaternary glaciation in East Siberia was stronger when compared with Europe, but because of the insufficient amount of moisture in air masses over Central Asia, a permanent snow cover persisted only on the mountain ranges and on their arms. This factor is responsible for a significant depth of freezing of sedimentary rocks ("eternal" frost, or permafrost), and for the occurrence of a smaller (in area and thickness) glacial shield advancing from the Arctic Ocean. Recent years saw the emergence of intensive studies into palaeoclimatic changes, based on analyzing core material from bottom sediments in Baikal. The history of the "Drilling on Baikal" or "Baikal-Drilling" program is widely known from publications of such outstanding scientists as M.I. Kuzmin, G.K. Khursevich, A.A. Prokopenko, S.A. Fedenya, and E.B. Karabanov (Kuzmin et al., 2009); S.M. Colman, J.A. Peck, E.B. Karabanov S. J. Carter, J. P. Bradbury, J.W. King and D.F. Williams (Colman et al., 1995); T.G. Moore, K.D. Klitgord, A.Ya. Golmstok, and E. Weber (Moore et al., 1997), and others. The program has opened up a new era in the palaeoclimatic investigations into the Baikal hollow. The point is that as early as 1989 D.F. Williams, professor of South Carolina University, suggested to their Russian colleagues that a cooperative project be started aimed at studying global changes in the natural environment and climate of Central Asia based on deep-water drilling on Baikal. In 1992, Japanese scientists headed by Professor Sh. Horie joined the "Baikal-Drilling" project. As part of the drilling operations, several continuous palaeoclimatic records covering a time span of several million years were obtained. The question as to how these palaeoclimatic records are correlated with intra-continental deposits is a currently challenging palaeoclimatic problems of Central Asia which will remain one of the "hottest" areas for many years to come.

2. Background and formulation of the problems

Prince P.A. Kropotkin was the first to investigate the glacial deposits of the Baikal mountainous region surrounding Lake Baikal. Drawing on the characteristic glacial striation on rocks and on the numerous erratic boulders of granitoid composition, P.A. Kropotkin (1876) found for the first time that the entire northern part of Pribaikalie and the Patoma

Upland underwent glaciation at some time in the past. He suggested that the glaciation was two-fold. P.A. Kropotkin's conclusions were sharply objected by I.D. Chersky (1877) and A.I. Voyeikov (1952), who argued that formation of blanket glaciers is simply impossible in a dry climate. Later, the ancient glacial forms in the Baikal mountainous region were studied by Academician V.A. Obruchev (1931), who found that the permafrost in this region is more than a hundred meters in depth and that a relatively local occurrence of a significant amount of products of glacial activity points to the fact that the glaciation was discontinuous (patchy) rather than continuous. Subsequent researchers concerned with Quaternary deposits at Lake Baikal (Yatsenko, 1950; Gurulev, 1959; Voskresensky, 1959; Lamakin, 1961; 1963; Olyunin, 1969) suggested that this region experienced two glaciations that manifested themselves during the Mid- and Late Pleistocene. To substantiate the Mid-Late Pleistocene Interglacial in the southwestern Pribaikalie, S.S. Voskresensky (1959) used deposits from the 35- and 22-meter terraces of the Angara characterized by pollen from heat-loving vegetation. The eminent glaciologist M.G. Grosvald (1965), who studied in detail the deposits from the glacial age on the western slopes of Eastern Sayan, reconstructed the pattern of thrice-repeated glaciation. Around the same time E.I. Ravsky and collaborators (Ravsky et al., (Rayevsky et al., 1964) and S.M. Tseitlin (1964), based on analyzing the terrace deposits along the valley of the Angara and Lower-Tunguska rivers, suggested that there occurred a fourfold glaciation. Somewhat later, this same conclusion was arrived at by O.M. Adamenko, A.A. Kul'chitsky and R.S. Adamenko (Adamenko et al., 1974) as well as by N.A. Logachev and collaborators (Logachev et al., 1974) and S. Back with R.M. Strecker (Back, Strecker, 1998). After many years of investigations into the morphology of the glacier relief in the mountains of the northern part of the Baikal mountainous region, an outstanding geomorphologists A.G. Zolotarev (1961) identified traces of the six-fold movement of the valley glaciers. However, because of lack of exhaustive geologo-palaeogeomorphological factual evidence, he was unable to discriminate between traces corresponding to independent glaciations and traces corresponding to stages or oscillatory movements. On the other hand, among the geomorphologists studying the Baikal mountainous region there appeared adherent to the concept of thrice-repeated glaciation. According to data reported by N.V. Dumitrashko (1952), there occurs no complete thawing away of the glaciers throughout the Pleistocene. She considers the entire glacial age to be a common glaciation and attributes the terminal moraines to the phase changes in the glaciers. Among the four phases identified, she assigns the oldest phase to the Pliocene, the second (maximal) phase to the Early Pleistocene, and the last two phases to the end of the glacial age (Late Pleistocene). The possible existence of a single glaciation in the past was also supported by N.P. Ladokhin (1959) and V.V. Zamoruyev (1971, 1978). According to the conclusion drawn by V.V. Zamoruyev (1971), the age of the morainic deposits that were stripped at the foot of the southern part of Khamar-Daban does not exceed the length of the Mid-Pleistocene, but they became of widespread occurrence at the end of the Late Pleistocene (Sartanian time). Over the course of many years, N.A. Logachev (Logachev et al., 1964; 1974), A.A. Kul'chitsky (1973; 1985; 1993), S.S. Osadchy (1982), D.B. Bazarov (1986), A.B. Imetkhenov (1987), V.D. Mats (Mats et al., 1982; 2001), and other workers were concerned with investigations into the geology of Quaternary deposits and, accordingly, into glacial formations at Lake Baikal. Their geological research was instrumental in substantially complementing and systematizing the data on Quaternary deposits of Lake Baikal. The results obtained through the implementation of the international "Baikal-Drilling" project aroused considerable interest in reconstructing the climatic changes at Lake Baikal during the Pleistocene (Team ..., 1998). Deep-water drilling operations on Baikal provided continuous palaeoclimatic records for Central Asia for the last 10 mln years. It was found that at the period of global glaciations there disappeared diatom algae, and the lake was drainless. Also, its water level was 30-50 m below the contemporary elevation. Among the recent investigations into the Pleistocene-Holocene deposits of the Baikal mountainous country, the research done by S.K. Krivonogov (2010) deserves mention. Drawing on geomorphological data, he found that the Tyiskaya phase of technogenesis that occurred in the Mid-Late Pleistocene brought about a significant differentiation of the relief. The emergence of high mountains caused the cold climatic belt to expand, so that at the time of the last global cooling during the Late Pleistocene the mountains surrounding the hollows became a region of intense glaciation which determined, to a significant extent, the pattern of sedimentation. According to his data, the maximum of Pleistocene glaciation in the mountains of the inner parts of Eurasia corresponds to the first half of the Late Pleistocene (100–70 ths years ago), which differs drastically from the glaciation in the mountains of oceanic areas, with their maximum occurring about 20 ths years ago. The publications cited above contributed to a general understanding of the complicated history of Quaternary glaciation evolution in East Siberia. Almost all the references cited above point out an involved glaciation degradation process and identified a number of stadial terminal moraines. On the other hand, there is no consensus among researchers even regarding the key issues relating to ancient glaciation. It is well known that global changes in atmospheric circulation, humidification, hydrographic network and in the regime of river runoff, continental sedimentogenesis and soil-formation processes as well as migration of flora, fauna and ancient man were associated with the evolution of glaciation. The geological activity of glaciers and associated processes was responsible for the peculiar kind of landscapes of regions occupied by ancient glaciers. Besides, they played an important role for the preservation or destruction of Pleistocene placers of gold and other metals (Obruchev, 1931; Kazakevich and Vashko, 1965). Therefore, the resolution of the ancient glaciation problems has not only a scientific-technological but also important practical significance. The primary objectives of this study are: 1) to identify the main types of glacial deposits and study their matter composition, and 2) to carry out a correlation of the products of activity of Quaternary glaciation in the continental part of the lake, with an attempt to correlate them with the time frames of the glaciations as identified using deep-water drilling cores of bottom sediments from lake Baikal.

3. Materials and methods of investigation

This paper draws on results from the expedition-based research done in the arms of the Baikalsky, Akitkansky, Barguzinsky and Khamar-Daban Ranges surrounding Lake Baikal (Fig. 1). Furthermore, this investigation used the results from a lithological study of deep-water drilling cores BDP-97 and BDP-99, under the Baikal Drilling Project (Team ..., 1998). Thus, to achieve the objectives of this study, use was made of the field and laboratory investigation techniques for Quaternary deposits. Field investigations involved a local survey of glacial, fluvioglacial and morainic landforms. A study was made of the lithologo-facial composition of Pleistocene deposits and their relationship with the deposits of conjugate terraces on the shores of Lake Baikal. The lithological and palynological investigations were carried out in the Laboratory of Stratigraphy and Lithogenesis at the Institute of the Earth's Crust SB RAS (Irkutsk). A total of 356 samples were subjected to granulometric and mineralogical analyses, and an X-ray diffraction analysis was made of clay fraction. The particle-size composition of subaerial sediments was determined through sieving (fractions of >1.0; 1.0-0.5; 0.5-0.25; 0.25-0.1, and 0.1-0.05 mm) and elutriation (fractions of 0.05-0.001 and <0.001 mm) after 20 min and 24 h, respectively (Strakhov, 1957). All mineral grains 0.05 to 0.25 mm in size were separated with bromoform into light and heavy fractions to be thoroughly studied in immersion preparations. A mineral-based classification of sand-silty deposits was performed according to a systematization proposed by Shutov (1972), and a classification of terrigenous material and determination of its degree of roundness were made following Rukhin (1969). The analyses were made by S.P. Sumkina (particle-size composition), I.A. Kalashnikova & E.G. Polyakova (mineral composition), O.N. Shestakova & N.V. Kulagina (palynological composition), and by T.S. Fileva & M.N. Rubtsova (clay fraction).

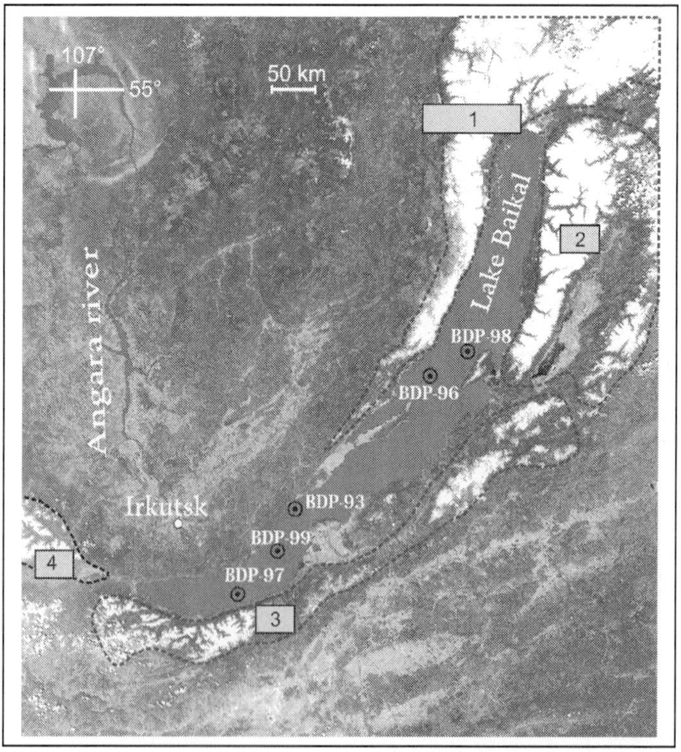

Fig. 1. Satellite image of the study region with the dashes showing the main Quaternary glaciations units at lake Baikal: 1 – Northwestern (Primorsky, Baikalsky, Akitkansky and other mountain ranges); 2 - Eastern (Barguzinsky and Ikatsky ranges); 3 – Southeastern (Khamar-Daban and Ulan-Burgasy mountain ranges); 4 – Southwestern (mountain structures of Eastern Sayan). BDP - deep-water boreholes under the "Baikal Drilling Project". The spring 2010 image of Lake Baikal is used as the basis (Ministry of Natural Resources of Russia, Baikal Information Computer Center, www.geol.irk.ru). Snow-covered mountain summits which were experiencing glaciations appear white.

4. Results

Within the Baikal mountainous country we identify four u7nits of Pleistocene glaciation: Northwestern (the Primorsky, Baikalsky, Akitkansky, Delyun-Uransky and Synnyrsky Ranges), Eastern (the Barguzinsky and Ikatsky Ranges), Southeastern (the Khamar-Daban and Ulan-Burgas Ranges), and Southwestern (the mountain structures of Eastern Sayan) (Fig. 1). The northwestern unit is the most immense region of glaciation. It was an extensive province feeding the numerous mountain-valley glaciers that descended at different times to the Baikal hollow, and to the Prebaikalian foredeep. A geological mapping of the glacial deposits that was carried out in one of the areas of this unit was instrumental in reconstructing the paths of the individual glaciers. They are quite well traceable from ridges and hills of aqueous-glacial layers (oses and kames) and the numerous sediments of the lake-bog type and end with banks of morainic deposits (Figs. 2 and 3).

Fig. 2. Sand-boulder-block morainic deposits of the Del'bichinda rock unit (mouth of the Del'bichinda River; Northwestern glaciations unit, see Fig. 1.

4.1 Morainic deposits

Morainic deposits occur mainly in the foothill area. Only occasional moraines penetrated deep into the Cisbaikalian basin, to a distance of up to 20 km (the Minya-Okunaika interfluve). Most likely, they were brought by thick glaciers from the Baikalsky and Akitkan Ranges. We have studied the sections of morainic deposits in arms of the Baikalsky Range at an absolute altitude of 767 m (N: 55°44' 05,1"; E: 108°44' 22,6"). Here, across the wide valley of the Kunerma River, near the mouth of its right tributary, the Del'bichinda River, morainic deposits (morainic bar) are outcropping, with their visible thickness being 21 m. The depth of the valley increases from 625 m in the Davan mountain pass to 950 m at the place where the Kunerma River leaves the mountains; the slope dip of the valley increases correspondingly from 20° to 30-35°, and the width of the area between its brows, on the contrary, decreases from 7 to 3.5-4 km. The valley bottom narrows from 1.7 to 1-0.5 km when approaching the ridge margin and drastically widens only at the river exit to the Central Siberian upland. In the Kunerma valley, the arms of the Baikalsky Range are virtually at the same absolute altitude, 1400-1700 m. The absolute altitudes of the valley bottom decrease from 980 m in the Davan mountain pass to 519 m at the river exit from the mountains. The downstream slope of the valley bottom averages 9 m/km. Near the mouth of the Del'bichinda River (in the area of the village of Granitnyi), the reconstructed valley along which a glacier moved has a bottom slope of 20 m/km and a slope dip of 20°. The slopes are complicated by tectonic benches, especially well expressed over a 4 km stretch of the Del'bichinda River. Owing to these benches, the slopes locally look as a staircase. Similar benches exist on mountain slopes along the Kunerma valley as well as in areas north of the Medvezhii Brook and between the latter and the Dikii Brook. The benches are composed mainly of fractured bedrocks separated into blocks and plates. The loose sedimentary cover of the slopes is thin and broken. Below the slope piedmonts, it occurs as separate relics mainly in relief sinks, overlying Proterozoic granodiorites and orthogneisses. The relics are unstratified sand-boulder-block (Fig. 2) and block-boulder-sand (Fig. 3) deposits with debris and gruss, which form a thick uniform rock unit called by us "Del'bichinda" Akulov et al., 2008). Gray silty inequigranular sand is the main filling material occurring in the form of a polymict glacier milk. All sediments are intensely milled but absolutely unsorted, because

Fig. 3. Boulder-pebble-sand morainic deposits on the northwestern slope of the Baikalsky Range. Block-boulder-sand morainic deposits (Granitnyi Village region; Northwestern glaciations unit).

the glacial meltwaters did not participate in their lithologic formation. Rock fragments that occur in these deposits are of the first and second roundness grades. In places, elongate boulders and pebbles among the morainic deposits are oriented so that their long axes are directed along the glacier movement, i.e., along the strike of the ancient valleys. We measured the rock block sizes: 150x90x90, 100x64x55, 85x80x70, 85x80x70, and 65x60x50 cm. The blocks, boulders, and pebbles have the same petrographic composition: granitoids, quartzites, felsitic and quartz porphyry, amphibolites, etc. We also discovered glacier scores on the surface of some boulders and large pebbles from the base of the studied morainic deposits (Fig. 4). Erratic blocks are chaotically scattered across the territory o glaciation, and only the upper part of the largest of them outcrops to the day surface (Fig. 5). Counting of the rows of morainic bars showed that there were no less than seven mountain glaciers that slipped down to the Cisbaikalian basin from the side of the Akitkan Ridge, whereas in the Baikal Ridge area we have revealed only five rows of terminal moraines. Probably, part of the bars is the result of oscillations of the same glacier. In plan, the terminal moraines are segments, 3 to 12 km long, arranged in a fan. As seen in the schematic geologic section, these segment-like morainic bars support each other. In the Minya-Okunaika interfluve, terminal morainic bars reach 1.5 km in length, 600 m in width, and 30 m in height. They are composed of sand-gravel deposits and pebbles with some boulders. The sand-gravel deposits are unstratified; only locally there are inequigranular sands with horizontal and oblique layering. Large obliquely layered rock series usually pinch out over a distance of 12 to 25 m, giving way to laterally layered gravel-pebble deposits. Our laboratory studies of the Del'bichinda unit deposits showed that the granulometric composition of sand-silty fraction gradually changes from base to roof of the moraine (Fig. 6). The lower part of the unit is

Fig. 4. Glacial scores on boulder (Del'bichinda rock unit; Northwestern glaciations unit).

Fig. 5. Erratic blocks and boulders were scattered by the glaciers across the taiga plains of the western Pribaikalie, and only the upper part of the largest blocks outcrops to the day surface. Northwestern glaciation unit.

dominated by inequigranular sands containing 22–30% coarse-grained (fine pebble-gravelstone) (> 2 mm), 40–50% large-grained (2–0.5 mm), 10% medium-grained (0.5–0.25 mm), and 15–24% fine-grained (0.25–0.05 mm) fractions. The content of silt fraction (0.05-0.005 mm) does not exceed 6.5%, and that of clay fraction, 8%. In the roof of the unit, the content of coarse-grained fraction drastically increases to 52%, that of large-grained fraction remains the same (40%), and the contents of medium-grained, fine-grained, silt, and clay fractions decrease to 4.6, 1, 1, and 1.3%, respectively. The content of quartz in the light fraction slightly increases from base to roof (up to 50%) of the unit, whereas other minerals: plagioclases (20-28%) and K-feldspars (25-40%) are evenly distributed throughout the analyzed moraine section. The fraction also contains rock fragments (3-6%). Deposits with this mineralogical composition are referred to as arkoses. Mineral composition of the heavy fraction almost does not change throughout the rock unit (magnetite-amphibole association); the contents of minerals are variable: magnetite 3-30%, amphiboles 55-80%, and ilmenite, rutile, epidote, and sphene 2-5% each. Moreover, occasional shows of garnets, zircon, pyroxenes, siderite, goethite, hematite, leucoxene, fluorite, biotite, tourmaline, staurolite, and sillimanite (portions of percent) are present. All mineral grains are angular or angular-rounded. The above data on the lithology of the Del'bichinda unit overlying granitoids of the Upper Proterozoic Irel' complex indicate that the unit resulted from the exaration of mainly amphibolites of the Lower Proterozoic Ilikta Formation widespread on the Baikal Ridge peaks. Similar deposits were found in terminal moraines 10 km from the mouth of the Minya River and in other areas, but the most complex section of morainic deposits was revealed in the Okunaika River valley. Here, the steep scarp is composed of

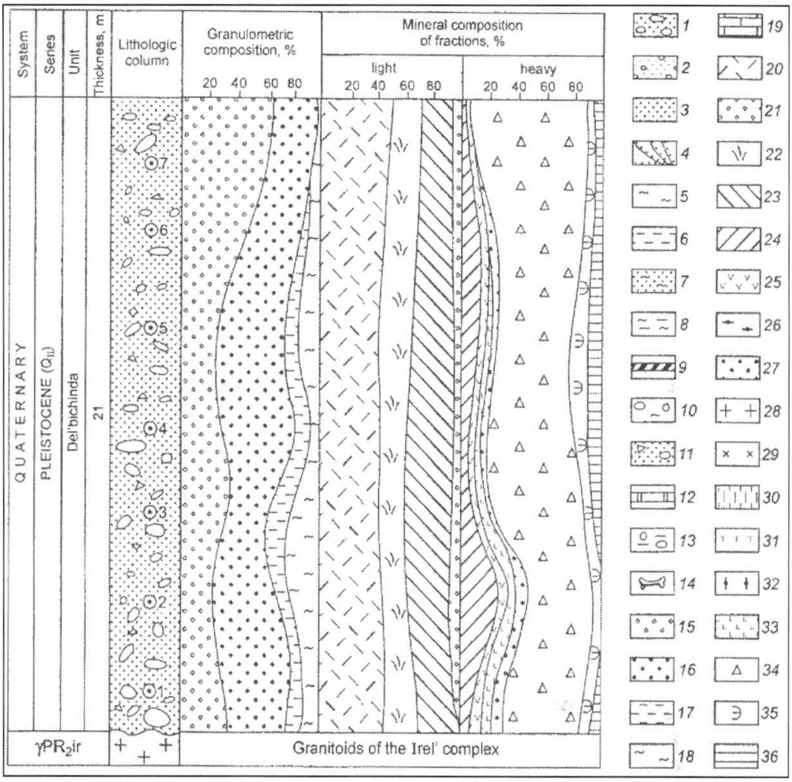

Fig. 6. Lithologo-stratigraphic section of morainic deposits (Del'bichinda rock unit; Northwestern glaciation unit).Circles with points and digits in lithologic columns mark sampling localities. *1-13* - *deposits*: *1* - sand-boulder-pebble, *2* - sand-pebble, *3* - sands, *4* - layered sands, *5* - clays, *6* - silts, *7* - sandy loams, *8* - loams, *9* - cuirasse, *10* - clays with boulders and pebbles, *11* - sand-debris-pebble, *12* - buried soils, *13* - loams with gruss and pebble; *14* - bones of a large mammal; *15-19* - *granulometric composition* (fractions, mm): *15* - >2, *16* - 2- 0.5, *17* - 0.5-0.25, *18* - 0.25-0.1, *19* - <0.1; *20-23* - *mineral composition of light fraction*: *20* - quartz, *21* - rock fragments, *22* - plagioclases, *23* - K-feldspars; *24-36* - *mineral composition of heavy fraction*: *24* - magnetite, *25* - ilmenite, *26* - zircon, *27* - rutile, 28 - leucoxene, 29 - siderite, *30* - sphene, *31* - tourmaline, *32* - psilomelane, *33* - goethite, *34* - amphiboles, *35* - epidotites, *36* - apatite.

boulders and pebbles with yellowish-gray sandy loam filling all the space around the psephitic material. In petrographic composition (granitoids, felsitic porphyry, granite-porphyry, quartzites, limestones, dolomites) these deposits are identical to Lower Devonian pebbles and boulders of the Ornochekan Formation. Exaration and subsequent redeposition of boulder-hosting red-colored poorly cemented conglomerates of the Ornochekan Formation took place over a vast area covering the northwestern part of the Cisbaikalian basin. The area of preserved relics of these conglomerates reaches 25 km^2. The absence of scores, furrows, and other traces of glacier activity from rock blocks and boulders from morainic deposits is due to the fact that the creeping mountain glaciers eroded exposing red-colored boulders and pebbles that were well rounded in the Baikal paleobasin in the Early Devonian (Akulov, 2004). Owing to the good roundness, the boulders and pebbles easily rolled and were less subject to abrasion. Possibly, part of boulders and pebbles with glacier scores is buried in the lower moraine horizons. We investigated the morainic and fluvioglacial deposits also in the southern part of the Baikal depression (the southern block of glaciation at Baikal) on the northwestern arms of the Khamar-Daban Range. The ground moraine from the Mid-Pleistocene glaciation up to 12 m in visible thickness is detected here under the layer of fluvioglacial pebbles. It is traceable as far as nearly the littoral zone of Lake Baikal (Figs. 7–9). Exaration caused by glaciers moving along mountain slopes and ancient river valleys led to the formation of ice-dressed rocks and glacier furrows and polishing on Proterozoic magmatic and metamorphic bedrocks. Results of this process are seen almost everywhere on the exposed steep (≥ 40°) slopes of the Khamar-Daban, Akitkan and Baikal Ridges. In the rear of the terminal-moraine bars in the middle course of the Minya River, upper reaches of the Domutka River, and other areas glacier-dammed lakes appeared, which are now filled with terrigenous sediments and are vast bogs. A series of lakes, though much smaller, also formed along the leading edge of segment-like moraine bars. It is here that fluvioglacial deposits begin, extending from terminal-moraine bars for almost 20 km to the west; they occupy an area of more than 200 km^2. Farther westward from the Akitkan and Baikal Ridges, the thickness of fluvioglacial deposits gradually decreases up to their vanishing near the Kirenga River valley (Fig. 10–11).

Fig. 7. Glacial deposits of the middle part of the morainic complex. Southeastern glaciations unit; mining open cast in the area of the settlement of Vydrino.

4.2 Aqueoglacial deposits

Aqueoglacial deposits are subdivided into river glacial (fluvioglacial), mud-fluvioglacial, and glaciolacustrine. By the type of formation, glaciolacustrine deposits are subdivided into periglacial and intraglacial. In the case of periglacial deposits, the relief after the end of glaciation and disappearance of glaciers is formed by glaciolacustrine plains (outwash plains) and small, extending for 5-6 km along the glacier path, elongate-oval hills (radial oses). In the case of intraglacial deposits, the relief is formed by rounded conical domes with flat tops - kames. In radial oses, the long axis of hills coincides with the direction of the movement of fluvioglacial streams (250°SW). Their relics are localized in watersheds of rivers that deeply separated the postglacial relief, which gave grounds to refer them to "watershed pebbles". One of such sites is the mouth of the Okunaika (right tributary of the Kirenga River), where morainic deposits rise above the river level to almost 60 m. Adamenko et al. (1974), who studied such deposits near the mouth of the Baldakhin'ya River (left tributary of the Kirenga), called them the Ul'kan fluvioglacial rock unit. They separated the unit into three members (from bottom to top):

	Thickness, m
1. Reddish-gray sands with horizontal and oblique layering	10.0
2. Red-brown clays with banded layering	18.0
3. Red-cherry clays with boulders and rock blocks 0.2-0.3 tons in weight	15.0
Soil-vegetation layer	0.07

Fig. 8. Well-rounded erratic block 1.6x3.2 m in size as part of morainic deposits (Southeasetrn glaciations unit; 3.2 km to the south of the settlement of Vydrino).

Our research has shown that only deposits of member 3 are properly fluvioglacial. The other two members formed in the preglacial warm epoch. This is evidenced from their lithology and spore-and-pollen complexes discovered in varved clays of member 2. Palynological analyses carried out by L.M. Shestakova and L.A. Filimonova showed that the clays contain pollen of broad-leaved trees (oak, hazel, etc.). Moreover, red-brown loams of the same member have preserved bone fragments, which, according to L.N. Ivan'ev, are parts of radius and pelvic bones of a large mammal Elephas sp. (s. 1) inhabiting this area in the Pleistocene preglacial epoch (Adamenko et al., 1974; Kul'chitsky, 1973). Therefore, only boulder-block clays (member 3) should be referred to as fluvioglacial deposits of the Ul'kan rock unit. On geological mapping (scale 1 : 200,000) all Cenozoic preglacial deposits to the Chingor rock unit (N_2-Q_1 - Pliocene-Lower Quaternary deposits). The deposits consist of yellowish-gray gravel, brown clays, gray sands, and grayish-brown loams. Thus, deposits of members 2 and 3 are similar to the Chingor unit. In some quarries, gravel-pebble deposits of the Chingor unit resulted from the erosion and redeposition of earlier weathered gravel and pebble and products of Cretaceous-Paleogene crust of weathering. This is evidenced from the fact that the unit contains almost evenly distributed (about 30% of the total volume of terrigenous material) white kaolinitized pebble, lenses of monomict quartz sands, and white kaolin, which imparts a whitish tint to its grayish-yellow deposits. Preglacial deposits of the upper member of the Chingor unit, considered by Adamenko et al. (1974) as glacial, have quite different lithologic properties. We comprehensively studied them in quarry 1 located 1.5 km from the left bank of the Baldakhin'ya River and 2.5 km from its meeting with the Kirenga River. These deposits are eroded and overlie the Lower Ordovician limestones of the Ust'kut Formation (Fig. 12) in the sequence (from bottom to top):

Fig. 9. Bedding conditions of morainic deposits (via plucking) on granitoids of Khamar-Daban. On the GPS image, they lie on bedrocks represented by granitoids (southwestern glaciations unit; right bank of the Osinovka River, 6 km to the south-east of the settlement of Tankhoi).

		Thickness, m
1.	Well-washed and well-sorted brownish-gray gravel with low-angle cross bedding	2.1
2.	Light-gray massive loam	0.3
3.	Well-washed and well-sorted light-yellow fine-grained polymict obliquely laminated sand	0.2
4.	Poorly washed gray massive gravel	0.3
5.	Light-gray fine-grained quartz sand with low-angle cross bedding	1.2
6.	Poorly washed gray massive gravel	0.18
7.	Gray inequigranular (large- and medium-grained) sand	0.12
8.	Whitish-yellow horizontally laminated clay	1.1
9.	Dark-yellow horizontally laminated fine-grained sand with a thin (4 cm) cuirasse parting in the bed roof	0.8
Total thickness of stripped deposits		6.3

The low-angle cross bedding and high degree of grading and washing of the psammitic material evidence its accumulation in lake-delta environments with warm humid climate. The age of the rock unit was determined from analysis of the spore-and-pollen complex, which includes *Abies* sp., *Tsuga* sp., *T. diversifolia*, *Picea* sp., *Larix* sp., *Pinus diploxylon*, small-leaved pollen *Betula* sp. and *Alnus* sp., occasional broad-leaved pollen *Corylus* sp., *Tuglans* sp. and *Quercus* sp. pollen. Grass pollen is abundant (*Artemisia* sp., *Gramineae* sp., etc.). Also, exotic coniferous *Pinus haploxylon* and *P. strobes* were found. The spore-and-pollen complex is dominated by pollen of the family Pinecede (55–69%). The composition of this complex gives no grounds to date the hosting deposits of the Chingor unit to the Pliocene-Early Eopleistocene. Granulometric analysis of the gravel deposits showed that the content of coarse-grained material in them gradually decreases from base (sample 13; 40%) to roof (sample 4; 9%) of the rock unit, whereas the contents of large-, medium-, and fine-grained fractions increase. The main mineral of the light fraction of gravel and sandy deposits is quartz (55–90%); plagioclases and K-feldspars amount to no more than 2%. The remainder of the fraction is rock fragments. The mineral composition of the fraction permits us to refer the deposits to siliceous clastites and quartzy rocks. The mineral composition of the heavy fraction is the same almost over the entire section: zircon (22–52%), rutile (4–16%), and leucoxene (23–38%). The lower beds of the section also contain psilomelane, whose content decreases from basal horizon (42%) to middle part (2.8%) of the Chingor unit. The upper beds bear tourmaline (up to 8%) and ilmenite (up to 7%). Magnetite, garnets, sphene, brookite, amphiboles (hornblende, tremolite, actinolite), pyroxenes, disthene, staurolite, siderite, goethite, and pyrite are minor (no more than portions of percent). Note that all types of studied deposits bear a rare mineral - florencite, whose content increases from occasional shows in the roof to 1.4% in the base of the section. The grains of all minerals are well rounded; only quartz grains and rock fragments of the light fraction are angular-rounded. It is important to emphasize that the above-mentioned deposits are overlain with erosion by rather uniform massive reddish-brown fluvioglacial clays with boulders (boulder clays) amounting to about 10 vol. % of the rock. The boulders are perfectly rounded; these are granitoids and felsitic porphyry. They are of equal size, mainly no larger than 29x27x19 cm. These clays reach 1.6 m in thickness. They are similar to the above-described fluvioglacial deposits of the Ul'kan Formation. Mineralogical analysis showed that the clays differ considerably from the underlying deposits in the mineral composition of heavy fraction (goethite-zircon assemblage): goethite (23.8%), siderite (10.2%), sphene (8.4%), florencite (0.2%), zircon (26.8%), magnetite (8.6%), ilmenite (8.8%), rutile (7%), epidote (1.6%), tourmaline (1.6%), and hornblende (0.8%). As in the underlying deposits, the light fraction of the above clays is dominated by quartz (84.4%), the remainder being rock fragments (9.6%). Similar composition of the light fraction was observed in the underlying siliceous clastites and quartzy rocks of the Chingor unit. In addition, the fraction contains plagioclases (5%), K-feldspars (0.6%), muscovite (0.3%), and carbonized plant remains (0.1%0. The mineral grains are angular-rounded. Granulometric analysis of the rock fragments separated through the water washing of the clay fraction showed the following composition of the terrigenous material: coarse-grained ~ 20%, large-grained - 32%, medium-grained - 14%, and fine-grained – the balance. The boulder-clayey deposits seem to have formed during the intense thawing of glaciers, which gave rise to strong mud flows in vast areas.

Fig. 10. Morainic boulder-block-pebble deposits in clayey cement. Northwestern glaciations unit; pit No. 64 is located in the mouth zone of the Minya River (the right tributary of the Kirenga River).

Fig. 11. Marginal part of glaciations or boundary of glacial rafting of boulders and pebbles occurring on products of physical weathering of dolomites of the Ust'-Kut formation. Northwestern glaciations unit; pit No. 79 is located on the left bank of the Kirenga River in the area of the settlement of Kazachenskoye.

4.3 Mud-fluvioglacial deposits

Mud-fluvioglacial deposits about 1.5 m thick lie over the Chingor unit or the Middle-Upper Cambrian red-colored mudstones of the Upper Lena Formation. Such deposits were also stripped on the left bank of the Kirenga River. They are composed of red-colored or light-brown supersticky clay hosing mainly well-rounded boulders and occasional pebbles. The red color of the deposits is due to the presence of erosion products of the underlying bedrocks of the Upper Lena Formation. Obviously, some fluvioglacial streams crossed the Kirenga River bed. In the pit on the slope of the left Kirenga bank (opposite the Okunaika mouth), at a height of 150 m above the river level, the following deposits were stripped (from base to top):

	Thickness, m
1. Red-colored mudstones with horizontal layering	>0.35
2. Brown-gray boulder-debris-clayey deposits	0.3
3. Light-brown boulder-debris-loam deposits	0.15
4. Brownish-gray loams with gruss and pebble	0.65
5. Light-gray sandy loam	0.15
Soil-vegetation layer	0.05

Proper mud-fluvioglacial deposits occur in beds 2 and 3. Angular fragments among them are local limestones and sandstones of the Ust'kut Formation. The deposits also abound in boulders of different sizes (cm): 25x13x14, 20x12x11, 29x27x17, 21x16x11, 25x14x9, and

24x17x13. They are mainly of granitoid composition: granite-porphyry, sheared granites, granodiorite-porphyry, and blastomylonite granites. Effusive rocks are scarce. The rock grains are well rounded. Granite-porphyry is of holocrystalline texture. Some large (2-4 mm) feldspar phenocrysts contain altered plagioclase. The groundmass consists of plagioclase, feldspar, quartz, amphibole, chlorite, epidote, apatite, and ore mineral; the latter, like amphibole, forms large (up to 2.5 mm) accumulations. Mineralogical analysis of thin sections showed the following composition of granodiorite-porphyry: plagioclase (40-45%), K-feldspar (20-25%), quartz (10-20%), hornblende (10-15%); accessory minerals are apatite (up to 2%), sphene, zircon, titanomagnetite, and, less often, orthite. Boulders in effusive rocks belong to rhyolite-porphyry (liparites) according to their petrographic composition. They have a distinct porphyritic texture, with the groundmass composed of granophyric microgranite. Porphyry phenocrysts consist of plagioclase, feldspar, and quartz. The latter occurs as a short-prismatic crystals, in places intensely corroded. Feldspars and plagioclase are partly replaced by chlorite and carbonate. Chlorite is of radiolith structure. Ore inclusions reach 0.3 mm in size. The liparite groundmass is of cryptocrystalline texture and, locally, of distinct fluidal structure. The liparite contains inclusions of quartz, K-Na-feldspar, and plagioclase. The high degree of roundness of the rock blocks and boulders unambiguously points to their redeposition from more ancient deposits during the transportation of the latter by waters. The boulders were supplied mainly from the Irel' intrusive complex and Devonian conglomerates of the Ornochekan Formation. Mineralogical analysis of sandy material in boulders and debris of the Okunaika unit showed that the light fraction contains quartz (up to 42%), plagioclases (16%), K-feldspars (8%), and minerals in rock fragments (up to 45%) (Fig. 13). Psammitic deposits with a similar mineral composition of light fraction are usually called feldspathic-quartz graywackes. Minerals of the heavy fraction are magnetite (37%), amphiboles (28%), ilmenite (23%), zircon (12%), and epidote (~8%). Also, minor (no more than portions of percent) garnets, sphene, pyroxenes, apatite, tourmaline, florencite, goethite, and hematite are present. All grains are rounded or angular-rounded. We performed a spore-and-pollen analysis of samples taken from bed 2 of the above-described section of loose deposits. Three samples are dominated by pollen of tree species (52–87%), mainly coniferous: pine (*Pinus sylvestris*) - 30-55%, cedar (*Pinus sibirica*) - 23%, spruce (*Picea obovata*) - 4–7%, and fir (*Abies sibirica*). Deciduous trees are represented by occasional pollen of alder (*Alnus*) and birch (*Betula secs.* Albae). Pollen of bushes is minor and includes cedar elfin wood (*Pinus pumila*), shrubby birch (*Betula secs.* Nanae), and Duschekia (*Duschekia*). Two samples from the lower beds (depths 0.6-0.9 m) abound in pollen of grasses, particularly wormwood (*Artemisia*), and the third sample from the upper bed, in moss spores (Lycopodium sp.), including coniferous species *L. clavatum*, *L. annotinum*, and *L. complanatum*, as well as grape fern (*Botrychium*, growing on taluses and debris substrates) and ferns (Polypodiaceae). The lower samples (0.6–0.9 m) exhibit a considerable role played by pollen of grasses, especially wormwood *(Artemisia)*, while club-moss (*Lycopodium sp.*) spores are dominant in the upper samples. The spore-and-pollen spectra show that the area was earlier inhabited by cedar-pine forests including spruce, fir, and, seldom, alder and birch. Bushes and grasses were scarce. There were also open sites such as taluses and slightly turf-covered debris slopes. The great number of forest moss species and decrease in the portion of tree species (52%) in the upper part of the section might indicate the presence of larch in the forests, whose pollen has been poorly preserved. The plants revealed in this section evolved, most likely, in the Late Pleistocene or early Early Holocene. Thus, the studied mud-fluvioglacial deposits correspond in the time of formation to the Sartan (Q^4_{III}) Glaciation in West Siberia, i.e., to the Late Würm (Würm III) according to the Alpine Stratigraphic Chart. It is important to note that a bone of large mammal was found in brownish-gray loams of bed 4. In similar deposits stripped during archaeological excavation in the Baldakhin'ya-Tala watershed (exc. 1), a broken bone, fragments of ceramic vessels, and abundant flakes were discovered (Fig. 14).

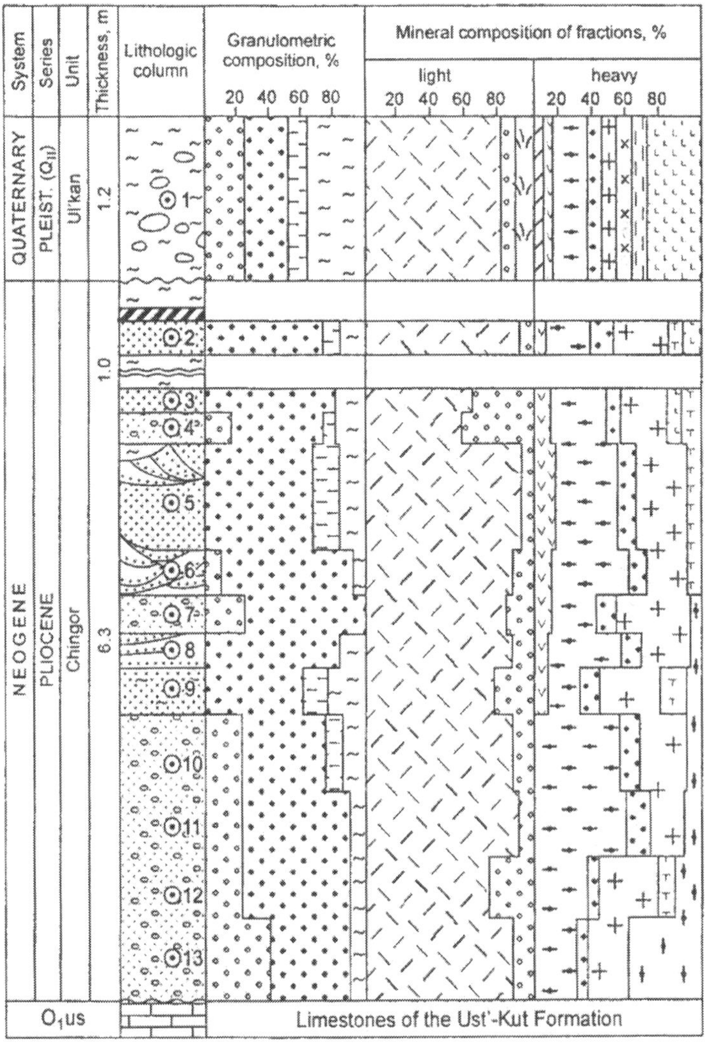

Fig. 12. Lithologo-stratigraphic section of preglacial (Chingor unit) and fluvioglacial (Ul'kan unit) deposits (quarry 1). The designations are the same as in Fig. 6. Northwestern glaciations unit.

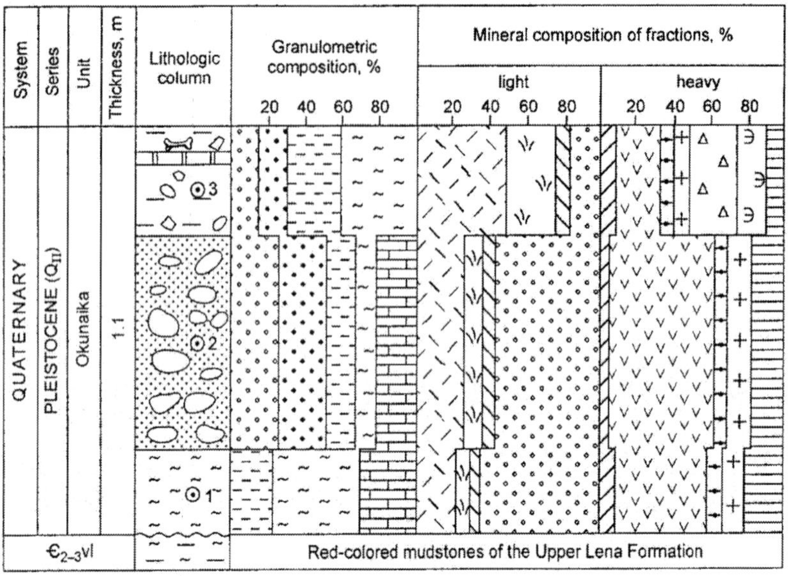

Fig. 13. Lithologo-stratigraphic section of the Okunaika unit, pit No. 62. The designations are the same as in Fig. 6. Northwestern glaciations unit.

Fig. 14. Fragments of ceramics from the first layer having preserved archaeological cultural specimens in excavation 1 (vicinity of the Okunaika mouth). Northwestern glaciations unit.

4.4 Fluvioglacial deposits

Fluvioglacial deposits are widespread on the right bank of the Kirenga River, in the vicinity of the mouths of its right tributaries Okunaika, Umbella, Minya, and Kutima (Fig. 15). These are sandy-pebble deposits containing well-rounded boulders and rock blocks. The blocks reach 2.1 m in size along the long axis. The average size of boulders is 23x20x16 cm, and that of pebbles, 5x3x2 cm. Blocks amount to ~5 vol. %; boulders, 10 vol. %; pebble, up to 40 vol. %; the remainder is inequigranular sand. The boulders and blocks are composed mainly of granitoids. The deposits are yellowish-gray, poorly sorted, with horizontal and low-angle cross bedding. The inequigranular sands are quartz graywackes: They consist of quartz (27%), plagioclases (6%), and rock fragments (77%). Minerals of the heavy fraction are ilmenite (58%), apatite (21%), leucoxene (12%), zircon (6%), and the rest (3%) is amphiboles, magnetite, Cr-spinellide, siderite, florencite, tourmaline, rutile, hematite, and goethite. Fluvioglacial deposits are best exposed near the mouth of the Umbella River. Zamaraev et al. (1976) reported on exposed horizontally layered pebbles alternating with clay-silty deposits and sands formed during the thawing of glaciers in the terminal-moraine area. These exposures were observed in a 50-55 m high scarp beneath yellowish-gray 1.5 m thick loess-like loams. The researchers suggested that the glaciers formed at the peak of the Samarovo Glaciation (Q^2_{III}). The morainic deposits are block-boulder accumulations with psammite-clay filler. Sand-clayey fluvioglacial deposits contain spores and pollen typical ofpine-birch forests including alder, willow, and minor herbs (data provided by M.V. Litvinova). Before we proceed further upon considering our identified genetic types of glaciation-associated deposits, it would be of interest to at least briefly run through the glaciodislocations caused by the thawing out of blocks of "dead" ice and by subsequent deformations having a subsidence and landsliding character. At the present time, such deposits usually form at the edge of retreating glaciers in the periglacial zone (Fig. 16). Of particular interest are the glacial dikes or glacial injections resulting from pressure of the ice on deposits of the palaeolake. In such cases there is taking place the squeezing-out of lacustrine deposits into cracks and weakened portions of the lower layers of dead ice (Fig. 17). Furthermore, the glaciations were accompanied by an aridization of the climate, and by an intensification of wind activity. The climate aridization was proceeding through the freezing out of moisture from air currents and deposition of hoarfrost on glaciers. An intensification in the activity of cold air flows, mixed with warm air flows, caused an enhancement in winds and their constancy thereby contributing to the widespread occurrence of aeolian deposits. It should be noted that the term "cryoaridization" was coined by V.S. Sheinkman (2002a), who maintains that the glaciation process creates the conditions for progressive cryoaridization accompanied by climate drying due to the freezing out of air masses and the transition of moisture at a solid state.

Thus there was taking place a reduction in the supplies of moisture transported to the areas of Siberian glaciation, as it was intercepted by glacial shields situated in northwestern Europe (Sheinkman, 2002). According to data reported in the reference just cited, unlike the northern part of Western Siberia, deep within Siberia the most favorable conditions for the advancement of glaciers existed at the beginning of the Late Pleistocene, and they were enhanced toward the end of the Late Pleistocene. It should be stressed that the climatic fluctuations under consideration are quite well recorded on the isotope-oxygen curve that was constructed on the basis of analyzing the section of bottom sediments from Baikal

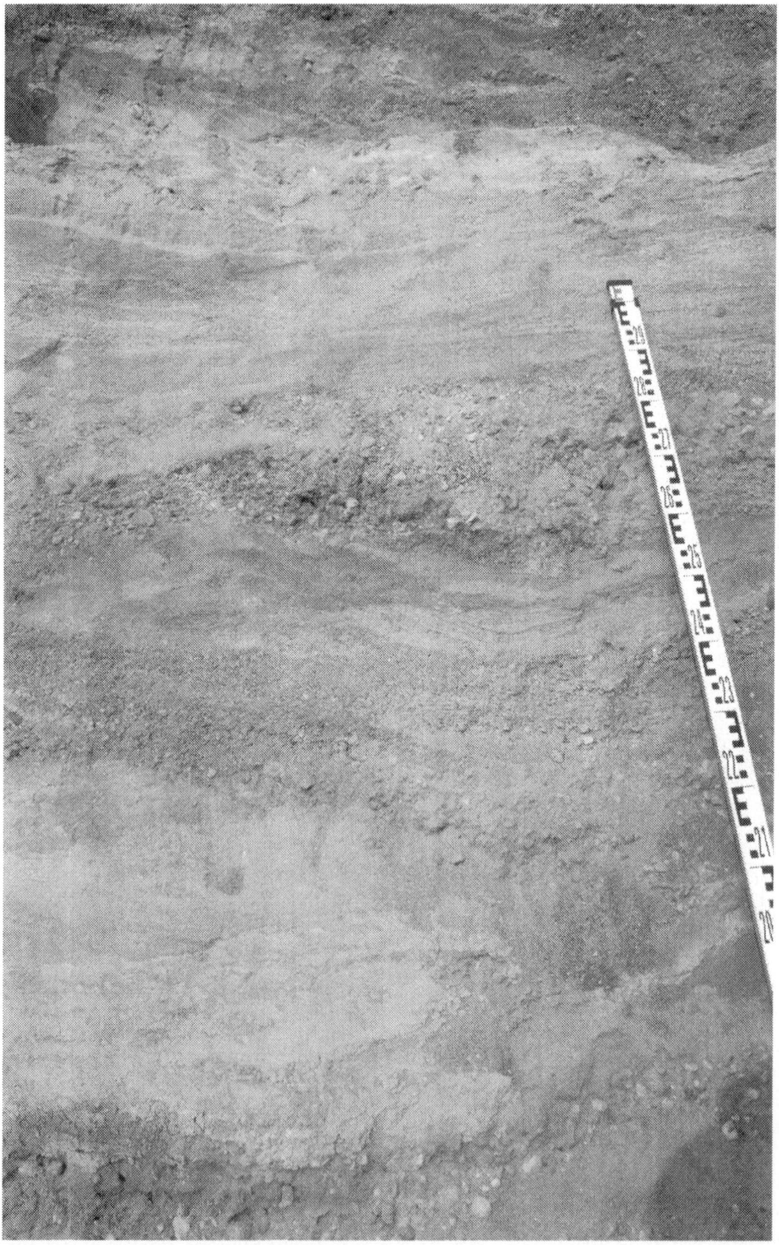

Fig. 15. Gravel-clay-sand fluvioglacial deposits. Northwestern glaciation unit, mining open cast on the right bank of the Umbella River.

Fig. 16. Glaciodislocations are represented by deformations of subsidence and landslide character. On the fragments: a - perhaps, it is a large layered glacial cotun; b - glacial diapirism or injective form that was produced in the process of squeezing out plastic clay-limestone depositions into cracks of the degrading ice. Northwestern glaciations unit; mining open cast in the upper reaches of the Kunerma River.

Fig. 17. Glaciodikes or glacial injections resulted from pressure of the ice on deposits of the glacial palaeolake. Northwestern glaciations unit, right bank of the Del'bychindy River.

(Karabanov et al., 1998; 2000). Storm winds are characteristic for the present-day Arctic and Antarctic. Their periphery can include fields of aeolian sands whose formation is possible if the following three conditions are satisfied: 1) dry climate, 2) strong wind, and 3) existence of a sand-producing "factory". The main sand "factories" on Earth are provided by water bodies: oceans, seas, lakes, and rivers. In the littoral parts of water bodies where sand material is being removed to the beach zone, sand undergoes drying к freezing out; after that, it is captured by storm winds to be transported deep into the continent where it is deposited to form fields of aeolian sands and loessal formations (Akulov et al., 2008; Akulov, Rubtsova, 2011). The widespread occurrence of the subaerial sedimentary formation in regions of manifestation of cryoaridization was pointed out by S.K. Krivonogov (2010). According to data reported by him, a considerable part of sandy massifs in the Charskaya hollow of the Baikal rift system has an aeolian genesis. Previously, there was also no doubt that the sand dunes occurring in the Charskaya hollow have an aeolian genesis. The situation is different with the question: "From where did they originate, given the surrounding mountains only?" It was thought that they were produced by deflation of fluvioglacial and lacustrine-alluvial sediments. According to our data, the sands of the Charskaya hollow emerged in the large barrier palaeolake (Charsky) during the Pleistocene glaciation. It disappeared within several tens of millennia, and the sands that were deflated by it and by the wind have been traveling across the hollow till the present (Akulov, Rubtsova, 2011). Loessal (silty-fine) aeolian sands are of widespread occurrence on the mountain arms of the Baikalsky Range (Fig. 18). Their layer is as thick as three meters. They occur on the exaration surface of granite gneisses, and it is only in the saddles of the relief where they are supported by deluvial-proluvial formations very small in thickness (as small as 0.2 m). The aeolian silty sands are overlain by a soil-vegetation layer (as thick as 0.2 m).

Fig. 18. Loessal whitish-yellow aeolian sands are well traceable in the form of a sand ribbon lying on hard rocks of the arms of the Baikalsky Range and clearly outline the mountain relief (Northwestern glaciations unit; Baikal-Amur Railroad; territory adjacent to the western entry into the Baikalsky tunnel 6.7 km in length.

4.5 The matter composition of bottom sediments

The matter composition of bottom sediments of Lake Baikal was studied by using cores from boreholes BDP-96 and BDP-98 which we collected with permission from the Baikal Drilling Project managers (Kuzmin et al., 2001). It should be noted at this point that the multichannel seismic research into bottom sediments in the middle part of Lake Baikal (in the area of boreholes BDP-96 and BDP-98) permitted T.G. Moore and collaborators (Moore et al., 1997) to divide them into two seismocomplexes. According to their data, the formation of Pleistocene-Holocene complex "A" is associated with the Angara development phase of Baikal, and its thickness is about 200 meters. The lithological investigations of core material from boreholes BDP-96-1; 2 and BDP-98 which exposed seismocomplex "A", showed that the matter composition of Pleistocene deposits is highly uniform (Kashik and Lomonosova, 2006; Kashik and Akulov, 2008). They are typical, horizontally occurring thin-layered polymictic graywackes that are represented by an alternation of aleurite clays and aleurite clay silts. Their mineralogical analysis intimated that light fraction consists of quartz, plagioclases, potassium feldspars, biotite, muscovite, chlorite, graphite, calcite, and vivianite as well as fragments of rocks, carbonificated plant tissues, remains of diatom loricae and spicules of sponges. Quartz content exceeds total percentage of feldspars, reaching 58%. There occur both pure, colorless, transparent varieties and turbid, semitransparent grains.

The amount of plagioclases varies from 11.6% to 34%. The group of micas is represented by biotite, muscovite and chlorite. They are dominated by biotite whose amount varies from 0.4 to 39%. Graphite and vivianite in the section of sediments reach 2.4%. Vivianite is present both in the form of oxidized differences of grayish-blue color and in an unaltered form. The content of diatoms and loricae fluctuates over a wide range from trace amounts to 92%; in this case, they produce interlayers of diatomites. Abrupt changes in concentration of diatom loricae are nearly coincident with intervals of increased contents of vivianite and sponge spicules. The yield of heavy fraction in analyte samples is very small, 17 g/kg. The highest contents of heavy minerals occur at the lower boundary of layer "A" (199 m). Hornblende is universally dominant (about 43%), which has a prismatic angular-rounded shape of grains. Pyroxenes (up to 4%) revealed the presence of the monoclinic as well as the rhombic differences. In the group of epidote, the main component is represented by epidote proper. Zoisite and clinozoisite are present in small amounts. The contents of the minerals from the group of epidote show a rather dramatic fluctuation, from total absence to 26%, with the content averaging 10.5%. Heavy fraction always includes ilmenite, magnetite, garnets, sphene, zircon, and apatite whose total content does not exceed 25%. Sillimanite, kyanite, spinel, chromspinellids, anatase, staurolite, tourmaline, brookite, leucoxene, chloritoid, biotite, and chlorite occur in a few percent or some fractions of percent. An investigation into clay minerals in bottom sediments of Lake Baikal showed that their contents fluctuate from 16 to 52% of the sample volume. They are characterized by a low degree of perfection of structure, a high dispersion, and by the presence of mixed-layered hydromica-smectite and chlorite-smectite phases with variable contents of swelling packets. According to the degree of perfection of structure and dispersion, they are entirely identical to the varieties characteristic for even-aged deposits in the various hollows of the lake (Kashik et al., 2001).

5. Discussion

It is quite obvious that this overview of the research done is not exhaustive for the whole variety of glacial deposits in the Baikal mountainous region. This notwithstanding, we managed to relatively thoroughly reconstruct the activity of the Northwestern unit of Pleistocene glaciation. It encompassed the mountain structures of the northwestern part of the Baikal mountainous region and had a substantial influence on many other natural processes and phenomena at Lake Baikal.

5.1 Glaciations and climatic records as deduced from bottom sediments of Baikal

The Pleistocene climate was characterized by an alternation of the epochs of cold (glacial) and warm (interglacial) climate. A reconstruction of the palaeoclimate for the glacial ages is difficult because the deposits having origins at that time are not supported by palaeontological and spore-pollen evidence. The investigations of the glaciation areas which we and our predecessors have made indicate that mountain-valley glaciers were dominant during the Pleistocene. Furthermore, subsurface glaciation, or "permafrost", came to be of widespread occurrence. Thus, as early as V.A. Obruchev's epoch, mining of Bodaibo gold-bearing alluvium buried under glacial deposits, by means of adits established that productive horizons are in the permafrost zone. According to data reported by V.A. Kudryavtsev (1978), the thickness of the permafrost zone within the region under investigation varies from 150 to 900 m. The existence of permafrost zone provides a whole variety of advantages in commercial mining of placer gold. Firstly, permafrost protected the ancient gold-bearing alluvium against erosion due to the activity of glaciers. Secondly, since gold-bearing materials are frozen, they can be mined using adits, obviating the need to reinforce the walls. Thirdly, mining workings in permafrost are entirely devoid of water influx; therefore, they are relatively dry and do not require any forced water drain. In cases where the frozen alluvial complex is being mined through open-cast operations by means of hydraulic giants, the overlying fluvioglacial deposits are removed by means of a bulldozer. After that, the site thus prepared is thawing out for several months under the hot summer sun, so that hydraulic giants can easily flush out the thawed-out gold-bearing alluvium. In the mountains surrounding Baikal, products of activity from up to four glaciations are identifiable. On the other hand, it has not yet been possible to reveal any traces of glaciations older than the Sartanian glaciation in the lake's mountain surroundings because of their total scouring or redeposition. It should also be remembered that the manifestation of two or three glaciations would suffice to destroy the traces of all the previous ones. Hence it comes as no surprise that the records of bottom deposits from Baikal include traces of a significantly larger number of glaciations than revealed in the mountainous areas surrounding it. In documenting the deep-water drilling core, the composition of bottom sediments from the Pliocene revealed thin interlayers of glacial deposits that were produced by ice- and iceberg-rafting (Kuzmin et al., 2009). They are represented by fine glacial clays with the inclusion of sand and sparse gravel grains as well as coarse sand lenses. This suggests the existence of earlier glaciations than the Pleistocene glaciations. The presence of clays in deposits dating back to 2.5–2.6 mln years points to a sharp cooling in the Baikal mountainous region as early as the Late Pliocene. A significant cooling, with the age of 2.2 mln years, was recorded in buried soil horizons on Olkhon Island (Vorobyeva et al., 1995). On the other hand, according to data reported by E.B. Karabanov (1999), a record of biogenic silicon for the Brunhes epoch in the core for one of the boreholes contains 19 stadials (10 warm and 9 glacial). Comparison of the Baikal curve for biogenic silicon with the maritime isotopic curve permitted him to identify not only the correspondence of the number of peaks and minima but also their remarkable similarity. The degree of correlation of the Baikal records with maritime records is 0.71–0.84 (Peck et al., 1994; Colman et al., 1995). All this indicates that changes in the Pleistocene climate in Siberia were following the general planetary behavior of its development. In estimating the climate existing in the area of the lake over the course of the last glacial age, E.I. Ravsky (1972), and subsequently V.A. Belova (1985) pointed out that it was characterized y a decrease in annual mean temperatures by 8–12°C, and by a considerable drying. Natural landscapes were dominated by cold glacial steppes. The mountain ranges surrounding the lake were covered by mountain-valley glaciers. All this led to a sharp decrease in production of the lake's phytoplankton, primarily the diatom phytoplankton. According to data reported by G.K. Khursevich (Kuzmin et al., 2009), the distribution of diatom algae in Pleistocene sediments of Baikal (about 1.6 mln years ago) helped to identify 18 boundaries of extinction of diatom species. They attribute the cause for their extinction to the climatic minima of phases of a sharp climate cooling in the region. About 13.7 ths years ago, a warming set in, because the section of bottom sediments in Lake Baikal begins to incorporate valves of diatom algae forming a small peak on the interval of 13.7–11.3 ths years. This peak corresponds in terms of age to the warming of the final stage of the last glacial age. This warming has a complicated structure and includes two minima. A relative warming was followed by a cooling which is more pronounced in the opal record. The age of this cooling in the Baikal record is 11.3–9.5 ths years. A detailed analysis of the entire spore-pollen complex, obtained from the section of Baikal's bottom sediments, intimates two major restructuring boundaries of plant communities that occurred on land surface for the last 5 mln years (Belova, 1985; Bezrukova et al., 1999). The first restructuring took place at the boundary of 2.8–2.5 mln years, and the second restructuring occurred at the boundary of 1.6 mln years. The first boundary was distinguished by a considerable increase in steppe and forest-steppe plant communities, and by a significant reduction in forest communities. An abrupt change of this kind is indicative of a sharp cooling and drying of the climate over the range of 2.8–2.5 mln. Years, which is supported by data on the first glaciation of Asia during that time span. After that cooling, the plant communities did not restore their structure in full measure, and larch, Siberian stone pine, spruce, and fir acquired a predominant significance, whereas the representatives of the moderately thermophilic flora remained in the form of a rare inclusion. Vegetation of the forest-tundra type shaped itself during the second cooling. A restructuring at the boundary of 1.6 mln years led to an almost total disappearance of moderately thermophilic species of the dendroflora, and to the formation of the type of vegetation similar to the contemporary one (cold-enduring boreal-taiga vegetation). It should be emphasized that the previously obtained results from lithologo-facial and other analyses for the boreholes under the Baikal Drilling Project (Team ..., 2004) permitted the project participants to project about 30 climatic minima. It is beyond reason to say whether they were glaciations or short-lasting cooling phases because of lack of adequate supporting evidence. The surprisingly large number of Pleistocene glaciations in Siberia or in Central Asia was beyond the imagination. Hence, not all of the reconstructed climatic minima must be viewed as a glaciation; instead, they should be interpreted as local climatic fluctuations within the Baikal hollow. A particular climatic minimum can only be associated with glaciation, if it has been ascertained that the deposits that had formed under the conditions of the interglacial climate preceding and following the glaciation will be detected in river and lake terraces of the study region. In any case, interpretation of the results of palaeoclimatic records for Baikal's bottom sediment core demands further elaboration.

5.2 Number of glaciations and their datings

The question as to the number of glaciations in the Baikal mountainous region was posed for the first time by P.A. Kropotkin (1876) and has remained open till the present. Perhaps, the answer to this question is simple. It implies that the glacial deposits were studied in different areas that were retreating during degradation of blanket glaciations and mountain-valley glaciations. In the areas at the largest distance from the center of glaciation, the remaining complexes of terminal morainic deposits provide evidence in support of the manifestation of a whole variety of glaciations. On the other hand, in areas which are at the shortest distance from its center, the remaining products of activity of the glaciers correspond with a smaller number of glaciations. This is complicated further by the fact that, orographically, the middle part of the Siberian Platform from which the glaciation was advancing toward Lake Baikal constitutes the Middle-Siberian Upland. It consists of a large number of leveled (by the Quaternary glaciation) plateaus with lake-bog landscape and poor exposure. In all likelihood, from the Pliocene to the Holocene the upland was under a thick blanket glaciation which was descending southward to the present-day location of the Lower-Tunguska River. According to data reported by T.A. Burashnikova and collaborators (Burashnikova et al., 1978), the Arctic glacier mantle of the last glaciation that, in East

Siberia, almost reached the mouth of the Lower-Tunguska River, was 1–2 km in thickness. At the same time, as pointed out above, the glaciations within the Baikal mountainous region had a mountain-valley pattern. It is important to note that the Arctic blanket glaciation was separated from deposits that had been formed by the largest terminal moraine descending the Baikal mountains, by the off-glacier zone about a thousand kilometers in width. Its length was about two thousand kilometers (from the Vekhoyansk mountains on the east to the Yenisei mountain-ridge and Eastern Sayan on the west). A buried soil of the steppe type was discovered on huge expanses of the off-glacial zone which we call the Angara-Tunguska zone (Ravsky et al., 1964). The Angara-Tunguska off-glacial zone was an immense pasture land with succulent grass irrigated by thawing glaciers providing habitat to mammoths, woolly rhinoceros, and other mammals, including primitive man. This zone was the home to the numerous camp sites of ancient man in East Siberia, including in the city of Irkutsk. The morainic deposits which we investigated on the Minya-Okunaika interfluve were termed the Kunermian glaciation by A.A. Kul'chitsky (1973). The Kunermian glaciation was the oldest among the visually recorded glaciations. It was one of the most severe glaciations in East Siberia, time-coincident with the Samarian glaciation of West Siberia, as well as to the Riss-I glaciation, according to the Alpine Stratigraphic Chart, and is dated back to the end of the first half of the Mid-Pleistocene (Q_{II}^2). In Europe, the Riss glaciation started about 250 ths years ago. According to data reported by V.M. Gavshin, S.A. Arkhipov, V.A. Bobrov and collaborators (Gavshin et al., 1998), by analyzing the distribution of natural radioactive elements in Holocene-Pleistocene deep-water deposits of Lake Baikal, they were able to assign the time of the Samarian glaciation to the boundary of 276–247 ths years, and the subsequent Shirtian Interglacial, to the interval of 247–190 ths years. According to their data, the Tazovian glaciation manifested itself 190–127 ths years ago, the Kazantsevian Interglacial - 127–73 ths years ago, and the Yermakovskoye and Sartanian glaciations, respectively, 73–38 ths years ago and 28–11 ths years ago. The last two glaciations are separated by the Kargian Interglacial (38–28 ths years ago). Thus the first wave of Pleistocene glaciation was responsible fro the fact that the Kunerma glaciers caused a maximum possible advance of the morainic deposits which, subsequently, experienced multiply occurring intense erosion processes thereby leading to their significant dwindling. With the distance from the main front of the moraines, the amount of coarse-debris material decreases and its roundedness increases. The second wave of substantial cooling corresponds to the Tazovian glaciation of West Siberia as well as to the Riss-II glaciation according to the Alpine Stratigraphic Chart, and dates back to the end of the Mid-Pleistocene (Q_{II}^4). Morainic deposits of this phase of glaciation were discovered on the Minya-Okunaika watershed and in the area of urochishche Toka-Makit. The Kazantsevian warming was followed by one of the strongest glaciations, Yermakovian (Q_{III}^2), which was time-coincident with the Würm-I glaciation (Alpine Stratigraphic Chart). Products of the activity of this glaciation were revealed on the Minya-Okunaika watershed. It was time-coincident with the formation of terrace complexes of the Ulkan (river mouth, 8–10-meter terrace), the Kirenga (10–12-meter terrace), and the Munok (8–10-meter terrace). The Kargian Interglacial set in, which was replaced by the fourth phase of glaciation, the Sartanian, at the end of the Late-Pleistocene (Q_{III}^4) and, according to the Alpine Stratigraphic Chart, in time coincidence with the Late Würm (Würm-III). Thus the Late Pleistocene sedimentary complex is comprised of deposits of two interglacial and two glacial horizons. The first interglacial horizon is represented by lacustrine sands in the straths of the low Baikalian terraces, overlain in places by morainic debris deposits. The glacial horizons as such were opened in the composition of four Baikalian terraces which are composed of morainic and fluvioglacial deposits. For the more detailed solution of the problem regarding the number of glaciations at Baikal, in addition to invoking conventional techniques and the aforementioned methods, some researchers used absolute geochronology data (thermoluminescence and radiocarbon). As yet, thee number of resulting datings is small; nevertheless, the currently available C14 and TYa-dates provided some insight into the solution of a number of questions. More specifically, in the most complete section of Pleistocene deposits in the northwestern part of Baikal, in the valley of the Rel River, on the semi-scoured moraine of maximum advance there occur Rel deluvial-proluvial interglacial sands with Coelodonta antiquitatus with C14 - 25880 ± 350 SOAN-829, which correlated with (occurring in this same area) palaeosoils of the Kurlinskaya camp site, C14 - 24060 ± 570 SOAN-1397 (Mats et al., 2001). Based on this, it was possible to infer the lower boundary (the beginning) of Late-Pleistocene glaciation. The above-lying sand-pebble deposits of the first Baikalian terrace formed during the first optimum of the Holocene, with deposits of wave-cut grottoes being level with the terrace, with a dating C14 -7890 ±235 SOAN-580. It important to notice that the above-lying deposits contain strata with Neolitic culture (C14 - 4470 ±65 SOAN-830). The research done by E.E. Kononov (2009) ascertained that the formation of the glacial barrier-lacustrine basin in the Muisko-Kuandinskaya hollow (the northern part of Baikal) occurred about 40 еры years ago. The entire complex of sandy deposits occurring in this basin is combined into the composition of the Kobylinskaya formation. The base of the formation was dated from stump wood: 38320±775, 40500±930, and 36500±2500 (KI-3951) (Kul'chitsky, 1995). The upper boundary of the formation was established at borehole 14a drilled on the left-bank terrace of the Mudirikan River. A dating of 22 300 years was obtained from peaty layers at a depth of 95 km (SOAN-2484) (Filippov, 1997). The formation start of lacustrine deposits is time-coincident with burial of tree trunks, and its cessation coincides with the termination of lacustrine sedimentation as a result of the discharge of the barrier palaeolake at the interglacial period. Thus the lifetime of the barrier lake fits in the interval ranging from 38 to 22 thousand years and corresponds to the Karginian Interglacial. Nowadays it is generally recognized that the Baikal hollow includes deposits corresponding to the Karginian interglacial, or Megainterstadial. The Karginian warming in Siberia was documented sufficiently clearly by many investigators (Rayevsky, 1972; Belova, 1985; Vorobyeva, 1994).

5.3 Embodiment of Pleistocene glaciations in the deposits of river floodplains and terraces

According to present views, the valleys of the Lena, Angara, Lower Tunguska and Vilyui rivers had formed as early as the end of the Pliocene. The Late Pleistocene glaciation encompassed largely their upper portions which have retained a more-or-less clearly pronounced shape of troughs till the present. The smoothed shape of the benches and buttresses on the valley slopes is well preserved on the outcrops of hard rocks. Gentle platforms of the smoothed benches often are the home to erratic boulders and exotic pebble gravel witnessing the former filling of the valleys with the ice. Degradation of glaciation is represented mainly by various accumulative formations having a paragenetic connection with it. The ground moraine is composed of debris material mostly not larger than pebble in size. Only rarely does it include small boulders. Even though the ground moraine is a uniform mechanical mixture without any obvious glacio-dynamical textures, it drastically differs precisely in this feature from the ablation moraine saturated with boulders and blocks. Such a difference in the lithology of the ground and ablation moraines is also conserved when terrace complexes form on them. Morainic material that is present in the deposits of the river terraces is readily identified and diagnosed. Beyond the edge of the Late-Pleistocene moraines, traces of glacial activity are observed in the eastern and northern halves of the Lake Baikal hollow along the valleys of the Lena, Vitim and Patoma rivers, but such attributes are lost in places where the rivers enter planate areas. This notwithstanding, indirect signs of, at least, Sartanian glaciation can be found even in floodplain deposits of the off-glacial zone. They are represented by typical varieties of cryogenic formations in the form of frost wedges and fissures, cryoturbations, sinkholes, etc. (Fig. 19). Nevertheless, given the present stage of understanding the friable deposits in the mouth area of the Angara River, one cannot draw any definite conclusions regarding the glacial processes in this area. The high terraces are weakly exposed; therefore, the composition and structure of the deposits composing them have to be inferred from rather disconnected outcrops. Our survey of these terraces suggests that 1) the high terraces are composed of sediments of a common, complex-structured layer which is characterized by clearly pronounced facial changes both in section and in strike; 2) from the base of the layer forming part of the high terraces, upward the section there is a decrease in cementation density of the deposits; 3) the deposits of the high terraces contain varying amounts of large erratic boulders and blocks concentrated largely at the upper termination of the terrace; 4) the high terraces are largely composed of stratified, poorly rounded, small-size, free-flowing pebbles, gravel and debris

Fig. 19. Cryogenics deformations in floodplain deposits of the Angara-Tunguska off-glacial zone (Northwestern glaciations unit; left bank of the Kirenga River in the vicinity of the settlement of Magistralny, pit No. 5).

which do almost not contain any clay material; in some places, however, they include large erratic boulders and blocks, and 5) pebbles of the low terraces are fluvioglacial and contain a relatively large amount of clay material. In closing this Section, it is worth noting that the loessal aeolian sands that were opened on the arms of the Baikal mountains (see Fig. 18) constitute a reflection of the powerful Aeolian processes occurring widely after the Glacial Age. The huge exposed planate area of the Prebaikalian trough following the thawing out of the ice and snow was covered with morainic, aqueous-glacial, fluvioglacial-debris flow and other derivative glacial deposits and immediately became the scene of Aeolian activity. Northwesterly storm winds were producing huge amounts of dust, transporting it to considerable distances. As the wind was losing energy, aleuro-pelitic material was uniformly deposited on the surfaces of the mountains, and on the mountain slopes.

6. Conclusions

The research reported here suggests the following conclusions: 1) the morainic complex of the Baikal region is composed by sandy-boulder-block and block-boulder-sandy unstratified deposits, with the occasional inclusion of scree and gravel, the main filling material in which comprises arkose sands with magnetite-amphibole assemblage of heavy fraction; 2) fluvioglacial deposits are of the most widespread occurrence among all the types of glacial formations, and 3) the aqueous-glacial layers are overlain by loamy sands and loams which often contain scree and buried soils and are culture-host ones. Their composition revealed all archaeological finds; 4) during the Pleistocene, blanket glaciation that had been advancing from the Arctic Ocean did not reach Lake Baikal and stopped about one thousand kilometers from it. It was separated from the glacial formations of the Baikal mountainous region by a vast Angara-Tunguska off-glacial zone with steppe landscape; 5) the problem of a correlation of the climatic minima as identified in the bottom sediment core from Lake Baikal, with classical glaciations of Europe involves a difficulty in detecting time-coincident continental glacial deposits in the arms of the mountains or in their valleys, and in the terraces of rivers or lakes; 6) during the Pleistocene, at Lake Baikal there occurred four mountain-valley glaciations: Kunermian (Samarian; Riss-I; 276–247 ths years ago), Tazovskian (Riss-II; 190–127 ths years ago), Yermakovian (Zyryanian, Würm-I; 73-38 ths years ago), and Sartanian (Würm-III; 28–11 ths years ago), and 7) the distinctive characteristics of the glaciations of the Baikal mountainous region were a dramatic aridization of the climate, and a widespread occurrence of aeolian processes.

7. References

Adamenko, O.M.; Kul'chitskii, A.A. & Adamenko, R.S. (1974). The Stratigraphy of the Quaternary Deposits in the Cisbaikalian Basin, *Geologiya i Geofizika (Soviet Geology and Geophysics)*, Vol.15, No.8, pp. 34-42, ISSN 0016-7886

Akulov, N.I., Agafonov, B.P. & Krasnoshchekov, V.V. (2008). Aeolian Deposits Covering Camp Sites of Ancient Man, *Litologiya i poleznye iskopaemye*, No.2, pp. 209-222, ISSN 0002-3337

Akulov, N.I. & Rubtsova, M.N. (2011). Aeolian Deposits of Rift Zones, *Quaternary International*, No.234, pp. 190-201, ISSN 1040-6182

Akulov, N.I., Agafonov B.P. & Rubtsova M.N. (2008). Glacial Deposits and "Watershed Pebbles" in Western Baikal Area, *Russian Geology and Geophysics*, Vol.49, No.1, pp. 28-39, ISSN 0016-7886

Akulov, N.I. (2004). Paleogeography and Conditions of Accumulation of Devonian Sediments in the Southern Siberian Platform, *Stratigrafiya. Geologicheskaya Korrelyatsiya*, Vol.12, No.3, pp. 26-36, ISSN 0869-592X

Back, S. & Strecker, R.M. (1998). Asymmetric Late Pleistocene Glaciations in the North Basin, of the Baikal Rift, Russia, *J. of the Geological Society*, Vol.155, pp.61-69, ISSN 0016-7649

Bazarov, D.B. (1986). *The Cainozoic of the Prebaikalia and Western Transbaikalia*, Nauka, Novosibirsk, Russia

Bezrukova, E.V., Kulugina, N.V., Letunova, P.P. & Shestakova, O.N. (1999). The Directedness of Vegetation and Climate Changes in the Baikal Region for the Last 5 Million Years (as Deduced From data of Palynological Investigations of Lake Baikal), *Geologiya i geofizika*, Vol.40, No.5, pp. 739-749, ISSN 0016-7886

Belova, V.A. (1985). *Vegetation and Climate of the Late Cainozoic in the South of East Siberia*, Nauka, Novosibirsk, Russia

Burashnikova, T.A., Grosvald, M.G. & Suyetova, I.A. (1978). The Volume of Arctic Glacial Cover at the Epoch of the Last Glaciation of the Earth, *DAN SSSR*, Vol.238, No.5, pp. 1169-1172, ISSN 0869-5652

Chersky, I.D. (1877). Opinions Regarding the Former (in the Post-Glacial Period) Highly Significant Occurrence of Waters of the Arctic Ocean in Siberia, *Izvestiya Sibirskogo otdeleniya Russkogo geograficheskogo obshchestva*, Vol.III, pp. 86-113.

Colman, S.M., Peck, J.A., Karabanov, E.B., Carter, S. J., Bradbury, J. P., King, J.W. & Williams, D.F. (1995). Continental Climate Response to Orbital Forcing From Biogenic Silica records in Lake Baikal, *Nature*, Vol.378, (30 November 1995) pp. 769–771, ISSN 0028-0836

Dumitrashko, N.V. (1952). Geomorphology and Palaeogeography of the Baikal Mountainous Region, In: *Materials on Geomorphology and Palaeogeography of the USSR* (ed. by V.A. Obruchev), Izd-vo AN SSSR, Moscow,191 p.

Filippov, A.G. (1997). *Detailing of the Local Litho- and Biostratigraphic Differentiation of Quaternary Deposits on the Basis of Studying Reference Sections for Improving StratigraphicPatterns of the Muiskaya Series and the Angara-Lena Block of the South of E.Siberia*, Irkutsk, Russia

Florensov, N.A. (1960). *Mesozoic and Cenozoic Depressions in the Baikal Area*, Izd. AN SSSR, Moscow.

Grossvald, M.G. (1965). *Development of Topography of the Sayan-Tuva Upland*, Nauka, Moscow.

Grossvald, M.G. (2002). P.A. Kropotkin and the Problem of Ancient Glaciation of Siberia, In: *P.A. Kropotkin's Ideas and Natural Science. On P.A. Kropotkin's Biography*, Moscow, pp. 17-36.

Gurulev, A. (1959). On the Age of the Quaternary Glaciation in the Northern Baikal Region, *Trudy VSGI, seriya geologicheskaya*, Issue 2, Materials on Geology of East Siberia, Irkutsk, pp. 175-186.

Imbrie, J. & Imbrie, K.P. (1979). *Ice Ages Solving the Mystery*, New Jersey, Hillside.

Imetkhenov, A.B. (1987). *Late Cainozoic Deposits in the Shore Area of Lake Baikal*, Novosibirsk: Nauka, Russia

Karabanov, E.B., Prokopenko, A.A., Williams, D.F. & Colman, S.M. (1998). The Link Between Insolation, North Atlantic Circulation and Intense Glaciations in Siberia During Interglacial Periods of Late Pleistocene, *Quaternary Research*, Vol.50, pp. 46-55, ISSN 0033-5894

Karabanov, E.B. (1999). Geological Structure of Sedimentary Layer of Lake Baikal and Reconstruction of Climate Change of Central Asia in the Late Cainozoic, In: *Abstract of Doctor of Geologic-Mineralogical Sciences Degree Dissertation*, Institute of Lithosphere RAS, Moscow.

Karabanov, E. B., Prokopenko, A. A., Williams, D. F. & Khursevich, G. K. (2000). Evidence for Mid-Eemian Cooling in Continental Climatic Record From Lake Baikal, *Journal of Paleolimnology*, Vol.23, pp. 365–371, ISSN 0921-2728

Kashik, S.A. & Akulov, N.I. (2008). Pleistocene–Miocenic Sedimentation in Baikal Lake. *The 7th International Symposium on Environmental Changes in East Eurasia and Adjacent Areas - High Resolution Records of Terrestrial Sediments*, pp. 83-85, Ulaanbaatar-Hatgal, Mongolia, August-September, 28-3, 2008

Kashik, S.A., Lomonosova, T.K. & Fileva, T.S. (2001). Genetic Types of Clay Minerals in Bottom Sediments of the Southern Depression of Lake Baikal, *Geologiya i geofizika*, Vol.42, Nos.1-2, pp. 164-174, ISSN 0016-7886

Kashik, S.A. & Lomonosova, T.K. (2006). Cainozoic Sediments of the Underwater Akademichesky Range in Lake Baikal, *Litologiya i poleznyye iskopayemye*, No.4, pp. 339-353, ISSN 0002-3337

Kazakevich, Yu.P. & Vashko, N.A. (1965). Role of Glacial Processes in the Preservation and Destruction of Gold-Bearing Placers as Exemplified by Some Areas of Siberia, In: *Geology of Placers*, Nauka, Moscow, pp. 157-164.

Kul'chitsky, A.A. (1973). Deposits and Paleontology of the Epoch of the Maximum Glaciation in the Cisbaikalian Basin, *Geologiya i geofizika*, No.9, pp. 60-66, ISSN 0016-7886

Kul'chitsky, A.A. (1985). Pleistocene Glaciations of the Mountains of the Northwestern Baikal Region in the BAM Zone (Exemplified by the Kunerma River Basin), *Geologiya i geofizika*, No. 2, pp. 3-9, ISSN 0016-7886

Kul'chitsky, A.A. (1993). Deposits and Palaeogeography of the Epoch of the Maximum Glaciation of the Baikal Region, *Geologiya i geofizika*, No.9, pp. 60-67, ISSN 0016-7886

Kul'chitsky, A.A. (1995). Deformation of Cainozoic Deposits in the Muisko-Kuandinskaya Depression of the Baikal Rift Zone, In: *RFBR in the Siberian Region*, Irkutsk, pp. 35-36.

Kononov, E.E. (2009). On the Origin of Sandy layers in the Northern Baikal Region, *Vestnik IrGTU*, No.4(40), pp. 23-27, ISSN 1814-3520

Krivonogov, S.K. (2010). *Sedimentation in the Hollows of the Baikal Rift Zone in the Late Pleistocene and Holocene. Author's Abstract of Doctor of Geological-Mineralogical Sciences Dissertation*, Izd-vo IGM SO RAN, Novosibirsk, Russia

Kropotkin, P.A. (1876). *Studies on the Glacial Period*, St. Petersburg, Russia

Kudryavtseva, V.A. (1978). *General Permafrost Science*, Moscow: Izd-vo MGU, Russia

Kuzmin, M.I., Karabanov, E.V., Kawai, T., Williams, D., Bychinsky, V.A., Kerber, E.V., Kravchinsky, V.A., Bezrukova, E.V., Prokopenko, A.A., Gelety, V.F., Kalmychkov,

G.V., Goreglyad, A.V., Antipin, V.S., Khomutova, M.Yu., Soshina, N.M., Ivanov, E.V., Khursevich, G.K., Tkachenko, L.L., Solotchina, E.P., Yoshida, N. & Gvozdkov, A.N. (2001). Deep Drilling on Baikal - Main Results, *Geologiya i geofizika*, Vol.42, Nos.1-2, pp. 8-34, ISSN 0016-7886

Kuzmin, M.I., Khursevich, G.K., Prokopenko, A.A., Fedenya, S.A. & Karabanov, E.B. (2009). *Centric Diatoms in Lake Baikal During the Late Cenonzoic: Morphology, Systematics, Stratigraphy and Stages of Development (Based on the Deep Cores of the Baikal Drilling Project),* Editor-in-Chief Professor A.M. Spiridonov, Academic Publishing House "GEO", Novosibirsk, ISBN 978-5-9747-0137-5

Ladokhin, N.P. (1959). Toward the Ancient Glaciation of the Baikal Region, In: *Materials on the Geology of East Siberia. Seriya geologicheskaya,* Issue 2, Irkutsk, pp. 153-173.

Lamakin, V.V. (1961). The Quaternary Geology of the Baikal Hollow and of Its Surrounding Mountains, In: *Some Questions of the Geology of the Anthropogene (to the 4th Congress of the INQUA to Be Held in Warsaw in 1961),* Moscow: Izd-vo AN SSSR, pp. 152-165.

Lamakin, V.V. (1963). Glacial Deposits in the Littoral Strip of Baikal, *Trudy komissii po izucheniyu chetvertichnogo perioda AN SSSR,* issue 21, pp. 126-147.

Logachev, N.A., Antoshchenko-Olenev, I.V., Bazarov, D.B. & Galkin, V.I. (1974). *The Uplands of the Prebaikalia and Transbaikalia,* Nauka, Moscow, Russia

Mats, V.D., Ufimtsev, G.F. & Mandelbaum, M.M. (2001). *The Cainozoic of the Baikal Rift Hollow: Structure and Geological History,* Novosibirsk: Izd-vo "Geo", Russia

Mats, V.D., Pokatilov, A.G., Popova, S.M., Kravchinsky, A.Ya., Kulagina, N.V. & Shimarayeva, M.K. (1982). *The Pliocene and Pleistocene of Middle Baikal,* Novosibirsk: Nauka, Russia

Moore, T.G., Klitgord, K.D., Golmstok, A.Ya. & Weber, E. (1997). The Central and North Basins of Lake Baikal:The Early Phase of Basin Formation, *Geol. Soc. Amer. Bull.,* Vol.9, No.6, pp. 746-766, ISSN 0016-7606

Obruchev, V.A. (1918). P.A. Kropotkin (On the 75th Anniversary of His Birth), *Priroda,* Nos.4-6, pp. 309-322.

Obruchev, V.A. (1931). Attributes of the Glacial Period in Northern and Central Asia, *Byulleten Komissii po izucheniyu chetvertichnogo perioda AN SSSR,* No.3, pp. 43-120.

Olyunin, V.N. (1969). Ancient Glaciation of Khamar-Daban, In: *Geography and Geomorphology of Asia,* Nauka, Moscow.

Osadchy, S.S. (1982). On the problem of the Relationship Between pluvial and Glacial Epochs on the Territory of the Transbaikalian North, In: *The Late Cainozoic History of Lakes in the USSR,* Nauka, Novosibirsk, pp. 61-71.

Peck, J., King., Colman, S.M. & Kravchinsky, V.A. (1994). A Rock-Magnetic Record From Lake Baikal, Siberia: Evidence for the Late Quaternary Climate Change, *Earth Planet Sci. Lett.,* Vol.122, pp. 221-238, ISSN 0012-821X

Ravsky, E.I., Alexandrova, L.P., Vangengeim, E.A., Gerbova, V.G. & Golubeva, L.V. (1964). *Anthropogenic Deposits in the South of East Siberia,* Nauka, Moscow.

Ravsky, E.I. (1972). *Sedimentation and Climates of Inner Asia in the Anthropogene,* Nauka, Moscow, Russia

Rukhin, L.B. (1969). *The Foundations of Lithology,* Nedra, Leningrad, Russia

Sheinkman, V.S. (2002a). Testing the S-S Technology of Thermoluminiscence Dating on the Sections Along the Shores of the Dead Sea, Its Implementation in Mountainous

Altai, and Palaeo Geographical Interpretation of Results, *Arkheologiya, etnografiya i antropologiya of Eurasia*, Vol.2, No.10, pp. 22-37, ISSN 1563-0102

Sheinkman, V.S. (2002b). Age Diagnostics of Glacial Deposits in Mountainous Altai and Their Testing Against the Sections of the Dead Sea, *Materialy glatseologicheskikh issledovanniy*, Vol.93, pp. 41-55, ISSN 0130-3686

Sheinkman, V.S. & Antipov, A.N. (2007). Baikal's Palaeoclimatic Records: Disputable Issues Relating to Its Possible Correlation With Ancient Glaciations of Siberia's Mountains, *Geografiya i prirodnye resursy*, No.1, pp. 6-13, ISSN 0206-1619

Shutov, V.D. (1972). Classification of Sandstones, In: *Graywackes*, Moscow: Nauka, pp. 21-24.

Strakhov, N.M. (Ed.) (1957). *Methods of Studying Sedimentary Rocks*, Moscow: Gosgeoltekhizdat.

Team of "Baikal-Drilling" Project Participants. Continuous Record of Climatic Changes in Sediments of Lake Baikal for the Last 5 Million years. (1998). *Geologiya i geofizika*, Vol.39, No.2, pp. 139-156, ISSN 0016-7886

Team of Project Participants. High-Resolution Sedimentary record from the New Core BDP-99 Deep-Water Drilling on the Posolskaya Bank in Lake Baikal. (2004). *Geologiya i geofizika*, Vol.45, No.2, pp. 163-193, ISSN 0016-7886

Tsetlin, S.M. (1964). *Comparison of Quaternary Deposits and Off-Glacial Zones of Central Siberia (Lower-Tunguska Basin)*, Nauka, Moscow.

Velichko, A.A. (1957). Kropotkin as the Originator of the Theory of the Glacial Period, *Izvestiya AN SSSR, seriya geograficheskaya*, No.1, pp. 122-126.

Vorobyeva, G.A. (1994). Palaeoclimates Around Baikal in the Plestocene and Holocene. In: *Baikal as a Natural Laboratory for the Study of Global Changes*, Izd-vo Lisna, Irkutsk, Vol. 2, pp. 54-55.

Vorobyeva, G.A., Mats, V.D. & Shimarayeva, M.K. (1995). Palaeoclimates of the Late Cainozoic of the Baikal Region, *Geologiya i geofizika*,Vol.36, No.8, pp. 82-96, ISSN 0016-7886

Voskreswensky, S.S. (1959). *Geomorphology of Siberia*, Izd-vo MGU, Moscow, Russia

Voyeikov, A.I. (1881/1952). Climatic Conditions of Glacial Phenomena in the Past and at Present, In: *Selected Works*, Izd-vo AN SSSR, Moscow, Vol.III, pp. 321-364.

Yatsenko, A.A. (1950). On the Glaciation of the Baikal Mountainous Region, *Voprosy geografii i geomorpfologii*, IGO, Moscow, pp. 179-188.

Zamoruyev, V.V. (1971). On the Character and Age of Quaternary Glaciation of the Mountains in the Southern Transbaikalia and Prebaikalia, In: *Chronology of the Glacial Age*, Izd-vo GO SSSR, Leningrad, pp. 92-100.

Zamoruyev, V.V. (1978). The Stadial Character of Glacial Retreat and the Position of the Snowline in the Khamar-Daban Range During the Late Quaternary Glaciation, *Izvestiya VGO*, Vol.II0, pp. 526-530, ISSN 0869-6071

Zamaraev, S.M., Adamenko, O.M., Ryazanov, G.V., Kul'chitsky, A.A., Adamenko, R.S. & Vikent'eva, N.M. (1976). *Structure and History of Evolution of the Cisbaikalian Piedmont Depression*, Nauka, Moscow, Russia

Zolotarev, V.G. (1961). Geomorphology and Geology of Quaternary Deposits in the Northern Part of the North-Baikal Upland, In: *Materials on Geology and Mineral Resources of Irkutsk Oblast*, Issue 1, No. 28), pp. 40-61.

15

Application of Illite- and Kaolinite-Rich Clays in the Synthesis of Zeolites for Wastewater Treatment

Carlos Alberto Ríos Reyes[1] and Luz Yolanda Vargas Fiallo[2]
[1]*School of Geology, Universidad Industrial de Santander*
[2]*School of Chemistry, Universidad Industrial de Santander*
Colombia

1. Introduction

Water is the source of life and is the basic condition of human survival. However, the severe water contamination and insufficient water source are nowadays two thorny problems. Industrial effluents are contaminated with highly toxic, non-biodegradble and cancerogenic heavy metals, which are generated by industries such as electroplating, mineral processing, galvanization plants, paints formulation, porcelain enameling, nonferrous metal and vegetable fat producing industries (Meena et al., 2005). Due to the discharge of large amounts of metal-contaminated wastewater, the electroplating industry is one of the most hazardous among the chemical-intensive industries (Pereira et al., 2010). If not carefully managed, however, wastewater may produce both short- and long-term effects on human health and the ecological system. In many developing countries there are deadly consequences associated with exposure to contaminated water, as many developing countries have increasing population densities, increasingly scarce water resources, and no water treatment utilities. Therefore, there are huge challenges all over the world regarding the handling of waste water for a sustainable future.

The processes of dissolution, transport and immobilization of heavy metal ions are very important in environmental science and technology. Many industrial processes involve solubilisation of heavy metal ions to aqueous solutions which then are released into the environment via wastewater; as heavy metal ions persist in the environment, an effective protection strategy requires the ions to be sequestered from the wastewater (Nestle, 2002). Several treatment technologies for wastewater treatment, including chemical precipitation, electrodeposition, ion exchange, membrane separation and adsorption, have been developed (Diz & Novak, 1998; Webster et al., 1998; Feng et al., 2000; Mohan & Chander, 2006; Chartrand & Bunce, 2003; Santos et al., 2004; Gibert et al., 2005; Johnson & Hallberg, 2005; Wattena et al., 2005; Wei et al., 2005; Kalin et al., 2006; Ríos et al., 2008), although adsorption has been the preferred method for heavy metal removal, because it is considered to be a particularly effective technique if it takes in consideration the use of suitable, cheap, and environmentally friendly sorbent materials. Heavy metal removal from electroplating wastewater have been investigated by several researchers (Algara et al., 2005; Sousa et al., 2009). Adsorption is usually quite a complex process, generally involving much more than simple ion exchange into the pore openings of the ion exchanger. Factors such as pH, nature and concentration of the counter ion (metal ion), ion hydration, varying metal solubilities, presence of competing and complexing ions, all affect the amount of metal ion to be adsorbed (Ikhsan et al., 1999) and therefore the sorbent selectivity.

One of the biggest advantages of clays that have to be used as raw materials in the synthesis of zeolites is their relatively low cost, which allows applications to be industrially feasible. Clays, such as clay minerals such as kaolinite (Breck, 1974; Barrer, 1982; Boukadir et al., 2002; Ríos, 2008; Ríos et al., 2009, 2011), halloysite (Klimkiewicz & Drąg 2004; Zhao et al., 2010), illite (Mezni et al., 2011), montmorillonite (Song & Guo, 1997; Ruiz et al., 1997; Cañizares et al., 2000; Boukadir et al., 2002), vermiculite (Johnson & Worrall, 2007), serpentine (Saada et al., 2009) and interstratified illite-smectite (Baccouche et al., 1998), have been used as the Al and Si sources for the synthesis of several types of zeolites.

Due to their exceptional properties, zeolites have been widely used as catalysts, adsorbents and ion exchangers (Breck, 1974). Numerous types of adsorbents such as organic and inorganic materials have been tested for their ability to remove heavy metals. Water researchers are seeking cheaper raw materials low-cost sorbents such as clay-based zeolites with application in the uptake of heavy metals from polluted effluents. Such adsorbents would be a viable replacement or supplement to chemicals, although they should be readily available, economically feasible, and should be regenerated with ease. The potential use of clay-based zeolites in the treatment of wastewater has been evaluated by a number of research groups (Bhattacharyya & Gupta, 2008; Jamil et al., 2010; Ibrahim, 2010).

The aims of this study are to combine two areas of expertise, water science and clay minerals and zeolites chemistry as well as to address the problem of environmental pollution by removal of heavy metal contaminants.

2. Experimental procedure and materials

2.1 Materials

The natural clays used as starting materials in zeolite synthesis corresponds to illite-rich clay from the Barroblanco mine, situated in the municipality of Oiba (Santander), and kaolinite-rich clay cropping out around the Sochagota Lake, Paipa (Boyacá). The raw materials were prepared prior to the synthesis process by drying during 24 h, and pulverized with an agate Mortar grinder RETSCH RM 100. Finally, the samples were sieved and particles of 63 μm selected for zeolite synthesis. Activating was done using the following chemical reagents: sodium hydroxide, NaOH, as pellets (99%, Aldrich) and distilled water. To determine the removal efficiency of Cr^{+3} and Ni^{+2} of the as-synthesized zeolite, a wastewater sample was collected from an electroplating industry located at Bucaramanga (Santander).

2.2 Hydrothermal transformation of clays into zeolites

The synthesis of faujasite-type zeolite from clays was conducted under hydrothermal conditions. An alkaline fusion step was introduced prior to hydrothermal treatment, because it plays an important role in enhancing the hydrothermal conditions for zeolite synthesis. On the other hand, this approach was adopted in this study because larger amounts of aluminosilicates can be dissolved employing this method. Raw and calcined at 900 oC materials were dry mixed with NaOH pellets (starting material/alkaline activator = 1/1.2 in weight) for 30 min and the resultant mixture was fused at 600 °C for 1 h. The alkaline fused product was ground in a mortar and then 4.40 g of this was dissolved in 21.50

ml of distilled water (ratio = 1/4.9) under stirring conditions for 30 min and then the reaction gel was aged for 24 h to form the amorphous precursors. The amount of reagents used for the preparation of the hydrogels was based on previous experimental work developed by Ríos and co-workers (Ríos, 2008; Ríos et al., 2009). Crystallization was carried out by hydrothermal synthesis under static conditions in PTFE vessels of 65 ml at 80°C for different reaction times (6, 24 and 96 h). At the end of the process the solid is separated by filtration, washed thoroughly several times with distilled water until the filtrate pH reduced to less than 10. The precipitated solid dried at 100° C overnight. The dried samples were weighed and kept in plastic bags for characterization.

2.3 Characterization of the raw materials and as-synthesized zeolites

Powder X-ray diffraction patterns of the raw materials and as-synthesized products were recorded with a Philips PW1710 diffractometer operating in Bragg–Brentano geometry with Cu-Kα radiation (40 kV and 40 mA) and secondary monochromation. Data collection was carried out in the 2θ range 3–50°, with a step size of 0.02°. Phase identification was performed by searching the ICDD powder diffraction file database, with the help of JCPDS (Joint Committee on Powder Diffraction Standards) files for inorganic compounds.

2.4 Sorption tests and water analyses

Clay-based faujasite was studied in laboratory batch experiments to determine its sorption of Cr^{+3} and Ni^{+2}, which was carried out at room temperature to investigate the efficiency of the as-synthesized zeolite as sorbent material for removing heavy metals from aqueous solution. A weighted amount of sorbent (0.25 and 0.5 g) was introduced in 180 g amber glass bottles, and then a volume of 50 ml of electroplating industry wastewater was added. Later, the sorbent:aqueous solution mixtures were continuously shaked for 24 h, and the temporal evolution of the solution pH and electrical conductivity was monitored. At several scheduled reaction times the bottles were removed from the shaker and the adsorbents were separated by filtration, while the filtrates were stored in a refrigerator for chemical analyses. All measurements were done according to the "Standard Methods for the Examination of Water and Wastewater" (APHA, AWWA, WEF, 2005). The pH and electrical conductivity of the original and treated aqueous solutions were measured using a pH Meter Lab 870 (Schott Instruments) and a 712 conductometer (Metrohm AG), respectively. The metal concentrations were determined using a Perkin-Elmer 372 atomic absorption spectrophotometer. The efficiency of treatment of the electroplating effluent using faujasite was then determined by the following equation:

Metal Removal Efficiency = $(C_1-C_2)/C_1 x100$,

Where C_1= initial metal concentration and C_2= metal concentration after treatment.

3. From clays to zeolites

Both clays and zeolites are aluminosilicates, which differ, however, in their crystalline structure. Clays have a layered crystalline structure and are subject to shrinking and swelling as water is absorbed and removed between the layers. Zeolites have a rigid, 3-dimensional crystalline structure consisting of a network of interconnected tunnels and cages. Water moves freely in and out of these pores but the zeolite framework remains rigid. Clays are characterized by two-dimensional sheets of corner sharing SiO_4 and AlO_4 tetrahedra. In these tetrahedral sheets, each tetrahedron shares 3 of its vertex oxygen atoms with other tetrahedra forming a hexagonal array in two-dimensions. The fourth vertex is not shared with another tetrahedron and all of the tetrahedra point in the same direction. The tetrahedral sheets are always bonded to octahedral sheets formed from small cations, such as Al^{+3} or Mg^{+2}, coordinated by six oxygen atoms. The unshared vertex from the tetrahedral sheet also form part of one side of the octahedral sheet but an additional oxygen atom is located above the gap in the tetrahedral sheet at the center of the six tetrahedra. This oxygen atom is bonded to a hydrogen atom forming an OH group in the clay structure. Clays can be categorized depending on the way that tetrahedral and octahedral sheets are packaged into layers, and they are commonly referred to as 1:1 and 2:1 clays with t-o and t-o-t layers, respectively. A 1:1 clay would consist of one tetrahedral sheet and one octahedral sheet (e.g., kaolinite, Figure 1a). A 2:1 clay consists of an octahedral sheet sandwiched between two tetrahedral sheets (e.g., illite, Figure 1b), which occurs due to two tetrahedral sheets with the unshared vertex of each sheet pointing towards each other and forming each side of the octahedral sheet. The crystal structure is formed from a stack of layers interspaced with the interlayers (spaces between the t-o or t-o-t layer packages). In the kaolinite structure, the layer will be electrically neutral (uncharged) and the t-o layers are bonded together only by weak intermolecular forces (van der Waals' bonds). In the illite structure the layer will have a net negative charge and K^+ ions will be attached themselves to clay surfaces in the so-called interlayer sites. In each case the interlayers can also contain H_2O molecules.

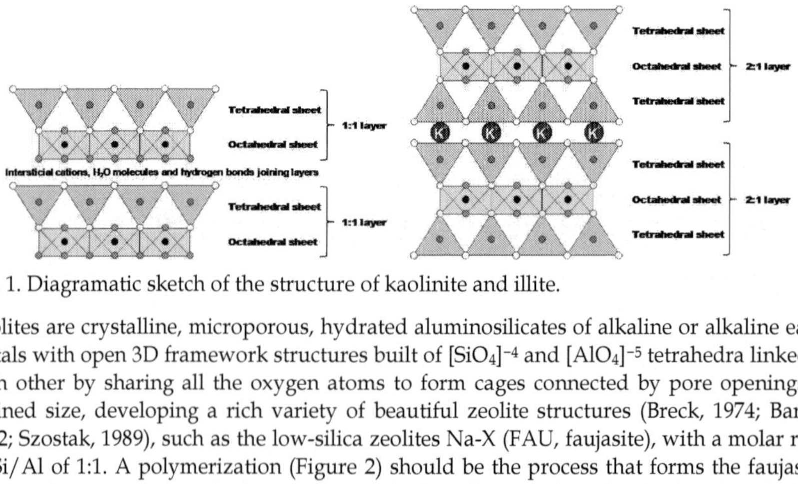

Fig. 1. Diagramatic sketch of the structure of kaolinite and illite.

Zeolites are crystalline, microporous, hydrated aluminosilicates of alkaline or alkaline earth metals with open 3D framework structures built of $[SiO_4]^{-4}$ and $[AlO_4]^{-5}$ tetrahedra linked to each other by sharing all the oxygen atoms to form cages connected by pore openings of defined size, developing a rich variety of beautiful zeolite structures (Breck, 1974; Barrer, 1982; Szostak, 1989), such as the low-silica zeolites Na-X (FAU, faujasite), with a molar ratio of Si/Al of 1:1. A polymerization (Figure 2) should be the process that forms the faujasite-type zeolite precursors, which contains tetrahedra of Si or Al randomly distributed along polymeric chains that are cross-linked so as to provide cavities sufficiently large to accommodate the charge balancing alkali ions. The faujasite-type zeolite is based on the primary building units (TO_4) where the central tetrahedrally bonded (T) atom is usually either Si^{4+} or Al^{3+}, surrounded by four O^{-2}. The primary TO_4 units can be linked to create secondary building units. In the faujasite-type zeolite, a combination of 4- and 6-rings promoted the formation of the β-cage. The secondary building units consist of n-ring structures, with each corner in the secondary building units representing the center of a tetrahedron. Secondary building units can be linked to form cages or channels within the faujasite structure. The aluminosilicate cages and the 6-rings connect to form a three dimensional net type structure. The framework of faujasite, consists of sodalite (SOD) cages composed of six 4-rings and eight 6-rings. Therefore, the framework of faujasite consists of β-cages (SOD) and α-cages (supercages); β-cages are linked together by double six-membered rings (D6R) and form the supercages of faujasite, which has a diameter of 13.0 Å, is surrounded by 10 β-cages and is interconnected to four other supercages by tetrahedrally disposed 12-membered-ring windows.

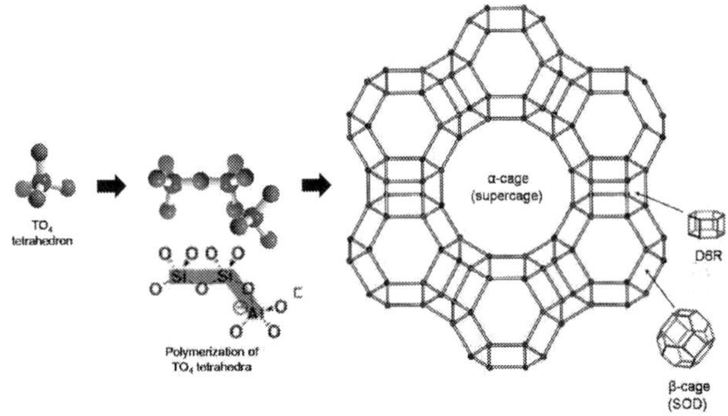

Fig. 2. Framework structure of zeolite Na-X (FAU), showing their characteristic cages and channels.

4. Results

4.1 Characterization of the raw clays

As shown in the XRD pattern of Fig. 3a, illite is the predominant mineral phase in the starting material and is identified by a series of basal reflections at 10.1 Å, 4.98–5.01 Å, 3.33 Å, and 2.89–2.92 Å. Similar results have been reported by Mezni et al. (2011). As shown in the XRD pattern of Fig. 3b, kaolinite is the predominant mineral phase, which can be identified by its distinctive reflections at 12.34° and 24.64° 2θ as reported by Zhao et al. (2004). In both clays quartz was identified by its distinctive reflections at 4.26 Å and 3.35 Å. The 3.35 Å peak of quartz was more intense than the other peaks.

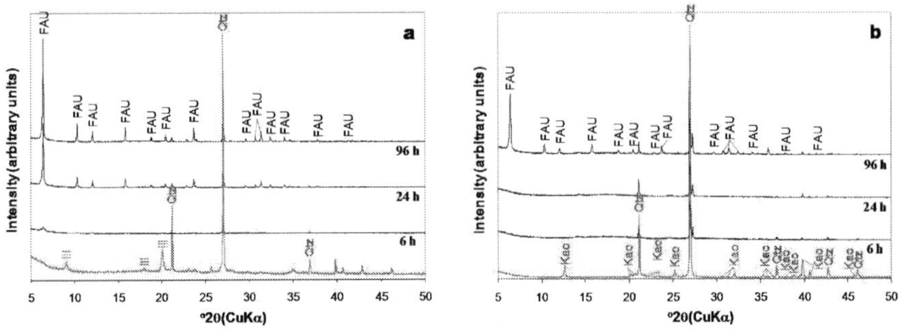

Fig. 3. X-ray diffraction patterns of the raw clays and as-synthesized faujasites.

4.2 Characterization of the clay-based faujasites

As shown in Figure 3, an almost complete transformation of the starting clay-rich materials into faujasite-type zeolite of high purity occurred. However, relictic quartz of the starting materials still remain in the synthesis products. Newly formed compounds dissolved in water more readily than the mineral phases in the starting materials, which have a low velocity of dissolution with the occurrence of some of them as relict phases in the synthesis products. The activation of the starting materials produced a rapid dissolution of the alkaline-fused products only after 6 h of reaction, and the complete disappearance of characteristic peaks of illite and kaolinite, accompanied by the gradual decrease in peak intensity of quartz that persisted in the products obtained. Moreover, the occurrence of faujasite was recorded after 24 h, showing an increase in the intensity of characteristic peaks between 24 and 96 h. Both samples produced a similar zeolite phase (faujasite), except for the fact that kaolinite-based faujasite showed less intense peaks that illite-based faujasite. Therefore, we decide to use the illite-based faujasite in the sorption tests.

5. Effectiveness of illite-based zeolites as sorbent material in heavy metal uptake

5.1 Kinetics of the neutralization reaction

The kinetics of the neutralization reaction was investigated by monitoring the pH and electrical conductivity of faujasite /aqueous solution mixtures (0.25 g/50 ml and 0.50 g/50 ml) over a period of 24 h. The effect of contact time on pH and EC during the sorption experiments for Cr^{+3} and Ni^{+2} is shown in Figure 4. Results reveal that the adsorption process of these heavy metals by faujasite was highly pH-dependent and increased with increasing pH conditions. pH increased rapidly within the first 5 min of contact between the solution and the sorbent (illite-based faujasite), and then it thereafter become stable (Figures 4a and 4b). According to Genç-Fuhrman et al. (2007), pH increases mainly due to dissolution of the sorbent in the process of shaking. Final pH values of 7.58-7.69 for Cr^{+3} and 8.42-8.53 for Ni^{+2} were observed in the batches. This significantly increased pH value during the experiments can be explained by the simultaneous uptake of hydrogen ions by faujasite, the hydrolysis of faujasite and the cationic exchange. Similar results are reported elsewhere with a remark that the pH increase is almost unavoidable in a removal of heavy metals by zeolite, taking into account its alkaline nature. On the other hand, results reveal that there is an increase in pH with sorbent dosage. A similar behavior was observed for EC as shown in Figures 4c and 4d.

5.2 Uptake of Cr^{+3} and Ni^{+2}

Removal of heavy metal ions such as Cr^{+3} and Ni^{+2} from electroplating wastewater has been investigated in order to determine the effectiveness of illite-based faujasite as sorbent material in the immobilization of Cr^{+3} and Ni^{+2}, exploiting the sorption capacity of this zeotype in order to evaluate its potential for the reduction of metal mobility and availability and its possible application for the remediation of wastewater.

The kinetics of adsorption process on porous materials such as zeolites is controlled by three consecutive steps (Mohan et al., 2001; Baniamerian et al., 2009): transport of the adsorbate from the bulk solution to the film surrounding the adsorbent, diffusion from the film to the proper surface of adsorbent, and diffusion from the surface to the internal sites followed by adsorption immobilization on the active sites.

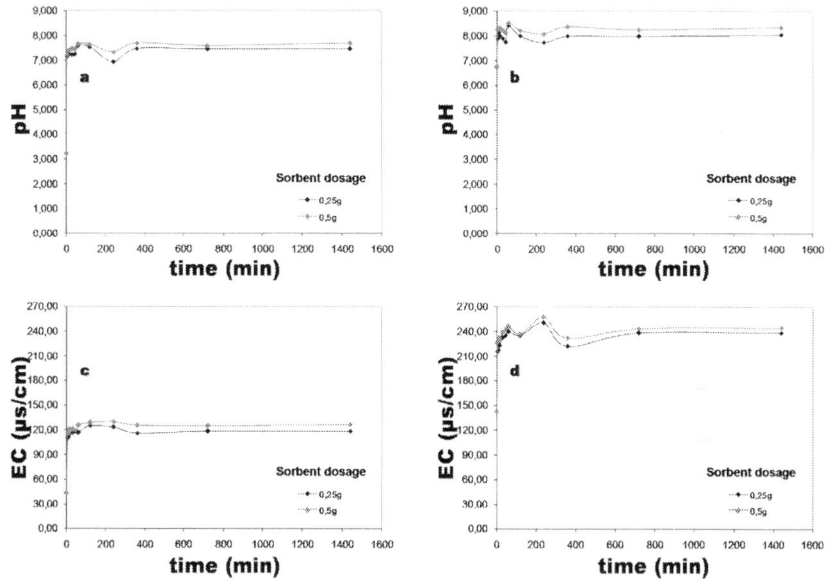

Fig. 4. Variation of pH and electrical conductivity as a function of time during the sorption tests for Cr^{+3} (a-c) and Ni^{+2} (b-d). Starting pH and EC of 3.20 and 43.24 µS/cm for Cr^{+3} and 6.75 and 142.80 µS/cm for Ni^{+2}.

Our study of adsorption kinetics of Cr^{+3} and Ni^{+2} ions was performed on faujasite-type zeolite at room temperature over 24 hours. Metal uptake trends as a function of contact time after batch reaction are illustrated in Figure 5. Results indicate that faujasite produced a steep decrease in Cr^{+3} concentration within the first 5 min, reaching very low residual concentrations. However, after 45 min plateau values were reached for the rest of the time intervals, indicating a complete removal. Cr^{+3} shows an abrupt decrease in concentration from 0 to 45 min and tends to stabilize at values between 0,922 and 1,695 mg/L (0.25g of faujasite) and between 0,946 and 1,513 mg/L (0.5g of faujasite). The Cr^{+3} exchange character is irreversible. Ni^{+2} showed an inconsistent variation of concentration between 0 and 360 min, which is revealed by the fluctuations observed during the batch experiments, and tends to stabilize at values between 35,614 and 38,763 mg/L (0.25g of faujasite) and between 47,963 and 50,184 mg/L (0.5g of faujasite). Therefore, a lower sorption of Ni^{+2} was observed which can be attributed to the higher selectivity to Cr^{+3} by illite-based faujasite. In general, the adsorption capacity increase as pH approaches neutral.

Several studies (e.g., Kannan & Rajakumar, 2003) report the apparent increase of the percentage of removal of heavy metals with increase in the dose of sorbent due to the active sites/surface area for the adsorption of metal ions, whereas, at lower sorbent dosage the number of metal ions was relatively higher, compared to availability of adsorption sites/surface area. However, results from this study reveal that there is not a strong difference in percentage of metal ion removal with sorbent dosage. Therefore, it is clear that it is very important to investigate in future studies the optimum dosage of faujasite-type zeolite tested as sorbent material.

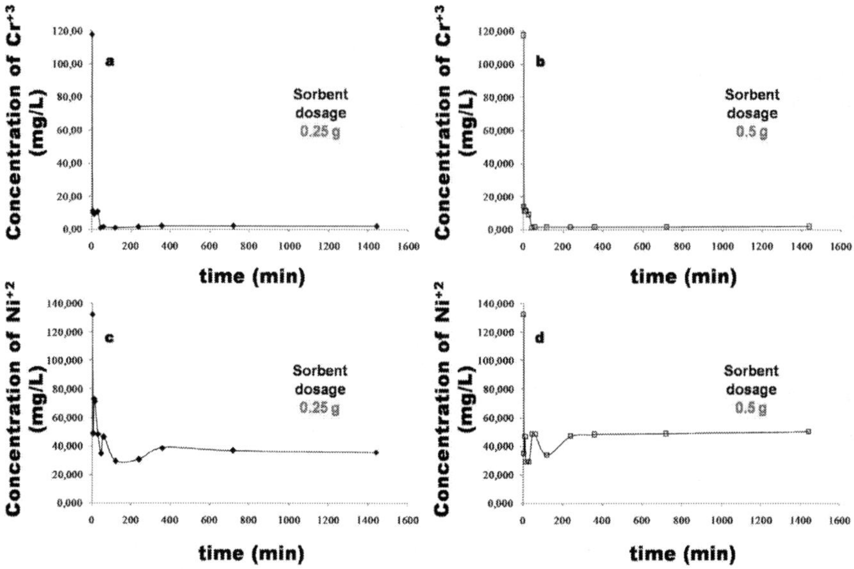

Fig. 5. Variation of concentration of Cr^{+3} and Ni^{+2} as a function of time during the sorption batch experiments. Starting concentration 117,300 mg/L for Cr^{+3} and 132,300 mg/L for Ni^{+2}.

According to Jenne (1998), heavy metal cations can be immobilized by zeolites by two mechanisms: ion exchange and chemisorption. Ion exchange involves substitution of ions present in zeolite crystalline lattice by metal ions from the solution (Inglezakis et al., 2002). The type of catión (the position of the cation in the selectivity series) as well as the cation concentration in the solution will determine the ion-exchange efficiency (Mozgawa & Bajda, 2005). Chemisorption results in the formation of stable inner-sphere complexes (Godelitsas, 1999). This is due to the fact that functional groups (mainly OH-) form strong chemical bonds with metal ions outside the hydration envelope (Jenne, 1998).

As reported in previous studies (e.g. Mozgawa & Bajda, 2005), after zeolite reaches the saturation level of a metal ion sorbed, further pH lowering causes the increase in the zeolite crystalline lattice positive charge, which reduces the zeolite ability of metal cations chemisorption. The proportion of chemisorption and ion-exchange processes depends on pH changes and inherent properties of the metal.

The retention efficiency (for the metals considered) depended, not only on the ionic exchange capacity of the as-synthesized zeolite, but also on the decrease of the acidity induced by the zeolitic product. The retention efficiency of faujasite produced the following ranges: Cr^{+3} (90.53-99.21%, 0.25 g of zeolite, and 88.31-99.19%, 0.5 g of zeolite), and Ni^{+2} (44.74-77.73%, 0.25 g of zeolite, and 62.07-78.30%, 0.5 g of zeolite). Sorption tests reveal that both metal ions were rapidly removed by faujasite within 45 min (Cr^{+3}) and 360 min (Ni^{+2}) with 88.31-90.53% and 63.24-73.54% of the metal removal achieved in the first 5 min for Cr^{+3} and Ni^{+2}, respectively. According to Mozgawa & Bajda (2005), the contribution of chemisorption and ion-exchange processes to the metal immobilization on zeolite depends on the metal type and the reaction time. In the case of Cr^{+3} and Ni^{+2}, after the first minutes of reaction, it is probably that the chemisorption process becomes more dominant than the ion-exchange process, similar to data reported by Mozgawa & Bajda (2005). Faujasite-type zeolite produced lower Ni^{+2} removal (88.31-99.21%) compared with that for Cr^{+3} (44.74-78.30%). However, the competition for sorbent adsorption sites in the presence of Cr^{+3} produced a decrease in the uptake of Ni^{+2}. No significant adsorption was observed after 45 min (Cr^{+3}) and 360 min (Ni^{+2}) of contact time.

According to Peric et al. (2004), the immobilization of heavy metals from aqueous media is a complex process, which consists of ion exchange and adsorption and is likely to be accompanied by precipitation of metal hydroxide complexes on active sites of the particle surface. On the other hand, the addition of an alkaline material such as faujasite to the electroplating wastewater increased the pH. (7.58-7.69 for Cr^{+3} and 8.42-8.53 for Ni^{+3}) and these metal ions could be hydrolyzed and precipitated as suggested by Evangelou and Zhang (1995). However, the efficiency of the tested sorbent with respect to metal retention and/or metal concentration control during its application for the treatment of metal-bearing aqueous media is governed by parameters like contact time, pH, temperature and sorbent nature as demonstrated in previous studies (Helquet et al., 2000). On the other hand, mechanisms of interactions, such as precipitation and adsorption, between Cr^{+3} and Ni^{+3} and illite-based faujasite are strongly influenced by pH.

6. Conclusions

Natural clays were successfully transformed into highly crystalline faujasite-type zeolite by fusion with NaOH powder followed by hydrothermal treatment. The adsorption studies showed rapid uptake in general for the first 45 min (Cr^{+3}) and 360 min (Ni^{+2}) with 88.31-90.53% and 63.24-73.54% of the metal removal achieved in the first 5 min for Cr^{+3} and Ni^{+2}, respectively. After this initial rapid period, the rate of adsorption decreases. The as-synthesized faujasite proved to have great potential in the immobilization of Cr^{3+} and Ni^{2+} from electroplating wastewater. Results of this work can be also used as a reference for future in depth studies considering alternative technologies applied to remediation of electroplating industry wastewater.

7. Acknowledgments

This research forms part of the background experience of the authors in the field of clay and zeolite technology and has benefited from research facilities provided by the Universidad Industrial de Santander, and the Instituto Zuliano de Investigaciones Tecnológicas. We thank to Miguel Ramos for assistance with XRD data acquisition.

8. References

Algarra, M.; Jiménez, M.V.; Rodríguez-Castellón, E.; Jiménez-López, A. & Jiménez-Jiménez, J. (2005). Heavy metals removal from electroplating wastewater by aminopropyl-Si MCM-41. *Chemosphere*, Vol. 59, No. 6, (May 2005), pp. 779-786, ISSN 0045-6535

American Public Health Association (APHA), American Water Works Association (AWWA), & Water Environment Federation (WEF). (2005). Standard Methods for the Examination of Water and Wastewater. A.D. Eaton, L.S. Clesceri, E.W. Rice, A.E. Greenberg and M.A.H. Franson (Eds.), 21st Ed., 1368p.

Baccouche, A.; Srasra, E. & Maaoui, M.E. (1998). Preparation of Na-P1 and sodalite octahydrate zeolites from interstratified illite-smectite. *Applied Clay Science*, Vol. 13, No. 4, (October 1998), pp. 255-273, ISSN 0169-1317

Baniamerian, M.J.; Moradi, S.E.; Noori, A. & Salahi, H. (2009). The effect of surface modification on heavy metal ion removal from water by carbon nanoporous adsorbent. *Applied Surface Science*, Vol. 256, No. 5, (December 2009), pp. 1347-1354, ISSN 0169-4332

Barrer, R.M. (1982). *Hydrothermal Chemistry of Zeolites*, Academic Press, ISBN 0120793601, New York, USA

Bhattacharyya, K.G. & Gupta, S.S. (2008). Adsorption of a few heavy metals on natural and modified kaolinite and montmorillonite: A review. *Advances in Colloid and Interface Science*, Vol. 140, No. 2, (August 2008), pp. 114-131, ISSN 1359-0294

Boukadir, D.; Bettahar, N. & Derriche, Z. (2002). Synthesis of zeolites 4A and HS from natural materials. *Annales de Chimie Science des Matériaux*, Vol. 27, No. 4, (July 2002), pp. 1-13, ISSN 0151-9107

Breck, D.W. (1974). *Zeolite Molecular Sieves: Structure, Chemistry and Use*, John Wiley, ISBN 0471099856, New York, USA.

Cañizares, P.; Durán, A.; Dorado, F. & Carmona, M. (2000). The role of sodium montmorillonite on bounded zeolite-type catalysts. *Applied Clay Science*, Vol. 16, No. 5-6, (May 2000), pp. 273–287, ISSN 0169-1317

Chartrand, M.M.G. & Bunce, N.J. (2003). Electrochemical remediation of acid mine drainage, *Journal of Applied Electrochemistry* Vol. 33, No. 3-4, (March 2003), pp. 259–264, ISSN 1572-8838

Diz, H.R. & Novak, J.T. (1998). Fluidized bed for the removing of iron and acidity from acid mine drainage. *Journal of Environment Engineering*, Vol. 124, No. 8, (August 1998), pp. 701-708, ISSN 1943-7870

Evangelou, V.P. & Zhang, Y.L. (1995). A review: Pyrite oxidation mechanisms and acid mine drainage prevention. *Critical Reviews in Environmental Science and Technology*, Vol. 25, No. 2, (November 1995) pp. 141-199, ISSN 1064-3389

Feng, D.; Aldrich, C. & Tan, H. (2000). Treatment of acid mine water by use of heavy metal precipitation and ion exchange. *Minerals Engineering*, Vol. 13, No. 6, (June 2000), 623–642, ISSN 0892-6875

Genç-Fuhrman, H.; Mikkelsen, P.S. & Ledin, A. (2007). Simultaneous removal of As, Cd, Cr, Cu, Ni and Zn from stormwater: Experimental comparison of 11 different sorbents. *Water Research*, Vol. 41, No. 3, (February 2007), pp. 591-602, ISSN 0043-1354

Gibert, O.; de Pablo, J.; Cortina, J.L. & Ayora, C. (2005). Municipal compost-based mixture for acid mine drainage bioremediation: Metal retention mechanisms. *Applied Geochemistry*, Vol. 20, No. 9, (September 2005), pp. 1648–1657, ISSN 0883-297

Godelitsas, A. (1999). Transition metal complexes supported on natural zeolitic materials: an overwiew. In: *Natural microporous materials in environmental technology*, P. Misaelides; F. Macásek; T.J. Pinnavaia & C. Colella (Eds.), 271–281, Springer, ISBN 97-8079-2358-88-6 Dordrecht, Germany

Hequet, V.; Ricou, P.; Lecuyer, I. & LeCloirec, P. (2000). Removal of Cu^{2+} and Zn^{2+} from aqueous solutions by sorption onto mixed fly ash. *Fuel*, Vol. 80, No. 6, (May 2001), pp. 851-856, ISSN 0016-2361

Ibrahim, H.S.; Jamil, T.S. & Hegazy, E.Z. (2010). Application of zeolite prepared from Egyptian kaolin for the removal of heavy metals: II. Isotherm models. *Journal of Hazardous Materials*, Vol. 182, No. 1-3, (October 2010), pp. 842-847, ISSN 0304-3894

Ikhsan, J.; Johnson, I.B.B. & Wells, J.D. (1999). A comparative study of the adsorption of transition metals on kaolinite. *Journal of Colloid and Interface Science*, Vol. 217, No. 2, (September 1999), pp. 403-410, ISSN 0021-9797

Inglezakis, V.J.; Loizidou, M.D. & Grigoropoulou, H.P. (2002). Equilibrium and kinetic ion exchange studies of Pb^{2+}, Cr^{3+}, Fe^{3+} and Cu^{2+} on natural clinoptilolite. *Water Research*, Vol. 36, No. 11, (June 2002), pp. 2784-2792, ISSN 0043-1354

Jamil, T.S.; Ibrahim, H.S., Abd El-Maksoud, I.H. & El-Wakeel, S.T. (2010). Application of zeolite prepared from Egyptian kaolin for removal of heavy metals: I. Optimum conditions. *Desalination*, Vol. 258, No. 1-3, (August 2010), pp. 34-40, ISSN 0011-9164

Jenne, E.A. (1998). Adsorption models. In: *Adsorption of metals by geomedia: variables, mechanism and model applications*, J.A. Jenne, (Ed.), 11-36, Academic, ISBN: 012-3842-45-X, San Diego, USA

Johnson, Ch.D. & Worrall, F. (2007). Novel low density granular adsorbents - Properties of a composite matrix from zeolitisation of vermiculite. *Chemosphere*, Vol. 68, No. 6, (June 2007), pp. 1153-1162, ISSN 0045-6535

Johnson, D.B. & Hallberg, K.B. (2005). Acid mine drainage remediation options: a review. *Science of the Total Environment*, Vol. 338, No. 1-2, (February 2005), pp. 3–14, ISSN: 0048- 9697

Kalin, M.; Fyson, A. & Wheeler, W.N. (2006). The chemistry of conventional and alternative treatment systems for the neutralization of acid mine drainage. *Science of the Total Environment*, Vol. 366, No. 2-3, (August 2006), pp. 395-408, ISSN: 0048- 9697

Kannan N. & Rajakumar, A. (2003). Suitability of various Indigenously Prepared Activated carbons for the adsorption of mercury(II). Toxicological and Environmental Chemistry, Vol. 84, No. 1-4, (December 2003), pp. 7-19, ISSN 0277-2248

Klimkiewicz, R. & Drąg, E.B. (2004). Catalytic activity of carbonaceous deposits in zeolite from halloysite in alcohol conversions. *Journal of Physics and Chemistry of Solids*, Vol. 65, No. 2-3, (March 2004), pp. 459–464, ISSN 0022-3697

Meena, A.K.; Mishra, G.K.; Rai, P.K.; Rajagopal, Ch. & Nagar, P.N. (2005). Removal of heavy metal ions from aqueous solutions using carbon aerogel as an adsorbent. *Journal of Hazardous Materials*, Vol. 122, No. 1-2, (June 2005), pp. 161-170, ISSN 0304-3894

Mezni, M.; Hamzaoui, A.; Hamdi, N. & Srasra, E. (2011). Synthesis of zeolites from the low-grade Tunisian natural illite by two different methods. *Applied Clay Science*, Vol. 52, No. 3, (May 2011), pp. 209-218, ISSN 0169-1317

Mohan, D., Gupta, V.K., Srivastava, S.K. & Chander, S. (2001). Kinetics of mercury adsorption from wastewater using activated carbon derived from fertilizer waste. *Colloids and Surfaces A-Physicochemical and Engineering Aspects*, Vol. 177, No. 2-3, (February 2000), pp. 169-181, ISSN 0927-7757

Mohan, D. & Chander, S. (2006). Removal and recovery of metal ions from acid mine drainage using lignite - A low cost sorbent. *Journal of Hazardous Materials*, Vol. 137, No. 3, (October 2006), pp. 1545–1553, ISSN 0304-3894

Mozgawa, W. & Bajda, T. (2005). Spectroscopic study of heavy metals sorption on clinoptilolite. *Physics and Chemistry of Minerals*, Vol. 31, (February 2004), pp. 706–713, ISSN (electronic): 1432-2021

Nestle, N. (2002). NMR studies on heavy metal immobilization in biosorbents and mineral matrices. *Reviews in Environmental Science and Biotechnology*, Vol. 1, No. 3, pp. 215-225, ISSN 1569-1705

Pereira, F.V.; Alves, L.V. & Gil, L.F. (2010). Removal of Zn^{2+} from aqueous single metal solutions and electroplating wastewater with wood sawdust and sugarcane bagasse modified with EDTA dianhydride (EDTAD). *Journal of Hazardous Materials*, Vol. 176, No. 1-3, (April 2010), pp. 856–863, ISSN 0304-3894

Peric, J.; Trigo, M. & Medvidovi´c, N.V. (2004). Removal of zinc, copper and lead by natural zeolite-a comparison of adsorption isotherms. *Water Research*, Vol. 38, No. 7, (April 2004), pp. 1893–1899. ISSN 0043-1354

Ríos, C.A. (2008). Synthesis of zeolites from geological materials and industrial wastes for potential application in environmental problems. PhD Thesis, University of Wolverhampton, Wolverhampton, England.

Ríos, C.A.; Williams, C.D. & Roberts, C.L. (2008). Removal of heavy metals from acid mine drainage (AMD) using fly ash, natural clinker and synthetic zeolites. *Journal of Hazardous Materials*, Vol. 156, No. 1-3, (August 2008) pp. 23-35, ISSN 0304-3894

Ríos, C.A.; Williams, C.D. & Fullen, M.A. (2009). Nucleation and growth history of zeolite LTA synthesized from kaolinite by two different methods. *Applied Clay Science*, Vol. 42, No. 3-4, (January 2009), pp. 446-454, ISSN 0169-1317

Ríos, C.A.; Williams, C.D. & Roberts, C. (2011). Synthesis and characterization of SOD-, CAN- and JBW-type structures by hydrothermal reaction of kaolinite at 200ºC. *Dyna*, No. 166, (April 2011), pp. 38-47, ISSN 0012-7353

Ruiz, R.; Blanco, C.; Pesquera, C.; Gonzalez, F.; Benito, I. & Lopez, J.L. (1997). Zeolitization of a bentonite and its application to the removal of ammonium ion from waste water. *Applied Clay Science*, Vol. 12, No. 1-2, (June 1997), pp. 73-83, ISSN 0169-1317

Saada, M.A.; Soulard, M.; Patarin, J. & Regis, R. (2009). Synthesis of zeolite materials from asbestos wastes: An economical approach. *Microporous and Mesoporous Materials*, Vol. 122, No. 1-3, (June 2009), pp. 275-282, ISSN 1387-1811

Santos, S.; Machado, R. & Correia, M.J.N. (2004). Treatment of acid mining waters. Minerals Engineering, Vol. 17, No. 2, (February 2004), pp. 225–232, ISSN 0892-6875

Song, S. & Guo, J. (1997). Synthesis of zeolite Y from bentonite. *Zeolites*, Vol. 18, No. 1, (January 1997), pp. 84, ISSN 1083-2718

Sousa, F.W.; Sousa, M.J.; Oliveira, I.R.N.; Oliveira, A.G.; Cavalcante, R.M.; Fechine, P.B.A.; Neto, V.O.S.; de Keukeleire, D. & Nascimento, R.F. (2009). Evaluation of a low-cost adsorbent for removal of toxic metal ions from wastewater of an electroplating factory. *Journal of Environmental Management*, Vol. 90, No. 11, (August 2009), pp. 3340-3344, ISSN 0301-4797

Szostak, R. (1998). *Molecular sieves*, Blackie Academic and Professional, ISBN 0751404802, London, England

Wattena, B.J.; Sibrella, P.L. & Schwartzb, M.F. (2005). Acid neutralization within limestone sand reactors receiving coal mine drainage. *Environmental Pollution*, Vol. 137, No. 2, (September 2005), pp. 295-304, ISSN 0269-7491

Webster, J.G.; Swedlund, P.J. & Webster, K.S. (1998). Trace metal adsorption onto an acid mine drainage iron (III) oxy hydroxy sulphate. *Environmental Science and Technology*, Vol. 32, No. 10, pp. 1361-1368, ISSN 0013-936X

Wei, X.; Viadero Jr., R.C. & Buzby, K.M. (2005). Recovery of iron and aluminium from acid mine drainage by selective precipitation. *Environmental Engineering Science*, Vol. 22, No. 6, (October 2005), pp. 745–755, ISSN 1092-8758

Zhao, H.; Deng, Y.; Harsh, J.B.; Flury, M. & Boyle, J.S. (2004). Alteration of kaolinite to cancrinite and sodalite by simulated hanford tank waste and its impact on cesium retention. *Clays and Clay Minerals*, Vol. 52, No. 1, (February 2004), pp. 1-13, ISSN 0009-8604

Zhao, Y.; Zhang, B.; Zhang, X.; Wang, J.; Liu, J. & Chen, R. (2010). Preparation of highly ordered cubic NaA zeolite from halloysite mineral for adsorption of ammonium ions. *Journal of Hazardous Materials*, Vol. 178, No. 1-3, (June 2010), pp. 658-664, ISSN 0304-3894

Part 5

Hydrogeology

16

Natural Radioactive Isotopes in Water in Relation with Geology and Hydrological Investigations in the Territory of Luxembourg

Zornitza Tosheva, Harald Hofmann and Antoine Kies
[1]University of Luxembourg,
[2]Monash University,
[1]Luxembourg
[2]Australia

1. Introduction

Luxembourg is situated in the middle of Europe and is surrounded by Belgium in the north-west, France in the south-west and south, and Germany in the east (Figure 1). It is one of the smaller countries in Europe with a surface area of 2.600 km^2 and a population of approximately 500.000 inhabitants. The northern third, known as 'Eislek', has a plateau character with an average elevation of 450 m, whereas the southern two-third are know as the 'Gutland', good land in English. This region has a more varied topography and an average elevation of 240 m.

Groundwater is the main water source for drinking water supply in the Grand Duchy of Luxembourg. About 60% of the country's drinking water comes from so called "Luxemburg Sandstone", which forms the most important aquifer. Most of the groundwater captures are natural springs or shallow wells and represent 80 % of the volume of this aquifer. The majority of the springs are distributed along the main river valleys. The unconfined part of the sandstone is located in the centre of the country and covers about 300 km^2 with a surface to water table depths ranging from 0 to 100 meters (Colbach, 2006).

Except from some bores in the north of Luxembourg and some in the south, natural springs are mainly used for public drinking purpose.

The radionuclide content of more than 300 springs and bores was determined as part of 2 monitoring campaigns, in a first step to relate the physico-chemical parameters with local geological conditions and in a second to obtain data on groundwater pollution regarding agricultural contaminants, such as pesticides, nitrate etc. The second campaign allows a general quality assessment of the groundwater resources in the country, which is also required by the European Water Framework Directive (EU, 2000).

Radio-nuclides are of interest for two reasons. Firstly, a high activity poses a health risk. An example is dissolved radon (222Rn) in groundwater that degasses when the water leaves the aquifer. Radon concentrations increase in the air in buildings where the water is trapped, such as spring trappings, or when used like in households. The inhalation of radon and the deposition of its daughter products in the lungs can cause lung cancer. Secondly, radio-nuclides can be useful tools as tracers in the hydrological / hydro-geological cycle.

Radioactive carbon and tritium are widely known and utilized tracers for groundwater dating and follows the diffusion and migration processes. The radioactive carbon and tritium are cosmogenic. Most of the cosmogenically produced radio-nuclides in the atmosphere are oxidized and become attached to aerosol particles. These particles act as condensation nuclei for the formation of cloud droplets and eventually coagulate to form precipitation. Similar is the mechanism of naturally produced ^{210}Pb, a long-lived decay product of the noble gas radon from the uranium decay series, which reach the atmosphere and are transported with the atmospheric circulations in form of particles until deposition on the surface with the precipitations. Other radioactive elements, besides these isotopes, such those of the uranium and thorium decay series, can be used as groundwater flow and mixing tracers.

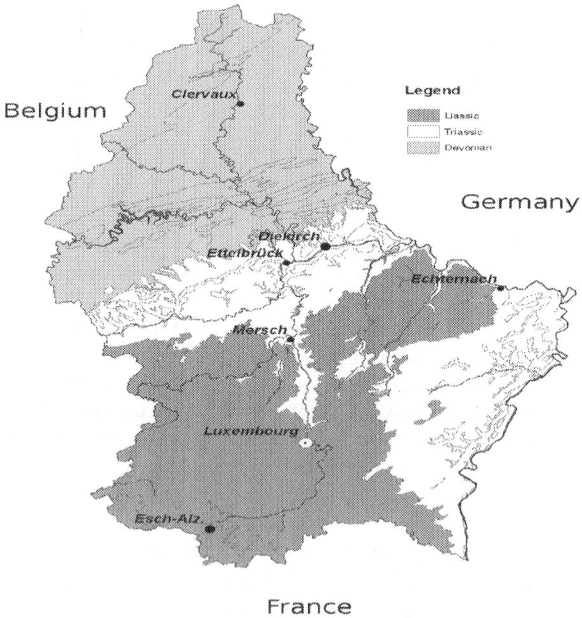

Fig. 1. Geological map of the main units in Luxembourg. The area of the Devonian formations in the North forms the 'Eislek', whereas the Southern 'Gutland' is represented by the Triassic and Jurassic units.

The uranium and thorium series isotopes derive from the release out of uranium / thorium bearing minerals in the soils and rocks. These naturally occurring radio-isotopes are supplementary tracers to conventional major ion chemistry and stable isotopes in hydrological applications, e.g. creates a particularly interest such as surface water/groundwater interactions.

There had been some research on the use of radio-nuclides in these fields but still only little was known about their distribution and the mobility over the country. If there presence is ubiquitous, there are physical and chemical aquifer characteristics, such as porosity, hydraulic conductivity or organic matter content that play major roles in their occurrence and migration in the environment.

This study does not aim to solve all questions, which are related to the distribution and mobility of radio-nuclides in the groundwater but it helps to get closer to some solutions.

In this work we present two different kind of investigations performed with the natural radioactive isotopes in relation with geological and hydrological studies.

The aim of the first part is to deliver a high resolved spatial radioactive isotope distribution maps and an analysis for the causes of activity variations. We rely on the major ion chemistry data and physical parameters that have been measured in Luxembourg's drinking water supplies. The relations of the radionuclide concentrations with the major ion chemistry were studied. Comparison of the obtained results versus local geology was followed in this approach.

Besides a dense spatial variability and chemical behavior along the hydro-geological regions, the second part of this work aims analyses of temporal variations of radio-isotopes, especially in the case of radon. These results are crucial for the use of radio-isotope data as groundwater and surface water tracers.

2. Sampling and methodology

The samples were taken in 2 litres, airtight glass bottles directly from the spring. Where possible they were taken right at the emergence of the water from the host rock. Water from wells and bores was pumped with the preinstalled pumps from the water suppliers. Purging was necessary because the sampled bores are production bores that are constantly in use. The bottles were filled completely without air space and stored in insulated coolers to avoid degassing during the transport to the laboratory.

The electrical conductivity and the temperature of the water was measured in-situ using a WTW i340 multi probe. The electrical conductivity was temperature corrected to 20°C.

Major ion chemistry was done at the National Water Laboratory of Luxembourg by ion chromatography for the anions and induced-coupled mass spectrometry for the cation content and trace elements.

The radio-nuclides were measured at the Radio-physics Laboratory of the University of Luxembourg. Radon (^{222}Rn) was measured a few hours after sampling by liquid scintillation counting (LSC) on a Perkin-Elmer "Guardian" Liquid Scintillation Counter, using 10 ml of Perkin-Elmer Ultima Gold F scintillator and 12 ml of sample. For some choosen samples 3H were measured by LSC after purification with Eichrom tritium columns. The other isotopes (^{238}U, ^{235}U, ^{234}U, ^{232}Th, ^{230}Th, ^{228}Th, ^{228}Ra, ^{226}Ra, ^{224}Ra, ^{210}Pb) were separated with sequential extraction (Tosheva et al. 2003, 2009) and ^{210}Po by electro-deposition on silver discs. After separation they were measured on Canberra alpha detector or LSC respectively.

3. Local geology and hydrogeology

3.1 Geology

In the official web page of Luxembourg geological survey (geology.lu) is written that Luxembourg shows on its small area of about 2500 km^2 a dense geological diversity.

As mentioned above, Luxembourg is orographicaly and geologically divided in two major natural regions, which are geologically different. On the one hand, there is the 'Eislek' in the

North, which is built up from Lower Devonian formations in Siegen facies (Figure 1). It belongs to the rhenohercynian block of the Eifel Mountains. The outcropping rocks consist mostly of folded and slated medium grained sediments with actual thickness of several thousand meters, transformed into schist, meta-sandstones, and bluish-grey low-grade methamorphosed shists sediments. Inter-beddings of quartzite sandstone occur in some parts of the formation. The geological substrate is covered by a thin soil layer of 0.5 to 1.5 m thickness.

On the other hand, there is the 'Gutland' in the South, which encompasses Mesozoic units forms the Perm-Triassic boundary to the Middle Jurassic. They belong to the units of the Paris Basin and extend about 150 km from the southern Eifel in Germany, via Luxemburg towards Belgium (Berners, 1983). The sequence starts with the Lower Triassic units of the 'Buntsandstein' with fluvial red bed sand-, siltstones and clays, followed by marls of the Upper Keuper (Rhät). However, the unit, which bears most drinking water springs in the country, is the 'Luxembourg Sandstone'. It is carbonate bound sandstone from the Middle Liassic (li), which can be assigned to a shallow marine shelf environment with wave influencing reworking and re-sedimentation (Berners, 1983). Heterogeneous layering boundaries can be observed within and in between the sandstone formation and the Psilonotas marls. It consists of light yellow, fine to medium grained sandstones with a low to medium carbonate content (15-50 %), which origins from the spartic cement and bioclastic components (Colbach, 2006). Some layers contain very well rounded pebbles (ca. 10 cm).

The Luxembourg sandstone is covered by younger formations in the South-West of the country. They encompass the iron Minette ore formation carbonate units of the Middle Dogger and carbonates of particularly the Upper Dogger. These south-west part ironstones are present in the form of limonitic and pisolithic surface formations.

3.2 Hydrography and hydrogeology

Luxembourg has a temperate climate with an average rainfall of 800 mm per year and an average temperature of 9°C. The land-use distribution is as follows: forested 34%, pastures 26%, agricultural-cultivated 23%, other 17%.

The majority of the samples come from the Luxembourg sandstone (grès de Luxembourg) aquifer, which is the most important aquifer for the drinking water supply in the country. It is generally limited on its bottom and top by the limestone-shale alternations of the Lorraine facies, although interbeddings with these exist (Colbach, 2005). The porosity of the aquifer is closely related to the carbonate content with approximately 5% in well cemented sandy limestones to 35% in poorly cemented sections, whereas water flow occurs in the well developed open fracture network. It is a dual-porosity aquifer and shows very heterogeneous permeabilites across its occurrence. The general flow pattern is dominated by the dip to the south-west of the unit.

The other hydrogeological units, which were encompassed in the campaigns, are Devonian shists and sandstone in Siegen facies, red-bed sandstones (Buntsandstein), Muschelcalc sandstone, dolomitic Keuper limestones, carbonate-bound sandstones from the middle Lias and iron-bearing limestones form the Dogger.

The Devonian schist and sandstones in Siegen and Ems build up the northern part of the country and belong to the rheno-hercynian block of the Rhenish Slate Mountains and the Ardennes Mountains. They are folded and fractured schist and sandstones with a low hydraulic conductivity. Flow and storage happens mostly in the fracture network.

The Buntsandstein in Luxembourg shows a sandy facies of fluvial environments with conglomerate and silt/clay interbeddings. Groundwater flow and storage happens in the fracture network as well. It has mixture permeability of fissures and pores. The dolomitic limestones of the Keuper are characterized by a more developed fracture network than the older formations. Carbonate dissolution has created a reasonable specific storage and hydraulic conductivity. The same hold true for the middle Lias carbonate-bound sandstones. The Dogger formations consist of oolithic iron ores from the Minette-type. The hydrogeology of the Minette rocks is complicated due to the mining activities in the past but general characteristics of the aquifer encompass a medium to high hydraulic conductivity. This aquifer is a dual porosity aquifer with a well developed open fracture network. The pore space in the matrix is high in the weathered areas and low in the non-weathered areas.

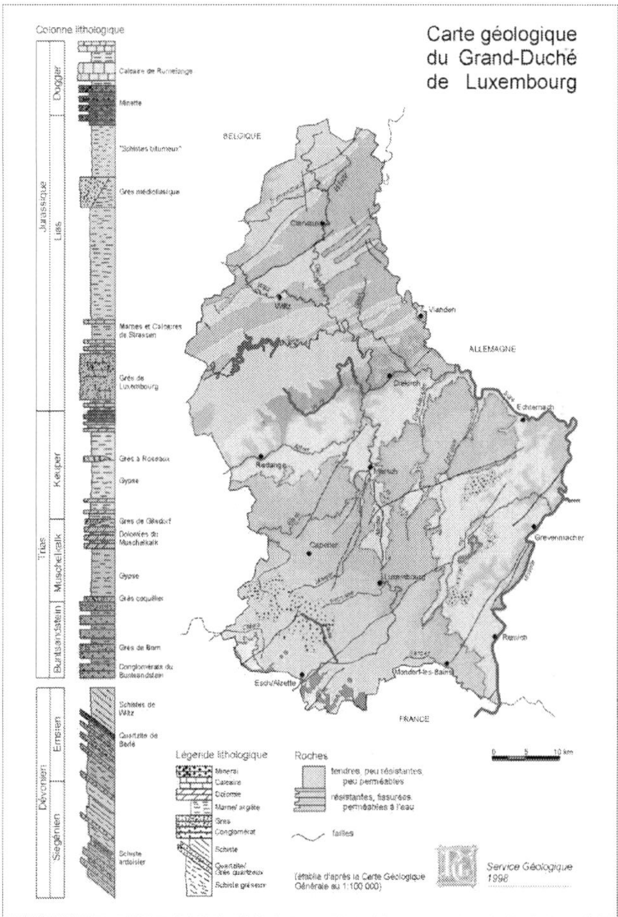

Fig. 2. Geo-morphological map of the main units in Luxembourg.

4. Results and discussion

The major ion chemistry of the water in Luxembourg is dominated by Ca^{2+}, Mg^{2+}, HCO_3^- and SO_4^{2-} and therefore the waters can be assigned to a calcium-bicarbonate type (figure 3). However Na^+, K^+ and Cl^- play a significant role, especially in places with higher permeability and fractures. Another not negligible contribution comes from NO_3^- ions, especially on the plateaus of Luxembourg Sandstones when agriculture affects the water table or organic matter is present in the rocks. The carbonate species are dominant in the carbonate units of the Luxembourg Sandstones and the younger Jurassic lime stones formations and carbonate units if Luxembourg's sandstones with carbonate cements and bioclastic components. Calcium values range from a few mg/l to over 200 mg/l, where values of the main aquifers, such as the Luxemburg sandstone, range from 80 to 200 mg/l. The bicarbonate follows the same trend with concentrations in between 1 to 40 mg/l.

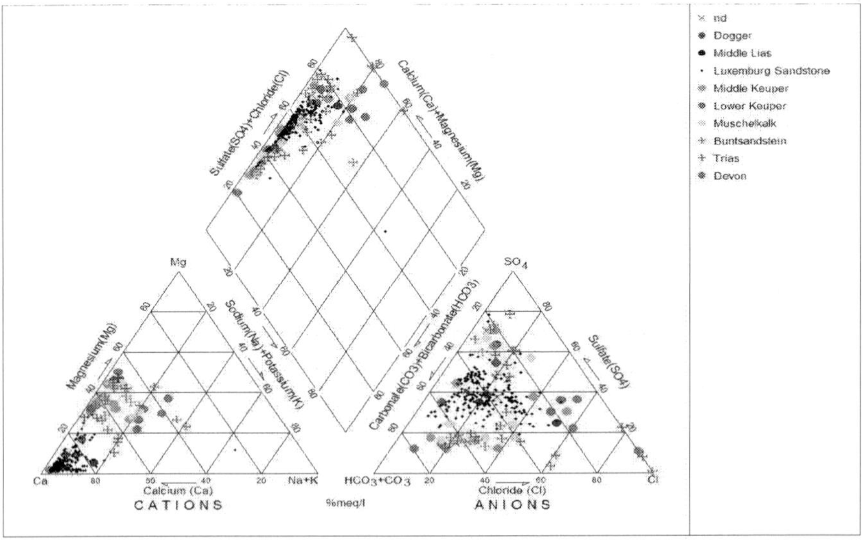

Fig. 3. Piper-diagram of the sampled water in the Spatial Monitoring campaign.

Radionuclide concentrations mainly vary with the local geology and in many cases are supported by different chemical mecanismus. Values for activity concentration of uranium vary from 0.2 to 74 mBq/L with predominant 10–30 mBq/L. They correspond to aquifers with similar mineralogy described by[Porceli et al., 2003].

Radium in natural water is derived mainly by interaction between the water and radium bearing materials, such as rock, soil and deposits due to its good solubility. As it forms sulfates, carbonates, nitrates and chlorides at different degrees and as such is a good indicator of the geochemical situation. In the analyzed samples the concentration varied from 3 to 35 mBq/L. Only five samples show significantly increased results, which can be explained with the different geological material in the aquifers (Devonian and Trias formations). Another data set with higher than median values was from a spring where the waters have a quite different chemistry and high salinity [Oliveira et al, 2006].

Lead mobility in water depends on the pH, hardness, sulfate and carbonate ion concentration and the presence of organic complexion agents [Oliveira et al, 2006]. In our study the spatial distribution of ^{210}Pb follows the same pattern as for ^{238}U and ^{230}Th (figure 4). These data prove that the investigated sources have almost no contact with the surface, mostly there has been no exchange between water horizons [Gonzalez-Labajo et al,2001] and the radio-nuclides originate from the surrounding rocks and soils.
The correlations among radio-nuclides and major ions were not significant, suggesting independence among the variables. However the analyzed waters were from a bicarbonate type; the main influence on the radionuclide distribution is due to chloride ions. Some explanation is related to the surprising fact that most of the waters contain non negligible iron and manganese concentrations.

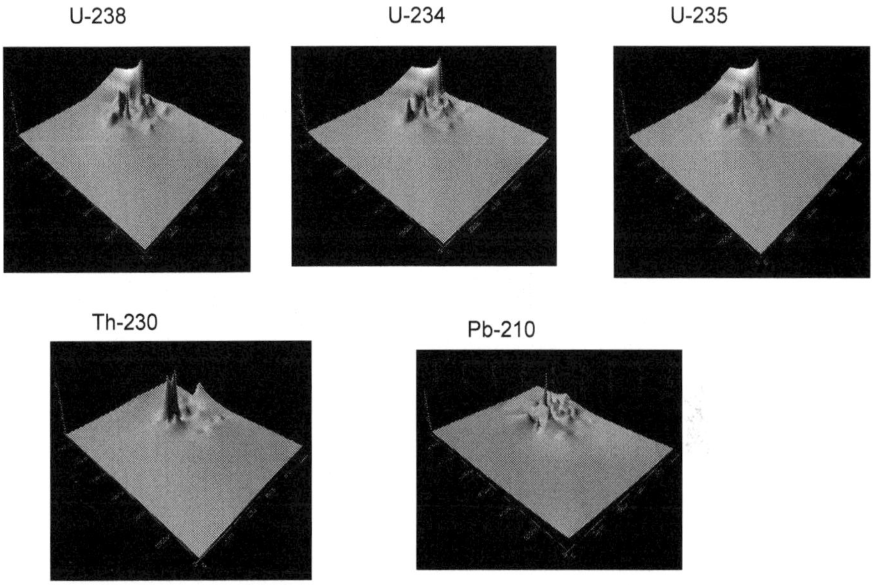

Fig. 4. Spatial distribution of uranium series isotopes over Luxembourg country. Activity concentration, represented on axe 'z' is in mBq/L

Correlation analysis show also no clear relation between calcium and magnesium content, however calcium is substituted well by strontium in almost all geological formations (figure 5). We obtain good correlation of strontium and radium radio-nuclides in Buntsandstein and Luxembourg sandstones formation where these form sulfates. In both formations reducing conditions exist in the deep wells.
In figure 6 are represented average values of main ions influencing radio-nuclide distribution over geological formations as well as the mean of different radioisotopes distributions over the same formations. The values are in 'ppm', except where for representation facility, they are converted in 'mg'. Radioactive elements are in 'mBq/L except radon that is in Bq/L. In Figure 6a are represented main cations and anions that may affect the solution of natural radio-nuclides, following the fact that generally Luxembourg

waters are bicarbonate. In Figure 6b, after detailed analysis of the data are represented the main influencing radioactive distribution ions. Strontium and sulfates affect the radium distribution, as discussed previously; whereas iron and its supported copper and manganese are responsible for uranium and thorium distributions.

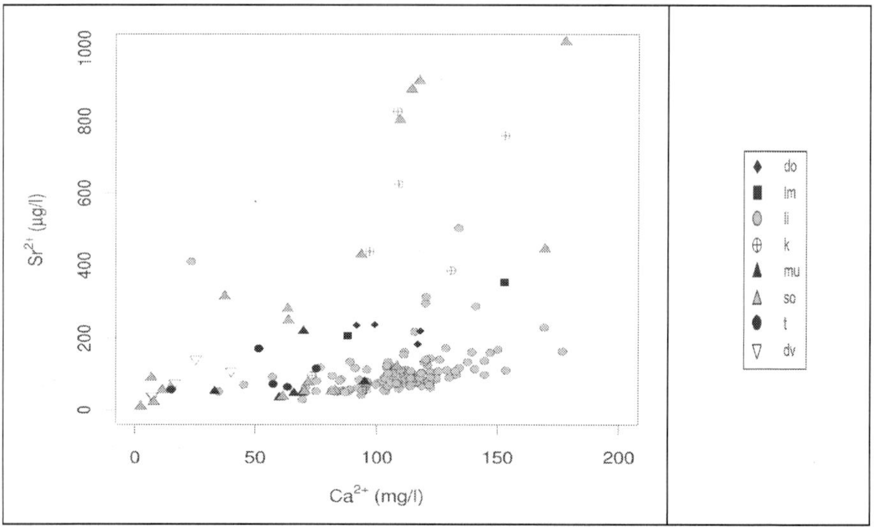

Fig. 5. Correlation between calcium and strontium content in analysed water vs. geological formation. The abbreviations are: do (dogger), km (keuper), li (liasic), mu (muschelcalc), so (buntsandstein 1), t (buntsandstein 2), dv (devonien)

In Figure 6c and 6d are represented the uranium and thorium series. The geological formations are also separated as for the Buntsandstein, where we observe a mixture of fissures and pores and permeability; besides an increasing mineralization from north to north-east. In the first ones we distinguish more sulfate and bicarbonate, while in the second more chlorides. This explains the low radium radio-isotopes and polonium contents in the bi-carbonate clay-sand and gypsum part, as well the presence of lead sulfate, and its dissolutions in mobile chlorides. Noticed is also the clear influence of iron and strontium in reducing conditions for the uranium and thorium presence. For similar reasons the sandstones of Luxembourg are presented into two different sub-formations. In the widely represented Luxembourg sandstones water circulation is favoured along vertically formed diaclases; the water table is charged continuously. Fractures are enlarged by water passages and by carbonate dissolutions; whereas the dissolved residue fraction in the sands constitutes a natural filter. As a consequence the reducing/oxidizing conditions change and induce the formation of sulfates where stable and radioactive thorium (80%) correlates with sodium, copper and manganese.

The $^{226}Ra/^{238}U$ ratios over the country range from the unique very low value of 0.3 to the values of 20, having a mean value not significantly different from 1. Similar is the range of ratios between ^{210}Po and ^{210}Pb. The highest observed values can be explained with the high solubility of radium and with polonium mobility in anoxic water.

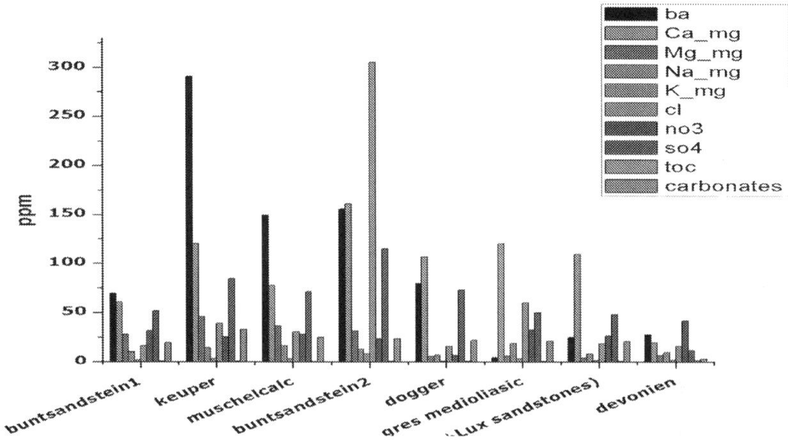

Fig. 6a. Distribution of major ions versus geological formation

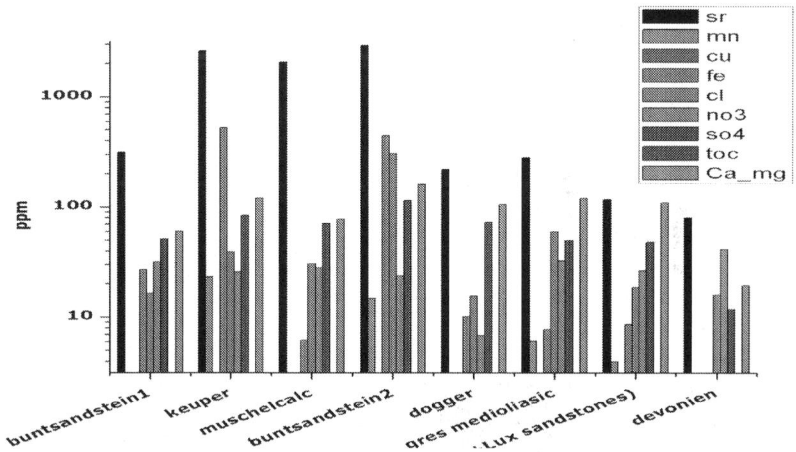

Fig. 6b. Spatial distribution of some influencing radio-nuclides occurrence ions versus geological formations

Analyzing the cross ratios $^{238}U/^{232}Th$ and $^{230}Th/^{232}Th$, it is observed that the first ones are higher than the second ones, but follow the same surface distribution. This fact again is explained with the specific chemistry of these radio-nuclides and should be attributed to the very low thorium solubility.

One can see that in many cases over the country, an opposite distribution between radon and radium is present. An analysis of the relation between radium and oxygen content shows higher radium associated with lower oxygen content. The initially anoxic water, when enriched in oxygen, deposits radium; this radium will produce elevated concentrations of radon in the source.

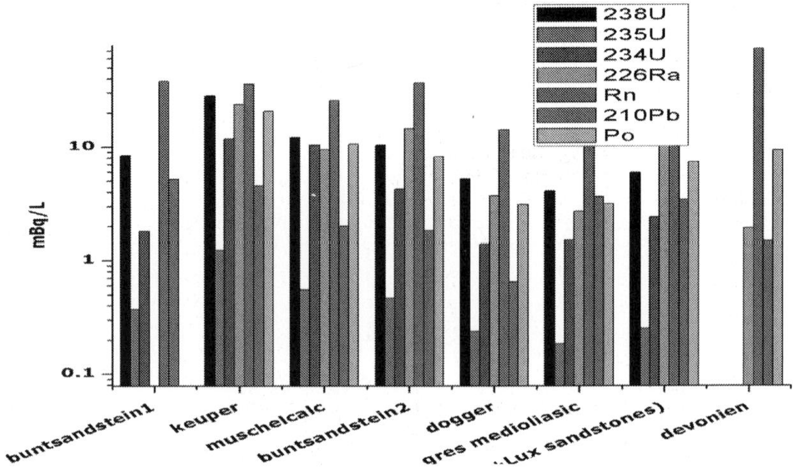

Fig. 6c. Uranium series isotopes distribution versus geology

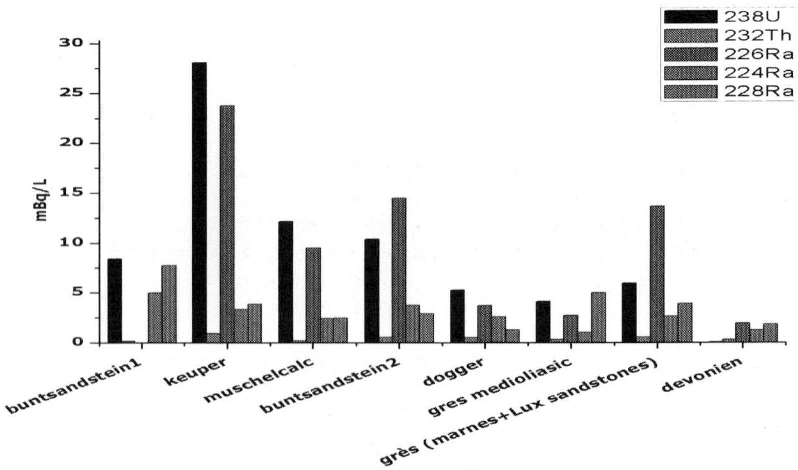

Fig. 6d. Distribution of main radio-nuclides versus geology

Another feature is the relatively high concentrations of ^{210}Po compared to the parent ^{226}Ra in the middle part of the country. One explanation is that, polonium product of ^{210}Pb is supplied to body waters from atmospheric input, run-off and the "in situ" decay of ^{226}Ra, its precursor in the water column. As the lead distribution follows the uranium one and it is 10 times lower than polonium the dependence of the ^{210}Po/^{226}Ra ratio with the radon should be investigated in location with high values. Naturally produced 210Pb can exhale from the soils and can get transported through the lower atmosphere. It is a long-lived decay product of the noble gas radon from the uranium decay series, which reaches the atmosphere and is transported with the atmospheric circulations in form of particles until deposition on the surface with the precipitations. Generally in keuper, muschelcalc and sandstone formations we observed some equilibrium between radon and radium. In these cases the 226Ra/210Po ratio is upper or near to 1. In the escape cases we had some spots of stable lead and it is related to Ba/Ca sulfates formations. Polonium values were high due to re-deposition of aerosols too.

In the devonien formation radon concentrations were higher than radium ones and 226Ra/210Po ratios lower than 1. In this case possible explanation of disequilibrium between polonium and lead were the exhalation to atmosphere processes.

For every analysed water sample the activity ratios $^{238}U/^{232}Th$ and $^{230}Th/^{232}Th$ are calculated. From the results is we obtain that these ratios may have significant values. In most cases, they can be attributed to the very low ^{232}Th concentrations and not to high ^{238}U. The mean ratio value of $^{238}U/^{232}Th$ for surface and soil infiltrated water is 2.4, whereas it exceeds 12 for underground water. It should be mentioned that in the investigated underground water with high $^{238}U/^{232}Th$ ratios, we remarked that the ^{210}Pb content is very low, and the ^{230}Th concentrations are twice as high as those of ^{232}Th [5]. This is associated with the higher solubility of uranium isotopes compared with thorium isotopes.

The obtained data also help to study the radioactive equilibrium disturbance within the uranium series. For each sample, the ratios $^{210}Pb/^{238}U$, $^{234}U/^{238}U$ are calculated. As expected for surface waters, the $^{238}U/^{234}U$ ratio is close to 1 and reveals a secular equilibrium (normal background). In the same samples, the lead content exceeds by more than 30 times the uranium content related to the normal hydrological cycle and atmospheric precipitations.

In all drilled wells the $^{234}U/^{238}U$ ratios range from the very low value of 1.3 to a value of 5, with a mean value of 1.8. A correlation with increasing depth or fault zones is presumed, which causes the irregularities.

As above -mentioned, the majority of the natural springs occurs in the outcropping region of the Luxembourg sandstone in the central part of the country. The normal background radon concentrations in water vary in the range of 2-60 Bq/L with an average value of 17 Bq/L. The distribution of the obtained data follows a positively skewed log- -normal dependence with the maximum at 15 Bq/L that is characteristic for deeper groundwater in this area. Only 3% of the obtained results are close to 100 Bq/L and are well explained with the high mineralization of these sources.

The comparison between indoor radon concentrations over the country and water radon concentrations show no direct relation. The reasons for this lack of relationship are not clear because soil air radon and, therefore, indoor radon concentrations are dependent on the geological substrates and their U-bearing minerals in general. It is supposed that the radon propagation in the water is more dependent of the local hydrology processes than from the geological structures.

The geographical distribution of radon shows two axes of elevated concentrations. One is located in the north-western region, followed by the second one in the central East of the country. They both are mainly situated on Lower Jurassic and Triassic formations. A radiation profile map of Luxembourg, prepared with aero-gamma measurements in 1995 [Geo-service of Luxembourg, 1997], also reveals that these parts of the country have increased radiation levels as a consequence of high geochemical distribution of 238U in the rocks of these formations. The test of radon dependency on main physical parameters, such as EC, temperature and pH, did not show any correlation for all sampled springs. The temperature of the waters varied in-between 8-16°C with average values in the range of 10-12°C. An anti-correlation of the radon and temperature in 65% of the cases is known from the literature, but this could not be observed in this case due to the small temperature fluctuations of the groundwater [Huyadi et al, 1999]. The same situation was found for the EC.

Radon concentrations in the north-central part show some differences in the 2003 and the 2007 campaigns. The latter campaign coincides with the recent construction of the northern highway in the time of this investigation. A hydro-geological and hydro-geochemical study of the Geological Survey of Luxembourg showed that the construction works influenced the flow regime and the hydro-geochemistry. A decreasing radon concentration could be observed in the same period. It is assumed that drilling and blasting works changed the flow path within the aquifer and, therefore, also changed the water chemistry. This was correlated with degreasing discharges in most of the monitored springs.

The spatial variations can be led back to the different geological materials and structural/tectonic features. Figure 6c shows that the radon activity changes with different geological units where the sources are located. Most aquifers in the region are double porosity aquifers, with proportion of matrix storage and fracture flow. Radon concentrations are influenced by the contact time between the water and the aquifer rock. Water from fast fracture flow is usually in contact with the rock matrix only for a short period of time and, therefore, cannot accumulate so much radon. On the other hand, radon with its short half-life reaches secular equilibrium after approximately 30 days. Hence, the groundwater flow should not influence its concentrations because groundwater flow is usually by magnitudes slower. Nevertheless, hydraulic conductivities of up to 100 m/d have been found along open fracture zones through tracer tests [Pistre et al, 2005], which support the results of this study.

The relation between radon and radium is of particular interest. Radon is a tracer, which can be used in a variety of groundwater studies, such as hydrograph separation, or any other type of groundwater/surface water interaction [Schmidt et al, 2009; Wu et al, 2004]. It is a daughter isotope of radium. Reasonable amounts of radium get into the hydrological cycle mainly through the dissolution of silicates and carbonates. It is not very mobile and can precipitate with other bivalent cation, such as Ba^{2+} or Ca^{2+}, as sulfates or carbonates. Nevertheless, considerable amounts of radium were measured in the sampled sources.

It is interesting to notice that the gamma airborne survey has shown a strong anti-correlation of radon and radium over Luxembourg, especially in the central-eastern part of the country. The radium concentrations in water are negatively correlated with radon. High levels of radium are associated with low levels of radon and vice versa. Radon activity in groundwater is controlled by the concentration of radium in the aquifer matrix [Vinson et al, 2009]. However, the amount of radon that can enter the groundwater depends on the emanating coefficient of the radium bearing material in the aquifer. This means that the radon emanation potential of the rocks is in some cases very low, especially for low permeable rocks. Then again, low radium content rock can nevertheless emit higher concentrations of radon, when their porosity and interconnected pore space is large. Both interpretations do not explain the negatively correlated behaviour in Luxembourg. The waters from the Devonian rocks, low porosity, show high concentrations in radon (Figure 6c), but low concentrations in radium, whereas the waters from the Lias and Buntsandstein, higher porosity, formations are low in radon, but high in radium. Hence, a possible interpretation is that in the first case the radon derives directly from emanating form the rock. The Devonian shits and sandstones are rich in uranium and radium, whereas the radon comes form dissolved radium in the water in the second case. The relation between radium and oxygen content shows higher radium concentrations associated with lower oxygen content. The initially anoxic water, poor in oxygen, mobilizes only a little amount of radium. If this water is deposited on the surface layer, it will produce elevated concentrations of radon in the springs.

Besides, a dense spatial resolution we analyze temporal variations of radon as well. The 316 springs from 2007-2008 campaign were sampled twice, once during the winter month, and once during the summer month and therefore during base flow conditions. Furthermore, five of the selected springs have been equipped with continuous flow and electrical conductivity (EC) devices and have been sampled for radon on monthly basis over a whole year (figure 7). Data from a previous campaign in 2003 are also used for comparison.

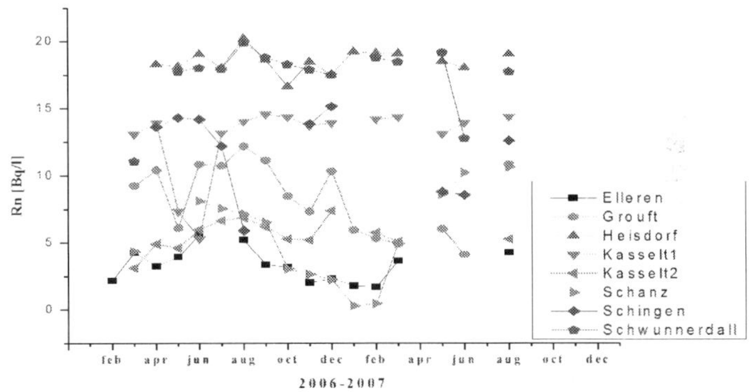

Fig. 7. Temporal variations of radon in sources of drinking water

Spatial heterogeneity was assumed in the rocks, with different correlation lengths. Long-scale effects were dominant for most of the sources.

The possible application of tritium is an essential tool for hydrological investigations. The most important applications are identifications of recharges in aquifers, estimation of hydraulic parameters related to pollutant transfer and determination of turnover time of groundwater. In the present work are analyzed for this isotope only 83 samples mainly for pollution investigation purposes this does not allow us to make significant conclusions. In all of them the concentrations are near to the capability of registration of used equipment. Analyses of all obtained data over these two campaigns are in progress of elaboration.

5. Conclusions

The spatial distributions of the radio-nuclides as well the major ions are driven by two kinds of forces, anthropogenic (agricultural land use, iron mining etc.) and geogenic Uranium / thorium series isotopes are integrated in the crystal lattice of minerals, e.g. as it is the case with uranium in zircons, or substitute positively charged major ions, e.g. the substitution of calcium and magnesium by radium. Depending of local site geo-morphology this substitution is guided by carbonates (trias, buntsandstein), sulphates and chlorides (deep water in keuper, muschelcalc, lias) or nitrates and iron oxide / hydroxides (Luxembourg sandstones diaclases, land-used trias part).

Radon (in this case ^{222}Rn) constitutes a particular case, as it is a noble gas and therefore inert. It is the daughter nuclide of Radium (^{226}Ra decays to ^{222}Rn) and accumulates in open crevasses in the rock matrix or gets dissolved in the groundwater.

As the distribution of the analysed sources is much more concentrated in the part of Luxembourg sandstones and alluvial formation, more of the results are reported to trias and inferior lias formations.
In medio-liasic sandstones and supra-liasic formations (Minette), due to iron mining, showed a marked uranium/thorium disequilibrium.
In Muschelcalc formation radon concentrations are low due to low permeability, increased levels are measured in sources due to the existence of fractures. In Luxembourg sandstones due to changing oxidizing/reducing conditions variable radon concentrations are observed.
Seasonal heterogeneities of radon are mainly relied to runoff production processes.
The response of radio-nuclides as tracers is more direct in sandstones compared to the schist in rain periods.
The investigation on spatial distribution dependency shows that the distribution in uranium series radio-nuclides, except of radium and polonium, are in correlation with local hydrogeology features.
Comparisons between two measurement campaigns are coherent and give us important information for the mechanism of distributions of the natural radio-nuclides in ground and surface water.

6. Acknowledgment

We want to thank the Water Service Department of Ministry of Interior and the Great Region of Luxembourg andespecially Dr. Denis Pittois form the Public Research Center Henri Tudor - CRTE for the chemical data and support. We would like to acknowledge the Service of Geology of Luxembourg for the guidelines and help during the analysis and whole project.

7. References

Colbach R (2005) Overview of the geology of the Luxem- bourg Sandstones. Scientific work of Museum of natural history of Luxembourg vol.44, pp. 155–160

EU Directive of drinking water (2000)

Tosheva Z., Taskaeva I., Kies A. (2006). International Journal of Environmental Analytical Chemistry, vol.86 No 9, pp.657-661

Tosheva Z., Hofmann H., Kies A. (2009). Journal of Radioanalytical Nuclear Chemistry, vol.282, pp.501-505

www.geology.lu

Berners (1983) Annales de la Société Geologique de Belgique 106:87-102

Porceli D., Swarzenski P (2003). Mineral Geochemistry, vol.52, pp.317-321

Oliveira J., Carvalho F (2006). Czechoslovak Journal of Physics, vol.56, No1, D545

Gonzales-Labajo J., Bolvar J.P., García-Tenorio R (2001). Radiation Physical Chemistry, vol.61, No3, pp.643-652

Geochemical map of G-D Luxembourg (1997). Geological Service Division of Luxembourg

Huyadi I., Csige I., Haki J., Baradacs E. (1999). Radiation and Measurements, vol.31, No116, pp.301-316

Pistre S., Marliak S., Joudre A., Bidoux P. (2005). Ground Watater, vol.40, pp.232-241

Schmidt A., Stringer Ch., Haferkorn U., Schubert M. (2009). Environmental Geology, vol. 56, No5, pp.855-863

Wu Y., Wen X., Zhang Y (2004). Environmental Geology, vol.45, No5, pp.647-653

Vinson D., Vengosh A., Hirschfeld D., Dwyer G. (2009). Chemical Geology, vol.260, No3/14, pp.159-171

Part 6

Minerology

17

Carbonate-Hosted Base Metal Deposits

Fred Kamona
University of Namibia
Namibia

1. Introduction

Carbonate-hosted base metal sulphide deposits are discussed with reference to ore deposit mineralogy and geology, isotope geochemistry, ore fluids and ore genesis of type examples. Three main carbonate-hosted deposit types are recognized: 1. Leadville-type mineralization (LTM); 2. Sedimentary-exhalative (SEDEX); and 3. Mississippi Valley-type (MVT) deposits. The three types constitute a distinct group of deposits characterized by orebodies within carbonate host-rocks containing Pb and Zn as the major metals with variable amounts of Cu and associated by-products, including Ag, Cd, Ge, V, Ga, As, Sb, Au and In.

LTM deposits form by magmatic-hydrothermal processes associated with igneous intrusions and correspond to the carbonate-hosted high-enthalpy type of Russell & Skauli (1991) and the chimney-manto type deposits of Hutchinson (1996). The SEDEX deposits are typically characterized by syngenetic to syndiagenetic processes of mineralization at the submarine surface within sedimentary basins prior to carbonate lithification and are spatially associated with faults located within, or at the margins of, the basins. They are also known as medium-enthalpy, Irish type or clastic dominated Pb-Zn deposits (Hitzman & Beaty, 1996; Russell & Skauli, 1991; R.D. Taylor et al., 2009). In contrast, typical MVT deposits (low-enthalpy type of Russell & Skauli, 1991) are epigenetic and form at the periphery of basins in lithified carbonate rocks spatially associated with faults, structural highs and facies changes. They are commonly referred to as MVT due to the occurrence of classic districts in the Mississippi River drainage basin of the USA (Leach et al., 2010). Both SEDEX and MVT deposits form from formation waters derived from sedimentary basins with high heat flows.

1.1 LTM deposits

LTM deposits are high temperature (>200°) carbonate-hosted Pb-Zn-Ag deposits distinctly different from MVT and SEDEX deposits by virtue of their temporal and spatial relationship to igneous intrusions. The fundamental diagnostic feature of this deposit type is its origin from magmatic-hydrothermal processes associated with felsic to intermediate magmatism and ore formation by replacement of carbonate rocks (Einaudi et al., 1981; Megaw et al., 1988; Thompson & Beaty, 1990). Although the type name originates in the Colorado mineral belt of the United States (Beaty et al., 1990), the type setting for these deposits is northern Mexico where the mineralization is hosted in carbonate-dominated portions of Jurassic-Cretaceous sedimentary sequences (Megaw et al., 1988).

The type of mineralization ranges from near-intrusion Cu and Zn-Pb skarns through distal Pb-Zn skarns to massive sulphide bodies hosted by limestone and dolostone cut by granite, quartz monzonite and other intermediate to felsic hypabyssal, porphyritic lithologies (Einaudi et al., 1981; Megaw et al., 1988). There is a continuous transition in most districts from skarn ores to massive orebodies beyond the skarn zones (Fig. 1).

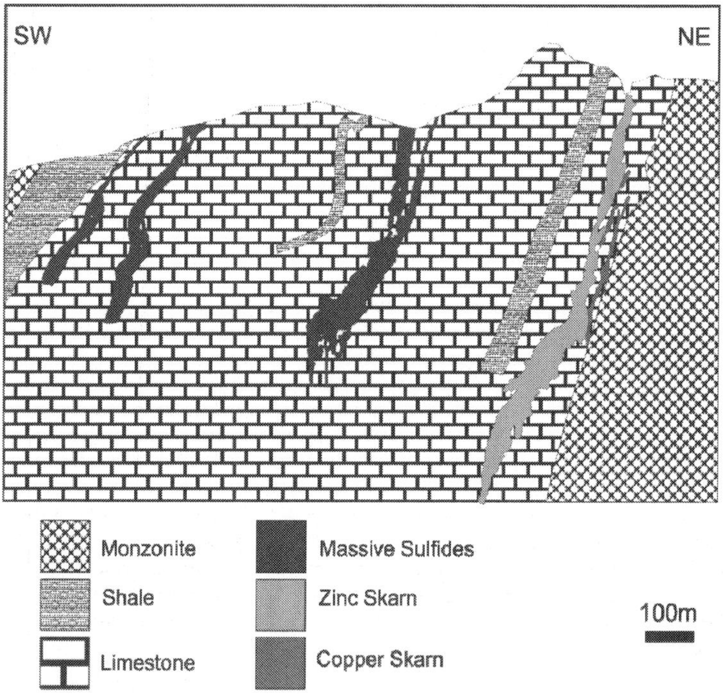

Fig. 1. Cross section of major orebodies of Providencia in Zacatecas, Mexico over a vertical interval of 1000m with concordant skarn and massive sulphide orebodies associated with a monzonite stock (modified from Megaw et al., 1988).

1.2 SEDEX deposits

Carbonate-hosted SEDEX Pb-Zn deposits are characterized by a stratiform morphology, an active tectonic setting during mineralization and low to moderate temperatures of formation (100° to 260° C). The Pb-Zn ores are associated with significant Ag and minor Cu (Hitzman & Large, 1986). The mode of occurrence of the mineralization includes cross-cutting and stratiform ores, such as in the classical Silvermines deposit in Ireland (Fig. 2), where epigenetic feeder zones occur below exhalative stratiform orebodies (Andrew, 1986a). The deposits are distinct from volcanogenic massive sulphide deposits because of a high Pb/Cu ratio and their lack of an intimate association with volcanic and volcaniclastic sequences (Turner & Einaudi, 1986). Some SEDEX deposits typically occur within marine shales or siltstones (Large, 1983) and are therefore distinct from carbonate-hosted Pb-Zn SEDEX deposits considered here.

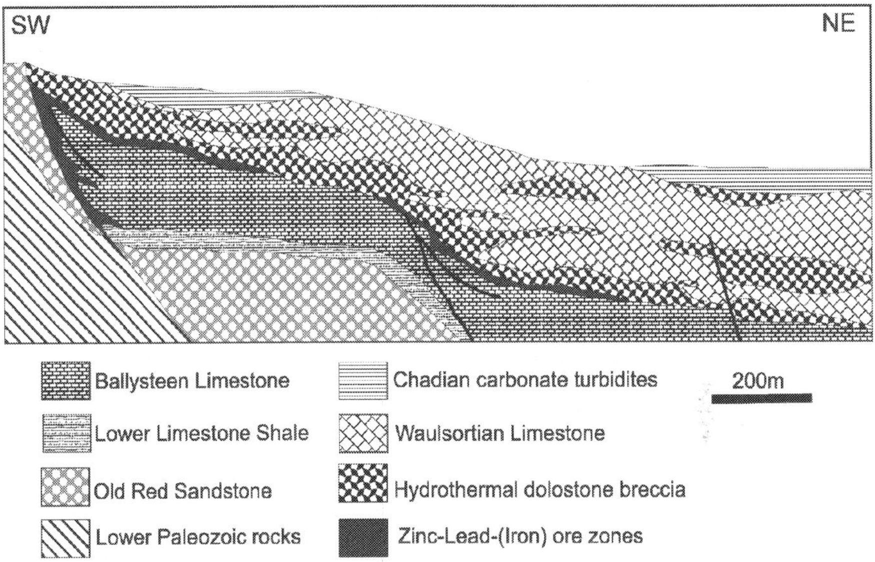

Fig. 2. Cross section of the Silvermines SEDEX deposit with an epigenetic feeder zone and stratiform ores (modified from Hitzman & Beaty, 1996).

The central Irish ore field may be considered as the type setting of carbonate-hosted SEDEX deposits. The ore field occurs in a first-order, intrashelf basin on the edge of the North Atlantic craton within Lower Carboniferous limestones (Andrew & Ashton, 1985; Hitzman & Large, 1986). The Alpine Triassic deposits of Central Europe are comparable to the Irish SEDEX deposits (Schneider, 1964; Sangster, 1976). They predominantly contain epigenetic replacement bodies and minor but extensive syngenetic stratiform orebodies in carbonate rocks (Schneider, 1964). Klau & Mostler (1986) regard the deposits as MVT based on the Pb-isotope data of Köppel (1983) which, according to them, shows no relation between magmatic rocks and mineralization. In addition, the ore lead is isotopically different from host-rock lead. However, according to Köppel (1983), basement rocks and upper Carboniferous to Permian magmatic rocks of the Southern Alps exhibit similar characteristics as the lead in the Triassic Pb-Zn deposits, whereas the lead of feldspars from Triassic volcanics shows significant differences. The syndiagenetic Pb-Zn deposits in Triassic carbonates could have obtained lead from lower crustal rocks, or material thereof, by leaching processes (Köppel, 1983).

1.3 MVT deposits

MVT deposits are stratabound, epigenetic orebodies that occur in clusters in carbonate formations of mineral districts distributed over large areas (Ohle, 1959, 1967; Heyl, 1967; Snyder, 1967). The orebodies are typically found at or near basin margins associated with domes and intrabasinal highs (Fig. 3). Like SEDEX deposits, MVT deposits are characterized by the absence of obviously associated igneous rocks, but the stratabound and often stratiform morphology of the former contrasts with the majority of MVT deposits whose morphology commonly crosscuts stratigraphy (Hitzman & Large, 1986; Sangster, 1990).

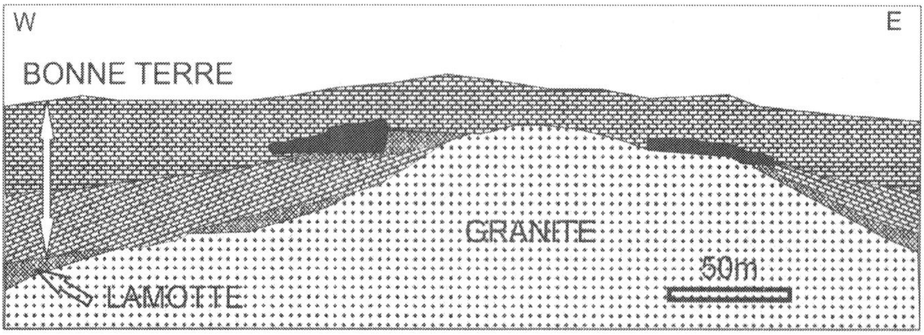

Fig. 3. Cross section through the Doe Run Mine, Southeast Missouri with ore zones (black) overlying a Precambrian granite "knob" (modified from Ohle, 1996).

Classical examples of MVT deposits are the districts of the mid-continent in the United States and the Pine Point district in Canada. However, they also occur in other districts, including the Appalachian in eastern United States, Silesia in Europe, southern and central Africa, and the Leonard shelf in Australia. They include districts varying from Zn- to Pb-dominant as well as Cu-rich deposits. With the exception of the Southeast Missouri district, which is dominantly Pb-rich, most districts are Zn-dominant and a few deposits (e.g. Kipushi and Tsumeb in Africa) contain significant amounts of Cu in addition to Zn and Pb.

2. Ore deposit geology

2.1 LTM deposits

This type of deposit is found on continental margins of orogenic belts and on the inner side of principal arcs where carbonate rocks representing miogeoclinal or stable platform depositional environments are widespread (Eunaudi et al., 1981; Sawkins, 1990). The deposits are associated with granodioritic to granitic magmatism which, together with the related metallogenesis in this environment, are believed to be a function of subduction (Einaudi et al., 1981; Bookstrom, 1990). In Mexico, the mineralized districts occur within or on the margins of a thrust belt with anticlinal ranges (the Sierra Madre Oriental). The thrust belt consists of Jurassic-Cretaceous sedimentary sequences of basal continental red beds with evaporites followed by a series of shale and carbonate facies which were deformed during the Laramide orogeny. The carbonate portions of these sequences are frequently more than 3 km thick and consist of limestone with minor dolomite (Megaw et al., 1988).

The mineralization-related intrusions range from batholiths and stocks to thin dikes and sills composed of diorite, granodiorite, quartz monzonite, monzogranite, quartz latite, and rhyolite (Einaudi et al., 1981; Megaw et al., 1988; Bookstrom, 1990). In Mexico the intrusions originated from the lower crust (Ruiz et al., 1988) and their age ranges from 47 to 26 Ma (Megaw et al., 1988). Mineralization occurred during a restricted interval of the Mid-Tertiary, well after evolution of the Laramide-aged Mexican thrust belt. Large deposits like Santa Eulalia and Naica grade outward from intrusion-associated mineralization to intrusion-free ores, suggesting that districts without intrusion relationships may not have been traced to their ends (Megaw et al., 1988).

The carbonate host rock may be predominantly limestone or dolomite and there are no consistent links between ore deposition and carbonate composition, facies, organic content, or insoluble components (Megaw et al., 1988). Many deposits contain mineralization in carbonate strata within or below relatively less permeable rocks such as cherts, shales, slates and volcanic rocks, and the formation of large orebodies appears to be related to structural enhancement of permeability in the carbonate rocks (Yun & Einaudi, 1982; Megaw, et al., 1988). Complex fold and fault structures are the dominant controls in strongly folded areas, whereas simple fault-related structures are more common on fold belt margins and radial and/or concentric structural patterns related to intrusions are important controls in domal areas (Megaw et al., 1988). The controlling structures apparently acted as conduits for channeling mineralizing fluids to sites of ore deposition.

The orebodies commonly display a combination of forms that include mantos, chimneys, pods and veins (Einaudi et al., 1981; Megaw et al., 1988; Beaty et al., 1990) and are temporally and spatially associated with igneous intrusions (Fig. 1). The change in orebody morphology is caused by local variations in stratigraphy, structural patterns, travel distance of hydrothermal fluids, and depth of intrusive emplacement (Megaw et al., 1988). Podiform skarn and sulphide bodies are the deepest modes of occurrence and are irregular or equant with no particular orientation or relationship to stratigraphy. They commonly occur along intrusive, fault, and lithological contacts (Megaw et al., 1988). Typical vein deposits include Bluebell in Canada and Uchucchacua in Peru (Einaudi et al., 1981). Vein mineralization is also dominant in some major skarn deposits such as San Martin (Rubin & Kyle, 1988).

Tonnage and grade data (Megaw et al., 1988) for skarn, chimney and manto ores in seventeen districts of Mexico show a wide range of values (Fig. 4). The ore districts have an average of 11.4 million tonnes (Mt) at 11.0 wt.% Pb, 8.5 wt.% Zn, 0.7 wt.% Cu, 243 ppm Ag and 1.8 ppm Au. Santa Eulalia is the largest district with 50 Mt of ore grading 5% wt. Pb, 6.7 wt.% Zn, 0.1 wt.% Cu and 242 ppm Ag. Other major ore districts include Providencia-Conception del Oro (25 Mt, 2 to 20 wt. % Pb+Zn, 30 to 500 ppm Ag) in Mexico (Megaw et al., 1988), and Leadville (24 Mt, 7 wt. % Pb+Zn, 320 ppm Ag) in Colorado (Beaty et al., 1990).

With the exception of the San Carlos and Los Lamentos manto districts which are Pb-rich and contain no Zn, the average Zn/(Zn+Pb) ratio of 0.5 in the other Mexican deposits indicates equal concentrations of Pb and Zn in most of the orebodies. However, Cu values are generally low (0.7 wt. %) and only range from 0.1 to 1.6 wt % Cu. The gold grades are also typically low and ranges from 0.2 to 6 ppm Au with an average of only 1.8 ppm Au for the 9 Mexican districts for which data is available (Megaw et al., 1988). In contrast, the Ag grades are high in the majority of deposits with an average of 243 ppm Ag and a range from 14 to 600 ppm Ag with most deposits (14 out of 17) having grades >100 ppm Ag.

2.2 SEDEX deposits

The central Irish ore field occurs in a first-order, northerly transgressive intrashelf basin on the edge of the North Atlantic craton (Andrew & Ashton, 1985). All the known major deposits lie along basin margins adjacent to historically active basement fault zones. The major stratiform deposits at Navan, Silvermines and Tynagh occur in early Carboniferous stratigraphic successions of shallow water carbonate sediments and argillites, and deep water Waulsortian limestones overlain by carbonates and argillites deposited in latest Courceyan to Chadian stages (Philips & Sevastopulo, 1986). The carbonate succession is underlain by basal Devonian sandstones (the Old Red Sandstones) which diachronously overlie a Precambrian-Lower Paleozoic basement.

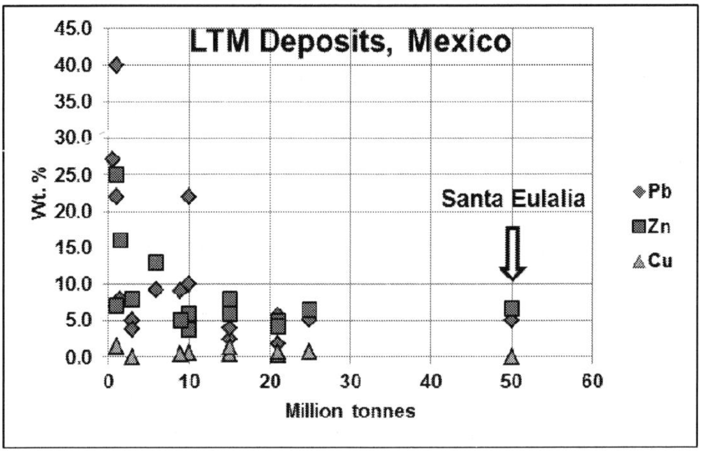

Fig. 4. Tonnage and grade data for 17 Mexican districts with skarn, chimney and manto ores based on data from Megaw et al. (1988).

There is no obvious spatial association between igneous rocks and ore deposits at Silvermines, Tynagh and Keel (Morrissey et al., 1971). A late Caledonian, post-tectonic syenite with a model lead isotopic age of 400 Ma intrudes Lower Paleozoic rocks beneath the Navan deposit, but it is not regarded as a mineralizing pluton as it is older than the mineralization age of 366 Ma (Andrew & Ashton, 1985; Ashton et al., 1986). Although thin pyroclastic beds occur at the same stratigraphic level as the most mineralized beds at Tynagh, the pyroclastic material is considered to have been derived from the Limerick volcanic centre 70 km away (Morrissey et al., 1971).

The mode of occurrence of SEDEX orebodies includes cross-cutting, epigenetic veins and breccia ores commonly regarded as feeders to overlying tabular stratiform lenses. Such a feeder-exhalative system is best developed at Silvermines (Fig. 2) where the upper stratiform ores occur at the base of and within Waulsortian carbonates (Andrew, 1986a). The mineralization is preferentially enriched immediately below siltstone and shale bands as well as in dolomitization voids and fracture systems. The thickness and distribution of the stratiform ore is related to the paleotopographic control of knolls of Waulsortian micrites (S. Taylor, 1984). Navan represents a different style of mineralization in that the orebody consists of five vertically superimposed stratiform lenses with cross-cutting veins and breccia zones within Lower Carboniferous limestones (Andrew & Ashton, 1985; Ashton et al., 1986). The mineralization is consistently restricted to non-argillaceous units, being generally best developed in micritic, oolitic, pelloidal or slightly arenaceous carbonate beds.

SEDEX deposits have a spatial relationship to fault structures active during mineralization, and, in the case of Silvermines and Tynagh, during host rock deposition (Hitzman & Large, 1986). All the ore zones at Silvermines are closely related to WNW or westerly faults (Andrew, 1986a). At Navan, the mineralization is grossly located adjacent to fault intersections on the flank of an anticline (Ashton, et al., 1986). The main orebody at Tynagh was a subhorizontal wedge that extended updip of the Tynagh fault in two elongated zones separated by a barren dolomitized reef (Boast et al., 1981a; Clifford et al., 1986). In addition, a residual deposit of oxidized and unoxidized mineralization occurred in the hanging wall of the fault in a deep post-karstification trench.

Tonnage and grade data for 28 carbonate-hosted SEDEX deposits (Fig. 5) from the data bank of Goodfellow & Lydon (2007) shows that the average deposit contains 8.9 Mt grading at 6.74 wt. % Zn, 2.51 wt. % Pb and 63 ppm Ag. The deposits are typically Zn-rich with a Zn/(Zn+Pb) ratio of 0.71.

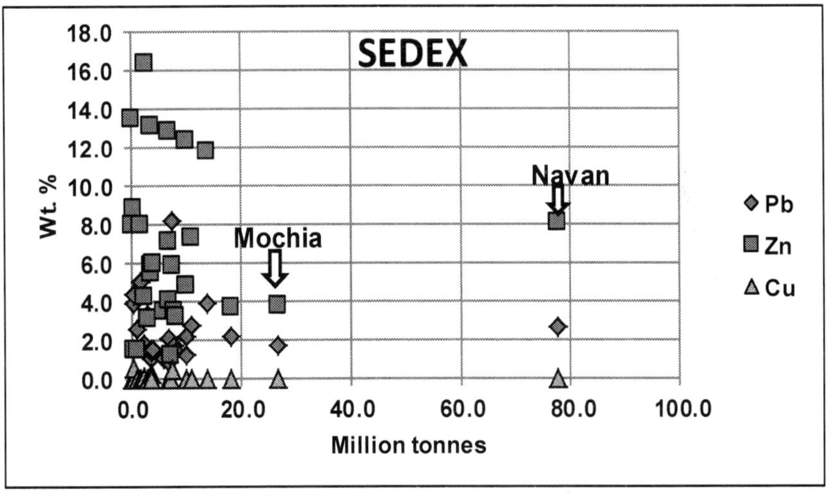

Fig. 5. Grade tonnage plot for carbonate-hosted SEDEX deposits based on data from Goodfellow & Lydon (2007).

With the exception of Navan (77.8 Mt at 8.04 wt. % Zn and 2.68 wt. % Pb) and the Palaeoproterozoic Mochia deposit in India (26.8 Mt at 3.79 wt. % Zn and 1.69 wt. % Pb) most carbonate-hosted SEDEX deposits contain less than 20 Mt of ore (Fig. 5). The Cu content of these deposits is also very low with an average of only 0.30 wt. % Cu in the four deposits for which Cu data is available. The Mehdiabad deposit in Cretaceous dolostone in Iran contains exceptionally large geological resources estimated at 218 Mt with 2.3 wt. % Pb, 7.2 wt. % Zn and 51 ppm Ag (Goodfellow & Lydon, 2007).

2.3 MVT deposits

MVT Pb-Zn deposits have low temperatures (100° to 150° C), uniform salinity, density and composition (Roedder, 1967, 1984) and typically occur in shallow-water platform carbonates peripheral to intracratonic basins (Sangster, 1990). The classical districts of the Upper Mississippi Valley, Tri-State and Southeast Missouri are found within the stable continental interior of the North American craton, whereas the Appalachian district occurs in an orogenic belt on the eastern margin of the continent (Hoagland, 1967, 1976; Thacker & K.H. Anderson, 1977). The mineral districts are distributed over thousands of square kilometres (Ohle, 1967) and the carbonate-hosted ore deposits occur over a stratigraphic interval of 200 million years ranging from Cambrian to Mississippian-Pennsylvanian (Ohle, 1980).

MVT orebodies are stratabound in essentially horizontal carbonates, chiefly dolomites and dolomitic limestones. They are epigenetic and typically occur within breccia bodies below unconformities (Callahan, 1967; Sangster, 1988). Zones of high permeability such as channels and troughs in algal reefs, collapse breccias, pinchouts, facies changes, bedding planes, faults and fractures are the focus of mineralization, particularly along flanks or on crests of basement highs (Fig. 3) or below unconformities. The ore-hosting breccias include discordant domes and columns that may be interconnected to concordant breccias. The breccia tops are vuggy and porous and contain mainly open-space filling mineralization while replacement mineralization is dominant near the base of the breccias (Rogers & Davis, 1977).

The regional distribution of the Southeast Missouri orebodies closely follow the pinchouts of the Lamotte Sandstone (Kisvarsanyi, 1977). Although algal reefs are also spatially associated with the mineralization, the reefs themselves are heavily mineralized only in some parts of the Viburnum Trend, and they are barren in many places. Faults seem to have been important in the Old Lead Belt where fault zone ores provided most of the lead (James, 1952). In the Tri-State district the ore occurs as elongate tabular bodies ("runs") of breccias up to 1 km or more in length which may form circular map patterns (Hagni, 1976). The dominant ore-bearing collapse breccias developed by solution enlargement of fractures along silicified dolomite-limestone contacts. In addition, flat "sheet" or blanket ore occurs in partly broken, stratified chert bodies. In the Upper Mississippi Valley the ore occurs dominantly as open-space filling in pitch-and-flat structures along bedding planes, joints and faults (McLimans et al., 1980; Heyl, 1983). The orebodies are linear, arcuate, or elliptical in plan and vary in length from 0.4 to 2 km. In the Appalachian district the ores are associated with karst breccias or reefs in dolomitic limestone at or adjacent to a dolomite-limestone interface (Hoagland, 1976).

The widespread regional distribution of MVT deposits is also observed in other districts, including the Otavi Mountainland in Namibia where regional scale ore fluid migration is indicated by the presence of Pb–Zn occurrences over 2500 km^2 within stratabound breccias of the Elandshoek Formation (Kamona & Günzel, 2007) as well as in central Africa (Fig. 6) where numerous prospects are associated with a few Neoproterozoic economic deposits such as Tsumeb, Kipushi, Kabwe and Berg Aukas (Kamona & Friedrich, 2007; Kampunzu et al., 2009).

Most MVT deposits are Zn-dominant as indicated by the Zn/(Zn+Pb) ratio of 0.7 in individual orebodies and districts. However, some deposits like the Viburnum No. 27 mine with 8 Mt grading 2.9 wt. % Pb and 0.2 wt. % Zn (Grundmann, 1977) and the Buick mine with >50 Mt, 8 wt. % Pb and 2 wt. % Zn (Rogers & Davis, 1977), indicate the Pb-rich nature of these ores. In contrast, sphalerite is the only ore mineral in Central Tennessee which contains 20 Mt with 5 wt. % Zn (Kyle, 1976).

Tonnage and grade data for 23 MVT deposits (Fig. 7) from Canada (Paradis et al., 2007) and Africa (Kamona & Günzel, 2007; Kamona & Friedrich, 2007; Kampunzu et al., 2009) indicate that the average MVT deposit contains 13 Mt with grades of 7.2 wt. % Zn and 3.4 wt. % Pb. The average Zn/(Zn+Pb) ratio of 0.69 is similar to that of carbonate-hosted SEDEX deposits (0.71). Copper averages 3.2 wt. % in nine of these deposits for which Cu grades are available with significant Cu grades in the Kipushi (8.0 wt. %), Tsumeb (4.42 wt. %) and Khusib Springs (10.06 wt. %) deposits. Silver contents can be considerable with a maximum of 584 ppm in the Khusib Springs deposit and an average of 96.8 ppm Ag in 11 of the deposits considered here.

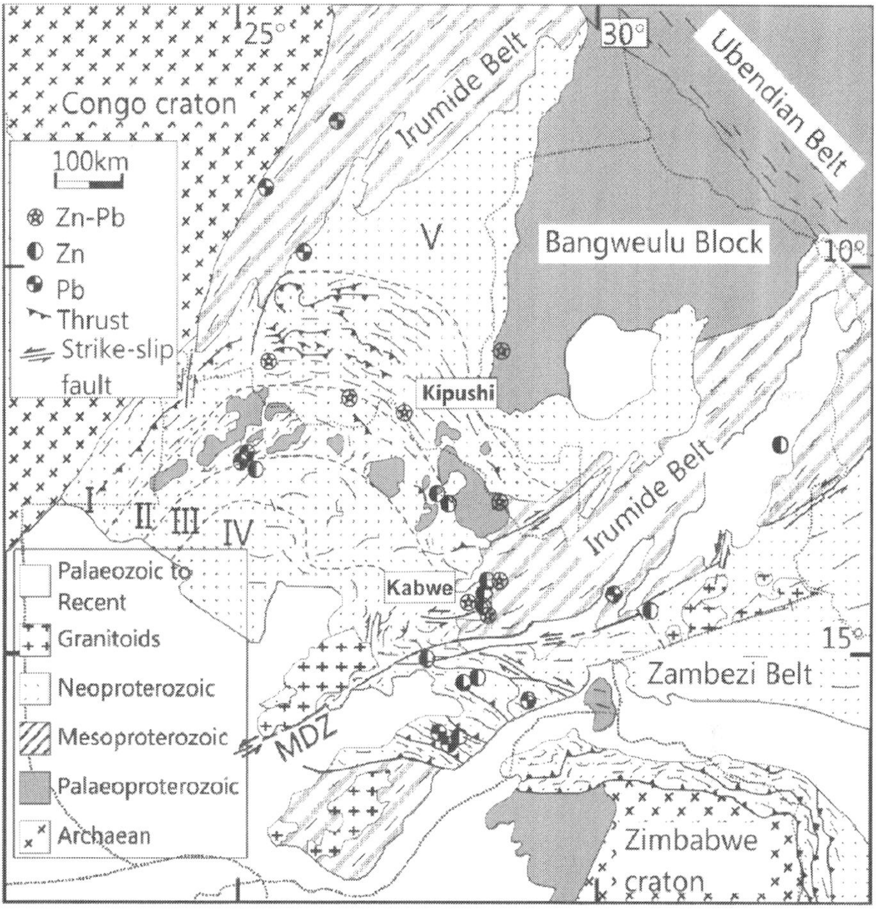

Fig. 6. Location of the Kipushi and Kabwe Pb-Zn deposits and Pb-Zn occurrences in the geotectonic framework of central Africa (modified from Kampunzu et al., 2009). I. External fold and thrust belt, II. Domes region, III. Synclinorial belt, IV. Katanga high, and V. Kundelungu aulacogen or palaeograben. MDZ (Mwembeshi Dislocation Zone).

The Kipushi (Democratic Republic of the Congo) and Gayna River (Canada) deposits are examples of large deposits of this type, each with 50 Mt of ore. The Zn-Cu dominant Kipushi deposit is characterised by zones of Cu, Zn+Cu and Zn+Pb along its length (Kampunzu et al. 2009). The 80 Mt Pine Point district, which contains 100 individual orebodies (Paradis et al., 2007) is included for comparison (Fig. 7). The data of Leach et al. (2005, 2010) and Paradis et al. (2007) indicate that MVT districts with carbonate-hosted orebodies in the USA and Canada contain an average of 233 Mt with grades of 3.3 wt. % Zn and 2.2 wt. % Pb with a Zn/(Zn+Pb) ratio of 0.72 (n=10).

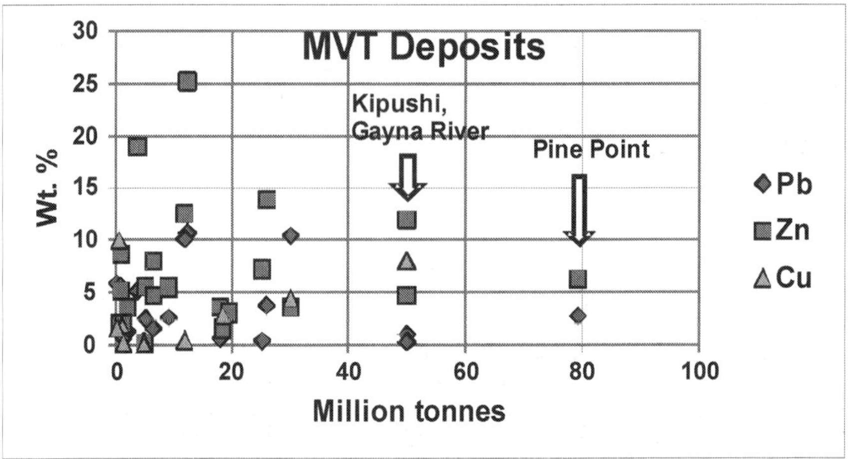

Fig. 7. Grade and tonnage plot for carbonate-hosted MVT deposits in Canada and Africa based on data from Kamona & Günzel (2007), Kamona & Friedrich (2007), Kampunzu et al. (2009) and Paradis et al. (2007).

3. Mineralogical and geochemical features

3.1 LTM deposits

The principal sulphides are galena and sphalerite with variable amounts of pyrite, pyrrhotite, marcasite, chalcopyrite and arsenopyrite (Einaudi et al., 1981; Megaw et al., 1988; Thompson & Arehart, 1990). The massive sulphide orebodies commonly exhibit banding due to textural, size or mineralogical differences between bands and they are composed of more than 65 wt. % sulphides. The sulphides usually exhibit a zonal distribution with galena being more abundant than sphalerite in mantos, but the latter is more abundant in chimneys. The chalcopyrite content may increase with depth, and districts that contain both pyrite and pyrrhotite show an increase of pyrrhotite with depth or an increase of pyrite with time.

Tetrahedrite-tennantite, chalcopyrite and arsenopyrite commonly occur near skarns at deep levels, and near mineralization-related intrusions. Acanthite, cinnabar, stibnite, realgar and silver sulfosalts are most common in the peripheral parts of districts. In addition, some skarns also contain bornite, covellite, molybdenite, scheelite, powellite, cassiterite, magnetite, and hematite. Silver occurs in solid solution in galena in most Mexican deposits (Megaw et al., 1988), whereas galena from the Leadville district is devoid of Ag, Sb and Bi, and the source of silver is tetrahedrite and electrum with the former occurring partly as disseminations within galena (Thompson & Arehart, 1990).

The gangue is dominated by carbonates, fluorite, or quartz with or without minor barite and anhydrite. Calc-silicates may be locally present and make up less than 5 % of the mineral content (Megaw et al., 1988). The main minerals of the skarns are pyroxene and garnet with associated wollastonite, bustamite, rhodonite, dannemorite, idocrase, olivine, ilvaite, chlorite, Mn-actinolite, epidote, fluorite and cummingtonite. Skarns are typically zoned from andraditic garnet to hedenbergitic-johannsenitic pyroxene to wollastonite-bustamite to marble (Einaudi et al., 1981; Shimizu & Iyama, 1982).

The contacts between mineralization and unaltered carbonate wall rocks are sharp, and where alteration exists it is variable in extent and may appear as zones of disseminated Mn-oxide mineralization, hydrothermal dolomitization and/or recrystallized carbonates (Megaw et al., 1988). Silicification or jasperoid formation varies from poorly to strongly developed peripheral to mineralization in some ore districts, but it may be totally lacking in other districts. Isotopic depletion halos of C and O surround orebodies at La Encantada and Santa Eulalia (Megaw et al., 1988). C and O isotopes in calcite veinlet stockworks above mineralization, and limestone wallrocks adjacent to mineralization show shifts to lighter values.

3.2 SEDEX deposits

Sphalerite is the major sulphide with lesser galena, pyrite and marcasite (S. Taylor & Andrew, 1978; Boast et al., 1981a; Andrew & Ashton, 1985; Andrew, 1986a; Ashton et al., 1986). In addition, tennantite is of economic importance at Tynagh (Boast et al., 1981a). Common accessories include chalcopyrite, bornite, pyrrhotite, arsenopyrite, freibergite, pyrargyrite, boulangerite, bournonite, semseyite, jordanite and cylindrite. Principal gangue minerals are barite (except at the Magcobar barite orebodies of Silvermines), calcite, dolomite, siderite, and quartz with fluorite as an accessory. In the upper stratiform ores at Silvermines Ag is directly proportional to Pb due to inclusions of boulangerite and jordanite in galena (S. Taylor & Andrew, 1978). In addition, an isolated lens of rich silver mineralization in barite contains prominent patches of discrete silver minerals such as proustite, xanthoconite, smithite, miargyrite, argentite-acanthite, and argyrodite as well as gersdorffite (S. Taylor, 1984).

The lateral zonation at Silvermines and Tynagh is Cu-Pb-Zn-Mn with Mn forming an extensive aureole of up to 7 km in the carbonates at Tynagh (Russell, 1978). The Magcobar barite body and associated pyrite-rich zones occur on the fringes of the stratiform sulphide ores, whereas the iron-oxide facies of the Iron Formation is lateral to the sulphide mineralization (Andrew, 1986a; Caulfield et al., 1986). Vertical zoning in feeder zones is characterized by increasing Pb/Zn ratios which decrease slightly in the upper parts of the ores as in the feeder zones at Silvermines (S. Taylor, 1984). Zoning patterns at Navan are complex due to variations in the metal content of deposited sulphides with time, resulting in superposition of differing zoning trends (Andrew & Ashton, 1985). Mn, As, Zn and Pb values are irregular, but they form an intensive halo over the mineralization at Navan.

Dolomitization and silicification of Waulsortian micrites is associated with the mineralization at Silvermines, Tynagh and Navan. At Silvermines, dolomitization occurred before silicification of the host rock and precipitation of laminated cherts which in turn preceded sulphide deposition (Andrew, 1986a). Dolomitization at Tynagh occurred before and during mineralization (Boast et al., 1981a; Clifford et al., 1986). In addition, alteration aureoles containing Na and K feldspars surround epigenetic Cu-Pb-Ba veins which represent the final stage of sulphide mineralization (Boast et al., 1981a). According to Clifford et al. (1986), silicification related to dolomitization post-dates the Cu-Pb-Ba veins. In rocks of the Navan Group, zones of complex dolomitization involving both ferroan and non-ferroan dolomite, as well as more coarsely crystalline dolomite, occur only in mineralized areas (Hitzman & Large, 1986).

Depletion in the heavier ^{13}C and ^{18}O isotopes was observed in mineralized limestones, ore-stage carbonates and post-ore carbonates with respect to unmineralized Waulsortian limestone at Tynagh (Boast et al., 1981a). The depletion in ^{13}C and ^{18}O is a result of isotopic exchange between host rock and mineralizing fluids. In general, the unmineralized micrites and diagenetic calcites define a heavy isotope end-member, whereas the post-ore carbonates tend to be most depleted in ^{13}C and ^{18}O. Carbonates coeval with sulphide mineralization have intermediate $\delta^{13}C$ and $\delta^{18}O$ values.

3.3 MVT deposits

Sphalerite is the main ore mineral of most MVT districts except Southeast Missouri where galena is more dominant. The gangue generally consists of dolomite, calcite, quartz, marcasite and pyrite (Hagni, 1976, 1983; Hoagland, 1976; Hagni & Trancynger, 1977; McLimans et al., 1980). Chalcopyrite and several Co-Ni sulphides such as siegenite, fletcherite, bravoite, vaesite and polydymite are abundant in Southeast Missouri (Jessey, 1983). Barite and/or flourite are associated with the mineralization in the Appalachian, Tri-State and Upper Mississippi Valley districts. Sphalerite in the upper Mississippi Valley is characterized by colour banding which can be correlated stratigraphically over several kilometers (Maclimans et al., 1980). Sphalerite in Appalachian deposits also shows fine laminar to single crystal banding with colour variations of light and dark zones (Craig et al., 1983). The dark zones contain numerous inclusions of fluid, vapour and solids, and the darkest bands are rich in hydrocarbons and Cd.

The paragenesis of the ores are complex and variable from deposit to deposit and usually consist of several generations due to repetitive sulphide formation over a long period of time (Hagni & Trancynger, 1977; McLimans et al., 1980; Sverjensky, 1981; Hagni, 1983). Pyrite and marcasite are generally early, followed by chalcopyrite, sphalerite and galena, in that order. According to Jessey (1983), the Ni-Co-Cu minerals in Southeast Missouri preceded Pb-Zn mineralization.

Dolomitization, silicification, recrystallization and associated isotopic changes in the host carbonates are the most significant alteration effects in MVT deposits. In the Southeast Missouri district, an early porosity network was enlarged through the corrosive action of the ore fluids with pre-ore dolomite (Type I) to produce Type II and Type III end-member dolomite compositions (Frank & Lohmann, 1986). The isotopic composition of the host dolomite exhibits trends from Type I dolomite to Type II (depletion of both $\delta^{13}C$ and $\delta^{18}O$ isotopic ratios) and Type III (depletion of $\delta^{13}C$ and enrichment of $\delta^{18}O$) dolomites. The isotopic trends suggest a fluid dominated system in which the host rock was pervasively altered in areas of fluid flow to compositions in equilibrium with mineralizing fluids.

The C and O isotopes of host rock limestones and dolomite as well as of gangue calcite in the Upper Mississippi Valley define a trend extending from the isotope ratios of unaltered dolomite, through those of limestone and dolomite with decreased $\delta^{13}C$ and $\delta^{18}O$ values, to the significantly lower values for calcite deposited late in the ore paragenesis (Sveryensky, 1981b). The significant decrease of $\delta^{13}C$ and $\delta^{18}O$ values toward the orebodies is attributed to recrystallization under conditions of progressively higher water to rock ratios near the orebodies.

In the Tri-State district the close association of ore with areas of coarsely crystalline dolomite constitutes the single most important guide to ore (Hagni, 1976). Most orebodies are located along one or both margins of dolomitic zones that are probably dolomitized bioherms (Hagni, 1982). Jasperoid with disseminated sulphides forms a zone around many dolomite zones and its lateral and vertical extent essentially coincides with sulphide mineralization. Recrystallization of limestone occurs for distances of 13 to 27 m beyond the margin of silicification.

Although minor local variations may exist, the following set of minerals is typical of massive Zn–Pb sulphide ores in the low grade metamorphic zone of the Copperbelt of Central Africa: sphalerite+pyrite+galena+chalcopyrite, with subsidiary ±arsenopyrite, ±gallite, ±tennantite, ±briartite, ±reniérite. Variations in the relative proportions of the three important sulphides (pyrite, galena and sphalerite) result in the identification of massive pyrite ore, galena-rich and sphalerite-rich ores, respectively (Kampunzu et al., 2009). In the Otavi Mountainland mineralogically simple Zn-Pb-dominated ores comprising sphalerite, galena, pyrite, minor chalcopyrite, bornite, colusite, reniérite (such as the Berg Aukas deposit) are distinguished from polymetallic ores of the Tsumeb-type which contain Cu, Pb, Zn and As, as well as trace elements such as Ag, Cd, Ga, Ge, Mo, and Sb. The sulphide assemblages in the latter ores are composed of variable amounts of galena, tennantite, chalcopyrite, sphalerite, chalcocite, enargite, bornite, pyrite, minor germanite, reniérite, briartite, Ge-bearing colusite, and Mo–W sulphides (e.g. Melcher et al., 2006; Kamona & Günzel, 2007).

4. Ore fluids

4.1 LTM deposits

These deposits are characterized by formation temperatures from 200° to 500° C (Megaw et al., 1988) as well as high chlorine concentrations of up to 21.2 wt. % Cl in fluid inclusions at Providencia (Rye & Haffty, 1969). In addition, Rye & O'Neil (1968) measured δD values in the range -68 and -83 ‰ and $\delta^{18}O$ from 5.8 to 6.2 ‰. At Leadville the ranges are from -45 to -70 ‰ δD and 5 to 8 ‰ $\delta^{18}O$, respectively (Thompson & Beaty, 1990). Similar values of δD= -45 to -60 ‰ and $\delta^{18}O$= 4 to 9 ‰ have been reported for the Gilman deposit (Beaty, ed. 1990). These values are comparable to those for waters which have equilibrated with magmas (Sheppard, 1986).

In general, the metasomatic fluid of skarn and massive sulphide ores has a low CO_2 content (X_{CO2} < 0.1) and moderate to high salinities (10 to 45 wt. % NaCl equivalent) (Einaudi et al., 1981). Fluid inclusion data from host minerals coeval with sulphide deposition or skarn formation (fluorite, sphalerite, quartz, pyroxene and garnet) indicate homogenization temperatures in the range 200° to 550° C and salinities with a total range from 1 to 60 wt. % NaCl equivalent (Einaudi et al., 1981; Megaw et al., 1988). The extreme variation in salinity may result from various mixtures of highly saline magmatic fluids and later circulating groundwaters (Roedder, 1984). Hotter (400° - 650° C), more saline (>26 wt. % NaCl equivalent) solutions are typical of the skarn zones, whereas the massive sulphide ores were deposited at temperatures between 200° and 450° C and variable salinities. Boiling appears to be characteristic of shallower environments (0.3 to 1 kb).

4.2 SEDEX deposits

Detailed fluid inclusion studies have been carried out at the Silvermines deposit (Samson & Russell, 1983, 1987). The mineralizing fluids have high Na concentrations with lesser, more variable K and Ca, and uniformly low Mg concentrations. Fluid inclusion waters in quartz, dolomite, sphalerite, galena and barite have δD values in the range of -23 to -58 ‰. The calculated $\delta^{18}O$ values of the mineralizing fluid, based on mineral values for quartz and dolomite, range from 1.1 to 7.7 ‰. The data represent aqueous fluids that equilibrated with the geosynclinal sequence and granites underlying the deposit (Samson & Russell, 1987).

The δD and $\delta^{18}O$ values of the ore fluid at Silvermines partly overlap that of both MVT and LTM ore fluids.

Homogenization temperatures and salinities in SEDEX deposits range from 100 to 260 °C and from 8 to 28 wt. % NaCl equivalent, respectively with deposition temperatures of Pb-Zn mineralization varying from 100° to 185° C at Navan, 150° to 195° C at Tynagh and 180° to 240° C at Silvermines (Andrew, 1986b). Ore stage fluids at Silvermines were more saline (12 to 22 wt. % NaCl equivalent) than those at Tynagh and Navan which had salinities ranging from 8 to 12 wt. % NaCl equivalent. According to Samson & Russell (1987), at Silvermines a negative correlation between the homogenization temperature and salinity of quartz-hosted fluid inclusions indicates mixing of high temperature, low salinity (8 to 12 wt. % NaCl equivalent) fluids with lower temperature, high salinity (18 to 22 wt. % NaCl equivalent) fluids.

4.3 MVT deposits

The ore fluids of MVT deposits have low temperatures (100° to 150° C), uniform salinity, density and composition (Roedder, 1967, 1984). The gross salinity of the fluid inclusions is usually >15 wt. % NaCl equivalent and frequently >20 % wt. % NaCl equivalent, but no daughter crystals are found, indicating the presence of cations other than Na (Roedder, 1984). Fluid inclusion investigations have shown that MVT deposits form from highly concentrated Na-Ca-Cl brines (Newhouse, 1932; Hall & Friedman, 1963; Roedder, 1967). Apart from Na and Ca chlorides, the solutions contain minor K and Mg with total sulphur (expressed as SO_4^{2-}) seldom exceeding a few thousand ppm. The density is always >1.0 g/cm³ and frequently >1.1 g/cm³ and the solute composition is uniform among various MVT districts. In addition, organic matter is frequently but not always observed as gases such as methane, as immiscible oil-like droplets, and as organic compounds in solution in the brines (Roedder, 1984).

Other regional fluid inclusion studies by Leach (1979), Leach & Rowan (1986) and Rowan & Leach (1989) have shown that sphalerite and associated sparry dolomite in the Ozark region have homogenization temperatures in the range 77° to 140° C with salinities from 14 to >23 wt. % NaCl equivalent. The narrow temperature range for sphalerite suggests that the ore fluid was near thermal equilibrium with rocks over a rather large geographic distribution. Even higher fluid inclusion temperatures have been found in sparry dolomite from the Reelfoot Rift Complex where values in the range of 150° to 280° C have been recorded (Leach et al., 1997). According to Leach et al. (1997), these high homogenization temperatures may be related to widespread igneous activity in the Reelfoot Rift Complex in Late Pennsylvanian to Permian times.

As discussed by Kampunzu et al. (2009), fluid inclusions studies in the Pb-Zn-Cu deposits of central Africa indicate at least two stages of mineralization: an earlier high temperature (220° to 390° C) and moderate to high salinity(15 to 43 wt. % NaCl equivalent) stage and a later stage associated with lower fluid temperature (≤80° to 180° C) and variable salinities (<12 to 31 wt. % NaCl equivalent).

The ore fluids have δD values ranging from 0 to -45 ‰ and $\delta^{18}O$ values from -3 to +4.7 ‰ (Hall & Friedman, 1963; Pinkney & Rye, 1972). Hall & Friedman (1963) established that the concentrations of deuterium in ore-stage fluids were relatively high in comparison to that in late gangue calcite which had low deuterium concentrations similar to the local groundwaters. In general MVT ore fluids are similar to SEDEX ore fluids in terms of salinity and homogenization temperatures as indicated above.

5. Sulphur and lead isotopes

5.1 LTM deposits

The isotope ratios of sulphides (including sphalerite, galena, pyrite, chalcopyrite and pyrrhotite) are homogeneous and cluster around 0 ±4 ‰ $\delta^{34}S$ in most LTM deposits (Sawkins, 1964; Rye, 1974; Gilmer et al., 1988; Thompson & Beaty, 1990; Beaty, 1990). Deposits with homogeneous isotope ratios clustering around 0 ±4 ‰ include Providencia, Velardeña, Leadville and Gilman. Other deposits have positive homogeneous values in the range +5.9 to +9.6 ‰ as at Temperino in Italy (Corsini et al., 1980) or from +18.9 to +24.1 ‰ as at Los Lamentos (Megaw et al., 1988). Isotopic equilibrium between sulphide species may or may not have been maintained during mineral deposition.

At Santa Eulalia, a wide spread is observed at the East Camp deposit with $\delta^{34}S$ ratios ranging from +4 to -11 ‰, whereas a much narrower spread occurs in the West Camp deposit where negative values from -8 to -15 ‰ are characteristic (Megaw et al., 1988).

The lead isotope ratios of galena from Mexico are relatively homogeneous within individual mineral districts, and deposits of similar type exhibit similar isotope ratios (Cumming et al., 1979). Lead isotope data for Cenozoic limestone replacement and vein deposits define a linear array which is regarded as a secondary isochron resulting from a two-stage process. The linear array partly coincides with the field of mature arc volcanics of Doe & Zartman (1979), but extends to more radiogenic compositions characterized by pelagic sediments and/or upper crustal material. Isotopic data for these carbonate replacement deposits are characterized by intermediate values between the less radiogenic SEDEX deposits and the more radiogenic MVT Appalachian subdistrict of Ivanhoe-Austinville.

5.2 SEDEX deposits

The $\delta^{34}S$ ratios of sulphide minerals from the epigenetic feeder zones at Silvermines and Keel cluster around 0 ± 10 ‰ (Coomer & Robinson, 1976; Caulfield et al., 1986). Other sulphides, particularly from the stratiform zones at Silvermines and most of the sulphides from Tynagh, are characterized by isotopically light sulphur generally ranging from -42.5 to -4 ‰. Sulphides with relatively narrower spreads occur within this wide range of $\delta^{34}S$ values, such as most of the stage one pyrite (-3.1 to -7.4 ‰) at Tynagh, stratiform ore sphalerite and galena (-20 and -18 ‰) at Silvermines, and sedimentary sphalerite (-15 and -22 ‰) from Navan (Coomer & Robinson, 1976; Boast et al., 1981a; I.K. Anderson et al., 1986). Pyrite from hydrothermal vents at Silvermines also exhibits relatively homogeneous $\delta^{34}S$ values in the range -42.5 to -31.3 ‰ (Type I pyrite vents) and -23.7 to -18.4 ‰ (Type II pyrite vents) (Boyce et al., 1983).

Sulphides with heavy $\delta^{34}S$ values of up to +14 ‰ include post-sedimentary galena at Navan (0 to +14 ‰), and sphalerite and galena from Tatestown (-3 to +14 ‰) (I.K. Anderson et al., 1986; Caulfield et al, 1986). Barite from the Irish deposits forms a fourth distinct isotopic group characterized by positive $\delta^{34}S$ values in the range 14.2 to 22.6 ‰ (Boyce et al., 1983; Andrew & Ashton, 1985; Caulfield et al., 1986).

Caulfield et al. (1986) observed that the lead isotope ratios of the Irish deposits depend on geographic location rather than on the mode of occurrence or age of host rocks, indicating direct involvement of local basement rocks. The $^{206/204}Pb$ isotope ratios are parallel to basement trends of Caledonide inheritance and closely follow geophysically identified basement structures. The isotope data defines a linear array of both Lower Paleozoic and Carboniferous-hosted mineralization on the lead-lead plot which reflects mixing between a

less radiogenic source with a mantle type lead signature and a more radiogenic source with upper crustal type characteristics (O'Keefe, 1986).
With the exception of the Navan deposit, the lead isotope ratios of galena from Tynagh and Silvermines (Boast et al., 1981b; Boast, 1983), as well as minor deposits such as Keel, Tatestown, Ballinalack and Moyvoughly (Caulfield et al., 1986), are characterized by uniform values within a given deposit. All the lenses except the stratigraphically highest number 1 lens at Navan have uniform isotope ratios on a lens scale (Mills et al., 1987). In general, the average $^{206/204}Pb$ ratio decreases from the stratigraphically lowest number 5 lens to the number 1 lens. The number 1 lens contains a less radiogenic component in addition to the more radiogenic lead characteristic of the other lenses.

5.3 MVT deposits

The isotope ratios of sulphur in MVT deposits are generally positive and uniform as in the Upper Mississippi Valley (Pinckney & Rafter, 1972; Maclimans, 1975), or positive with wide ranges as in Southeast Missouri (Sverjensky, 1981). Sulphur with negative isotope ratios has been reported from the Tri-State district (Ault & Kulp, 1960; Deloule et al., 1986).
The data of Pinckney & Rafter (1972) for the stratigraphically correlated bands in ores from four deposits in the Upper Mississippi Valley indicate that the total $\delta^{34}S$ range of the main stage sulphides is from 6.3 to 15.9 ‰, but individual sulphides have narrower spreads. The bands include early (bands 1 to 7), intermediate (bands 8 to 10), and late (bands 11 to 18) paragenetic stages. The latest stage is represented by post-sphalerite deposition of galena. The isotopic variations as deposition proceeded are small and largely due to equilibrium isotopic fractionation between sphalerite and galena with decreasing temperature from an ore fluid with essentially constant $\delta^{34}S_{H2S}$ (Pinckney & Rafter, 1972; Maclimans, 1975).
An example of isotopic disequilibrium during sulphide precipitation is provided by sulphides from the Buick mine, Southeast Missouri (Sverjensky, 1981) where the successive stages include early (dark sphalerite), intermediate (cuboctahedral galena), late (cubic galena) and latest (pale sphalerite). Early sphalerite ($\delta^{34}S$ = 17.54 ‰) and intermediate galena ($\delta^{34}S$ = 16.53 ‰) have similar average values, but galena has a wider spread ranging from 13.0 to 21.1 ‰. Cubic galena of the late stage is isotopically distinct from the earlier phases as indicated by a spread from 0.2 to 9.9 ‰. The latest stage of sphalerite is a minor phase with isotope ratios overlapping those for cuboctahedral and cubic galena.
In the Tri-State district Deloule et al. (1986) found a total range of -6 to -12 ‰ $\delta^{34}S$ in a single galena crystal from the Pitcher orebody. The inner part of the crystal had a constant $\delta^{34}S$ ratio of -9 ‰ $\delta^{34}S$, whereas the outer part had a value of -12 ‰ $\delta^{34}S$ with local variations. The main variation is spatially related to a major change in the lead isotope ratio of the crystal, indicating a common source for the lead and sulphur.
In southern and central Africa, the sulphur isotopes of MVT deposits fall into three categories: (1) heavy and homogeneous $\delta^{34}S$ values (e.g. Tsumeb $\delta^{34}S$ = 20.5 ‰, n= 35; Kamona & Günzel (2007)), (2) highly variable with a wide range of values (e.g. Kombat $\delta^{34}S$ = −11.3‰ to +26.2‰ with an average value of +5.2‰, n=19; Hughes (1987)) and (3) negative and homogeneous $\delta^{34}S$ values (e.g. Kabwe $\delta^{34}S$ = −15.8‰, n= 43; Kamona & Friedrich (2007)). The negative and homogeneous sulphur isotope ratios of ore sulphides (−18 to −12‰ $\delta^{34}S$) from Kabwe are typical of sedimentary sulphides produced through bacterial reduction of seawater sulphate and suggest a sedimentary source for the sulphur, whereas the heavy $\delta^{34}S$ isotopes at Tsumeb and other deposits (Berg Aukas, Khusib Springs) indicate

seawater as the ultimate source of the sulphur (Kamona & Günzel, 2007). A seawater sulphate source is also suggested for the Kipushi deposit (Kampunzu et al., 2009).

The seawater sulphate may have been initially concentrated in basinal brines and evaporite beds that later provided heavy sulphur to ore-forming hydrothermal solutions that eventually deposited the base metals in the carbonate host rocks. The large variation in the sulphur isotope ratios at Kombat is more enigmatic and could be due to thermochemical sulphate reduction (Innes & Chaplin, 1986), mixing of seawater with a fluid containing isotopically light sulphur derived by bacteriogenic reduction of seawater sulphate, variations in oxygen fugacity and/or pH with decreasing temperature (Hughes, 1987) or the involvement of a magmatic ore fluid (Innes & Chaplin, 1986).

Mississippi Valley type deposits of the United States are characterized by anomalous radiogenic linear trends that do not fit standard lead evolution models (Heyl et al., 1966). This anomalously radiogenic lead is traditionally termed "J-type" after the Joplin galena in the Tri-state district. Galena from the Appalachian district is the least radiogenic of all the MVT Pb-Zn districts (Heyl et al., 1966).

In contrast, the lead isotopes of galena from the Austinville-Ivanhoe subdistrict of the Appalachians (Foley et al., 1981) have a relatively narrow spread. According to Foley et al. (1981), galenas of different generations have distinctive lead isotope ratios, with galena becoming increasingly more radiogenic as ore deposition proceeded. In addition, smaller orebodies have a wider range in lead ratios and are more radiogenic than larger orebodies due to contamination by more radiogenic crustal lead.

Lead isotopic variations have also been observed in single galena crystals from MVT deposits (e.g. Sverjensky, 1981; Deloule et al., 1986; Crocetti et al., 1988). The data indicate frequent and rapid isotopic changes that show increasing radiogenic lead with time. In addition, the data shows several isotopic domains that suggest mixing between at least three end members with distinct isotope ratios such as different sedimentary layers and basement rocks (Deloule et al., 1986).

Sulphides from the Zn–Pb orebodies in southern and central Africa show small variations in their Pb isotopic compositions with values typical of the upper continental crust (Hughes et al., 1984; Kamona et al., 1999).

6. Discussion

6.1 LTM deposits

The δD and $\delta^{18}O$ ratios for waters in fluid inclusions at Providencia and Leadville (Rye & O'Neil, 1968; Thompson & Beaty, 1990), as well as the close spatial and temporal relationship between ores and intrusives, suggest a magmatic origin for the ore fluids and metals (Shimazaki, 1980; Einaudi et al., 1981). The deposits may be associated with subduction-related, magnetite-series granitoids (Einaudi et al., 1981) or with magnetite- and ilmenite-series granitoids as in Japan (Shimazaki, 1980).

The lead isotope data of Cumming et al. (1979) indicates that in the case of the Mexican deposits, the lead was derived from the Cenozoic mantle as well as from the Precambrian crust by magma contamination. The sulphur isotope values do not indicate a unique magmatic source for all the deposits since the $\delta^{34}S$ values of granitoid source-magmas are controlled by geographic location and may therefore reflect sulphur isotope ratios of local country rocks (Ohmoto, 1986). Nevertheless, a predominantly magmatic sulphur source is indicated in deposits such as Providencia (Rye, 1974; Sawkins, 1964), Velardeña (Gilmer et al., 1988), and Leadville (Thompson & Beaty, 1990).

The high concentrations of chlorine in fluid inclusions at Providencia (Rye & Haffty, 1969) indicate that the major metals are transported as chloride complexes to the deposition site. The general lack of sulphates suggests that H_2S is the dominant sulphur species (Ohmoto, 1986). The solutions were moving up fast enough to maintain uniformly heated wall rocks adjacent to the fluid conduits as indicated by the low vertical temperature gradient of less than 50° C/km at Providencia (Sawkins, 1964). Initial skarn formation by infiltrational metasomatism and diffusion in limestones occurs between 400° and 650° C and precedes sulphide deposition (Einaudi et al., 1981; Einaudi & Burt, 1982). Sulphide deposition takes place as a consequence of declining temperature, local oxidation-reduction reactions, or neutralization of the ore fluid at the marble contact. A generalized reaction between metal chlorides and hydrogen sulphide leading to sulphide deposition is (Barnes, 1979):

$$MeCl_{2(aq)} + H_2S_{(aq)} = MeS + 2H^+ + 2Cl^- \qquad (1)$$

where Me is metal (e.g. Zn or Pb). From reaction (1) deposition of sulphides could be caused by an increase in reduced sulphur (due to sulphate reduction or leaching of previously formed sulphide minerals), an increase in pH (e.g. acid neutralization by carbonate dissolution), and a decrease in chloride concentration (by removal of Cl^- ions through pairing by ions such as Ca^{2+} and Mg^{2+}). A decrease in temperature of the chloride-sulphide bearing fluid stabilizes H_2S relative to SO_2 and thus promotes deposition of metal sulphides (Rubin & Kyle, 1988).

6.2 SEDEX deposits

In the case of SEDEX deposits there is no general consensus about fluid and metal sources (White, 1986). The problem is caused by the fact that although the lead isotope data indicates the involvement of basement rocks, it does not distinguish between lead derived directly from the basement and lead from disaggregated basement rocks. O'Keefe (1986) established that epigenetic lead veins hosted in Lower Paleozoic rocks in Ireland are of the same age and have been derived from the same lead sources as the surrounding Carboniferous-hosted deposits, suggesting a pre-Carboniferous source for the mineralization. The relative abundance of Pb over Cu indicates silicic continental-derived rocks and sediments which are relatively rich in Pb rather than deep ocean rifts and basaltic rocks with high Cu and Zn but low Pb contents (White, 1986).

A basement source for the metals in the Irish SEDEX deposits is postulated by the deep convection cell model of Russell (1978) and Russell et al. (1981) in which fluid convection within deep fault structures in upper crustal rocks leads to ore fluid generation. In the model, the ore solution evolves from seawater into a brine capable of leaching metals from underlying rocks during fluid-rock interaction with increasing depth down to the base of the brittle zone in the upper crust where temperatures approach 250° C.

Objections raised against the deep fluid convection cell model concern the lack of an intimate water-rock contact (Cathles, 1986) and the absence of suitable rock types such as basalts susceptible to mineral:solution disequilibrium (Lydon, 1986). Lydon (1986) proposed a stratal aquifer model in which the ore fluid is generated from connate waters within an arenaceous aquifer. The metals are derived mostly from clays and iron-oxide pigments in the aquifer itself with minor contributions from faulted basement rocks during reflux convection to depths of 1-3 km and not up to 15 km as in the Russell model. Dense chloride rich brines were produced by evaporitic conditions during the early part of a marine transgression and collected as connate pore waters in Old Red Sandstone lithologies within a basinal structure. The brines became geopressured and heated to over 200° C due to hydraulic sealing and thermal insulation by about 500 m of semi-compacted Lower Carboniferous shale and argillaceous limestone. Heat to the stratal aquifer is supplied directly by conduction from basement rocks and by reflux convection of brines along dilatant fault zones to depths of 3 km.

The sources of sulphur are less problematic. The sulphur isotope ratios of sulphides from the epigenetic feeder zones of the Irish SEDEX deposits are similar to those of diagenetic and vein sulphides in underlying Lower Paleozoic strata from which the sulphur was presumably leached (I.K. Anderson et al., 1989). An ultimate source for the sulphur may be Caledonian granites in the age range 390 to 435 Ma which have a mean $\delta^{34}S$ value of +0.7 ± 2.6 ‰ (Laouar, et al., 1990). Isotopically light sulphur with ratios from -42.5 to -4 ‰ is considered to have originated through bacteriogenic reduction of Lower Carboniferous seawater sulphate and was the dominant sulphur source in the Pb-Zn deposits (Hitzman & Large, 1986; I.K. Anderson et al., 1989). A third group of sulphides in the Irish SEDEX ores is characterized by $\delta^{34}S$ values in the range -3 to +14 ‰. The heavy sulphur component in these sulphides is believed to be of evaporitic origin (Caulfield et al., 1986), but it may represent deep-seated sulphur or low temperature chemical reduction of seawater at the deposition site (I.K. Anderson et al., 1986). Barite formed by reaction of exhaled Ba^{2+} with Lower Carboniferous seawater sulphate (Boyce et al., 1983).

It is generally agreed that stratiform sediment-hosted ores are deposited with or soon after the enclosing sediments and are therefore syngenetic or syndiagenetic (White, 1986). In the deep fluid convection model, circulation of seawater is initiated during rifting and is driven by a high geothermal gradient as indicated by local volcanic flows and/or shallow intrusives in the rock successions at Navan, Tynagh and Silvermines (Russell et al., 1981). The modified seawater with metals and H_2S in solution returns to surface along convective updrafts such as intersecting major fractures. Ore deposition takes place as a result of fluid mixing with alkaline groundwaters and seawater brines. In contrast, in the stratal aquifer model, geopressured formation waters are released intermittently to the surface along fault zones during periods of tectonic activity (Lydon, 1983) and ore precipitation takes place from a brine pool in a third-order basin due to cooling and/or bacteriogenic reduction of sulphate to sulphide within the water column.

6.3 MVT deposits

The origin of MVT deposits is regarded as a normal part of sedimentary basin evolution (Jackson & Beales, 1967). According to the basinal brine model of Beales & Jackson (1966, 1967), ore-forming brines are produced during burial of sediments. Heat, metals and other solutes are acquired during brine migration along aquifers. The similarity of the fluid inclusion data in minerals from MVT deposits to typical oil-field brines with respect to element compositions, high salinities, and hydrogen and oxygen isotopes has been noted (Hall & Friedman, 1963; Roedder, 1967). The main difference is the much lower Na/K ratio of MVT fluids (Sawkins, 1968; White, 1968). This difference may be due to water-rock interactions in aquifers during fluid migration (Sverjensky, 1984; Roedder, 1984), or to contributions of interstitial fluids from evaporite beds enriched in residual potassium (G.M. Anderson & Macqueen, 1982). According to Crocetti & Holland (1989), the lower Na/K ratio of MVT fluids results from equilibration with K-feldspar-albite assemblages in basement and/or sedimentary rocks, whereas higher ratios are due to equilibrium reactions with clay minerals.

The much higher salinities of the brines compared to seawater result from dissolution of salt beds by surface-derived groundwaters or by expulsion of interstitial fluids from evaporite beds (G.M. Anderson & Macqueen, 1982). The chloride-rich brines are able to leach metals from rocks during fluid migration by desorption of loosely bound metals, or release of metals from metal-organic complexes through thermal alteration or destruction of such complexes.

The isotope ratios of sulphur are generally positive and may be uniform or have wide ranges, indicating a crustal, ultimately seawater origin involving sulphate reduction (G.M. Anderson & Macqueen, 1982; Sverjensky, 1986). Multiple sources for the lead have been suggested, including a radiogenic lead component from the basement rocks, lead from carbonate cements and feldspars in sandstones, and normal lead from oil-field brines (Brown, 1967; Doe & Delevaux, 1972). The isotopic studies of Deloule et al. (1986) indicate that numerous and discontinuous inputs of Pb and S from isotopically distinct sources such as different sedimentary layers and basement rocks in the same basin are involved.

In the basinal-brine model, fluids are expelled from strata by sediment compaction during early stages of diagenesis. However, Bethke (1986) observes that fluids displaced from deep basins by compaction-driven flow move too slowly to avoid conductive cooling and cannot account for the deposition temperatures of 100° to 150° C. Episodic dewatering from overpressured aquifers (Cathles & Smith, 1983) also appears to be incapable of adequate heat transfer because of the small volumes required and a narrowly defined aquifer (Leach & Rowan, 1986). In addition, significant overpressures did not develop in basins such as the Illinois basin in the Upper Mississippi Valley (Bethke, 1986). A gravity-driven groundwater flow system due to topographic differences across basins with high heat flows and adequate aquifer permeabilities is considered more likely for the mid-continent districts (Bethke, 1986; Leach & Rowan, 1986).

The fluids are expelled from basins that have undergone deformation or uplift as indicated by the proximity of MVT deposits to tectonically deformed or uplifted basin margins, a coincidence which may be an important unifying factor for MVT deposits in diverse geologic provinces (Leach & Rowan, 1986). Other mechanisms include topography-driven, sediment compaction and overloading, overpressured gas reservoirs, and thermal and density reflux drives (e.g. Leach et al., 2010; Vearncombe et al., 1996)

Due to the low solubilities of sphalerite and galena in the presence of significant H_2S in solutions at 100° to 150° C, the metals and H_2S are not transported together (G.M. Anderson, 1983). The sulphide is instead supplied at the deposition site by reduction of sulphate already in the brine or by adding H_2S to the brine, resulting in sulphide deposition. The source of reduced sulphur is sulphate, either from locally available sulphate minerals or dissolved brine-transported sulphate. This is the so-called "mixing-model" which has been applied with success at the Pine Point deposit where H_2S was derived locally through thermochemical reactions involving sulphate, bitumen and H_2S (Macqueen & Powell, 1986). Alternatively, transport of metals and H_2S in the order of 1 ppm is possible under conditions of acidic pH and high total dissolved CO_2 concentrations (Sverjensky, 1986), as indicated by the correlation of Pb and S isotopes in galena in the Southeast Missouri district and the district wide sphalerite color banding in the Upper Mississippi Valley. This is the "non-mixing" or "reduced-sulphur" model in which deposition of metal sulphides may be caused by pH change, cooling or dilution with groundwater.

The Zn–Pb–(Cu) deposits of southern and central Africa formed from basinal brines during two main mineralizing events that characterize syntectonic deposits (e.g., Kabwe, Tsumeb) and post-tectonic deposits (e.g., Kipushi). These deposits exhibit many of the typical characteristics of Mississipi Valley-type ore-forming systems, including their classical stratabound, epigenetic nature and occurrence in clusters within platform marine carbonates overlying continental crust (e.g. Frimmel et al., 1996; Kampunzu et al., 2009). A sialic crustal origin of the metals from basinal sediments and basement rocks is supported by lead and strontium isotope data, whereas the sulphur isotope compositions are compatible with a sedimentary origin of sulphur from seawater via evaporites and/or diagenetic sulphides. The main ore-forming fluids were saline with moderate to high temperatures and could have been produced by normal geothermal gradients during basin evolution. Ore deposition occurred in carbonate rocks with favourable permeable structures including faults, veins, breccias and hydrothermal karsts as a result of cooling, fluid mixing, pH change or addition of H_2S.

Although the one universal and fundamental characteristic of MVT deposits is the absence of igneous rocks that are potential sources of the ore solutions (Ohle, 1959), intermittent igneous activity occurred in the Southeast Missouri district from Cambrian to Cretaceous time (Thacker & K.H. Anderson, 1977). In particular, Paleozoic explosive igneous activity and intrusion of dikes occurred during Upper Cambrian and Devonian times in the district (Gerdemann & Myers, 1972). In addition, a volcanic tuff bed of up to 1 m or more in thickness occurs over a limited area near the top of a sandy transition zone at the Lamotte-Bonneterre contact. Although the igneous activity was localized and short-lived to have had much direct bearing on the mineralization, it provides evidence that the region had a high geothermal gradient (Gerdemann & Myers, 1972).

7. Conclusions

Carbonate-hosted Pb-Zn deposits range from high temperature, intrusion-related LTM ores through moderate temperature, syngenetic to syndiagenetic SEDEX ores to low and moderate temperature MVT ores characterized by a lack of associated igneous rocks. Metal ratios expressed as Zn/(Zn+Pb) vary from 0.5 in LTM districts to 0.7 in both SEDEX and MVT deposits. Pyrite, sphalerite and galena are the major sulphides, but the mineral paragenesis and relative proportions of these sulphides vary from deposit to deposit, even within the same district. On average Ag grades are relatively high in LTM (>200 ppm) compared to SEDEX (63 ppm) and MVT (96.8 ppm) ore deposits. Copper may be significant in some MVT deposits, especially those in which sulphosalts of tennantite and tetrahedrite form part of the mineral paragenesis as at the Tsumeb and Kipushi deposits.

There are numerous mechanisms that may be responsible for MVT ore fluid migration in sedimentary basins, including topography-driven, sediment compaction, orogenic squeezing, overpressured gas reservoirs, as well as thermal and density reflux mechanisms. However, no single mechanism can be applied to all districts. In the case of the LTM deposits metal and sulphur sources include mantle and crustal rocks, whereas for SEDEX and MVT deposits the metals may have been derived from basement rocks as well as basin sediments during fluid-rock interactions over a prolonged period of time in sedimentary basins with a high heat flow. In all three cases the formation of large deposits requires focusing of ore fluids along conduits that include faults and folds, stratigraphic aquifers, discontinuities and contacts, breccias, pinch-outs and reefs in carbonate host rocks.

8. References

Anderson, G.M. (1983). Some geochemical aspects of sulphide precipitation in carbonate rocks. In: *International Conference on Mississippi Valley-type lead-zinc deposits*. Kisvarsanyi, G., Grant, S.K., Pratt, W.P. & Koenig, J.W. (Eds.), Proceedings Vol., pp. 61-76, University of Missouri, Rolla.

Anderson, G.M. & Macqueen, R.W. (1982). Mississippi Valley-type lead zinc deposits. *Geoscience Canada*, Vol. 9, No. 2, pp. 108-117, ISSN 0315-0941.

Anderson, I.K., Andrew, C.J., Ashton, J.H., Boyce, A.J., Caulfield, J.B.D., Fallick, A.E. & Russell, M.J. (1989). Preliminary sulphur isotope data of diagenetic and vein sulphides in the Lower Paleozoic strata of Ireland and northern Scotland: implications for Zn + Pb + Ba mineralization. *Journal of the Geological Society of London*, Vol. 146, No. 4, pp. 715-720.

Anderson, I.K., Boyce, A.J., Russell, M.J., Fallick, A.E., Hall, A.J. & Ashton, J. (1986). Textural and sulphur isotopic support for Lower Carboniferous sedimentary exhalative base metal deposition at Navan, Ireland. *Terra Cognita*, Vol. 6, p.133.

Andrew, C.J. (1986a). The tectono-stratigraphic controls to mineralization in the Silvermines area, County Tipperary, Ireland. In: *Geology and genesis of mineral deposits in Ireland*, Andrew, C.J., Crowe, R.W.A., Finlay, S., Pennell, W.M. & Pyne, J.F. (Eds.), pp. 377-417, Irish Association for Economic Geology, Dublin.

Andrew, C.J. (1986b). Sedimentation, tectonism, and mineralization in the Irish orefield. In: In: *The genesis of stratiform sediment-hosted lead and zinc deposits: conference proceedings*, Turner, R.J.W. & Einaudi, M.T. (eds.), Vol. XX, pp. 44-56, Stanford University Publications.

Andrew, C.J. & Ashton, J.H. (1985). Regional setting, geology and metal distribution patterns of the Navan orebody, Ireland. *Transactions of the Institute of Mining and Metallurgy*, Vol. 94, B66-B93, ISSN 0371-7453.

Ashton, J.H., Downing, D.T. & Finlay, S. (1986). The geology of the Navan Zn-Pb orebody. In: *Geology and genesis of mineral deposits in Ireland*, Andrew, C.J., Crowe, R.W.A., Finlay, S., Pennell, W.M. & Pyne, J.F. (Eds.), pp. 243-280, Irish Association for Economic Geology.

Barnes, H.L. (1979). Solubilities of ore minerals. In: *Geochemistry of hydrothermal ore deposits*, Barnes, H.L. (Ed.), pp. 404-460, Wiley-Interscience, 2nd. ed., New York.

Beales, F.W. & Jackson, S.A. (1966). Precipitation of lead-zinc ores in carbonate rocks as illustrated by Pine Point ore field. *Transactions of the Canadian Institution of Mining and Metallurgy*, Vol. 75, pp. B278-B285, ISSN 0371-5701.

Beaty, D.W., Landis, G.P. & Thompson, T.B. (1990). Carbonate-hosted sulfide deposits of the Central Colorado Mineral Belt: Introduction, general discussion and summary. *Economic Geology Monograph 7*, pp. 1-18.

Beaty, D.W. (Ed.) (1990). Origin of the ore deposits at Gilman, Colorado. *Economic Geology Monograph 7*, pp. 193-265.

Bethke, C.M. (1986). Hydrologic constraints on the genesis of the Upper Mississippi Valley mineral district from Illinois basin brines. *Economic Geology*, Vol. 81, No. 2, pp. 233-249, ISSN 0361-0128.

Boast, A.M. (1983). Discussions and contributions. *Transactions of the Institute of Mining and Metallurgy*, Vol. 92, p. B101, ISSN 0371-7453.

Boast, A.M., Coleman, M.L. & Halls, C. (1981a). Textural and stable isotopic evidence for the genesis of the Tynagh base metal deposit, Ireland. *Economic Geology*, Vol. 76, No. 1, pp. 27-55, ISSN 0361-0128.

Boast, A.M., Swainbank, I.G., Coleman, M.L. & Halls, C. (1981b). Lead isotope variation in the Tynagh, Silvermines and Navan base-metal deposits, Ireland. *Transactions of the Institute of Mining and Metallurgy*,Vol. 90, pp. B115-B119, ISSN 0371-7453.

Bookstrom, A.A. (1990). Igneous rocks and carbonate-hosted ore deposits of the Central Colorado Mineral Belt. *Economic Geology Monograph 7*, pp. 45-65.

Boyce, A.J., Coleman, M.L. & RusselL, M.J. (1983). Formation of fossil hydrothermal chimneys and mounds from Silvermines, Ireland. *Nature*, Vol. 306, pp. 545-550.

Brown, J.S. (1967). Isotopic zoning of lead and sulfur in Southeast Missouri. In: *Genesis of stratiform lead-zinc-barite-fluorite deposits in carbonate rocks*, Brown, J.S. (Ed.),. Economic Geology Monograph 3, pp. 410-426.

Callahan, W.H. (1967). Some spatial and temporal aspects of the localization of Mississippi Valley-Appalachian type ore deposits. In: *Genesis of stratiform lead-zinc-barite-fluorite deposits in carbonate rocks*, Brown, J.S. (Ed.), Economic Geology Monograph 3, pp. 14-19.

Cathles, L.M. (1986). A tectonic/hydrodynamic view of basin-related mineral deposits. In: *The genesis of stratiform sediment-hosted lead and zinc deposits: conference proceedings*, Turner, R.J.W. & Einaudi, M.T. (Eds.), Vol. XX, pp. 171-176, Stanford University Publications.

Cathles, L.M. & Smith, A.T. (1983). Thermal constraints on the formation of Mississippi Valley-type lead-zinc deposits and their implications for episodic basin dewatering and deposit genesis. *Economic Geology*, Vol. 78, No. 5, pp. 933-1002, ISSN 0361-0128.

Caulfield, J.B.D., LeHuray, A.P. & Rye, D.M. (1986). A review of lead and sulphur isotope investigations of Irish sediment-hosted base metal deposits with new data from the Keel, Ballinalack, Moyvoughly and Tatestown deposits. In: *Geology and genesis of mineral deposits in Ireland.* Andrew, C.J., Crowe, R.W.A., Finlay, S., Pennell, W.M. & Pyne, J.F. (Eds.), pp. 591-615, Irish Association for Economic Geology, Dublin.

Clifford, J.A., Ryan, P. & Kucha, H. (1986). A review of the geological setting of the Tynagh orebody, County Galway. In: *Geology and genesis of mineral deposits in Ireland. Irish Association forEconomic Geology*, Andrew, C.J., Crowe, R.W.A., Finlay, S., Pennell, W.M. & Pyne, J.F. (Eds.), pp. 419-439, Irish Association for Economic Geology, Dublin.

Coomer, P.G. & Robinson, B.W. (1976). Sulphur and sulphate-oxygen isotopes and the origin of the Silvermines deposits, Ireland. *Mineralium Deposita*, Vol. 11, No. 2, pp. 155-169, ISSN 0026-4598.

Corsini, F., Cortecci, G., Leone, G. & Tanelli G. (1980). Sulfur isotope study of the skarn-(Cu-Pb-Zn) sulfide deposit of Valle del Temperino, Campiglia Marittima, Tuscany, Italy. *Economic Geology*, Vol. 75, No. 1, pp. 83-96, ISSN 0361-0128.

Craig, J.R., Solberg, T.N. & Vaughan, D.J. (1983). Growth characteristics of sphalerites in Appalachian zinc deposits. In: *International Conference on Mississippi Valley type lead-zinc deposits*, Proceedings Vol., Kisvarsanyi, G., Grant, S.K., Pratt, W.P. & Koenig, J.W. (Eds.), pp. 317-327, University of Missouri, Rolla.

Crocetti, C.A. & Holland, H.D. (1989). Sulfur-lead isotope systematics and the composition of fluid inclusions in galena from the Viburnum Trend, Missouri. *Economic Geology*, Vol. 84, No. 8, pp. 2196-2216, ISSN 0361-0128.

Crocetti, C.A., Holland, H.D. & Mckenna, L.W. (1988). Isotopic composition of lead in galenas from the Viburnum Trend, Missouri. *Economic Geology*, Vol. 83, No. 2, pp. 355-376, ISSN 0361-0128.

Cumming, G.L., Kesler S.E. & Krstic D. (1979). Isotopic composition of lead in Mexican mineral deposits. *Economic Geology*, Vol. 74, No. 6, 1395-1407, ISSN 0361-0128.

Deloule, E., Allegre, C. & Doe, B. (1986). Lead and sulfur isotope microstratigraphy in galena crystals from Mississippi Valley-type deposits. *Economic Geology*, Vol. 81, No. 6, pp. 1307-1321, ISSN 0361-0128.

Doe, B.R. & Delevaux, M.H. (1972). Source of lead in southeast Missouri galena ores. *Economic Geology*, Vol. 67, No. 4, pp. 409-425, ISSN 0361-0128.

Doe, B.R. & Zartman, R.E. (1979). Plumbotectonics, the Phanerozoic. In: *Geochemistry of hydrothermal ore deposits*, Barnes, H.L. (Ed.), 2nd. ed., pp. 22-70, Wiley-Interscience, New York.

Einaudi, M.T., Meinert, L.D. & Newberry, R.J. (1981). Skarn deposits. In: *Economic Geology 75th Anniversary Volume 1905-1980*, Skinner, B.J. (Ed.), pp. 317-391, Economic Geology Publishing Company, Catalog Card No. 81-84611, New Haven.

Einaudi, M.T. & Burt, D.M. (1982). Introduction - Terminology, classification, and composition of skarn deposits. *Economic Geology*, Vol. 77, No. 4, pp. 745-754, ISSN 0361-0128.

Foley, N.K., Sinha, A.K. & Craig, J.R. (1981). Isotopic composition of lead in the Austinville-Ivanhoe Pb-Zn district, Virginia. *Economic Geology*, Vol. 76, pp. 2012-2017, ISSN 0361-0128.

Frank, M.H. & Lohmann, K.C. (1986). Textural and chemical alteration of dolomite: interaction of mineralizing fluids and host rock in a Mississippi Valley-type deposit, Bonneterre Formation, Viburnum Trend. In: *Process Mineralogy VI: conference proceedings, Metallurgical Society Annual Meeting*, pp. 103-116, Louisiana, March 2-6, 1986, 103-116.

Frimmel, H.E., Deane, J.G. & Chadwick, P.J. (1996). Pan-African tectonism and the genesis of base metal sulphide deposits in the northern foreland of the Damara Orogen, Namibia. In: *Carbonate-hosted lead-zinc deposits*, Sangster, D.F. (Ed.), pp. 204-217, Society of Economic Geologists, 1-887483-95-0, Littleton.

Gerdemann, P.E. & Myers, H.E. (1972). Relationships of carbonate facies patterns to ore distribution and to ore genesis in the southeast Missouri lead district. *Economic Geology*, Vol. 67, No. 4, 426-433, ISSN 0361-0128.

Gilmer, A.L., Clark, K.F., Conde, J.C., Hernandez, I.C., Figueroa, J.I.S. & Porter, E.W. (1988). Sierra de Santa Maria, Velardeña Mining District, Durango, Mexico. *Economic Geology*, Vol. 83, No. 8, pp. 1802-1829, ISSN 0361-0128.

Goodfellow, W.D. & Lydon, J.W. (2007). Sedimentary exhalative (SEDEX) deposits. In: *Mineral Deposits of Canada: A Synthesis of Major Deposit Types, District Metallogeny, the Evolution of Geological Provinces, and Exploration Methods.* Goodfellow, W.D., (Ed.), Geological Association of Canada, Mineral Deposits Division, Special Publication No. 5, pp. 163-183.

Grundmann, W.H. Jr. (1977). Geology of the Viburnum No. 27 mine, Viburnum Trend, Southeast Missouri. *Economic Geology*, Vol. 72, No. 3, pp. 349-364, ISSN 0.61-0128.

Hagni, R.D. (1976). Tri-State ore deposits: the character of their host rocks and their genesis. In: *Handbook of stratabound and stratiform ore deposits*, Wolf, K.H. (Ed.), Vol. 6, pp. 457-494. Elsevier, New York.

Hagni, R.D. (1982). The influence of original host rock character upon alteration and mineralization in the Tri-State district of Missouri, Kansas, and Oklahoma, U.S.A. In:

Ore genesis: the state of the art, Amstutz, G.C. et al., (Eds.), pp. 97-107, Springer-Verlag, Heidelberg.

Hagni, R.D. (1983). Ore microscopy, paragenetic sequence, trace element content, and fluid inclusion studies of the copper-lead-zinc- deposits of the Southeast Missouri lead district. In: *International Conference on Mississippi Valley type lead-zinc deposits.* Kisvarsanyi, G., Grant, S.K., Pratt, W.P. & Koenig, J.W. (Eds.), Proceedings Vol., pp. 243-256, University of Missouri, Rolla.

Hagni, R.D. & Trancynger T.C. (1977). Sequence of deposition of the ore minerals at the Magmont mine, Viburnum Trend, Southeast Missouri. *Economic Geology,* Vol. 72, No. 3, pp. 451-464, ISSN 0361-0128.

Hall, W.E. & Friedman, I. (1963). Composition of fluid inclusions, Cave-in-Rock fluorite district, Illinois, and Upper Mississippi Valley zinc-lead district. *Economic Geology,* Vol. 58, No. 6, pp. 886-911, ISSN 0361-0128.

Heyl, A.V. (1967). Some aspects of genesis of stratiform zinc-lead-barite-fluorite deposits in the United States. In: *Genesis of stratiform lead-zinc-barite-fluorite deposits in carbonate rocks.* Brown, J.S. (Ed.), pp. 20-32, Economic Geology Monograph 3.

Heyl, A.V. (1983). Geologic characteristics of three major Mississippi Valley districts. In: *International Conference on Mississippi Valley type lead-zinc deposits,* Kisvarsanyi, G., Grant, S.K., Pratt, W.P. & Koenig, J.W. (Eds.), Proceedings Vol., pp. 27-60, University of Missouri, Rolla.

Heyl, A.V., Delevaux, M.H., Zartman, R.E. & Brock, M.R. (1966). Isotopic study of galenas from the Upper Mississipppi Valley, the Illinois-Kentucky, and some Appalachian Valley mineral districts. *Economic Geology,* Vol. 61, No. 5, 933-961, ISSN 0361-0128.

Hitzman, M. W. & Beaty, D.W. (1996). The Irish Zn-Pb-(Ba) orefield. In: *Carbonate-hosted lead-zinc deposits,* Sangster, D.F. (Ed.), pp. 112-143, Society of Economic Geologists, 1-887483-95-0, Littleton.

Hitzman, M. W. & Large, D. (1986). A review and classification of the Irish carbonate-hosted base metal deposits. In: *Geology and genesis of mineral deposits in Ireland,* Andrew, C.J., Crowe, R.W.A., Finlay, S., Pennell, W.M. & Pyne, J.F. (Eds.), pp. 217-238, Irish Association for Economic Geology, Dublin.

Hoagland, A.D. (1967). Interpretations relating to the genesis of East Tennessee zinc deposits. In: *Genesis of stratiform lead-zinc-barite-fluorite deposits in carbonate rocks,* Brown, J.S. (Ed.), pp. 52-58, Economic Geology Monograph 3.

Hoagland, A.D. (1976). Appalachian zinc-lead deposits. In: *Handbook of stratabound and stratiform ore deposits,* Wolf, K.H. (Ed.), Vol. 6, pp. 495-534, Elsevier, New York.

Hutchinson, R.W. (1996). Regional metallogeny of carbonate-hosted ores by comparison of field relationships. In: *Carbonate-hosted lead-zinc deposits,* Sangster, D.F. (Ed.), pp. 08-17, Society of Economic Geologists, 1-887483-95-0, Littleton.

Jackson, S.A. & Beales, F.W. (1967). An aspect of sedimentary basin evolution: the concentration of Mississippi Valley-type ores during the late stages of diagenesis. Bulletin of the Canadian Society of Petroleum Geologists, Vol. 15, No. 4, pp. 393-433.

James, J.A. (1952). Structural environments of the lead deposits in the southeastern Missouri mining district. *Economic Geology,* Vol. 47, No. 6, pp. 650-660, ISSN 0361-0128.

Jessey, D.R. (1983). The occurrence of nickel and cobalt in the southeast Missouri mining district. In: *International Conference on Mississippi Valley type lead-zinc deposits,*

Kisvarsanyi, G., Grant, S.K., Pratt, W.P. & Koenig, J.W. (Eds.), Proceedings Vol., pp. 145-154, University of Missouri, Rolla.

Kisvarsanyi, G. (1977). The role of the Precambrian igneous basement in the formation of the stratabound lead-zinc-copper deposits in Southeast Missouri. *Economic Geology,* Vol. 72, No. 3, pp. 435-442, ISSN 0361-0128.

Kamona, F. & Friedrich, G.H. (2007). Geology, mineralogy and stable isotopes geochemistry of the Kabwe carbonate-hosted Pb–Zn deposit, Central Zambia. *Ore Geology Reviews,* Vol. 30, Nos. 3-4, pp. 217–243, ISSN 0169-1368.

Kamona, A.F. & Günzel, A. (2007). Stratigraphy and base metal mineralization in the Otavi Mountain Land, Northern Namibia. *Gondwana Research,* Vol. 11, No. 3, pp. 396–413, ISSN 1342-937X.

Kamona, A.F., Lévêque, J., Friedrich, G., Haack, U. (1999). Lead isotopes of the carbonate hosted Kabwe, Tsumeb, and Kipushi Pb–Zn–Cu sulphide deposits in relation to Pan African orogenesis in the Damaran–Lufilian fold belt of Central Africa. *Mineralium Deposita,* Vol. 34, No. 3, pp. 273–283, ISSN 0026-4598.

Kampunzu, A.B., Cailteux, J.L.H., Kamona, A.F., Intiomale, M.M. & Melcher, F. (2009). Sediment-hosted Zn–Pb–Cu deposits in the Central African Copperbelt. *Ore Geology Reviews,* Vol. 35, Nos. 3-4, pp. 263-297, ISSN 0169-1368.

Klau, W. & Mostler, H. (1986). On the formation of Alpine Middle and Upper Triassic Pb-Zn deposits, with some remarks on Irish carbonate-hosted base metal deposits. In: *Geology and genesis of mineral deposits in Ireland,* Andrew, C.J., Crowe, R.W.A., Finlay, S., Pennell, W.M. & Pyne, J.F. (Eds), pp. 663-675, Irish Association for Economic Geology, Dublin.

Köppel, V. (1983). Summary of lead isotope data from ore deposits of the Eastern and Southern Alps: some metallogenic and geotectonic implications. In: *Mineral deposits of the Alps and of the Alpine Epoch in Europe,* Schneider, H.-J. (Ed.), pp. 162-168, Springer-Verlag, 0387122311, Heidelberg.

Kyle, J.R. (1976). Brecciation, alteration, and mineralization in the Central Tennessee zinc district. - *Economic Geology,* Vol. 71, No. 5, pp. 892-903, ISSN 0361-0128..

Laouar, R., Boyce, A.J., Fallick, A.E. & Leake, B.E. (1990). A sulphur isotope study on selected Caledonian granites of Britain and Ireland. *Geology Journal,* Vol. 25, pp. 359-369.

Large, D. E. (1983). Sediment-hosted massive sulphide lead-zinc deposits: an empirical model. In: *Sediment-hosted stratiform lead-zinc deposits,* Sangster, D.F. (Ed.), Short Course Handbook, Vol. 8, SC9, pp. 1-29, Mineralogical Association of Canada.

Leach, D.L. (1979). Temperature and salinity of the fluids responsible for minor occurrences of sphalerite in the Ozark region of Missouri. *Economic Geology,* Vol. 74, No. 4, pp. 931-937, ISSN 0361-0128.

Leach, D.L. & Rowan, E.L. (1986). Genetic link between Ouachita foldbelt tectonism and the Mississippi Valley-type lead-zinc deposits of the Ozarks. *Geology,* Vol. 14, No.11, 931-935, ISSN 0091-7613.

Leach, D.L., Apodaca, L.E., Repetski, J.E., Powell, J.W. & Rowan, E.L. (1997). Evidence for hot Mississippi Valley-type brines in the Reelfoot rift complex, south-central United States, in late Pennsylvanian-Early Permian. U.S. Geological Survey professional paper 1577, Accessed 09/04/2003, Available from: http://www.pubs.usgs.gov/puprod.

Leach, D.L., Sangster, D.F., Kelley, K.D., Large, R.R., Garven, G., Allen, C.R., Gutzmer, J., & Walters, S. (2005). Sediment-hosted lead-zinc deposits: a global perspective. Economic Geology One Hundredth Anniversary Volume 1905–2005, pp. 561–607, *Society of Economic Geologists*, ISSN 0361-0128.

Leach, D.L., Taylor, R.D., Fey, D.L., Diehl, S.F., & Saltus, R.W. (2010). A deposit model for Mississippi Valley-Type lead-zinc ores. In: *Mineral deposit models for resource assessment.* Accessed 15 January 2011, Available from; http://www.usgs.gov/pubprod.

Lydon, J.W. (1983). Chemical parameters controlling the origin and deposition of sediment-hosted stratiform lead-zinc deposits. In: *Sediment-hosted stratiform lead-zinc deposits*, Sangster, D.F. (Ed.), Short Course Handbook, Vol. 8, SC9, pp. 175-250, Mineralogical Association of Canada.

Lydon, J.W. (1986). Models for the generation of metalliferous hydrothermal systems within sedimentary rocks and their applicability to the Irish Carboniferous Zn-Pb deposits. In: *Geology and genesis of mineral deposits in Ireland*, Andrew, C.J., Crowe, R.W.A., Finlay, S., Pennell, W.M. & Pyne, J.F. (Eds.), p. 555-577, Irish Association for Economic Geology, Dublin.

Macqueen, R.W. & Powell, T.G. (1986). Origin of the Pine Point lead-zinc deposits, N.W.T., Canada: organic geochemical and isotopic evidence. *Terra Cognita*, Vol. 6, p. 134.

McLimans, R.K. (1975). Systematic fluid inclusion and sulfide isotope studies of the Upper Mississippi valley Pb-Zn deposits. Economic Geology, Vol. 70, No. 7, p. 1324, ISSN 0361-0128.

McLimans, R.K., Barnes, H.L. & Ohmoto, H. (1980). Sphalerite stratigraphy of the Upper Mississippi Valley zinc-lead district, Southwest Wisconsin. *Economic Geology*, Vol. 75, No. 3, pp. 351-361, ISSN 0361-0128.

Megaw, P.K.M., Ruiz, J. & Titley, S.R. (1988). High-temperature, carbonate-hosted Ag-Pb-Zn(Cu) deposits of Northern Mexico. *Economic Geology*, Vol. 83, No. 8, pp. 1856-1885, ISSN 0361-0128.

Melcher, F., Oberthur, T & Rammlmair, D. (2006). Geochemical and mineralogical distribution of germanium in the Khusib Springs Cu-Zn-Pb-Ag sulphide deposit, Otavi Mountain Land, Namibia. *Ore Geology Reviews*, Vol. 28, No. 1, pp. 32-56, ISSN 0169-1368.

Mills, H., Halliday, A.N. Ashton, J.H. Anderson, I.K. Russell, M.J. (1987). Origin of a giant orebody at Navan, Ireland. *Nature*, Vol. 327, No. 6119, pp. 223-226, ISSN 0028-0836.

Morrissey, C.J., Davis, G.R., & Steed, G.M. (1971). Mineralization in the Lower Carboniferous of Central Ireland. *Transactions of the Institute of Mining and Metallurgy*, Vol. 80, pp. B174-B184, ISSN 0371-7453.

Mouat, M.M. & Clendenin, C.W. (1977). Geology of the Ozark Lead Company mine Viburnum Trend, Southeast Missouri. *Economic Geology*, Vol. 72, No 3, pp. 398-407, ISSN 0361-0128.

Newhouse, W.H. (1932). The composition of vein solutions as shown by liquid inclusions in minerals. *Economic Geology*, Vol. 27, No. 5, pp. 419-436, ISSN 0361-0128.

Ohle, E.L. (1959). Some considerations in determining the origin of ores of the Mississippi Valley-type. *Economic Geology*, Vol. 54, No. 5, pp. 769-789, ISSN 0361-0128.

Ohle, E.L. (1967). The origin of ore deposits of the Mississippi Valley type. In: *Genesis of stratiform lead-zinc-barite-fluorite deposits in carbonate rocks*, Brown, J.S. (Ed.), pp. 33-39, Economic Geology Monograph 3.

Ohle, E.L. (1980). Some considerations in determining the origin of ores of the Mississippi Valley-type, Part II. *Economic Geology*, Vol. 75, No. 2, pp. 161-172, ISSN 0361-0128.

Ohle, E.L. (1996). Significant events in the geological understanding of the southeast Missouri district. In: *Carbonate-hosted lead-zinc deposits,* Sangster, D.F., pp. 1-7, Society of Economic Geologists, 1-887483-95-0, Littleton.

Ohmoto, H. (1986). Stable isotope geochemistry of ore deposits. In: *Stable isotopes in high temperature geological processes,* Valley, J.W., Taylor, H.P. & O'Neil, J.R. (Eds.), Reviews in Mineralalogy, Vol. 16, pp. 491-559 Mineralogical Society of America.

O'Keefe, W.G. (1986). Age and postulated source rocks for mineralization in Central Ireland, as indicated by lead isotopes. In: *Geology and genesis of mineral deposits in Ireland,* Andrew, C.J., Crowe, R.W.A., Finlay, S., Pennell, W.M. & Pyne, J.F. (Eds.), pp. 617-624, Irish Association for Economic Geology, Dublin.

Paradis, S., Hannigan, P. & Dewing, K. (2007). Mississippi Valley-type lead-zinc deposits. In: *Mineral Deposits of Canada: A Synthesis of Major Deposit-Types, District Metallogeny, the Evolution of Geological Provinces, and Exploration Methods,* Goodfellow, W.D., (Ed.), pp. 185-203, Geological Association of Canada, Mineral Deposits Division, Special Publication No. 5.

Phillips, W.E.A. & Sevastopulo, G.D. (1986). The stratigraphic and structural setting of Irish mineral deposits. In: *Geology and genesis of mineral deposits in Ireland,* Andrew, C.J., Crowe, R.W.A., Finlay, S., Pennell, W.M. & Pyne, J.F. (Eds.), pp. 1-30, Irish Association for Economic Geology, Dublin.

Pinckney, D.M. & Rafter, T.A. (1972). Fractionation of sulfur isotopes during ore deposition in the Upper Mississippi Valley zinc-lead district. *Economic Geology,* Vol. 67, No. 3, 315-328, ISSN 0361-0128.

Roedder, E. (1967). Environment of deposition of stratiform (Mississippi Valley-type) ore deposits, from studies of ore inclusions. In: *Genesis of stratiform lead-zinc-barite-fluorite deposits in carbonate rocks,* Brown, J.S. (Ed.), pp. 349-362, Economic Geology Monograph 3.

Roedder, E. (1984). Fluid inclusions. Mineralogical Society of America, Reviews in Mineralogy, Vol. 12, 644 p.

Rogers, R.K. & Davis, J.H. (1977). Geology of the Buick mine Viburnum Trend, Southeast Missouri. *Economic Geology,* Vol. 72, No.3, 372-380, ISSN 0361-0128.

Rowan, E.L. & Leach, D.L. (1989). Constraints from fluid inclusions on sulphide precipitation mechanisms and ore fluid migration in the Virburnum Trend Lead District, Missouri. *Economic Geology,* Vol. 84, No. 7, pp. 1948-1965, ISSN 0361-0128.

Rubin, J.N. & Kyle, J.R. (1988). Mineralogy and geochemistry of the San Martin skarn deposit, Zacatecas, Mexico. *Economic Geology,* Vol. 83, No. 8, pp. 1782-1801, ISSN 0361-0128.

Ruiz, J., Patchett, P.J. & Ortega-Gutierrez, F. (1988). Proterozoic and phanerozoic basement terranes of Mexico from Nd isotopic studies. Geological Society of America Bulletin, Vol. 100, pp. 274-281.

Russell, M.J. (1978). Downward-excavating hydrothermal cells and Irish-type ore deposits: importance of an underlying thick Caledonian prism. *Transactions of the Institute for Mining and Metallurgy,* Vol. 87, pp. B168-B171, ISSN 0371-7453.

Russell, M.J. & Skauli, H. (1991). A history of theoretical development in carbonate-hosted base metal deposits and a new tri-level enthalpy classification. In: *Historical perspectives of genetic concepts and case histories of famous discoveries, Economic Geology*

Monograph 8, Hutchinson, R.W. & Grauch, R.I., (eds.), pp. 96-116, Economic Geology Publishing Company, No. 91-073091, New Haven.

Russell, M.J., Solomon, M. & Walshe, J.L: (1981). The genesis of sediment-hosted, exhalative zinc + lead deposits. *Mineralium Deposita*, Vol. 16, No. 1, 113-127, ISSN 0026-4598.

Rye, R.O. (1974). A comparison of sphalerite-galena sulfur isotope temperatures with filling temperatures of fluid inclusions. *Economic Geology*, Vol. 69, No. 1, pp. 26-32, ISSN 0361-0128.

Rye, R.O. & Haffty, J. (1969). Chemical composition of the hydrothermal fluids responsible for the lead-zinc deposits at Providencia, Zacatecas, Mexico. *Economic Geology*, Vol.64, No. 6, pp. 629-643, ISSN 0361-0128.

Rye, R.O. & O'Neil, J.R. (1968). The O^{18} content of water in primary fluid inclusions from Providencia, North-Central Mexico. *Economic Geology*, Vol. 68, No. 3, pp. 232-238, ISSN 0361-0128.

Samson, I.M. & Russell, M.J. (1983). Fluid inclusion data from Silvermines base metal-baryte deposits, Ireland. Transactions of the Institute for Mining and Metallurgy, Vol. 92, pp. B67-B71, ISSN 0371-7453.

Samson, I.M. & Russell, M.J. (1987). Genesis of the Silvermines zinc-lead-barite deposit, Ireland: Fluid inclusion and stable isotope evidence. *Economic Geology*, Vol. 82, No. 2, pp. 371-394, ISSN 0361-0128.

Sangster, D.F. (1976). Carbonate-hosted lead-zinc deposits. In: *Handbook of stratabound and stratiform ore deposits*, Wolf, K.H. (Ed.), Vol. 6, pp. 447-456, Elsevier, New York.

Sangster, D.F. (1988). Breccia-hosted lead-zinc deposits in carbonate rocks. In: *Paleokarst*, James, N.P. & Choquette (Eds.), pp. 102-116, Springer-Verlag, New York.

Sangster, D.F. (1990). Mississippi Valley-type and sedex lead-zinc deposits: a comparative examination. *Transactions of the Institution of Mining and Metallurgy*, Vol. 99, pp. B21-B42, ISSN 0371-7453.

Sawkins, F.J. (1964). Lead-zinc ore deposition in the light of fluid inclusion studies, Providencia Mine, Zacatecas, Mexico. Economic Geology, Vol. 59, 883-919.

Sawkins, F.J. (1968). The significance of Na/K and Cl/SO_4 ratios in fluid inclusions and subsurface waters, with respect to the genesis of Mississippi Valley-type ore deposits. - Economic Geology, Vol. 63, 935-942.

Sawkins, F.J. (1990). *Metal deposits in relation to plate tectonics* (2nd edition), Springer-Verlag, 0-387-50920-8, Heidelberg.

Schneider, H.-J. (1964). Facies differentiation and controlling factors for the depositional lead-zinc concentration in the Ladinian geosyncline of the eastern Alps. In *Sedimentology and ore genesis*, Amstutz, G.C. (Ed.), pp. 29-45, Elsevier, Amsterdam.

Sheppard, S.M.F. (1986). Characterization and isotopic variations in natural waters. In: *Stable isotopes in high temperature geological processes*, Valley, J.W., Taylor, H.P. Jr. & O'Neil, J.R. (Eds.), Reviews in Mineralogy, Vol. 16, pp. 165-183, Mineralogical Society of America.

Shimazaki, H. (1980). Characteristics of skarn deposits and related acid magmatism in Japan. *Economic Geology*, Vol. 75, No. 2, pp. 173-183, ISSN 0361-0128.

Shimizu, M. & Iiyama, J.T. (1982). Zinc-lead skarn deposits of the Nakatasu Mine, Central Japan. *Economic Geology*, Vol. 77, No. 4, pp. 1000-1012, ISSN 0361-0128.

Snyder, F.G. (1967). Criteria for origin of stratiform ore bodies with application to Southeast Missouri. In: *Genesis of stratiform lead-zinc-barite-fluorite deposits in carbonate rocks*, Brown, J.S. (Ed.), pp. 1-12, Economic Geology Monograph.

Sverjensky, D.A. (1981). The origin of a Mississippi Valley-type deposit in the Viburnum Trend, Southeast Missouri. *Economic Geology*, Vol. 76, No. 7, pp. 1848-1872, ISSN 0361-0128.

Sverjensky, D.A. (1984). Oil field brines as ore-forming solutions. *Economic Geology*, Vol. 79, No. 1, pp. 23-37, ISSN 0361-0128.

Sverjensky, D.A. (1986). Genesis of Mississippi Valley-type lead-zinc deposits. *Annual Reviews of Earth and Planetary Science*, Vol. 14, pp. 177-199.

Sweeney, P.H., Harrison, E.D. & Bradley, M. (1977). Geology of the Magmont mine, Viburnum Trend, Southeast Missouri. *Economic Geology*, Vol. 72, No. 3, pp. 365-371, ISSN 0361-0128.

Taylor, R.D., Leach, D.L., Bradley, D.C., & Pisarevsky, S.A. (2009). Compilation of mineral resource data for Mississippi Valley-type and clastic-dominated sediment-hosted lead-zinc deposits: U.S. Geological Survey Open-File Report 2009–1297, pp. 42, U.S. Geological Survey, Reston, Virginia.

Taylor, S. (1984). Structural and paleotopographical controls of lead-zinc mineralization in the Silvermines ore bodies, Republic of Ireland. - Economic Geology, Vol. 79, 529-548.

Taylor, S. & Andrew, C.J. (1978). Silvermines orebodies, County Tipperary, Ireland. - Trans. Inst. Min. Met., Vol. 87, B111-B124.

Thacker, J.L. & Anderson, K.H. (1977). The geologic setting of the Southeast Missouri lead district - regional geologic history, structure and stratigraphy. *Economic Geology*, Vol. 72, No. 3, 339-348, ISSN 0361-0128.

Thompson, T.M. & Arehart, G.B. (1990). Geology and the origin of ore deposits in the Leadville District, Colorado: Part I. Geologic studies of orebodies and wall rocks. *Economic Geology Monograph 7*, pp. 130-155.

Thompson, T.M. & Beaty, D.W. (1990). Geology and the origin of ore deposits in the Leadville District, Colorado: Part II. Oxygen, hydrogen, carbon, sulfur, and lead isotope data and development of a genetic model. Economic Geology Monograph 7, pp. 156-179.

Turner, R.J.W. & Einaudi, M.T. (1986). Introduction. In: *The genesis of stratiform sediment-hosted lead and zinc deposits: conference proceedings*, Turner, R.J.W. & Einaudi, M.T. (Eds.), Vol. XX, pp. 1-2, Stanford University Publications.

Vearncombe, J.R., Chisnall, A.W., Dentith, M.C., Dorling, S.L., Rayner, M.J., & Holyland, P.W. (1996). Structural controls on Mississippi Valley-type mineralization, the southeast Lennard Shelf, Western Australia. In: *Carbonate-hosted lead-zinc deposits*, Sangster, D.F. (Ed.), pp. 74-95, Society of Economic Geologists, 1-887483-95-0, Littleton.

White, D.E. (1968). Environments of some base-metal ore deposits. *Economic Geology*, Vol. 63, No. 4, pp. 301-335, ISSN 0361-0128.

White, D.E. (1986). Environments favorable for generating sediment-hosted Pb-Zn deposits. In: *The genesis of stratiform sediment-hosted lead and zinc deposits: conference proceedings*, Turner, R.J.W. & Einaudi, M.T. (Eds.), Vol. XX, pp. 177-180, Stanford University Publications.

Yun, S. & Einaudi, M.T. (1982). Zinc-lead skarns of the Yeonhwa-Ulchin district, South Korea. *Economic Geology*, Vol. 77, No 4, pp. 1013-1032, ISSN 0361-0128.

18

Simulation of Tunnel Surrounding Rock Mass in Porous Medium with Hydraulic Conductivity Tensor

Lin-Chong Huang and Cui-Ying Zhou
Sun Yat-sen University
China

1. Introduction

The deformation of surrounding rock in tunnel is a comparatively complex process, because of the heterogeneous and discontinuous characters in deformation, which belongs to a highly nonlinear problem.

In recent years, many researchers have paid a lot of attention to the study of deformation in the soft surrounding rock. Sulem(1987) and Stille(1989) got the analytic solution of displacement under the hydrostatic pressure state. In fact, this result is based the linear yield criterion in the condition of small deformation. According the experiments, Lade(1977), Agar(1985), and Santarelli(1987) pointed out that, in soft surrounding rock, especially in the soft soil, the relationship between the maximum principle and the minimum stresses is nonlinear, and the linear relationship is only just the special case. In 1966, Hobbs proposed the Power law nonlinear criterion for the fist time, and then Ladanyi suggested a new nonlinear criterion from the crack theory of Griffith in 1974. Kennedy and Linderg studied it using the segment linear theory in 1978, and Brown got the Hoek-Brown nonlinear failure criterion based on predecessors.

With the rapid development of computing power, extensive research has been done on the 3D modeling of tunnel construction. In addition to the special issue on tunneling mentioned earlier, Shahrour and Mroueh(1997) performed a full 3D FEM simulation to study the interaction between tunneling in soft soils and adjacent surface buildings. Their analysis indicated that the tunneling-induced forces largely depended on the presence of the adjacent building and neglecting of the building stiffness in the tunneling-structure analysis yielded significant over-estimation of internal forces in the building members. Tsuchiyama et al. (1988) analysed the deformation behaviour of the rock mass around an unsupported tunnel intersection in the construction of a new access tunnel to the existing main tunnel using 3D linear elastic FEM and found that the influence area along the main tunnel was on the order of one tunnel diameter on the obtuse angle side and about three times the tunnel diameter on the acute angle side from the point of intersection. Rowe and Kack(1983) carried out a numerical analysis based on the finite element method and compared the results with case histories for predicting and designing the settlement above tunnels constructed in a soft ground. Kasper and Meschke(2004) developed a 3D finite element model for a shield-driven tunnel excavation in a soft ground and reproduced settlement, pore pressure distribution, stress levels, and deformations in the lining and in the soil.

Constitutive modeling is an important aspect for the analysis of the displacement and stress fields. A review of existing literature suggests that elastoplastic theory seems to be the most popular framework for constitutive modeling. Borja et al. (1997) formulated the problem of elastoplastic consolidation at finite strain. They showed that for saturated soil media with incompressible solid grains and fluids, balance of energy suggests that Terzaghi's effective stress was the appropriate measure of stress for describing the constitutive response of the soil skeleton. Consequently, the formulation had the advantage of being able to accommodate a majority of the effective stress-based models developed in geotechnical engineering for describing the deformation behaviour of compressible clays. Elastoplastic analyses of circular tunnels excavated in Mohr-Coulomb media have been performed by numerous investigators. For elastic brittle plastic case, Brown et al. (1983) presented the closed-form solution for stress and radial displacement in the plastic zone. However, they did not consider the variation of elastic strain, resulting in the neglect of the influence of the unloading in the plastic zone. Recently, several improved solutions were provided. An analytical solution for Hoek - Brown rock mass given by Sharan(2005) was not exact in calculating displacements in plastic zone, as it was assumed that the elastic strain field in the plastic zone was the same as that of thickwall cylinder problem. Solutions by Park and Kim (2006) offered an exact expression for displacement in the plastic zone.

To sum up, most of the research efforts on tunneling to date have focused on the assessment of the ground surface settlement although some have begun to pay attention to the interaction between tunneling and existing surface structures such as adjacent buildings. Relatively, little research work can be found in the literature on the particular constitutive model for soft soil, especially for the tunnel in porous medium. Also, in the numerical modeling, the Finite Difference Method (FDM) is seldom used in the simulation. The purpose of this paper is to present an elastoplastic model with hydraulic conductivity tensor perform utilizing FLAC3D, and monitor the mechanical behaviour of the tunnel deformation response in the porous medium tunnel during construction.

2. Constitutive model framework

There is the need to establish a link between the state of stresses and the deformations, also a link between the flow vector and the fluid pressure in soft argillaceous shale tunnel. In these links, the stresses are assumed to be a nonlinear function of the deformation via an elastoplastic constitutive response.

The deformation of the surrounding rock includes elastic and plastic deformation in elastoplastic model theory, thus we should set up the constitutive model by combining elastic and plastic theory. In the constitutive relation, there are Mohr-Coulomb and Drucker-Prager yield criterions for the surrounding rock material.

As showed in Fig. 1, when friction angle φ>0, the yield surface of the Drucker-Prager criterion is a conical surface in the principal stress space, which is inscribed at the Mohr-Coulomb yield surface; while if φ=0, the Drucker-Prager criterion is exactly the Mises criterion.

3. Deformation numerical simulation in soft argillaceous shale tunnel

The Guan Kouya tunnel is taken as the research objection here, which is a typical soft argillaceous tunnel located at Hunan province in China. It is a four-lane bidirectional separated tunnel with 880m in length. The geologic investigation data show that the rock is mainly soft argillaceous shale with distinct stratification structure, and the surrounding rock is classified as IV and V. The structure is supervised by NATM, and the support obtains the composite lining, with anchor and sprayed concrete to be the primary support together with the reinforced concrete as the secondary lining.

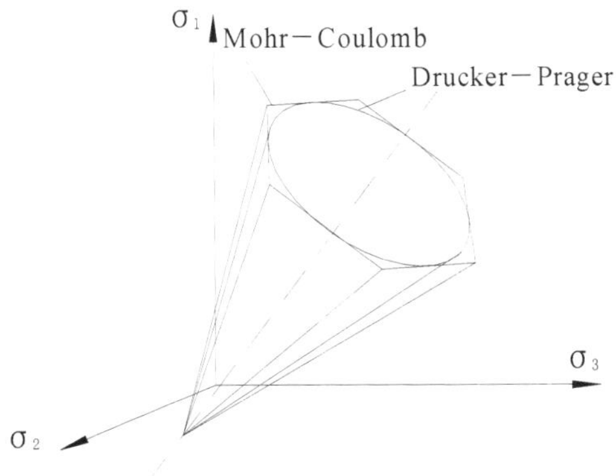

Fig. 1. Mohr−Coulomb & Drucker−Prager yield surface

3.1 Computational model

In order to obtain the deformation characters and compare them with the measurement results in site, using the software FLAC3D, the typical cross section (YK73+839.5, classified as V) is simulated and computated, which is exactly the section measured in site.

3.1.1 Computational bound

The boundary of the computational model is more than three times of the cavity width in each direction, so that the adverse influnce caused by the boundary constraint condition can be reduced sharply in the process of computation. To be specific, the computational zone includes 100 m in the horizontal direction, 30 m from the arch head to the ground in the vertical direction, and 30 m along with the route direction, which can be clearly observed in Fig. 2. The water height is 5.6m over the arch crown.

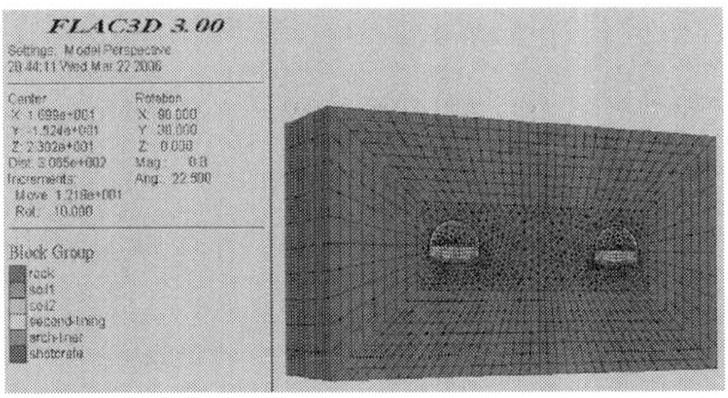

Fig. 2. Numerical computation model

3.1.2 Anchor simulation

It is unrealistic to simulate each anchor in computation, because there are so many anchors in the three-dimensional model. There is an effective method to solve this problem shown in [Huang, 2009] by enhancing the cohesive force instead of the effection of anchoring. The equation of the cohesive force in the anchoring rock is shown as:

$$C = C_0\left[1+\frac{\eta}{9.8}\cdot\frac{\tau S_m}{ab}\times 10^4\right] \tag{1}$$

Where C_0 and C are the cohesive forces before and after adding the anchor(MPa), respectively; τ is the maximum shear stress of the anchor(MPa); S_m is the acreage of the anchor $\left(m^2\right)$; a and s are the distances between each other in the longitudinal and lateral directions, respectively (m); η is a empirical coefficient, and it equals 4.0 in this project.
The material parameters utilized in the simulation is summarized in Table 1.

Items	R (kN/m^3)	E (GPa)	μ	C (kPa)	ϕ (°)
soft argillaceous shale	2.0	1	0.45	10	31
C25 concrete	23	29.5	0.2	—	—
C30 concrete	23	31	0.2	—	—

Table 1. Summary of simulation papameters

3.1.3 Construction procedure

This structure is simulated as two saparate sigle cavities. We obtain the "bench method" in computational simulation, just as the method used in the pratical construction phases, and the procedures are followed as:
step 1→excavating 5 m in the upper bench and setting primary support; Step 2→excavating 5 m in the upper bench additionally and setting primary support, at the same time, excavating 5 m in the lower bench and setting primary support; Step 3→setting the secondary support; Step 4→excavating 5 m in the upper bench and setting primary support, at the same time, excavating 5 m in the lower bench and setting primary support. The construction simulation is done according to this flow operation.
In order to compare the computational results with the measurement data in site conveniently, some typical locations are chosen to be computated and analyzed, and these locations are the same as those measured in site. The construction procedure and these typical locations are shown in Fig. 3.

3.2 Deformation analysis

The maximum computational displacement results of the typical locations around the cavity are summarized in Table 2, and the displacement convergence of the arch crown at cross section YK73+839.5 is shown in Fig. 4.

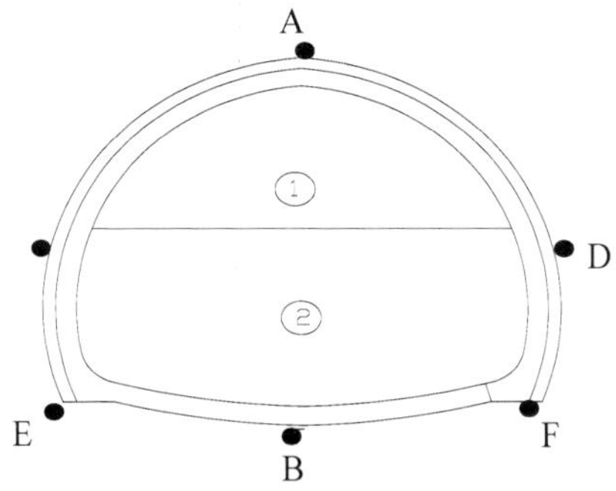

Fig. 3. Typical location & construction procedure

Construction phase	Vertical displacement of the typical locations (mm)					
	A	C	E	B	F	D
Step 1	-37.853	-22.046	-13.091	11.276	-13.581	-24.482
Step 2	-38.518	-12.091	-13.528	11.143	-13.170	-27.905
Step 3	-39.175	-22.975	-14.039	10.791	-14.786	-31.044
Note: minus means the displacement direction downwards.						

Table 2. Displacement computation results of the typical locations around the cavity

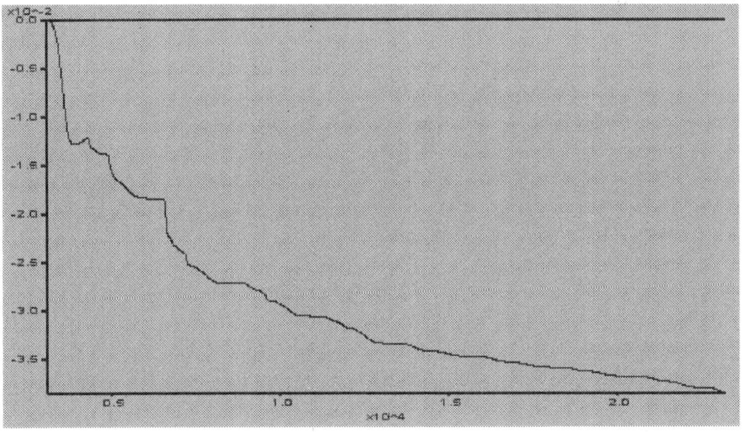

Fig. 4. Displacement convergence of arch crown (Unit:m)

The relative displacements at the arch crown (point A) in section YK73+839.5 are plotted in Figure 5.

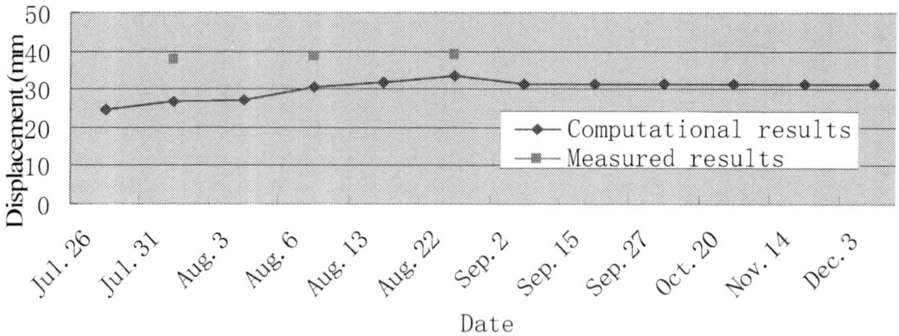

Fig. 5. Graph of the relative displacement change at the arch crown (point A)

While the relative displacements at the right of arch springing (point F) in section YK73+839.5 are plotted in Figure 6.

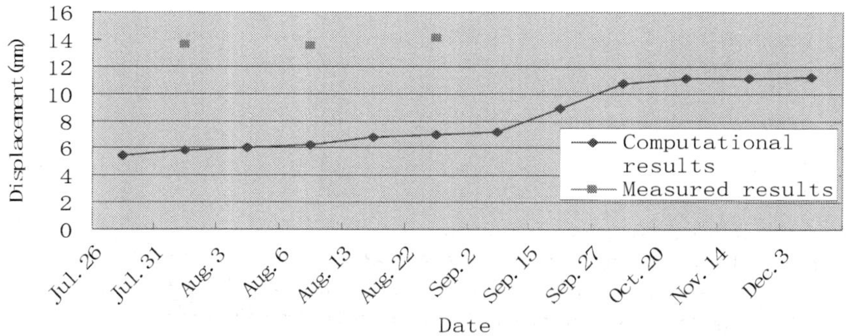

Fig. 6. Graph of the relative displacement change at the arch spinging (point F)

The computational results and the measured results of some typical points are comparatively summarized in Table 3.(Noting that the Graph of the relative displacements at the left arch springing (point E) is elided)

Result	Left arch springing（E）	Arch crown (A)	Right arch springing (F)
Computational value	14.039	39.175	14.786
Measured value	10.02	31.16	11.19

Table 3. Internal displacement value of the section YK73+839.5 (Unit：mm)

The comparative analysis about the computational and the measured results are followed below.

Computational results analysis:

1. The deformaion velocity is much high in the beginning, compared with the late phase, while the time that the deformation get steady is much long. These laws are coincident with the data measured in site.

2. The displacement convergence at the arch crown is 0.51576% after the construction of the secondary lining, while it is 0.6504% at the arch. Spingring. Both of the convergence values are much large, but are not beyond the allowable values, which are less than 0.4%~1.2%. At this point, evaluating with the displacement convergence, this tunnel is steady under constructrue howbeit the convergence values are large. These laws are also coincident with the data measured in site.

Measured results analysis:

1. The deformation velocity rises rapidly in the first 6 days, and goes to steady state after 40 days.
2. The displacement at the arch crown get steady after 30 days, which shows that the deformaion velocity is much high, and the time that the deformation get steady is also much long
3. The displacement at the arch spinging goes to steady state after 40 days, at which point the time is longer than the arch crown.
4. The vertical displacement of the arch spinging increases at all time, and the displacement at the arch crown are always larger than the arch spinging. This rule is also coincident with the computational results.

3.3 Plastic zone analysis

Figure 7 and 8 show the plastic zone of the computational section after excavating the upper bench and the lower bench, respectively, where "shear-n" means failure, while "shear-p" means yielding but no failure.

The plastic zone is small after step 2, in which tension and shearing yield appears in few zone. The whole initial support is almost in a state of "yield" before the construction of the secondary lining, which takes the form of the so-called "pulling yield" from the middle to the bottom of the arch, whereas it appears as the so-called "shear yield" in the 3~4 *m* near the excavating surface and the arch spinging. These suggest that the secondary lining and the Invert should be set in time during the construction of soft surrounding rock tunnel, so that the closed support can be formed in time, which is especially important to insure the tunnel safety, which are shown in Figure 7 and 8.

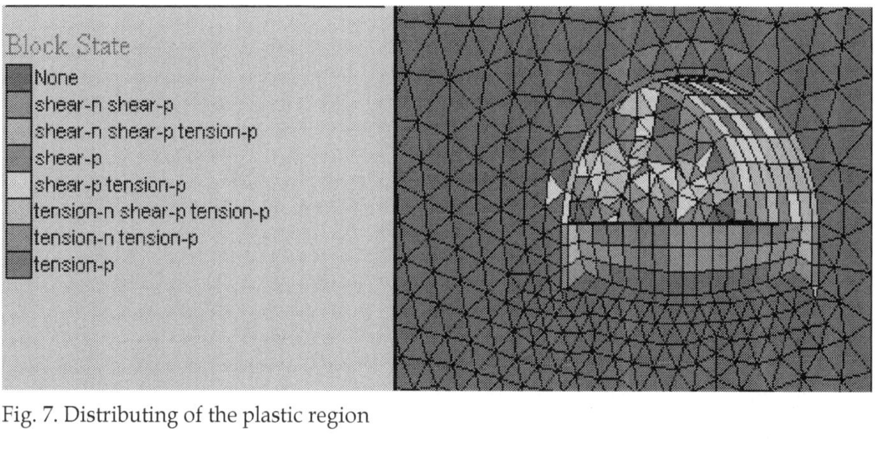

Fig. 7. Distributing of the plastic region

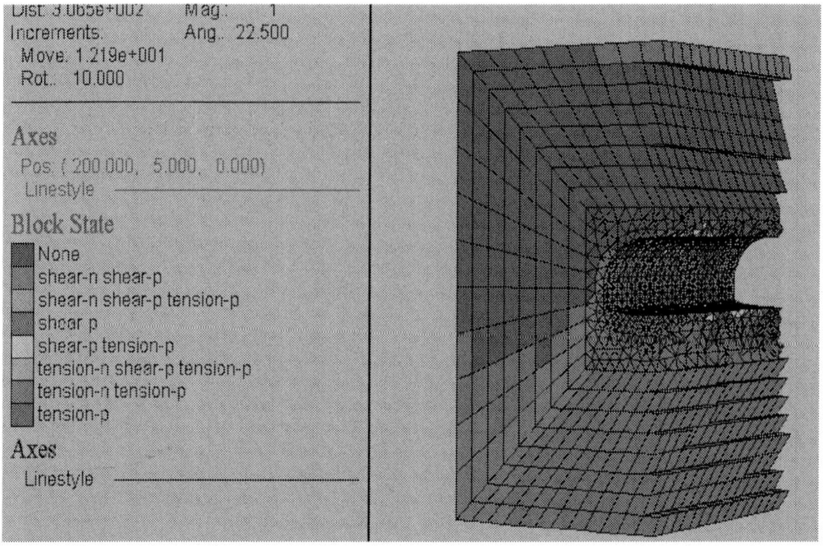

Fig. 8. Portrait distributing of the plastic region

3.4 Secondary lining stress analysis

Figure 9 and 10 show the maximum and minimum principle stresses of the computational values, respectively, while the the maximum and minimum principle stresses of the typical points can be clearly observed in Table 4.

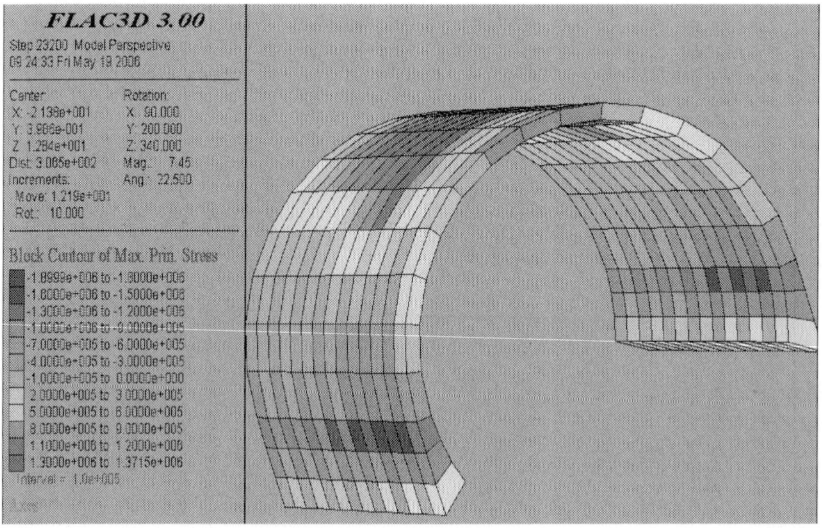

Fig. 9. Maximum main stress of the 2ed lining (Unit:Pa)

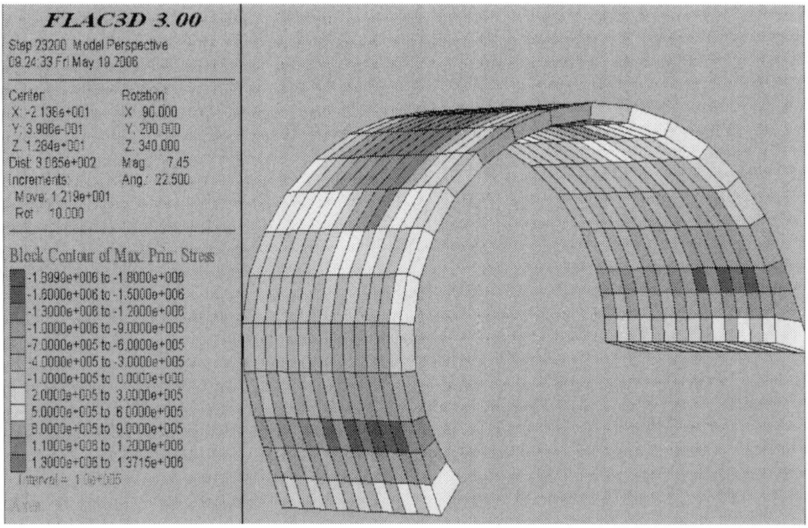

Fig. 10. Maximum main stress of the 2ed lining (Unit:Pa)

Item	A	C	E	B	F	D
Maximum principle stress（MPa）	6.50	4.16	9.91	None	10.76	10.06
Minimum principle stress (MPa)	-0.11	-0.19	-0.39	None	-1.34	-1.76
Note: The minus means tension stress.						

Table 4. The max. & min. principle stress of the typical locations at the secondary lining

Some results can be obtained from the computation such that: There are mainly pression stress in secondary lining, and the maximum pression stress is 10.76 *MPa*, which is located at the middle of the arch, at the same time, which is 44.46% of the designed concrete compressive strength. At this point, the secondary lining is steady. However, there are few tension stresses near the arch spinging, and the maximum is 1.76 *MPa*, 88% of the designed concrete tensile strength, which suggests some reinforcing steel bars should be configured.

Table 5 shows the internal force values of the locations at secondary lining according to the measrued data, from which, we get the shematic diagram of the axial force and moment, shown as Figure 11 and 12, respectively.

Some results can also be suggested from the measured data, i.e.:

1. In the secondary lining, the initial stresses are mostly the tension, because of the contractive stress of the concrete in the prophase. However, they will trun to be the compressive stresses about one month later, in particular, these compressive stresses increace slowly with the time, which will be steady about two months later.
2. The secondary lining moment is much high at the arch crown and the arch spinging, which is basiclly coincident with the mechanical characteristic of the mold casting lining.

location	axial force (KN)	moment (KN·m.)
A	542	11
C	262	4.6
E	678	-9.5
D	433	1.3
F	981	-5.3

Table 5. Internal force results of the measure locations at the secondary lining

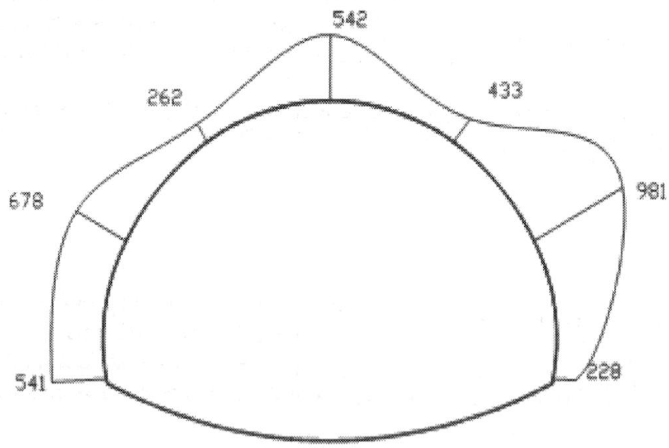

Fig. 11. Sketch map of axial force in 2ed lining (unit: KN)

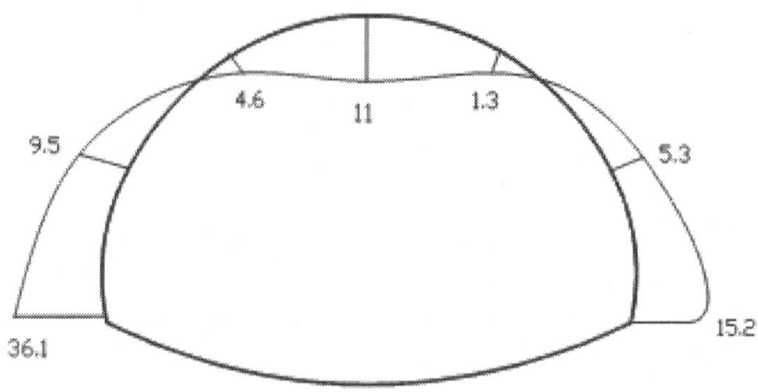

Fig. 12. Sketch map of moment in 2ed lining (unit: KN· m)

4. Conclusion

We have presented a nonlinear finite element model for the simulation of the tunnel in porous medium with hydraulic conductivity tensor. Using the FLAC3D code, the numerical simulation analysis is done to the Guan Kouya soft argillaceous shale tunnel .Some results are obtained, compared with the measurement data in site, such that:

1. The deformation velocity is fast during the prophase of the tunnel excavating, compared with the later phase, but the steady time remains long;
2. Regarding the vertical displacement, the displacement of the arch crown appears bigger than that of the middle, while the displacement of the arch springing is the least obvious, and the horizontal displacement always remains small;
3. The whole initial support is almost in a state of "yield" before the construction of the secondary lining, which takes the form of the so-called "pulling yield" from the middle to the bottom of the arch, whereas it appears as the so-called "shear yield" in the 3~4m near the excavating surface and the arch. These suggest that the secondary lining and the Invert should be set in time during the construction of soft surrounding rock tunnel, so that the closed support can be formed in time, which is especially important to insure the tunnel safety;
4. The surrounding rock stress releases rapidly after the excavating of the cavity, at this point, some assistant methods, especially the lock-foot anchor should be applied to enhance the steady stability due to the big compressive stress located at the arch of the two sides in the tunnel.

5. Acknowledgment

This work has been supported by China Postdoctoral Science Foundation (No. 2009046082724 & 201003386), the National Natural Science Foundation of China (No.51108472),the Natural Science Foundation of Guangdong Province, China (No.S2011040005172),and the State Key Program of National Natural Science Foundation of China(No.41030747), these supports are gratefully acknowledged.

6. References

Sulem, J. ; Panet, M. & Guenot, A (1987). An analytical solution for time-dependent displacement in a circular tunnel. *Int J. Rock Mech Mi Sci & Geomech Abstr*, Vol.24, No.3,pp. 155-164.

Stille, H. ; Holmoery, M. & Mord, G(1989). Support of week rock with grouted bolts and shoterete. *Int J Rock Mech Mi Sci & Geomech Abstr*, Vol.26, No.1,pp. 99-103.

Lade, P. V. (1977). Elasto plastic stress strain theory for cohesionless soil with curved yield surface. *Int J Skilds Struct*, Vol.13, pp. 1019-1035.

Agar, J. G. ; Morgensteren, N. R. & Scott, J. Shear strength and stress strain behaviour of Athabasca oil sand at elevated temperatures and pressure (1985). *Can. Geotech. J.*, Vol.24, No.1, pp. 1-10.

Santarelli, F. Theoretical and experimental investigation of the stability of the axisy metric borehole(1987). University of London.

Shahrour, I. & Mroueh, H.(1997). Three-dimensional non linear analysis of a closely twin tunnels. *Sixth International Symposium on Numerical Models in Geomechanics*, Montreal, pp. 481-487.

Tsuchiyama, S. ; Hayakawa, M. ; Shinokawa, T. & Konno, H.(1988). Deformation behaviour of the tunnel under the excavation of crossing tunnel. *Proceedings of the 6th International Conference on Numerical Methods in Geomechanics*, Innsbruck, pp. 1591-1596.

Rowe, R.K. & Kack, G.J(1983). A theoretical examination of the settlements induced by tunneling four case histories. *Can. Geotech. J.*, Vol.20, pp, 299-314.

Kasper, T. & Meschke, G.(2004). A 3D finite element simulation model for TBM tunneling in soft ground. *Int. J. Numer. Anal. Meth. Geomech.*, Vol.28, pp, 1441-1460.

Borja, R.I. & Andrade, J.E.(2006). Critical state plasticity, Part VI: Meso-scale finite element simulation of strain localization in discrete granular materials. *Computer Methods in Applied Mechanics and Engineering*, Vol.195, pp, 5115-5140.

Borja, R.I.; Tamagnini, C. & Amorosi, A.(1997). Coupling plasticity and energy conserving elasticity models for clays, *J. Geotech. Geoenviron. Engrg.*, Vol.123, pp, 948-957.

Brown, E.T. ; Bray, J.W ; Ladanyi, B. & Hoek, E.(1983). Ground response curves for rock tunnels. *J. Geotech. Eng., ASCE,* Vol.109, pp, 15 - 39.

Huang, L.C, Xu, Z.S, Wang, L.C. Constitutive equations and finite element implementation of strain localization in sand deformation(2009). Journal of Central South University of Technology. Vol.16,No.3, pp, 482-487

Sharan, S.K.(2005). Exact and approximate solutions for displacements around circular openings in elastic - brittle - plastic Hoek - Brown rock. *Int. J. Rock Mech. Min. Sci.* Vol.42, pp, 542-549.

Park, K.H. & Kim, Y.J.(2006). Analytical solution for a circular opening in an elasto-brittle-plastic rock. *Int. J. Rock Mech. Min. Sci.* Vol.43, pp, 616-622.

Kumar, P. (2000). Infinite Elements for Numerical Analysis of Underground Excavations[J]. Tunneling and Underground Space Technology, Vol.15, No.1, pp, 117-124.

Part 7

Soil

19

Soil Contamination by Trace Metals: Geochemical Behaviour as an Element of Risk Assessment

Monika Zovko and Marija Romić
University of Zagreb, Faculty of Agriculture
Croatia

1. Introduction

Trace metals occur naturally in rocks and soils, but increasingly higher quantities of metals are being released into the environment by anthropogenic activities. Metals are chemically very reactive in the environment, which results in their mobility and bioavailability to living organisms. People can be exposed to high levels of toxic metals by breathing air, drinking water, or eating food that contains them. As a consequence, metals get into the human body by different routes - by inhaling, through skin, and via ingestion of contaminated food.

Every decision on the application of any measures in the environment related to soil quality and management, whether statutory regulations or practical actions, must be based on reliable and comparative data on the status of this part of environment in the given area. Various aspects have to be considered by the society to provide a sustainable environment, including a soil clean of heavy metal pollution. The first step is to identify environments (or areas) in which anthropogenic loading of heavy metals puts ecosystems and their inhabitants at a health risk.

Long-term and extensive use of land for agriculture with frequent application of agri-chemicals is one of the major causes of trace metal, such as copper, nickel, zinc and cadmium, accumulation in soil. Accumulation of Cu in agricultural soil is a consequence of the century - old practice of using copper-sulphate (Bordeaux mixture) and other copper containing fungicides to control vine downy mildew. It is estimated that every time vines are sprayed with copper-containing solutions, some 2 to 5 kg ha^{-1} of copper enter the soil (Romić & Romić, 2003).

Widespread distribution of Cd and its high mobility makes it a potential contaminant in a wide range of natural environments. Generally, soil Cd concentrations exceeding 0.5 mg kg^{-1} are considered evidence of soil pollution (McBride, 1994). Phosphatic fertilizers are one of the most ubiquitous sources of Cd contamination in agricultural soils throughout the world. Total Cd inputs to soils through fertilizers in the countries of the European Union have been estimated to be around 334 t yr^{-1} (Alloway & Steinnes, 1999). Long-term investigations worldwide have shown that application of phosphate fertilizers has resulted in soil enrichment with Cd (0.3-4.4 g ha^{-1} year^{-1}), depending on the rates and kinds of fertilizers applied (Singh, 1994). Andrews et al. (1996) and Gray et al. (1999) also determined a highly significant correlation between total concentrations of Cd and phosphorus in agricultural soils of New Zealand, which they attributed to long-term application of phosphate fertilizers. Chen et al. (2008) observed significant correlations between Cd and Pb and soil phosphorus in California vegetable croplands, indicating the application of P-fertilizers contributes significantly to the accumulation of Cd and Pb in soils. Zinc belongs to a group of trace metals that are potentially most dangerous for the biosphere. The main sources of the pollution are industry and use of liquid manure, composted materials and agrochemicals such as fertilizers and pesticides in agriculture.

Beside anthropogenic sources, trace metals can be also found in the parent material from which the soils developed. Whether the said inputs will become toxic and to what degree mobile depends on a number of factors: specific chemical and physical trace metal characteristics, soil type, land use, geomorphological characteristics within the soil type and exposure to emission sources. Processes that control the mobility, transformation and toxicity of metals in soil are of special importance in the soil root developing zone - the rhizosphere. For this reason, there is a considerable interest in understanding trace metals behaviour in soil, with special emphasis on the way they build-up in soil and on processes of by which plants take up metals.

2. Factors controlling trace metals behaviour in soil

2.1 Trace metal characteristics

Two thirds of all elements found in nature are metals. According to their chemical definition, metals are elements and as such cannot be synthesized or degraded by biological or chemical processes, though these processes can change chemical forms of metals. Metals are contained in the Earth's crust and in parent rocks, by whose weathering soils are formed, so their presence differs in different geographic regions. Terms like heavy metals, metalloids and microelements are the most commonly encountered in ecological studies. Among the 96 known metals, 17 are semimetals or metalloids (e.g., B, Si, Ge, As, Sn, Te, Po ...). The term heavy metal refers to a group of 53 metals with density higher than 5 g/cm^3. From the geochemical point of view, trace elements are metals whose percentage in rock composition does not exceed 0.1% (e.g., Cu, Cr, F, Fe, Mo, Ni, Se, Zn, As, Cd, Hg, Pb). In very small amounts, some of these elements are essential for normal growth and development of living organisms and they are, from the physiological point of view, called micronutrients or microelements (e.g., Fe, Mn, Zn, Cu, Mo, Ni, Se), while others are toxic even in small concentrations. The issue of toxicity is usually merely a matter of quantity, with the range varying for each element.

Regardless of whether soil metals originate from a natural source or are a consequence of anthropogenic activity, metals appear in groups; one element by itself is rarely the source of contamination. Hence, synergistic and antagonistic interactions of metals should not be disregarded in assessing bioavailability. For example, Zn and Cd are usually present together in ores and have similar physical and chemical properties.

Distribution, mobility, bioavailability and toxicity of metals depend not only on metal concentration but also on the form in which metal exist. Full understanding and prediction of chemical behaviour of an element in the environment is possible only by identification of all forms in which that element can be found under different environmental conditions. Metal speciation is one of the most important properties that determine the behaviour and toxicity of metals in the environment. Chemical speciation of an element refers to its specific form characterized by a different isotopic composition, molecular structure, and electronic or oxidation state (Manouchehri et al., 2006). Speciation is the process of identification and determination of different chemical and physical forms of elements present in a sample (Wang et al., 2006). Metals that occur in cationic forms have a higher ability of binding to negatively charged soil colloids, and are thus less bioavailable, but more easily accumulate in soil, unlike the anionic forms that are mainly present in soil solution and are more bioavailable, but are more readily leached from the soil.

2.2 Soil properties

Because of their very low abundance, trace elements are particularly sensitive to surrounding environmental conditions, which influence their physico-chemical speciation and their behaviour in the ecosystems (N'guessan et al., 2009). In agricultural environment soil is the main sink and source of trace metals. Not all soil properties have equal influence on the mobility and availability of a particular metal. For each metal it is therefore important to know the dominant soil property that will control the behaviour of that metal in that particular soil. With regard to bioavailability, the following metal fractions have an important role in soil: metals in soil solution; precipitated metals; metals bound to clay minerals, oxides and hydroxides, organic matter, and metals in the soil mineral matrix. All fractions of metals are in dynamic equilibrium, and only metals in aqueous soil solution are directly available to plants. Soil solution is in direct contact with the soil solid phase and transformations going on in it are a consequence of mineral equilibrium, exchange processes and sorption processes in the soil mineral phase and organic matter, as well as complexation with organic matter in the solid phase and in solution (figure 1). Major soil properties that affect changes in metal speciation, and thereby also their fractionation, are soil reaction, redox-potential, and existence of different organic and inorganic reactants - ligands.

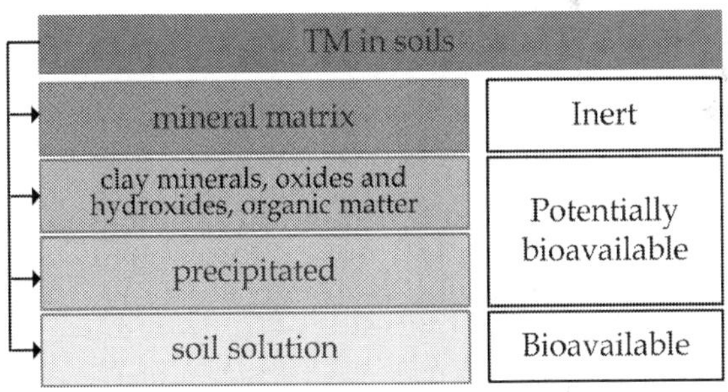

Fig. 1. Simplified schematic presentation of different trace metal (TM) fractions in soil

Whether soil metals will form complexes with organic matter, mineral colloids, inorganic complexes or exist as free ions depends on the soil solution reaction. Constant changes in soil pH are a consequence of cation/anion balance, leaching of organic acids, root respiration and oxidation-reduction reactions. Each change in pH alters the adsorption capacity of minerals and organic colloids. Positive charge prevails at low pH conditions under which anion exchange capacity is dominant, while at high pH negative charge prevails and cation exchange capacity predominates. Solubility of Mn and Zn is strongly dependent on soil reaction. The lower the pH, the stronger the Mn and Zn competition with H^+ and Al^{3+} ions for places in the soil cation exchange complex, and higher the Mn and Zn solubility and proportion of readily available free ions in soil solution. Soil reaction will affect Cu speciation, its solubility and adsorption; however, a weak correlation was determined between soil reaction and Cu concentration in soil solution (Wang et al., 2002) since dissolved Cu has high affinity for organic matter, that is, for phenolic groups and groups of weak acids of humic compounds (Cornu et al., 2007). Gummuluru et al. (2002) showed that Cu complexes with dissolved organic matter are the most dominant species, averaging 97.1, in the neutral and mildly acidic soils.

The presence of carbonate, high contents of organic matter and percentages of clay my suggest that trace metals could be retained in these soils, as these properties increase the adsorption capacity of metals by soil (Peris et al., 2008). Organic substances and oxyhydroxides have the highest surface charge density and are thus the most important and strongest sorbents of trace metals. Carboxylic groups (-COO-) of organic compounds can form stable complexes with metal ions, which leads to changes in the metal fraction available to biota. Surface charge of organic matter and oxyhydroxides is strongly dependent on soil pH, contrary to surface charge of clay (except kaolinite), which does not depend on pH (Reichman, 2002). Organic matter affects the soil redox potential. Fe and Mn hydroxides are especially susceptible to changes in soil redox potential. Electrons are released through organic matter decomposition, which results in a decrease in redox potential and reduction of Fe(III) into the easily soluble Fe(II) form. Manganese ions display similar behaviour. Organic matter can also influence temporary immobilization of Zn. Available Fe and Mn ions inhibit Zn uptake, which can affect precipitation of dissolved Zn into the mineral franklinite ($ZnFe_2O_4$).

Selective sequential extraction was applied to investigate the potential of mobilizing trace elements in agricultural soils of northwestern Croatia (Romić & Romić, 2003). In alluvial soils developed on Quaternary (Upper Pliocene to Holocene) deposits, extraction with 1M $Mg(NO_3)_2$ (pH 7) (Shuman, 1985) indicated possible remobilization of elements from the solid phase into soil solution, particularly in the case of copper. There are several possible mechanisms that increase solubility of metals in the surface layer: 1) the soil mineral component is more susceptible to weathering in shallower than in deeper soil layers owing to faster infiltration of precipitation, higher biological activity and greater changes in temperature; 2) shallower soil horizons are richer in organic matter, which can stimulate metal desorption by formation of soluble organic complexes; and 3) exchangeable complex of shallower soil layers contains more basic cations (Na^+, K^+, Ca^{2+}, Mg^{2+}), which can also reduce sorption of metals by increasing the competition for exchange sites. All these processes are even more pronounced in the anthropogenic horizon of arable soils.

For trace metals to be translocated by water through the soil profile, they have either to be in the soluble phase or bound to mobile particles. Metals can form complexes with particles of organic matter in topsoil and as such can be translocated vertically along the profile depth. It is generally taken that the water-soluble and exchangeable fractions, and sometimes also the organic fraction, are bioavailble forms. Investigations dealing with the leaching of metals from soil revealed a marked vertical distribution of organometallic compounds that can be leached up to 3 times deeper compared to metals not complexed with organic matter. This phenomenon is particularly pronounced in application of organic soil amendments. The observed enhanced mobility of metals in soils improved by different organic conditioners (Al-Wabel et al., 2002) is a consequence of metal binding to dissolved organic carbon (DOC) formed through decomposition of such conditioners. An increase in DOC concentration increases metal complexation with organic matter, that is, the proportion of metals in the soil liquid phase, and thereby also their mobility. In their investigations, Al-Wabel et al.

(2002) assessed that more than 99% of Cu and Zn in soil solution and about 90% of Pb were complexed with dissolved organic carbon or mineral colloids without being in the form of free ions or inorganic complexes. In contrast to Cu and Zn, speciation of Cd in soil solution was not strongly influenced by organic matter, but Cd existed in solution as a free divalent cation (Cd^{2+}) or with inorganic complexes Cl^-, SO_4^{2-} or HCO_3. It has been well documented in the literature that complex formation between metals and inorganic ligands Cl^- and SO_4^{2-} inhibits the adsorption of Cd on soil and soil constituents due to the formation of cadmium complexes that were not strongly adsorbed by soil (McLean & Bledsoe, 1992). In many arid and semiarid regions, including European Mediterranean region, saline irrigation water containing high level of chloride might aggravate trace metals pollution problem, as a result of metals mobilisation due to the formation of metals-chloride complexes (Ghallab & Usman, 2007). Soil salinity strongly increased the concentration of dissolved chloride (Cl^-) ligands, and significantly influenced Cd solubility and thereby its bioavailability and phytoaccumulation. Exposure to increasing NaCl salinity in the rhizosphere environment increased accumulation of Cd in muskmelon and radish leaves (Ondrašek et al., 2009).

The mobilization potential of trace metals, like Cd, Cu, Pb and Zn, may get increased in dredged (oxidised) sediments compared to submerged (reduced) sediments (Tack et al. 1998; Vandecasteele et al., 2007). Indeed, an important factor in the metal mobilization is changing redox conditions. In that sense, the Mn behavior is particularly varying. In oxic environment Mn precipitates as oxides with large mineral surfaces entering into the reactions with both trace metals and soil organic matter (Kaiser & Guggenberger, 2003).

3. Assessment of soil contamination by trace metals – geochemical methods

Various chemical methods, geochemical models and biotests are used for assessment of the bioavailable metal fraction in soil. However, these methods are not universally applicable for all elements and different soil characteristics. Chemical methods for assessment of metal bioavailability are commonly grouped within methods for assessment of total metal content in soil, methods for assessment of currently available and potentially available fractions, methods for assessment of metal speciation in soil solution.

Determination of the total content of metals in soils is an important step in estimating the hazards to the vital roles of soil in the ecosystem, and also in comparison with the quality standards in terms of the effects of pollution and sustainability of the system. Methods for assessment of total metal content are based on soil digestion with strong acids such as HNO_3, HF, $HClO_4$ and aqua regia. Although the total metal content in soil does not show a good correlation with the bioavailable fraction, it is still used in most countries as a statutory regulation for assessment of soil contamination.

From the ecotoxicological aspect, however, it is equally important to determine the bioavailabilty of trace metals accumulated in agricultural soils. Single and sequential extraction methods are applied for assessment of currently available and potentially available metal fractions in soil. There are several kinds of extraction solutions, the most commonly used being 0.001 to 1 M salt solutions ($CaCl_2$, $Ca(NO_3)_2$, $NaNO_3$), weak acids (acetic acid, citric acid) and strong complexes (EDTA, DTPA). Extraction methods are based on complexometric reactions between extractants and metals. In complexometric titrations use is made of suitable indicators, commonly compounds that can produce a less stable coloured product with a free (hydratized) cation. Extraction methods are extensively applied in bioavailability investigations, but are not acceptable for all kinds of metals; for example, EDTA is not suitable for metal bioavailability assessment in contaminated soils, particularly for Cu assessment (Brun et al., 2008). Available Cu in topsoils of the wine-growing regions in north-western Croatia was evaluated with DTPA extraction and calcium chloride extraction methods. Highly significant positive correlation was determined between total copper content and DTPA-extractable copper contents. Such strong correlation indicates that the DTPA extraction method is not suitable for assessing copper availability to plants (Romić et al., 2003).

There are two approaches to determining metal speciation in solution: analytical determination and chemical balance models. Direct measurements of metal ion speciation in soil solution are rare. Measurements are mostly conducted in solutions extracted with dilute salts (0.005 M HNO_3) of soil metals. Precise measurement of different forms of metals present in aqueous solution is the most demanding procedure, since it requires analytical methods of high selectivity and sensitivity; hence, different computer models (GEOCHEM, SOILCHEM) are mostly used for metal speciation determinations. These models are based on geochemical thermodynamic principles (Peijnenburg et al., 2003). Despite their high efficiency, computer models still have shortcomings that must not be disregarded in interpretation of the results obtained. The problem in the application of geochemical models for calculation of element speciation is the modelling of organic matter-metal complexation, since there are no reliable values for stability constants of dissolved organic matter, the properties of which can vary considerably in dependence on environmental conditions.

The main flaw of all chemical methods for bioavailability assessment is that they invalidate all complexity of the mechanism of metal uptake by plants and neglect the very important role of root metabolism and of the microorganisms that surround it. For this reason, these tests are not adequate for assessment of plant available metals in soil (Chaignon et al., 2003). An ideal method for metal phytoavailability assessment should simulate soil-plant interactions as closely as possible (Fang et al., 2007). Further, such a method should extract the amount of metal that corresponds to metal concentration taken up by the plant. Along these lines, the RHIZO - method (rhizosphere based method) was developed. It is based on the application of extraction of 0.01 M solution of organic acids of low molecular mass in wet rhizosphere soil. In laboratory investigations, Feng (2005) demonstrated the efficiency of the RHIZO method application in acid, neutral and alkaline soils, which makes this method more suitable than other bioavailability assessment methods. However, the results were satisfactory only for Cr, Cu, Zn, and Cd, while the method was not a good indicator of Pb and Ni phytoavailability.

Application of bioassays with plants offers a different approach to assessment of metal bioavailability in contaminated soils. Biotests can be done in nutrient solutions or in soil, depending on the research goal. Nutrient solutions can be considered as models of the soil system, more precisely, soil solution. Whether a nutrient solution will be a good soil solution model depends on the experimental design, but it should still be pointed out that this is a simplified soil system and that metal-plant interactions are different in solution and in soil. Thus, the interaction between Zn and Cu is of synergistic character in soil, and of antagonistic character in solution. It is also important to know metal rhizotoxicity because some metals, e.g., Cu, accumulate much more in roots than in stalks (Chaingon et al., 2003). Therefore it is essential for bioavailability assessment to develop biotests that will enable an unobstructed approach to root and rhizosphere studies.

4. Assessment of soil contamination by trace metals: A case of NW Croatia

4.1 Environmental soil functions

Soil plays many important roles in the environment. As being situated at the interface between the atmosphere and the lithosphere it acts as a filter and a buffer: it may weaken and degrade environmentally harmful compounds protecting the air quality. It also has an interface with hydrosphere and therewith it affects surface and groundwater quality. Furthermore, soil, as a part of biosphere, provides nutrient-bearing environment that sustains the growth of plant and animals. As a habitat and protecting media of flora and fauna it contributes to the maintaining of the global nutrient cycling as well as biomass production, whether by natural vegetation growing or plant cultivation. Beside these ecological functions, soil is ground to build and live on, raw material and reserve of cultural heritage.

Because soil quality and its utilization are directly linked, each of above mentioned functions or use mode requires a certain soil quality level. Otherwise, any change of soil quality may affect its utilization potential. Soil is a natural resource essential for the food production and global economy. The way and rate of soil degradation on the global scale point out the importance of the sustainable land use. Harris et al. (1996) define the soil quality as a capacity of the certain soil volume in given conditions (land use, relief, and climate) to protect water and air quality, to sustain plant and animal growth, promoting thus the human health. Out of total degraded land on the global scale that are estimated to 1,965 mha, about 55% was water eroded, about 28% wind eroded, and about 12% is polluted by chemicals (Adriano et al., 1995). Land degradation caused by physical, biological and chemical processes runs up the changes of the key soil properties that have a pivotal role in geochemical cycling.

From the standpoint of soil degradation, the presence of some trace elements in a toxic concentration may be due to both natural and anthropogenic factors. Therefore, it may become quite difficult to discriminate among the different causes. The parent material largely influences trace metals content in many soil types, with concentration sometimes exceeding the critical values (Palumbo et al., 2000; Romić & Romić, 2003; Salonen & Korkka-Niemi, 2007). Some metals, such as Ni, Cr and Mn, are contained as trace elements in some rock types of volcanic and metamorphic origin (Alloway, 1995). During weathering processes the primary crystalline structures of some rock minerals are completely broken, relevant chemical elements may be thus either adsorbed in the topsoil or transported towards surface water or ground water targets. The transformation of metals and metalloids is influenced significantly by adsorption-desorption reaction in soil environments; these reactions are affected by physicochemical and biological interfacial interactions, which should be especially important in rhizosphere (Huang, 2008). Soil buffer capacity may be defined as its ability to postpone the negative effects of more or less continuous input of toxic substances by inactivation of contaminants (Moolenar & Lexmond, 1999). This inactivation can generally be reached by effective binding of contaminant and soil particles, or by forming of insoluble complexes. When the contaminant input exceeds the level of so-called «critical content», their buffer capacity is getting overcome as well and then the soil is characterized as polluted (de Haan, 1996). So that, buffer capacity of the diverse soil types defers considerably regarding the soil characteristics reflecting thus its vulnerability or resistance.

Generally, two main types of pollution may be distinguished: diffuse pollution or non-point source, and point source (O'Shea, 2002). The example of the non-point source is atmospheric deposition as a result of urban, transport and construction activities, as well as mineral fertilizer or sewage sludge application in agriculture. Diffuse sources of pollution are not easy to control, and the best methods for soil pollution control often depend on the legal regulations and management strategies. It becomes easier to control point sources of pollution, because it usually refers to the single source that is easy to identify (local pollution caused by chance, accidentally or undertaking prohibited activities). Sources of agricultural land contamination, especially if places near urban or industrial area, certainly have a diffuse nature.

Geochemical maps are good visual demonstration of contaminant changes in the space, and enable the identification of the areas that are likely to contain harmful substances (Goodchild et al., 1993). The knowledge on spatial variability becomes equally important both for the assessment of the study site and for the prediction of the possible risks. The procedure of the geochemical surveys, including exhausting field work, long-lasting and expensive chemical analysis, require the sampling scheme optimization for the efficient interpolation and mapping.

4.2 Soil pollution assessment

A soil pollution assessment becomes very difficult to carry out when different sources of contamination are present and their products are variably distributed. In these cases the spatial variability of the trace metal concentrations in soils is basic information for identifying the possible sources of contamination and to delineate the strategies of site remediation. An approach was described that interpolate sampled trace metal concentrations using numerous environmental predictors and then represent the overall pollution by using the continuous limitation scores, as proposed by Romić et al. (2007). Such visualizations can supplement maps of separate trace metals so that the areas of high overall pollution can be more easily delineate presenting the basis for further studies on risk assessment or decision making.

In Europe, decision makers and spatial planners more and more require information on soil quality for different purposes: to locate areas suitable for organic (ecologically clean) farming and agro-tourism; to select sites suitable for conversion of agricultural to non-agricultural land, particularly for urbanization; setting up protection zones for groundwater pumped for drinking water; to estimate costs of remediation of contaminated areas and similar. Every decision on the application of any measures in the environment relating to soil quality and management, whether statutory regulations or practical actions, must be based on reliable and comparable data on the status of this part of environment in the given area. Various aspects must be considered by the society to provide a sustainable environment, including a soil clean of heavy metal pollution. The first among them is to identify environments (or areas) in which anthropogenic loading of heavy metals puts ecosystems and their inhabitants at health risk. Maps indicating areas with pollution risks can provide decision-makers or local authorities with critical information for delineating areas suitable for the planned land use or soil clean up (Van der Gaast et al., 1998; Broos et al., 1999). Maximum permissible concentrations of trace metals in soil are now regulated by law in many countries.

In urban and industrial environments there are numerous potential sources of contamination with harmful substances, including trace metals, mainly combustion processes in industry and transportation. Lead and cadmium are the main trace elements arising from combustion and are often associated with zinc owing to tyre wear on the roads. Waste water from industrial processes may contain an important load of zinc, copper, chromium and nickel. Moreover, mining activities for extraction and manufacturing of metal products may result in a large amount of pollutants to be released into the atmosphere and, secondly, in the adjoining soils and waters.

Long-term and extensive use of agricultural land with frequent application of growing practices and use of pesticides (Nicholson et al., 2003) may cause heavy metals such as copper, nickel, zinc and cadmium to be strongly accumulated in the topsoil.

The estimation of the total trace metals content and spatial variability of these elements in soil is the main indicator of the degree of contamination, but is not sufficient for establishing the relevant guidelines or decision making. GIS-based mapping techniques in conjunction with statistical and geostatistical analysis of the data are widely used to highlight the influence of human activities on the trace metals content of topsoils in urban and sub-urban areas (Kelly et al., 1996) and to assess the transport pathways, sinks, and impact of particulate associated trace metals in the various spheres making up the urban environment (Charlesworth et al., 2011). Intensive urbanisation of the Croatian capital of Zagreb has led to a situation where very good agricultural soils, particularly for vegetable production, are entrapped within urban and suburban areas. On the example of the Zagreb region (Northwest Croatia), different approaches to the assessment of the soil trace metals spatial variability and level of contamination will be demonstrated.

4.2.1 Data acquisition

The research on pollution in agricultural soils of the Zagreb region and establishment of the monitoring on a regional scale has started in 1997. At the beginning, the survey was carried out on approximately 860 km^2 of urban and suburban areas of the city of Zagreb. Later on, the research was extended for about 3000 km^2, covering agricultural soils of the City of Zagreb and the surrounding Zagreb County.

A total of 916 topsoil (0-20 cm) samples were collected using a systematic sampling on a 2-km grid, with sampling density increasing to 1 km within the industrial and residential area and wine-growing areas (Fig. 2). The observation sites were spatially referenced using GPS and data were stored in different GIS layers. Site survey, carried out during sampling process, provided site-specific information related to land use and other human activities near the sampling points.

For the determination of soil properties, the surface soil samples (average weight of 2 kg) were air-dried and mixed well. A subsample of about 1 kg soil was sieved through a 0.5-mm mesh. Digestion in aqua regia (HRN ISO11466, 2004) was done by the microwave technique on a PerkinElmer Multiwave 6MF 100 (1000W) apparatus in closed TFM vessels and with automatic pressure and temperature regulation. Heavy metal concentrations in soil digests were determined by inductively coupled plasma optical emission spectroscopy (ICP-OES) on a VistaMPX AX (Varian).

Before any solution for the problem of soil heavy metal pollution can be suggested, a distinction needs to be made between natural anomalies and those resulting from human activities. Namely, it often happens that also natural concentrations and distribution of potentially toxic metals could present health problems, like in the case of chromium, cobalt, and particularly nickel in ultramafic soils (Proctor & Baker, 1994). Rock type and geological-geochemical processes can change markedly in a relatively small area, resulting in great spatial variability in the soil content of elements. The region exhibits a variety of soils developed on diverse lithologies described in detail by Sollito et al. (2010). The oldest stratigraphic units are represented by the Paleozoic magmatic and metamorphic complex of the deep earth crust (Fig.

3a), mostly comprising diabases, gabbros, greenshist and blueshist facies rocks outcropping at Mt Medvednica (Belak & Tibljaš, 1998). These types of rocks are known to have a chemical composition characterized by high content of heavy metals, such as Ni and Cr, which are accumulated during the weathering processes in the soil (Alloway, 1995). Mesozoic calcareous rocks (mostly dolomite and limestone) outcrop at NW. Paleozoic-Mesozoic massif is rimmed by Miocene sediments due to the presence of tectonic structures. These deposits are composed by a transgressive sequence of calcareous breccias and conglomerates, marls, clays, sands and silts (Vrsaljko et al., 2005). The main geo-lithological features are the Pliocene and Quaternary alluvial sediments of the Sava River basin in the central and southern parts of the studied area. Terraced sediments outcrop mainly in the western and southwestern sectors of the region and consist of gravel and sands, and secondarily of sandy and silty clays. The sediments in the floodplain area and in the recent stream beds consist of coarse grained sandy-clayey silts and silty clays, with thin layers of charcoal that were flooded from the Slovenian coal mines. Moreover, Pleistocene deposits are made of pond sediments and non-carbonate loess mixed with sand and gravel. Quartz is the main component of the light mineral fraction of these sediments.

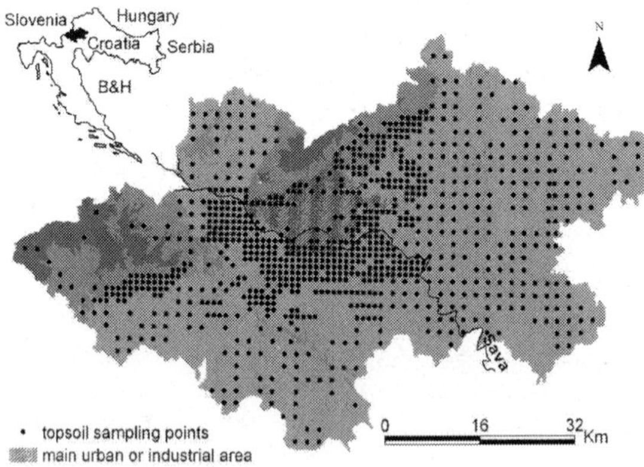

Fig. 2. Study area and sampling locations

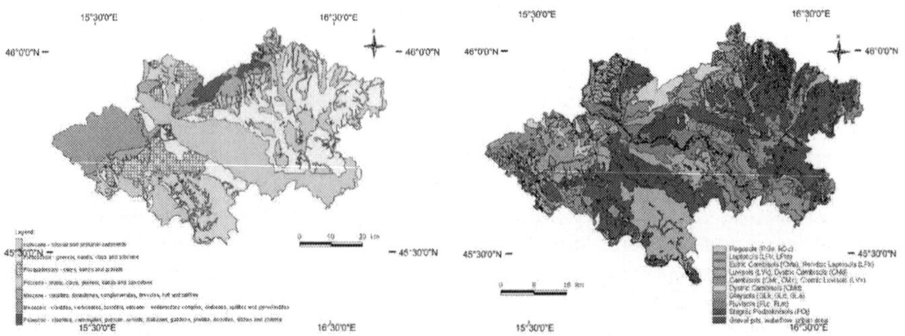

a. b.

Fig. 3. a) Simplified geologic map of the studied area; b) Simplified soil map of the study area

The large heterogeneity of the parent material, combined with the climate and geomorphology of the Zagreb region, has caused the development of a wide variety of soil types (Fig. 3b). Since the floodplain soils are mixture of the material eroded from the geological units present in the catchment basin, the mineralogical and chemical composition as well as the textural properties of the sedimentary deposits in the alluvial region are largely influenced by the dynamic of the Sava River flow and its solid transportation. Hydromorphic soils prevail in the Sava River valley, where the recent deposits form fluvial terraces. Molic Fluvisols, Calcaric Fluvisols, Eutric Cambisols, Eutric and Calcic Greysols are developed on the Holocene deposits. In the area of Pleistocene terraces, Stagnic Podzoluvisols prevail on the plateau, whereas Stagnic Podzoluvisols and Glayic Podzoluvisols are prevalent on the slopes (FAO, 1998). More than 50% of the land is used for agriculture: soils developed on loamy aeolian materials are mostly used for intensive cultivation of field crops (like cereals), but some of them are under permanent grassland. Anthropogenic vineyard soils, classified as Aric Anthrosols (FAO, 1998), of the Mt Medvednica foothills were formed on Tertiary carbonate deposits of marl and limestone. Moreover, as the Žumberačka Gora piedmont spreads perpendicularly to the mountains, some slopes are firmly interlinked by ridges, forming well protected, amphitheatre-shaped vineyard areas. The presence of intense agricultural activities imposes the risk of soil contamination due to the use of pesticide, which sometimes may constitute a diffuse source over large regions. In addition to agricultural land use other main usage classes are forest, covering the mountain areas and some parts of the terraces, pasture and orchard. Local source of pollution may be related to industrial areas and urban networks, which are scattered distributed over the floodplain, as well as to mining activities which are mainly located in the upstream regions.

Soils in the vicinity of urban areas and industry are exposed to input of potentially toxic elements, and the situation of agricultural soils gets additionally complicated due to continuous application of agrochemicals.

In practice, soil pollution by heavy metals is commonly assessed by interpolating concentrations of heavy metals sampled at point locations, so that each heavy metal is represented in a separate map (Fig. 4).

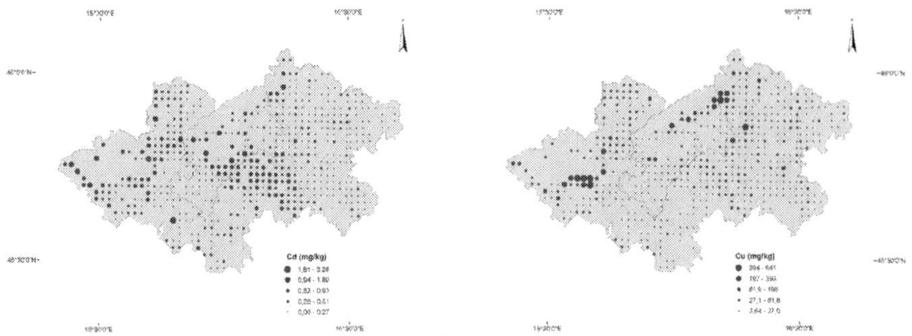

Fig. 4. Cadmium and copper contents at the point locations of the study area

4.2.2 Assessing spatial variability

Spatial variation of soil properties within a field, including trace element contents, is extremely complex, even in a small scale. Therefore it is important to apply methods that may describe this variability. First step in the data processing is the exploratory analysis in spatial data processing. Summary statistics calculation on the dataset (shown in Table 1) provided information about the frequency distribution on the concentrations of chemical elements in the topsoil; the results were compared with some reference values (Table 1).

Variable	Mean	Median	SD	Min	Max	Skewness	Kurtosis	Soils worldwide[a]	Upper continental crust[b]	MPCc
Ba ($mg\ kg^{-1}$)	224	220	68.2	12.7	421	0.19	3.09	500	584	-
Ca ($g\ kg^{-1}$)	18.4	5.20	37.3	0.67	214	3.25	13.4	14	29.5	-
Cd ($mg\ kg^{-1}$)	0.40	0.31	0.34	0.02	4.94	5.58	57.7	0.3	0.102	2
Co ($mg\ kg^{-1}$)	10.9	10.8	3.62	2.74	47.2	1.96	16.7	10	24	50
Cr ($mg\ kg^{-1}$)	54.6	51.2	21.1	11.5	400	5.86	91.2	80	126	100
Cu ($mg\ kg^{-1}$)	56.1	23.5	117	3.64	1335	4.56	29.3	25	25	100
Fe ($g\ kg^{-1}$)	29.7	28.9	7.77	5.85	59.1	0.26	2.91	35	30.9	-
Mg ($g\ kg^{-1}$)	8.04	6.51	5.43	0.77	36.6	2.98	12.7	9	13.5	-
Mn ($mg\ kg^{1}$)	597	556	266	79.2	4537	4.28	56.0	530	527	-
Ni ($mg\ kg^{-1}$)	35.2	29.7	23.8	0.70	488	8.51	148	20	56	60
P ($mg\ kg^{-1}$)	722	675	300	213	3023	2.36	13.6	750	665	-
Pb ($mg\ kg^{-1}$)	23.2	19.6	14.4	1.00	216	5.24	54.2	17	14.8	150
S ($mg\ kg^{-1}$)	407	349	249	47.7	2683	2.52	16.7	800	-	-
Sr ($mg\ kg^{-1}$)	92.9	55.5	136	17.2	1846	6.17	56.5	240	333	-
Zn ($mg\ kg^{-1}$)	77.9	70.7	33.6	27.1	479	4.28	39.6	70	65	300

[a] Median, after Reimann and de Caritat (1998)).
[b] Mean, after Wedepohl (1995).
[c] Maximal permissible concentrations defined by the Croatian government

Table 1. Summary statistics of element concentrations in soils of the study area

The background concentrations of the elements in soils were mostly lower than the average element concentrations in the upper continental crust (Wedepohl, 1995) and were similar to the worldwide median values in soils (Reimann & de Caritat, 1998). Except for the cobalt, the concentrations of all other elements exceeded the maximum permissible concentration (MPC) defined by the Croatian government, with critical conditions recorded for copper, nickel and chromium. For these elements, the maximum concentrations exceed the MPC by up to ten times. Nevertheless, the mean values for all samples were lower than the critical thresholds, meaning that the high values are unevenly distributed in the region. Only a relatively small proportion of samples had concentrations of Ni, Cu and Cr exceeding the MPC (9%, 8% and 1.4% of all samples, respectively). As confirmed by the skewness values (Table 1), the concentrations of elements are characterized by large variability, with positively skewed frequency distributions. This is common for heavy metals, because they usually have low concentrations in the environment, so that the presence of a point source of contamination may cause a sharp increase of local concentration, so exceeding the thresholds. The concentrations of Ni and Cr higher than the MPC were mainly recorded near the urban area of Zagreb and on the northern edge of the city, whereas the Cu concentration exceeding MPC was found all over the piedmont belt surrounding the northern Paleozoic and Mesozoic relieves. Only the Fe and Ba data distributions are nearly normal, with small differences between the mean and median values and the skewness values close to 0. In contrast, the Mg and Ca concentrations in the topsoil showed a skewed distribution with high values where calcareous rocks outcrop.

As confirmed by the skewness values (Table 1), the concentrations of elements are characterized by large variability, with positively skewed frequency distributions. This is common for trace metals, because they usually have low concentrations in the environment, so that the presence of a point source of contamination may cause a sharp increase of local concentration, so exceeding the thresholds. Skewnes measures the asymmetry of the observations. Normal distribution is symmetrical and its mean, mode and median coincide at its centre. When the distribution is skewed than the mean does not represent the central data value that causes the unreliability of the statistics. The problem of processing geochemical and environmental data sets is elaborated in details elsewhere (Zang & Selinus, 1998; Reinman & Filmoser, 2000; Webster & Oliver, 2001). Data transformation enables approaching to the normal distribution, reduces the influence of high values, stabilizes variance and thereby enables the next data processing, as shown on Figure 5.

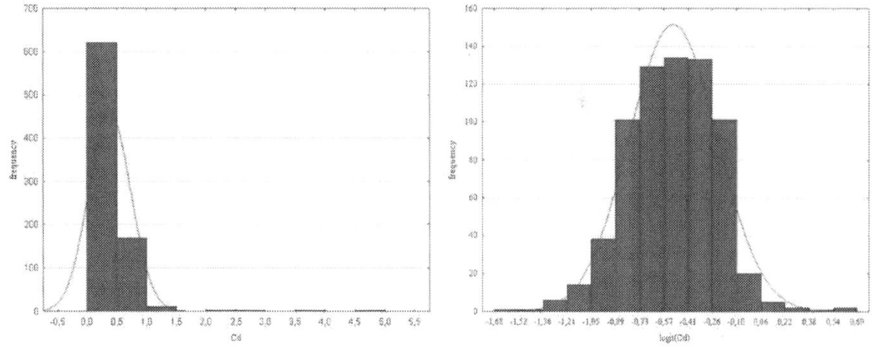

Fig. 5. Histogram of cadmium concentrations before and after logit transformation

Classic geostatistics methods based on univariate analysis can provide distribution maps for each pollutant and delineate the areas where its concentration exceeds the threshold values (Fig. 6). Although the concentration patterns of the elements could suggest probable sources of contamination (Atteia et al., 1995) each chemical element or pollutant might have its own distinctive spatial distribution making difficult to get an overall picture of the contamination.

To solve a problem of presenting overall polluted areas, Romic & Romic (2003) applied factor analysis prior to interpolation and then interpolated factor scores (Fig. 7). As a multivariate method, factor analysis (FA) facilitates the reduction, transformation and organization of the original data by the use of intricate mathematical techniques, which eventually results in a sample form of factor model. Factor analysis creates a new set of uncorrelated variables, which are the linear combination of the original ones with the same amount of information. Since the FA is conducted if the original variables have significant linear intercorrelations, the first few factors will include the largest part of the total variance. The interpretation of dominant factors was made by taking into account the highest factor loadings on chemical elements. The theoretical details of the FA are given by Johnson (1998).

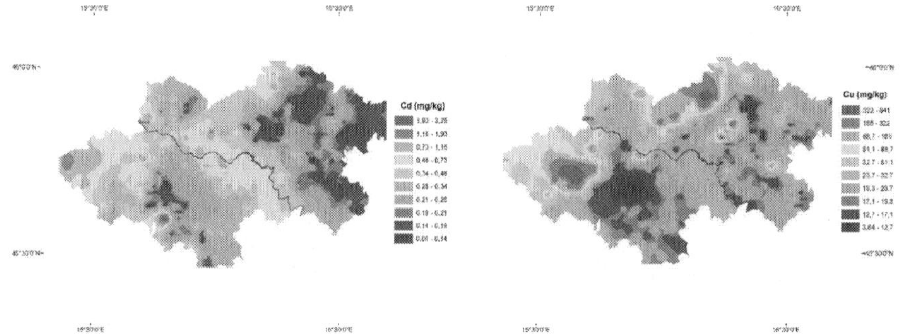

Fig. 6. Interpolated maps of cadmium and copper contents in soils of the study area

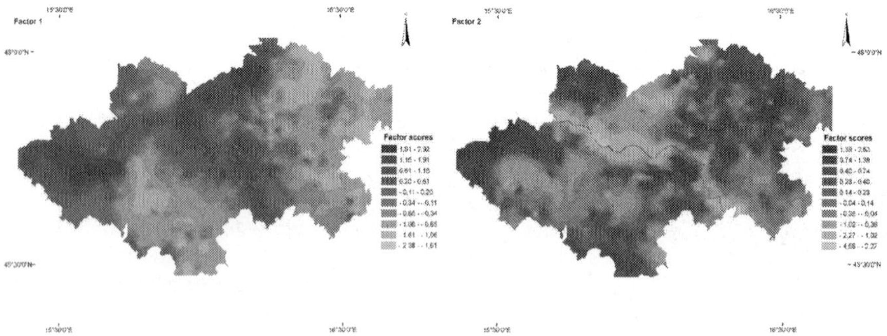

Fig. 7. Contour maps of the distribution of factor scores

Van der Gaast et al. (1998) used maps of background values of soil contaminants focusing on the 90-percentiles. Hanesch et al. (2001) tested fuzzy classification algorithms to distinguish different sources of pollution. Amini et al. (2004) classified HMCs using unsupervised fuzzy k-means to partition the values optimally. The final outputs are maps of memberships to each cluster, which commonly reflect the combination of most correlated heavy metals. In all these examples the procedures are statistically valid, but the meaning of such factors and continuous memberships is hard to interpret. In practice, decision makers usually only wish to see the areas that are polluted without any training in (geo)statistics. Legislative regulations on maximal permitted soil concentrations of potentially toxic elements from practical reasons are setting sharp boundaries. As additional criteria for the soil pollution assessment other soil properties that control metal behaviour in soil are usually set up. Therefore, geochemical data have to be integrated with the detailed soil characterization. Spatial variation of soil properties within a field is extremely complex, even in a small scale. Main questions interrogating in so doing are how we can measure and model this variability and how this variability affects land use decision-making and environmental quality.

For the study area of NW Croatia, it was assumed that the distribution of trace element contents is systematic, i.e. controlled by natural and anthropogenic factors. The problem appears when the complex interactions between these factors allows for local yield of high natural concentrations of potentially toxic metals in soils, which may exceed the threshold limits designated for contaminated areas (Myers & Thorbjornsen, 2004, Amorosi & Sammartino, 2007). Therefore, statistical and spatial analysis tools were utilised in order to combine the quantitative information obtained from the chemical analysis of the soil samples with the area-specific qualitative information. Romić et al. (2007) applied the regression block kriging for spatial interpolation of heavy metals. A list of potential predictors was used as auxiliary data in the RK model (geological and land cover map, NCVI, water table depth, slope, distance to urban area and roads and wind exposition. After the heavy metal concentrations had been interpolated, they were converted to limitation scores. This study was shown that in the Zagreb region, only 7.2% of the total area is critically polluted by one or more heavy metals.

The applied procedure for geostatistical analysis of heavy metal concentration data successfully identified a number of contamination hotspots in the studied region. The limitation of using the scores is that the high overall pollution can be due to very high values of single element, or due to a cumulative effect of large number of contaminants. But the methodological framework of cumulative limitation scores (CLS) opens several perspectives, i.e. to relate the CLS directly with the remediation costs (Broos et al., 1999), or to observe how heavy metal concentrations change at different scale. In addition, one might consider the methodology of error propagation (Heuvelink, 1998) to derive the composite uncertainty of the final soil pollution map. Geostatistical simulations would help us to get an idea about the propagated uncertainty, but can also be used as an input to a more complex environmental data modelling.

The maps of cumulative limitation scores might be advantageously used to target sampling and/or delineating contaminated zones with lower costs. However, the procedure is suitable in many other decision situations, especially for making decisions about soil remediation, health risk assessment or contaminated land management in general.

5. Conclusion

Risk assessment is the basic element of sustainable management of the agricultural environment because it provides answers to the question about how safe is a studied medium for a population in a given time period and under defined conditions of that medium and population. Risk assessment is the basis for selecting sustainable solutions for allocation of agricultural soils as well as for undertaking efficient measures for remediation of contaminated soils. In the case of trace metals, risk assessment is a demanding and complex task. Metals have a very complex chemical behaviour in the environment and, unlike organic compounds, are not subject to degradation processes. The problem of metals in agricultural environments is enhanced by their potential for mobilization under specific soil physical and chemical conditions. A direct effect of this mobilization is a faster translocation of metals into deeper soil layers, and thereby possible contamination of groundwater. Increased mobility of metals in soil is a frequent effect of fertilization or application of composts and other organic amendments in a solid form. Therefore, metals risk assessment is an important tool for establishing protective levels in different environmental media. The critical scientific issues that need addressing are the various properties of metals environmental chemistry, bioavailability and bioaccumulation. Bioavailability assessment methods need to be elaborated on as well.

Chemical speciation of metals, their bioavailability, bioaccumulation and toxicity are the key elements in assessing the potential harmful impacts of metals upon the environment and human health. Metal bioavailability is a complex issue that depends on a series of properties related to soil matrix, plant characteristics, and environmental conditions. An interdisciplinary approach is therefore required in risk assessment studies. Bioavailability can be used as an element of risk assessment only if all risk assessors apply the same approach. Such a uniform approach to bioavailability assessment will enable establishment of standard methods for soil analyses and contribute towards developing new legislation on soil protection, particularly those related to persistent immobile substances such as metals and polycyclic aromatic hydrocarbons.

Climate changes are currently the focus of numerous studies because their indirect consequences can affect the geochemical and biological cycling of metals as well. The main aim of these studies is to define the indicators that best describe bioavailability of a metal under given environmental conditions. Reliable information on soil contamination by trace metals is needed, as well as the application of geostatistical methods for determining trace metal variation in soil and identifying possible sources of contamination.

6. References

Adriano, D.C., Chlopecka, A., Kaplan, D.I., Clijsters, H., Vangronsveld, J. (1995). Soil contamination and remediation: philosophy, science and technology. In: Prost R., ed. *Contaminated Soils: Third International Conference on the Biogeochemistry of Trace Elements*, Paris, 15-19 May, 1995. (INRA editions, Paris, France, CD-room, full text)

Alloway, B.J. (1995). The mobilisation of trace elements in soils, In: *Prost R., ed. Contaminated Soils: Third International Conference on the Biogeochemistry of Trace Elements*, Paris, 15 - 19 May 1995. (INRA editions, Paris, France, CD-room, full text)

Alloway, B.J. & Steinnes, E. (1999). Anthropogenic Additions of Cadmium to Soils, In: *Cadmium in Soils and Plants, McLaughlin, M.J. and B.R. Singh (Eds.). Kluwer Academic Publishers,* Dordrecht, Netherlands, pp: 97- 123

Al-Wabel M.A, Heil D.M, Westfall D.G, Barbarick K.A. (2002). Solution chemistry Influence on metal mobility in biosolids-amended soils. *J. Environ. Qual.* 31: 1157-1165

Amini, M., Afyuni, M., Fathianpour, N., Khademi, H., Fluhler,H. (2004). Continuous soil pollution mapping using fuzzy logic and spatial interpolation. Geoderma 124 (3–4), 223–233

Amorosi, A., Sammartino, I. (2007). Influence of sediment provenance on background values of potentially toxic metals from near-surface sediments of Po coastal plain (Italy), *International journal of earth sciences*, 96, 2 , 389-396

Andrews, P., Town, R.M., Hedley, M.J., Loganathan, P. (1996). Measurement of plant-available cadmium in New Zealand soils. *Australian Journal of Soil Research*, 34 (3): 441-452.

Atteia, O., Thelin, P.H., Pfeifer, H.R., Dubois, J.P., Hunziker, J.C. (1995). A search for the origin of cadmium in the soil of the Swiss Jura. *Geoderma, 68*, 149–172

Belak, M., Tibljas, D. (1998). Discovery of blueschists in the Medvednica Mountain (northern Croatia) and their significance for the interpretation of the geotectonic evolution of the area, *Geologia Croatica* 51 (1), 27–32

Bross, M.J, Aarts, L., van Tooren, C.F., Stein, A. (1999). Continous soil pollution mapping using fuzzy logic and spatial interpolation, *Geoderma*, 124 (3-4), 223-233

Bross, M.J., Aarts, L., van Tooren, C.F., Stein, A. (1999). Quantification of the effects of spatially varying environmental contaminants into a cost model for soil remediation, *Journal of Environmental Management*, 56 (2), 133–145

Brun, L.A, Maillet J, Richarte J, Herrmann P, Remy J.C. (1998). Relationships between extractable copper, soil properties and copper uptake by wild plants in vineyard soils, *Environmental Pollution* , 102: 151-161

Chaignon V, Hinsinger P. (2003). Heavy Metals in the Environment: A Biotest for Evaluating Copper Bioavailability to Plants in a Contaminated Soil, *J. Environ. Qual.* ,32:824–833

Charlesworth, S., De Miguel, E., Ordóñez, A. (2011). A review of the distribution of particulate trace elements in urban terrestrial environments and its application to considerations of risk, *Environmental geochemistry and health,* 33, 2 , 103-123

Chen, W., Krage, N., Wu, L., Pan, G., Khosrivafard, M., Chang, A.C. (2008). Arsenic, cadmium, and lead in California cropland soils: role of phosphate and micronutrient fertilizers. Journal of Environmental Quality, 37(2), 689-95.

Cornu, J. Y., Staunton, S., Hinsinger, P. (2007). Copper concentration in plants and in the rhizosphere as influenced by the iron status of tomato (Lycopersicon esculentum L.). *Plant Soil* , 292:63–77

De Haan, F.A.M. (1996). Soil quality evaluation. In: *Soil pollution and soil protection.* Ed. F.A.M. de Haan, M.I. Visser-Reyneveld. Wageningen Agricultural University and International Training Centre (PHLO). Wageningen. pp. 1 -17

Fang J, Wen B, Shan X, Lin J, Owens G. (2007). Is an adjusted rhizosphere-based method valid for field assessment of metal phytoavailability? application of non-contaminated soils, *Environmental Pollution* 150: 209-217

FAO/ISRIC/ISSS (1998). World reference base for soil resources, *World Soil Resources Report*, 84, FAO, Rome

Feng M. H, Shan X. Q, Zhang S, Wen B. (2005). A comparison of the rhizosphere-based method with DTPA, EDTA, $CaCl_2$, and $NaNO_3$ extraction methods for prediction of bioavailability of metals in soil to barley, *Environmental Pollution* 137: 231-240

Ghallab, A., Usman, A.R.A (2007). Effect of Sodium Chloride-induced Salinity on Phyto-availability and Speciation of Cd in Soil Solution, *Water, Air, and Soil Pollution*, 185 (1-4). p.43

Gray, C.W., McLaren, R.G., Roberts, A.H.C., Condron, L.M. (1999). Cadmium phytoavailability in some New Zeland soils. *Australian Journal of Soil Research*, 37: 461-477

Goodchild, M.F., Parks, B.O., Steyaret, L.T. (Eds.) (1993). Environmental modeling with GIS, *Oxford University Press*, New York, USA

Gummuluru, S.R., Krishnamurti, G.S.R., Ravendra, N. (2002). Solid-Solution Speciation and Phytoavailability of Copper and Zinc in Soils. *Environmental Science & Tecnology*, 36 (12) 2645-2651

Hanesch, M., Scholger, R., Dekkers, M.J., (2001). The application of fuzzy c-means cluster analysis and non-linear mapping to a soil data set for the detection of polluted sites, *Physics and Chemistry of the Earth, Part A: Solid Earth and Geodesy*, 26 (11–12), 885–891

Harris, R.F.D., Karlen, D.L., Mulla, D.J. (1996). A conceptual framework for assessment and management of soil quality and health, *Methods for assessing soil quality*. Ed. Doran J.W., A.J. Jones, Soil Science Society of America, Special Publication, 49: 61-82

Hengl T., Heuvelink GBM., Stein A. (2004). A generic framework for spatial prediction of soil variables based on regression-kriging, *Geoderma*, 122(1-2), 75-93

Heuvelink, G. (1998). *Error Propagation in Environmental Modelling with GIS*. Taylor & Francis, London, UK, pp 127

Johnson, D.E. (1998). Applied multivariate methods for data analysts. Duxbury Press, Pacific Grove, CA, USA

Kaiser, K. & Guggenberger, G. (2003). Mineral surfaces and soil organic matter, *European Journal of Soil Science*, 54, 219-236.

Kelly, J., Thornton, I., Simpson, PR. (1996). Urban Geochemistry: A study of the influence of anthropogenic activity on the heavy metal content of soils in traditionally industrial and nonindustrial areas of Britain. *Applied geochemistry*, 11 , 1-2, 363-370

Manouchehri, N., Besancon, S., Bermond, A. (2006). Major and trace metal extraction from soil by EDTA: Equilibrium and kinetic studies. *Analytica Chimica Acta* 559: 105-112

McBride M.B. (1994). *Environmental chemistry of soils*, Oxford University Press, New York

McLean & Bledsoe, (1992). *EPA Ground Water Issue*, EPA 540-S-92-018:25 pp

Moolenaar, S.W., Lexmond, T.M. (1999). General aspects of cadmium, copper, zinc, and lead balance of agro-ecosystems, *Journal of Industrial Ecology*, 2 (4): 45-60

Myers, J; Thorbjornsen, K. (2004). Identifying metals contamination in soil: A geochemical approach, *Soil & sediment contamination*, 13 , 1 , 1-16

Nicholson, FA., Smith Sr., Alloway, BJ., Carlton-Smith, C., Chambers, BJ. (2003). An inventory of heavy metals inputs to agricultural soils in England and Wales, *Science of the total environment*, 311 , 1-3, 205-219

N'guessan, Y.M., Probst, J.L., Bur, T., Probst, A. (2009). Trace elements in stream bed sediments from agricultural catchments (Gascogne region, S-W France): Where do they come from? Science of the total environment 407, 2939– 2952

Odeh I, McBratney A, Chittleborough D. (1995). Further results on prediction of soil properties from terrain attributes: heterotopic cokriging and regression kriging, *Geoderma*, 67 (3-4) 215-226

Ondrašek, G., Romić, D., Rengel, Z., Romić, M., Zovko, M., (2009). Cadmium accumulation by muskmelon under salt stress in contaminated organic soil. *Science of the Total Environment*, *407*(7), 2175-2182

O'Shea, L. (2002). An economic approach to reducing water pollution: point and diffuse sources, *Science of the Total Environment*, 282:49-63

Palumbo, B., Angelone, M., Bellanca, A., (2000). Influence of inheritance and pedogenesis on heavy metal distribution in soils of Sicily, Italy, *Geoderma*, 95 (3–4), 247–266

Peijnenburg, W.J.G.M, Jager, T. (2003). Monitoring approaches to assess bioaccessibility and bioavailability of metals: Matrix issues, *Ecotoxicology and Environmental Safety*, 56: 63–77

Peris, M., Recatalá L., Micó, C. , Sánchez, R., Sánchez, J. (2007). Increasing the Knowledge of Heavy Metal Contents and Sources in Agricultural Soils of the European Mediterranean Region, *Water Air Soil Pollut.*, 192: 25–37

Proctor, J., Baker, A.J.M. (1994). The importance of nickel for plant growth in ultramafic (serpentine) soils, In: *Toxic Metals in Soil-Plant System. Ed. S.M. Ross. John Wiley & Sons Ltd.* pp. 417-432

Reichman S.M (2002). The responses of Plants to Metal Toxicity: A review on Copper, Manganese and Zinc. *Australian Minerals & Energy Environment Foundation.* www.ameef.com.au

Reimann, C., de Caritat, P. (1998). *Chemical Elements in the Environment.* Springer.

Reimann, C., Filzmoser, P. (2000). Normal and lognormal data distribution in geochemistry: death of a myth. Consequences for statistical treatment of geochemical and environmental dana, *Environmental Geology,* 39 (9): 1001-1014

Romić M., Hengel T., Romić D., Husnjak S. (2007). Representing soil pollution by heavy metals using continuous limitation scores, *Computers & Geosciences,* 33(10): 1316-1326

Romić M. and Romić, D. (2003). Heavy metals distribution in agricultural topsoils in urban area. *Environmental Geology, 43*(7), 795-805

Salonen, V., Korkka-Niemi, K. (2007). Influence of parent sediments on the concentration of heavy metals in urban and suburban soils in Turku, Finland, *Applied Geochemistry,* 22, 906–918

Shuman, L.M. (1985). Fractionation method for soil microelements, *Soil science,* 140 (1): 11-22

Singh, B.R. (1994). Trace element availability to plants in agricultural soils, with special emphasis on fertilizer inputs, *Environmental Reviews,* 2: 133-146

Sollitto, D., Romić, M., Castrignano, A., Romić, D., Bakić, H. (2010). Assessing heavy metal contamination in soils of the Zagreb region (Northwest Croatia) using multivariate geostatistics. *Catena, 80*(3), 182-194

Tack, F.M.G., Singh, S.P., Verloo, M.G. (1998). Heavy metal concentrations in consecutive saturation extracts of dredged sediment derived surface soils. *Environmental Pollution, 103*(1), 109-115.

Vandecasteele, B., Quataert, P. and Tack, F.M.G. (2007). Uptake of Cd, Zn and Mn by willow increases during terrestrialisation of initially ponded polluted sediments. *Science of the Total Environment,* 380, 133-143.

Van der Gaast, N., Leenaers, H., Zegwaard, J. (1998). The grey areas in soil pollution risk mapping the distinction between cases of soil pollution and increased background levels, *Journal of Hazardous Materials,* 61 (1–3), 249–255

Vrsaljko, D., Pavelić, D., Bajraktarević, Z. (2005). Stratigraphy and palaeogeography of Miocene deposits from the marginal area of Žumberak Mt. and the Samoborsko Gorje Mts. (Northwestern Croatia), *Geologia Croatica* 58 (2), 133–150

Wang G., Su, M., Chen, Y., Lin, F., Luo, D., Gao, S. (2006). Transfer characteristics of cadmium and lead from soil to the edible parts of six vegetable species in southeastern China. *Environmental Pollution* , 144: 127–135

Wang, Z., Shan, X., Zhang S. (2002). Comparison between fractionation and bioavailability of trace elements in rhizosphere and bulk soils. *Chemosphere,* 46: 1163–1171

Webster, R., Oliver, M. (2001). *Geostatistics for Environmental Scientists,* John Wiley & Sons, LTD.

Wedepohl, K.H. (1995). The composition of the continental-crust, *Geochimica et Cosmochimica Acta,* 59(7): 1217-1232

Yanai, J., Yabutani, M., Kang, YM., Huang, B., Luo, GB., Kosaki, T. (1998). Heavy metal pollution of agricultural soils and sediments in Liaoning Province, China, *Soil science and plant nutrition,* 44 , 3, 367-375

Zhang, C., Selinus, O. (1998). Statistics and GIS in environmental geochemistry, some problems and solutions, *Journal of Geochemical Exploration,* 64: 339-354

20

Evaluation of Soil Hydraulic Parameters in Soils and Land Use Change

Fereshte Haghighi[1], Mirmasoud Kheirkhah[2] and Bahram Saghafian[2]
[1]*Soil Conservation and Watershed Management Institute, Tehran,*
[2]*Soil Conservation and Watershed Management Research Institute, Tehran,*
Iran

1. Introduction

The knowledge of soil water properties and land-use effects on these properties are important for efficient soil and water management. Furthermore, the use of the pedotransfer functions (PTFs) to estimate soil water content (θ_h) is important to assess. The loosening effect of dryland farming on soil water retention is known. In this chapter we review soil water content, pedotransfer functions and some infiltration models applicability for two land-use types. The land-use effect on soil water retention may be significant at water potentials of -33 *kPa* and 0 *kPa* in the soil. At the -1500 *kPa* pressure head, water content may not be affected by cultivation of rangeland at different soil depths. In addition, pedotransfer functions can be used as a physically based model for soil water retention characterization in the various areas. Moreover, it is essential to evaluate the infiltration models applicability for different soils and various land-uses.

1.1 Definition

1.1.1 Soil hydraulic properties

Soil hydraulic properties govern transport processes and water balance in soils. Water retention capacity, infiltration rate, and saturated hydraulic conductivity are important soil hydraulic properties. Soil water retention and saturated hydraulic conductivity (K_s) are necessary input data for the simulations of water flow in soil and water engineering. Characterizing hydrological behavior of catchments requires knowledge of hydraulic parameters.

1.1.2 Soil water retention

Soil water retention at field capacity (*FC*) and permanent wilting point (*PWP*) are used to estimate the water depth applied by irrigation (Hansen et al., 1980), and to calculate water availability, as a crucial factor to assess the land area suitability for crop producing (Sys et al., 1991).

2. Soil water retention capacity and land use

One important soil hydraulic property is water retention capacity, which affects soil productivity and management. Soil water content (θ_h) governs the transport characteristics of water and solutes in soils. The knowledge of water retention capacity and land use effects on this property is important for efficient soil and water management. Upon conversion of natural lands to cultivated fields, water retention capacity is strongly influenced (Schwartz *et al.*, 2000; Bormann and Klaassen, 2008; Zhou *et al.*, 2008). Soil water retention at field capacity (*FC*) and permanent wilting point (*PWP*) are important to estimate the irrigation water depth which may be affected by land use change. Soil water retention characteristic, is affected by soil organic matter (SOM) content and porosity, which are significantly influenced by land use type (Zhou *et al.*, 2008).

We conducted a study to evaluate, document, and quantify the effect of cultivation of rangeland on soil water retention in field capacity (*FC*), permanent wilting point (*PWP*), and to test the use of the van Genuchten equation to estimate θ_h in cultivated and natural lands in the same soils of the Taleghan watershed in Iran.

Significant differences in the OM and bulk density (BD) were observed between dryland farming and rangeland at both depths of 0 cm - 15 cm and 15 cm - 30 cm. Soil sample water contents at different pressure heads under both land use types are presented in Figures 1. The overall measured and fitted soil water retention curves did not show significant difference within the selected water potentials for both land use types in this study. However, measured θ_s (0 *kPa*) values were found to be significantly lower for dryland farming when compared with rangeland at depths of 0 cm - 15 cm and 15 cm - 30 cm, respectively. Moreover, the land use effect on soil water retention was significant at a water potential of -33 *kPa* (*FC*) based on laboratory measurements only at the top (15 cm depth). The results indicated that the conversion of rangeland to dryland farming led to a significant decrease (16.56% on average) in the *FC* at a depth of 0 cm - 15 cm. The mean -1500 *kPa* (*PWP*) water content was not affected by the land-use type. Figure 1 indicates that the mean total field capacity (*FC*) was significantly greater in rangeland when compared with dryland farming at a depth of 0 cm - 15 cm. In this study, there were not statistically significant differences in water content at other potentials (-50 *kPa*, -100 *kPa*, -500 *kPa*, and -1000 *kPa* pressures) between the two types of land use presented in Figure 1. At those pressure heads and at a -1500 *kPa* water content, the amount of micropores were not affected by cultivation of rangelands (Fig. 1). Overall, the results showed that the soil pore system and reduced total porosity under dryland farming can decrease water storage capacity at water potentials of -33 *kPa* and 0 *kPa*. Ndiaye et al., (2007) has shown that improper soil management decreases the soil macroporosity in the long-term affecting the θ_s. The data obtained in our study demonstrated the loosening effect of dryland farming on soil water retention. Previous studies on the effect of land use have demonstrated clear changes in soil physical properties, such as soil porosity, SOM, and BD, in relation to hydraulic properties (Bormann and Klassen 2008; Haghighi et al., 2010b).

3. Pedotransfer functions (PTFs)

Determination of soil water properties required as input data for simulation models is time consuming and relatively costly (Wösten *et al.*, 1995). Thus, indirect estimation of these characteristics has been proposed as one alternative to direct estimation of the soil hydraulic parameters based on the measured water retention data. Pedotransfer functions (PTFs) are emerged as the relationship between soil hydraulic and other more available measured properties (Bouma, 1989) which can be used to estimate hydraulic parameters. PTFs are useful tools for modeling applications.

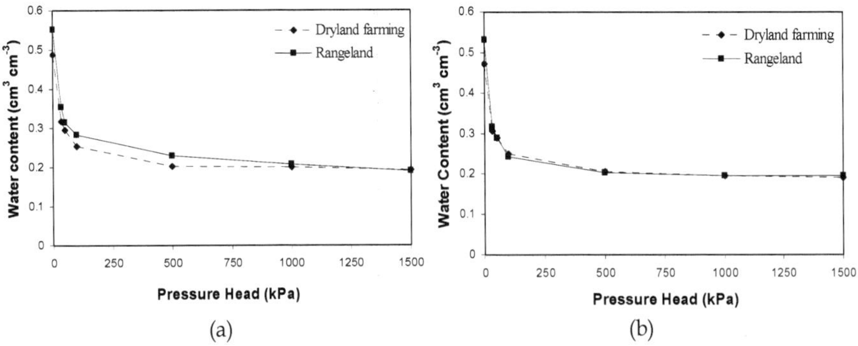

Fig. 1. Soil water content as a function of the pressure head for two landuse types at depths of (a) 0 cm -15 cm and (b) 15 cm - 30 cm.

4. Pedotransfer functions (PTFs) and different land uses

To estimate the land use effects on soil water retention, van Genuchten model (Van Genuchten, 1980) may be applied. Some researches have correlated van Genuchten parameters with soil organic matter, bulk density (BD), and soil particle size distribution and many researchers have estimated the water retention curve using soil texture, bulk density, and porosity.

Many statistical equations (pedotransfer functions) characterizing the water retention curve have been presented (Kutilek and Nielsen., 1994). PTFs are useful tools for modeling applications. Such analytical functions are derived involving various soil data. Such data are measured in the field and laboratory analysis. Soil hydraulic parameters derived through PTFs can be used to express soil hydraulic properties and water retention (Brooks and Corey, 1964). Consequently, physically based models such as van Genuchten representing a pedotransfer function may be considered as a valuable tool to simulate the soil water properties in different land uses.

The $\theta_{(h)}$ data may be fitted to van Genuchten equation to derive retention curves and parameters (a, n, and θ_r), using the RETC (RETention curve) optimization computer code (Van Genuchten *et al.*, 1991). The van Genuchten model (van Genuchten, 1980) is defined as:

$$\theta(h) = \theta_r + \frac{\theta_s - \theta_r}{(1+|\alpha h|^n)^m} \qquad \theta_{(h)} = \theta_s \quad h \geq 0 \tag{1}$$

Where $\theta_{(h)}$ (cm^3 cm^{-3}) is the volumetric water content (for h<0), θ_r (cm^3 cm^{-3}) is the residual water content, and θ_s (cm^3 cm^{-3}) is the saturated water content. Here, m is 1-(1/n) with n>1. a (cm^{-1}) and n are empirical parameters determining the shape of the curve which were obtained for each core. Parameter n is related to steepness of the water retention curve.

$$K(S_e) = K_s S_e[1-(1-S_e^{(1/m)})^m] \tag{2}$$

Where K_s (mm/h) is saturated hydraulic conductivity and S_e is the effective saturation expressed as:

$$S_e = \frac{\theta - \theta_r}{\theta_s - \theta_r} \tag{3}$$

The effect of landuse type on soil water retention and PTF applications have not been documented for different land-uses to the best of our knowledge. In developing countries, there is a lack of large databases that are needed to develop PTFs. Thus, in many developing countries, the use of available PTFs can cause errors for estimating soil hydraulic properties. This encourages further investigations of the model applications and development of suitable point and parametric PTFs for estimating soil hydraulic properties in the studied area. The selection of more suitable PTFs for application where there are not developed PTFs caused by a lack of large databases is difficult. Consequently, it is essential to evaluate the model applicability and to develop point and parametric PTFs for estimating soil hydraulic properties for the soils in various sites. Thus, the estimates may be improved by comprehensive local studies.

Location (site)	Depth: 10–20 cm					Depth: 25–50 cm				
	K_{sat} (cm/day)	θ_{sat} (–)	α (1/cm)	l (–)	n (–)	K_{sat} (cm/day)	θ_{sat} (–)	α (1/cm)	l (–)	n (–)
A	16.0	0.43	0.007	0.944	1.37	23.1	0.43	0.014	–0.350	1.41
B	16.0	0.42	0.007	0.331	1.36	15.8	0.40	0.009	–0.448	1.33
C	17.4	0.40	0.008	0.093	1.35	14.7	0.38	0.010	–0.482	1.33

Table 1. Mualem-van Genuchten parameters calculated for old grassland (site A), recently reseeded grassland (site B) and previous maize cultivated land (site C) (Sonneveld et al, 2003)

5. Evaluation of common infiltration models for different land-uses

The evaluation of infiltration characteristics as a hydrologic process in soils is necessary in agricultural studies. The knowledge of final steady infiltration rate is important for irrigation water efficiency, designing desirable irrigation systems, and loss of water. Thus, infiltration rate is important factor in sustainable agriculture, effective watershed management, surface runoff, and retaining water and soil resources. Since measuring the final infiltration rate is time consuming, several physical and empirical models have proposed to determine it. The empirical models such as Kostiakov (1932) and Horton (1940), and physical model such as Philip (1957) are the most common models to estimate infiltration rate of the soils.

5.1 Kostiakov-Lewis model

The model of Kostiakov modified for long times as follows:

$$f = at^{-b} + f_c \tag{4}$$

Where a and b are the equation's parameters ($a>0$ and $0<b<1$). i_c is the steady infiltration rate (LT^{-1}).

5.2 Horton's model

The Horton's infiltration model (Horton, 1940) is expressed as follows:

$$f = (f_0 - f_c)e^{-kt} + f_c \qquad (5)$$

Where i_c is the presumed final infiltration rate (LT^{-1}), i_0 is the initial infiltration rate (LT^{-1}) and t is time (T). k is the infiltration decay factor.

5.3 Philip two-parameter model

The Philip two-term model is expressed as (Philip, 1957):

$$f = \frac{1}{2}St^{-0.5} + A \qquad (6)$$

Where f is the infiltration rate (LT^{-1}) as a function of time.
A= Transmissivity factor (LT^{-1}) as a function of soil properties and water contents, S = Sorptivity that is function of soil matric suction ($LT^{-0.5}$).
t= time (T)
Singh (1992) expressed that the various models can estimate different values of the final infiltration rate in a soil which seems to be uncorrect, because of the final infiltration rate is a soil-dependent factor. Compared to the previous investigations on soil infiltration properties and models, studies on soil infiltration modelling depending on land use are scarce. Nevertheless, it can be assumed that landuse type have a significant impact on soil infiltration and infiltration models performance. Machiwal et al (2006) observed the infiltration process was well described by the Philip's model in a wasteland of Kharagpur, India. However, different soil management that influences the final infiltration rate is a major reason for different applicability of these models. Long-term effects of land use changes on soil infiltration and infiltration models (e.g., Horton, Kostiakov, and Philip models) can be observed (Navar and Synnott, 2000; Shukla et al, 2003).
Thus, the variability of soil infiltration characteristics and goodness of fit of the infiltration models for different land-uses should be considered during infiltration modelling studies helping on correct predictions of final infiltration for different land uses. Ability of these models for estimating the infiltration rate in different land-uses and soil management has been examined by some researchers. Gifford (1976) observed among the Horton, Kostiakov and Philip's models, the Horton's model was the best model to fit the infiltration data in mostly semi-arid rangelands from the Australia, but only under specific conditions. Shukla et al (2003) evaluated some of the infiltration models at different soil management and land-use systems in Ohio and observed among infiltration models, the Swartzendruber model was the best ones and fitted the observed infiltration data with lower sum of squares and higher model efficiency. Davidoff and Selim (1986) examined the goodness of fit for eight infiltration models on a Norwood soil with four winter cover crop treatments and results of their study showed that the Philip, Kostiakov and Horton's models had best predictions than the other models. Haghighi et al (2010b) evaluated the effects of rangeland and dryland farming land uses on performance of some infiltration models to estimate the final infiltration rate of soils. The study was conducted on some soils of Taleghan watershed, Iran. According to reports (Taleghan watershed study report, 1993), investigated soils are calcareous and classified as Typic Xerorthents. Mean annual rainfall alters from 464 to 796 mm and lands slope is by 15 %. The soil texture varied from clay-loam to silty clay loam.

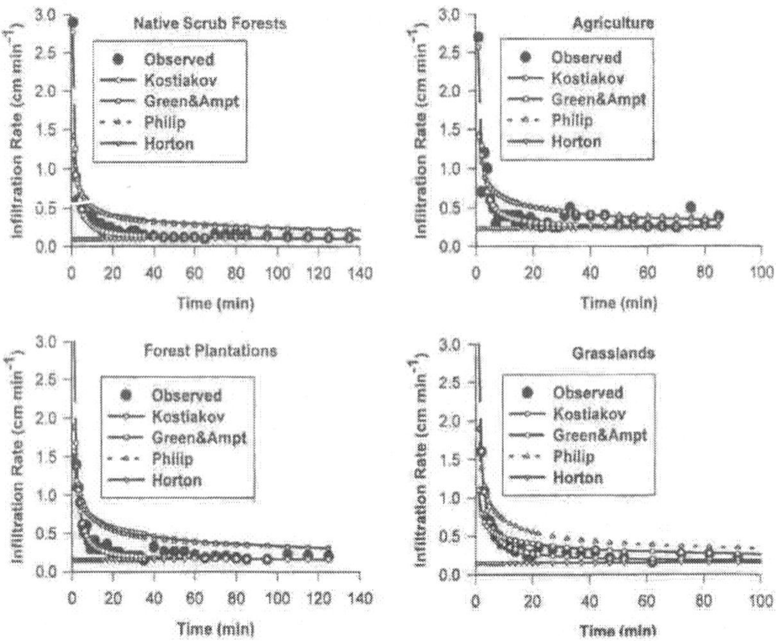

Fig. 2. Effects of land use changes on soil infiltration and infiltration models (Navar and Synnott, 2000)

In our study, the goodness of the fit of selected models and ability of them for estimating the final infiltration rate of rangeland and dryland farming soils was evaluated using the root mean squared errors (RMSE). The values of R^2 were determined high (0.99) and equal for all sites and land-uses, but the values of RMSE and the final steady infiltration showed that the estimated infiltration rates by the infiltration model of Horton, approached more closely to the measured ones at the selected area [Table 1]. The Horton's model was the best model selected for both of land-uses. It can be expressed that various models can suppose different final infiltration (f_c) values for a soil, which seems to be not practical, because f_c is a soil-dependent parameter, in general. Common changes in land-use negatively affect soil physical properties and decrease soil infiltration rate and could change modelling performance. Effect of land-use should be well documented aiming on good predictions in the studied areas and elsewhere.

The infiltration models can be used for estimating the infiltration rate in soils, well. But only, one or some of these models are better and appropriate for a specific site. Thus, the infiltration models should be analyzed for their ability to estimate the infiltration rate of each location. The investigation of Haghighi et al (2010a) showed that the Horton's model is the best ones selected for rangeland and dryland farming and land-use type is not an important factor to affect infiltration models efficiency. Due to a few number of investigation in this field of research, there is a need for further investigation on land-use effect on infiltration modelling and for the impact of land-use on soil infiltration characteristics, as well.

Land use	Kostiakov-Lewis model			Philip two-term model		Horton's model			Observed final infiltration rate (cm min^{-1})
	c	b	a	S	A	β	f_0	f_c	
Rangeland 1	0.2052	0.543	1.136	1.217	0.223	0.130	0.894	0.299	0.2803
Rangeland 2	0	0.770	1.058	1.322	0.235	0.021	0.596	0.190	0.2441
Rangeland 3	0.0870	0.606	0.839	0.970	0.127	0.090	0.556	0.183	0.1613
Rangeland 4	5.13×10^{-14}	0.855	0.5076	0.543	0.208	0.045	0.387	0.225	0.2285
Dryland farming 1	0.0585	0.636	1.475	1.741	0.162	0.085	0.909	0.258	0.2347
Dryland farming 2	0	0.522	2.863	2.984	0.018	0.120	1.601	0.200	0.1863
Dryland farming 3	1.37×10^{-12}	0.781	0.4857	0.592	0.119	0.029	0.290	0.120	0.1098
Dryland farming 4	0.0452	0.681	0.2555	0.315	0.073	0.071	0.198	0.089	0.0869

Table 2. Parameters of the selected infiltration models in both of land use types (Haghighi et al, 2010a)

6. Conclusions

Soil management and land use change may affect soil water retention at a -33 *kPa* (*FC*) potential in the soil based on laboratory measurements and model simulations. Lower water content at the -33 *kPa* potential would be expected upon conversion of natural lands to cultivated lands. In addition, the saturated soil water content (θ_s) may be affected by cultivation of rangeland. Moreover, because cultivation of natural lands affects soil macroporosity, we suggest measuring soil water retention at higher suction heads to document the land use effect on soil water retention properties in relation to soil macropores. Appropriate technology for dryland farming and suitable measures are necessary to improve soil water retention where cropping is required.

The findings show that the van Genuchten model is useful in describing soil water retention. Thus, use of this model may be considered as a valuable tool to gain more knowledge of hydraulic properties for various soil types. The effect of land use type on soil water retention and PTF applications have not been documented for dryland farming to the best of our knowledge. In many developing countries, such as Iran, the use of available PTFs can cause errors for estimating soil hydraulic properties. This review encourages further investigations of the model applications and development of suitable point and parametric PTFs for estimating soil hydraulic properties. The selection of more suitable PTFs for application where there are not developed PTFs caused by a lack of large databases is difficult. Consequently, it is essential to evaluate the model applicability and to develop point and parametric PTFs for estimating soil hydraulic properties for different land uses.

7. References

[1] Bormann, H., Klaassen, K., 2008. "Seasonal and land use dependent variability of soil hydraulic and soil hydrological properties of two Northern German soils." Geoderma. 145, 295-302.

[2] Brooks, R.H., Corey, A.T., 1964. "Hydraulic properties of porous media." Civil Engineering Dept., Colorado State University, Fort Collins, CO.

[3] Davidoff, B. and Selim, H.M., 1986. Goodness of fit for eight water infiltration models, Soil Sci. Soc. Am. J. 50, 759-764

[4] Gifford, GF. 1976. Applicability of some infiltration formulae to rangeland infiltrometer data. Journal of Hydrology. 28, 1-11.

[5] Haghighi F, Gorji M, Shorafa M, Sarmadian F, Mohammadi MH. 2010. "Evaluation of some infiltration models and hydraulic parameters." *Spanish Journal of Agricultural Research (INIA)*, 8 (1), 210-217.

[6] Haghighi F, Gorji M, Shorafa M. 2010. "A study of the effects of land use change on soil physical properties and organic matter." Land Degradation & Development Journal. *In press* 10.1002/ldr.999.

[7] Horton, R.E. 1940. An approach toward a physical interpretation of infiltration-capacity. Soil Sci. Soc. Am. Proc.5, 339-417.

[8] Kostiakov, A. N. 1932. On the dynamics of the coefficient of water percolation in soils and on the necessity of studying it from a dynamic point of view for the purposes of amelioration.

[9] Kutílek M. 2004. Soil hydraulic properties as related to soil structure, *Soil Till. Re* 79, 175–184.

[10] Machiwal, D., K.J.., and B.C. Mal. 2006. Modelling infiltration and quantifying spatial soil variability in a watershed of Kharagpur, India. Biosystems Engineering, 95(4), 569-582.

[11] Mishra, s.k, J.V.Tyagi, and V.P.Singh. 2003. Comparison of infiltration models. Hydrological processes. 17, 2629-2652 .

[12] Navar.J, and T.J. Synnote, 2000. Soil infiltration and land use in Linares, N.L., Mexico, Terra Latinoamericana. 18 (3), 255-262.

[13] Ndiaye, B., Molenat, J., Hallaire, V., Hamon, C.G.Y., 2007. "Effects of agricultural practices on hydraulic properties and water movement in soils in Brittany (France)". J. Soil & Tillage Res. 93, 251-263.

[14] Philip, J.R. 1957. The theory of infiltration: 4. Sorptivity and algebric infiltration equations. Soil Sci. 84, 257-264.

[15] Schwartz, R.C., Unger, P.W., Evett S.R., 2000. "Land use effects on soil hydraulic properties." ISTRO.

[16] Shukla, M.K., R.Lal., and P.Unkefer. 2003. Experimental evaluation of infiltration models for different land use and soil management systems. Soil Sci. 168 (3), 178-191.

[17] Singh V.P. 1992. Elementary Hydrology. Prentice Hall: Englewood Cliffs, NJ.

[18] Sonneveld MPW., Bachx MAHM., Bouma.J. 2003. Simulation of soil water regimes including pedotransfer functions and land use related preferential flow. *Geoderma*, 112, p: 97-110.

[19] University of Tehran. "Taleghan Watershed study report"., 1993. Irrigation Engineering Department, Iran.

[20] Van Genuchten, M.Th., 1980. "A closed-form equation for predicting the hydraulic conductivity of unsaturated soils." Soil Sci. Soc. Am. J. 44, 892–898.

[21] Van Genuchten, M.Th., Leij, F. J., Yates, S.R., 1991. "The RETC code for quantifying the hydraulic functions of unsaturated soils." EPA/600/2-91/065. R. S. Kerr, Environmental Research Laboratory. U. S. Environmental Protection Agency, Ada, OK. p: 83.

[22] Wösten, J.H.M., Finke, P.A., Jansen, M.J.W., 1995. "Comparison of class and continuous pedotransfer functions to generate soil hydraulic characteristics." Geoderma. 66, 227–237.

[23] Zhou, X., Lin, H.S., White, E.A., 2008. "Surface soil hydraulic properties in four soil series under different land uses and their temporal changes." Catena. 73, 180-188.

Part 8

Remote Sensing

21

Integration of Satellite Imagery, Geology and Geophysical Data

Andreas Laake
WesternGeco Cairo
Egypt

1. Introduction

Satellite imagery is a large scale surface geological mapping too, which offers the unique opportunity to investigate the geological characteristics of remote areas of the earth surface without the need to access the area on the ground. The resolution of the technique is limited by the resolution of the imagery. This chapter explains how geological information can be extracted from satellite imagery and how this information can be merged with geological and geophysical data to build consistent geological models for the surface and subsurface. On the one hand, the interpretation of satellite imagery can generate start models prior to the start of geophysical surveys. On the other hand, geological and geophysical data can calibrate models derived from satellite imagery.

2. Methodology

When studying the shape of the earth surface in connection with the rock layers and their deformation by tectonic forces, we often notice a correlation between shapes and structures at the surface and in the subsurface (Short and Blair 1986). This opens the opportunity to map the characteristics of the surface and infer characteristics of the subsurface. We can describe the surface by its shape and by its structure. The surface shape depends on topography, terrain gradient and surface lithology, which we call **geomorphological properties**. The surface structure is determined by lithological boundaries and fracture zones outcropping at the surface, which we call **litho-structural properties**. Fracture zones can also be inferred from the characteristics of recent or paleo-drainage (Short and Blair 1986). Geomorphology and litho-structure allow building a **static geological model**. If information is available about the elevation change with time, then the statics model can be expanded into a **dynamic geological model**. Figure 1 gives an over of the building blocks for geological model building.

We illustrate the methodology at a simple layer-cake geological model which is deformed by a vertical fault (fig. 2). The surface is formed by a soft sandstone layer resting on a hard limestone layer. These two layers form the near-surface. We call the layers between the bottom of the near-surface and the top of the basement subsurface. Prior to the deformation by the fault only the soft sandstone was visible at the surface. The fault has lifted part of the layer package and exposed the near-surface sandstone and limestone layers at the fault plane and made them accessible for mapping by satellites.

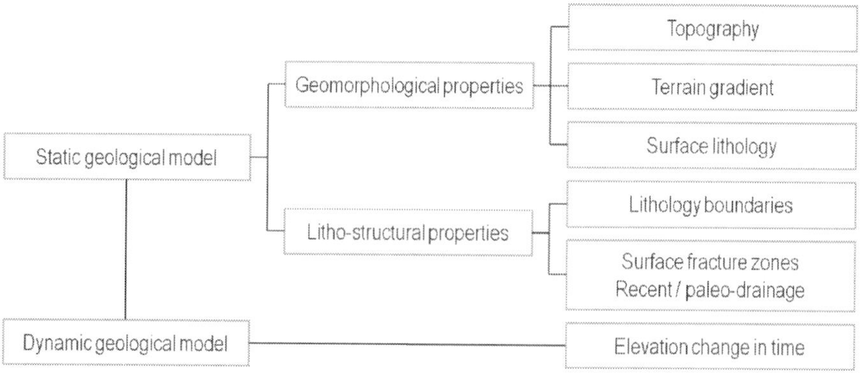

Fig. 1. Building blocks of near-surface and subsurface geological models

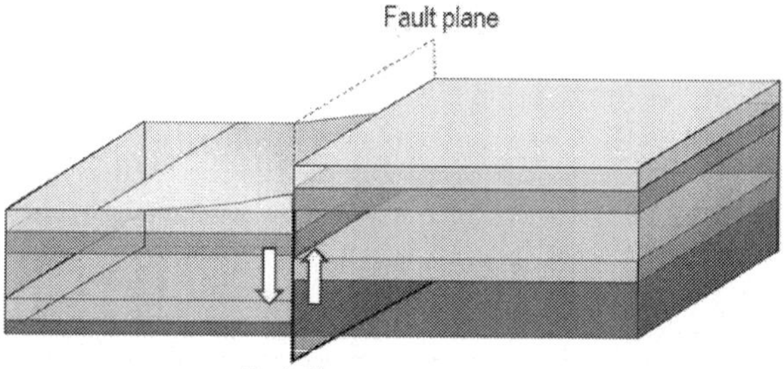

Fig. 2. Correlation of surface shape and subsurface geology

3. Satellite imagery

Earth observation satellites map the physical properties of the earth surface and near-surface. In the context of geological mapping we distinguish two types of electro-magnetic methods (see figure 3) :

- **passive optical methods**: use the sunlight as the source and measure the reflectance of the earth surface in the visible and infrared spectral bands. We used Landsat 7 ETM+ and the ASTER instrument from the Terra satellite.
- **active microwave radar methods**: use a microwave source onboard of the satellite and measure the back-scatter from the earth. We used Radarsat-1 and the radar sensor from the Shuttle Radar Tomographic Mission (SRTM).

Details about the acquisition and processing are provided among others by the USGS (2011) and Short (2010). For an introduction into the interpretation of satellite imagery see Sabins (1996). The **visible imagery (VIS)** covers the colors blue, green and red and provides information about water features, infrastructure and landuse as well as limited information about selected rock types. Infrared imagery is split into three classes : **very near infrared (VNIR)**, which detects specifically vegetation; **short wave infrared (SWIR)** which is the best option for the discrimination of sedimentary rocks; and finally **thermal infrared** (TIR). The thermal infrared radiation from the earth surface represents the property of the surface material to convert the solar spectrum into heat radiation. We distinguish between a warm response from dark materials such as non-sedimentary rocks and cool response from ground moisture or voids, where evaporation absorbs energy. In general optical imagery does not penetrate the earth surface.

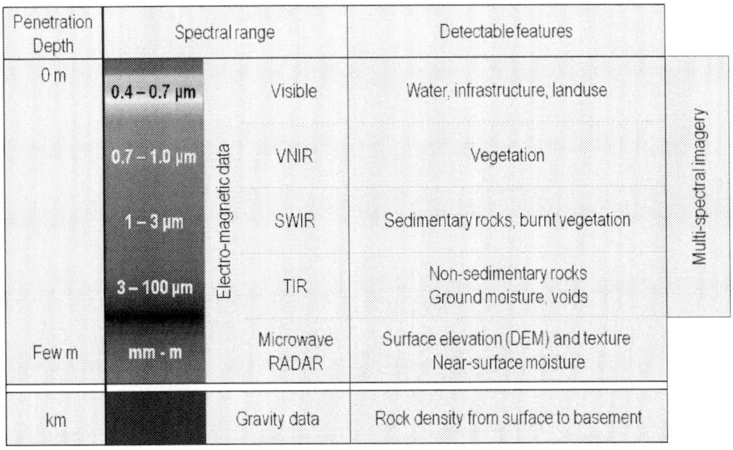

Fig. 3. Spectral overview of electro-magnetic satellite imagery

Microwave radar uses electro-magnetic waves in the mm to m range. At hard surfaces microwaves are almost completely back-scattered. Their travel time can be used to determine the distance between the satellite and the surface, which is used to generate digital elevation models (DEM). For soft, non-conductive surfaces the microwaves penetrate into the subsurface; the back-scattered signal is generated from volume scatter in the subsurface.

A special application of microwave radar is the estimation of **gravity anomalies**. Sandwell and Smith (2009) have studied anomalies in the orbits of radar satellites and inferred Bouguer gravity anomalies using a geoid model. These gravity anomalies can be interpreted for the thickness of the sedimentary cover above the crystalline basement.

The following examples illustrate the information obtainable from the various satellite imagery sets. Figure 4 shows an example of multi-spectral data from ASTER for each range : visible (fig. 4.a), short wave infrared (fig. 4.b) and thermal infrared bands (fig. 4.c) of the island of Bahrain with a colour scale from blue (low reflection intensities) via green and yellow to red (high intensities). The visible green band penetrates water for a few meters (blue shades). It is relatively insensitive to different rock types (red shades). The short wave infrared band does not penetrate water and therefore shows uniform blue color, whereas it allows the discrimination of the different rock types in the oval anticlinal structure covering most of the island (yellow and red shades). The thermal infrared band shows the thermal properties such as warm response from the built-up areas (dark red). Cool response (greenish areas in the coastal areas) is observed for wet coastal salt flats called sabkha.

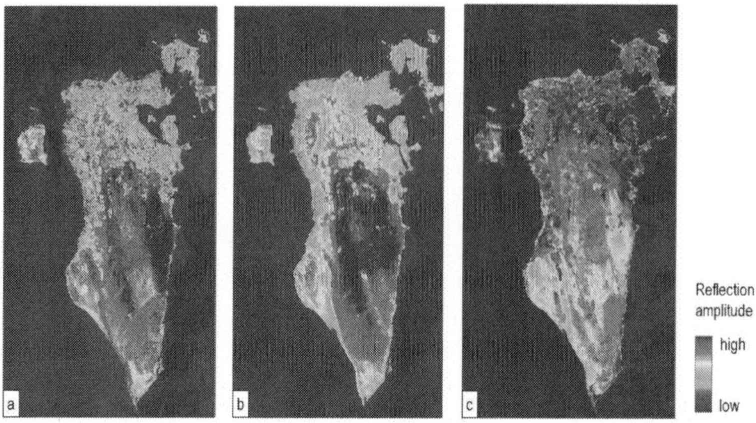

Fig. 4. Examples for optical satellite imagery from ASTER: a - visible blue, b - short wave infrared, c - thermal infrared [after Laake et al. 2006]

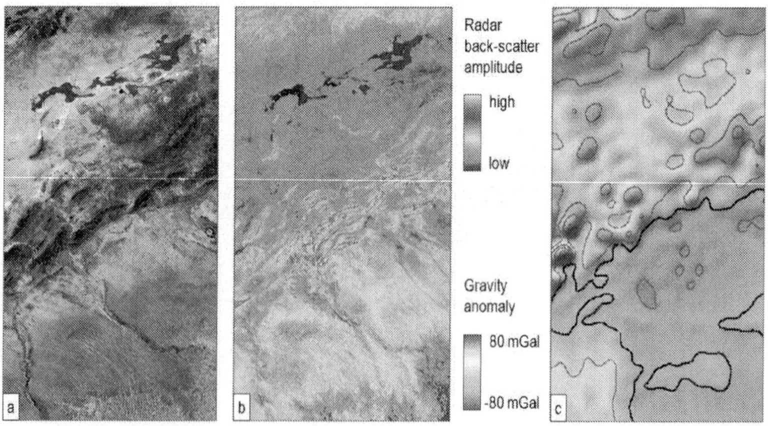

Fig. 5. Examples for optical and radar satellite imagery: a - optical data [Landsat 742 RGB] for comparison, b - radar scatter intensity [Radarsat-1], c - gravity anomaly derived from radar

The examples for microwave radar data show the Atlas area in Algeria. For orientation we have supplied a Landsat 742 RGB image (fig. 5.a), which shows mountain ranges (dark purple), gravel planes (yellow-gray), salt lakes (blue) and vegetation (green). The radar back-scatter intensity from Radarsat-1 (fig. 5.b) is displayed from blue (total absorption in conductive salt brine) via intermediate volume back-scatter (green and yellow) to strong surface back-scatter from hard rock (brown and white).

Figure 5.c shows the gravity anomaly inferred from radar satellite imagery. Negative anomalies correspond to low density rocks as from thick sedimentary cover at the foot of the

Atlas mountains whereas positive anomalies correlate with dense metamorphic and basement rocks in the Atlas ranges.

4. Extraction of information from satellite imagery

Satellite imagery is provided as sets of digital images, one image for each spectral band. Each image displays the measured values as intensity. The information contained in the satellite imagery can be extracted using either single bands or combinations of bands. **Single bands** are usually displayed as maps coding measured amplitude as colour.

Figure 6 shows radar data from the very dry desert in south-west Egypt : radar data from the shuttle radar tomographic mission (SRTM, Jarvis et al., 2008) interpreted for a DEM using topographic colour coding, i.e. green for low elevations, yellow to brown for high ground (fig. 6.a). Raw radar data from Radarsat-1 reveal high amplitudes from the back-scatter at hard sandstone (white and brown colors in fig. 6.b). Low radar back-scatter (fig. 6.c) correspond to volume back-scatter from microwaves penetrating sand sheets. They reveal buried paleo-rivers (blue colors in fig. 6.c) following the interpretation by El-Baz and Robinson (1997) and Robinson et al. (2007).

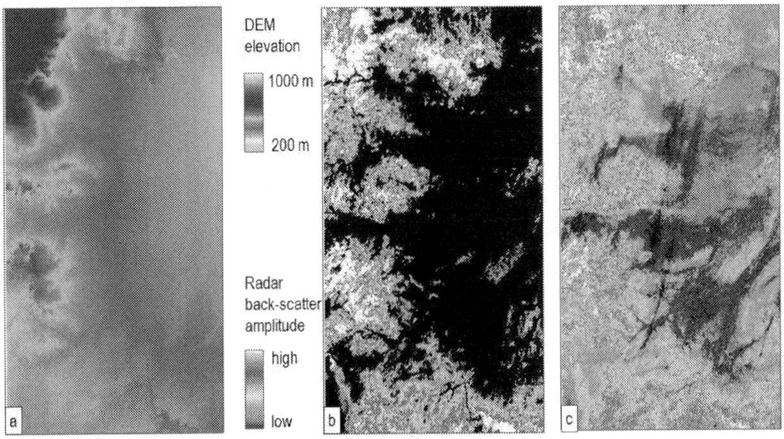

Fig. 6. Single band data examples : a - radar based digital elevation model [SRTM DEM], b - high intensity radar data [Radarsat-1], c - low intensity radar data [Radarsat-1]

Dual band images use combinations of two bands from Landsat for the north-western desert in Egypt (fig. 7) thereby enhancing subtle features in the data that would not be imaged by a single band alone. The ratio of the infrared bands 7 and 4 (fig. 7.a) highlights for example clay minerals which fill karst holes (red and cyan) in an otherwise homogeneous limestone plateau (dark blue). The band difference of the very near infrared band 4 and the green band 2 for the same area highlights difference in lithology between the pure limestone (yellow to red) and the more sandy cover (blue tones) towards the top of the image (fig. 7.b).

Multi-band images use three or more bands combined in continuous colour or red-green-blue (RGB) images. RGB images provide significantly more shades than single or dual band images: for 8 bit imagery an arithmetic combination of 3 bands provides 256 shades whereas an RGB image offers 16.8 million colors (Guo et al., 2008).

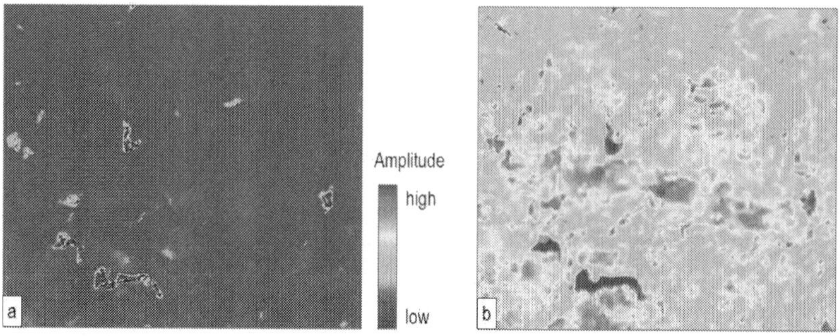

Fig. 7. Dual band data examples: a - band ratio [Landsat 7/4], b - band difference [Landsat 4-2] for the same area

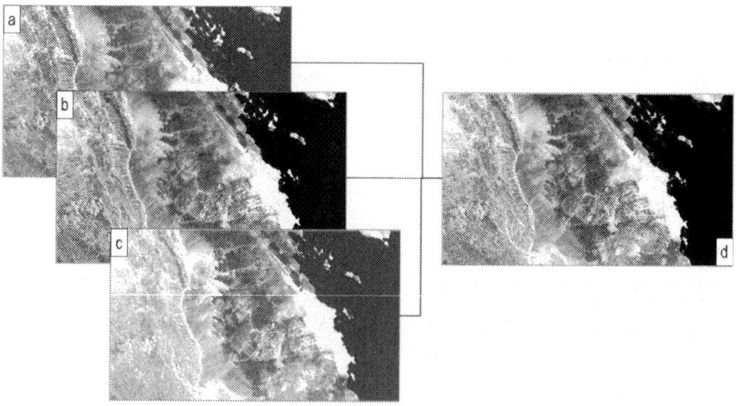

Fig. 8. Multi-band data example : merge of single band data (a to c) into RGB image (d) [Landsat 742 RGB]

To illustrate this we study an example from the southern Red Sea Mountains and the adjacent coastal area in south-east Egypt (fig. 8). The raw input bands (fig. 8 a-c) show only marginally different signatures for the very different rock types, whereas the multi-band RGB image (fig. 8.d)distinguishes clearly between the basement rocks (dark), the Mesozoic clastic sedimentary rocks (yellow tones), coastal carbonates (white) and the sea (black).

5. Integration of satellite imagery and geology

In this section we describe techniques to extract geological information from satellite imagery and how to integrate this information with geological data. Satellite imagery is interpreted for surface topography and lithology as well as for surface and subsurface structure with the goal of generating geological models for the near-surface and subsurface (Laake and Insley, 2007, and Laake et al., 2008). The techniques are illustrated through a series of case histories starting with simple layer-cake geology.

5.1 Layer-cake geology (Qattara Depression, Egypt)

The surface north of the Qattara depression in Egypt is dominated by flat layering of hard and soft rocks at the surface (fig. 9). The raw DEM (fig. 9.a) shows a platform (yellow), which is located between rough hills (brown tones) and a sharp escarpment towards the Qattara Depression (green to blue). The terrain classification map (fig. 10.c) allocates the different elevations to three classes. The lithological analysis of the multi-spectral image (fig. 9.b) allows a clear separation into rock types : two types of limestone (blue tones), two types of sandstone (yellow to orange) and evaporites of the sabkha at the bottom of the depression (cyan).
Ideally the surface lithology interpretation is validated in the field (see fig. 10) using GPS tracked lithological analysis. The combination of terrain classification and lithology suggests that the plane represents the top of the sandstone formation, which continues also below the limestone layers, which form the higher ground. We will use this geological model to estimate statics for seismic data processing in section 5.

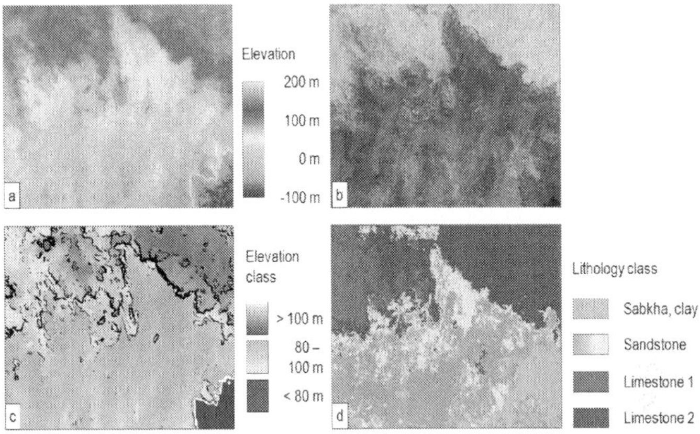

Fig. 9. Layer-cake geology near Qattara Depression, Egypt : a - digital elevation model [SRTM DEM], b - multi-band satellite image [ASTER 631 RGB], c - terrain classification map, d - lithology map [Cutts and Laake 2009]

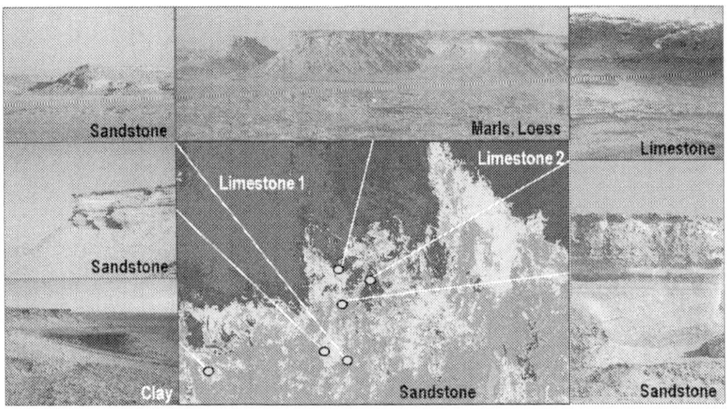

Fig. 10. Validation of lithology map in the field [after Coulson et al. 2009]

5.2 Anticline (Awali, Bahrain)

The outline of the island of Bahrain is determined by the topographical and lithological structure of the Awali anticline (fig. 11). We use satellite imagery from the ASTER sensor for the discrimination of clastic and carbonatic rocks (Laake et al., 2006).

The continuous colour image from short wave infrared and visible bands (ASTER 631 RGB in fig. 11.a) allows discriminating the coastal farmland in the north (green) from the carbonates of the anticline (purple tints) and coastal sabkha (cyan). The structure of the anticline can be delineated using the difference of visible and short wave infrared data (fig. 11.b), which can be traced even in the built-up area of Manama city in the north. The outer contour of the anticline appears highlighted in the west through the strong signature of the coastal sabkha (dark red). Draping the continuous colour image over the vertically exaggerated DEM generates a strong structural impression, which is useful for structural interpretation (fig. 11.c).

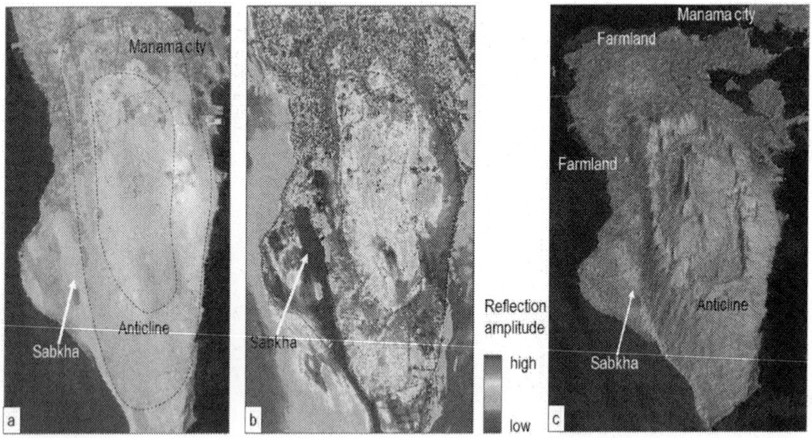

Fig. 11. Anticlinal structure of Awali, Bahrain : a - multi-spectral image [ASTER 631 RGB], geological structure from band difference image [ASTER visible minus short wave infrared], c - rendering of multi-spectral image on DEM [after Laake et al. 2006]

5.3 Mapping of basins from gravity and radar data (Illizi Basin, Algeria)

The mapping of basins from satellite imagery is the only remote sensing technique which infers deep geological structures. Bouguer gravity anomalies inferred from satellite radar data give an indication of the thickness of the sedimentary cover above the basement. However, this interpretation requires additional information for example from magnetic data to constrain the model. Following the concept of geomorphology, which correlates surface and subsurface geology and litho-structure, surface structural lineaments can be used to infer subsurface structures.

In figure 12 we show structural maps from satellite imagery for the Illizi Basin, which is indicated by the black continuous line in each figure. The lithology image (Landsat 742 RGB, fig. mapping 12.a) distinguishes dark paleozoic limestone (blue tones) and Mesozoic clastics (brown tones), which surround recent sand dunes (yellow). The geomorphological map (fig. 12.b) highlights the litho-structural boundaries, which are co-located with topographic ridges. Satellite gravity data (fig. 12.c) reveal a basin (blue). The combination of surface geomorphology and gravity (fig. 12.d) gives an indication of the basin outline.

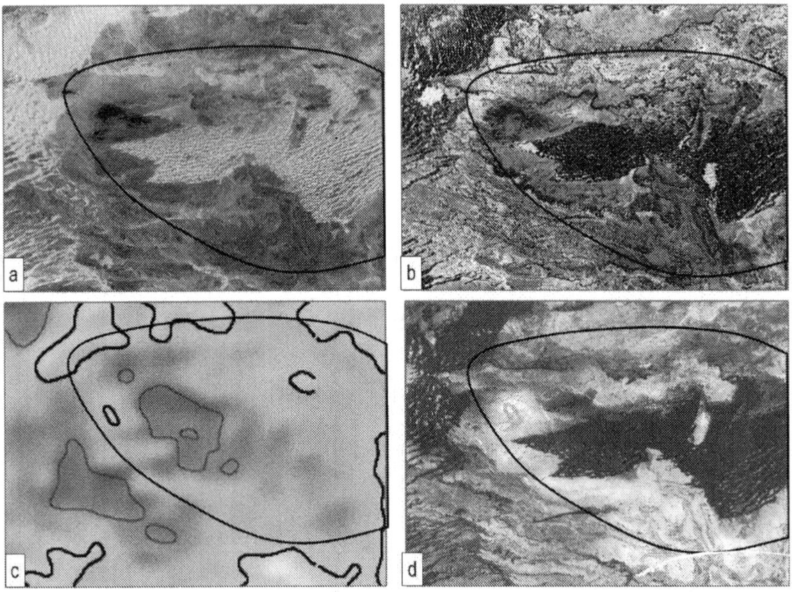

Fig. 12. Basin mapping of Illizi basin, Algeria: a - multi-spectral image (Landsat 742 RGB], b - geomorphological map [SRTM DEM, Landsat 742 RGB], c - gravity anomaly derived from radar, d - overlay of geomorphology on gravity data

5.4 Mapping of fracture zones below sand cover (Gilf Kebir, Egypt)

When sand dunes cover fracture zones in the near-surface, satellite radar data can map ancient river courses, from which fracture related weak zones in the near-surface can be derived (fig. 13). We start with the outcrops of faults in the hard rock surrounding the sand dune field mapping straight valleys in the hard sand stone. The topography of the study area (fig. 13.a) is composed of the flat Gilf Kebir plateau (brown tones) with gentle slopes (yellow) and a large plane (green to white). This corresponds to a surface lithology (fig. 13.b) of hard sandstones (greenish-brown) and basement rocks (dark brown) as well as belts of sand dunes (white), which are locally discoloured by hematitic iron (purple). The shape of the escarpment and the valleys intersecting the Gilf Kebir plateau reveal fairly straight fault lines (black lines). Below sand neither topographical nor optical satellite imagery can reveal buried fracture zones. Therefore we use satellite radar which can penetrate dry sand for up to 20 m to map the clay contained in buried paleo-river beds. The radar data (fig. 13.c) do not only map the fracture zones in the sandstone of the plateau (brown lineaments), it also shows W-E and SW-NE trends in the paleo-river courses (low radar intensities in blue) which continue from the lineaments in the rock outcrops. The overlay of the paleo-river courses from radar data on the geomorphology delineates the outcropping fracture zones across all terrains (fig. 13.d).

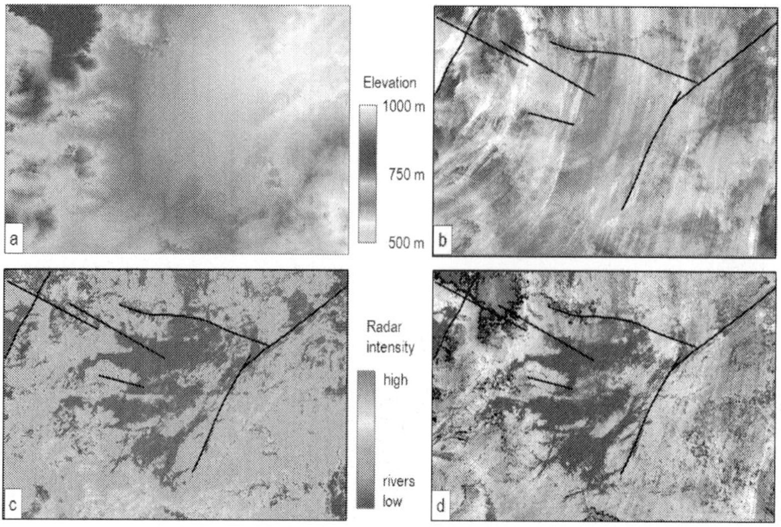

Fig. 13. Structural delineation from radar data close to Gilf Kebir, Egypt: a - DEM, b - multi-spectral image [Landsat 742 RGB], c - low intensity radar data [Radarsat1], d - merge of radar data with geomorphology. Faults indicated by lines.

5.5 Mapping of glacial moraine structures (Pechora Basin, Russia)

The last case study concerns the mapping of post-glacial structures in the Pechora basin in northern Russia using vegetation and water features as indicators; the lithology is not exposed in the study area. The structures studied comprise different types of moraines as well as drainage features in front of the glaciers (Laake 2009, Astakhov et al. 1999). We distinguish terminal and lateral moraines, which are composed of gravel and glacial till, from ground moraines, which are characterized by undulating terrain. Moraine ridges are often drier than the surrounding terrain because the elevated gravels cannot support a shallow water table. In contrast to this, ground moraines deposit more glacial till, which provides the basis for a shallow water table with very wet terrain and lakes. Significant differences in the ground moisture attract different plant species, which can be distinguished by their different response in the very near infrared and visible bands of satellite imagery. The topography allows delineating of the lateral and terminal moraines as long as they are still elevated above the surrounding plane (fig. 14.a).

Rendering the geomorphological map on the DEM in 3D improves the detection of these moraines. The location of lakes, which is indicative of ground moraines, is obtained from the landuse analysis of short wave infrared and visible green data. (fig. 14.c). The combination of all maps yields a clear delineation of the glacial moraines particularly when rendered in 3D (fig. 14.d). In this case four glacial stages can be interpreted (numbers in fig. 14) : an initial stage where the entire area was covered by a thick ice shield (1), which deposited ground moraines over the entire study area. The second (2) and the third (3) phase comprise two distinguishable glaciers, where the glacier from phase two created the partly broken moraine wall in the west. The glacier in phase three covered the centre of the study area thereby partly levelling the lateral moraines from phase two. After melting, this glacier left the huge Lake Komi between the lateral and terminal moraines, which forms an extensive swamp today. The final glacier advance (4) covered only the northern third of the study area, leaving an irregular line of terminal moraines and extensive planes of ground moraines behind.

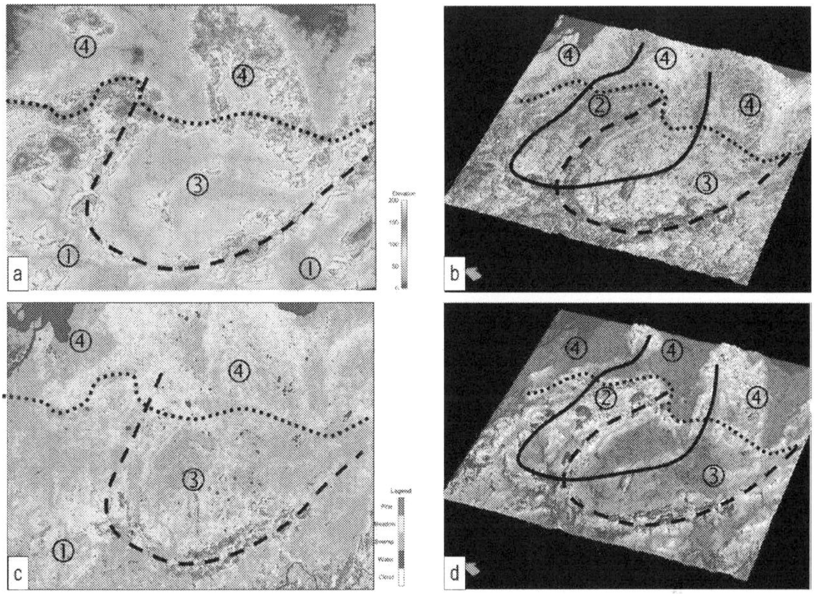

Fig. 14. Identification of glacial structures in Pechora basin, Russia: a - DEM [SRTM DEM], b - rendering of geomorphological map in 3D, c - landuse classification from Landsat, d - interpretation of geomorphology for glaciers and moraines [after Laake 2009]

6. Integration of satellite imagery and geophysics

The geological information extracted from satellite imagery can be used to build geological models from basin scale (several 10^5 sqkm) to survey scale (few 10^3 sqkm) well before any geophysical data are acquired on the ground. These models can be used for the estimation of logistic risks for personnel and vehicles as well as for the estimation of the quality of the acquired geophysical data (Laake and Insley (2004a and b, Laake and Cutts 2007, Coulson et al. 2009). In turn, geophysical data can be employed to calibrate geophysical characteristics of near-surface layers inferred from satellite imagery.

We distinguish the following types of geophysical surveys :

1. **Frontier exploration** aiming at identification of new basins in large unexplored areas. Satellite imagery can assist defining the outline of potential basins and the definition of scouting surveys.
2. **Structural imaging** focuses on potential structures once the outline and character of a basin has been identified. Satellite imagery can support the design and logistic planning

of geophysical surveys on the ground and can provide estimates for the quality of the acquired data before the data are acquired.

3. **Reservoir characterization** targets the most comprehensive study of the subsurface geological structure and fluids for individual reservoirs and therefore requires the best data quality. Satellite imagery can provide detailed models of the surface and near-surface which provide input to data quality estimation before and during acquisition. For data processing satellite imagery can supply input to processes that correct for noise related to near-surface properties.

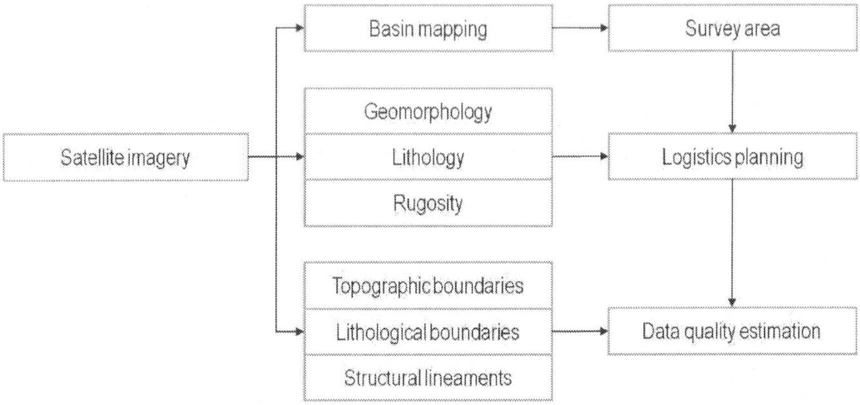

Fig. 15. Workflow for application of satellite imagery to geophysical methods

An outline of the information satellite imagery can provide for geophysical methods and the parameters studied is shown in figure 15. This section focuses on the aspects of survey design and data quality estimation.

The design of geophysical surveys requires an understanding of the logistics and data quality aspects of the target area to correctly estimate the effort required to provide the desired quality of the final subsurface geological product. Satellite imagery can provide the information about the surface and near-surface before the start of the data acquisition (Cutts and Laake, 2009a and b). The method uses satellite imagery to generate geomorphological and litho-structural models of the surface and near-surface as described in the previous section to derive logistic planning and data quality estimation maps.

Figure 16 shows the result of the method for a survey in the Western Desert of Egypt. DEM and multi-spectral data provide topography (fig. 16.a) and lithology class maps (fig. 16.b). The planning of the logistics requires information about limitations for access and maneuver for personnel and vehicles. The terrain classification provides the locations and steepness of escarpments as well as the surface roughness related to hard limestone. The combination of these terrain attributes defines the logistics risk (fig. 16.c). A major obstacle for data quality for this survey is the scattering of seismic waves at the numerous escarpments resulting from the different weathering of limestone and sandstone. Fig. 16.d shows an estimate map for the scatter risk associated with escarpments.

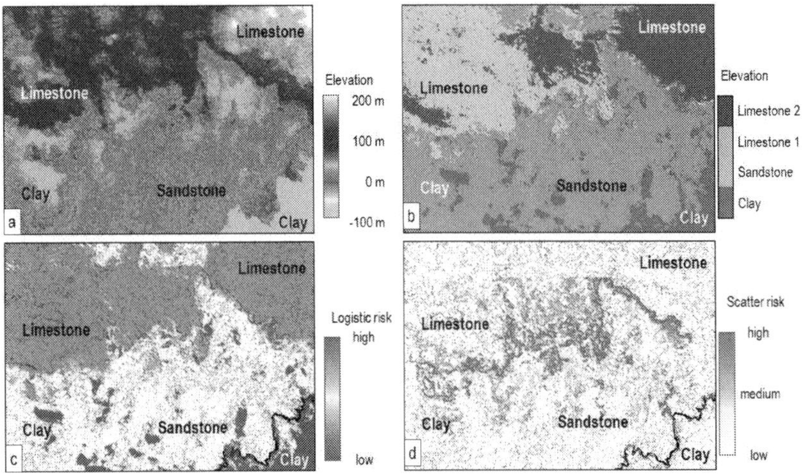

Fig. 16. Survey design based on satellite imagery (Qattara Depression, Egypt) : a - terrain class map based on DEM, b - lithology map, c - logistic risk estimate, d - data quality risk estimate [after Cutts and Laake 2008]

6.1 Logistic planning in volcanic terrain

Vegetation may obstruct the assessment of the logistic risk (Laake 2005b). The basalt plateau close to the Payun Volcano in the Andean foothills of Argentina exposes the risk of big basalt blocks obstructing the access to large areas whereas in other parts of the survey basalt grit exposes no maneuver risk at all (fig. 17.a to c). The analysis of the multi-spectral imagery from ASTER for big basalt blocks is challenging because in both cases - big basalt blocks and bush as well as basalt grit and grass - the ratio of basalt and vegetation is very similar. The resolution of the satellite image of 15 m does not allow the direct mapping of the basalt blocks (fig. 17.d). However, the thermal properties of big basalt blocks are sufficiently different from basalt grit, which allows the discrimination of the basalt block size using the thermal infrared bands from ASTER imagery (fig. 17.e), which is supported by the intermediate back scatter intensity recorded by radar data from Radarsat-1 (fig. 17.f).

6.2 Planning of safe operation in sand dunes

In high sand dunes, both logistics for wheeled vehicles and high absorption for seismic waves may impact the survey design severely (Laake and Insley, 2004a). The analysis of the topography can assist in outlining the dunes and characterizing their shape (fig. 18). Digital elevation models derived from radar or stereo optical data (ASTER in our case) provide the slope index, which can be represented as an index for safe operation. In seismic operation surface gradients above 15 degrees are considered unsafe for operation, whereas in areas with 10 to 15 degrees slope an inspection of the terrain would be advised. Figure 18 shows a composite image of seismic line crossing a dune field. The visible band ASTER image (fig. 18.b) gives a photographic impression of the surface, whereas the DEM (fig. 18.c) provides an impression of the elevations. The slope risk map (fig. 18.d) is obtained from the gradient of the DEM.

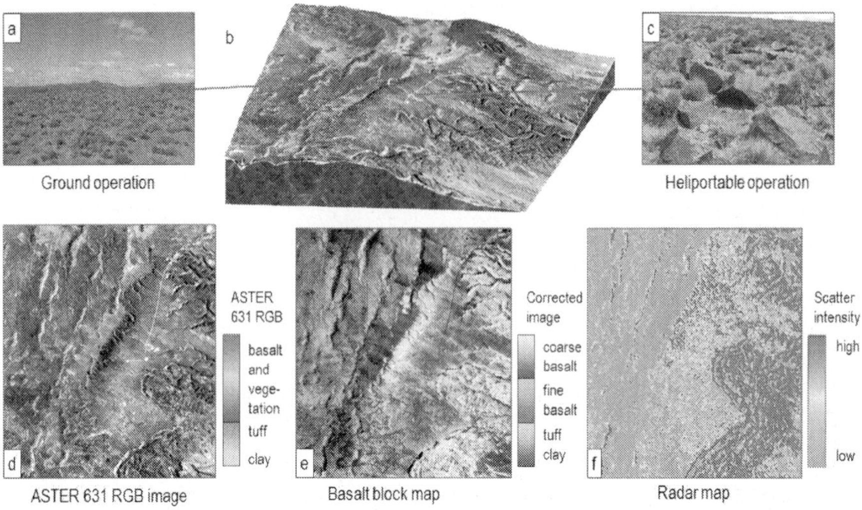

Fig. 17. Satellite imagery based logistic planning in Andean foothills, Argentina : a - basalt gravel plane, b - rendering of ASTER image, c - basalt blocks, d - landuse image [(ASTER 631 RGB], e - basalt block map, f - surface rugosity from Radarsat-1 [after Laake 2005b]

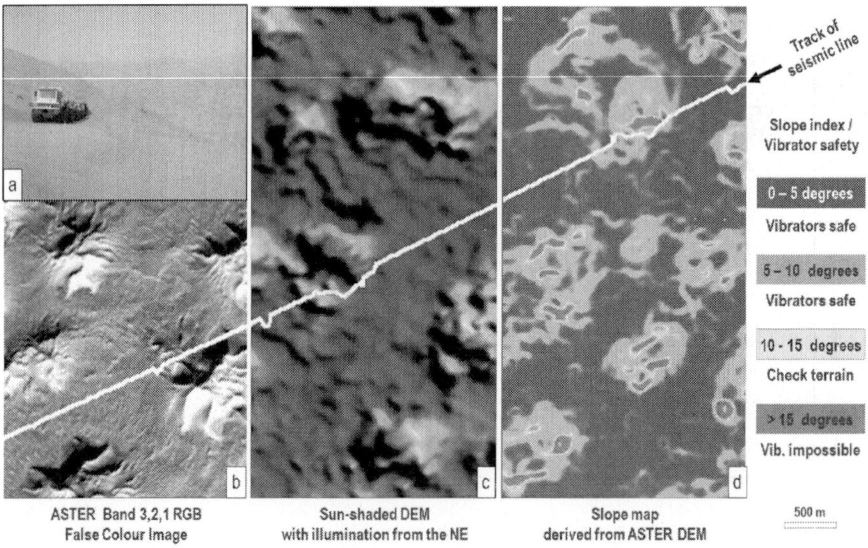

Fig. 18. Planning of safe vibroseis operation in high sand dunes, Berkine basin, Algeria: a - vibrator array in high dunes, b - multi-spectral satellite image [ASTER 321 RGB], c - sun-shaded DEM, d - Slope map derived from DEM [ASTER DEM] [Laake and Insley, 2004a]

6.3 Prediction of accessibility and data quality in sabkha

In arid coastal areas often sabkha called salt flats pose a severe threat to personnel and equipment (Cutts and Laake 2009b). The salt flats may be subject to seasonal inundation, which may change their accessibility significantly. Analysis of the multi-spectral satellite Landsat imagery can allow the detection of halite minerals found in the surface crusts of sabkha. Thermal infrared imagery provides information about the different thermal properties of wet and dry sabkha. The visible band Landsat image in figure 19.b gives an indication of the location of the wet sabkha. The risk that the sabkha surface would not support heavy vehicles (see fig. 19.a) is directly correlated with presence of sabkha and can be derived from a sabkha detection map (fig. 19.c). In addition to its impact on the logistics sabkha also affects the vibrator data quality. High distortion of the vibrator signal, an undesirable data property, is directly related to the presence of sabkha (fig. 19.d).

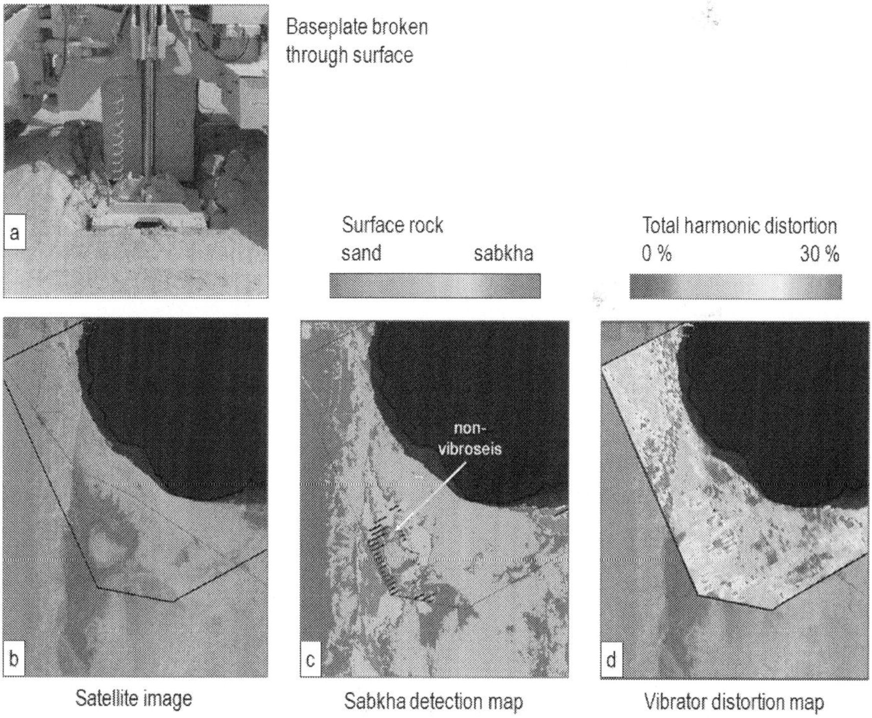

Fig. 19. Prediction of accessibility and data quality of vibroseis operations in sabkha, Sabkha Matti, UAE: a - vibrator baseplate broken through surface crust, b - satellite image [Landsat 321 RGB], c - sabkha detection map, d - map of vibrator total harmonic distortion [after Cutts and Laake 2008]

6.4 Estimation of the vibrator baseplate-to-ground coupling

Hard rock terrain affects the coupling of the vibrator baseplate to the ground (Laake and Tewekesbury 2005). Good ground coupling is achieved when the entire baseplate is in contact with the ground for example on gravel (fig. 20.c). If the baseplate rests on a big boulder, the coupling surface from the baseplate to the ground may be reduced to a very small area resulting in substantial distortion of the transmitted signal and in severe damage to the baseplate (fig. 20.a).

In this case the so-called point loading risk is correlated with hard limestone. The limestone prediction map draped over the DEM is shown in figure 20.b. But even for gravel the ground coupling may be compromised if the force level of the vibrator shaker and the frequency exceed the levels at which the ground supports the weight of the vibrator. The time signal of the baseplate amplitude (fig. 20 bottom) shows the increasing frequency of the shaker force with time along with measured signal distortion. At three moments of the sweep photos of the shaker were taken. As long as the coupling of the baseplate to the ground is perfect the signal distortion is low (fig. 20.d and e). When the ground surface breaks dust is whirled up by the baseplate motion and the distortion goes up (fig. 20.f). Usually frequencies higher than the frequency at which the baseplate breaks through the surface show high phase distortion and deteriorate the data quality. The high distortion for very low frequencies (corresponding to fig. 20.d) is associated with motions of the entire vibrator and are not considered here.

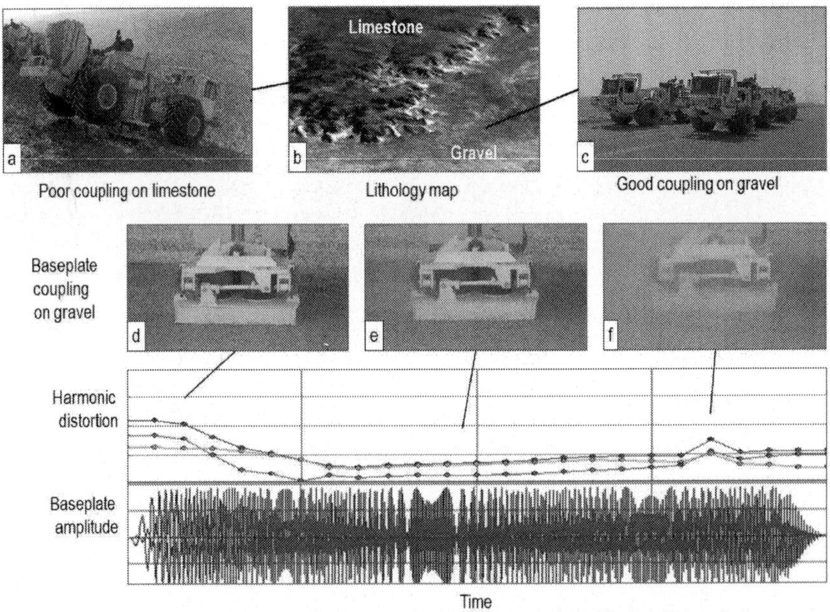

Fig. 20. Prediction of accessibility and data quality for vibroseis operations in hard rocky terrain, Tademait, Algeria: a, c – operations photos, b – virtual 3D lithology map, d - f – photos from baseplate coupling, bottom – baseplate distortion and amplitude time signal [after Laake and Tewkesbury 2005]

Soft terrain may have an impact on both the source coupling as well as on the receiver signal output (Laake 2005a). Due to higher absorption in soft sand, the signal level on sandy surface (fig. 21.b and d) is generally lower than on hard gravel plain (fig. 21.a and c). Satellite imagery can discriminate gravel with shallow evaporite pans on the higher ground from sandy terrain on the lower ground (fig. 21.e). The quality of the source signal also correlates with the terrain character : signals from the sandy terrain show higher distortion (fig. 21.f) and lower ground stiffness (fig. 21.g) than signals from the gravel plain. Interpretation of satellite imagery can provide estimates for the quality of seismic signals obtained from the terrain characterization.

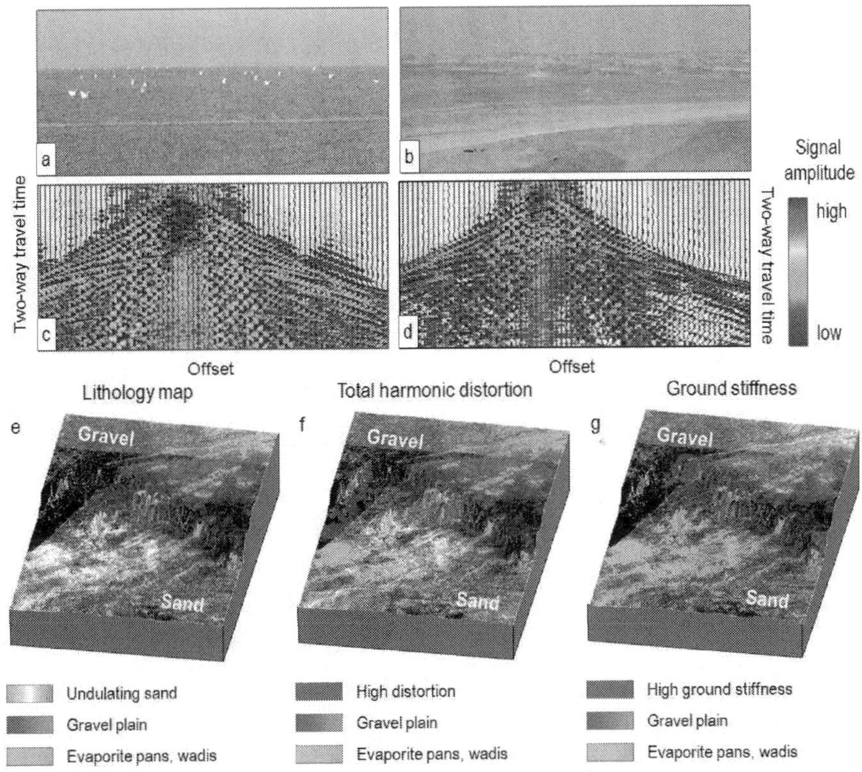

Fig. 21. Correlation of source and receiver data quality with terrain, Kuwait: a - gravel plain, b - undulating sandy surface, c, d - shot record displays, e - surface lithology, f - total harmonic distortion, g - ground stiffness [after Laake 2005a]

6.5 Improvement of static corrections

Near-surface geological models can be generated from the interpretation of satellite imagery for geomorphology and lithology (Laake et al. 2008). Using standardized seismic velocities the geological model is converted into an elastic model, from which estimates for statics corrections can be computed (Laake and Zaghloul 2009). The analysis of the DEM delivers the topographic classification into three layers (see fig. 22): the platform (sandstone), which occupies most of the study area, as well as the lower platform of the Qattara Depression and the rough higher layer (limestone). The interpretation of the shortwave infrared data reveals that the platform is composed of sandstone, whereas the higher ground consists of limestone (fig. 22 top). When choosing standardized P-wave velocities of 2200 m/s for sandstone and 3300 m/s for limestone we can compute estimated statics values (fig. 22 bottom). The estimated values are compared against refraction static corrections from a 3D seismic survey (fig. 23).

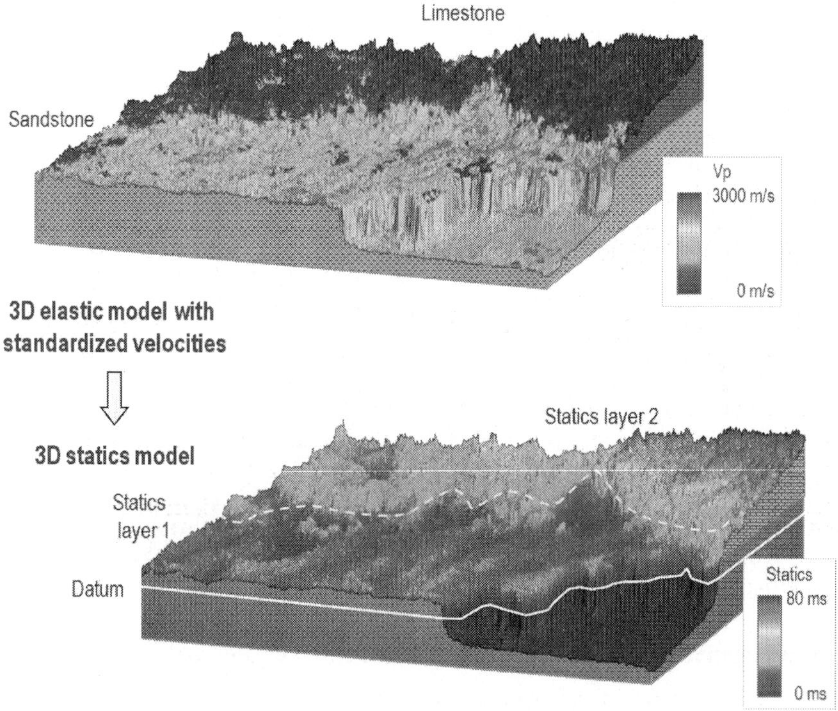

Fig. 22. Estimation of statics corrections from shallow geological model, near Qattara depression, Egypt : top - near-surface geological model from satellite imagery, bottom - 3D statics model [after Laake and Zaghloul 2009]

The static corrections for model 1 (fig. 23.a) use the geological model generated from satellite imagery. The statics contain the lithological details of the sandstone plateau and retain the sharp velocity contrast at the edge of the rough limestone plateau (see fig. 23.c for comparison with the lithology). The statics estimates for model 2 are based on the first break picks from a 3D seismic survey, which are remarkably smoother than the statics from model 1 (fig. 23.d). First break picking involves velocity smoothing over offsets up to 1500 m, which results in a spatial low pass filter. This may result in attenuation of sharp structural lineaments such as fracture zones outcropping at the surface. The difference between the two models (fig. 23.d) reveals another effect of the spatial smoothing: local velocity anomalies may not be corrected properly and might remain in the data as an artefact, which may lead to the generation of structural artefacts in the deeper seismic data.

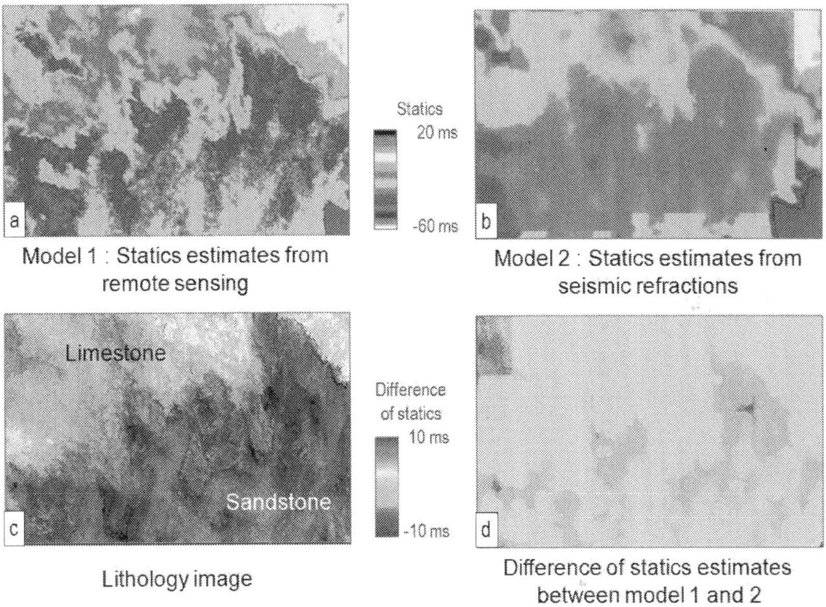

Fig. 23. Comparison of statics from near-surface geological model and from seismic refractions, Qattara depression, Egypt : a - statics corrections from satellite imagery, b - statics from seismic refractions, c - lithology image, d - difference between statics from both models [after Laake and Zaghloul 2009]

7. Consistent deep geological models from data integration

In the final section we will show the benefits of merging interpreted satellite imagery with geological and geophysical data on the western shore of the Gulf of Suez in Egypt by demonstrating the correlation of the geomorphology with the subsurface litho-structure . The processing of multi-spectral imagery from Landsat for lithology and drainage reveals fault structures which are masked by recent wadi deposits (Laake 2010, Laake et al. 2011). The approach follows the idea that tectonic movements deviate the courses of wadis in a characteristic way.

7.1 Mapping of fault outcrops at surface from satellite imagery

The geological setting of the study area is determined by rift faulting in NW - SE direction and transform faults in the perpendicular direction (Darwish and El-Araby 1993, Alsharhan and Salah 1993). The natural colour Landsat 321 RGB image (fig. 24.a) shows the almost featureless beige gravel plain between the Red Sea Mountains in the west and the Gulf of

Suez in the East. The only visible features are SW - NE running wadis delineated through their light sediments. The inverted natural colour image directs the eye to the wadi courses, which reveal several anomalously straight sections (fig. 24.b). When we use all data from Landsat and spectrally enhance the resulting image boundaries along the main tectonic directions are imaged (fig. 24.c): parallel to the coast NW - SE trending linear anomalies point to outcropping parallel fracture zones (dashed lines). The dominating wadi in the southern part of the study area is confined between two straight SW - NE trending lineaments which point to faults running perpendicular to the rift orientation (dotted lines). This map also reveals differences in the mineral composition of the bedrock of the Red Sea Mountain granites. The erosion fans from these two granites may actually be used as tracers to highlight the structural boundaries on the gravel plain. The lineaments are also highlighted in the drainage map, which results from the ratio of the thermal and pan-chromatic bands (fig. 24.d).

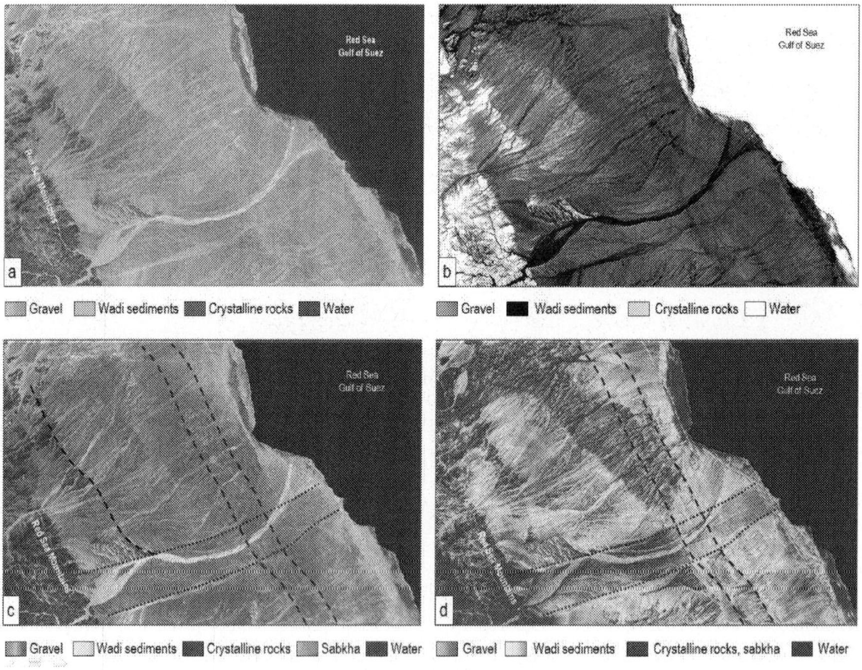

Fig. 24. Surface and near-surface litho-structural mapping from Landsat : a - natural colour image [Landsat 321 RGB], b - inverted natural colour image, c - high-discrimination lithology image, d - drainage map [Landsat bands 6 - 8] [after Laake et al. 2011]

7.2 Mapping of faults in near-surface from shallow geophysical data

In the study area, the fault outcrop pattern mapped from satellite imagery is calibrated with surface features and shallow seismic data from a 3D seismic survey. The only tree in the study area has been found at the intersection of two fault outcrops (fig. 25.a) indicated by the arrow in the lithology map (fig. 25.b). The surface structure becomes evident when draping the litho-structural map over the vertically exaggerated DEM (fig. 25.c). Arrows indicate the outcrop of faults at the surface. The correlation with faults in the near-surface is achieved by extracting weak zones of low surface wave velocity from seismic 3D data (fig. 25.d). Surface wave velocity analysis provides an iso-velocity horizon which corresponds to a shallow lithologic horizon. This horizon shows very low velocities along the rift parallel fault outcrop zones mapped by remote sensing. In the direction orthogonal to the rift faults a high-velocity structure below the wadi is revealed but no sharp boundaries. This may be due to the transform character of these faults which does not provide an impedance contrast that could be detected from the seismic.

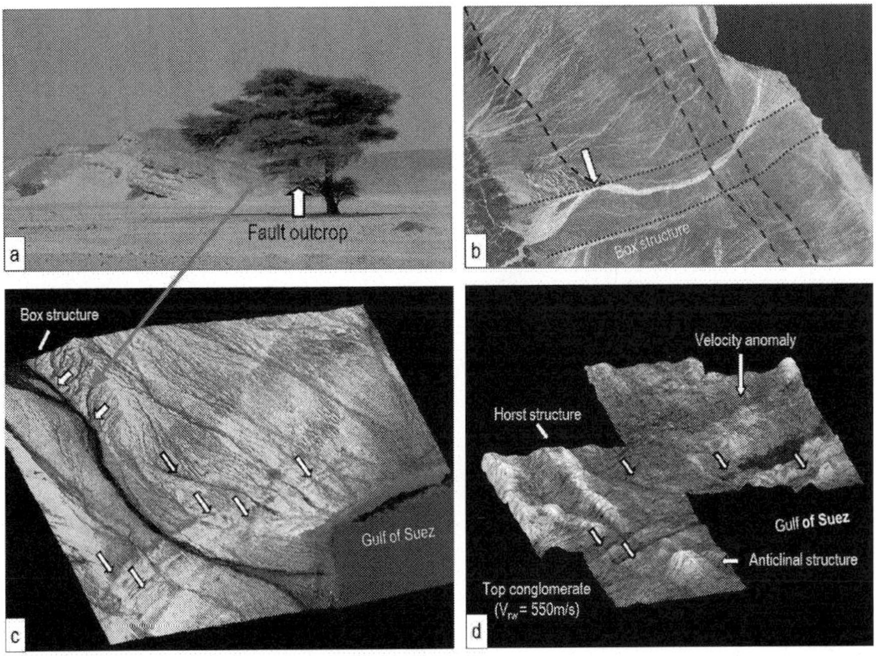

Fig. 25. Calibration of satellite imagery results with near-surface seismic data: a - tree located at a fault intersection, b - high-discrimination lithology map, c - virtual 3D rendering of the litho-structural map - arrows indicate fault outcrops, d - near-surface formation surface obtained from the velocity analysis of seismic surface waves [after Laake et al. 2011]

7.3 Mapping of faults in the subsurface using surface tamplates

The characteristics of drainage patterns delineated from satellite imagery may also be used to generate shape templates for the detection and extraction of similar structures in subsurface seismic data (fig. 26). We use the drainage map to define the template for a wadi (fig. 26.a) consisting of the braided stream and the fan delta. In our case the braided stream is confined by the SW - NE trending faults perpendicular to the rift orientation, whereas the fan delta part shows so little elevation change that the fan delta crosses the boundary faults (fig. 26.b).

The shape template is used in the processing of the shallow seismic data to search for the correct seismic attribute suited to map the faults which are hidden in most other seismic attributes. In our case the instantaneous frequency attributes were identified for mapping the SE - NE faults because this attribute highlights locally high spectral amplitudes which were interpreted as resonances within the palao-wadi structures (fig. 26.c). Once the shallow paleo-wadi has been identified the same methodology is applied to the deeper data; a total of four paleo-wadis spanning the entire rift history were mapped (fig. 26.d).

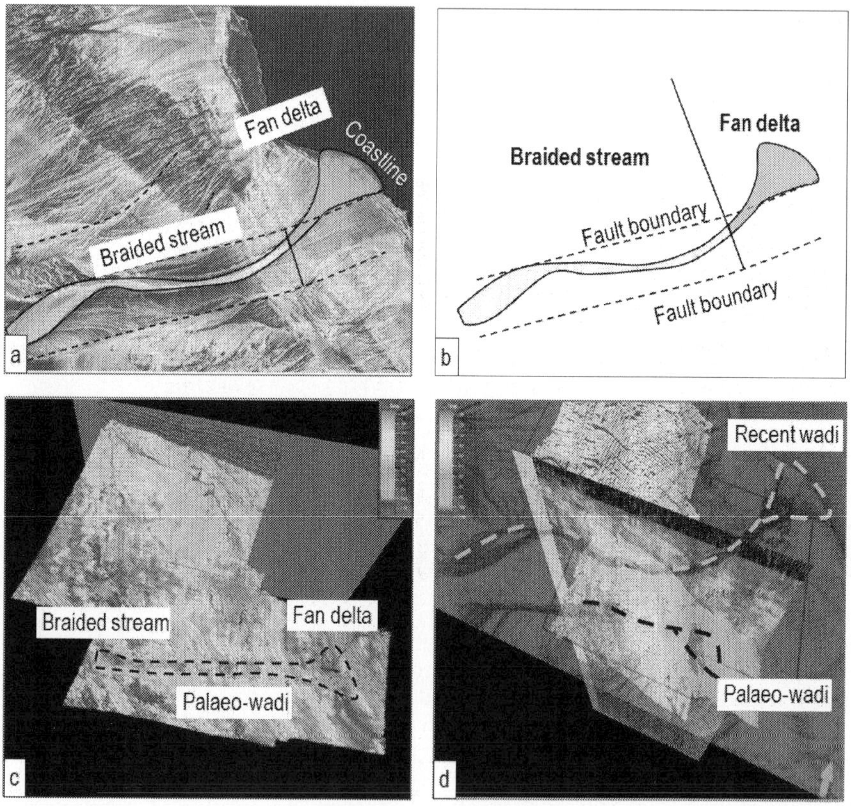

Fig. 26. Satellite based generation of templates for subsurface geobody extraction: a - drainage map, b - extracted wadi pattern, c - detection of similar pattern in shallow seismic data, d - delineation of geobodies in deep seismic data [after Laake et al. 2011]

7.4 Mapping of shallow drilling risk related to faults

Finally the surface and shallow subsurface mapping can be merged into a shallow structural map with indications of shallow drilling risks related to outcropping fault zones (fig. 27). The map basis is a landuse map (Landsat 742 RGB) onto which the near-surface topography is projected. The fault lines delineate areas of shallow drilling risk related to fault induced weak zones, which might lead to collapsing boreholes and/or loss of circulation.

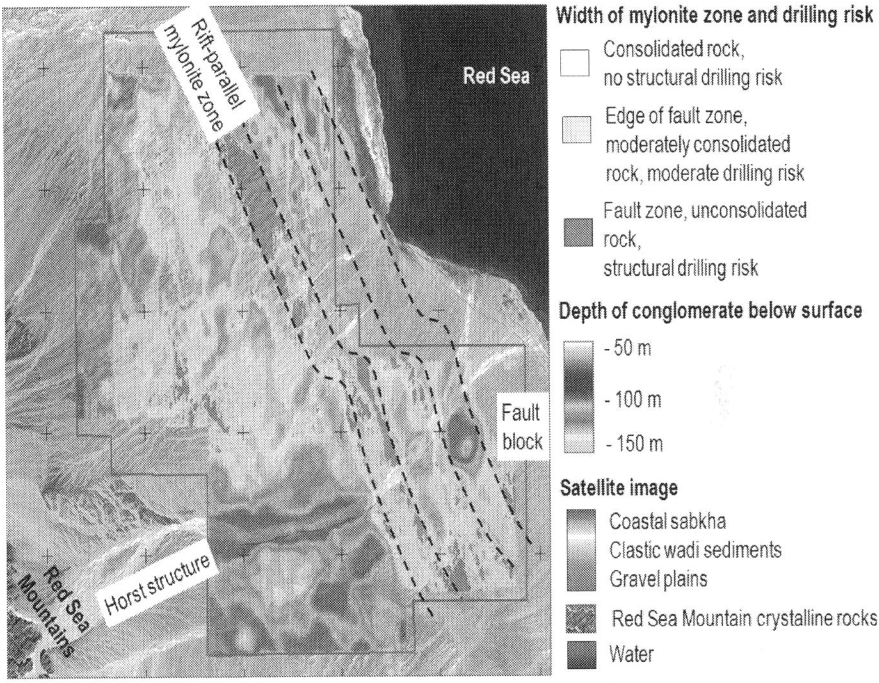

Fig. 27. Shallow structural and drilling risk mapping guided by interpretation of satellite imagery [after Laake et al. 2011]

8. Conclusion

Satellite imagery enables the investigation of the properties of the earth surface in remote areas and over large areas through the mapping of physical properties from satellite based sensors. Optical imagery delivers information about land-use, water features and surface rocks. Microwave radar imagery maps surface roughness when back-scattered at hard surfaces and paleo-drainage when penetrating dry sand cover. Microwave radar distance measurements can be converted into surface topography maps or digital elevation models. The joint interpretation of radar elevation measurements and the geoid shape delivers satellite based gravity anomaly maps, which reflect the thickness of the sedimentary layers above the crystalline basement.

The interpretation of satellite imagery can assist in the planning of surface seismic surveys through assessment of logistic and data quality risks early in the planning specifically when exploring in frontier areas. Interpretation of satellite imagery can provide estimates of the source and receiver data quality and static corrections. Consistent geological models can be generated from the interpretation of satellite imagery for geomorphological and litho-structural properties integrated with geological and geophysical data.

Fig. 28 and table 1 give an overview of the geological features detectable from satellite imagery and their impact on seismic data quality.

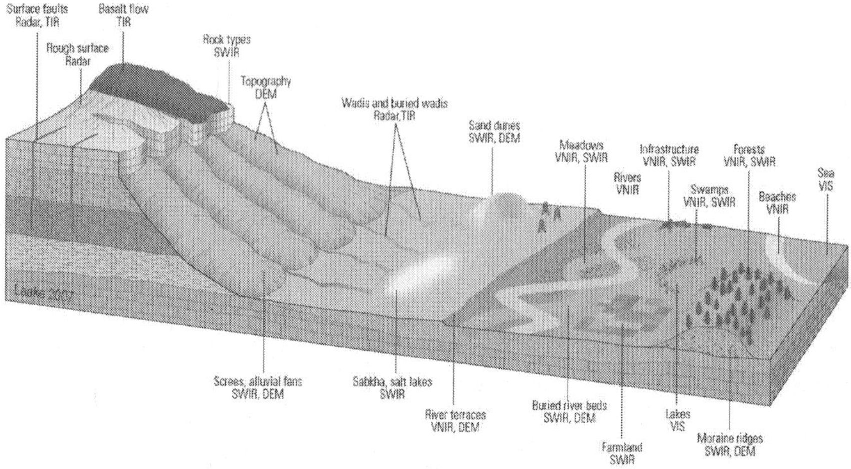

Fig. 28. Characterization of surface and near-surface properties from satellite imagery

Surface Class	Surface Feature	Satellite Imagery	Impact on Data Quality	Impact on Logistics
Topography and texture	Escarpments, river terraces	DEM, radar	Scattering noise	Severe risk for 15 - 25% slope, no access above
	Rough surfaces	Radar	Poor source / receiver coupling	Severe risk of tire damage for vehicles
	Surface faults	TIR, radar	Scattering noise	Only on escarpments
Land use	Swamps, marshes	VIS–NIR	Resonance; velocity statics	If wet, no access for vibrators and vehicles
	Water features	VIS	Special equipment	No vehicle access
Lithology	Basalt flows	TIR	Poor coupling; strong scattering	Often risk for vibrator and vehicle tires
	Claypans	DEM, NIR	Resonance	No access if wet
	Hard rock outcrops	NIR, SWIR, radar	Poor source / receiver coupling	Limited risk of access for vibrators
	Sabkhas, salt lakes	DEM, SWIR	Resonance; velocity statics	Severe risk for vibrator and vehicle access
	Buried rivers	TIR, radar	Velocity statics	No risk
Geo-morph-ology	Moraine ridges	SWIR, DEM	Low velocity, high attenuation	No risk
	Sand dunes	SWIR, DEM	Elevation statics; strong attenuation	Access for vibrators severely limited

Table 1. Detection of surface geological features from satellite imagery and impact on seismic data. For satellite imagery codes see section 3.

9. Acknowledgements

The author thanks ADNOC, Anadarko, Apache Oil Egypt, Bahrain Oil Company, Dara Petroleum Company, Egypt General Petroleum Corporation, Gaz de France, Kuwait Ministry of Oil, Kuwait Oil Company Ltd., Repsol YPF, Sonatrach, TransGlobe Energy, and WesternGeco for the permission to publish the data.

The original data for Landsat 7 ETM+ and MrSID is provided by NASA. The original data for the SRTM DEM is provided by NASA through the CIAT-CSI SRTM website (http://srtm.csi.cgiar.org), the Canadian Space Agency provided Radarsat microwave data. ASTER original data and ASTER GDEM are products of METI and NASA, the original data is property of METI and NASA. Radar based satellite gravity data are provided by the University of California, La Jolla Satellite.

The author also thanks Elisabeth DeTemple, Charles Woodward and Nick Moldoveanu for their reviews, Andrew Cutts for the support with GIS and Claudio Strobbia, Mohamed Sheneshen and Larry Velasco for their contributions to the data integration.

10. References

Alsharhan, A.S. and Salah, M.G. (1995). Geology and hydrocarbon habitat in rift setting: Northern and central Gulf of Suez, Egypt, Bulletin of Canadian Petroleum Geology, Vol. 43, 156–176.

Astakhov, V. I., Svendsen, J. I, Matiouchkov, A., Mangerud, J., Maslenikova, O. and Tveranger, J. (1999), Marginal formations of the last Kara and Barents ice sheets in northern European Russia. Boreas, Vol. 28, pp. 23–45. Oslo.

Coulson, S., Grabak, O., Cutts, A., Sweeney, D., Hinsch, R., Schachinger, M., Laake, A., Monk, D. and Towart, J. (2009). Satellite sensing : risk mapping for seismic surveys, Schlumberger Oilfield Review, Winter 2008/2009, pp. 40-51.

Cutts, A. and Laake, A. (2009). An Analysis of the Near Surface Using Remote Sensing for the Prediction of Logistics and Data Quality Risk, Tunis 2009 - 4th North African/Mediterranean Petroleum and Geosciences Conference & Exhibition, Tunis, March 2009, paper S30.

Darwish, M. and El-Araby, A.M. (1993). Petrography and diagenetic aspects of some siliciclastic hydrocarbon reservoirs in relation to the rifting of the Gulf of Suez. In: Philobbos, E.R. & Purser, B.H. (eds) Geodynamics and Sedimentation of the Red Sea-Gulf of Aden Rift System. Geological Society of Egypt, Special Publication, 1, pp. 155–187.

El-Baz, F. and Robinson, C. (1997), Paleo-channels revealed by SIR-C data in the Western Desert of Egypt : implications to sand dune accumulations, 12th International Conference and Workshops on Applied Geologic Remote Sensing, Denver November 1997.

Guo, H., Lewis, S. and Marfurt, K.J. (2008). Mapping multiple attributes to three- and four-component colour models - A tutorial. Geophysics, Vol. 73, pp. W7–W19.

Jarvis A., H.I. Reuter, A. Nelson, E. Guevara, 2008, Hole-filled seamless SRTM data V4, International Centre for Tropical Agriculture (CIAT), available from: http://srtm.csi.cgiar.org.

Laake, A. (2005a). Application of Landsat data to seismic exploration - Case study from Kuwait, Kuwait First International Remote Sensing Conference and Exhibition, Kuwait September 2005.

Laake, A. (2005b). Remote sensing application for vibroseis data quality estimation in the Neuquen Basin, Argentina, VI Congreso de Exploración y Desarrollo de Hidrocarburos, Mar del Plata, November 2005.

Laake, A. (2009). Hybrid near-surface modeling for seismic property estimation in arctic areas, 71st EAGE Conference & Exhibition, Amsterdam, June 2009, paper T003.

Laake, A. (2010). Enhancing the value of remote sensing data through integration with ground based data in 3D, ESA workshop on satellite earth observation for the oil and gas sector, Frascati, September 2010.

Laake, A. and Cutts., A. (2007). The role of remote sensing data in near-surface seismic characterization, First Break, Vol. 25, pp. 51 - 55.

Laake, A. and Insley, M. (2007). Near-surface characterization from remote sensing data, ENVISAT Symposium 2007, Montreux, April 2007, A457029.

Laake, A. and Tewkesbury, A. (2005). Vibroseis data quality estimation from remote sensing data, Proceedings of the 67th Conference and Exhibition, Madrid, June 2005, Expanded abstracts, paper G017.

Laake, A. and Zaghloul, A. (2009). Estimation of static corrections from geologic and remote-sensing data, The Leading Edge, February 2009, pp. 192-196.

Laake, A., Al-Alawi, H. and Gras, R. (2006). Integration of remote sensing data with geology and geophysics - Case study from Bahrain, GEO 2006, Manama, March 2006.

Laake, A., and Insley, M. (2004a). Applications of satellite imagery to seismic survey design, The Leading Edge, Vol. 23, No. 10, 1062-1064.

Laake, A., and Insley, M. (2004b). Satellite-based seismic technology, World Oil, Vol. 225, No. 9, pp. 27-33.

Laake, A., Sheneshen, M., Strobbia, C., Velasco, L. and Cutts, A. (2011). Integration of 4surface/subsurface techniques reveals faults in Gulf of Suez oilfields, Petroleum Geoscience, Vol. 17, 2011, pp. 165–179.

Laake, A., Strobbia, C. and Cutts, A. (2008). Integrated approach to 3D near-surface characterization, First Break, Vol. 26, pp. 109-112.

Robinson, C.A., Werwer, A., El-Baz, F., El-Shazly, M., Fritch, T. and Kusky, T. (2007), The Nubian aquifer in Southwest Egypt, Hydrogeology Journal, Vol. 15, 2007, pp. 33-45.

Sabins, F. (1996). Remote sensing, principle and interpretation (3rd ed.), Freeman, ISBN 0716724421, New York.

Sandwell, D.T. and Smith, W.H. (2009). Global marine gravity from retracked Geosat and ERS-1 altimetry: Ridge segmentation versus spreading rate, Journal of Geophysical, Vol. 114, B01411, pp. 1-18.

Short, N., (2010), The remote sensing tutorial, NASA 2010, Date of access : 15/06/2011, Available from : http://landsat.gsfc.nasa.gov/about/L7_td.html

Short, N.M.Sr., and Blair, R.W.Jr. (eds.) (1986), Geomorphology from Space, NASA 1986, Date of access : 19/09/2007, Available from : http://geoinfo.amu.edu.pl/wpk/geos/GEO_COMPLETE_TOC.html

USGS (2011), Landsat 7 science data users handbook, Date of access : 24/06/2011, Available from: http://landsathandbook.gsfc.nasa.gov/pdfs/Landsat7_Handbook.pdf

Part 9

Environmental Sciences

22

Climate History and Early Peopling of Siberia

Jiří Chlachula
Laboratory for Palaeoecology, Tomas Bata University in Zlín, Czech Republic

1. Introduction

Siberia is an extensive territory of 13.1 mil km^2 encompassing the northern part of Asia east of the Ural Mountains to the Pacific coast. The geographic diversity with vegetation zonality including the southern steppes and semi-deserts, vast boreal taiga forests and the northern Arctic tundra illustrates the variety of the present as well as past environments, with the most extreme seasonal temperature deviations in the World ranging from +45°C to -80°C. The major physiographical units - the continental basins of the Western Siberian Lowland, the Lena and Kolyma Basin; the southern depressions (the Kuznetsk, Minusinsk, Irkutsk and Transbaikal Basin); the Central Siberian Plateau; the mountain ranges in the South (Altai, Sayan, Baikal and Yablonovyy Range) and in the NE (Stavonoy, Verkhoyanskyy, Suntar-Hajata, Cherskego, Kolymskyy Range) constitute the relief of Siberia. The World-major rivers (the Ob, Yenisei, Lena, Kolyma River) drain the territory into the Arctic Ocean.

Siberia has major significance for understanding the evolutionary processes of past climates and climate change in the boreal and (circum-)polar regions of the Northern Hemisphere. Particularly the central continental areas in the transitional sub-Arctic zone between the northern Siberian lowlands south of the Arctic Ocean and the southern Siberian mountain system north of the Gobi Desert characterized by a strongly continental climate regime have have been in the focus of most intensive multidisciplinary Quaternary (palaeoclimate, environmental and geoarchaeological) investigations during the last decades. Siberia is also the principal area for trans-continental correlations of climate proxy records across Eurasia following the East-West and South-North geographic transects (Fig. 1). Among the terrestrial geological archives, loess (fine aeolian dust) represents the most significant source of palaeoclimatic and palaeoenvironmental data, together with Lake Baikal limnological records, with bearing for reconstruction of the past global climate history. The Siberian loess, being a part of the Eurasian loess-belt, has provided chronologically the most complete evidence of past climate change in the north-central Asia (Chlachula, 2003).

The Cenozoic neo-tectonic activity with the Pleistocene glaciations and interglacial geomorphic processes modeled the configuration of the present relief of Siberia and the adjoining Ural Mountains. During the cold Pleistocene periods, the vast extra-glacial regions of West Siberia south of the NW Arctic ice-sheet were transformed into a large periglacial super zone (Arkhipov, 1998), which became a major sedimentation area of aeolian (silty) deposits cyclically derived by winds from the continental ice-front ablation surfaces. Main palaeoenvironmental records, spanning over several hundred thousand years, have been preserved in deeply stratified sections within the major basins (Ob, Yenisei, Angara and

Lena River) exposed after progressive erosion triggered by constructions of large dams (Drozdov et al., 1990; Medvedev et al., 1990). Equally important sources of the Quaternary (geological and biotic) palaeoclimate proxy data originate from open-coalmines and other modern industrial surface disturbances (Zudin et al., 1982; Foronova, 1999).

Studies of Quaternary climates in Siberia, encompassing the last 2.5 million years, have advanced considerably in recent years, mainly because of increased awareness of the value of reconstructing geological and natural histories for understanding of the present-day ecosystems, and applied as a means of predicting the probable extent and consequences of future climatic changes. Because of the multi-factorial nature of a long-term climatic and environmental evolution, Siberian palaeoclimate-oriented studies have become increasingly interdisciplinary, integrating Quaternary geology and palaeogeography, palaeopedology, palaeontology, palaeobotany, Palaeolithic archaeology and other fields (Chlachula, 2001a-c, 2010a-b). Reconstructions of past environments in specific regions and time periods have been used to assess the effects of orbital variations on seasonal and latitudinal distribution of solar radiation and atmospheric circulation patterns, and the consequential changes in regional temperatures, precipitation and moisture balance. Some long-term models provide means of predicting future climatic evolution in the context of the global climate history and help in the assessment of the modern human factor in environmental change. Because of the pronounced climatic continentality of the territory, even minor variations in atmospheric humidity and temperature led to major transformations in local ecosystems, particularly in the open southern Siberian continental sedimentary basins and the upland depressions.

Principal information on past climates and climate change comes from the southern Siberian loess regions that have been intensively studied since the 1990's, following the initial field investigations and chronological interpretations of long-term sub-aerial sequences. The most recent studies have gradually shifted to more detailed and high-resolution Late Quaternary records and refinement of the regional loess-palaeosol chronostratigraphies, illustrating the landscape development and biota evolution particularly for the last 130 000 years by using advanced chronometric, geo-chemical and biostratigraphic marker analyses. The most complete Late Pleistocene palaeoclimate archives include four main stages (the Kazantsevo Interglacial, the Ermakovo /early Zyriansk/ Glacial, the Karginsk Interpleniglacial, and the Sartan /late Zyriansk/ Glacial) correlated with the Marine Isotope Stages (MIS) 5-2.

The broad Siberia also has the key relevance for elucidation of timing and conditions of environmental adaptation of the prehistoric and early historic people to high latitudes of Eurasia, as well as the initial colonization of the Pleistocene Beringia, including the north-western part of the American continent. The particular geographic location and the diversity in topographic configuration of regional landscape reliefs together with changing Quaternary environments governed by the past global climate change played the key role in this long and complex process. The spatial and contextual distribution of the documented archaeological sites reflects a climatic instability and a timely discontinuous inhabitability of particular geographical areas of Siberia delimited by the Central Asian mountain system in the south and the Arctic Ocean in the north. The cyclic nature of the glacial and interglacial stages led to periodic geomorphic transformations and generation of specific ecosystems adjusted to particular topographic settings and responding to acting atmospheric variations. Diversity of the present relief and environments (Fig. 1B), reflecting the past climate change, played the key role in the process of the initial peopling of the immense Siberian territory. Palaeoenvironmental databases (palynological, palaeontological as well as early cultural records) provide unique evidence of strongly fluctuating Pleistocene glacial and interglacial climates, corroborating the geological stratigraphic archives.

The human occupation of Siberia used to be traditionally associated with the Late Palaeolithic cultures. Systematic geoarchaeology investigations during the last 20 years across the entire Siberia (with the key research loci in the Tran-Ural region of West Siberia, the Altai region, the Upper Yenisei, Angara and Lena Basins, as well as at the easternmost margins of the Russian Far East in Primoriye and on the Sakhalin Island) revealed several hundred of Palaeolithic and Mesolithic sites (Serikov, 2007; Chlachula et al., 2003, 2004b; Derevianko & Markin,1999; Derevianko & Shunkov, 2009; Medvedev et al., 1990; Mochanov, 1992; Vasilevsky, 2008; Zenin, 2002). Particularly the discoveries of numerous Palaeolithic sites, some of potentially great antiquity (> 0.5 Ma), located in large-scale surface exposures (river erosions and open-pit mines) followed by systematic archaeological investigations within the major river basins of south and central Siberia between the Irtysh River in the west and the Lena River / Lake Baikal in the east (Fig. 1A), have provided overwhelming evidence of a much greater antiquity of human presence in broader Siberia and capability of early people to adjust to changing Pleistocene environments. Cultural remains are located in diverse geomorphic settings (i.e., lowland plains, mountain valleys, upland plateaus) and geological contexts (aeolian, fluvial, lacustrine, palustrine, alluvial, glacial and karstic), with the highest concentrations in the Pleistocene periglacial parkland steppe and the boreal tundra-forest foothill zone. Particularly the geographically extensive and deeply stratified loess-palaeosol sections in the southern Siberian open parklands have revealed a long and chronologically well-documented cultural sequence of human occupation. The variety of cultural finds provides witness to several principal stages of inhabitation of the Pleistocene Siberia, possibly encompassing the time interval close to 1 Ma with the earliest (Early and Middle Pleistocene stages) represented by typical "pebble tool" industries, followed by the Middle Palaeolithic complexes, including the (Neanderthal) traditions with the Levallois prepared-core stone-flaking technology, and the regionally diverse Late/Final Palaeolithic blade complexes eventually replaced by the microlithic Mesolithic cultural facies that developed in response to major natural transformations during the final Pleistocene.

A further northern geographic expansion of humans into the Arctic regions reflects a progressive cultural adaptation to extreme climatic conditions of (sub)polar Pleistocene environments (e.g., Mochanov & Fedoseeva, 1996, 2001). The occupation sites in the Polar Urals and North Siberia (Svendsen & Pavlov, 2003; Pitulko et al., 2004) provide eloquent evidence that people reached the Arctic coast already before the Last Glacial (>24 000 years ago). All these discoveries logically lead to revision of the traditional perceptions on a late peopling of northern Asia as well as the "late chronology" models of the initial human migrations across the exposed land-bridge of Beringia to the North American continent (Chlachula, 2003b). Geoarchaeology studies, particularly in the poorly explored and marginal geographic regions of northern and eastern Siberia (Pitulko et al., 2004; Vasilevsky, 2008), are of utmost importance for reconstruction of past climate change as well as the early human history in north Eurasia. Evolutionary processes in natural environments, and specific behavioral Palaeolithic adaptation patterns and material-technological conditions, as well as documentation of sequenced climatic events stored in geological records are the principal objectives of the current multidisciplinary Quaternary investigations in Siberia.

This contribution summarizes in a general overview the present evidence on the Quaternary climates and climate development in western, southern and eastern Siberia in respect to associated environmental transformation and the stages of early peopling of particular geographic areas of this extensive and fascinating territory.

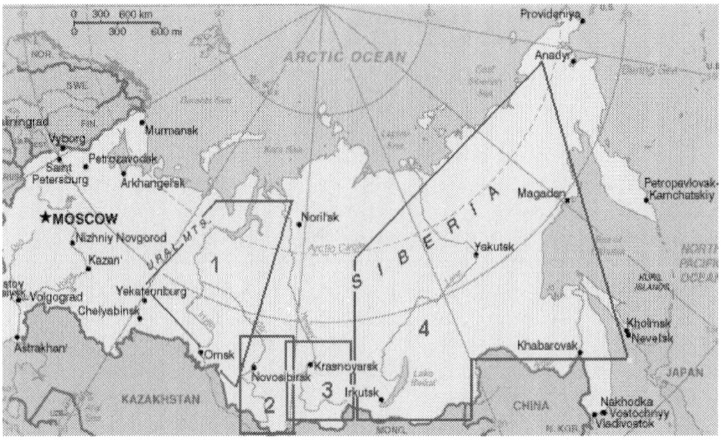

A

B-1

B-2

B-3

B-4

Fig. 1. A: Geographic map of Siberia. 1. Trans-Urals and West Siberia, 2. Altai, SW Siberia, 3. upper Yenisei Basin, 4. East Siberia (Lake Baikal, Lena Basin); B: Landscapes & ecosystems 1. parkland-steppe (West Siberian Lowland), 2. the Altai Mountain tundra; 3. semi-desert (Chuya Basin, southern Altai); 4. northern taiga with thermo-karst lakes (central Yakutia).

2. Pleistocene climates and natural environments of Siberia

Because of its vast territory and the geographical isolation between the major Central Asian (Himalayan) mountain systems and the Arctic Ocean, Siberia represents a unique place that is of major potential for mapping the climatic history of northern Eurasia, but also general pathways and rates of global change in the Northern Hemisphere. Tectonic uplift in the Miocene, continuing through to the late Pliocene, initiated formation of the present relief. Breaking up the original pre-Cenozoic continental Siberian Platform significantly influenced the Pleistocene atmospheric circulation by blocking a free influx of warm and humid air masses from the southeast. The dominance of the arctic atmospheric streams gradually led to the present strongly continental climatic regime, with dry winters with little snow cover, and warm to hot summers, of mean annual temperatures -0.5 to -3°C (with the recorded minimum -71.2°C at Oymyakon in central Yakutia). Permafrost underlies most of north and central Siberia (Fig. 1B-4), while perennial mountain ground ice is locally distributed on the upland plateaus north of Mongolia. Glaciers occur above 3000 m altitude in the Altai and Sayan Mountains. The NW-SE oriented southern Siberian mountain ranges (Fig. 1B-2) form a barrier between the northern sub-arctic uplands and Arctic lowlands, and the southern steppes and rocky deserts of Mongolia and north-west China. The regional geography and relief shape the pronounced latitudinal and altitudinal vegetation zonation, with (parkland)-steppe, taiga forest, alpine tundra, polar tundra-steppe and mountain semi-deserts (Fig. 1B). Steppe chernozems with thick humic (Ah) horizons, cryogenically distorted by frost wedges, are distributed in the southern loess zone, while podzols with brown forest soils and gleyed tundra soils prevail in the mountains and the northern regions, respectively.

The pre-Quaternary (>2.5 Ma) geology of the southern and eastern Siberian areas was controlled by a series of tectonic events. Both relief and the geological structure contributed to intensive geomorphic processes in the past. The Middle and Late Palaeozoic orogenesis formed uplands separated by isolated continental depressions. Later Hercynian and Oligocene tectonics modified the original configuration of the upper Lena, Angara, Kuznetsk, Minusinsk and other basins north of the Altai-Sayan Mountains and the Northern Baikal-Yablonovyy Mountain Range. Accumulation of extensive proluvia near the mountain fronts and lacustrine /alluvial formations in the principal sedimentary basins continued throughout the Miocene and Pliocene. During the Late Pliocene and Early Pleistocene, early fluvial systems were established, accompanied by a progressive regional uplift. In response to climate change in conjunction with orogenic activity, a series of river terraces gradually developed in the climatically most favorable and first inhabited basins of southern Siberia, with the earliest, Early and Middle Pleistocene (70-90 m, 110-130 m, 150-170 m-high) terraces preserved mostly in relics and buried by 10-50 m loess deposits. The Siberian river valleys (Fig. 1A) served as main migrations corridors of both Pleistocene fauna and early humans.

During glacial periods, the northward drainage of the Ob and Yenisei Rivers was diverted by the ice barrier of the northern inland ice-sheet southwest into the Aral and Caspian Sea (Arkhipov, 1998). The relief of southern Siberia with deep intermountain depressions preconditioned formation of major river dams blocked by mountain glaciers that periodically caused high-magnitude floods (Rudoy & Baker, 1993). These large-scale catastrophic events undoubtedly had a dramatic impact on the early human habitat. The geographical position close to the geographical centre of Asia also contributed to extreme continental climatic conditions with widespread permafrost developed during cold stadials. Siberia is thus of bearing significance for mapping Pleistocene changes in Arctic air-mass circulation above the central and eastern Eurasia, as reflected in textural and compositional variations in the aeolian deposits. Among the continental sediments used as climatically significant palaeo-archives, loess (fine aeolian dust) has attracted most attention because of its high environmental sensitivity and the long-term stratigraphic records it often provides. Increased rates of loess accumulation correspond to cold and dry (glacial) stages followed by soil development during warm (interglacial and interstadial) intervals. The Siberian loess and the associated palaeoclimate proxies contribute to the trans-Eurasian palaeoclimatic correlation, linking the European, Central Asian and Chinese loess provinces (Rutter et al., 2003; Bábek et al., 2011). Ultimately, the Siberian loess-palaeosol sections, rich in fossil micro- and macro-fauna and early cultural remains, provide a contextual and chronological framework for documentation of timing, natural conditions and processes acting during the early human peopling of northern Asia and the Pleistocene Beringia.

The principal distribution zone of loess and loess-like deposits lies in the southern part of the Siberian territory (50°-60° N and 66°-104° E) west of the Ural Mountains within a broad belt 500-1500 km wide (N to S), encompassing ca. 800 000 km^2 between the Ob and Angara River basins along the Altai, Salair and Sayan Mountains (Fig, 2A). The Siberian loess provides a detailed, high-resolution record of climatic shifts that may not be detectable in more uniform loess-palaeosol formations elsewhere on the Eurasian continent. Thickness of loess sections is from a few meters in the Lake Baikal area to 40 m in the Yenisei River valley and up to 150 m on the Ob River (Priobie) loess plateau and the North Altai Plains. The regional loess-palaeosol record spans throughout the Quaternary, yet it may be locally fragmentary, partly re-deposited, or may be completely absent for the earlier stages. The geomorphological setting of southern Siberia with the extensive open lowlands in the west and depressions towards the east, combined with dominantly westerly (NW-SW) winds, points to location of the major deflation surfaces (loess source areas) on the eastern/steppe Altai Plains between the Irtysh and Ob Rivers with accumulations of fine aeolian dust eastwards on the plains north of Altai-Sayan Mountains. Ice-sheet marginal areas, large alluvial floodplains, exposed valley floors and margins of (glacio-) lacustrine basins were important sources of the silty sediment. Local geological sources in the southern intramontane basins (Kuznetsk, Minusinsk, Irkutsk Basin), with aeolian dust discharge from the glaciated Altai and Sayan Mountains, the Kuznetskiy Alatau and the Baikal Range, played an important role in the regional input of wind-blown sediment in the adjacent river basins. The present westerly (SW/NW) winds likely prevailed throughout the Quaternary Period with several Late Pleistocene stages of sub-aerial deflation and sediment deposition.

The most continuous and high-resolution sections of aeolian aerosol dust deposits, interbedded with variously developed fossil soils from the Priobie Loess Plateau, the North Altai Plains, the upper Yenisei and Angara River basins, produced a consistent and unique evidence of the Late Quaternary climate evolution and the corresponding landscape development in parkland-steppe of southern Siberia. The temporarily preceding Early and Middle Pleistocene loess records are less complete with regional climatostratigraphic hiatuses. Climate-indicative proxy data (magnetic susceptibility, grain-size, % $CaCO_3$, % TOC mineral color parameters) supplemented by pedological, including thin-section studies from the most continuous loess-palaeosol reference sites (i.e., Iskitim, Biysk, Kurtak and Krasnogorskoye) (Fig. 2A, 2D) show marked and cyclic climatic variations spanning the last ca. 250 000 years (250 ka) corresponding to Marine Oxygen Isotope Stages (MIS) 7-1, most pronounced during the last interglacial-glacial cycle in congruence with the Lake Baikal detrital records and the globally indicative deep-sea isotope records (Fig. 2B) (Chlachula, 2003a; Evans et al., 2003; Bábek et al., 2011; Grygar et al., 2006; Prokopenko et al., 2006).

The loess stratigraphy shows aeolian dust deposition during dry and cold stages, and soil formation during warm intervals with surface stabilization and a subsequent cryogenic distortion by frost actions followed by colluviation due to climatic warming. The massive (late) Middle Pleistocene loess accumulation in the southern periglacial tundra-steppe zone is linked with a major Pleistocene glaciation of the Altai and Sayan Mountains during the penultimate glacial (MIS 6, 170-130 ka BP). An intensified pedogenic activity indicates a rapid onset of the last interglacial (MIS 5, 130-74 ka BP) (Fig. 2D) with a strongly continental warm climate during the interglacial climate optimum (MIS 5e, 125 ka BP). The palaeo-temperature data indicate an increase of temperature during the last interglacial peak in average by ca +3°C in the southern part of Siberia and up to +6°C in the arctic regions. This was succeeded by a dramatic interglacial cooling (MIS 5d) with a temperature minimum at ca. 115 ka BP evidenced by a major cryogenic event (with deformation of the MIS 5e chernozem by 1-3 m deep frost-wedges blanked by a thin loess cover) in congruence with the Lake Baikal stratigraphic record. Permafrost remained widely distributed in east Siberia. Cooler conditions during the following interglacial sub-stages (MIS 5c and 5a) correspond to the zonal vegetation shifts, with a gradual replacement of parkland-steppe and mixed southern taiga by boreal forests, interspersed by periglacial tundra-steppe (MIS 5d and 5b). The progressing cooling climatic trend of the early Last Glacial (MIS 4, 74-59 ka BP) culminated in several hyper-arid and cold stadials (MAT by ca. 10°C under present values) interrupted by intervals of climatic amelioration with (gleyed) periglacial forest-tundra soils formation and isolated brush and tree communities *(Betula nana, Salix polaris, Pinus sibirica)*. The following cycle with relatively short warm as well as very cold climate variations characterizes the mid-last glacial interstadial stage (MIS 3, 59-24 ka BP) with cryogenically distorted chernozemic soils, pointing to the existence of a high permafrost table for most of this (Karginsk) interstadial. This was ensued by a new stage of a massive loess accumulation during the late Last Glacial (MIS 2, 24-12 ka BP) with incipient (forest/steppe)-tundra gleysols formed in response to warming oscillations prior to the present Holocene surface stabilization (MIS 1, 12-0 ka BP). A marked drop in MAT by about 8°C (to -9/-10°C) and a decline of precipitation by 250 mm is assumed for the last glacial maximum (22-19 ka BP) (Velichko, 1993). The highest loess accumulation rates recorded during the Late Pleistocene glacial stages (Ermakovo / MIS 4 and Sartan /MIS 2) on the Altai Plains and in the upper Yenisei Basin indicate the most intensive aeolian dust deposition after the glacial maxima, with the most recent interval dated to ca. 19-15 ka BP (Chlachula, 2003a; Evans et al., 2003).

The interglacial and interstadial palaeosol markers in the loess series, characterized by an intensive syndepositional pedogesis, attest to a strongly fluctuating climate evolution and the corresponding landscape development with major biotic transformations in north-central Asia during the Late Quaternary. The past continental atmospheric shifts reflected by changes in the main vegetation zones display a cyclic pattern of interglacial/interstadial parkland-steppe and mixed taiga ecosystems replaced by boreal tundra-forest and arid periglacial tundra-steppe during cold stadial stages. The northern expansion of parkland-steppes during warm periods as well as periglacial steppes undoubtedly promoted a geographical enlargement of the Palaeolithic oikumene. Cold climatic stages furthermore stimulated human biological and cultural adaptation in the process of peopling of Siberia.

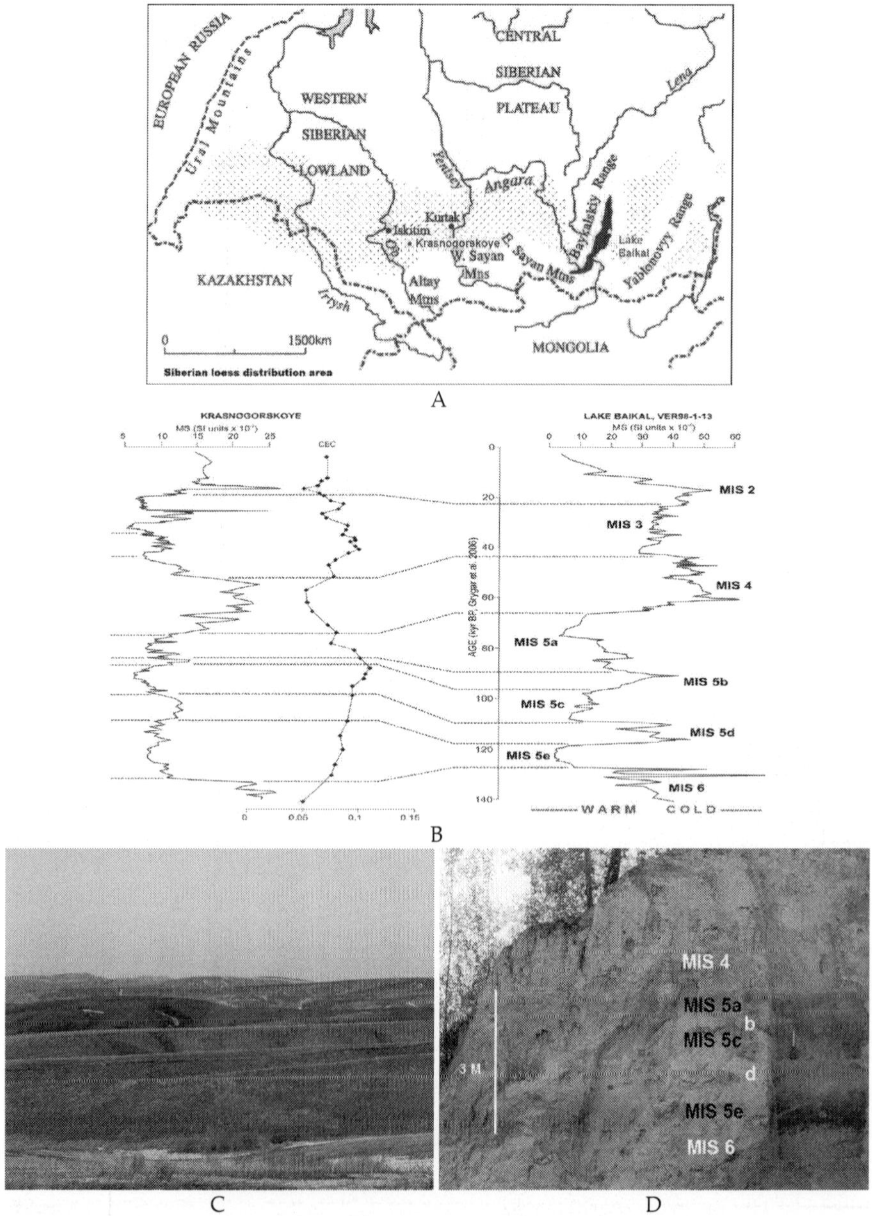

Fig. 2. A: Loess distribution in Siberia; B: Late Quaternary climate evolution in Siberia based on loess-palaeosol and Lake Baikal proxy records; C: the Altai loess landscape; D: the last interglacial (MIS 5) pedocomplex (130-74 ka BP) at Iskitim, with thee parkland chernozems.

3. Early human occupation and environmental adaptations

The vast territory of Siberia and the adjacent Trans-Ural region has principal bearing for elucidating the historical processes and environmental contexts of the Pleistocene expansion of people from the south-eastern parts of Europe into West-Central Siberia and the northern Russian Arctic areas. Interactions of past climate change and the regional relief modeling triggered by the neotectonic activity and reflected by natural transformations of ecosystems attest to the complexity of the Quaternary landscape development, ultimately affecting timing, intensity and adaptations of the earliest human occupation of north-central Asia.
The contextual geology, palaeoecology and paleontology records from the investigated archaeological sites and stratified geological sections provide evidence of pronounced (palaeo)environmental and biotic shifts triggered by the global climate evolution as well as the associated glacial and interglacial geomorphic processes. Quaternary climatic cycles regulated spatial and temporal movements of prehistoric people into the high latitudes of Eurasia. Integrated ecology multi-proxy databases document trajectories of the complex and long occupation history of this extensive, but still marginally known part of the World.

3.1 The Urals and the Trans-Ural regions of West Siberia

The Urals with the adjoining regions of the West Siberian Plain (*Zauralye* - the Trans-Urals) are of key relevance for multi-disciplinary Quaternary studies focusing on initial peopling of northern Eurasia in respect to the strategic geographical location of this mountain chain, forming the natural borderline between Europe and North Asia. Except for documenting the Pleistocene and Holocene climate evolution and related environmental transformations, the territory of the Urals has a major bearing for mapping the main time intervals of migrations of Pleistocene hominids / early humans into the high latitudes / sub-polar and polar regions of Eurasia, and reconstruction of adaptation strategies to past climate variations.

3.1.1 Geography and natural setting

The Urals, the principal mountain range separating the East European Plains and the West Siberian Lowland as the geographical milestone between Europe and Asia, extends for over 2000 km from the Arctic coast in the north to the arid steppes of western Kazakhstan in the south (Fig. 1A). It is characterized by a pronounced vegetation zonation with polar tundra-forest, taiga and open steppe being the dominant biotopes. The archaeologically most productive Central Urals (56-59°N) is one of the five physiogeographic zones of the Ural mountain range (1895 m asl.) , connecting the Southern Urals with the Northern, Sub-polar and Polar Urals (Fig. 3A). Most of the regional hilly relief (700-1500 m asl.) is transected by mountain valleys with the principal E-W oriented fluvial drainage discharge following the continental topo-gradient (Fig. 3C). The adjoining geographical area in the west *(Priuralye /* Fore-Urals) includes the loess plateau of the upper Kama River basin. The Trans-Ural region is delimited by the southern limits of the Sverdlovsk District, and the Tobol and Ob River basins from the (South-)East with the tributary valleys (Severnaya Sosva, Sosva, Tura, Tagil', Neiva, Pyshma) draining the eastern Ural foothills (Fig. 3D). Its western part is formed by small hills (300-400 m asl.) separated by shallow river valleys, lakes and bogs, extending east into the West Siberian Lowland. The present soil cover is characterized by well-developed parkland-steppe chernozems, mixed / coniferous taiga forest brunisols / gleysols and forest-tundra regosols, all with fossil analogues in the Quaternary pedostratigraphic record.

3.1.2 Palaeogeography and climate history

The topographically and biotically diverse territory of the broader Ural area attests to the complex palaeogeography and palaeoecology history reflecting large-scale landscape restructurings affected by orogenesis, and the Arctic and northern mountain glaciations (Arkhipov et al., 1986; Astakhov, 1997; 2001; Velichko, 1993; Mangerud et al., 2002), The regional Quaternary stratigraphy is based on geological, palaeontological, bio-stratigraphic and palaeomagnetic records related to the continental basin transgressions and regressions, and stages of loess deposition (Arkhipov et al., 2000; Vereshagina, 2001; Stefanovskiy, 2006).

The pre-Quaternary history of the broader Urals is closely linked with formation of the main hydrological systems, particularly in the Fore-Urals area, connected to the Pechora and Caspian Basins. This process was triggered by past global climate change in conjunction with the re-activated orogenesis of the Eurasian Plate, with dynamic sea-transgressions and prolonged periods of stands and regressions, shaping the continental relief and resulting in formation of the present northern (Pechora) and southern (Kama-Volga) drainage systems. In warm climatic stages, the basins were filled by alluvial and lake deposits superimposed by periglacial alluvial formations (Yakhimovich et al., 1987). The original N-S oriented drainage of the Oligocene-Miocene basins persisted until Pliocene. A new tectonic phase contributed to a palaeo-relief restructuring with shallow river basins. The Early Pleistocene interglacial biotic records document a forest-steppe habitat with presence of arboreal taxa absent in the area today *(Fagus, Acer, Fraxinus, Ulmus, Tilia, Castanea)* and brackish settings inhabited by *Archidiscodon meridionalis, Hipparion* sp., *Coelodonta* sp. (Stefanovskiy, 2006).

Climatic cooling at the end of the Early Pleistocene (ca. 740 ka BP) is manifested by glacial deposits distributed in the Pechora and Upper Kama basins, with regressions of the Caspian and Arctic seas. The reactivated tectonics at the beginning of the Middle Pleistocene in the broader Urals-Caucasus area accelerated restructuring of the former drainage system. Increased aridity put up to an intensified weathering in the mountain regions of the Urals. A reduced fluvial activity in the eastern (Trans-Ural) West Siberian lowlands contributed to genesis of closed lake and boggy basins. Floral and faunal communities indicate dry steppe environments replacing former woodlands with the Mediterranean vegetation (*Pterocarya fraxinifolia, Rhododendron* sp., *Taxus baccata, Castanea sativa*) (Stefanovskiy, 2006). The new hydrological network of the Tobol (MIS 9) Interglacial (390-270 ka) correlates with the 60 m-terrace, with the Tiraspol Fauna Complex of large herbivorous species (*Paleoloxodont* sp., *Mammuthus primigenius* Blum., *Bison priscus gigas, Alces alces, Megaloceros gigantheus, Cervus elaphus, Equus hemionus, Equus caballus, Camelus* sp. and *Ovis amon* sp.) indicative of warm semi-arid parklands. The following Samarovo (MIS 8) Glacial (270-244 ka) correlates with the maximum expansion of the Arctic ice to ca. the present line Perm-Nizhny Tagil, blocking the northern Ob / Irtysh River drainage of the West Siberian glacial waters diverted south into the Aral-Caspian Basin. Temperate parklands characterized the following the Shirta (MIS 7) Interglacial (244-170 ka). Recession of the ice-masses is assumed for the final Middle Pleistocene Tazov (MIS 6) Glacial (160-130 ka) with mountain glaciers confined to the Northern - Polar Urals (Arkhipov et al., 1986). The late Middle Pleistocene in the south-central Ural basins is associated with 20-30 m thick fluvial deposits with fossil fauna of the Khozarian Complex (*Mammuthus primigenius, Coelodonta antiquitatis, Bison priscus longicornis, Megaloceros gigantheus* and *Equus caballus chosaricus)* indicative of a moderately cold climate. A final Middle Pleistocene neotectotic stage accentuated the regional geomorphological relief restructuring triggered by continuing uplifts of the Central and Southern Ural ranges.

The beginning of the Late Pleistocene brought up a reduced regional topographic differentiation associated with lateral planation processes in the Trans-Ural area and the initial last interglacial (MIS 5e, 130-120 ka BP) transgression. The Ural foothills were covered by a mixed spruce and pine-dominated taiga; open parklands occupied lowlands inhabited by the "Mammoth Fauna Complex" (Bolikhovskaya & Molodkov, 2006).

The Late Pleistocene stratigraphy and glacial geomorphology reflects reducing geographical limits of the continental ice-masses during the Zyriansk (MIS 4) and the Sartan (MIS 2) glacial stages. Predisposed by the territorial topography, most of the present West Siberian Lowland was inundated by a large ice-dammed lake formed between the northern Arctic ice-sheet and the southern continental water divides during the early Last Glacial (ca. 90 ka BP) (Mangerud et al., 2001). Moraines in the Pechora Basin on the NE European Plains point to the expanding Arctic glaciations from the Kara and Barents Sea ice lobes, presumably reaching their Late Pleistocene maximum extent (Mangerud et al., 1999) in corroboration with the West Siberian glacial records (Astakhov, 1997, 2001). Dynamic climatic fluctuations of the second half of the glacial stage are linked with a gradual recession of the continental ice limits in the Northern Urals and activation of intensive periglacial and gravity-slope processes resulting in massive accumulations of up to 80 m thick polygenic colluvia in the Trans-Ural foothills. An intensified loess deposition formed a 20-25 m thick cover on the Fore-Ural plateaus, overlying periglacial alluvia of the Upper Kama, reflects the prevailing dry and cold climatic conditions of periglacial tundra-steppe.

The subsequent mid-last glacial (MIS 3) interstadial warming (59-24 ka) promoted expansion of spruce- and pine-dominated taiga. The new stage of glacial advance during the late Last Glacial /Sartan (MIS 2), probably confined to the Polar and Northern Urals with localized ice-caps, led to expansion of tundra-steppe and tundra-forests in the southern lowland and mountain areas, respectively (Velichko et al., 1997). Most of north-central Siberia presumably remained ice-free (Zemtsov, 1976, Mangerud et al., 2008), opposing the model of a more extensive ice-cover in the Russian Arctic (Grosswald & Hughes, 2002). The geographic distribution of glacial landforms and glacigenic deposits provides evidence of the Arctic ice expanding from the Kara Sea basin along the western slopes of the Urals, likely merging with the piedmont glaciers (Mangerud et al., 2002). Most of the eastern slopes of the Urals likely hosted ice-free environment possibly due to a rapid disintegration of the early Last Glacial (MIS 4) ice and dry climate. Accelerated loess accumulation in the Fore-Ural area marks onset of the full last glacial conditions associated with remains of large Pleistocene fauna *(Mammuthus primigenius, Coelodonta antiquitatis, Equus* sp., *Bison priscus)* and small species typical of periglacial tundra-steppe (Yakhimovich et al., 1987).

The late Pleistocene ecosystems in the Trans-Ural region are linked to alluvial formations of the fluvial basins draining the eastern slopes of the Urals, and incorporating the typical large and small periglacial fauna of periglacial tundra-steppe (mammoth, woolly rhinoceros, horse, bison, reindeer, elk, saiga) with smaller open-steppe and semi-desert taxa such as wolf, fox, hare, marmot, as well as of rodents and other small mammals indicative of mixed polar tundra - steppe biotopes (Stefanovskiy, 2006). Permafrost limits at the end of the Pleistocene (the Allerød warming) were still located at 55°N in the Trans-Urals / West Siberia (Velichko et al., 2002). Precipitation values during the late Last Glacial and the early Holocene (15-10 ka BP) amounted to ca. 60-65% on the East European Plains and ca. 80% in West Siberia comparing to the present-day values. The present geographic distribution of particular biotopes in the broader Ural territory reflects the Holocene climate development.

3.1.3 Pleistocene environments and early human peopling

The importance of the Trans-Urals lies in its specific geographic location at the westernmost limits of Siberia adjacent to the Ural Mountains in terms of study of prehistoric migration processes, adaptation strategies and cultural exchanges between Eastern Europe and Siberia. Interaction of the Pleistocene climatic variations with the regional relief modeling by the northern Arctic glaciations, interglacial sea-transgressions, and stages of river erosion and sediment weathering attest to the complexity of the Quaternary history of the broader Ural area. During most of the Pleistocene, the North and Central Trans-Urals, sloping east from the central mountain range, were ice-free. The bordering foothills and plains of West Siberia constituted parts of the periglacial super zone (Arkhipov, 1999) and the place of mass accumulation of aeolian deposits transported from the ablation zone of the NE European-Siberian ice-sheets. The geographic and topographic configuration of the Urals predisposed a long history of human occupation in this area, forming a natural gateway for the early human infiltration further east into the vast parklands and forest lowlands of West Siberia.

The initial peopling of the territory, presumably from the East European Plain and the northern Caucasus region, was a complex process governed by changing palaeoecology conditions in response to the Pleistocene climate variations. Warm and humid Middle Pleistocene interglacial, with rich biotic resources of forest-parklands, undoubtedly enabled early human migrations further north and east along the Ural range particularly through the major river systems (Volga, Kama, Pechora) draining the East European Plains. Mixed coniferous and broad-leaved forests with *Pinus silvestris, Betula pubescens, Quercus robur, Tilia cordata, Ulmus laevis* formed a mosaic vegetation cover during the late Middle Pleistocene (MIS 9 and MIS 7) interglacials (Bolikhovskaya & Molodkov, 2006). The oldest recorded (Early Palaeolithic) sites are mostly exposed along active river banks of the Kama reservoir by erosion of loess and loessic sediments overlying relics of the Middle Pleistocene (35-60 m) river terraces. The cultural material is represented by archaic and simply flaked core-and-flake stone industries (cores, retouched flakes, scrapers and other tools, partly bifacially worked) produced on cobbles from old river alluvia. Numerous fragmented animal bones with the anthropogenic working and use attributes (flaking, splitting, retouching) (Fig. 2B) indicate productive natural occupation habitats and complex behavioral activities. The formal technological uniformity of these Palaeolithic finds and their geological contexts related to the Middle Pleistocene alluvia suggests an intensive expansion of the Old Stone Age people across the middle latitudes of Eurasia prior to the last interglacial. (>130 ka BP) (Matyushin, 1994; Velichko et al., 1997; Chlachula, 2010a). Periglacial sub-Arctic tundra-steppe and open forest-tundra correlated with the late Middle Pleistocene (MIS 8 and MIS 6) glacial advances in the north-eastern part of Europe (the Pechora Urals) covered most of the non-glaciated territories occupied by periglacial fauna *(Mammuthus primigenius, Coelodonta antiquitatis* Blum., *Rangifer tarandus, Bison* sp., *etc.).* These prominent Middle Pleistocene glacial stages brought a major geographical reduction of the early human occupation areas.

The pronounced climatic warming at the beginning of the Last Interglacial (MIS 5), recorded in the loess-palaeosol sections from the East European Plain to southern Siberia (Little et al., 2002; Chlachula, 2003), may have created inundations and ground saturation of the formerly glaciated areas and thus temporarily impeded free movements of early people. Increased continentality and MAT led to expansion of parkland-steppe also facilitating new migrations of the Middle Palaeolithic (Neanderthal/early *Homo sapiens*?) communities. High summer temperature (increased by ca. 4-5°C in respect to present values) and mean annual

precipitation (by ca. 100 mm) led to the northward distribution of southern taiga along both sides of the Urals by 500-700 km beyond the present limits (Velichko et al., 1992; Velichko, 1993). Pine-spruce-birch forests with oak, hornbeam, lime, hazel and elm characterized the prevailing mosaic vegetation of the warm last interglacial climate sub-stages (MIS 5e, 5c).
Recession of human occupation is linked with the onset of the early Last Glacial stage (MIS 4, 74-59 ka BP) reducing the formerly inhabited areas of the periglacial steppe zone around the Southern Urals and repeatedly during the late Last Glacial stage (MIS 2, 24-12 ka BP). The most intensive late Middle / early Upper Palaeolithic geographic expansion occurred during the mid-Last Glacial interstadial interval (MIS 3, 59-24 ka BP) reaching the northernmost regions of the Urals and the adjacent territories of northern Europe and NW Siberia (Pavlov, 1986; Pavlov et al., 2001, 2004).The progressive biological and cultural adaptation enabled to retain the new lands prior and shortly after the LGM. The Central Urals forming the free geographic passage connecting the East European Plain with the vast West Siberian territories most likely played a key role in this historical evolutionary process.
The last glacial peopling of north-central Urals is principally associated with the cave complex in the Chusovaya River valley cutting E-W through the Ural mountain range. The multi-layer sites (e.g., Bezymyannaya Cave, Kotel Cave, Kamen Dyrovatyy, Bolshoy Glukhoy) distributed in a karstic formation over a 200 km distance encompass a long record spanning from the Late Pleistocene (Middle Palaeolithic) till the Iron Age (Fig, 3C). Apart of the main valley, the upper Chusovaya River tributaries (Koiva, Sylvitsa, Serebryanka) likely encompassed rich biotic niches even in glacial stages. A unique cultural record from the Kamen Dyrovatyy Cave treasured over 18,000 arrowheads made of various materials (stone, bone and metal) dating since the Late Pleistocene to Holocene, with the Late Palaeolithic level, dated to 13,757±250 yr BP (Serikov 2001). Other central Ural cave sites interpreted as cultic in view to archaeological inventories (Kumyshanskaya Cave, Ignatievskaya Cave, Grot Zotinskyy, Kotel Cave) as well as those in the southern Urals (Kapova Cave) (Bader, 1965; Matyushin, 1994) suggest a high socio-cultural and behavioral complexity. In the latter mountain zone, spruce and pine taiga forests persisted until the onset of the Last Glacial, ca. 22-21 ka BP (Danukalova & Yakovlev, 2006). A certain regional cultural non-uniformity across the Urals manifested in the lithic industry composition may indicate differences in environmental adaptations. The taxonomic diversity of hunted fauna is particularly evident at other Final Palaeolithic / Mesolithic sites, including reindeer- and mammoth-dominated sites in the Trans-Ural foothills and the eastern lowlands, respectively (Serikov, 1999; 2000). A multifunctional character and a variety of particular activities are well evident in the broad repertoire of stone and bone tools, including "hacking" instruments, barbed harpoons and arrowheads. A micro-lithization and stone tool polishing in the Mesolithic complexes distributed across the Urals and West Siberia indicate an adaptive adjustment and diversification of exploitation of the final Pleistocene/early Holocene environments.
Opposite to taiga forest of the Ural mountain valleys, the eastern Trans-Urals and the adjacent West Siberian Lowlands hosted a diversity of environments and settings related to seasonally water-saturated taiga forests with braided river streams, lakes and bogs. This vast and biotically diverse territory underwent a palaeogeographical and palaeoecological evolution reflecting large-scale environmental shifts, partly affected by the Arctic as well as mountain Pleistocene glaciations. The stratigraphic position of the mapped Pleistocene sites is determined by past geomorphic processes and biotic shifts of both the lowland and foothill areas with absence of thick loess deposits present on the eastern slopes of the Urals.

A B C D E F

Fig. 3. (Trans-)Urals. A: the Pleistocene glacial mountain topography of the Northern Urals; B: humanly flaked bones and ivory of the Middle Pleistocene fauna (the upper Kama basin); C: karstic formations of the Chusovaya River valley with multilayer prehistoric cave sites; D: open parkland taiga of the Sosva River valley, Trans-Urals of West Siberia. E: frost-wedge casts incorporating fossil wood and charcoal of the last glacial landscape diagnostic of low MAT and a periglacial setting. F: Gari Site 1, Late Palaeolithic occupation with concentration of mammoth bones indicating cultural exploitations of the last glacial Siberian ecosystem.

The Palaeolithic in the Trans-Ural regions, predating the mid-Last Glacial, is poorly known contrary to the western regions of the Urals (Pavlov et al., 1995). The archaic character of lithic implements from the local Pleistocene occupation sites must not imply antiquity and may reflects a time-transgressive character of expedient stone industries of the Palaeolithic and Mesolithic complexes pre-determined by local raw material sources (*e.g.* sandstone and siltstones at Gorbunovskiy Torfianik; quartzite tuffs at Golyy Kamen) (Serikov, 1999; 2000).
Geoarchaeology studies from the easternmost limits of the Trans-Ural area provide multiple lines of evidence of Palaeolithic peopling of this geographically marginal and still poorly explored territory of West Siberia following the mid-last glacial warming (35-24 ka). A specific cultural entity represents the Late Palaeolithic complex of open-air sites in the Sosva River basin on the periphery of the West Siberian Lowland contextually associated with massive concentrations of large Pleistocene fauna remains, previously interpreted as natural "mammoth cemeteries" known since the 19th century (Fig. 3D, F). Other fossil fauna with woolly rhinoceros, bison, horse, reindeer and other animals characteristic of periglacial tundra-steppe together with small (bird and fish) species indicate a diversity of exploitation of the local Late Pleistocene natural resources and adaptation to the last glacial sub-polar environments (Chlachula & Serikov, 2011). The skeletal remains from the main occupation localities suggest both active hunting and anthropogenic scavenging practices during the Last Glacial (with radiocarbon dates of ca. 25-12 ka BP). Human behavioral activities at the key Gari I site are also indicated by circular concentrations of mammoth and rhinoceros bones (including complete sculls, large bones and tusk fragments), some of them artificially modified, reminding primitive dwelling structures known in the Late Palaeolithic from the East European Plains (Pidoplichko,1998; Iakovleva & Djindjian, 2005).The geological context of the well-preserved fossil records, sealed in cryogenically distorted alluvial deposits of the meandering Sosva River, indicates complex taphonomic histories of the occupation sites buried in cold periglacial swampy-riverine settings. The palaeontological records show an extraordinary biotic potential of the Pleistocene periglacial West Siberian steppe marginally interspersed by the eastern Ural foothill ecosystems. The dominant exploitation of mammoth (98.5%) corroborates the faunal compositions from other Late Palaeolithic sites on both sides of the Urals (Byzovaya Cave, Mamontova Kurya in the Pechora basin, Volchiya Griva in west Siberia (Pavlov, 1996; Petrin, 1986; Savel'eva, 1997, Svendsen & Pavlov, 2003).
Due to the extreme environmental conditions (the high climate-continentality, surface saturation and natural water/stream barriers), the adjacent West Siberian Lowland is assumed to have been colonized rather late only after the Last Glacial Maximum at about 17-16 ka BP (Petrin, 1986). The Trans-Urals and probably the north Siberian Arctic regions remained largely ice-free even during the LGM and except for water-saturated and frozen boggy lands there were no major natural obstacles for free movement of Palaeolithic groups further east into NW Siberia prior and after that time interval. Fossil fauna displays a major biotic potential in the central northern Trans-Urals. The latitudinal zonal differentiation of the last glacial ecosystems of the broader Urals from the Arctic coast in the north to the Black/Caspian Sea was likely more uniform, largely covered by periglacial tundra-steppe in contrast to more mosaic interglacial parkland landscapes (Markova, 2007), facilitating a free migration of fauna as well as the Late Palaeolithic peoples particularly within the foothill corridors. Regional palaeoenvironmental deviations, differing from the overall territorial bio-geographic and climatic pattern (Panova et al., 2003), may have played a significant role in the process of early prehistoric peopling of the northern regions of Eurasia.

3.2 South-west Siberia (the Altai region and the Kuznetsk Basin)

The territory of SW Siberia shows a complex Quaternary history spanning the last 2.5 million years governed by climate changes leading to the present pronounced continentality of inner Eurasia in association with a regional Cenozoic orogenic activity. Dynamics of these processes is evident in formation of the relief of the southern Siberian mountain system. It is also witnessed in preserved geological records that may have a global relevance in view to the geographic location (such as deeply stratified loess-palaeosol sections indicative to high-resolution Pleistocene and Holocene atmospheric variations in the Northern Hemisphere). Reconstruction of the past climate dynamics that shaped the configuration of SW Siberian topography and local ecosystems is essential for understanding timing and adaptation of the initial peopling of the territory of northern Asia. Geoarchaeological investigations of the Palaeolithic and Mesolithic occupation have a long tradition particularly in the Altai region. The spatial distribution of the transitional geographic areas along the margin of SW Siberia believed to be the main gateway for early human migrations from the southern regions of Central Asia into Siberia is of utmost significance of multidisciplinary Quaternary studies.

3.2.1 Geography and natural setting

The territory of SW Siberia, including the Altai and the Kuzbass regions, is characterized by a rather diverse physiography with high mountain massifs in the south and east, and open lowlands in the north and west (Fig. 1A). The Altai Mountains (maximum elevation 4506 m asl.) and the Kuznetskiy Alatau (2171 m asl.) form a natural barrier from the South and East, respectively, connected by the Gornaya Shoria Mountains (1560 m asl.). The Salair Range (590 m) separates the adjacent continental depression in the north from the upper Ob Basin and the Kuznetsk Basin (150-300 m), representing the marginal parts of the Western Siberian Lowland. Hydrologically, the area belongs to the Ob River drainage system with the Katun', Biya, Irtysh and Tom' River being the main tributaries. The lower relief zone (>1200 m asl.) includes more then 50% of the mountain area and constitutes relics of old denudation surfaces covered by more recent Pleistocene deposits derived during past glaciations. The present climate is strongly continental with major seasonal temperature deviations between the northern lowlands and the southern mountains (Fig. 1B). In winter, climatic conditions in the montane zone are generally less severe than in the open northern steppes, and micro-climate prevails throughout the year in some protected locations in the Altai and West Sayan central valleys. Annual temperatures as well as the precipitation rate vary greatly according to particular topographic settings. Most of the precipitation falls on the W/NW slopes in the northern and central Altai, whereas the southern regions are more arid and increasingly continental with average July temperature +25°C. The average January temperature -33°C may drop to -60°C. Most of the area is underlain by perennial permafrost with the active thaw layer only 30-70 m thick. Vegetation is characterized by open steppe-parkland dominated by birch and pine. Mixed southern taiga (larch, spruce, pine, Siberian pine, fir and birch) is replaced by the alpine vegetation (with pine, larch and dwarf birch) at higher elevations. Semi-desert communities with admixture of taxa characteristic of the Mongolian steppes are established in upland depressions and on plateaus of the southern Altai. The present soil cover corresponds with the zonal vegetation distribution, with steppe chernozems in the northern lowland basins and plains, luvisolic soils at humid valley locations, brunisols and podzolic soils in the lower taiga forest zone, tundra regosols in the (sub)-alpine zone; and calcareous soils and solonets in the semi-desertic mountain basins.

3.2.2 Palaeogeography and climate history of SW Siberia

3.2.2.1 Southern mountain areas

Geological history of the broader territory of SW Siberia is linked with the formation of the southern Siberian mountain system initiated by the Miocene uplift of the Transbaikal region and reaching the Sayan-Altai area during the late Pliocene (ca. 3 Ma). This major orogenetic period that continued until the early Middle Pleistocene constructed a system of mountain ranges separated by deep depressions filled by large lakes (Fig. 1B-2). Following warm climatic conditions with landscape stability encompassing most of the Pliocene (5.3-2.5 Ma), the beginning of the Pleistocene period brought a major modification of the former relief as a result of dramatic climatic changes with progressive intra-continental cooling and aridity. The Pleistocene glaciations caused a regional topographic restructuring with intensive erosion in the glaciated alpine zones and accumulation of (pro)glacial, alluvial, proluvial, lacustrine and aeolian deposits in the intramontane depressions. Little is known about the earliest glacial events of the Altai, with evidence largely obliterated by erosional processes of subsequent glaciations. The earliest Quaternary (Early Pleistocene) non-glacial records are documented by gravelly alluvia and lacustrine sands in the Teleckoye Lake formation with pollen indicating a mixed taiga forest with warm broadleaf arboreal communities. A high-mountain forest-steppe stretched over the major southern Altai intermountain basins (the Chuya and Kuray Depressions) witnessed by pollen records with *Ephedra* and a large fossil fauna represented by *Hipparion* sp., *Coelodonta* sp., Elephantidae, Bovidae (Deviatkin, 1965).
The Middle Pleistocene glacial periods are evidenced by two glacial moraines correlated with the Kuyuss Glaciation and buried in the middle Katun' River basin. The pollen data from the lower till bear witness of a cold steppe environment (Markin, 1996). Warm interglacial conditions during the Middle Pleistocene from the Anui River valley, NW Altai, are evidenced by mixed pollen of pine, birch, oak, lime, maple and other arboreal taxa (Derevianko et al., 1992c). An analogous climatic development in the northern lowlands is documented in the Kuzbass Basin with a series of basal red soils indicative of warm and relatively humid forest-steppe followed by brown soils of cold tundra steppe (Zudin et al., 1982). Tectonic movements during the late Middle Pleistocene (0.3-0.2 Ma) initiated uplift of the mountain ranges to about the 1500-2000 m elevation, triggering intensive denudation processes of former unconsolidated deposits. The renewed orogenic activity caused a restructuring of the regional drainage system, with a dominant N-S oriented river-flow direction and a further deepening of the mountain river valleys by up to 100 m in the marginal areas (Anui, Charysh) and > 200 m in the central Katun River valley.
The regional climatic variations are documented by thick alluvial-proluvial sandy gravels of the Tobol (MIS 9) Interglacial separated by glacial till deposits (MIS 8) from the following Shirta (MIS 7) Interglacial (Markin, 1996). The warm periods are characterized by expansion of mixed taiga forests, and brown forest and forest-steppe soils in the Kuznetsk Basin (Zudin et al., 1982). The final Middle Pleistocene (MIS 6) glacial stage (Eshtykholskoye Glaciation) is believed to be the most extensive Pleistocene glacial event in the Altai and West Sayan area (Deviatkin, 1981) corroborated by the evidence from the Eastern Sayan Mountains (Nemchinov et al., 1999). The alpine glaciers presumably formed extensive coalescent ice fields supporting large glacial lakes in the central basins (Fig. 4A). Massive proglacial, ice-marginal deposits accumulated in the Chuya Basin with up to 50 m thick alluvial fans along the eastern flanks at the Kuray Range (4000 m asl.). In the foothills and the adjacent continental depressions, this climatic cooling is linked with appearance of

forest-tundra soils (Zykina, 1999). Cold and dry climatic conditions in the extra-glacial areas are indicated by accumulation of loess, incorporating the typical loessic mollusc fauna (*Succinea oblonga, Pupilla muscorum, Vallonia tenuilabris, Columella columella, Vertigo alpestris*) and rodent species characteristic of tundra-steppe. An extensive geographic expansion of southern taiga forests dominated by spruce and the Siberian pine into the former steppe and upland region characterizes the Last Interglacial (MIS 5) (Shunkov & Agadzhanyan, 2000).

During the Late Pleistocene, the Altai, as well as other southern Siberian mountain ranges experienced two glaciation events (MIS 4 and 2) represented by two moraines and tills separated by non-glacial interstadial deposits. The first (Ermakovo/Chibit) glaciation followed the same topographic ice-expansion pattern as the previous Middle Pleistocene glaciation (MIS 6). Large glaciofluvial ice-marginal lakes repeatedly filled the upper reaches of the Chuya and Kuray Basins, with icebergs released from the surging glaciers (Deviatkin, 1965). Accumulation of coarse and massive glacigenic sediments and formation of erosional surfaces in the lower reaches at relative elevations of 200 m, 80-100 m, 50 m, 30-40 m, 4-6 m and 1-1.5 m followed the glacial lake drainage cycles (Fig. 4B), indicating periodicity of these processes during the two glacial events (Rudoy et al., 2000). Pollen from the lower Katun River terraces dominated by dwarf birch *(Betula nana)* documents very cold conditions with tundra-steppe expanded within the northern lowlands and depressions of SW Siberia.

The mid-last glacial (MIS 3) warming (59-24 ka BP) is associated with accumulation of gravelly alluvial sediment facies in the mountains, resulting from the former ice ablation, and river sands and lacustrine clays in the upper Ob and the Kuznetsk Basins (the Bachatsk Formation) (Markin, 1996; Zudin et al., 1982). Geological records display a wide range of environments and climates. Pollen from the 30 m Katun River terrace shows expansion of taiga forests with appearance of warm broadleaf flora *(Quercus, Tilia, Ulmus, Juglans)* whose distribution range is well south beyond the present Altai limits. A gradual transformation to forest steppe and tundra steppe characterizes the later part of this stage (the Ust'-Karakol section), reflecting cooling of the Konoshelskoye Stadial (33-30/29 ka BP). Analogous climatic trends are indicated by pollen records from the pre-Altai Plain.

The second (MIS 2) Late Pleistocene (Sartan/Akkem) glaciation (24-17 ka BP) was less extensive than the preceding two (MIS 6 and 4) glaciations. In the Altai, it is associated with terminal moraines at the >2000 m elevations. Glacial basins formed in the intramontane depressions of the Biya, Chulyshman, Bashkaus, Katun, Chuya and Argut River basins with the synchronous glacier expansion followed by cataclysmic flooding (Butvilovskiy 1985, Okishev, 1982; Baker et al., 1993; Chlachula, 2001a, 2011b), with the latest around 13 ka BP. Large alluvial fans overlying or laterally merging with the highest alluvial terraces attest to dramatic geomorphic processes during the Last Glacial. Periglacial conditions with cold-adapted fauna prevailed in the northern extra-glacial zone. Pollen data illustrate distribution of cold periglacial steppe in the Altai foothills and on the adjacent plains with isolated tree (birch, pine, spruce, willow) communities in more humid settings (Zudin et al., 1982). The broken mountain relief differentiated the regional climatic pattern during the Last Glacial, with microclimate conditions in the protected locations along the northern Altai foothills and the central Katun basin allowing survival of warm Pleistocene flora and other biota until the Holocene. Large periglacial stony polygons on the Plateau Ukok, formed as a result of permafrost dynamics, attest to very severe post-glacial climatic variations. Dramatically wasting mountain glaciers in the southern Altai (the Tabon Bogdo Ula Range, 4120 m) are linked with a progressing global warming despite very low MAT (-10°C) (Rudoy et al. 2000).

3.2.2.2 Northern lowland areas

The Pleistocene climatic evolution in the lowland extra-glacial areas of SW Siberia and the continental depressions north of the Altai-West Sayans are characterized by accumulation of aeolian (silty and sandy) deposits during glacial periods and surface stabilization with soil formation during interglacial periods (Arkhipov et al., 2000). Particularly the loess-palaeosol formations widely distributed in southern Siberia provide the most detailed multi-proxy chronostratigraphic correlation of past climatic cycles (Chlachula, 2003; Bábek et al., 2011).

In the Kuznetsk Depression, being a major tectonic basin of SW Siberia bordered by the Salair Range and the Kuznetskiy Alatau Mountains, the Quaternary climate evolution is evidenced by the unique 30-60 m thick stratigraphic sequence of loess and lacustrine deposits intercalated by up to 16 Early, Middle and Late Pleistocene palaeosols above the Pliocene red soils (Zudin et al., 1982). The contextually associated fossil fauna, illustrates diversity of the local ecosystems (open steppe and parkland-forest environments) in terms of the regional climatic evolution and the geographic relief configuration (Foronova, 1999). Intensive aeolian dynamics over the south-west Siberian parkland-steppe, being a part of the Eurasian loess belt, is eloquently documented by up to 150 m high loess sections on the Priobie (Ob River) Loess Plateau, with a series of prominent interglacial pedocomplexes of chernozemic steppe soils and luvisolic / brown forest soils, spanning from the Early to Late Pleistocene. The stages of aeolian reactivation of silty dust derived from up to 100 m deep deflation surfaces in the lowland areas of SW Siberia (Zykina, 1999) correlate with the main glacial cycles in the mountain regions. Because of earlier stratigraphic hiatuses, best-documented are the Late Quaternary palaeoclimate loess archives spanning over the last ca. 300 000 years.

The prominent last interglacial (MIS 5) pedocomplex (Fig. 2D), with three parkland-steppe chernozems disturbed by cryogenesis under humid and cold conditions, shows complexity of the interglacial climate evolution (130-74 ka BP) (Chlachula & Little, 2011). The MAT during the interglacial climatic optimum (MIS 5e, 125 ka BP) in SW Siberia was by 1-3°C higher than at present, with about a 100 mm increase in annual precipitation. Environmental conditions were broadly similar to the present ones during the following interstadials. Increased precipitation and annual temperature contributed to a northward expansion of southern taiga forests by about 500-700 km beyond their present distribution limits due to raised summer temperature by 4-5°C relative to the present values (Velichko et al., 1992). Climate during the mid-last glacial (Karginsk/MIS 3) interstadial optimum was likely warmer than at the present as witnessed by chernozemic soils dated to 35-31 ka BP. The overlying podzolic forest soils (26-22 ka BP) show a gradual cooling towards the early last glacial stage (MIS 2) succeeded by a cold periglacial tundra-steppe with humic gleysols analogous to the modern soils of central Yakutia (with January temperatures below -27°C). Following the LGM, pronounced aridization at the western margin of southern Siberia correlates with loess accumulation (19-14 ka BP) and the time-equivalent formation of large, ice-marginal glaciolacustrine basins in the Irtysh and Ob River valleys (Arkhipov, 1980).

In sum, both glacial and sub-aerial formations provide evidence of cyclic nature of the Pleistocene climates of south-western Siberia with pronounced shifts in past ecosystems. The Early Pleistocene interglacials, characterized by a higher heat balance and atmospheric humidity, promoted distribution of meadow-forests and mixed parklands. The progressing climatic continentality during the Middle and particularly Late Pleistocene led to establishment of the present-type forest-steppe vegetation zones during interglacial periods and periglacial steppe during cold stages correlated with the Altai mountain glaciations.

3.2.3 Pleistocene environments and early human peopling

The Quaternary climatic changes and transformations in natural environments in south-western Siberia are well manifested by geological, biotic, but also the early cultural records. The past glacial dynamics, controlled by the regional atmospheric air-mass circulation flows in conjunction with the neoteotectonic activity and river erosion, shaped the relief to form biotically productive river valleys and depressions. The cyclic nature of the glacial and interglacial stages led to a restructuring of landscapes and generation of specific ecosystems adjusted to particular topographic settings and responding to ongoing climatic variations.

The spatial distribution of the Palaeolithic sites on the territory of SW Siberia shows a location of most sites within the transitional zone between the southern mountain ranges and the northern lowlands. These geomorphological zones, about 75-150 km wide, form a topographic relief belt of the 300-1000 m altitude, narrowing or expanding in respect to the particular physiographic conditions and the configuration of the relief (Baryshnikov, 1992). Particularly the Altai region shows the high density of Pleistocene occupation localities. This pattern of the (palaeo)geographic site location, reflecting specific environmental adaptation strategies to local settings, applies for both open-air localities buried in alluvial, colluvial or aeolian deposits, as well as cave sites concentrated in the NW Altai (the Anui valley) (Fig. 4E). Formation of the latter relates to a progressive down-cutting by fluvial erosion through the Devonian-Carboniferous limestone bedrock during the late Middle Pleistocene.

The periodic Palaeolithic migrations in the broader SW Siberia were principally dependent of the Pleistocene climatic development. Intensified orogenic uplifts, triggering large-scale erosions in the river valleys, re-shaped natural occupation habitats. Relatively stable environmental conditions seem to have persisted in the central and northern Altai due to increased regional precipitation and a tempering atmospheric effect of the mountain ranges. Limestone caves excavated by a fluvial bedrock erosion provided shelters for a more permanent human inhabitation of the Altai area, particularly in the northern foothills and the central intramontane depressions (Derevianko & Markin, 1992) characterized by warmer microclimate conditions. Cataclysmic drainage of glacier waters, periodically dammed by the mountain glaciers during glacial stages, had temporarily a dramatic impact to the local Pleistocene ecosystems. Enormous erosion processes associated with these major events significantly reduced the visibility potential of occupation sites in the flooded areas, with localities preserved only at high topographic elevations above the glacial basin waterlines.

The initial peopling of the Altai - West Sayan region likely occurred in some of the early Middle Pleistocene interglacials in the process of the northern expansion of warm biotic communities. Mixed coniferous and broadleaf forests became established in the tectonically active mountain areas with maximum elevations about 1500-2000 m. Parklands covered most of the adjacent plains of West Siberia with continental depressions filled by lakes and drained by meandering rivers. Rudimentary core and flake ("pebble tool") stone collections scattered on high river terraces and the former lakeshore margins of the present arid basins (Kuznetsk and Zaisan Basin) attest to several stages of early human inhabitation and a relative environmental stability (Chlachula 2001a, 2010b). There is limited evidence on persistence of the Early Palaeolithic occupation during glacial stages on the territory of SW Siberia, although some intermittent semi-continuity in the southernmost areas is assumed in view to finds of weathered lithic artifacts from the old periglacial alluvia in association with cold-adapted megafauna. Mastering the technique of fire making was clearly the main precondition for early human survival in cold tundra-steppe and tundra-forest habitats.

A B C D E F

Fig. 4. Altai, SW Siberia. A: former glacial valleys of central Altai; B: glacio-fluvial terraces (up to 200 m high) in the Chuya River basin formed by cataclysmic releases of glacial lakes dammed by valley glaciers; C: thick loess deposits along the Biya River, North Altai Plains; D: colluviated redsoils in NW Altai attesting to warm Middle Pleistocene climate conditions; E: the Anui River valley, NW Altai, with the Middle-Late Palaeolithic occupation cave sites; F: Denisova Cave, Anui valley. Mousterian (Neanderthal) tools of the Levallois tradition.

The last interglacial warming (130 ka BP), associated with the re-colonization of southern Siberia by coniferous taiga forests, is linked with the appearance of the Mousterian tradition. Changes in the relief configuration influenced a regional climate regime and opened new ranges of habitats for the Middle Palaeolithic population concentrated in the transitional zones (500-1000 m asl) in the karstic area of the NW Altai foothills (e.g., Ust'-Kanskaya, Strashnaya, Denisova, Okladnikova, Kaminnaya Caves) as well as at open-air sites in the central valleys (e.g., Kara-Bom, Ust'-Karakol, Tyumechin I and II) (Derevianko & Markin, 1999). The Middle Palaeolithic horizon, encompassing a time span of up to 140 000 years (180-40 ka BP), represents a marked cultural phenomenon in the Altai. Isolated teeth (2) from the Denisova Cave dated to MIS 5 and identified as *Homo neaderthalensis* support the model of biological evolution of pre-modern humans in Siberia (Derevianko & Shunkov, 2009). Major cooling during the early Last Glacial (MIS 4) led to establishment of full glacial conditions in the central and southern Altai, and a zonal geographic replacement of boreal forest by periglacial tundra-forest in the northern Altai and by arid tundra-steppe in the adjacent lowlands (the Ob River basin and the Kuznetsk Basin). Accentuated moderate climate fluctuations between cold stadials are evidenced by embryonic regosolic soils in the loess formations on the North Altai Plains and sparse cultural records in protected locations in the Altai foothills. Human occupation of the central and southern Altai during the early Last Glacial was impeded by harsh, ice-marginal environments and expansion of glaciers in the upper reaches of the Katun - Chuya valleys subsequently filled by large proglacial lakes.Progressive warming during the early mid-last glacial interstadial interval (MIS 3, 59-35 ka BP) caused a dramatic wasting of the ice fields accompanied by cataclysmic releases of ice-dammed lakes and large-scale mass-flow and slope erosional processes. The former valley glaciers either completely disappeared or receded to the highest elevations where they persisted as corie glaciers. The periodic outbursts of the glacial basins had a dramatic impact on the regional ecosystems, but also obliterating most of the earlier cultural records. Mixed forests dominated by birch, pine, spruce and fir invaded the former periglacial and ice-marginal landscape. The presence of broadleaf arboreal taxa (oak, lime, chestnut, maple) indicates a climate warmer than at the present time. Increased humidity and cooling during the later stage of the mid-last glacial interstadial interval (35-24 ka BP) initiated mass gravity slope processes and cryogenic deformations related to the Konoshelskoye Stadial (33-30 ka BP) followed by warmer oscillations with formation of podzolic forest gleysols, Appearance of the transitional early Late Palaeolithic cultural facies reflects human adaptation to mosaic interstadial habitats, including sub-alpine taiga, dark coniferous forest, mixed parklands and steppe with mixed non-analogue biotic communities. The identical geographical distribution of the Middle - Late Palaeolithic sites and the time-transgressive lithic technologies suggest a regional cultural (and biological?) continuity in the Altai area during the Late Pleistocene (Derevianko, 2010). A phalanx fragment from Denisova Cave dated to 40 000 years and interpreted on basis of DNA as an extinct human species (Dalton, 2010) reinforces this scenario. Re-establishment of cold tundra-steppe habitats correlates with dispersal of the developed Late Palaeolithic with blade-flaking techniques in stone tool production and associated with a periglacial megafauna that possibly survived in protected and milder microclimate locations in the northern Altai throughout the LGM (20-18 ka BP). Emergence of the microlithic stone tool assemblages with wedge-shape cores is linked with a new cultural adjustment in the final stage of the Palaeolithic development responding to natural transformations of the former periglacial ecosystems towards the end of Pleistocene.

3.3 South-central Siberia (the Yenisei Basin)

The upper Yenisei basin in the southern part of the Krasnoyarsk Region, south-central Siberia, is rather unique in view to its location near the geographical centre of Asia (Fig. 1A). It has been together with the upper Angara basin in the Irkutsk Region the key area of Quaternary stratigraphic studies mapping the Pleistocene climate evolution. The extensive loess cover with a suite of buried palaeosols from the Northern Minusinsk Basin, being a continuation of the southern Siberian loess belt, have provided the most complete, high-resolution Late Quaternary palaeoclimate record in the north-central Eurasia (Chlachula et al., 1997, 1998). Coupled with the pollen evidence, the continuous loess-palaeosol sequences document periglacial steppe-tundra established during cold stadial intervals, replaced by boreal forest and parkland steppe during the warm interstadials and interglacial stages. The associated Pleistocene landscape transformations, governed by marked climatic variations, are reflected by large diversity of fossil faunal species, including non-analogue communities to modern biota. Geoarchaeological investigations in the stratigraphic sections in the upper reaches of the Yenisei River basin, initiated after progressive erosion of the unconsolidated aeolian deposits, revealed a rich series of Early, Middle and Late Palaeolithic stone tools (industries) associated with an abundant Pleistocene fossil fauna (Drozdov et al., 1990, 1999; Chlachula, 2001b) (Fig.5). The main focus of the current studies is reconstruction of the Pleistocene ecology and chronology of the early human occupation of south-central Siberia. The unique loess-palaeosol records have a fundamental bearing for better understanding the past global climate and environmental history well beyond the limits of north-central Asia.

3.3.1 Geography and natural setting

The south-central Siberia, bordered by the western Mongolia from the South and the Central Siberian Plateau along the eastern margin of the West Siberian Lowland in the North, is a geographically extensive, and topographically and biotically diverse territory encompassing lowlands, high ranges of the Sayan Mountains and the transitional foothill areas transected by river valleys largely draining water discharge from the southern mountain massive. The broader southern Siberian continental basin is built by a system of regional tectonic depressions structured by the Hercynian orogenesis related to the formation of the central Asian mountain system. The upper Yenisei area, including the Northern and Southern Minusinsk Depression and the adjacent slopes of the Eastern and Western Sayan Mountains, lies in the southern Siberian loess zone along the upper reaches of the Yenisei River (52-56°N and 89-94°E). The central part of the Minusinsk Basin is structurally defined by a zone of tectonic breaks running in the north-south direction across the Batenevskiy Range (900 m asl.) in the north. From the east, it is bordered by ridges of the Eastern Sayan Mountains (1778 m), from the west by the Kuznetskiy Alatau (2178 m) and from the south by foothills of the Western Sayan Mountains (2735 m). In the northwest, the Yenisei basin broadens and joins a system of palaeo-valleys merging with the Nazarovskaya Depression of the West Siberian Lowland. The present continental climate (with MAT -0.4°C), with cold and dry winters with little snow cover and warm to hot summers, corresponds to the particular geographical location of the territory. Open steppe stretches over most of southern Central Siberia, with isolated mosaic mixed parkland forest-steppe communities dominated by birch and pine. A chernozemic soil cover underlies parklands, whereas brown forest and podzolic soils are developed under mixed and coniferous taiga forests, respectively. Sandy semi-deserts with saline lakes such as in Khakasia illustrate the marked regional climate diversity.

3.3.2 Palaeogeography and climate history of south-central Siberia

Pre-Quaternary geology of the broader southern Siberian territory was controlled by a series of tectonic events. Both relief and the geological structure encouraged intensive geomorphic processes in the past. The Cambrian and Proterozoic volcanism disturbed and subsequently dislocated the south Siberian pre-Cambrian crust composed of igneous and metamorphic rocks. The Middle and Late Palaeozoic orogenesis formed a system of high mountain ranges separated by deep intra-continental depressions. Later Hercynian and Oligocene tectonics in southern Siberia modified the original configuration of the Angara, Kuznetsk, Minusinsk and other basins north of the Altai, Salair and Sayan Mountains, which were filled by Devonian, Carboniferous, Jurassic and Palaeogene volcanic, lacustrine and alluvial deposits. Accumulation of extensive proluvia near the mountain fronts and lacustrine / alluvial formations in the main sedimentary basins continued throughout the Miocene and Pliocene. During the Late Pliocene and Early Pleistocene (ca. 3-1 Ma), an early fluvial system of the palaeo-Yenisei was established with relics old river terraces about 200 m above the present river level. A neotectonic movement during the early Middle Pleistocene divided the Minusinsk Basin into the northern and the southern parts. In response to climatic change in conjunction with the orogenic activity, a system of terraces gradually developed with the earlier, Early and Middle Pleistocene (70-90 m, 110-130 m, 150-170 m) preserved mostly in relics and buried by 10-50 m of aeolian loess as well as colluviated loessic deposits. The Late Pleistocene terraces (8-12 m, 15-20 m, 30-45 m) formed above river plains (Chlachula, 1999).

During glacial periods, the northward drainage of the Yenisei and the Ob River was diverted by the ice barrier of the northern inland (Arctic) ice-sheet to the southwest into the Aral and Caspian Sea (Arkhipov, 1998). The regional (palaeo)relief of southern Siberia with intermountain depressions preconditioned formation of large river dams blocked by glaciers periodically causing high-magnitude floods. These dramatic events resulted in an intensive syndepositional accumulation of alluvial and alluvio-lacustrine sediments in the river valleys adjacent to the Altai and Sayan Mountains (Rudoy & Baker, 1993; Yamskikh, 1996). Loess and loessic deposits, in average 10-20 m thick and derived from local alluvial plains during glacial periods, are distributed over most of the landscape, but mainly on the western, lee-side slopes where they reach up to 40 m (Chlachula, 2001b) (Fig. 5A). During warm interglacial and interstadial stages, the formation of alluvia as well as the aeolian sediment accumulation was interrupted by palaeo-surface stabilisation with formation of chernozemic and brunisolic palaeosols (Chlachula et al., 2004a). The geographical location of the Yenisei area close to the geographical centre of Eurasia contributed to extreme continental climatic conditions with widespread permafrost that developed during cold stadials. Permafrost was not preserved during the Last Interglacial (130-74 ka BP) in the southern part of Siberia, whereas it remained widely distributed farther north and east.

Detailed stratigraphic records of the last two glacial-interglacial cycles have been documented in the Kurtak area in the Northern Minusinsk Basin (Drozdov et al., 1990). Erosion of the Quaternary sedimentary cover overlying old terraces and igneous bedrock exposed the most complete loess - palaeosol sequences with a chronostratigraphic archive encompassing ca. the last 300 000 years. The unique geological record indicates pronounced and cyclic climatic variations during the Late Quaternary, displaying consistent correlation regularities with the deep-sea oxygen isotope records (Marine Oxygen Isotope Stages / MIS 7-1) also reflected by the corresponding patterned changes in the natural environments. Magnetic susceptibility of the fine aeolian dust has proven to be a very sensitive and useful proxy indicator of past climate change in this part of Siberia (Chlachula et al. 1997, 1998). A prolonged period of a relative environmental stability during the late Middle Pleistocene is manifested by a basal prominent chernozemic palaeosol (Kurtak 29 type-section) correlated with MIS 7a. The following cold period (MIS 6) experienced significant warm-cold climate fluctuations, expressed by an intensively colluviated loess unit traced across southern Siberia between two main cold and arid intervals defined by aeolian dust sedimentation. The last interglacial /(MIS 5) *sensu lato* (130-74 ka BP) is evidenced by a new period of landscape stabilization with intensified soil formation processes disrupted by intervals of loess deposition and periglacial surface deformation (Fig. 5B). The loess-palaeosol sequence attests to several relatively short, warm as well as very cold intervals (MIS 5e-5a), with a gradual shift from a strongly continental climate during the first half of the interglacial (130-100 ka BP) to cooler and more humid conditions. Reactivated loess sedimentation reflects a progressive cooling during the early stage of the last glacial (MIS 4), culminating in several cold and hyperarid stadials. Climatic amelioration during the early mid-last glacial (Karginsk) interval (MIS 3) is evidenced by the appearance of (gleyed) brunisolic soils. The interpleniglacial climatic optimum associated with the well-developed duplicate chernozem dated to ca. 31 ka BP was likely as warm, but more humid than the present (Holocene) interglacial. The secondary cryoturbation of the fossil chernozem shows a major drop of temperature and formation of underground permafrost. Gradual cooling leading to formation of gleyed brunisolic and later to regosolic horizons found in the upper Yenisei loess sections indicates that the transition towards the last glacial stage (MIS 2) was less dramatic than during the previous glacial interval (Chlachula, 1999). A further drop of annual temperatures associated with an intensive loess deposition marks the time around the Last Glacial Maximum / LGM (20 000 - 18 000 years BP). A progressive warming with several climatic oscillations expressed by an initial pedogenesis characterizes the Final Pleistocene climate development, eventually leading to establishment of the present (MIS 1) interglacial conditions. Frost wedges, filled with humus-rich sediments in the present chernozem, reflect cold winters with little snow cover and seasonally frozen ground.Pleistocene faunal remains are abundant in the larger southern Siberia and provide a significant line of proxy evidence on past natural habitats and biodiversity composition related to climate change. The most prolific and taxonomically rich large fauna collections originate from the Kuznetsk and Yenisei Basins (Fig. 5D). Taxonomically, more than 32 large fossil fauna species have been documented, with the evolutionary (Middle-Late Pleistocene) *Mammuthus/Elephas* lineage being the best-represented. Most frequent megafauna species include woolly rhinoceros *(Coelodonta antiquitatis* Blum.), mammoth *(Mammuthus chosaricus)*, horse (*Equus hemionus* Pall. and *Equus mosbachensis - germanicus*), bison (*Bison priscus* Boj.), roe deer *Capreolus capreolus* L.), giant deer (*Megaloceros giganteus* Blum.), reindeer (*Rangifer tarandus*), saiga (*Saiga tatarica)* and sheep *(Ovis* sp.) among other animals (Foronova, 1999). Malacofaunal (fossil mollusc) assemblages represent one of most abundant and indicative biotic records in the loess-palaeosols formations with both interglacial and glacial taxa.

In sum, the diversity of the Pleistocene faunal communities in the Minusinsk and Kuzbass Basins documents particular evolutionary stages reflecting regional environmental transformations and zonal biotic shifts triggered by climate change during Late Quaternary. The presence of cold-adapted species in the fossiliferous sedimentary beds indicates a very high biotic potential of periglacial steppe that expanded during the stadial intervals and was replaced by parklands and mixed forests during the warmer interglacial/interstadial stages.

3.3.3 Pleistocene environments and early human peopling

Until the 1980s, Siberia was generally believed to have been colonized by the Late Palaeolithic people (Tseitlin, 1979). Systematic geoarchaeological investigations particularly in the Yenisei valley revealed much earlier traces of human peopling. Progressing erosion of the Krasnoyarsk Lake have exposed the nearly complete stratigraphic Late Quaternary geological stratigraphic record, enclosing a series of various Palaeolithic stone industries accompanied by a rich Middle and Late Pleistocene fauna (Drozdov et al., 1999). Abundant archaeological finds represented by simply flaked, but diagnostic stone artifacts originate, together with the taxonomically diverse fossil fauna remains, from basal alluvial gravels of the 60-70 m terrace beneath 20-40 m thick loess deposits (Chlachula, 2001b) (Fig. 5C).

The archaeological and palaeontological records incorporated in the fluvial and sub-aerial formations provide new insights on timing, processes and conditions of the initial peopling of this territory, as well as on biodiversity of the particular Pleistocene habitats. Abundant "pebble tool" assemblages bear witness of human occupation of the Yenisei area prior to the last interglacial (>130 ka BP) in agreement with other regions of southern and eastern Siberia (Fig. 1A). These earliest cultural records, classified on the basis of stone flaking technological attributes as the Early Palaeolithic, are age-corroborated by the contextual geological position (the Kamennyy Log, Sukhoy Log, Verkhnyy Kamen and Razlog sites). These archaic stone implements were largely redeposited by erosional processes, but are also documented *in situ* in the original geological context of the Middle Pleistocene alluvia. The formal variability of the Early Palaeolithic industries, displaying a differential degree of patination and aeolian abrasion, indicates several stages of the initial peopling of this area. These cultural finds, in conjunction with the Middle Pleistocene cold-adapted fossil fauna, suggest an early human adjustment to local periglacial environments. The archaeological records from the Minusinsk Basin, supported by analogous finds from the Kuzbass and the upper Angara River basins (Medvedev et al., 1990), show that parts of southern Siberia were occupied repeatedly at several stages during the Middle Pleistocene.

Evidence on the Middle Palaeolithic peopling of central Siberia is principally from the foothill areas of the Altai and Sayan Mountains and the local tributary river valleys (Derevianko & Markin, 1999). In the Minusinsk Basins, the Middle Palaeolithic is found in the Late Pleistocene loess strata and the interbedded soils, particularly of the last interglacial pedocomplex, but may be exposed on the present surface in low-sedimentation-rate areas of Khakhasia, Tuva, Gorno Altai, East Kazakhstan (Astakhov, 1986; Chlachula, 2001b, 2010b). Favorable climatic conditions promoted expansion of the Mousterian (Levallois) tradition during the Last Interglacial (MIS 5) that persisted until the early last glacial stage (MIS 4).

An exceptional Middle Palaeolithic site producing a unique archaeological as well as palaeontological locality is Ust'-Izhul' exposed by erosion at the Krasnoyarsk lakeshore (Chlachula et al., 2003). Numerous concentrated fossil skeletal remains were incorporated *in situ* on top of the last interglacial chernozem IRSL dated of 125 ka±5 ka (Sib-1) and overlying the 65 m Middle Pleistocene Yenisei terrace (Fig. 5E) The fauna included an early form of mammoth (*Mammuthus primigenius* Blum.), woolly rhinoceros (*Coelodonta antiquitatis*), bison (*Bison priscus*), horse (*Equus mosbachensis*), elk (*Cervus elaphus*), as well as small mammals such as marmot (*Marmota* cf. *baibacina*), beaver (*Castor sp.*) and badger (*Meles meles*). The most abundant species -- mammoth -- was represented by several hundreds bone fragments and 42 molar teeth of at least 12 individuals, mostly juveniles, parts of which were recorded in anatomical position. The associated mollusc fauna shows a xerotheric grassland setting. A perfectly preserved stone tool assemblage (ca. 200 pcs) was found in association with the

Fig. 5. Central Siberia. A: eroded late Quaternary loess sections along the Krasnoyarsk Lake; B: the key section (Kurtak 29) with the interglacial (MIS 7, 244-170 ka; MIS 5, 130-74 ka BP) and interstadial (MIS 3 59-24 ka BP) pedocomplexes; C: exposed gravel alluvia of the Middle Pleistocene (65 m) Yenisei River terrace incorporating fossil fauna and archaic pebble-tools; D: Kuznetsk Basin, mammoth teeth (left: *M. trogonterii, M.* sp., *M. primigenius*); E: the Middle Palaeolithic site Ust'-Izhul' with remains of early *Mammuthus primigenius* and stone artifacts (F) *c.* 125 000 year old, representing a most unique and complete early occupation in Siberia.

fossil fauna and three fireplaces with a fir-wood charcoal radiocarbon dated to >40 ka BP. The stone artifacts mostly represented by simple flakes (Fig. 5F) were used for processing the slaughtered animals. Human activity is also manifested by flaked and cut bones of mammoth, rhinoceros, bison and elk, and mammoth tusk fragments. The palaeogeographic site configuration with the concentrations of the skeletal remains suggest that the game was hunted by people in the nearby area across the 100-110 m terrace over a cliff onto the present 65-70 m terrace (which formed the floodplain) and transported in dissected pieces to the habitation place. The composition of the inventories reinforces the interpretation that it was a short-term occupation / processing campsite. In respect to the high age and the contextual completeness, the Ust'-Izhul' locality is at present without parallel in Siberia. Together with other Pleistocene megafaunal assemblages from the Minusinsk Basin, the associated fauna record bears witness to the rich biological potential of the area for early human occupation. There is no consensus if the Middle Palaeolithic tradition in southern Siberia can be associated with the European and Near Eastern cultural milieu, although some "classical" Mousterian influences in the Altai cave sites are evident (Derevianko & Markin 1992) (Fig. 4F). The Middle Palaeolithic stone flaking technology, especially the Levallois technique, is still reminiscent in the Late Palaeolithic traditions, suggesting a certain continuity of the cultural and possibly biological human evolution in Siberia during the Late Pleistocene.

The Late Palaeolithic occupation in the southern Central Siberia is documented at both open-air and cave sites. Warming during the mid-last glacial interstadial stage (MIS 3) accelerated formation of the Late Palaeolithic cultures characterized by the developed blade flaking techniques. The upper Yenisei basin is one of the major loci of the Late Palaeolithic sites in Siberia (the Krasnoyarsk - Kanskaya forest-steppe, the Northern and Southern Minusinsk Basin, and the Western Sayan foothills) (Astaknov, 1986; Drozdov et al., 1990; Larichev et al., 1990; Vasiliev, 1992; Yamskikh, 1990). The earliest occupation sites have been found buried in the Karginsk (MIS 3) Pedocomplex (31-20 ka BP), in the early Sartan loess and the intercalated, weakly developed, humic loamy interstadial forest-tundra soils (dated to 25-22 ka BP). Intervals of significant climatic deterioration and onset of full glacial (MIS 2) conditions are manifested by the absence of any archaeological record except for biotic refugia in the northern foothills of the Western Sayan Mountains. During the LGM (22 000 - 19 000 yr BP), the Yenisei area, as most of southern Central Siberia, seems to have been vacated due to very cold and arid environmental conditions with the biologically productive mid-last glacial parkland-steppe gradually replaced by a harsh and inhospitable steppe-tundra.

More recent Final Palaeolithic finds from the steppe and foothill areas (Abramova, 1979a, 1979b; Drozdov et al., 1990; Vasiliev et al., 1999) provide evidence of re-colonization of the upper Yenisei region during warm climatic oscillations at the end of the Pleistocene. A pre-dominant smaller fauna (reindeer, red deer, argali sheep) hunting, composite bone tools (spears/harpoons) and a micro-blade stone flaking illustrate major shifts in adaptation strategies in response to changing environments during the warm interstadial (16-13 ka BP), preceding the last cold climatic interval of the Final Pleistocene (Younger Dryas, 12.9 ka BP). In sum, the present archaeological records from the larger southern Central Siberia provide eloquent evidence that this territory was occupied repeatedly by early people at several stages during the Pleistocene, including glacial stages with harsh periglacial natural conditions. Human occupation may have persisted throughout the Last Glacial in protected southernmost locations despite severe hyperarid climates over most of the northern plains.

3.4 East Siberia (the Angara, Lake Baikal and the Lena Basin)

The palaeogeographical and palaeoenvironmental evolution over East Siberia during the Quaternary Period is characterized by a dynamic interaction of past global climatic changes and a regional neotectonic relief modeling. Sub-rifting orogenic regimes of the Mongolian-Siberian mountain zone gave rise to a diversity of geomorphic settings throughout the Pleistocene. The Cenozoic tectonics in conjunction with mountain glaciations continuously shaped the topography of Eastern Siberia structured by a system of mountain ranges (North Baikalskyy, Verkhoyanskyy, Cherskego, Kolymskyy) separated by the major (Angara, Lena, Viluy, Aldan, Indigirka, Yana, Kolyma) river valleys and their tributaries. The geographical isolation with a partly reduced atmospheric circulation regime contributed to a pronounced climatic continentality of this vast territory. Pleistocene glaciations over the East Siberian mountain ranges, though largely localized to isolated ice-caps, played a significant role in the landscape development and related environmental shifts in the extraglacial depressions. Increased accumulation rates of sub-aerial (sandy and loessic) sediments and periglacial palaeosol surface deformations point to a gradually progressing climate cooling during the Late Quaternary. Fossil pollen, fauna and early cultural records from stratified geological contexts provide evidence of pronounced palaeoecology changes in natural habitats. Study of the palaeogeographical processes is essential for reconstruction of the Quaternary climate history, geomorphic and biotic transformations of ecosystems as well as human occupation. The key regions for the Quaternary studies are located in the southern part of East Siberia.

3.4.1 Geography and natural setting

The south-eastern Siberia (Fig. 1A),, encompassing the Pribaikal area (the upper Angara and Lena basins) and the SE Transbaikal area (the Selenga River basin), is a territory of contact of the major tectonic structures of the Siberian Platform and the adjacent mountain massifs. The southern part of the territory is formed by the Irkutsk Depression (800-1200 m asl.), geographically confined by the Eastern Sayan Mnts. and the Baikal Range (2500-3000 m asl.). Lake Baikal (-1620 m max. depth) occupies a tectonic basin of 636 x 26-79 km or 31,500 km^2. A system of palaeolakes existed in this area since the late Oligocene (25 Ma ago), although formation of the present lake (Fig. 6A) relates to a major orogenic activity initiated during the Pliocene/Pleistocene (the Baikal Tectonic Phase spanning for the last 3.5 Ma). The Baikal rift zone, 150 km wide in average, extends for about 2,500 km from Lake Khubsugul in Mongolia to the upper Aldan River in south-east Yakutia, and includes a series of tectonic depressions (Khubsugul, Tunkin, Baikal, Barzugin, and the upper Angara depression).

Most of the relief is shaped by smooth mountain ranges (800-1300 m asl.) separated by river valleys and shallow basins. Taiga forests occupy the northern lowland areas, while open steppe grasslands (Pribaikalye) and semi-deserts (Transbaikalye) broadly extend in the southern regions. Present climate is strongly continental, reflecting the geomorphological configuration of East Siberia, forming an orographic atmospheric barrier strengthening the influence of the Siberian Anticyclone. Average temperatures -20°C (January) and +23°C (July), and low annual precipitation values (250 mm) characterize most of the territory. Maximum precipitation falls on the north-exposed slopes (700-800 mm), covered by coniferous (mostly pine) forests. More humid regional climate prevails along the margins of the Irkutsk and upper Lena basins due to a warming effect of the Lake Baikal (350-450 mm). Mixed southern taiga is largely represented by spruce, larch and birch. Most arid interior valleys (250-300 mm *per* year) include xenotheric steppe and mosaic parkland communities.

3.4.2 Palaeogeography and climate history of East Siberia

Both orogenesis and climate have shaped the present relief of East Siberia. The pre-Cenozoic geological formations are represented by relics of the Palaeogene palaeo-relief preserved on plateaus tectonically raised to 300-500 m asl above the basal igneous and metamorphic rocks and sedimentary, partly lithified deposits. The Oligocene and Miocene uplift of the marginal regions of the Irkutsk Basin (including the upper Lena area) initiated formation of the united morphostructural province of the broader Baikal region. Tectonic processes, significantly reactivated during the Cenozoic Period, generated massive colluvial deposits that filled the former riverine basins and shallow depressions originating by the Late Miocene pre-rifting (Rezanov, 1988).The orogenic movement around the Pliocene/Pleistocene boundary (2.5 Ma ago), opened the Baikal rift zone and caused a major uplift of the Pribaikal and West Transbaikal regions. The raised topographic gradient lead to intensive river erosion followed by the establishment of the present drainage of the major East Siberian rivers (Angara, Lena, Selenga). Formation of the surrounding mountain ranges precipitated the accumulation of fluvial deposits and a gradual filling of the Pliocene rift graben (Bazarov, 1986). In the Lake Baikal surrounding, the geomorphic dynamics is evidenced by 6-10 raised lake terraces related the former lake water-stands and elevated up to 200 m above the present lake level. During the Early-Middle Pleistocene, the Lake Baikal discharge followed a passage through the Irkut River valley into the former Yenisei basin. The present regime was established during the Late Pleistocene as a result of a new tectonic movement opening an outlet into the present Angara River drainage system (Fig. 6B).

The neotectonic movements during the early Middle Pleistocene (after 750 ka BP) contributed to an accumulation of thick deltaic, fluvio-lacustrine and slope deposits filling the major river valleys and intermountain depressions, particularly in the Selenga River basin (Rezanov, 1988, 1995). A reactivation of the orogenic activity during the initial stage of the late Middle Pleistocene resulted in the Lake Baikal transgression. A system of 40-50 m-high river and lake terraces and adjacent alluvial fans formed during the Tobol (MIS 9) interglacial and continued in the following Samarovo (MIS 8) glacial period. During this maximum Middle Pleistocene glaciation, initiated by global climate cooling coupled by a regional uplift, mountain glaciers extended to 1000-1300 m elevation with terminal moraines along the foothills of the Baikal, Barguzin and Angarsk Ridges (Rezanov & Kalmikov, 1999). A renewed regional tectonic activity of the Baikal rift at the beginning of the Late Pleistocene (130 ka BP) triggered an accelerated erosion of the Selenga River, and the upper Angara and Lena River (Bazarov, 1986). Marked climatic shifts characterize the last glacial-interglacial cycle, with ice-cap formations in the East Siberian mountains during the cold stages (MIS 4 and 2). The Ermakovo (MIS 4) glacial stage, with a major glaciation in the Baikal Range and the East Sayans, is evidenced by moraine ridges and altiplanation terraces at >1000 m elevations (Nemchinov et al., 1999). Increased aridity and decrease of annual temperatures during the cold stadials together with strong NW winds stimulated formation of deflation and drifting sandy surfaces on top of the Early and Middle Pleistocene terraces, and aeolian dust deposition in main river valleys and depressions (Medvedev et al., 1990). Climate fluctuations of the early Last Glacial (MIS 4) caused intensive solifluction processes. The mid-last glacial (MIS 3) optimum (dated to ca. 31 ka BP) was thermally approaching the last interglacial (MIS 5e) climate conditions. A reduction of the glaciated areas is observed for the following Sartan (MIS 2) glacial stage. The present geomorphological configuration and a regional environmental zonation became gradually established during the Holocene.

3.4.2.1 Quaternary climate change and environments of the glacial mountain areas

Pleistocene glaciations in the East Siberian mountains, bordering the Angara and Lena basin from the south and east played a major role in the landscape development and related shifts in ecosystems of the extra-glacial depressions (Fig. 6E). The main glacial events documented in the Eastern Sayan Mountains occurred during the late Middle Pleistocene and Late Pleistocene, with the penultimate (Ermakovo, MIS 4) glaciation (74-50 ka) been the most extensive (Nemchinov et al., 1999). This evidence corroborates the glacial records from the northern part of the Baikal Depression and the Baikal Range as witnessed by relics of terminal moraines assigned to the Samarovo (MIS 8) glacial stage (Bazarov, 1986; Rezanov & Kalmikov, 1999). Warmer interglacial / interstadial periods correlate with a mountain ice retreat and expansion of thermophilous biotic communities over large geographical areas.
The Late Pleistocene climate variations caused a major landscape restructuring with a deep bedrock frost weathering, sediment deposition, solifluction and cryogenesis during cold stages, and pedogenic development during warm interstadials, particularly in the loess area of the Irkutsk Basin. Renewed tectonic activity contributed to intensive river incision, leading to formation of the present river valleys, but also of passages for the last glacial ice expansion. The initial Late Pleistocene interglacial (MIS 5) promoted spread of pine-birch forests including some warm arboreal broad-leaved taxa *(Tsuga, Tilia, Corylus).* A progressing climatic cooling at the beginning of the early Last Glacial (Zyriansk Glaciation, MIS 4) gave rise to major montane glaciers in the Baikal Range and the most intensive Pleistocene glaciation in the Sayan and Altai Mountains. Mountain valley glaciers, expanding from the north-eastern slopes of the Bolshoy Sayan Ridge, formed mighty piedmont glaciers penetrating up to100 km north into the extra-glacial zone. Accumulation of massive colluvia (up to 20 m thick) was apparently triggered by the reactivated regional neotectonic movements towards the end of the early last glacial stage. A climatic warming during the following (Karginsk) mid-last glacial interval (MIS 3) is evidenced by pollen records form alluvial and lacustrine sedimentary formations. Apart of coniferous species *(Pinus silvestris, Pinus sibirica, Abies, Picea* and *Tsuga),* deciduous trees (*Quercus, Ulmus, Corylus* and *Alnus*) and shrubs (*Salix, Alnaster*) were present. The East Siberian mountain glaciers largely wasted or retreated into high topographic locations to form isolated ice-caps. The late Last Glacial (Sartan Glaciations, MIS 2) in the Eastern Sayan Mountains was largely confined to active corrie glaciers above the 2700 m elevation (Nemchinov et al., 1999). On the contrary, the Pribaikal Highlands likely experienced the most extensive Pleistocene glaciation during the LGM (22-19 ka BP), with piedmont glaciers advancing far down into the foothill valleys (Rezanov & Kalmikov, 1999). Only isolated ice-caps formed on the NE Siberian mountain ranges (Verkhoyanskyy, Cherskogo, Kolymskyy) despite very low temperature, but due to a very high aridity and a lack of winter precipitations. Cold Late Pleistocene climatic intervals are also documented by cryogenic surface deformations, with the most prominent (up to 5 m deep) at the contact of the Pleistocene/Holocene formations assigned to Younger Dryas (12.9-11.5 ka BP). Cold periglacial tundra with isolated shrub and dwarf trees occupied the adjacent northern foothills until the warm climatic oscillations sparked off a complete zonal vegetation rearrangement with the approaching end of the last glacial stage. Cold and strongly continental climatic conditions prevail in the East Siberian mountain region at the present time, with MAT of -4°C. A considerable part of the territory is covered by mountain tundra with larch *(Larix* sp.), Siberian pine *(Pinus sibirica)* and dwarf birch *(Betula nana).* Most of the area is underlain by active perennial permafrost.

3.4.2.2 Quaternary climate change and environments of the extra-glacial areas

Quaternary geology of the extra-glacial areas of the Baikal territory (the Irkutsk Depression, the Transbaikal and upper Lena River areas) shows a complex structure resulting from the interaction of past climatic changes and a topographic modeling by the neotectonic activity. This encouraged mass deposition of alluvial, aeolian, colluvial and other clastic materials. The iron-stained gravels and reddish strongly weathered lacustrine clays attest to warm climatic conditions during the late Pliocene in corroboration with the incorporated pollen of thermophylous taxa *(Quercus, Carya, Tsuga)* of mixed coniferous-broadleaved forests.

The Pliocene/Pleistocene transition around 2.5 Ma BP defines the first major cooling evidenced by prominent cryogenic features. The Early Pleistocene climate trend of progressing continentality caused a complete zonal palaeogeographic rearrangement (Medvedev et al., 1990). The Early and Middle Pleistocene climatic variations are witnessed by a wide range of palaeosols (kashtanozems, chernozems, luvisols, brunisols, gleysols), indicating a mosaic vegetation and relief zonation. Stages of climatic deterioration correlate with weakly developed initial soils (regosols), solifluction horizons, cryogenic deformations and relic periglacial landforms. A seasonal strengthening of temperate and humid climate promoted a latitudinal zonal vegetation distribution. The Middle Pleistocene interglacial periods were warm and humid, allowing a northern expansion of a mixed pine-spruce-fir taiga with broad-leaved species *(Quercus, Fagus, Ulmus, Corylus, Pterocarya, Juglans* and *Tilia amurensis)* with up to a 50% increase in precipitation in the warmest Tobol Interglacial (MIS 9) (Rezanov & Kalmikov, 1999). The following climate deterioration during the late Middle Pleistocene is manifested by progressive fossil fauna forms of open periglacial parklands.

Diversity of soils developed during the Last Interglacial on alluvial and aeolian deposits (tundra gleysols, gleyed podzolic soils of northern taiga, podzols of central taiga, brown soils of southern taiga and chernozems of parkland-steppe). Polygons with up to 3-4 m deep frost wedges on top of the last interglacial (MIS 5e) chernozem indicate a dramatic cooling during the following stage (MIS 5d) with temperature minimum around ca. 115 ka BP, suggesting a major glaciation in eastern Siberia corroborated by the record from Lake Baikal and the loess-palaeosol sequences from other parts of southern Siberia (Karabanov et al., 1998; 2000; Prokopenko et al., 1999, Chlachula 2003, Chlachula et al., 2004b) (Fig. 2B).

Onset of the early Last Glacial stage (Ermakovo Glaciation) is associated with an intensive erosion of the exposed land surfaces leading to a massive transfer of sediments, particularly fine sands from the major denudation areas /deflation surfaces along the margin of the Irkutsk Depression. Solifluction of loessic deposits underwent during the following time intervals as a result of a moderate warming. The mid-last glacial Karginsk interstadial stage (MIS 3, 59-24 ka BP) included an earlier (cold and arid) phase of loessic sediment deposition succeeded by a later (warm and humid) phase of progressive soil formation (the Osinsk Pedocomplex in the Angara - upper Lena region). The first phase of the late Last Glacial (MIS 2), radiocarbon dated to 24-17 ka BP, is linked with marked climatic oscillations evidenced by solifluction and cryoturbation of the preceding (MIS 3) soils, succeeded by loess sedimentation with the approaching Last Glacial Maximum (22 ka BP). Cold climates corroborate appearance of cold-adapted fauna characteristic of periglacial steppe *(Saiga tatarica, Alces alces, Equus hemionus*). A warming interval after the LGM (17-16 ka BP) is manifested by an intensified gleying and degradation of permafrost in the southern part of the territory due to a global climatic amelioration also reflected by the expansion of coniferous taiga forests, and by the Palaeolithic re-colonization of south-eastern Siberia.

3.4.3 Pleistocene environments and early human peopling

Present cultural evidence attest to several stages of human occupation of Eastern Siberia during the Pleistocene, with the oldest represented by the Early (?) and Middle Pleistocene records, implying very early hominid migrations into the mid- and high-latitudes of Asia. Systematic investigations at the occupation sites in the upper Angara, Lena, Aldan, Viluy and Selenga River basins, along the Baikal Range and Eastern Sayan foothills, contextually associated with diverse palaeo-geomorphic zones and geological (alluvial, colluvial, aeolian, karstic) settings, deliver detailed multi-proxy information on the Pleistocene climate change and associated palaeoenvironmental processes in the particular areas (e.g., Medvedev et al., 1990; Konstantinov, 1994; Mochanov, 1992; Mochanov & Fedoseeva, 2002; Chlachula, 2001c). Principal Quaternary research in the Pribaikal and Trensbaikal regions (the upper Angara, Lena and Selenga basins and their tributary valleys) reflects a limited transport accessibility of the north-eastern territories. Archaeological sites may be partly obliterated or poorly preserved around the Lake Baikal due to neotectonic activity triggering erosional processes. The cultural finds, chronologically defined by the stratigraphic geological position and diagnostic technological attributes of stone flaking, include: 1) the Early Palaeolithic from Early(?) / Middle Pleistocene alluvial deposits (>130 ka BP) (Fig. 6B); 2) the Middle Palaeolithic buried in the last interglacial (MIS 5) pedocomplex and the early last glacial (MIS 4) gleysol horizons; 3) the early Late Palaeolithic (42-30 ka BP) from the mid-last glacial (MIS 3) soils; 4) the "classical" Late Palaeolithic from the late mid- and early last glacial (MIS 2) gleyed soil horizons (30-17 ka BP); 5) the final Palaeolithic (17-12 ka); and f) Mesolithic (12-8 ka BP) from diverse geo-contexts (Medvedev et al., 1990; Chlachula et al., 2004b).

The earliest Pleistocene sites with definite cultural occurrences are associated with old alluvial formations in the Lena and Angara River basins. Presently one of the oldest sites in East Siberia was found at the Diring Uriah within coarse sandy deposits of a 200 m-high terrace of the Lena River near Yakutsk. Despite its unclear chronological assignment ranging from 2.5 Ma to 400 ka and the postulated model of the extra-tropical origin of humans in NE Siberia (Mochanov, 1992) this site eloquently demonstrates a very early Pleistocene peopling of northern parts of Asia in the principal river valleys much earlier than previously believed probably during some of the Early or Middle Pleistocene interglacials (Waters et al., 1995). Temperate and drier conditions of the late Middle Pleistocene interglacial interval (MIS 7) promoted expansion of mixed taiga (*Pinus, Picea, Alnus, Salix, Betula*) and open parklands. A Middle Palaeolithic occupation is documented at the Mungkharyma site (64°N) located on the 70 m alluvial terrace of the middle Viluy River with a diagnostic stone industry, including elaborated bifaces, bifacial knives and side-scrapers, found in association with Pleistocene fauna (mammoth, woolly rhinoceros). Luminescence dates 150±38 ka BP (RTL-958) in the overlying sandy-silt layer and 600±150 ka BP (RTL-957) in the basal part suggest a (late) Middle Pleistocene age of this unique locality (Mochanov & Fedoseeva, 2001) (Fig. 6C-D). These as well as other pre-last glacial (>74 ka BP) sites from the central and southern Lena basin and its major tributaries (Viluy, Aldan) also have the principal implications in respect to the initial prehistoric colonization of the American continent (Chlachula, 2003b).

The archaeological records from the upper Lena and Angara River areas dated to the Middle Pleistocene, represented by analogous stone artifacts made on quartz / quartzite cobbles with the typical archaic tool forms (choppers, bifaces, scrapers on flakes, polyhedral cores), display strongly abraded surfaces as a result of aeolian weathering. The age of these earliest cultural assemblages from the Angara region exposed along the Bratsk Lake (Igetei locality), referred to as of the Acheulian-Mousterian tradition, is estimated to ca. 200 ka BP, corresponding to the Shirta Interglacial /MIS 7 (244-170 ka BP) (Medvedev et al., 1990). Climatic cooling, leading to expansion of mosaic steppe and pine-larch parklands, is linked with the onset of the Samarovo (270-244 ka BP) glacial period. A drop of annual temperature and increase in aridity during the Tazov Glacial (170-130 ka BP) hastened a degradation of interglacial forests and extension of open periglacial tundra landscapes in the broader Baikal region (Nemchinov et al., 1999). Human survival in south-east Siberia during the Samarovo (MIS 8) and Tazov (MIS 6) glacial stages presumes knowledge of fire-making. The marked final Middle Pleistocene environmental deterioration is indicated by records of cold-adapted fossil fauna from river terrace formations, including more progressive taxa typical of open periglacial landscapes, such as mammoth, wooly rhinoceros, bison, horse, kulan, giant deer, argali, and some species specific to the Trans-Baikal area - Kiakhta antelope *(Spirocerus kiakthensis),* the Baikal yak *(Poephagus baikalensis),* dzeren/gazelle *(Procapra gutturosa),* camel (*Camelus* sp.) (Kalmikov, 1990), in congruence with the preserved glacial landscape relics in the Northern Baikal Range and the Eastern Sayan Mountains (Rezanov & Nemchinov, 1999). During the Last Interglacial (MIS 5), mixed taiga forests with Siberian pine *(Pinus sibirica),* spruce, larch, birch, willow and hemlock were widely distributed in the mountain areas, indicating a temperate continental climate similar to the present one (Nemchinov et al., 1999). Brown fossil chernozems at the Mal'ta and Igetei Middle Palaeolithic sites (dated to MIS 5e and 5c, respectively) illustrate open parkland-steppe settings of the Angara basin. Marked cooling, increased aridity accompanied by a strong aeolian activity contributed to accumulations of extensive sandy deposits derived from drying up river beds in the Irkutsk Depression and Lena Basin during the early last glacial (MIS 4), also witnessed by an intense abrasion of the contextually associated stone industries. In the western Baikal region, this stimulated drifting sandy surfaces and accumulation of fine, aeolian dust on lee slopes in the form of a loess-like series at the eastern margin of the Eurasian loess belt. Cold intervals with increased atmospheric humidity correlate with cryoturbation processes, solifluction horizons and pollen records of invading periglacial grasslands and open pine-birch tundra. A local Middle Palaeolithic (Mousterian?) occupation may have survived under such harsh environmental conditions in the upper Angara at the Igetei Site III (Medvedev et al., 1990).

The mid-last glacial interstadial interval (MIS 3), with an early cold and arid loess sedimentation phase (59-40 ka BP) followed by a warm and humid pedogenic phase (40-24 ka BP) brought a major change in the distribution of vegetation zones in East Siberia, with a northern expansion of mixed taiga forests and pine-birch dominated parklands. Broad-leaved arboreal taxy (oak, beech, elm, hazel) distributed in more humid river valleys of the Baikal region testify to mosaic habitats with climate conditions possibly warmer than today (Rezanov & Kalmikov, 1999). The pedogenic record in the Transbaikal region (the lower hor. Kamenka site) indicates very high MAT of +8/10°C, thus considerably exceeding the present day temperature values (Dergachova et al., 1995). This warming correlates with a transgression of Lake Baikal, triggering deposition of thick deltaic and estuarine sediments. Cultural finds from the principal occupation sites (Ust'-Kova in the Angara basin, Mezin in the Kana valley, Kamenka in the Selenga basin, the Aldan River complex) dated to 30-24 ka BP display a poor preservation due to cryogenesis and solifluction processes persisting until the onset of late Last Glacial (Sartan) stage (MIS 2). Remains of exploited Pleistocene fauna (horse, antelope, woolly rhinoceros, mammoth, bison, sheep-argali and camel) similarly as rodent taxa are indicative of an open steppe and parkland habitat. At the Kamenka site in

Fig. 6. East-Central Siberia. A: the central west coast of Lake Baikal with marked orogenic structures; B: Upper Angara Basin (Igetei Site), eroded fossiliferous Late Quaternary loess-palaeosol sections overlying a Middle Pleistocene alluvium incorporating cultural remains; C: the Mungkharyma Site (64°N) on the 70 m Viluy River terrace, east-central Yakutia; D: Middle Palaeolithic quartzite tools (Mungkharyma II Site; courtesy Yu. A. Mochanov); E: coarse gravelly alluvial fans in the formerly glaciated Verkhoyanskyy Range, NE Yakutia; F: the meandering Yana River valley was one of corridors for peopling of the Arctic Siberia.

southern Buryatia, the most represented animal species, Mongolian gazelle *(Procapra gutturosa),* lives today in dry grassland steppes and semi-deserts in eastern Altai, Mongolia and Inner Mongolia in gently rolling landscapes with a reduced snowfall. Several specific site functional complexes related to game processing, woodworking, stone and bone tool production and manufacturing of mineral paints, and other behavioral cultural (ritual) activities were documented (Lbova 1996). The taxonomic fossil fauna variety from different (mountain) ecotones shows a wide (>100km) mobility range of the local Palaeolithic hunters.

Expansion of the Late Palaeolithic occupation ambit into the extreme parts of the East Siberian Arctic is documented at the Yana RHS site located 100 km south of the Laptev Sea coast (70°43'N, 123°25'E) (Pitulko et al., 2004). The cultural evidence (stone and bone industry and fossil fauna), sealed in frozen and cryogenically distorted silt blocs on a 18 m Yana River terrace radiocarbon dated to ca. 28-26 ka BP, illustrates warm interstadial (MIS 3) climates and a relative environmental stability of floodplain meadows of the Yana River delta (Fig. 6F). This exceptional site provides eloquent evidence of humans migrating along the ice-free northern coast of Siberia (the exposed continental Arctic shelf) across the exposed Bering land-bridge during the mid-last gladial interstadial (50-24 ka BP), opposing the model of "initial" peopling of Alaska by the East Siberian Diyuktai Culture at 12 ka BP.

The late Last Glacial stage (MIS 2) - the Sartan Glaciation (24-12 ka BP) - is the best-studied Pleistocene time interval in eastern Siberia in view to contextual cultural records. Tundra-steppe covered most of the territory occupied by the Late / Final Palaeolithic people with the famous sites Mal'ta and Buret' in the upper Angara basin (Tseitlin, 1979). Climate deterioration with sparse Arctic vegetation and progressive loess accumulation around the LGM (18 ka BP) caused a major decline in the population density in East Siberia, although some adaptation to extreme periglacial environments was suggested at the Krasnyy Yar Site, with animal bones and fossil coal used as fuel (Medvedev et al., 1990). A periglacial fauna (horse, wooly rhinoceros, mammoth, bison, giant deer, elk, saiga) implies cold periglacial tundra steppe across vast areas across the Angara, upper Lena and the Trans-Baikal basins.

Climate amelioration after the LGM (18 ka BP) is best-evidenced by micro-mammal (rodent) faunas from the Baikal-Angara-Lena Palaeolithic sites (Buret', Krasnyy Yar, Igetei, Mal'ta, Bolshoi Jakor) that indicate a gradual transition from tundra-steppe and meadow-steppe to forest-steppe landscapes corresponding to shifts from a cool and dry climate to milder and humid conditions (Medvedev et al., 1990, Khenzykhenova, 1999). This warming trend fostered dispersal of the Final Palaeolithic complexes with micro-blade stone and bone technologies during the late Last Glacial (18-12 ka BP). Re-colonization of broad areas of East Siberia regions reached the marginal sub-polar regions of the NE Russian Arctic (e.g, Berelekh site in northern Yakutia at 70°N, dated to 14-13 ka BP) (Vereshagin, 1977; Vasil'ev, 2001). The broad geographical distribution of new technologies to the most distant parts of NE Siberia and the Russian Far East Islands (Vasilevskyy, 2008) documents a successful prehistoric adaptation to the Final Pleistocene environments. This process culminated in a spatial spreading and presumably a major population increase by the end of the Pleistocene, represented by the early Mesolithic (11,000-10,000 yr BP) hunters and gatherers.

In sum, palaeoenvironmental (geological and biotic) proxy records document well the evolutionary development of Quaternary climate change in East Siberia as reflected in the gradual cultural adaptation of people and the geographic expansion of occupied areas. A great diversity of the early Quaternary biotic communities became reduced as a result of climatic cooling, leading to a relative taxonomic faunal uniformity in the Late Pleistocene.

4. Conclusion

Geological and palaeoecological evidence across western, south-central and eastern Siberia, including the Ob, Irtysh, Kuznetsk, Yenisei, Angara and upper Lena basins and the adjacent mountain regions of the Trans-Urals, Altai, Western and Eastern Sayans Mountains, Baikal Range and the NE Siberian mountains, shows patterned cyclic climatic changes during the Quaternary leading to establishment of the present natural conditions. A progressing global trend towards a strongly continental climate, with increased aridity and high seasonal temperature fluctuations, can be traced since the late Pliocene. At that time, northern and north-eastern Siberia experienced the first major regional glaciations (Laukhin et al., 1999) in response to changes in solar radiation and atmospheric circulation over high latitudes of the northern Hemisphere. The subsequent development during the Quaternary Period (last 2.5 Ma) is evidenced by zonal geographic shifts in the vegetation distribution, with expansion of boreal (taiga) forests northward during the interglacial periods and warmer interstadials, succeeded by sub-arctic periglacial forest-steppe, and tundra-steppe during cold glacial intervals, when the tree cover became much reduced and spatially limited to biotic refugia in the southernmost portions of central Siberia (the Altai and Sayan Mountain foothills).

Complex palaeoenvironmental evolution archives are stored in the high-resolution loess-palaeosol sequences, pollen records and fossil fauna remains from deeply stratified alluvial and loessic formations. The loess-palaeosol sections on the Northern Altai Plains and in the upper Yenisei basin have provided most complete information on the past climatic variations, the landscape development and the associated changes in the Pleistocene biotic communities on the territory of southern Siberia. The high-resolution stratigraphic records coupled with pollen and palaeontology data indicate marked Pleistocene ecosystem transformations, with arctic tundra and forest-tundra during cold stadial intervals, being replaced by boreal forest and later by parkland - steppe during warm interstadial intervals.

The Early and Middle Pleistocene climates brought a major modification of natural habitats, facilitating the northward dispersal of Palaeolithic people from the southern areas of Central Asia and Mongolia, and their environmental adaptation to the Siberian habitats and regional settings. The earliest unequivocally documented Middle Pleistocene (Early Palaeolithic) occupation centered in the southern continental basins and river valleys north of the Altai-Sayan Mountains. The human dispersal into East Siberia is assumed to have principally occurred during warm interglacials in the processes of the northern expansion of mixed parkland forests and associated fauna communities, whereas only local movements of early human groups are envisaged during cold stages. The Tobol (MIS 9) Interglacial (390-270 ka BP), when the MAT was by ca. 3-4°C higher than at present, is likely to have been (one of) the most favorable time period for the main initial migration to northern Asia through the major Siberian river valleys, reaching as far north as 60°N latitude. The Early and Middle Palaeolithic finds bear witness of repeated inhabitation of the Irtysh, Ob, upper Yenisei, Angara, Viluy, Aldan and the upper Lena River basins prior to the Last Interglacial. During the Middle Pleistocene glacial periods, glaciers in the Western and Eastern Sayan ranges expanded into the foothills to about 300-400 m altitude preceded by a down-slope retreat of dark coniferous taiga forests. In the Northern and Southern Minusinsk Depression, in the Kuznetsk Basin and some extraglacial locations in the northern Altai valleys protected by mountains from the west and east, and separated by high ranges from the frigid arctic tundra in the north, propitious (although periglacial) conditions with great biomass concentrations of steppe-parklands are likely to have persisted during the glacial stages. This is evidenced by the abundant fossil remains from the old (60-80 m) Yenisei and Angara alluvia, as well as the colluvial reworking of the older (late Middle Pleistocene) loess cover, indicative of a relatively humid, moderately cold and fluctuating climate regime.

Within the Last Interglacial (MIS 5; 130-74 ka BP), most of Siberia was covered by coniferous or mixed forests, with forest-steppe distributed at lower elevations and in river valleys. At that time, the Middle Palaeolithic (Neanderthal or early *Homo sapiens*) people entered the territory from Central Asia and/or the East European Plains. Expansion of the occupation habitat into the mountain areas, following the last interglacial climatic optimum, likely occurred in the later (MIS 5c and 5a) interstadials. During the early Last Glacial (Zyriansk) stage (MIS 4; 74-59 ka BP), cold periglacial tundra or tundra-steppe expanded across Siberia. The approaching glacial maximum disrupted human occupation, although this may have persisted in some protected southern locations. Following the interval of intensive loess deposition at the end of the glacial, renewed warm climate pulses during the mid-glacial (Karginsk) interval (MIS 3; 59-24 ka) preconditioned formation of zonal soils contextually associated with the transitional Middle/early Late Palaeolithic stone industries, suggesting a certain regional cultural (and possibly biological) evolutionary continuity in the Late Pleistocene Siberia. Moderately cold and stable environments during the second half of the interstadial interval (30-24 ka BP) promoted a major enlargement of habitats marking a climax of the Palaeolithic peopling in Siberia associated with the emergence of the "classical" Late Palaeolithic cultures. Productive interstadial ecosystems with mixed parkland-forest vegetation were gradually transformed into periglacial tundra with the approaching last (Sartan) glacial stage (MIS 2; 24-12 ka BP). A reduced population density is assumed around the LGM (22-19 ka BP) hindered by extremely harsh climate conditions. Some occupation continuity persisting until the end of the Pleistocene may have applied just for biotic refugia in the protected southernmost locations along the Altai-Sayan foothills.

Overall, the spatial and temporal distribution of the cultural records documents climatic instability over large parts of Siberia during the Quaternary Period (the last 2.5 Ma). Specific geographical and contextual locations of early sites indicate that environmental conditions during the earlier periods were generally more stable and favorable for peopling than during the later periods. On the other hand, increased continentality and gradual shifts towards cold and arid conditions accelerated adaptation of Palaeolithic populations to harsh periglacial climates promoting progressive development of sophisticated survival strategies. Timing and evolutionary processes related to the initial colonization of northern Asia are still insufficiently mapped, although ongoing archaeological investigations supply new and often surprising evidence about particularities and general trajectories of this evolutionary process. The traditional views, assuming only a Late Pleistocene inhabitation of Siberia and Beringia, have been definitely challenged. The archaeological discoveries disprove the long-held assumption of a late penetration (by Late Palaeolithic people) into the middle and high latitudes of northern Asia. Instead, glacial-interglacial and stadial-interstadial climate cycles regulated a geographic movement of early people northwards, predetermining the inhabitability of particular geographical areas. During glacial maxima, most of Siberia seems to have been vacated, especially during the earlier periods, because of the expansion of continental glaciers in the north, and harsh and inhospitable environments in the southern extra-glacial regions. Gradual adaptation to cold natural habitats accelerated during the Late Pleistocene in connection with the advanced cultural and biological adjustment, enabling people to establish permanently in the vast and geographically diverse Siberian territory.

5. References

Abramova, Z.A. (1979a). *The Palaeolithic of the Yenisei. The Kororevo Culture.* Nauka, Novosibirsk, 199 p. (in Russian).

Abramova, Z.A. (1979b). *The Palaeolithic of the Yenisei. The Afontovskaya Culture.* Nauka, Novosibirsk, 156 p. (in Russian).

Arkhipov, S.A., Ed. (1980). *Paleogeography of West-Siberian Plain at the Late Zyryanka Glaciation Maximum.* INQUA Project: Quaternary Glaciations of the Northern Hemisphere. Nauka, Novosibirsk, 108 p. (in Russian).

Arkhipov, S.A. (1998). Stratigraphy and paleogeography of the Sartan glaciation in West Siberia. *Quaternary International,*Vol.45/46, pp. 29-42.

Arkhipov, S.A. (1999). Natural habitat of early man in Siberia. In: Chlachula, J., Kemp, R.A., Tyráček, J. (Eds), *Quaternary of Siberia, Anthropozoikum,* Vol. 23, pp. 133-140.

Arkhipov, S.A.; Bespaly, V.G.; Faustova, M.A.; Glushkova, O.Yu.; Isaeva, L.L., Velichko, A. (1986). Ice-sheet reconstructions. Quaternary Science Reviews, Vol. 5, pp. 475-482.

Arkhipov, S.A.; Gnibidenko, Z.N. & Shelkoplyas, V.N. (2000). Correlation and paleomagnetism of glacial and loess-paleosol sequences on the West Siberian Plain.*Quaternary International,* Vol. 68-71, pp. 13-27.

Astakhov, S. N. (1986). *The Palaeolithic of Tuva. Nauka,* Novosibirsk, 174 p. (in Russian).

Astakhov, V. (1997). Late glacial events in the central Russian Arctic, *Quaternary International,* Vol. 41-42, pp.17-25.

Astakhov, V. (2001). The stratigraphic framework for the Upper Pleistocene of the glaciated Russian Arctic: changing paradigms, *Global and Planetary Change,* V.31, pp. 283- 295.

Bader, O.N. (1965). *The Kapova Cave – Palaeolithic Paintings.* Nauka, Moskva.

Baker, V.C.; Benito, G. & Rudoy, A.N. (1993). Paleohydrology of Late Pleistocene super-flooding, Altai Mountains, Siberia, *Nature* ,Vol. 259, pp. 348-350.

Baryshnikov, G.Ya., (1992). *Relief Evolution in Transitional Zones of Mountain Areas during the Cenozoic Era,* University of Tomsk Press, Tomsk, 182 p. (in Russian).

Bazarov, D.B.(1986). *Pribaikal and West Transbaikal Cenozoic.* Nauka, Novosibirsk (in Russian).

Bolikhovskaya, N.S. & Molodkov, A.N. (2006). East European loess-palaeosol sequence: Palynology, stratigraphy and correlation, *Quaternary International,* Vol. 149, pp. 24-36.

Butvilovskiy, V.V. (1985). Catastrophic releases of waters of glacial lakes of the south-eastern Altai and their traces in relief, *Geomorphology,* Vol. 1985, No. 1, pp. 65-74. Chlachula, J. (1999). Loess-palaesol stratigraphy in the Yenisei Basin, southern Siberia. In: Chlachula, J.; Kemp, R.A. & Tyráček, J. (Eds.), *Quaternary of Siberia. Anthropozoic* , Vol. 23, pp. 55-70.

Chlachula J. (2001a). Pleistocene climates, natural environments and palaeolithic occupation of the Altai area, west Central Siberia, *Lake Baikal and Surrounding Regions* (S. Prokopenko, N. Catto and J. Chlachula, Eds.), *Quaternary International,* Vol. 80-81, pp. 131-167.

Chlachula J. (2001b). Pleistocene climates, natural environments and palaeolithic occupation of the upper Yenisei area, south Central Siberia. *Lake Baikal and Surrounding Regions,* (S. Prokopenko, N. Catto and J. Chlachula, Eds.), *Quaternary International,* Vol. 80-81, pp. 101-130.

Chlachula J. (2001c). Pleistocene climates, natural environments and palaeolithic occupation of the Angara-Baikal area, east Central Siberia.*Lake Baikal and Surrounding Regions*

(S. Prokopenko, N. Catto and J. Chlachula, Eds.), *Quaternary International,* Vol. 80-81, pp. 69-92.

Chlachula J. (2003a). The Siberian loess record and its significance for reconstruction of the Pleistocene climate change in north-central Asia, *Dust Indicators and Records of Terrestrial and Marine Palaeoenvironments (DIRTMAP)* (Edward Derbyshire, Editor). *Quaternary Science Reviews,* Vol. 22. No. 18-19, pp. 1879-1906.

Chlachula J. (2003b). Controverse sur l'age des premiers Américains, *La Recherche* (Paris), Vol. 370, pp. 42-46.

Chlachula, J.; Evans, M.E. & Rutter, N.W. (1998). A magnetic investigation of a late Quaternary loess/palaeosol record in Siberia., *Geophysical Journal International,* Vol. 132, pp. 399-404.

Chlachula, J. (2010a). Environmental context and human adaptation of Palaeolithic peopling of the Central Urals, *Eurasian Perspectives of Environmental Archaeology* (J.Chlachula & N.Catto, Editors). *Quaternary International.,* Vol. 220, pp. 47-63. Chlachula. J. (2010b). Pleistocene climate change, natural environments and Palaeolithic peopling of East Kazakhstan, *Eurasian Perspectives of Environmental Archaeology* (J.Chlachula and N.Catto, Editors). *Quaternary International,* Vol. 220, pp. 64-87.

Chlachula, J.; Rutter, N.W. & Evans, M.E. (1997). A late Quaternary loess-palaeosol record at Kurtak, southern Siberia. *Canadian Journal of Earth Sciences* , Vol. 34, pp. 679-686. Chlachula J.; Drozdov N.I. & Ovodov N.D. (2003). Last interglacial peopling of Siberia: the Middle Palaeolithic Site Ust'-Izhul', the Upper Yenisei area. *Boreas,* Vol 32 (2003), pp. 506-520.

Chlachula J.; Kemp R.A.; Jessen .; Palmer A.P. &Toms P.S. (2004) (Chlachula et al. 2004a). Landscape development in response to climate change during the Oxygen Isotope Stage 5 in the southern Siberian loess region, *Boreas,* Vol. 33, pp. 164-180.

Chlachula J.; Medvedev G.I. & Vorobyova , G.I., 2004 (Chlachula et al. 2004b). Palaeolithic occupation in the context of the Pleistocene climate change in the Angara region, Central Siberia. *Anthropozoikum,* Vol. 25, pp. 31-49.

Chlachula, J. & Serikov, Yu. (2011). Last glacial ecology and geoarchaeology of the Trans-Ural area: the Sosva River Upper Palaeolithic complex, West Siberia, *Boreas,* Vol 40. No. 1, pp. 146-160.

Babek, O.; Chlachula, J. & Grygar, J. (2011). Non-magnetic indicators of pedogenesis related to loess magnetic enhancement and depletion from two contrasting loess-paleosol sections of Czech Republic and Central Siberia during the last glacial-interglacial cycle, *Quaternary Science Reviews,* Vol. 30, pp. 967-979.

Baryshnikov, G.Ya. (1992). *Relief Evolution in Transitional Zones of Mountain Areas during the Cenozoic Era,* University of Tomsk Press, Tomsk, 182 p. (in Russian).

Butvilovskiy, V.V. (1985). Catastrophic releases of waters of glacial lakes of the SE Altai and their traces in relief, Geomorphology, Vol.1985, No. 1, pp. 65-74 (in Russian).

Chlachula, J. & Little, E. (2011). A Late Quaternary high-resolution loess-palaeosol record from Iskitim, south-western Siberia, *The Second Loessfest (2009),* (Markowic, S.B., Catto, N., Smalley, I.J., Zoeller, L), *Quaternary International,* Vol. 240, pp. 139-149. Dalton, R. (2010). Fossil finger points to new human species. DNA analysis reveals lost relative from 40.000 years ago, *Nature,* Vol. 464, pp. 472-473.

Derevianko, A.P. (1990). *Palaeolithic of North Asia and the Problem of Ancient Migrations,* SB AS SSSR, Novosibirsk, 122 p.

Derevianko A.P. (2010). Three Scenarios of the Middle to Upper Paleolithic Transition: Scenario 1: The Middle to Upper Paleolithic Transition in Northern Asia. *Archaeology, Ethnology and Anthropology of Eurasia,* Vol. 38, No. 3, pp. 2-32.

Derevianko, A.P. & Markin, S.V. (1992). The Mousterian of the Gorno Altai, Nauka, Novosibirsk, 224 p. (in Russian).

Derevianko, A.P. & Markin, S.V. (1999). The Middle and Upper Palaeolithic of the Altai, *Quaternary of Siberia* (Chlachula, J., Kemp, R.A. & Tyráček, J., Eds), *Anthropozoikum,* Vol. 23, pp. 157-166.

Derevianko, A.P. & Shunkov, M.V. (2009). Development of Early Human Culture in Northern Asia, *Paleontological Journal,* Vol. 43, No. 8, pp. 881-889.

Dergacheva, M.I., Fedeneva, I.K. & Lbova, L.V. (1995). A pedogenic evidence of the Upper Palaeolithic occupation of the Varvarina Mountain, *Ecology and Culture of the peoples of Siberia and America, horizons of complex research,* Chita's Pedological Institute, Chita, pp. 88-89.

Deviatkin, E.V. (1965). Cenozoic Deposits and Neotectonics of the South-Eastern Altai, *Proceedings GIN AN SSSR,* Vol. 126, 244 p. (in Russian).

Deviatkin, E.V. (1981). *The Cenozoic of the Inner Asia,* Nauka, Moskva, 196 p. (in Russian).

Drozdov; N.I., Laukhin, S.A.; Chekha, V.P.; Kol'tsova, B.G.; Bokarev, A.A. & Vikulov, A.A. 1990 (Drozdov et al., 1990). The Kurtak Archaeological Region, *Chronostratigraphy of Palaeolithic of North, Central and Eastern Asia, and America,* SB RAS, Krasnoyarsk.

Drozdov, N.I.; Chlachula, J. & Chekha, V.P. (1999). Pleistocene environments and Palaeolithic occupation of the Northern Minusinsk Basin, Krasnoyarsk Region *Quaternary of Siberia* (Chlachula, J., Kemp, R.A. & Tyráček, J., Eds), *Anthropozoikum,* Vol. 23, pp. 141-155.

Evans, M.E.; Rutter, N.W.; Catto, N.; Chlachula, J. & Nyvlt, D. (2003). Magnetoclimatology: teleconnection between the Siberian loess record and North Atlantic Heinrich events. *Geology,* Vol. 31, No. 6, pp. 537-540.

Foronova, I.V., 1999. Quaternary mammals and stratigraphy of the Kuznetsk Basin (southwestern Siberia). *Quaternary of Siberia* (Chlachula, J., Kemp, R.A. & Tyráček, J., Eds), *Anthropozoikum,* Vol. 23, pp. 71-97.

Grosswald, M.G. & Hughes, T.J. (2002). The Russian component of an Arctic ice-sheet during the Last Glacial Maximum, *Quaternary Science Reviews,* Vol. 21, pp. 121-146.

Grygar; T., Kadlec, J.; Pruner, P.; Swann, G.; Bezdička, P.; Hradil, D.; Lang, K.; Novotná, K. & Oberhansli, H. (2006). Paleoenvironmental record in Lake Baikal sediments: Environmental changes in the last 160 ky, *Palaeogeography Palaeoclimatology Palaeoecology,* Vol. 237, pp. 240-254.

Kalmikov, N.P. (1990). *Large mammalian fauna of the Pribaikal and West Transbaikal Pleistocene.* Proceedings of the Buryat Scientific Centre, Ulan-Ude, 116 p (in Russian).

Khenzykhenova, F.I.. (1999). Pleistocene disharmonius faunas of the Baikal regiona (Russia, Siberia) and their implication for palaeogeography, *Quaternary of Siberia* (Chlachula, J., Kemp, R.A. & Tyráček, J., Eds), *Anthropozoikum,* Vol. 23, pp. 119-124.

Konstantinov, M.V. (1994). *L'age de Pierre de la Région est de l'Asie de Baikal.* Ulan-Ude, 180 p.

Larichev, V.; Khol'ushkin, U. & Laricheva I., (1990). The Upper Palaeolithic of Northern Asia: Achievements, Problems, and Perspectives. II. Central and Eastern Siberia. *Journal of World Prehistory,* Vol. 4, No. 3, pp. 347-385.

Laukhin, S.A.; Klimanov, V.A. & Belaya, B.V. (1999). Late Pliocene and Pleistocene palaeoclimates in northeastern Chukotka, *Quaternary of Siberia* (Chlachula, J., Kemp, R.A. & Tyráček, J., Eds), *Anthropozoikum*, Vol. 23, pp. 17-24.

Lbova, L. (1996). The Palaeolithic site Kamenka, *New Palaeolithic Sites of the Zabaikalye*(Konstantinov, M., Ed.), Nauka, Chita, pp. 24-47 (in Russian).

Little, E.C.; Lian, O.B.; Velichko, A.A.; Morozova T.D.; Nechaev, V.P.; Dlussky, K.G. & Rutter, N.W. (2002). Quaternary stratigraphy and optical dating of loess from the East European Plain (Russia), *Quaternary Science Review,* Vol. 21, pp. 1745-1762.

Mangerud, J.; Svendsen, J.I.. & Astakhov, V.I. (1999). Age and extent of the Barents and Kara Sea ice-sheets in northern Russia, *Boreas, Vol.* 28, No. 1, pp. 46-80.

Mangerud, J.; Astakhov, V.I. & Svedsen, J.I. (2002). The extent of the Barents-Kara ice sheet during the Last Glacial Maximum. *Quaternary Science Reviews,* Vol. 21, pp. 111-119.

Mangerud, J., Gosse, J., Matiouchkov, A. & Dolvik, T. (2008). Glaciers in the Polar Urals, Russia, were not much larger during the Last Global Glacial Maximum than today, *Quaternary Science Reviews,* vol. 27, pp. 1047-1057.

Markin, S.V. (1996). *The Palaeolithic of the North-West of the Altai-Sayan Mountain Region,* SB RAS, Novosibirsk, 58 p. (in Russian).

Markova, A.K. (2007). Pleistocene mammal faunas of Eastern Europe, *Quaternary International,* Vol. 160, pp. 100-111.

Matyushin, G.N. (1994). *Stone Age of the Southern Urals.* Nauka, Moskva (in Russian).

Medvedev, G.I.; Savel'ev, N.A. & Svinin, V.V. (1990). *Stratigraphy, Palaeogeography and Archaeology of Southern Central Siberia,* AN SSSR, Siberian Branch. Irkutsk, 164 p.

Mochanov, Y.A. (1992). *The Early Palaeolithic at Diring and the Question of Extra-Tropical Origin of Man,* SB RAS, Novosibirsk.

Mochanov, Y.A. & Fedoseeva, S.A. (1996). Berelekh, Allakhovsk Region, *American Beginnings: The Prehistory and Palaeoecology of Beringia* (F. H. West, Ed.), University of Chicago Press, Chicago., pp. 218-222.

Mochanov, Yu.A. & Fedoseeva, S.A. (2001). *Neosphere and Archaeology,* Nauka i Tekhnika v Yakutii, Vol. 2001, No. 1, pp. 28-33 (in Russian).

Mochanov, Yu.A. & Fedoseeva, S.A. (2002). Archeology, Palaeolithic of North-East Asia, Ex-Tropical Homeland of Humankind and Early Human Occupation Stages of America, Polyrnyj Krug, Yakutsk, 59 p.

Nemchinov, V.G.; Budaev, R.T. & Rezanov, I.N. (1999). Pleistocene glaciations of the eastern Sayan Mountains), *Quaternary of Siberia* (Chlachula, J., Kemp, R.A. & Tyráček, J., Eds), *Anthropozoikum,* Vol. 23, pp. 11-15.

Okishev, P.A. (1982). *Dynamics of glaciations in the Late Pleistocene and Holocene,* Tomsk State University Press, Tomsk. 209 p. (in Russian).

Panova, N.K.; Jankovska, V.; Korona, O.M. & Zinov'ev, E.V. (2003). The Holocene dynamics of vegetation and ecological conditions in the Polar Urals. *Russian Journal of Ecology,* Vol. 3, No. 4, pp. 219-230.

Pavlov, P.Yu. (1996). *Palaeolithic Monuments of the North-Eastern European Part of Russia,* Ural Department RAS, Syktyvkar, 194 pp. (in Russian).

Pavlov, P.; Svendsen, J.I. & Indrelid, S. (2001). Human presence in the European Arctic nearly 40,000 years ago, *Nature,* Vol. 413, pp. 64-67.

Pavlov, P.; Roebroeks, W. & Svendsen, J.I. (2004). The Pleistocene colonization of north--eastern Europe, *Journal of Human Evolution,* Vol. 47, No. 1-2, pp. 3-17.

Petrin, V.T. (1986). *Palaeolithic sites of the West Siberian Lowland,* Siberian Branch RAS, Novosibirsk, 142 p. (in Russian).

Pitulko, V.V.; Nikolsky, P.A.; Girya, E. Yu.; Basilyan, A.E.; Tumskoy, V.E.; Koulakov, S.A.; Astakhov, S.N.; Pavlova, E. Yu. & Anisimov, M.A. (2004). The Yana RHS Site: Humans in the Arctic before the Last Glacial Maximum, *Science, Vol.* 303, pp. 52-56.

Prokopenko, A.A.; Hinnov, L.A.; Williams, D.F. & Kuzmin, M.I. (2006). Orbital forcing of continental climate during the Pleistocene: a complete astronomically tuned climatic record from Lake Baikal, SE Siberia, *Quaternary Science Reviews,* Vol. 25, pp. 3431-3457.

Rezanov, I.N. (1988). *Cenozoic Deposits and Morphostructure of East Pribaikal.* Nauka, Novosibirsk, 128 p. (in Russian). Rezanov, I.N. (1995). The role of the neotectonic support in the formation of sandy series of the West Transbaikal valley systems, *Geomorphology,* Vol. 1, pp. 80-87 (in Russian).

Rezanov, I.N. & Kalmikov, N.P. (1999). Palaeogeography of the Pribaikal and Transbaikal Quaternary), *Quaternary of Siberia* (Chlachula, J., Kemp, R.A. & Tyráček, J., Eds), *Anthropozoikum,* Vol. 23, pp. 43-48.

Rudoy, A.N. & Baker, V.R. (1993), Sedimentary effects of cataclysmic Late Pleistocene glacial outburst flooding, Altay Mountains, Siberia, *Sedimentary Geology,* Vol. 85, pp. 53-62.

Rudoy, A.N., Lysenkova, Z.V., Rudskiy, V.B. & Shishin, M.Yu. (2000). Ukok. The Past, Present and Future. Altai State University Press, Barnaul. 174 p. (in Russian).

Rutter N.W.; Rokosh C.E.; Evans M.E.; Little C.E.; Chlachula J. & Velichko A.A. (2003). Correlation and interpretation of paleosols and loess across European Russia and Asia over the last interglacial cycle, *Quaternary Research,* Vol. 60, No. 1, pp. 101-109.

Savel'eva, E.A., Ed. (1997). *Archaeology of the Komi Republic.* DiK, Moskva, 758 p. (in Russian).

Serikov Yu.B. (1999). *Palaeolithic of the Central Urals,* Nizhniy Tagil University Press, Nizhniy Tagil, 103 p. (in Russian).

Serikov Yu.B. (2000). *Palaeolithic and Mesolithic of the Central Trans-Urals,* University Press, Nizhniy Tagil', 271 p. (in Russian).

Serikov Yu.B. (2001). The Cave Palaeolithic of the Chusovaya River valley and problems of the initial peopling of the Central Urals, *Problems of Prehistoric Culture,* Gilem Press, Ufa, pp. 117-135 (in Russian).

Serikov, Yu.B. (2007). *The Palaeolithic Occupation Site Gari and Some Problems of the Ural Palaeolithic,* Nizhniy Tagil' University Press, Nizhniy Tagil', 137p. (in Russian).

Stefanovskiy, V.V. (2006). *Pliocene and Quaternary of the Eastern Slope of the Urals and Trans-Urals,* Ural Branch RAS, Ekaterinburg, 223 p. (in Russian).

Shun'kov, M. V. & Agadzhanyan, A.K. (2000). Palaegeography of Palaeolithic of Denisova Cave, *Archaeology, Ethography and Anthropology of Eurasia,* Vol. 2000, No. 2, pp. 2-20.

Svendsen, I.G. & Pavlov, P.Yu. (2003). Mammontova Kurya: an enigmatic, nearly 40 000 year-old Paleolithic site in the Russian Arctic, *Chronology of the Aurignacian and the Transitional Technologies, Dating Stratigraphies, Cultural Implications* (Zilhao, J. & d'Errico, F., Eds), *Trabalhos de Arquelogia,* Vol. 33, pp. 109-120.

Tseitlin, S.M. (1979). *Geology of Palaeolithic of Northern Asia,* Nauka, Moskva, 286 p

Vasiliev, S.A. (1992). The Late Paleolithic of the Yenisey: a new outline, *Journal of World Prehistory,* Vol. 6, pp. 337-383.

Vasiliev, S.A.; Yamskikh, A.F.; Yamskikh, G.Y.; Svezhentsev, Y.S. & Kasparov, A.K. (1999). Stratigraphy and palaeoecology of the Upper Palaeolithic sites near the Maima village (Upper Yenisei valley, Siberia), *Quaternary of Siberia* (Chlachula, J., Kemp,

R.A. & Tyráček, J., Eds), *Anthropozoikum*, Vol. 23, pp. 29-35. Vasil'ev, S.A. (2001), Man and mammoth in the Pleistocene Siberia, The World of Elephants. Proceedings of the First International Congress. Consiglio Nazionale delle Ricerche, Rome. (Cavarretta, G., Gioia, p., Mussi, M., Palombo, M.R., Eds.), pp. 363-366.

Vasilevskyy, A.A. (2008). Stone Age of the Sakhalin Island, Yuzno Sakhalinsk Press, 412 p. Velichko, A.A., Ed. (1993). Evolution of Landscapes and Climates of Northern Eurasia. Late Pleistocene - Holocene. Elements of Prognosis. 1. Regional Palaeogeography. RAN, Institute of Geography RAN, Nauka, Moskva, 102p. (in Russian).

Velichko, A.A., Ed. (1993). Evolution of Landscapes and Climates of Northern Eurasia. Late Pleistocene - Holocene. Elements of Prognosis, 1. Regional Palaeogeography. RAN, Institute of Geography. Nauka, Moscow.

Velichko, A.A.; Catto, N.; Drenova, A.N.; Klimaov, V.A.; Kremenetski, K.V. & Nechaev, V.P. (2002). Climate changes in East Europe and Siberia at the Late glacial-Holocene transition, *Quaternary International*, Vol. 90, pp. 75-99.

Velichko, A.A.; Grichuk, V.P.; Gurtovaya, E.E.; Zelikson, E.M.; Borisova, O.K. & Barash, M.S. (1992). Climates during the last interglacial, *Atlas of Paleoclimates and Paleoenvironments of the Northern Hemisphere, Late Pleistocene – Holocene* (Frenzel, B., Péczi, M. & Velichko, A.A., Eds.), Geographical Institute, Hungarian Academy of Sciences and Gustav Fisher Verlag, Budapest - Stuttgart, pp. 86-89.

Velichko, A.A.; Kononov, Yu.M. & Faustova, M. (1997). The last glaciation of the Earth: Size and volume of ice-sheets, *Quaternary International,* Vol. 41/42, pp. 43-51.

Vereschagin, N.K. (1977). The Belerekh mammoth „cemetery". *Trudy Zoologicheskogo Instituta,* Vol. 72, pp. 5-50 (in Russian).

Vereschagina, V.S. (2001). Alluvial deposits of the high antropogenic terraces of the Urals, *Antropogen Urala,* Perm State University Press, Perm, pp. 74-95 (in Russian).

Yakhimovich, V.L.; Nemkova, V.K. & Sidnev, A.V. (1987). *Pleistocene of the Fore-Urals,* Nauka, Moskva, 113 p. (in Russian).

Waters, M.R., Forman, S. &. Pierson, J. (1995). Diring Yuriakh: A Lower Palaeolithic Site in Central Siberia and its implications to the Pleistocene peopling of the Americas. - *Abstracts,* Annual Meeting of the Society for American Archeology, Minneapolis, May 3-7, SAA, p. 194.

Yamskikh, A.F. (1990). Late Paleolithic ecology of the Upper and Middle Yenisei valley, *Early Man News,* Vol. 15, pp. 39-41.

Yamskikh, A.F. (1996). Late Quaternary intra-continental river palaeohydrology and polycyclic terrace formation: the example of south Siberian river valleys, *Global Continental Changes: the Context of Palaeohydrology* (Branson, J., Brown, A.G. & Gregory, K.J., Eds.), Geological Society, London, Vol. 115, pp. 181-190.

Zenin, V.N., (2002). Principal stages of peopling of the West-Siberian Plain by Palaeolithic man, *Archaeology, Ethnography and Anthropology of Eurasia,* Vol. 12, pp. 22-44.

Zudin, A.N.; Nikolaev, S.V.; Galkina L.I.; Butkeeva, O.Yu.; Efimova, L.I.; Panychev, V.A. & Ponomareva, V.A. (1982). Stratigraphic scheme of the Neogene and Quaternary deposits of the Kuznetsk Basin, *Problems of Stratigraphy and Palaeogeography of the Pleistocene Siberia,* Nauka, Novosibirsk, pp. 133-149 (in Russian).

Zykina, V. S. (1999). Pleistocene pedogenesis and climate history of western Siberia. *Quaternary of Siberia* ((Chlachula, J., Kemp, R.A. & Tyráček, J., Eds), *Anthropozoikum,* Vol. 23, pp. 49-54.

23

Multivariate Analysis Techniques in Environmental Science

Mohammad Ali Zare Chahouki
Department of Rehabilitation of Arid and Mountainous Regions, University of Tehran, Iran

1. Introduction

One of the characteristics of environmental data, many of them and the complex relationships between them. To reduce the number variables, different statistical methods exist. Multivariate statistics is used extensively in environmental science. It helps ecologists discover structure and previous relatively objective summary of the primary features of the data for easier comprehension. However, it is complicated in theorical structure and in operational methodology.

In this chapter some important statistical methods such as Principal component analysis (PCA), Canonical correspondence analysis (CCA), Redundancy analysis (RDA), Cluster analysis, and Discriminate function analysis will be explained briefly.

This chapter too cover the statistical analysis of assemblage data (species by samples matrices of abundance, area cover etc) and/or multi variable environmental data which arise in a wide range of applications in ecology and environmental science, from basic ecological studies (e.g. of dietary composition or population size-structure), through community-based field studies, environmental impact assessments and monitoring of large-scale biodiversity change, to purely physical or chemical analyses.

The use of multivariate analysis has been extended much more widely over the past 20 years. Much more is included on techniques such as Canonical Correspondence Analysis (CCA) and Non-metric Multidimensional Scaling (NMS) and another technique to include organisms and organism-environment relationships other than vegetation. Spatially constrained data analysis will be introduced and the importance of accounting for spatial autocorrelation will be emphasized. Use of the methods within ecology and environmental reconstruction will also be covered. A study and review of the application of multivariate analysis in biogeography and ecology is provided in: Kent, M. (2006).

2. Landscape ecology

Landscape is simply an area of land (at any scale) containing an interesting pattern that affects and is affected by an ecological process of interest. Landscape ecology, then, involves the study of these landscape patterns, the interactions among the elements of this pattern, and how these patterns and interactions change over time. In addition, landscape ecology involves the application of these principles in the formulation and solving of real-world problems (Turner et al, 2001).

Landscape ecology is perhaps best distinguished by its focus on: 1) spatial heterogeneity, 2) broader spatial extents than those traditionally studied in ecology, and 3) the role of humans in creating and affecting landscape patterns and process (Turner et al, 2001).

In effect the role of ecology, and especially that of vegetation science, has been mainly restricted to the evaluation of the landscape with respect to particular demands: either evaluation as an assessment of the qualities of the ecosystem or evaluation as a socio-economic procedure intended to estimate the functions the natural environment fulfills for human societ (Van der Ploeg and Vlijm, 1978).

Landscape ecology theory stresses the role of human impacts on landscape structures and functions. It also proposes ways for restoring degraded landscapes. Landscape ecology explicitly includes humans as entities that cause functional changes on the landscape (Mielke& Berry, 2001).

Landscape ecology theory includes the landscape stability principle, which emphasizes the importance of landscape structural heterogeneity in developing resistance to disturbances, recovery from disturbances, and promoting total system stability (Mielke& Berry, 2001). This principle is a major contribution to general ecological theories which highlight the importance of relationships among the various components of the landscape. Integrity of landscape components helps maintain resistance to external threats, including development and land transformation by human activity. Analysis of land use change has included a strongly geographical approach which has led to the acceptance of the idea of multifunctional properties of landscapes (Mielke et al, 1976). There are still calls for a more unified theory of landscape ecology due to differences in professional opinion among ecologists and its interdisciplinary approach (Bastian 2001).

Landscape ecology is distinguished by its focus on broader spatial extents than those traditionally studied in ecology. This stems from the anthropocentric origins of the discipline.

Despite early attention to the effects of sample area on measurements, such as species-area relationships, the importance of scale was not widely recognized until the 1980's. Recognition that pattern-process relationships vary with scale demanded that ecologist give explicit consideration to scale in designing experiments and interpreting results.

It became evident that different problems require different scales of study, and that most problems require multiple scales of study. The theory of scale and hierarchy emerged as a framework for dealing with scale. The emergence of scale and hierarchy theory provided a partial theoretical framework for understanding pattern-process relationships, which became the basis for the emergence of landscape ecology as a discipline (Turner MG. 1989).

3. Sampling methods

Sampling design varies considerably with habitat type and specific taxonomic groups. Sampling design begins with a clear statement of the question(s) being asked. This may be the most difficult part of the procedure because the quality of the results is dependent on the nature of the original design.

If the sampling is for densities of organisms then at least five replicate samples per sampling site are needed because many statistical tests require that minimal number. Better yet, consider 20 replicates per sampling site and in some cases 50 or more. If sample replicates are less than five then bootstrapping techniques can be used to analyze the data (name). Some type of random sampling should be attempted (e.g., stratified random sampling) or a line intercept method used to estimate densities (e.g., Strong Method).

Measurements of important physical-chemical variables should be made (e.g.,temperature, salinity, sediment grain size, etc...). Field experiments need to be carried out with carefully designed controls. The correct spatial scale needs to be considered when planning experiments (Stiling, 2002).
If vegetation is correlated with geomorphologic landforms, for this reason stratified random sampling method is better and this method employ (Greig-Smith, 1983; Ludwig and Reynold, 1988)
Environmental impact assessments ideally attempt to compare before and after studies. There were differences in sampling sites, sampling dates, effort, replication, taxonomic categories, and recovery data. (Ferson et al., 1986).
Gotelli and Ellison (2004) and Odum and Barrett (2005) studied on sampling design. Diserud and Aagaard (2002) present a method that tests for changes in community structure based on repeated sampling. This may be the plot method of sampling generally consists of three major types:
(1) Simple random or random sampling without replacement, (2) stratified random, and (3) systematic (Cochran, 1977)
Simple random sampling with replacement is inherently less efficient than simple random sampling without replacement (Thompson, 2002). It is important not to have to determine whether any unit in the data is included more than once. Simple random sampling consists of using a grid or a series of coordinate lines (transects) and a table of random numbers to select several plots (quadrats), the size depending on the dimensions and densities of the organisms present.

4. Classification methods

4.1 Measures of similarity and difference (similarity and dissimilarity)

Dissimilarity or distance measures can be categorized as metric, semi-metric, or nonmetric (McCune et al., 2002). Symmetric distances are extremely useful in community ecology. There are different Similarity coefficients based on binary data. Two methods that are simple and give good results are Jaccard and Czekanowski (Sorenson) coefficients range 0 to 1.o.
Following equations are for Jaccard (1901) and Sorenson similarity coefficients

$$J = \frac{c}{a+b+c}$$

Where J= Jcacard coeffiecient, a= number of occurrences of species a alone, b= number of occurrences of b alone, c= number of co-occurences of two species (a and b).

$$C = \frac{2c}{2a+b+c}$$

Where: (same as in Jaccard)
There are different Dissimilarity coefficients based on meristic or metric data. Coefficient that are widely used in ecology studies are Bray-Curtis (1975) dissimilarity coefficient and Morisita' s Index.
Bray-Curtis is recommended by Clarke and Warwick (2001) and others as the most appropriate dissimilarity coefficient for community studies.

$$BC = \frac{\sum_{j=1}^{n} \left| X_{1j} - X_{2j} \right|}{(X_{1j} + X_{2j})}$$

Where
Σ= sum (from 1 to n)
X1j= # organisms of species j (attribute) collected at site 1 (entity).
X2j= # organisms of species j collected at site 2.
BC = Bray - Curtis coefficient of distance
| | = absolute value
J= 1 to n
N= number of species
Krebs (1999) considers Morisita' s index as the best similarity index, as follows:

$$C_\lambda = \frac{2\sum X_{ij}X_{ik}}{(\lambda_1 + \lambda_2)N_jN_k}$$

$$\lambda_1 = \frac{\sum\left[X_{ij}(X_{ij} - 1)\right]}{N_i(N_i - 1)}$$

$$\lambda_2 = \frac{\sum[X_{ik}(X_{ik} - 1)]}{N_k(N_k - 1)}$$

Where
Σ= sum
Cm= Morisita' s Index of Similarity
Xij= No. individuals of species I in sample j
Xik= No. individuals of species I in sample k
Nj= Total No. individuals in sample j
Nk= Total No. individuals in sample k
And Coefficients of Association are two types, ranging from -1 to +1 (e.g., Pearson Correlation Coefficient) and from 0 to X (e.g., χ^2) and applicable to binary and continuous data. Pearson coefficient moment correlation uses conjoint absences, the use of which is inappropriate for comparing sites and appropriate for comparing species (Clarke and Warwick, 2001)
Euclidian Distance Euclidean distance is another measure of distance that can be applied to a site by species matrix. It has been widely used in the past because it is compatible with virtually all cluster techniques.

4.2 Cluster analysis

Clustering is a straightforward method to show association data, however, the confidence of the nodes are highly dependent on data quality, and levels of similarity for cluster nodes is dependent on the similarity index used. Krebs (1989) shows that mean linkage is superior to single and complete linkage methods for ecological purposes because the other two are extremes, either producing long or tight, compact clusters respectively. There are, however, no guidelines as to which mean-linkage method is the best (Swan, 1970).

The purpose of two-way clustering (also known as biclustering) is to graphically expose the relationship between cluster analyses and your individual data points. The resulting graph makes it easy to see similarities and differences between rows in the same group, rows in different groups, columns in the same group, and columns in different groups. You can see graphically how groups of rows and columns relate to each other. Two-way clustering refers to doing a cluster analysis on both the rows and columns of your matrix, followed by graphing the two dendrograms simultaneously, adjacent to a representation of your main matrix. Rows and columns of your main matrix are re-ordered to match the order of items in your dendrogram.

Group Linkage Methods	
1. Nearest Neighbor	5. Centroid
2. Farthest Neighbor	6. Ward's Method
3. Median	7. Flexible Beta
4. Group Average	8. McQuitty's Method
Ward's is also know as Orloci's and Minimum Variance Method	

Table 1. Major types of hierarchical, agglomerative, polythetic clustering strategies

Cluster analysis can be performed using either presence–absence or quantitative data. Each pair of sites is evaluated on the degree of similarity, and then combined sequentially into clusters to form a dendrogram with the branching point representing the measure of similarity.

4.3 TWINSPAN

The TWINSPAN method (from Two Way Indicator Species Analysis, Hill 1979; Hill et al. 1975) is a very popular method among community ecologists and it was partially inspired by the classificatory methods of classical phytosociology (use of indicators for the definition of vegetation types). Two popular agglomerative polythetic techniques are Group Average and Flexible. McCune et al. (2002) recommend Ward' s method in addition. Gauch (1982) preferred to use divisive polythetic techniques such as TWINSPAN.

This method works with qualitative data only. In order not to lose the information about the species abundances, the concepts of pseudo-species and pseudo-species cut levels were introduced. Each species can be represented by several pseudo-species, depending on its quantity in the sample. A pseudo-species is present if the species quantity exceeds the corresponding cut level.

TWINSPAN is a program for classifying species and samples, producing an ordered two-way table of their occurrence. The process of classification is hierarchical; samples are successively divided into categories, and species are then divided into categories on the basis of the sample classification. TWINSPAN, like DECORANA, has been widely used by ecologists.

4.4 Indicator species analysis

Indicator species analysis is a divisive polythetic method of numerical classification applicable to large sets of qualitative or quantitative data.

This method provides a simple, intuitive solution to the problem of evaluating species associated with groups of sample units Dufrêne and Legendre's (1997). It combines information on the concentration of species abundance in a particular group and the faithfulness of occurrence of a species in a particular group. This method produces indicator values for each species in each group. These are tested for statistical significance using a Monte Carlo technique. It requires data from two or more sample units. Indicator values (range 0 for no indication to 100 for perfect indication) are presented for each species. The statistical significance of the maximum indicator value recorded for a given species is generated by a Monte Carlo test. There are many types of indicator species ranging from individual species (Dufrêne & Legendre's ,1997).

The identification of characteristic or indicator species is traditional in ecology and biogeography. Field studies describing sites or habitats usually mention one or several species that characterize each habitat. There is clearly a need for the identification of characteristic or indicator species in the fields of monitoring, conservation, and management. Indicator species add ecological meaning to groups of sites discovered by clustering, they can compare typologies derived from data analysis, identify where to stop dividing clusters into subsets, and point out the main levels in a hierarchical classification of sites.

Indicator species differ from species associations in that they are indicative of particular groups of sites. Good indicator species should be found mostly in a single group of a typology and be present at most of the sites belonging to that group (Hill, 1975).

With Classical problem in community ecology and biogeography, species are the best indicators we have for particular environmental conditions. And in long-term environmental follow-up, conservation, ecological management, researchers are looking for bioindicators of habitat types to preserve or rehabilitate.

McGeoch & Chown (1998) found the indicator value method important to conservation ecosystem because it is conceptually straightforward and allows researchers to identify bioindicators for any combination of habitat types or areas of interest, e.g. existing conservation areas, or groups of sites based on the outcome of a classification procedure.

Indicator species are species that, due to their niche preferences, can be used as ecological indicators of community types, habitat conditions, or environmental changes (McGeoch 1998, Carignan and Villard 2002, Niemi and McDonald 2004).

They are usually determined using an analysis of the relationship between the observed species presence–absence or abundance values in a set of sampled sites and a classification of the same sites (Dufrêne and Legendre 1997).

Finally this method may represent the groups of sites in the classification, in different qualitative characteristics of the ecosystem, such as habitat or community types, environmental or succession states, or the levels of controlled experimental designs.

Since indicator species analysis relates two elements, the species and the groups of sites, it can be used for gaining information on either or both. Indeed, indicator species analysis allows the characterization of the qualitative environmental preferences of the target species (for instance, when the groups are habitat types), and identifyes indicators of particular groups of sites, which can be used in further surveys.

The applications of indicator species analysis are many, including conservation, land management, landscape mapping, or design of natural reserves. Indicator species are commonly referred to as 'diagnostic species' in vegetation studies (Chytr et al. 2002).

4.5 Multi-response permutation procedures (MRPP)

Multi-response Permutation Procedure (MRPP) was introduced by Mielke, Berry, and Johnson (1976) as a technique for detecting the difference between a priori classified groups.

It turned out to be an extremely versatile data-analytic framework from which a number of applications fall out, such as the measurement of agreement, multivariate correlation and association coefcients, and the detection of autocorrelation (see Mielke & Berry, 2001 for a complete coverage of applications of the MRPP framework).

MRPP is a non-parametric method for testing the hypothesis of no difference between two or more groups of entities. The groups must be a priori. For example, one could compare species composition between burned and unburned plots to test the hypothesis of no treatment effect. Discriminant analysis is a parametric method that can be used on the same general class of questions. However, MRPP has the advantage of not requiring assumptions (such as multivariate normality and homogeneity of variances) that are seldom met with ecological community data (Bakus, 2007).

Multiple Response Permutation Procedure (MRPP) provides a test of whether there is a significant difference between two or more groups of sampling units. This difference may be one of location (differences in mean) or one of spread (differences in within-group distance). Function MRPP operates on a data frame matrix where rows are observations and responses data matrix. The response(s) may be uni or multivariate. The method is mathematically allied with analysis of variance, in that it compares dissimilarities within and among groups. If two groups of sampling units are really different (e.g. in their species composition), then average of the within-group compositional dissimilarities ought to be less than the average of the dissimilarities between two random collection of sampling units drawn from the entire population.

The MRPP method is simply the overall weighted mean of within-group means of the pairwise dissimilarities among sampling units.

The MRPP algorithm first calculates all pairwise distances in the entire dataset, then calculates delta. It then permutes the sampling units and their associated pairwise distances, and recalculates a delta based on the permuted data. It repeats the permutation step permutations times.

The function also calculates the change-corrected within-group agreement. And it also calculates classification strength which is defined as the difference between average between group dissimilarities and within group dissimilarities (Van Sickle 1997).

If the first argument data can be interpreted as dissimilarities, they will be used directly. In other cases the function treats datas observations, and uses vegdist to find the dissimilarities. The default distance is Euclidean as in the traditional use of the method, but other dissimilarities in vegdist also are available.

Function meandist calculates a matrix of mean within-cluster dissimilarities (diagonal) and between-cluster dissimilarities (off-diagonal elements), and an attribute n of grouping counts. Function summary finds the within-class, between-class and overall means of these dissimilarities, and the MRPP statistics with all weight type options and the classification strength.

The MRPP a robust alternative to the traditional normal theory based parametric tests, such as the t test and the analysis of variance. However, MRPP is not widely known to researchers, and part of the reason is that it has not been incorporated into major statistical packages.

4.6 Mantel test

The Mantel test evaluates the null hypothesis of no relationship between two dissimilarity (distance) or similarity matrices. The Mantel test is an alternative to regressing distance matrices that circumvents the problem of partial dependence in these matrices.

For example evaluating the correspondence between two groups of organisms from the same set of sample units or comparing community structure before and after a disturbance is evaluate by mantel test.

In Consequency, Two methods are available in PC-ORD: Mantel's asymptotic approximation and a randomization (Monte Carlo) method.

The Mantel Test (Mantel, 1967) compares two dissimilarity (distance) matrices using Pearson correlation. The matrices that use in this test must be of the same size (i.e., same number of rows). It would appear that the Mantel Test could be used to evaluate the internal structure in two sets of samples by comparing the two (dis)similarity matrices.

McCune and Mefford (1999) have a computer program for performing the Mantel test. After the Mantel statistic has been calculated, the statistical significance of the relationship is tested by a permutation test or by using an asymptotic t-approximation test.

The minimum number of permutations (randomizations) recommended by Manly (1997) is 1000. The Mantel and ANOSIM procedures produce similar probabilities (Legendre and Legendre, 1998).

Some of the problems associated with the Mantel test include (1) weakness in detecting spatial autocorrelation where the spatial pattern is complex and not easily modeled with distance matrices, (2) a larger number of data points may be needed for field experiments than is usually obtained, and (3) multivariate data are summarized into a single distance or dissimilarity and it is not possible to identify which variable(s) contributed the most (Fortin and Gurevitch, 2001).

The use of Partial Mantel tests can distinguish the relative contributions of the factors of a third matrix considered as covariables.

Mantel and partial Mantel tests can produce complementary information that other methods such ANOVA cannot provide and may do a better job than other method, in detecting block effects (Fortin and Gurevitch, 2001).

Thus, it is possible to distinguish the effects of spatial pattern from those of experimentally imposed treatment effects with these Mantel tests (Bakus, 2007).

4.7 The best clustering strategies

The similarity or dissimilarity measures, forming the backbone of most multivariate clustering and ordination techniques, which are favored by many ecologists, are Jaccard and Bray-Curtis.

The most popular clustering techniques in ecological studies are Group Average and Flexible group average clustering. For many of these techniques, it is impossible to choose a "best method" because of the heuristic nature of the methods (Jongman et al., 1995; Anderson, 2001).

The choice of an index must be made based on the investigator's experience, the type of data collected, and the ecological question to be answered. When comparisons are made, the Jaccard index is among the least sensitive of the similarity (or dissimilarity) indices (Jongman et al. 1995). Similarly, hierarchical agglomerative techniques (e.g., particularly Group Average) have proven to be very useful to ecologists in the construction of dendrograms.

As part of a study is needed information on plant communities and their distribution. Then the objective of best method of multivariate analysis is to:

1. Identify and describe plant communities on the ecosystem
2. Map these communities to provide a tool for ecological studies and monitoring of ecological community;

3. Characterize the interrelationships between plant communities, soil particle size, moisture availability, grazing pressure and elevation.

5. Ordination methods

Ordination is a collective term for multivariate techniques which adapt a multi-dimensional swarm of data points in such a way that when it is projected onto a two dimensional space any intrinsic pattern the data may possess becomes apparent upon visual inspection (Pielou, 1984). As mentioned in the introduction, ordination is the arrangement of samples along gradients. Indeed, ordination can be considered a synonym for multivariate gradient analysis.

Basically, ordination serves to summarize community data by producing a low-dimensional ordination space in which similar species and samples are plotted close together, and dissimilar species and samples are placed far apart.

Indirect gradient analysis

- Distance-based approaches
- Polar ordination, PO (Bray-Curtis ordination)
- Principal Coordinates Analysis, PCoA (Metric multidimensional scaling)
- Nonmetric Multidimensional Scaling, NMDS

Eigenanalysis-based approaches

- Linear model
- Principal Components Analysis, PCA
- Unimodal model
- Correspondence Analysis, CA (Reciprocal Averaging)
- Detrended Correspondence Analysis, DCA

Direct gradient analysis

- Linear model
- Redundancy Analysis, RDA
- Unimodal model
- Canonical Correspondence Analysis, CCA
- Detrended Canonical Correspondence Analysis, DCCA

5.1 Polar ordination (Bray- Curtis)

(Polar Ordination) arranges samples with respect to "poles" (also termed end points or reference points) according to a distance matrix (Bray and Curtis 1957).. These endpoints are two samples with the highest ecological distance between them (objective approach), OR two samples suspected of being at opposite ends of an important gradient (subjective approach). This procedure is especially useful for investigating ecological change (e.g., succession, recovery).

Advantages of this method is Ideal for evaluating problems with discrete endpoints.

Polar Ordination ideal for testing specific hypotheses (e.g., reference condition or experimental design) by subjectively selecting the end points Disadvantages:

This technique does not provide a general-purpose description of the community (perspective is biased) Very sensitive to outliers (by definition - "end points") Select a distance measure (usually Sorensen Index) and calculate matrix of distances (D) between all pairs of points (Beals, 1984).

In the earliest versions of PO, these endpoints were the two samples with the highest ecological distance between them, or two samples which are suspected of being at opposite ends of an important gradient (thus introducing a degree of subjectivity).

Beals (1984) extended Bray-Curtis ordination and discussed its variants, and is thus a useful reference. The polar ordination, simplest method is to choose the pair of samples, not including the previous endpoints, with the maximum distance of separation.

5.2 Principal component aanalysis

PCA was invented in 1901 by Karl Pearson (Dunn, et al, 1987) Now it is mostly used as a tool in exploratory data analysis and for making predictive models.

Principal Components Analysis is a method that reduces data dimensionality by performing a covariance analysis between factors.

It can be done by eigenvalue decomposition of a data covariance matrix or singular value decomposition of a data matrix, usually after mean centering the data for each attribute.

The results of a PCA are usually discussed in terms of component scores (the transformed variable values corresponding to a particular case in the data) and loadings (the weight by which each standarized original variable should be multiplied to get the component score (Feoli and Orl¢ci. 1992).

Principal components analysis is the basic eigenanalysis technique. It maximizes the variance explained by each successive axis.

PCA was one of the earliest ordination techniques applied to ecological data. PCA uses a rigid rotation to derive orthogonal axes, which maximize the variance in the data set.

Both species and sample ordinations result from a single analysis. Computationally, PCA is basically an eigenanalysis. The sum of the eigenvalues will equal the sum of the variance of all variables in the data set. PCA is relatively objective and provides a reasonable but crude indication of relationships.

This method is a mathematical procedure that uses an orthogonal transformation to convert a set of observations of possibly correlated variables into a set of values of uncorrelated variables called principal components.

The number of principal components is less than or equal to the number of original variables. This transformation is defined in such a way that the first principal component has as high a variance as possible (that is, accounts for as much of the variability in the data as possible), and each succeeding component in turn has the highest variance possible under the constraint that it be orthogonal to (uncorrelated with) the preceding components.

Principal components are guaranteed to be independent only if the data set is jointly normally distributed. It is sensitive to the relative scaling of the original variables.

Depending on the field of application, it is also named the discrete Karhunen–Loève transform (KLT), the Hotelling transform or proper orthogonal decomposition (POD).

Broken-stick eigenvalues are provided to help evaluating statistical significance. Principal component analysis (PCA) (ter Braak and Sˇ milauer, 1998) was used to determine the association between plant communities and environmental variables, i.e. in an indirect non-canonical way (ter Braak and Loomans, 1987).

While PCA finds the mathematically optimal method (as in minimizing the squared error), it is sensitive to outliers in the data that produce large errors PCA tries to avoid. It therefore is common practice to remove outliers before computing PCA.

However, in some contexts, outliers can be difficult to identify. For example in data mining algorithms like correlation clustering, the assignment of points to clusters and outliers is not known beforehand.

A recently proposed generalization of PCA based on Weighted PCA increases robustness by assigning different weights to data objects based on their estimated relevancy.

Although it has severe faults with many community data sets, it is probably the best technique to use when a data set approximates multivariate normality. PCA is usually a poor method for community data, but it is the best method for many other kinds of multivariate (Bakus, 2007).

5.3 Principal coordinate analysis (PCoA)

Principal Coordinate Analysis (PCoA) is a method to represent on a 2 or 3 dimensional chart objects described by a square matrix containing that contains resemblance indices between these objects.

This method is due to Gower (1966). It is sometimes called metric MDS (MDS: Mutidimensional scaling) as opposed to the MDS (or non-metric MDS). Both methods have the same objective and produce similar results if the similarity matrix f square distances are metric and if the dimensionality is sufficient.

Principle coordinates are similar to principal components in concept. The advantage of PCoA is that it may be used with all types of variables (Legendre and Legendre, 1998). Because of this, PCoA is an ordination method of considerable interest to ecologists.

Most metric (PCoA) and nonmetric MDS plots are very similar or even identical, provided that a similar distance measure is used. The occurrence of negative eigenvalues, lack of emphasis on distance preservation, and other problems are discussed in detail by Legendre and Legendre (1998) and Clarke and Warwick (2001).

One of the biggest differences between PCA and PCoA is that the variables (i.e. species) representing the original axes are projected as biplot arrows. In the bryophyte communities, these biplot arrows greatly aid in interpretation (Bakus, 2007).

In most applications of PCA (e.g. as a factor analysis technique), variables are often measured in different units. For example, PCA of taxonomic data may include measures of size, shape, color, age, numbers, and chemical concentrations. For such data, the data must be standardized to zero mean and unit variance (the typical default for most computer programs).

For ordination of ecological communities, however, all species are measured in the same units, and data should not be standardized. In matrix algebra terms, most PCAs are eigenanalyses of the correlation matrix, but for ordination they should be PCAs of the covariance matrix.

In contrast to Correspondence Analysis and related methods, species are represented by arrows. This implies that the abundance of the species is continuously increasing in the direction of the arrow, and decreasing in the opposite direction. Thus PCA is a 'linear method'.

Although the discussion above implies that PCA is distinctly different from PCoA, the two techniques end up being identical, if the distance metric is Euclidean.

Unfortunately, this linear assumption causes PCA to suffer from a serious problem, the horseshoe effect, which makes it unsuitable for most ecological data sets (Gauch 1982).

Principal Coordinates Analysis (PCoA, = Multidimensional scaling, MDS) is a method to explore and to visualize imilarities or dissimilarities of data. It starts with a similarity matrix or dissimilarity matrix (= distance matrix) and assigns for each item a location in a low-dimensional space, e.g. as a 3D graphics.

PCOA tries to find the main axes through a matrix. It is a kind of eigenanalysis (sometimes referred as "singular value decomposition") and calculates a series of eigenvalues and eigenvectors. Each eigenvalue has an eigenvector, and there are as many eigenvectors and eigenvalues as there are rows in the initial matrix.

Eigenvalues are usually ranked from the greatest to the least. The first eigenvalue is often called the "dominant" or leading" eigenvalue. Using the eigenvectors we can visualize the main axes through the initial distance matrix. Eigenvalues are also often called "latent values".

The result is a rotation of the data matrix: it does not change the positions of points relative to each other but it just changes the coordinate systems!

By using PCoA we can visualize individual and/or group differences. Individual differences can be used to show outliers.

There is also a method called 'Principal Component Analysis' (PCA, sometimes also misleadingly abbreviated as 'PCoA') which is different from PCOA.

4. Principal Coordinates Analysis (Principal Coordinates Analysis (PCoA PCoA or PCO) or PCO)
5. Maximizes the linear correlation between distance measures and distance in the ordination.
6. useful if one has only a distance (or similarity) matrix
7. The underlying model is that there a fixed number of explanatoryoriginal variables. In contrast, PCA, RA, and DCA assume that there are potentially many variables, but of declining importance.
8. One cannot easily put new points in a PCoA.
9. For Euclidean distance, PCoA= PCA
10. PCoA can be expressed as an eigenanalysis.

5.4 Factor analysis (FA)

FA and PCA (principal components analysis) are methods of data reduction. Take many variables and explain them with a few "factors" or "components".

Correlated variables are grouped together and separated from other variables with low or no correlation. Patterns of correlations are identified and either used as descriptives (PCA) or as indicative of underlying theory (FA). It process of providing an operational definition for latent construct (through regression equation).

FA and PCA are not much different than canonical correlation in terms of generating canonical variates from linear combinations of variables. Although there are now no "sides" of the equation.

For calcuting this method we should:

1. Selecting and Measuring a set of variables in a given domain
2. Data screening in order to prepare the correlation matrix
3. Factor Extraction
4. Factor Rotation to increase interpretability
5. Interpretation

Factor analysis is seldom used in ecology. Several statisticians state that it should not be used because it is based on a special statistical model.

Estimating the unique variance is the most difficult and ambiguous task in FA (McGarigal et al., 2000).

5.5 Redundancy analysis

RDA is a linear method and since it is a linear method, species as well as environmental variables are represented by arrows. In most cases, it is best to represent the two sets of arrows

in two figures for ease of display. Thus, if have a gradient along which all species are positively correlated, RDA will detect such a gradient while CCA will not. RDA can use 'species' that are measured in different units. If so, the data must be centered and standardized. But in fact, as an ordination technique, the species should not be standardized.
Redundancy analysis is the linear method of direct ordination. It also goes under the name of principal components analysis with respect to instrumental variables, which in our case are environmental variables (Sabatier et al., 1989). It is also called least-squares reduced rank regression so as to emphasize its link with multivariate regression (ter Braak & Prentice, 1988; ter Braak & Looman, 1994). In statistical textbooks on multivariate analysis, redundancy analysis is usually neglected.
RDA is useful when gradients are short. In particular, RDA may be the method of choice in a short-term experimental study. In such cases, the treatments are the explanatory variables (and are usually dummy variables). The sample ID or block might be a covariable in a partial RDA, if one wishes to factor out local effects (Bakus, 2007).
'variance explained' is actually a variance explained, and not merely inertia. Thus, variance partitioning, and interpretation of eigenvalues, are more straightforward than for CCA (Lepš & Šmilauer, 1999).
The relatively short space devoted to RDA should not be given as an indication that it is less valuable than, or inferior to, other method. It is simply used for different purposes.
Consequently, the linear method of direct ordination, redundancy analysis, can often efficiently display the interesting effects, and there is no need for methods that also work when the range of community variation is larger (such as canonical correspondence analysis) nor for unconstrained nonmetric multidimensional scaling.

5.6 Correspondence analysis (CA) or reciprocal averaging (RA)

Reciprocal averaging is also known as correspondence analysis (CA) because one algorithm for finding the solution involves the repeated averaging of sample scores and species scores (citations). It is a graphical display ordination technique which simultaneously displays the rows (sites) and columns (species) of a data matrix in low dimensional space (Gittins, 1985). Row identifiers (species) plotted close together are similar in their relative profiles, and column identifiers plotted close together are correlated, enabling one to interpret not only which of the taxa are clustered, but also why they are clustered (Bakus,2007).
Reciprocal averaging (RA) yields both normal and transpose ordinations automatically. Like DCA, RA ordinates both species and samples simultaneously. Instead of maximizing 'variance explained', CA maximizes the correspondence between species scores and sample scores.
If species scores are standardized to zero mean and unit variance, the eigenvalues also represent the variance in the sample scores (but not, as is often misunderstood, the variance in species abundance).
Since CA is a unimodal model, species are represented by a point rather than an arrow. This is (under some choices of scaling; see ter Braak and Šmilauer 1998) the weighted average of the samples in which that species occurs. With some simplifying assumptions (ter Braak and Looman 1986), the species score can be considered an estimate of the location of the peak of the species response curve.
The CA distortion is called the arch effect, which is not as serious as the horseshoe effect of PCA because the ends of the gradients are not incurved. Nevertheless, the distortion is prominent enough to seriously impair ecological interpretation (Bakus, 2007).

In other words, the spacing of samples along an axis may not affect true differences in species composition.
Gradient compression can be quite blatant in simulated data sets. The problems of gradient compression and the arch effect led to the development of Detrended Correspondence Analysis (Bakus, 2007).

5.7 Detrended correspondence analysis (DCA)

Detrended Correspondence Analysis (DCA) eliminates the arch effect by detrending (Hill and Gauch 1982). It's a series of rules that are used to reshape data to make it friendlier for analysis. Once again, primarily used for ecological data, but can be extended to anything (data simply can't contain negative values).
The reason that this technique is used is to over come the arch effect (the horseshoe effect). Found in data whenever "PCA or other distance conserving ordination techniques are applied to data which follow a continuous gradient, along which there is a progressive turnover of dominant variables." Such as in ecological succession
After ordination by a distance conserving technique and the first two axes are plotted against each other, one would find an arch shape.
DCA is another eigenanalysis ordination technique that based on reciprocal averaging (RA; Hill 1979). DCA is geared to ecological sets, is based on samples and species. DCA ordinates both species and samples simultaneously.
There are two basic approaches to detrending: by polynomials and by segments (ter Braak and Šmilauer 1998). Detrending by polynomials is the more elegant of the two: a regression is performed in which the second axis is a polynomial function of the first axis, after which the second axis is replaced by the residuals from this regression. Similar procedures are followed for the third and higher axes.
The compression of the ends of the gradients is corrected by nonlinear rescaling. Rescaling shifts sample scores along each axis such that the average width is equal to 1.
Rescaling has a beneficial consequence: the axes are scaled in units of beta diversity (SD units, or units of species standard deviations). Thus if the underlying gradient is important well known, it is possible to plot the DCA scores as a function of the gradient, and there by determine whether the species 'perceive' the gradient differently than we measure it.
The shape of the species response curves may change if axes are rescaled. Thus, skewness and kurtosis are largely artifacts of the units of measurement for which we choose to measure the environment.

5.8 Nonmetric multimentional scaling (MDS, NMDS, NMS, NMMDS)

Nonmetric Multidimensional Scaling (NMDS) rectifies this by maximizing the rank order correlation. For this proceeds at first the user selects the number of dimensions (N) for the solution, and chooses an appropriate distance metric and then The distance matrix is calculated. And initial configuration of samples in N dimensions is selected. This configuration can be random, though the chances of reaching the correct solution are enhanced if the configuration is derived from another ordination method.
And finally, the final configuration of points represents your ordination solution. The configuration is dependent on the number of dimensions selected; e.g. the first two axes of a 3-dimensional solution does not necessarily resemble a 2-dimensional solution. The stress will typically decrease as a function of the number of dimensions chosen; this function can aid in the selection of the results (Bakus, 2007).

This is why it is sometimes useful to rotate the solution (such as by the Varimax method) - although there is no theory that states that the final solution will represent a 'gradient' Other problems and advantages of NMDS will be discussed later, when comparing it to Detrended Correspondence Analysis (Bakus, 2007).

5.9 MANOVA and MANCOVA

A factorial MANOVA may be used to determine whether or not two or more categorical grouping variables (and their interactions) significantly affect optimally weighted linear combinations of two or more normally distributed outcome variables.

These parametric multivariate techniques (Multivariate Analysis of Variance and Multivariate Analysis of Covariance) are similar to ANOVA and ANCOVA MANOVA (Wilks' Lambda) and ANCOVA are advantageous in that performing multiple univariate tests can inflate the a value, leading to false conclusions (Scheiner, 2001).

MANOVA seeks differences in the dependent variables among the groups (McCune et al, 2002). Assumptions of MANOVA include multivariate normality (error effects included), independent observations, and equality of variance-covariance matrices (Paukert and Wittig, 2002). Because of these assumptions, among others, MANOVA is not often used in ecology although its use in increasing.

The power of traditional MANOVA declines with an increase in the number of response variables (Scheiner, 2001). Unequal sample sizes are not a large problem for MANOVA, but may bias the results for factorial or nested designs. Before ANCOVA is run, tests of the assumption of homogeneity of slopes need to be performed (Petratis et al., 2001).

Early attempts to develop nonparametric multivariate analysis include those of Mantel and Valand (1970). They are more complex as they handle three or more variables simultaneously. They are frequently used with the analysis of experimental studies, especially laboratory experiments.

More recently, the Analysis of Similarities was developed to compare communities or changes in communities because of pollution (Clarke, 1993). For typical species abundance matrices, an Analysis of Similarities (ANOSIM) permutation procedure is recommended over MANOVA.

Multiple analysis of variance (MANOVA) is used to see the main and interaction effects of categorical variables on multiple dependent interval variables. MANOVA uses one or more categorical independents as predictors, like ANOVA, but unlike ANOVA, there is more than one dependent variable. Where ANOVA tests the differences in means of the interval dependent for various categories of the independent(s), MANOVA tests the differences in the centroid (vector) of means of the multiple interval dependents, for various categories of the independent(s). One may also perform planned comparison or post hoc comparisons to see which values of a factor contribute most to the explanation of the dependents.

There are multiple potential purposes for MANOVA.

To compare groups formed by categorical independent variables on group differences in a set of interval dependent variables.

To use lack of difference for a set of dependent variables as a criterion for reducing a set of independent variables to a smaller, more easily modeled number of variables.

Multiple analysis of covariance (MANCOVA) is similar to MANOVA, but interval independents may be added as "covariates." These covariates serve as control variables for the independent factors, serving to reduce the error term in the model. Like other control procedures, MANCOVA can be seen as a form of "what if" analysis, asking what would happen if all cases scored equally on the covariates, so that the effect of the factors over and beyond the covariates can be isolated. The discussion of concepts in the ANOVA section also applies, including the discussion of assumptions.

5.10 Discriminate analysis

Discriminate Analysis (DA) is a powerful tool that can be used with both clusters of species data and environmental variables. It determines which variables discriminate between two or more groups, that is, independent variables are used as predictors of group membership (McCune et al., 2002). It is very similar to MANOVA and multiple regression analysis (Statsoft, Inc., 1995; McGarigal et al., 2000). Clusters can be identified by several methods using raw data: (1) constructing a dendrogram, (2) using PCA (even if you have field data) for initial visual identification of clusters, and (3) point rotation in space by rotating ordinations (i.e., rotating axes - see McCune and Mefford, 1999). If any method indicates groups or clusters of data then DA can be used. However, the number of groups is set before the DA analysis. DA finds a transform for the minimum ratio of difference between pairs of multivariate means and variances in which two clusters are separated the most and inflated the least.

DA produces two functions: (1) classification function consisting of 2 groups or clusters of points (this information can be used for prediction with probabilities) and (2) discriminate function containing environmental variables that can be used to discriminate differences among the groups.

DA differs from PCA and Factor Analysis in that no standardization of data is needed (PCA and FA need standardization because of scaling problems) and the position of the axes distinguishes the maximum distance between clusters. (Davis, 1986).

DA assumes a multivariate normal distribution, homogeneity of variances, and independent samples (Paukert and Wittig, 2002). Violations of normality are usually not fatal (i.e., somewhat non-normal data can be used). A description of the procedure to use DA with Statistica is given in the Appendix. Multiple Discriminate Analysis (MDA) is the term often used when three or more clusters of data are processed simultaneously. MDA is particularly susceptible to rounding error. Calculations in double precision for at least the eigenvalue-eigenvector routines are advisable (Green, 1979). Limitations of Discriminate Analysis are discussed by McGarigal et al. (2000) Dytham (1999).

5.11 Canonical correspondence analysis (CCA)

In ecology studies, the ordination of samples and species is constrained by their relationships to environmental variables. When species responses are unimodal (hump-shaped), and by measuring the important underlying environmental variables, CCA is most likely to be useful.

It was used to examine the relationships between the measured variables and the distribution of plant communities (Ter Braak, 1986). CCA expresses species relationships as linear combinations of environmental variables and combines the features of CA with canonical correlation analysis (Green, 1989). This provides a graphical representation of the relationships between species and environmental factors.

Canonical Correlation Analysis is presented as the standard method to relate two sets of variables (Gittins, 1985). However, the latter method is useless if there are many species compared to sites, as in many ecological studies, because its ordination axes are very unstable in such cases.

The best weight for CCA describes environment variables with the first axis shows. Species information structure using a reply CCA Nonlinear with the linear combination of variables will consider environmental characteristics of acceptable behavior characteristics of species with environment shows. CCA analysis combined with non-linear species and environmental factors shows the most important environmental variable in connection with the axes shows.

In Canonical Correspondence Analysis, the sample scores are constrained to be linear combinations of explanatory variables. CCA focuses more on species composition, i.e. relative abundance.

When a combination of environmental variables is highly related to species composition, this method, will create an axis from these variables that makes the species response curves most distinct. The second and higher axes will also maximize the dispersion of species, subject to the constraints that these higher axes are linear combinations of the explanatory variables, and that they are orthogonal to all previous axis.

Monte Carlo permutation tests were subsequently used within canonical correspondence analysis (CCA) to determine the significance of relations between species composition and environmental variables (ter Braak, 1987)

The outcome of CCA is highly dependent on the scaling of the explanatory variables. Unfortunately, we cannot know a priori what the best transformation of the data will be, and it would be arrogant to assume that our measurement scale is the same scale used by plants and animals. Nevertheless, we must make intelligent guesses (Bakus, 2007).

In CCA possible that patterns result from the combination of several explanatory variables. And many extensions of multiple regression (e.g. stepwise analysis and partial analysis) also apply to CCA.

It is possible to test hypotheses (though in CCA, hypothesis testing is based on randomization procedures rather than distributional assumptions). Explanatory variables can be of many types (e.g. continuous, ratio scale, nominal) and do not need to meet distributional assumptions.

Another advantage of CCA lies in the intuitive nature of its ordination diagram, or triplot. It is called a triplot because it simultaneously displays three pieces of information: samples as points, species as points, and environmental variables as arrows (or points).

If data sets are few, CCA triplots can get very crowded then should be separate the parts of the triplot into biplots or scatterplots (e.g. plotting the arrows in a different panel of the same figure) or rescaling the arrows so that the species and sample scores are more spread out. And we can only plotting the most abundant species (but by all means, keep the rare species in the analysis).

Noise in the species abundance data set is not much of a problem for CCA (Palmer, 1988). However, it has been argued that noise in the environmental data can be a problem (McCune 1999). It is not at all surprising that noise in the predictor variables will cause noise in the sample scores, since the latter are linear combinations of the former.

It is probably obvious that the choice of variables in CCA is crucial for the output. Meaningless variables will produce meaningless results. However, a meaningful variable that is not necessarily related to the most important gradient may still yield meaningful results (Palmer, 1988).

If many variables are included in an analysis, much of the inertia becomes 'explained'. Any linear transformation of variables (e.g. kilograms to grams, meters to inches, Fahrenheit to Centigrade) will not affect the outcome of CCA whatsoever.

There are as many constrained axes as there are explanatory variables. The total 'explained inertia' is the sum of the eigenvalues of the constrained axes. The remaining axes are unconstrained, and can be considered 'residual'. The total inertia in the species data is the sum of eigenvalues of the constrained and the unconstrained axes, and is equivalent to the sum of eigenvalues, or total inertia, of CA. Thus, explained inertia, compared to total inertia, can be used as a measure of how well species composition is explained by the variables. Unfortunately, a strict measure of 'goodness of fit' for CCA is elusive, because the arch effect itself has some inertia associated with it (Bakus, 2007).

The ordination diagrams of canonical correlation analysis and redundancy analysis display the same data tables; the difference lies in the precise weighing of the species (ter Braak & Looman, 1994; Van der Myer, 1991)

One of limitations to CCA is that correlation does not imply causation, and a variable that appears to be strong may merely be related to an unmeasured but 'true' gradient. As with any technique, results should be interpreted in light of these limitations (McCune, 1999).

5.12 Multiple regression (MR) (multiple linear regression)

The mechanics of testing the "significance" of a multiple regression model is basically the same as testing the significance of a simple regression model, we will consider an F-test, a t-test (multiple t's) and R-sqrd. However, unlike simple regression where the F & t tests tested the same hypothesis, in multiple regression these two tests have different purposes. R-sqrd is still the percent of variance explained but is no longer the correlation squared (as it was with in simple linear regression) and we will also introduce adjusted R-sqrd. When considering a multiple regression (MR) model the most common order to interpret things consists of first looking at the R-sqrd, then testing the entire model by looking at the F-test, and finally looking at each individual coefficient individually using the t-tests. NOTE: The term "significance" is a nice convenience but is very ambiguous in definition if not properly specified. Thus when taking this class you should avoid simply saying something is significant without explaining (1) how you made that determination, and (2) what that specifically means in this case. You will see from the examples that those two things are always done. If you cannot do that then any time you use the word "significant" you are potentially hurting yourself in two ways; (1) you won't do well on the quizzes or exams where you have to be able to be more explicit than simply throwing out the word "significant", and (2) you will look like a fool in the business world when somebody asks you to explain what you mean by "significant" and you are stumped. Remember if you can't explain your results in managerial terms than you do not really understand what you are doing.

In order to show the relationship between biotic (principal component axes) and abiotic factors (environmental factors), a multiple regression type analysis is used. Multiple regression solves simultaneously normal equations and produces partial regression coefficients. Partial regression coefficients each give the rate of change or slope in the dependent variable for a unit of change in a particular independent variable, assuming all other independent variables are held constant. MR is not considered by some statisticians as a multivariate procedure because it includes only one dependent variable (Paukert and Wittig, 2002).

The objective of multiple regression is to determine the influence of independent variables on a dependent variable, for example, the effect of depth, sediment grain size, salinity, temperature, and predator density on the population density of species.

The parameters are estimated by the least-squares method, that is, minimizing the sum of squares of the differences between the observed and expected response (Jongman et al., 1995)

Normally component loadings (e.g., scores in PCA) suggest which variables are most important. However, with species abundances the only variable in some ordinations, one must use other techniques to attempt to suggest what may have produced the gradients for Axis 1, 2, 3, and so forth.

Univariate analyses such as the Spearman rank correlation coefficient are not as ecologically realistic as multivariate analyses such as multiple regression because some variables are correlated and there are interaction effects between variables (Jongman et al., 1995).

The highest standardized partial regression coefficients (positive or negative) suggest the most important factors (e.g., sediment size, predator density, etc.) in controlling the population density of species X. Significance tests and standard errors then can be calculated from the data.

Multiple regression has many potential problems such as the type of response curve, error distribution, and outliers that may unduly influence the results (Jongman et al., 1995). Multiple regression variables may be highly correlated, therefore, examine the correlation coefficients first (i.e., run a multiple correlation analysis between variables) to exclude some of them before doing multiple regressions.

Multiple regression generally should employ a maximum of 6 variables. Legendre and Legendre (1998) suggest a stepwise procedure for reducing numerous variables. This involves a process of alternating between forward selection and backward elimination (Kutner et al., 1996).

Lee and Sampson (2000) took ordination scores, representing a gradient in fish communities, and regressed them against a group of environmental variables and time. Many of scientist study on multiple regression and computer program (Such as Davis, 1986; and Sokal and Rohlf, 1995)

In multiple regression, it is typical to include quadratic terms for explanatory variables. For example, if you expect a response variable to reach a maximum at an intermediate value of an explanatory variable, including this explanatory variable AND the square of the explanatory variable may allow a concave-down parabola to provide a reasonable fit.

This is an analogous situation to multiple regression: the multiple r^2 or 'variance explained' increases as a function of the number of variables included.

Both multiple regression and CCA find the best linear combination of explanatory variables, they are not guaranteed to find the true underlying gradient (which may be related to unmeasured or unmeasurable factors), nor are they guaranteed to explain a large portion of variation in the data.

5.13 Path analysis

In statistics, path analysis is used to describe the directed dependencies among a set of variables. This includes models equivalent to any form of multiple regression analysis, factor analysis, canonical correlation analysis, discriminate analysis, as well as more general families of models in the multivariate analysis of variance and covariance analyses (MANOVA, ANOVA, ANCOVA).

Path analysis is a straightforward extension of multiple regression. Its aim is to provide estimates of the magnitude and significance of hypothesised causal connections between sets of variables. This is best explained by considering a path diagram.
Path analysis is an extension of multiple linear regression, allowing interpretation of linear relationships among a small number of descriptors (Legendre and Legendre, 1998).
This method was originally developed by Sewall Wright in which he introduced the concept of a path diagram. It handles more than one dependent variable and the effects of dependent variables on one another (Mitchell, 2001).
Path analysis assumes a normal distribution of residuals, additive and linear effects, inclusion of all important variables, that residual errors are uncorrelated, and that there is no measurement error.
Path analysis was developed as a method of decomposing correlations into different pieces for interpretation of effects (e.g., how does parental education influence children's income 40 years later?). Path analysis is closely related to multiple regression; you might say that regression is a special case of path analysis.
Some people call this stuff (path analysis and related techniques) causal modeling. The reason for this name is that the techniques allow us to test theoretical propositions about cause and effect without manipulating variables. However, the causal in causal modeling refers to an assumption of the model rather than a property of the output or consequence of the technique. That is, people assume some variables are causally related, and test propositions about them using the techniques.

5.14 Canonical correlation analysis (CVA)

Canonical variate analysis (CVA) is a widely used method for analyzing group structure in multivariate data. It is mathematically equivalent to a one-way multivariate analysis of variance and often goes by the name of canonical discriminate analysis. Change over time is a central feature of many phenomena of interest to researchers. This dissertation extends CVA to longitudinal data. It develops models whose purpose is to determine what is changing and what is not changing in the group structure. Three approaches are taken: a maximum likelihood approach, a least squares approach, and a covariance structure analysis approach. All methods have in common that they hypothesize canonical variates which are stable over time.
The maximum likelihood approach models the positions of the group means in the subspace of the canonical variates. It also requires modeling the structure of the within-groups covariance matrix, which is assumed to be constant or proportional over time.
In addition to hypothesizing stable variates over time, one can also hypothesize canonical variates that change over time. Hypothesis tests and confidence intervals are developed. The least squares methods are exploratory. They are based on three-mode PCA methods such as the Tucker2 and parallel factor analysis. Graphical methods are developed to display the relationships between the variables over time.
Stable variates over time imply a particular structure for the between-groups covariance matrix. This structure is modeled using covariance structure analysis, which is available in the SAS package Proc Calis.
Canonical Variate Analysis is a special case of Canonical Correlation Analysis (Jongman et al., 1995). It is also described as a type of linear discriminate analysis (McGarigal et al., 2000). The set of environmental variables consists of a single nominal variable defining the classes.

CVA is usable only if the number of sites is much greater than the number of species and the number of classes. Many ecological data sets cannot be analyzed by CVA without dropping many species, thus CVA is not used much in ecology.
Instead, they usually give the impression that there is only one such hypothesis, and therefore only one statistical technique is needed-Hotelling's canonical variate analysis (CVA).
Most discussions of CVA are restricted almost entirely to a description of the underlying mathematical theory, computing directions, and perhaps an example of the computations. Very little is usually said about the logic of the method, so that the reader is unable to judge for himself whether the method described can actually be used to test the hypothesis of interest to him.
Even when CVA is appropriate, several prominent sources have recommended misleading interpretations of the statistics computed in CVA.
Perhaps because of these deficiencies, many behavioral scientists have concluded incorrectly that CVA has few or no valid and important uses in the behavioral sciences. The use originally proposed by Hotelling has been rejected by most behavioral scientists.

6. Ordination and classification methods in various ecology studies

The more applied an ecological study is, the more the emphasis is on the effects on ecological communities of particular environmental factors, for example pollutants, management regimes, and other human-induced changes in the environments. (ter Braak, 1994).
Correspondingly, the statistical analysis should not 'just' show the major variation in the species assemblage, but focus on the effects on the variables of prime interest. Applied studies thus call for direct methods of ordination, typically with a very limited number of (qualitative or quantitative) environmental variables (ter Braak, 1994)
The range of community variation in an applied study tends to be quite small compared to that in the early ordination studies (e.g. Whittaker, 1965; Hill & Gauch, 1980).
Determining which factors control the distribution patterns of plant communities remains a central goal in ecology Classification assumes from the outset that the species assemblages fall into discontinuous groups, whereas ordination starts from the idea that such assemblages vary gradually.
Ordination compares sites on their degree of similarity, and then plots them in Euclidian space, with the distance between points representing their degree of Similarity. Ordination techniques include principal components analysis (PCA), detrended correspondence analyses (DCA), and nonmetric multidimensional scaling (NMS)
Ordination (or inertia) methods, like principal component and correspondence analysis, and clustering and classification methods are currently used in many ecological studies (Zare Chahouki et al., 2009, Anderson, 2002; Gauch et al., 1977; Orloci, 1975; Whittaker (ed.), 1967; Legendre & Legendre, 1988).
Ordination methods can be divided in two main groups, direct and indirect methods. Direct methods use species and environment data in a single, integrated analysis. Indirect methods use the species data only (Jongman et al., 1987). In contrast, if a unimodal response model is assumed, the relationships are unimodal. Unimodal relations are usefully summarized by their modes or - more conveniently - weighted averages (ter Braak & Looman, 1986), so that a sensible coefficient for the species $\times$ environment table is the weighted average. The other

way round, if a correlation coefficient is chosen, the implied response model is linear or approximately linear and if the chosen coefficient is the weighted average than the implied response model is unimodal (i.e. if the true model is bimodal, the ordination will fail, and if the true model is linear, the ordination will be inefficient). Assumptions about the response model and the choice of coefficients to use in secondary tables are thus interrelated.

New methods of exploring differences among groups include the nonparametric, recursive classification, and regression tree (CART). It is used to classify habitats or vegetation types and their environmental variables (McCune et al., 2002). It produces a top-to-bottom visual classification tree that undergoes a "pruning" or optimization process. CART is used to generate community maps, wildlife habitats, and land cover types in conjunction with a GIS (see p. 195 in Chapter 3). Another multivariate technique is Structural Equation Modeling (unfortunately termed SEM), a merger of factor analysis and path analysis (McCune et al. (2002). It is a method of evaluating complex hypotheses (e.g., effects of abiotic factors on plant species richness).

With multiple causal pathways among variables. It requires the initial development of a path diagram. It is an analysis of covariance relationships, effectively limited to about 10 variables. See Shipley (2000), McCune et al. (2002), Pugesek et al. (2002) and Bakus (2007).

7. References

Anderson, M.J. (2001). A new method for non-parametric multivariate analysis of variance. Austral. Ecol. 26:32-46.

Alisauskas, R.T. (1998). Winter range expansion and relationships between landscape and morphometrics of midcontinent Lesser Snow Geese. Auk 115(4):851-862.

Anderson, C.W., Barnett, V., Chatwin, P.C. and El-Shaarawi A.H. (2002). Quantitative Methods for Current EnvironmentalIssues. Springer-Verlag, New York.

C.T. Bastian, S.R. Koontz and Menkhaus D.J. (2001). The Impact of Forward Contract Information on the Fed Cattle Market: An Experimental Investigation into Mandatory Price Reporting, UW Agricultural and Applied Economics Seminar. August 31, 2001. (Presented by Bastian).

Bakus Gerald J. (2007). Quantitative Analysis of Marine Biological Communities Field Biology and Environment. WILEY-INTERSCIENCE, A John Wiley & Sons, Inc., Publication, 453p.

Beals, E.W. (1973). Ordination: Mathematical elegance and ecological naivete. J. Ecol. 61:23–35.

Bray, J.R. and Curtis J.T. (1957). An ordination of the upland forest communities of southern Wisconsin.Ecol. Monogr. 27:325-349.

Carignan, V. and Villard M. (2002). Selecting indicator species to monitor ecological integrity: a review. Environ. Monitor. Assess. 78: 45-61.

Chytr, M. (2002). Determination of diagnostic species with statistical fidelity measures. J. Veg. Sci. 13: 79–90.

Clarke, K.R. (1993). Non-parametric multivariate analyses of changes in community structure. Aust. J. Ecol.18:117-143.

Clarke, K.R. and Warwick R.M. (2001). Change in Marine Communities: An Approach to Statistical Analysis and Interpretation. 2nd edition. PRIMER- E, Plymouth Marine Laboratory, Plymouth, U.K.

Clifford, H.T. and Stephenson W. (1975). An introduction to numerical classification. Academic Press, New York, pp. 229.

Cochran, W.G. (1977). Sampling Techniques. Wiley, New York.

Davis, J.C. (1986). Statistics and Data Analysis in Geology. Wiley, New York.

Diserud, O.H. and Aagaard K. (2002). Testing for changes in community structure based on repeated sampling. Ecology, 83(8): 2271-2277.

Dufrene, M. and Legendre P. (1997). Species assemblages and indicator species: the need for a flexible asymmetrical approach. Ecol. Monogr. 67:345-366.

Dunn, C. P., and Stearns F. (1987). Relationship of vegetation layers to soils in southeastern Wisconsin forested wetlands. Am. Midl. Nat. 118:366-74.

Dytham, C. (1999). Choosing and Using Statistics: A Biologist's Guide. Blackwell Publishing, Williston, VT.

Ferson, S., Downey P., Klerks P., Weissburg M., Kroot S.I., S. Jacquez O., Ssemakula J., Malenky R. and Anderson K. (1986). Competing reviews, or why do Connell and Schoener disagree? Am. Nat., 127: 571-576.

Feoli, E., and Orl¢ci L. (1992). Thre properties and interpretation of observations in vegetation study. Coenoses 6:61-70.

Fortin, M-J. and Gurevitch J. (2001). Mantel tests: Spatial structure in field experiments. pp. 308-326 in: Scheiner S.M. and J. Gurevitch (eds.). Design and Analysis of Ecological Experiments. Oxford University Press, Oxford.

Gauch, H.G. (1977). ORDIFLIX—A flexible computer program for four ordination techniques: weighted averages, polar ordination, principal components analysis, and reciprocal averaging. In: Ecology and Systematics, Cornell University, Ithaca, N.Y.

Gauch, H.G., J. (1982). Multivariate Analysis in Community Ecology. Cambridge University Press, New York.

Gittins, R. (1985). Canonical analysis: a review with applications in ecology. Springer-Verlag, Berlin.

Gotelli, N.J. and Ellison A.M. (2004). A Primer of Ecological Statistics. Sinauer Associates, Sunderland, Maine

Gower, J. C. (1966). Some Distance Properties of Latent Root and Vector Methods used in Multivariate Analysis. Biometrika 53, 325-338.

Green, R.H. (1979). Sampling Design and Statistical Methods for Environmental Biologists. John Wiley and Sons, New York.

Greig-Smith, P. (1983). Quantitative Plant Ecology, 3rd Edition. Blackwell Scientific Publications, London, 359 pp.

Green, R.H. 1989. Power analysis and practical strategies for environmental monitoring. Environ. Res. 50:195-205.

Hill, M.O., Bunce R.G.H. & Shaw M.V. (1975). Indicator species analysis, a divisive polythetic method of classification, and its application to survey of native pinewoods in Scotland. Journal of Ecology, 63: 597-613

Hill, M.O. (1979). TWINSPAN - a FORTRAN Program for Arranging Multivariate Data in an Ordered Two-way Table by Classification of the Individuals and Attributes. Ithaca: Section of Ecology and Systematic, Cornell University.

Hill M.O. & Gauch H.G. (1980). Detrended correspondence analysiss, an improved ordination technique. Vegetatio, 42: 47-58

Jaccard, P. (1901): Etude comparative de la distribution florale dans une portion des Alpes et du Jura. Bulletin de la Socie´ te´ Vaudoisedes SciencesNaturelles, 37: 547–579.

Jongman, R.H.G., ter Braak C.J.F. and Van Tongeren O.F.R. (1987). Data analysis in community and landscape ecology. Cambridge University Press, Cambridge, UK.

Jongman, R.H.G., Ter Braak, C.J.F. and van Tongeren O.F.R. (eds.) (1995). Data Analysis in Community and Landscape Ecology. Cambridge University Press, Cambridge.

Kent, M. (2006). Numerical classification and ordination methods in biogeography. Progress in Physical Geography, 30: 399-408

Krebs, C.J. (1989). Ecol. Method. Harper Collins, New York.

Krebs, C.J. (1999). Ecological Methodology. Harper & Row, New York.

Kutner, M.H., Nachtscheim, C.J., Wasserman, W. and Neter J. (1996). Applied linear statistical models. WCB/McGraw-Hill, New York.

Lee, Y.W. and Sampson D.B. (2000). Spatial and temporal stability of commercial ground fish assemblages off Oregon andWashington as inferred from Oregon travel logbooks. Canadian J. Fish. Aqua. Sci., 57:2443-2454.

Legendre, P. & Gallagher E.D. (2001). Ecologically meaningful transformations for ordination of species data. Oecologia, 129: 271–280

Legendre, P. and Legendre L. (1998). Numerical Ecology. 2nd Edition.Elsevier, Amsterdam.

Lepš, Jan & Šmilauer P. (1999). Multivariate Analysis of Ecological Data Faculty of Biological Sciences, University of South Bohemia Ceské Budejovice,110pp

Ludwig, J.A., Reynold, J.F. (1988). Statistical Ecology. Wiley, New York, 337pp

Niemi, G.J. and McDonald M.E. (2004). Application of ecological indicators. - Annu. Rev. Ecol. Evol. Syst. 35: 89–111.

Manly, B.F.G. (1997). Randomization, Bootstrap and Monte Carlo Methods in Biology. Chapman and Hall, London.

Mantel, N. 1967. The detection of disease clustering and generalized regression approach. Cancer Res. 27:209-220.

Mantel, N. and Valand R.S. (1970). A technique of nonparametric multivariate analysis. Biometrics 26:547–558.

McCune, B. and Mefford M.J. (1999). PCORD, Multivariate Analysis of Ecological Data, Version 4. MjM Software Design, Gleneden Beach, Oregon, USA.

McCune, B., Grace J.B. and Urban D.L. (2002). Analysis of Ecological Communities. MjM Software Design, Gleneden Beach, Oregon.

McGarigal, K., Cushman, S. and Stafford S. (2000). Multivariate Statistics for Wildlife and Ecology Research. Springer-Verlag, New York

McGeogh, M.A. (1998). The selection, testing and application of terrestrial insects as bioindicators. Biol. Rev. 73: 181–201.

McGeoch, M.A. and Chown. S.L. (1998). Scaling up the value of bioindicators. Trends Ecol. Evol. 13: 46-47.

Mielke, P.W. & Berry K.J. (2001). Permutation methods: A distance function approach. New York: Springer-Verlag.

Mielke, P. W., Berry K.J. & Johnson E. S. (1976). Multi-response permutation procedures for a priori classications. Communications in Statistics- Theory and Methods, 5: 1409-1424

Mitchell, R.J. (2001). Path analysis: Pollination. pp. 217–234 in: Scheiner S.M. and J. Gurevitch (eds.). Design and Analysis of Ecological Experiments. Oxford University Press.

Petratis, P.S., Beaupre S.J. and Dunham A.E. (2001). ANCOVA: Nonparametric and Randomization Approaches. pp. 116–133 in: Scheiner S.M. and Gurevitch (eds.). Design and Analysis of Ecological Experiments. Oxford University Press, Oxford.

Pugesek, B., Tomer A. and von Eye A. (eds.) (2002). Structural Equations Modeling: Applications in Ecological and Evolutionary Biology Research. Cambridge University Press, Cambridge, U.K.

Paukert, C.P. and Wittig T.A. (2002). Applications of multivariate statistical methods in fi sheries. Fisheries 27(9):16-22

Palmer, M.W. (1988). Fractal geometry: a tool for describing spatial patterns of plant communities. Vegetatio 75:91–102.

Odum, E.P. and Barrett G.W. (2005). Fundamentals of Ecology. Thomas Brooks/Cole, Belmont, CA.

Orloci, L. (1975). Multivariate Analysis inVegetation Research. Junk, The Hague.

Rohlf, F.J. (1995). BIOM: A Package of Statistical Programs to Accompany the Text Biometry. Exeter Software, Setauket, New York.

Scheiner, S.M. (2001). Theories, hypotheses, and statistics. pp. 3-13 in: Scheiner S.M. and J. Gurevitch (eds.). Design and Analysis of Ecological Experiments. Oxford University Press, Oxford.

Shipley, B. (2000). Cause and Correlation in Biology. Cambridge University Press, Cambridge, U.K.

Statsoft, Inc. 1995. STATISTICA for Windows. 2nd edition. Tulsa, OK.

Stiling, P. (2002). Ecology: Theories and Applications. Prentice Hall, Upper Saddle River, N.J.

Swan, J.M.A. (1970). An examination of some ordination problems by use of simulatedvegetational data. Ecology 51: 89–102.

Ter Braak C.J.F. & Looman C.W.N. (1986). Weighted averaging, logistic regression and the Gaussian response model. Vegetatio, 65: 3-11

Ter Braak, C.J.F. (1987). The analysis of vegetation-environment relationships by canonical correspondence analysis. Vegetatio, 69:69-77.

Ter Braak, C. J. F. (1994). Canonical community ordination. Part I: Basic theory and linear methods. Ecoscience 1 (2), 127-140.

Ter Braak C.J.F. & Šmilauer P. (1998). CANOCO Reference Manual and User's Guide to Canoco for Windows. Microcomputer Power, Ithaca, USA. 352 pp.

Thompson, S.K. (2002). Sampling. John Wiley & Sons, New York. Second Edition.

Turner MG. (1989). Landscape ecology: the effect of pattern on process. Ann. Rev. Ecol. Syst. 20:171-197.

Turner MG, RH Gardner and RV O'Neill (2001). Landscape Ecology in Theory and Practice: Pattern and Process. Springer, New York.

van der Meer, J. Heip C.H., Herman P.J.M., Moens T. and van Oevelen D. (2005). Measuring the Flow of Energy and Matter in Marine Benthic Animal Populations. pp. 326–408 in: Eleftheriou. A. and A. McIntyre (eds.). 2005. Methods for Study of Marine Benthos. Blackwell Science Ltd., Oxford, UK.

Van der Plogeg, S.W.F. & Vlijm L. (1978). Ecological evaluation, nature conservation and land use planning with particular reference to the methods used in the Netherlands.Biol.Consero.14:197-221.

Van Sickle, J. (1997). Using mean similarity dendrograms to evaluate classifications. Journal of Agricultural, Biological, and Environmental Statistics, 2:370-388.

Whittaker R.H. (1965). Dominance and diversity in land plant communities. Science 147: 250–260.

Whittaker, R.H. (1967). Gradient analysis of vegetation. Biol. Rev. 42:207-264.

Zare Chahouki, M. A. Azarnivand H., Jafari M. & Tavili A. (2009). Multivariate Statistical Methods as a Tool for Model_Based Prediction of Vegetation Types, Russian Journal of Ecology, 41(1): 84-94.

Zare Chahouki, M.A. (2006). Modeling the spatial distribution of plant species in arid and semi-arid rangelands. PhD Thesis in Range management, Faculty of Natural Resources, University of Tehran, 180 p. (In Persian).

Schluter. D. and Grant P.R. (1982). The distribution of *Geospiza difficilis* on Galapagos islands: test of three hypotheses. Evolution 36:1213-1226

24

Accumulation of Bio Debris and Its Relation with the Underwater Environment in the Estuary of Itanhaém River, São Paulo State

Fresia Ricardi Branco[1], Sueli Yoshinaga Pereira[1], Fabio Cardinale Branco[2] and Paulo R. Brum Pereira[3]
[1]*D. Geologia e Recursos Naturais, Instituto de Geociências, Universidade Estadual de Campinas -UNICAMP, Campinas, SP*
[2]*Environmentality, São Paulo, SP*
[3]*Instituto Florestal, SP*
Brazil

1. Introduction

The following chapter presents some considerations regard to the interrelation between the bioclastic accumulations (plants and animals) in a coastal estuarine environment, and the influence of sub surface water in its preservation in the geological record. To it, the plants and zoo debris accumulations and the aquifers in the Itanhaém River basin, south coast of São Paulo State, Brazil were studied and characterized.

The estuary Itanhaém River is located on the southern coast of São Paulo State as part of Santos - Peruibe Quaternary Coastal Plain. It has characteristic of tropical environments, mainly by the exuberance of the Atlantic Rain Forest (Brazilian biome characterized by its high biodiversity which grows associated with the Brazilian coastal portion), largely in primary conditions of conservation, by the Forest of Restinga or dense ombrophyle forest of low altitudes (forest characterized by a high biodiversity that develops in sandy soils of the coastal regions of Brazil) also mostly in primary conditions for conservation and by the Mangrove, which is considered one of the most well protected area of the state.

The Coast of São Paulo has two climatic determiners that influence the dynamic of atmospheric water: (1) the line of the Tropic of Capricorn crosses the region, and (2) the presence of the Serra do Mar with altitudes (maximum of about 1000 meters) very close to the coast of the Atlantic Ocean, and covered by dense tropical vegetation, the Atlantic Forest. Thus, polar air masses and tropical, have interactivity with elements of the relief. The altitudes near the coast, the moisture of ocean evaporation, the evapotranspiration, the slope orientation are the factors responsible to make this region as one of the most humid in Brazil.

The characterization of climate dynamics is based on information from weather stations used in the regional analysis presented in Table 1.

The weather stations of São Paulo city was taken as a reference to the availability of climate information for long period to high altitudes. The station of the Research Center of the Itanhaém river (CPeRio), despite the short period of observation, is a complete station, situated in the Itanhaém river basin and was considered as a local reference.

Stations (name)	Period (years)	Latitude (degrees and Min.)	Longitude (degrees and Min.)	Altitude (m)
Santos	1961-1990	23°56′	45°20′	14
São Paulo	1961-1990	23°30′	46°37′	792
Ubatuba	1961-1990	24°26′	45° 06′	08
Paranaguá	1961-1990	25°31′	48°31′	05
Iguape	1977-1986	24°42′	47°33′	03
CPeRIO*	2002-2004	24°10′	46°47′	02

Table 1. Weather stations used for the study (the south coastal region of the state). * Climate Station of the Research Center of Itanhaém River.

The Central and Southern coast region of São Paulo has average annual temperatures between 19.6 and 21.8 ° C. The highest values are found during January, February and March, while the lowest values are during June, July and August. The absolute maximum temperatures on record occurred in the summer and their values ranged between 39.0 and 40.0 ° C. The absolute minimum temperatures were recorded in the winter months and ranged from 2.4 to 6.4 ° C.

The rates of average annual total rainfall range between 1,932.2 and 2,080.8 mm. The rainiest quarter is January, February and March with rates accrued between 741.6 and 816.2 mm, making a percentage in relation to the total annual rates between 35.9 and 42.2%. The less rainy quarter is in June, July and August with combined rates between 271.9 and 283.0 mm, present in relation to the total annual rates ranging between 13.6 and 14.1%. The maximum daily rainfall recorded ranged between 206.7 and 289.9 mm.

The average values of atmospheric pressure range between 867.7 and 968.6 hPa. The highest values are observed in July (range between 917.2 and 1,018.8 hPa), while the lowest values are found in November (776.0 hPa) and February (1,009.9 hPa).

In the area, the total annual evaporation ranges from 736.9mm 968.6 mm. The maximum values occur in the summer with rates ranging between 83.4 and 93.3mm (January), while the lowest values are observed in winter, where rates range from 45.6 (June) and 70.4mm (July).

Annual insolation values range between 1,227.6 and 1,494.1 hours. Each month, the highest values are found in summer and range between 142.7 and 149.6 hours (January), while the lowest values are found in the spring and ranges between 73.0 and 88.7 hours (September).

The monthly average rates of cloudiness range between 6.3 and 6.4 tenths. The highest rates are found in late winter and spring, ranging between 6.1 and 7.3 tenths, while the lowest values are found in autumn / winter, ranging between 5.2 and 6.0 tenths.

The average annual rates of relative humidity range between 80.0 and 84.0%. The highest rates are observed in late summer with rates between 83.0 and autumn winter further south in the area, with rates of 86%, however, the lower values in the central portion, in winter, with a rate of 75%.

The hydrographic basin of the Itanhaém River is the second largest on the coast of the state of São Paulo, with an area of 930 km², mostly located in the municipality of Itanhaém, in the southern coast of the state (Camargo et al., 2002; Ricardi-Branco et al., 2009, 2011). The Quaternary coastal plain of the Itanhaém is limited by the city of Peruibe to the southwest and the city of Mongaguá to the northeast, extending for some 50 km, it stretches 15-16 km to the west of the base of the Serra do Mar mountain range (Suguio and Martin, 1978;

Camargo et al., 2002; Amaral et al., 2006). The upper portion of the basin is contained within the Serra do Mar State Park with its sources well preserved. In the middle and lower portions of the basin, the anthropogenic influence is more intense. The main variations in the basin over time are thus the consequence of the patterns of variation of the tides and the intensity of the rain (Camargo et al., 2002). The tidal range is relatively low (less than 1 m, Giannini et al., 2009), but these microtides influence all of the river flowing across the narrow plain.

The main rivers draining the basin are the Itanhaém, formed by the junction of Branco and Preto rivers, and the Mambu, Aguapeú and Guaraú rivers, all tributaries of the Branco River. Studies have been conducted in Itanhaém, Preto and Branco rivers as described next.

The Itanhaém river is formed by the confluence of Branco and Preto rivers, presenting estuarine features and drainage area of approximately 26 km^2 (Camargo et al., 2002). On its banks are still present the Restinga Forest and Mangrove Forest, as well as ornamental species introduced in recent years (e.g. *Terminalia*). Its course was modified in the first half of the twentieth century (the 40s) by opening a channel (artificial cut levees, joining two different points of the river after a meander), then about 1 m, now more than 100 meters wide. The opening of this artificial channel caused the abandonment of a portion of the meander, less active today and known as Acima River.

The Branco River originates in the Serra do Mar and runs much of its extension in the Precambrian terrains (Camargo et al., 2002) until it reaches the coastal plain. It is the largest river of the basin with 68 km in length. At its mouth there is a mixture of salt and fresh water under the influence of saline. In its margins are present the Atlantic Forest and the Restinga Forest, both considerably well preserved, as well as plantations (banana, cassava, etc.). Finally the Preto River has about 40 km long, originating in the Serra do Mar, although its greatest extent is located in the Quaternary plain (Camargo et al., 2002). It has little influence of salt water, mostly in its estuary with the Itanhaém River and its banks covered by the Atlantic and the Restinga Forests.

2. Bio debris taphonomy

2.1 Plant debris accumulations

Besides being a classical sedimentary environment of fossilization, the macro plant debris accumulations were studied in the middle and coastal portion for being the ones with best conditions for the biodebris accumulations. These accumulations were studied in detail by Ricardi-Branco et al. (2009, 2011) employing the original analytical methodologies for taphonomic studies. In addition, a systematic survey of the vegetation along the river basin was developed to draw comparisons with the leaves found in the accumulations. Despite the accumulation of plant debris are to be composed of leaves, twigs and fronds, preference was given to the study of angiosperm leaves, since they are the most abundant elements of the accumulations. The branches were small, often in the larger petioles. Finally, in the accumulations studied, none skeletal remains of invertebrates or vertebrates were found.

The survey was conducted during the years of 2002-2009, and were preferentially studied and / or accompanied with nine sites with leaves accumulations (Fig. 1), selected because they are in facies with the best potential for preservation and therefore fossilization. These accumulations are spatially distributed on the Quaternary plain as follows:

- in the levees in the interior of the plain on the banks of the Branco River (P1);
- in the middle portion of the plain in point bars - Inclined heterolithic structure desposits (IHS deposits) of the Preto River (P2 - P5);

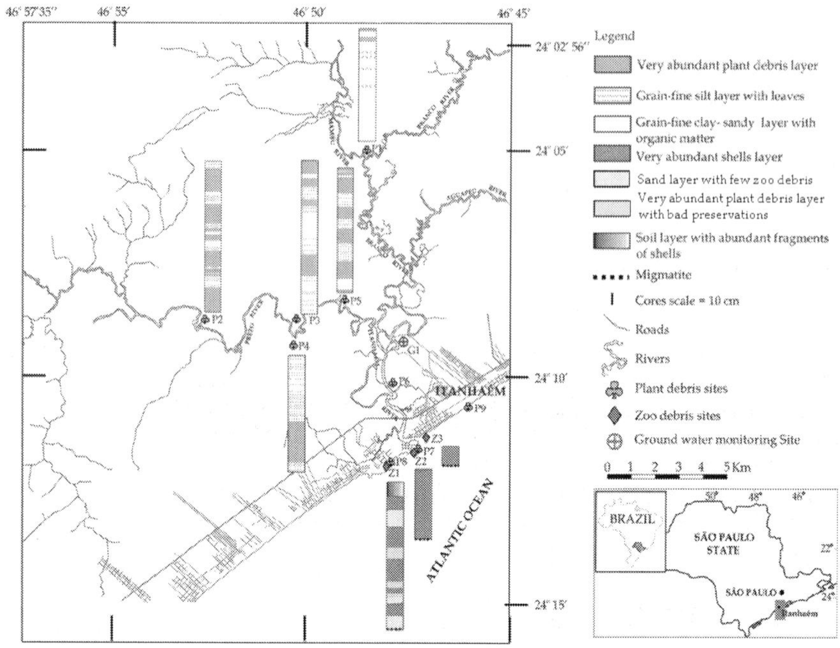

Fig. 1. Location map of the sites studied to plant debris (P1-P8) and to zoo debris (Z1-Z3). It was also plotted the cores developed to P1 to P5, in order to illustrate the constant presence of leaves packages in the margins of those rivers that drain the Itanhaém plain. The cores developed in Sites Z1 - Z3 show the studies to the shell accumulations collected in the coast of the basin.

- in the lower portion of the plain within the Mangrove of the Itanhaém River (P6) and
- in the coastal portion (Sibratel beach - Poço de Anchieta - P8, Mouth of the Itanhaém River - P7 and Itanhaém - Suarão beach -P9). These were followed up and did not have a detailed study of their components, because they are very scattered and still, quickly fragmented by wave action.

With the purpose of carrying out the identification of angiosperm leaves, the methodology described by Ricardi-Branco et al. (2009) was applied. First, the leaves collect were included in morphotypes (Figs. 3, 4 and 5) and later compared to the leaves of plants collected on the banks of the study sites. The results of plant debris analysis clearly showed their differential distribution in the quaternary plain of the Itanhaém River. So within the plain (Branco River, Site P1), in the middle portion (Preto River, Sites P2-P5), and the low portion (Itanhaém River Site P6) the accumulation of leaves were characterized as parautochthonous, primarily reflecting the surrounding vegetation of the study sites. In the Preto and Branco rivers the species corresponded to the Restinga Forest, and in the Mangrove of the Itanhaém River the leaves corresponded to that forest, the latter being the least biodiverse accumulation of those studied. In the coastal portion (Itanhaém River Mouth, Site P7; Sibratel Beach, Site P8 and Suarão Beach/Itanhaém, Site P9), plant debris were characterized as allochthonous composed by various elements from the mainland and from the ocean as well, and these were the only ones with plant and animal origins Ricardi-Branco et al. (2009).

Fig. 2. Sites of plant debris studies in the Itanhaém River Basin. 1, panoramic view of the Itanhaém River near its mouth; the arrow indicates the location of the Mangrove; 2 and 3, Site P1 overview and hand collection of leaves for study, 4 and 5, Preto River, meanders where layers of plant debris are deposited; 4 low tide and 5 high tide. 6 and 7, outcropping level of plant debris and 7 level of plant debris in core developed at P2.

Morphotipe	C1 - B1	C2 - B2	C5 - B3	C6 - B4	Ac1 - B5	Ac2 - B6	P1 - B7
Venation	Campt	Campt	Campt	Campt	Acro	Acro	Parallelo
Leaf shape	Obovate	Elliptic	Elliptic	Oblong	Obovate	Elliptic / Ovate	Ovate / Oblong
Base Shape	Cuneate	Cuneate	Cuneate	Convex	Cuneate	Cuneate/Convex	Convex
Apex shape	Acuminate	Acuminate	Convex	Rounded	Convex	Straight	Straight
Margin shape	Entire	Entire	Entire	Entire	Entire	Entire	Entire
Quantity	160	121	7	140	25	72	5
Related family	Fabaceae *(Inga)*	Lauraceae	Myrtaceae	Fabaceae *(Senna)*	Melastomataceae	Melastomataceae	Poaceae
Image	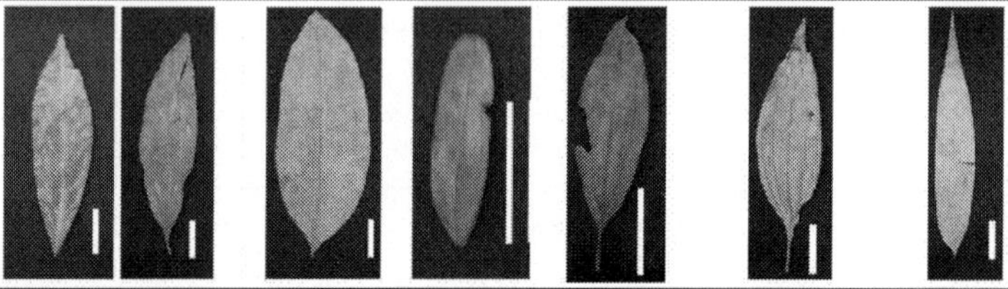						

Fig. 3. Morphotypes of the Branco River and their botanical affinity. Site P1. Campt = camptodromous; Acro= Acrodromous and Parallelo = Parallelodromous.

Morphotipe	C1-P1	C2-P2	C7-P3	C8-P4	Cl1-P5	Cr3-P6	P1 -
Venation	Campt	Campt	Campt	Campt	Campt	Craspedo	Parallelo
Leaf shape	Elliptic	Elliptic	Oblong	Ovate	Elliptic	Elliptic	Elliptic
Base Shape	Complex	Cuneate	Cuneate	Complex	Convex	Convex	Cuneate
Apex shape	Straight	Convex	Convex	Acuminate	Convex	Acuminate	Straight
Margin shape	Entire	Entire	Entire	Entire	Entire	Crenate	Entire
Quantity	4	48	25	81	19	73	3
Related family	Lauraceae	Fabaceae *(Inga)*	Rubiaceae	Fabaceae *(Inga)*	Clusiaceae	Sapindaceae	Poaceae
Image	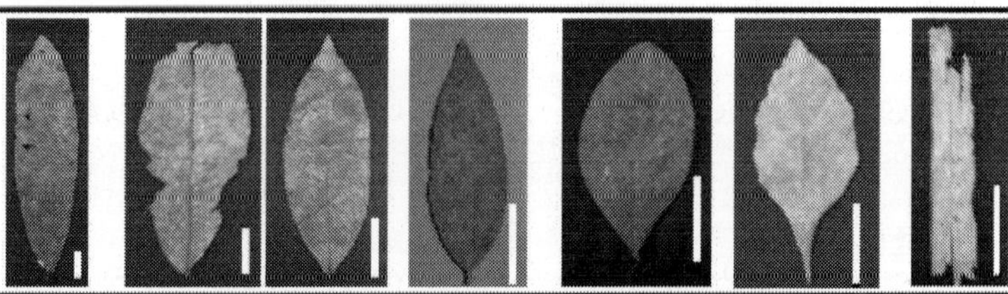						

Fig. 4. Morphotype of the Preto River and its botanical affinity. Site P2. Campt = camptodromous; and Parallelo = Parallelodromous.

At the same time, shallow cores were conducted with aluminum tubes of 7cm, manually buried during low tide up to 2m depth (Ricardi-Branco et al., 2009 , 2011), in three of the sites monitored for assembly of plant debris to follow the changes and evolution of the accumulations selected after the burying. The localities selected were: one on the margin of the Branco River inside the quaternary plain (P1), four on the margins of the Preto River in IHS deposits, in the middle portion of the basin (P2-P5) and the third in the mangroves in the lower portion of the plain. The best results regarding the preservation of the leaves were found in the cores of the Preto River (Fig. 1). The accumulations of plant debris along the meandering course of the Preto River to be characterized in greater detail with the aid of geophysical methods, were very frequent and associated with virtually all the meanders of the river (Fig. 4), featuring IHS deposits with an estuarine system with micro seas, where the monthly variations of the tide deposit thin laminated layers or tidal coupplets the cause of variations in river level. So the more thick layers (around 20 cm) of plant debris observed in the study correspond to climatic events that defoliated the trees of the Restinga Forest (Ricardi-Branco et al., 2011).

Morphotipe	M1	M2
Venation	Campt	Campt
Leaf shape	Obovate	Elliptic
Base Shape	Cuneate	Convex
Apex shape	Convex	Convex
Margin shape	Entire	Entire
Quantity	8	30
Related family	Rizophoraceae (*Rizophora*)	Combretacaea (*Laguncularia*)
Image	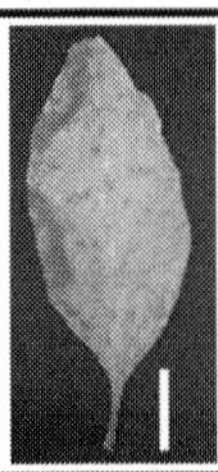	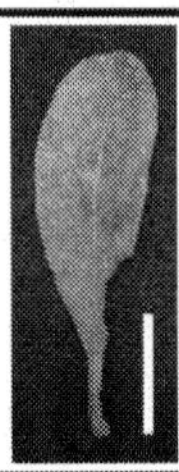

Fig. 5. Morphotype of the Itanhaém river and its botanical affinity. Site P6. Campt = camptodromous.

2.2 Zoo debris accumulations

From the second half of 2005 until the first half of 2006, natural deposits of zoo debris were characterized and described in the coastline portion of the Itanhaém River Basin (Zincone et al., 2007), showing the change in relative sea level during the late Holocene. These are the deposits where the presence of shell accumulations were detected (Fig. 1):

- Cibratel beach – beachrock of Poço de Anchieta (Site - Z1);
- Conchas beach in Costão de Paranambuco (Site - Z2) and
- the beach of Cabras Island located at a distance of approximately 140 m from the mainland (Site - Z3).

Pits were made in every site of the studied area up to the contact with the accumulations with the basement (Migmatite). The profiles were described and collected approximately 5 kg of each level has been taken to the laboratory for study and dating by C^{14}.

Site - Z1. Cibratel beach - beachrock of Poço de Anchieta (24 ° 12 '05.1 "S / 46 ° 48' 41.3" W): consisting of composed zoo debris cemented accumulation with approximately 1.30 m depth (Table 2). This kind of accumulation can be considered a beachrock, chenier or conglomerates of beach, usually consisting of quartz grains bound by carbonate cement, located in the intertidal zones and containing biogenic intraclasts (Suguio, 1992, 2004).

Site - Z2. Conchas beach, (24 ° 11 '46.7 "S / 46 ° 48' 06.01"): the accumulation is associated with a small shore stuck on the left margin of the coast, situated between the walls of migmatites, that features 13 meters wide in low tide and 22.5 meters wide at the posterior border with the coast. During the sampling collect, excavations were conducted at three different points of this beach. The deposit of bioclasts (mollusc shells) has 60 cm depth up to the base where it borders the migmatite.

Site Z3- Cabras Island (24 ° 11 '38.1 "S / 46 ° 47' 33.1" W): as the site is located on the Cabras Island at approximately 140m far from the Pescadores beach, the access occurs during low tide when the tide allows crossing the canal. In this place, three trenches were made in order to know the thickness of the shells deposit (Fig. 1 and tables 2 and 3).

The systematic composition of the deposits studied can be seen in Table 2. As the composition of sites Z1, Z2 and Z3 is very similar (Fig. 6).

Sites Bioclasts	Site - Z1 Beachrock (Anchieta beach) (24°12'05.1" S. / 46°48'41.3" W)	Site -Z2 Conchas beach (24°11'46.7" S / 46°48'06.1" W)	Site Z3 Cabras island (24° 11'38.1" S / 46°47'33.1"W)
Molusca - Bivalvia	*Anachis* sp., *Perna perna*, *Tellina sp.*, *Tivela* sp. and *Trachycardium* sp.	*Anadara* sp., *Macoma* sp., *Tivela* sp. and *Trachycardium* sp.	*Anadara* sp., *Brachidonte* sp., *Mactra* sp., *Mytella* sp., *Ostrea* sp., *Perna perna*, *Pisania* sp., *Tivela* sp. and *Trachycardium* sp.
Molusca - Gastropoda	*Colissella* sp., *Diodora* sp., *Dorsanum* sp., *Hastula* sp., *Littorina* sp., *Olivancillaria* sp., *Thais* sp. and microgastropds	*Colissella* sp., *Crepidula* sp., *Hastula* sp., *Janthina* sp., *Littorina* sp., *Pisania* sp. e *Thais* sp. and microgastropods	*Colissella* sp., *Diodora* sp., *Hastula* sp., *Olivancillaria* sp., *Pisania* sp. and *Thais* sp.
Polychaeta -	Tubes	Tubes	Tubes
Artropoda	Chelae of crabs and barnacles	Chelae of crabs and barnacles	Chelae of crabs and barnacles
Vertebrata			Fish vertebrae
Fragments	Shells and sea urchin spines	Shells and sea urchin spines	Shells and sea urchin spines

Table 2. Systematic composition of the studied deposits with zooclasts.

Fig. 6. Zoo debris found in points Z1, Z2 and Z3. 1. *Trachycardium* sp.; 2, *Tivela* sp.; 3, *Perna perna*; 4, *Thais* sp.; 5, fragments of gastropods, 6, chelicerae crab, 7, tubes of polychaetes and 8, fragments of spines of sea urchins. Scale bar= 1, 2, 3, 4, 5 and 6 = 1 cm. 7 = 5mm e 8 = 0.5 mm

Regard to the taphonomy in points Z1, Z2 and Z3 the bioclasts (shells) are disconnected, densely packed and poorly chosen, since they vary in size and polymodal distribution, which does not observe a preferential orientation of bioclasts. The polymodal distribution indicates activity a turbulent flow during the formation of fossiliferous assemblage as in the case studied here where the waves remove the accumulations daily, the paleoecological fashion is polytypic (Holz & Simões, 2002). Regarding the presence of taphonomic signatures, the most obvious ones found were drilling marks by predation, mainly gastropods, and fragmented valves. These signatures were observed in the valves of mollusks collected at all study sites.

The age obtained for the beachrock (Table 3) indicates that although it is a deposit partially eroded and emerged, it reflects the environmental conditions during the deposition in intertidal zone (Angulo et al., 2002). On the other hand it is interesting to mention that as stated above, the composition in terms of bioclasts among sites Z1, Z2 and Z3, despite having different age composition is very similar, showing that at 1,185 cal yr BP the malacofauna of the Itanhaém coastal region was very similar to that currently inhabits the coast.

Age Sample/Depth (cm)	Conventional age (^{14}C years AP)	Calibrated Age (2 σ) (cal yr BP)	Historical Age (yr AD/BC)	Samples Numbers ^{14} Cena - USP
Z2 -Conchas beach 30 - 50	570+/- 70	510 - 665	766 – 1,024 AD	975 - CENA 547
Z1 - Anchieta Beach 92 - 98	1,130 +/- 70	930 – 1,185	1,286 – 1,442 AD	976 – CENA 549

Table 3. Ages obtained from deposits of shells Z1 and Z2.

3. River – Ground waters

3.1 Regional aspects

The aquatic environment presents a great sensitivity in regions where the contacts between seawater and fresh water outflow are related. This estuarine environment that sustains unique living beings is a region that has developed under a dynamic that involves relief, tides, river flows and climate dynamics of masses of tropical and polar fronts.

The river belongs to the Southeast Atlantic Hydrographic Macro-region of Santos Coastline and has an area of 102.57 km^2. The formation rivers, Branco River and Preto River, have areas of 411.66 km^2 and 426.46 km^2, respectively. Yet, this report estimates of average flows for long period of 4.10 m^3/s, 12.9 m^3/s, 18.9 m^3/s, and minimum flows ($Q_{7,10}$) of 1.0 m^3/s, 3.15 m^3/s and 4.63 m^3/s for the Itanhaém , Preto and Branco rivers, respectively (Comitê de Bacias Hidrográficas/ Baixada Santista [CBH-BS], 2000).

The region is influenced by the tide and consequent saline intrusion. According to Camargo et al. (2002) seawater at high tide penetrates through the river channel of the Itanhaém and Branco rivers, occurring predominantly in the lower river basin. The average values reported are of 2,252.6 ppm salinity in the Itanhaém River. The Branco River has a small influence of saline water with maximum records of 200.0 ppm salinity and no influence on the Preto River (average of 4.3 ppm and maximum value of 20.0 ppm). However, the influence of pressure of the tide is noticed in the entire length of the rivers mentioned, with the lifting and lowering of the water level. It is interesting to note the differences in the aquatic environments between rivers Itanhaém and Branco, and the Preto River. In the latter one, its water is dark attributed by the presence of organic matter. The clear water of the other rivers may be attributed to poor soils, the large slope of the terrain, turbulence and the small number of primary producers, consumers and decomposers (Camargo et al., 2002).

The contrast of the steep topography of the Serra do Mar and the coastal plain provide different aquifer systems with dynamics of complex flow and interactivity of groundwater. The Crystalline Aquifer System is mainly composed of granites, gneisses, migmatites, schists and phyllites, Açungui Group (granites and schists varied), the Embu Complex and Coastal Complex (heterogeneous migmatites), and eluvial - colluvium continental sediments, bordering such rocks and being their alteration products. It is characterized by its large regional extension, fractured, free to semi-confined, heterogeneous, discontinuous and anisotropic. In the study area are characterized by the Serra do Mar, the hills that exist in the plains, and in the subsurface with basement of the sediments deposited which makes up the Cenozoic Coastal System.

The Cenozoic Coastal System is composed by sands interbedding with clay and siltstone and characterized by having a limited extension, granular, free and semi-confined, discontinuous, heterogeneous and anisotropic, usually with shallow water level. It is composed by Holocene sediments of Mangrove and swamp (sand and clay) by Santos Formation of fluvial sediments - lagoon (sand and clay), coastal marine sediments (sands), and reworked coastal marine sediments and coastal marine sediments (sands) of Cananéia Formation.

3.2 Groundwater monitoring

An area of six thousand square meters (6,000 m^2) was selected for installation of eight monitoring wells for groundwater, located in the meander on the left bank of the Itanhaém River (Fig. 7), in order to understand the influence of tides on the river and in groundwater. Information from local people revealed the existence of at least three aquifers in this area: the first, superior, unconfined, with 7 meters deep; the second, confined between two clay layers, between 8 and 20 meters deep; and the third, below the clay layer from 20 meters deep.

Thus, monitoring wells were installed in the first two aquifers, four in the shallower aquifer and three in the second aquifer, whose main characteristics are shown in Table 4. Conductivity Hydraulic essays (slug tests) were made and the results and hydraulic conductivity (Hvorslev, 1951, as cited in Fetter, 2001) and transmissivity are presented in Table 4 and Fig. 7 shows the study area and location of the monitoring wells.

Monitoring wells	Coordinates Latitude/ Longitude	Topographic elevation (m)	Well depth (m)	K (m/s)	T (m^2/s)	Aquifer
P-1	24° 8' 44" 46° 47' 10"	2.020	6	1.6×10^{-4}	8.8×10^{-4}	unconfined
P-2	24° 8' 43" 46° 47' 9"	1.854	6	1.6×10^{-4}	8.8×10^{-4}	unconfined
P-3	24° 8' 44" 46° 47' 9"	2.109	6	1.2×10^{-4}	6.6×10^{-4}	unconfined
P-4	24° 8' 44" 46° 47' 9"	2.005	6	1.3×10^{-4}	7.2×10^{-4}	unconfined
PII-1	24° 8' 44" 46° 47' 8"	2.261	21	1.8×10^{-6}	4.6×10^{-6}	confined
PII-2	24° 8' 45" 46° 47' 7"	1.926	21	5.1×10^{-6}	1.2×10^{-5}	confined
PII-3	24° 8' 45" 46° 47' 10"	1.452	21	1.8×10^{-4}	4.4×10^{-4}	(semi) confined

Table 4. Localization of monitoring wells. K - hydraulic conductivity (m/s); T - transmissivity (m^2/s)

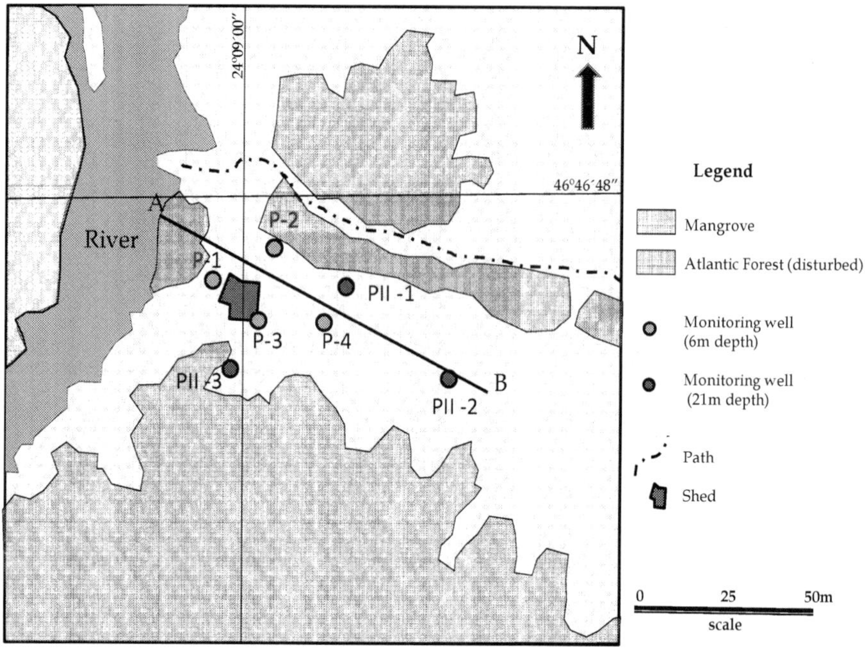

Fig. 7. Localization of monitoring wells and cross section (A – B) in the study area

The study area shows organic sandy soil in the upper portion, with fine-grained, and black color due to the presence of organic matter (up to half a meter deep). Then the soil becomes sandy with fine texture and a yellowish color (up to 1 meter deep), becoming dark gray after 2.5 meters deep. In the depth of 3.5 meters it presents light gray sand with the presence of plant debris to a depth of 7.0 meters. A 1- meter- thickness layer of silty clay is found in this interval, and between depths of 8 to 18.5 meters is found fine sand, with light gray color, but without the presence of plant debris. In the interval of 18.5 to 20 meters deep is found light gray fine sand with remains of calcareous shells. Finally, below 20 meters was found a silty clay layer, compact, with dark gray color (Fig. 8).

Two aquifers were identified: (a) the shallowest, whose bottom is impervious is at 7 meters deep (silty clay) and (b) the confined aquifer located between two silty clay layers, and with a 12-meter- thickness. So the silty clay layers correspond to the aquifuges of the site.

The shallow aquifer, unconfined and homogeneous, presents hydraulic conductivity between 1.6 x 10-4 to 1.2 x 10-4 m/s (transmissivity between 8.8 x 10-4 and 6.6 x 10-4 m2/s). The deeper aquifer presents lower hydraulic conductivity, between 4.6 x 10-6 and 1.2 x 10-5 m/s (transmissivity between 4.6 x 10-6 and 1.2 x 10-5 m2/s). The PII-3 well presented values of K and T intermediate (1.8 x 10-4 m/s to 4.4 x 10-4 m2/s) indicating the possibility of semi-confinement.

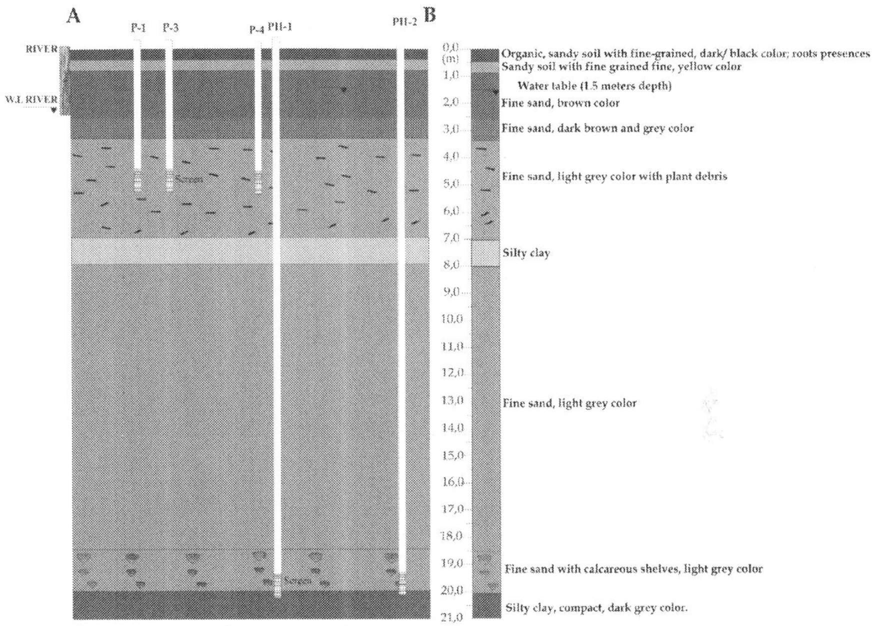

Fig. 8. Cross section (A - B) and geologic profile description (modified from Batista Filho, 2006)

3.3 Groundwater dynamics under tidal fluctuations

Two field campaigns were conducted to monitoring the dynamics of groundwater during the daily fluctuations of the tide. The first campaign was conducted between March-April 2005 (03/28 to 03/04/2005 - rainy season), and the second in August 2005 (08/20 to 08/26/2005 - less rainy season). The monitoring consisted of continuous measurements (every 30 minutes) of water levels from monitoring wells, pH and electrical conductivity (E.C.) during the fluctuation of the tide.

Table 5 shows the average results of monitoring of the river water (water level, pH, Eh and E.C.). The waters sampled during the rainy season are quite distinct from the waters of the less rainy period, as in the differences in tidal height variation as the values of pH, Eh and E.C.

The fluctuation measured of the level of river water is related to the location point of the tide gauge, located four kilometers from the Itanhaém river, downstream of the study area. The other physical-chemical parameters were measured in the study area, where the monitoring was conducted. The tide measured in CPeRio takes 1 hour and 30 minutes to reach the study area, through the river channel.

The differences of the variation in height of water level indicates a higher or lower influence of fresh water from the river (higher flow), combined with the fluctuation of the tide; in the rainy season there is a decrease in the difference between the maximum and minimum values of height. During the dry season, there is a reduction of river flow and consequently greater influence of the tides and greater height variation of the water level monitored.

RIVER	Water level (m) *		pH		Eh (mV)		E.C. (mS/m)	
	03/28 to 03/04/05 Rainy season	08/20 to 08/26/05 Less rainy season	03/28 to 04/04/05 Rainy season	08/20 to 08/26/05 Less rainy season	03/28 to 04/04/05 Rainy season	08/20 to 08/26/05 Less rainy season	03/28 to 04/04/05 Rainy season	08/20 to 08/26/05 Less rainy season
Average	0.51	0.74	7.5	6.4	224	260	27	827
Minimum	-0.19	-0.20	6.5	3.4	-169	182	10	186
Maximum	1.08	1.22	8.2	6.9	112	357	278	1100

Table 5. Average results (maximum and minimum) found in the monitoring of water level parameters, pH, Eh and E.C. of the river, from 03/28 to 03/04/2005 and 08/20 to 08/26/2005. * Change the water level in CPeRio (located four kilometers from the study area).

During the rainy season (January to March), the waters are less acidic, reducing environment and lower concentrations of salts and less influence of seawater in the study area.

During the less rainy season (June to August), the waters become more acidic due to the increasing concentration of humic acids. There is the increase of the concentration of dissolved salts, the lower river flow and higher facility of inflow of sea water in the channel at great distances.

In both monitoring periods the accumulated rainfall was 18.8 mm in 03/28 to 03/04/2005 and 3.0 mm in 08/20 to 08/26/2005. The rains in the first monitoring occurred on 03/28/2005 (7.4 mm) and on 03/04/2005 (11.4 mm); in the second monitoring, the rainfall occurred on 08/21/2005 (0.2 mm), 08/23/2005 (0.2 mm), 08/24/2005 (0.8 mm), 08/25/2005 (1.6 mm) and 08/26/2005 (0.2 mm).

Potentiometric surface maps for the two aquifers, shallow and confined, were prepared at various times of the rise and fall of the tide. Monitoring the values of pH, Eh and E.C. were made in conjunction with measurements of water level.

Figures 9 to 12 presents three moments each monitored day representing the dynamics of groundwater in accordance with the fluctuations of the tide. The days selected, representative of each period (wet and dry) were 03/30/2005 and 08/20/2005, at the same times of water sampling for physical and chemical analysis.

The potentiometric surface maps of the day 03/30/2005 of the shallow aquifer at high tide presents S-N direction, SE- NW and W-E, with hydraulic head higher in PM 3, followed by a hydraulic head of PM 1 (Fig. 9). This flow pattern seems to indicate the pressure of high tide in the aquifer near the river in the opposite movement of freshwater into the river (effluent). During the falling tide and low tide the flow of groundwater toward the river has lower hydraulic gradients, but realize also that the flow directions of S-N and SW-NE are predominant. In the deep aquifer (Fig. 10) the direction of groundwater flow is the opposite (NE-SW and NE-SE), however, the distribution of hydraulic head at various times of tide fluctuations indicates vertical oscillations of the aquifer potentiometry. At high tide, the hydraulic gradient is higher and decreases with low tide.

On 08/20/2005 (Figs. 11 and 12) the movement of groundwater in the shallow aquifer has lower hydraulic gradient and the predominant SW-NE direction. The influence of the tides (high and low) in the groundwater is noted with the increase in hydraulic head at high tide. In the deep aquifer groundwater, the direction of water flow is NE-SW (toward the river).

The higher hydraulic head moves at high tide and at the beginning of the fall (Fig.12). In the deep aquifer, the groundwater flows toward the river (NE-SW), the hydraulic head increases with the rising tide and decrease during the fall of the tide.

The dynamics of groundwater flow in the shallow aquifer has a direct influence of fluctuations of the tide. In March 2005, the potentiometric maps presented a strong northeast flow direction that is intensified at high tide, with greater hydraulic gradients than at low tide. The same behavior is observed in August 2005. Yet the prevailing flow direction to the northeast, but with lower gradients. The hydraulic heads, however, present higher values in August than in March.

In the deep aquifer, the groundwater flow has a less variable behavior, flowing to the southwest toward the river. However, there is a shift in the area of higher hydraulic head from north to northeast.

The major direction northeast flow of groundwater in the shallow aquifer is consistent with the situation of the river channel of the Itanhaém river, which greatly influences the movement of groundwater. In the case of deep aquifer the behavior is different and has SW direction toward the river. There is influence of the tide, but to a lesser extent than the shallow aquifer.

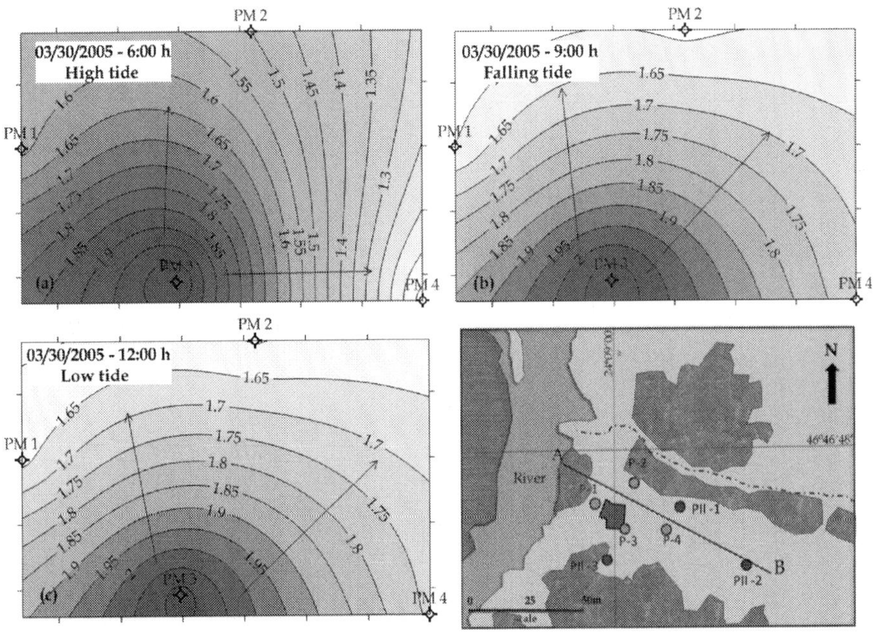

Fig. 9. Potentiometric surface maps representing groundwater flows in the following fluctuations: (a) high tide; (b) falling tide and (c) low tide - Shallow Aquifer (03/30/2005).

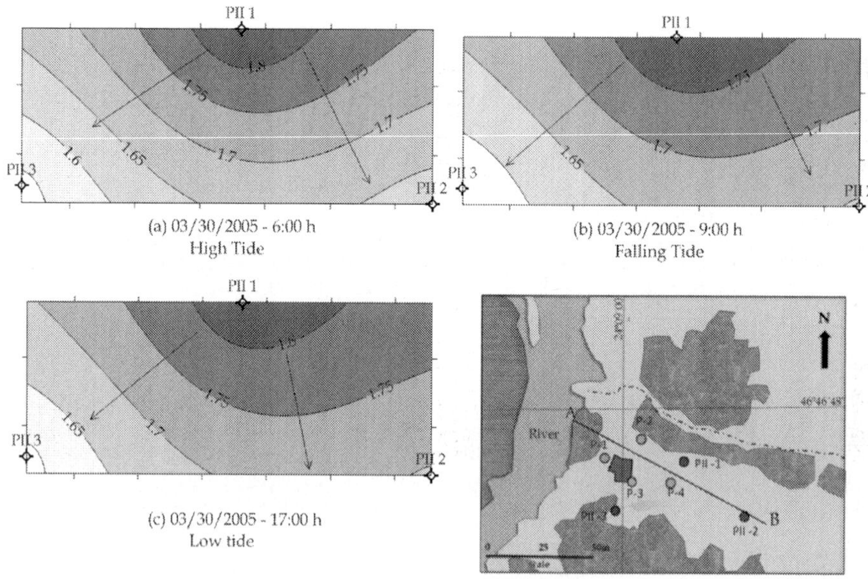

Fig. 10. Potentiometric surface maps representing groundwater flows in the following fluctuations: (a) high tide; (b) falling tide and (c) low tide - Deep Aquifer (03/30/2005).

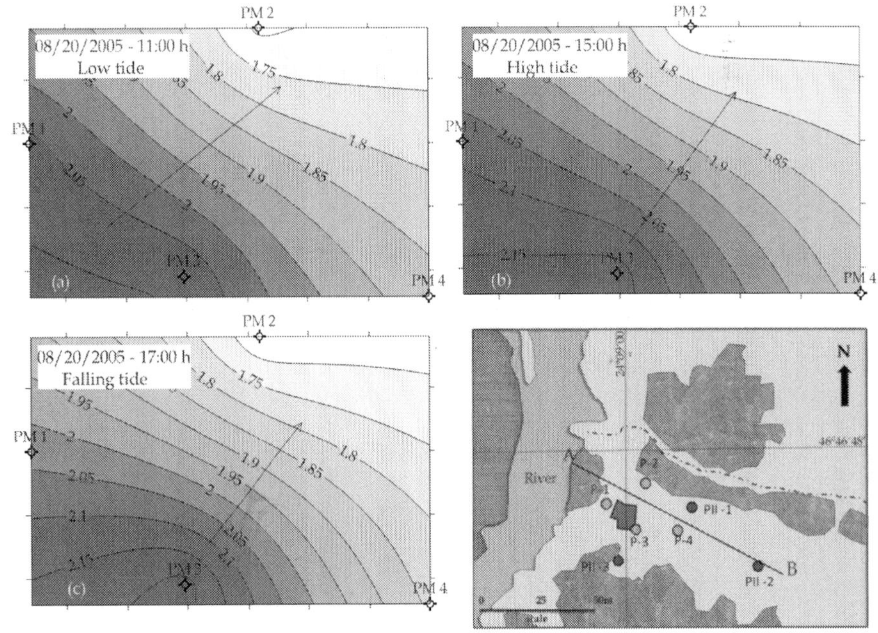

Fig. 11. Potentiometric surface maps representing groundwater flows in the following fluctuations: (a) low tide; (b) high tide and (c) falling tide - Shallow Aquifer (08/20/2005).

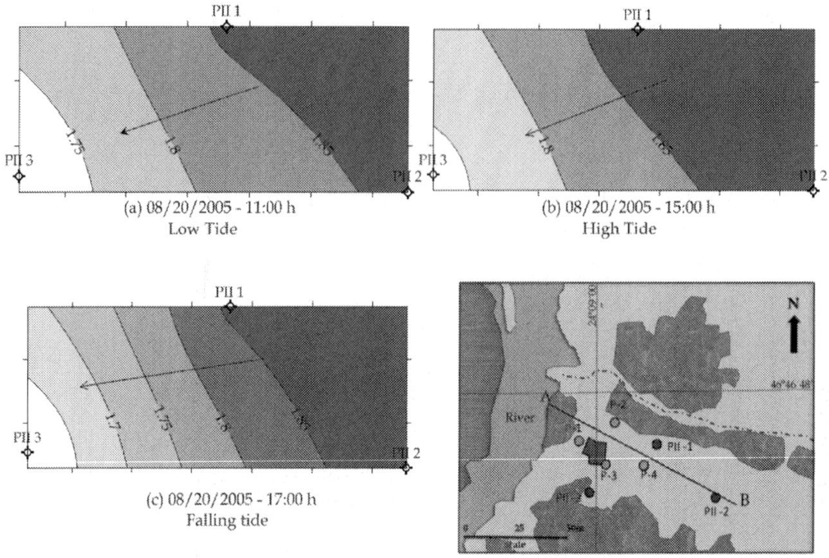

Fig. 12. Potentiometric surface maps representing groundwater flows in the following fluctuations: (a) high tide; (b) falling tide and (c) low tide - Deep Aquifer (03/30/2005).

3.4 Hydrochemistry

Samples of river water and groundwater were collected at different times of oscillation of the diary tide to study the variations of physical and chemical composition. The parameters pH, Eh and E.C. were determined by portable field equipment *in loco*, and HCO_3^- by titrations. The chemical analysis F^-, Cl^-, Br^-, N-NO_3^-, N- NO_2^- ,SO_4^{2-} e PO_4^{2-} and PO_4^{2-} were performed by ion chromatography and the elements Na, K, Si, Mg, Ca, Sr, Co, Mn, Cu, Zn, Pb, Al, Ba, Cd, Ni, Fe and Cr by ICP-OES (Inductively Coupled Plasma Optical Emission Spectrometry).

The test results are presented in tables 6 and 7. The elements Cd, Co, Cr, Cu, Ni and Pb showed no concentrations above the detection limit of the equipment.

In 03/30/2005 the river presented alkaline pH (7.5 to 7.8) and Eh between 227 and 259 mV. The values of E.C. in this period ranged between 12 and 13 mS/m. In general, low concentrations are presented for all elements analyzed. The element zinc is not present in these waters, and the concentrations of phosphate and barium are very low. The waters have low concentrations of calcium (average of 4.4 mg/L), magnesium (2.2 mg/L), sodium (13.4 mg/L), potassium (1.8 mg/L), fluoride (0.05 mg/L) chloride (23.5 mg/L), sulfate (3.78 mg/L), nitrate (0.26 mg/L) and bicarbonate (16.9 mg/L). The concentrations of elements between the different moments of the tide almost did not change with the fluctuation of the tide in the river in the study area.

The waters of the shallow aquifer had acid pH (between 4.8 and 7.3, average 5.4), Eh between 222 and 271 mV and low E.C. values (between 7 and 8 mS/m). The concentrations of all elements of water from each well hardly varied over the period monitored, however the well PM-2 showed higher concentrations of calcium ions and bicarbonate and lower concentrations of potassium and sulfate in relation to other wells (PM-1, PM-3 and PM-4). In general, calcium ions, magnesium, sodium and potassium ranged from 1.1 to 4.0 mg/L, 0.4 to 0.8 mg/L, 5.5 to 8.3 mg/L and 0.8 to 2.8 mg/L, respectively. Bicarbonate concentrations ranged from 6.7 to 10.1 mg/L, sulfate between 4.8 and 8.3 mg/L, and chloride between 11 and 13 mg/L. There is presence of fluoride (up to 0.06 mg/L), bromide (from 0.46 to 0.88 mg/L), aluminum (0.082-0.400 mg/L), barium (0.017-0.042 mg/L), iron (0.20 to 0.93 mg/L), manganese (0.009 to 0.017 mg/L), strontium (0.016-0.044 mg/L) and zinc (up to 0.009 mg/L).

The water of the deep aquifer is more mineralized with E.C. between 57 and 270 mS/m, alkaline pH (between 6.6 to 7.5), and lower Eh (from 129 to 187 mV). However, there is significant variation in concentrations among the water of the wells (PMII-1, and PMII-2 and PMII--3). The waters of the well-PMII-1 had E.C. of 270 mS/m, followed by the waters of PMII-2, E.C. between 79 and 95 mS/m, and PMII-3 with E.C. of 58 mS/m. It is notorious the concentration of sodium and bicarbonate in these waters (between 58 and 501 mg/L, and 322 and 1,152 mg/L, respectively), as well as chloride (40 to 260 mg/L). Among the elements of lower concentrations are the fluoride (between 0.17 and 0.45 mg/L), bromide (between 0.10 and 2.23 mg/L). The concentrations of calcium, magnesium and potassium ranged from 25.0 to 39.9 mg/L, 10.3 to 41.5 mg/L, and 8.4 to 41.2 mg/L, respectively. Sulfate, nitrate and phosphate presented low concentrations of up to 2.8 mg/L, from 3.30 to 21.90 mg/L and up to 2.61 mg/L respectively. Other elements like aluminum and barium showed negligible concentrations; iron, manganese, strontium and zinc presented variations between 0.02 to 0.43 mg/L, 0.006 to 0.091 mg/L, and absent to 0.056 mg/L, respectively. The three wells in the deep aquifer presented distinct compositional aspects, while the lowest mineralized water was in PMII-3 and the one with highest concentration water was in PMII-1.

The water of the shallowest aquifer had lower concentrations of chemical elements and higher compositional homogeneity in relation to the deepest aquifer. The river water had low concentrations of elements and less reduced environment. The monitoring in this period showed a predominance of fresh water in the estuary at that location, since the river flow had a strong contribution of rainwater from the entire drain area of the basin.

According to the monitoring of August (08/20/2005), the groundwater analyzed presented low compositional variation in relation to the monitoring results of March (03/30/2005). However, the river water presented high mineralization, with E.C. between 860 to 1,100 mS/m, acidic pH (6.5 and 6.8) and Eh from 240 to 289 mV. The surface water had low calcium concentration (between 5.6 and 7.4 mg/L), magnesium (23.26 and 16.90 mg/L) and potassium (between 6.93 and 9.76 mg/L), and high concentrations of sodium (between 151.5 and 192.4 mg/L). Regarding the anions, the concentrations vary as follows: bicarbonate, between 35.58 and 39.04 mg/L, sulfate, 362.4 and 470.6 mg/L, chloride, between 2,654.4 and 3,446.3 mg/L. There is an absence of phosphate, nitrite, fluoride, barium and zinc. Nitrate occurs at concentrations between 0.986 and 1.243 mg/L.

The waters of the shallow aquifer presented more acidic pH than in the first monitoring (4.2 to 4.8), with Eh values between 254 to 341 mV and E.C. between 6 and 10 mS/m. Elements of calcium, magnesium, sodium and potassium concentrations presented between 1.5 and 3.8 mg/L, 0.71 to 1.19 mg/L, 6.4 to 9.2 mg/L and 0.84 and 4.14 mg/L. Bicarbonate (between 4.9 and 17.1 mg/L), sulfate (4.5 to 13.2 mg/L), chloride (9.1 to 15.0 mg/L) presented at low concentrations, and absence of nitrate and nitrite. Fluoride (0.02 mg/L) occurs only in the first and second sampling of PM-1. The other elements such as bromide, aluminum, barium, iron, manganese, zinc and strontium in their respective concentration ranges: 0.35 to 0.52 mg/L, 0.248 to 0.513 mg/L, 0.021 to 0.062 mg/L, 0.421 to 0.934 mg/L, 0.011 to 0.025 mg/L, 0.023 to 0.047 mg/L and 0.009 to 0.025 mg/L.

The deep aquifer presented slightly acidic to neutral water (6.7 and 7.2), Eh between 109 and 196 mV, and EC 49 and 230 mV. As in the shallow aquifer, the waters of the deep aquifer did not show significant compositional differences between one period and another. The concentrations of calcium, magnesium, sodium and potassium presented respective values of 23.6 and 49.0 mg/L, 12.28 and 44.49 mg/L, 54.3 to 442.4 mg/L, 302.56 and 1,073.6 mg/L. High concentrations of bicarbonate (302.6 and 1,073.6 mg/L) and chloride (38.0 and 262.0 mg/L) are also present, as well as phosphate (2.86 and 30.90 mg/L). There is no presence of barium, and low concentrations of nitrate and zinc (occurring on average concentration of 0.045 mg/L in PMII-1). There is nitrite concentration up to 60.0 mg/L. Other elements such as fluoride, bromide, aluminum, iron, manganese, strontium and zinc occurred in the respective concentrations: 0.17 and 0.41 mg/L, 0.85 and 4.16 mg/L, 0.008 and 0.017 mg/L, 0.267 and 0.519 mg/L , from 0.037 to 0.145 mg/L, 0.278 and 0.287 mg/L, and absent at 0.049 mg/L.

The Piper Plot (Fig. 13) shows the compositions of the water analyzed and monitored in the river in the shallow aquifer, and in the deep aquifer. The chemical groundwater type does not change with the fluctuations of the tide.

The river water presents Na-Cl water type, typical of marine waters, although in the sample of 03/30/2005 waters have mixed behavior in relation to the sampling results of 08/20/2005, which are predominantly Na-Cl water type.

The waters of the shallow aquifer had mixed composition, Na-Cl-SO_4 water types.

The waters of the deep aquifer had Na-HCO_3 water type, and high concentrations. However, there are samples of Ca-HCO_3 water type. The presence of bicarbonate may result from the calcite dissolution, present in the organic matter but also on the shells found in this aquifer. The increased presence of sodium in these coastal environments over calcium may indicate exchange of bases where calcium can be adsorbed and sodium released, resulting Na-HCO_3 water type. It happens when fresh water flushes salts that are in the aquifer (Appelo & Postma, 2007).

The fraction of seawater (*f*sea) was calculated using the chloride concentration of the water sample. The chloride molar relation from the sample with the chloride from the sea water results in the contribution of salinity in fresh water (Appelo & Postma, 2007).

It is assumed that chloride is a conservative parameter and its only contribution is the sea water. Thus, Equation 1 was used.

$$fsea = m_{CL-,\,sample} / \; 566\ (\mathrm{mmol/L}) \tag{1}$$

*f*sea - fration of seawater

$m_{CL-,\,sample}$ - concentration of Cl^- (mmol/L)

566 mmol/L - concentration of 35‰ (grams of salt per kilogram) seawater

The results of *f*sea of the waters from the shallow aquifer ranged between 0.05 and 0.07%. *f*sea of the waters from the deep aquifer ranged from 0.20 to 1.29%, and the waters of PMII-1 are those with the highest rate.

There was no variation in these rates in groundwater in the daily tidal fluctuations.

The river water presented variations in the fraction. The waters on 03/30/2005 showed low values, 0.11 and 0.12. On 08/20/2005, the waters had rates of 13.2, 17.2 and 16.1%, indicating influence of seawater into the river in the study area.

03/30/2005	pH	E.C. (mS/m)	Eh (mV)	Ca (mg/L)	Mg (mg/L)	Na (mg/L)	K (mg/L)	HCO_3 (mg/L)	$SO_4{}^{2}$ (mg/L)	(mg/L)	$PO_4{}^{3-}$ (mg/L)	NO_3 (mg/L)	(mg/L)	(mg/L)	Al (mg/L)	Ba (mg/L)	Fe (mg/L)	Mn (mg/L)	Sr (mg/L)	Zn (mg/L)
PM-1 (6:00 h)	5.0	8	252	1.6	0.8	6.8	2.7	6.7	7.6	13	< 0.05	<0.013	0.02	0.62	0.230	0.041	0.60	0.017	0.021	0.005
PM-1 (9:00 h)	4.9	8	271	1.6	0.8	6.6	2.8	6.7	7.8	15	< 0.05	0,043	0.02	0.88	0.210	0.042	0.53	0.015	0.020	< 0.005
PM-1 (12:00 h)	5.0	8	269	1.5	0.8	6.0	2.7	6.7	8.2	13	< 0.05	<0.013	0.06	0.52	0.180	0.034	0.42	0.013	0.018	0.005
PM-2 (6:00 h)	5.3	7	233	4.0	0.7	6.0	0.8	10.1	5.0	11	< 0.05	<0.013	0.01	0.47	0.190	0.017	0.47	0.009	0.043	< 0.005
PM-2 (9:00 h)	5.1	8	245	3.9	0.7	5.5	0.9	10.1	5.1	11	< 0.05	<0.013	0.02	0.46	0.082	0.014	0.20	0.008	0.044	< 0.005
PM-2 (12:00 h)	5.3	8	256	4.1	0.8	6.8	0.8	10.1	4.8	11	< 0.05	<0.013	0.03	0.49	0.220	0.017	0.41	0.008	0.044	0.005
PM-3 (6:00 h)	4.9	8	248	1.5	0.7	6.6	2.0	6.7	8.3	12	< 0.05	<0.013	<0.001	0.57	0.270	0.039	0.65	0.015	0.021	0.007
PM-3 (9:00 h)	4.8	8	256	1.6	0.7	6.8	1.9	6.7	7.8	12	< 0.05	<0.013	0.02	0.52	0.270	0.040	0.59	0.013	0.021	< 0.005
PM-3 (12:00 h)	5.2	7	280	1.6	0.6	6.1	1.7	6.7	6.2	11	< 0.05	<0.013	<0.001	0.59	0.270	0.028	0.54	0.012	0.020	< 0.005
PM-4 (6:00 h)	7.3	8	222	1.2	0.4	7.3	1.3	6.7	6.4	13	< 0.05	<0.013	0.02	0.69	0.360	0.038	0.92	0.011	0.016	0.005
PM-4 (9:00 h)	5.7	8	227	1.1	0.4	8.3	1.3	6.7	6.5	13	< 0.05	<0.013	0.03	0.76	0.400	0.041	0.93	0.011	0.016	0.009
PM-4 (12:00 h)	6.5	8	228	1.2	0.5	7.7	1.3	6.7	6.4	13	< 0.05	<0.013	<0.001	0.74	0.390	0.039	0.92	0.010	0.016	0.005
PMII-1 (6:00 h)	7.2	270	170	25.1	39.0	440.0	32.2	1102.9	<0.01	245	18.60	<0.013	0.37	0.65	< 0.010	<0.0005	0.43	0.033	0.240	0.056
PMII-1 (9:00 h)	7.1	270	175	25.0	41.1	501.0	41.2	1151.7	0.9	259	19.50	<0.013	0.45	1.01	< 0.010	<0.0005	0.41	0.036	0.240	0.049
PMII-1 (12:00 h)	7.2	270	187	24.4	41.5	473.0	35.8	1132.2	2.8	260	21.90	2.61	0.33	2.23	< 0.010	<0.0005	0.38	0.032	0.240	0.044
PMII-2 (6:00 h)	7.5	81	147	22.6	14.7	121.0	14.0	409.9	1.3	64	3.60	0.09	0.29	1.01	< 0.010	<0.0005	0.02	0.006	0.170	< 0.005
PMII-2 (9:00 h)	7.5	79	163	25.9	18.4	118.0	14.3	429.4	<0.01	67	5.40	<0.013	0.30	0.18	< 0.010	<0.0005	0.04	0.012	0.180	< 0.005
PMII-2 (12:00 h)	7.5	95	161	28.1	21.1	165.0	15.6	507.5	1.8	85	8.10	0,043	0.31	2.31	< 0.010	<0.0005	0.05	0.018	0.200	< 0.005
PMII-3 (6:00 h)	6.7	58	129	39.9	10.3	59.4	8.5	327.0	0.4	40	3.30	<0.013	0.17	<0.005	< 0.010	<0.0005	0.31	0.077	0.230	< 0.005
PMII-3 (9:00 h)	6.7	58	138	39.5	10.6	58.9	8.4	322.1	1.6	40	3.60	<0.013	0.17	0.10	< 0.010	<0.0005	0.32	0.091	0.220	< 0.005
PMII-3 (12:00 h)	6.6	57	147	39.0	10.4	58.0	8.5	331.8	1.6	40	3.30	<0.013	0.17	0.10	< 0.010	<0.0005	0.30	0.074	0.220	< 0.005
River (6:00 h)	7.6	12	259	4.7	2.2	12.6	1.8	16.8	3.8	21	0.17	0.26	0.05	0.24	0.260	0.006	1.09	0.058	0.037	< 0.005
River (9:00 h)	7.8	13	233	3.8	2.2	13.9	1.8	16.8	3.5	25	< 0.05	0.26	0.05	0.23	0.230	0.008	0.77	0.032	0.032	< 0.005
River (12:00 h)	7.5	12	227	4.7	2.3	13.7	1.9	16.8	4.0	24	< 0.05	0.26	0.05	0.35	0.150	0.004	0.89	0.046	0.037	< 0.005

Table 6. River and ground water compositions under tidal fluctuations in the study area (03/30/2005)

08/20/2005	pH	E.C. (mS/m)	Eh (mV)	Ca (mg/L)	Mg (mg/L)	Na (mg/L)	K (mg/L)	HCO_3 (mg/L)	SO_4^{-2} (mg/L)	Cl^- (mg/L)	PO_4^{3-} (mg/L)	NO_3^- (mg/L)	NO_2^- (mg/L)	F^- (mg/L)	Br^- (mg/L)	Al (mg/L)	Ba (mg/L)	Fe (mg/L)	Mn (mg/L)	Sr (mg/L)	Zn (mg/L)
PM1 (11:00 h)	4.8	10	287	1.8	1.19	7.8	2.96	8.5	6.9	13.8	< 0.05	<0.013	<0.017	0.02	0.36	0.285	0.062	0.934	0.025	0.026	0.022
PM1 (15:00 h)	4.5	7	308	1.8	1.16	7.7	4.14	7.3	8.5	13.0	< 0.05	<0.013	<0.017	0.02	0.47	0.304	0.055	0.748	0.021	0.025	0.020
PM1 (17:00 h)	4.4	8	315	1.6	1.00	7.4	2.88	7.3	8.8	13.1	< 0.05	<0.013	<0.017	<0.001	0.35	0.308	0.053	0.737	0.020	0.024	0.015
PM2 (11:00 h)	4.7	7	342	3.5	0.88	6.4	0.95	14.6	4.6	9.1	0.22	<0.013	<0.017	<0.001	0.37	0.248	0.023	0.588	0.013	0.047	0.013
PM2 (15:00 h)	4.8	6	341	3.6	0.85	6.4	0.84	17.1	4.5	9.5	< 0.05	<0.013	<0.017	<0.001	0.37	0.251	0.021	0.421	0.011	0.046	0.009
PM2 (17:00 h)	4.8	6	299	3.8	0.85	6.5	0.87	17.1	4.6	9.6	< 0.05	<0.013	<0.017	<0.001	0.39	0.251	0.021	0.430	0.011	0.047	0.011
PM4 (11:00 h)	4.2	9	266	1.5	0.71	9.2	1.50	4.9	13.2	14.4	< 0.05	<0.013	<0.017	<0.001	0.50	0.533	0.058	1.367	0.018	0.023	0.025
PM4 (15:00 h)	4.3	8	265	1.5	0.72	9.1	1.45	5.5	12.5	14.8	< 0.05	<0.013	<0.017	<0.001	0.52	0.513	0.054	1.357	0.018	0.023	0.019
PM4 (17:00 h)	4.4	9	254	1.5	0.74	9.2	1.45	5.5	12.4	15.0	< 0.05	<0.013	<0.017	<0.001	0.51	0.496	0.054	1.360	0.017	0.023	0.017
PMII-1 (11:00 h)	7.1	230	135	23.9	44.49	439.0	29.93	1073.6	<0.01	258.6	30.21	<0.013	58.62	0.41	3.50	0.012	<0.006	0.519	0.039	0.278	0.049
PMII-1 (15:00 h)	7.1	230	134	23.6	44.30	435.4	29.99	1073.6	<0.01	260.1	29.94	<0.013	<0.017	0.41	4.16	0.017	<0.006	0.509	0.037	0.280	0.042
PMII-1 (17:00 h)	7.2	230	109	24.1	43.08	442.4	30.20	1073.6	<0.01	262.0	30.90	<0.013	60.00	0.34	3.99	0.015	<0.006	0.500	0.038	0.285	0.044
PMII-3 (11:00 h)	6.7	49	196	49.0	12.28	55.5	9.36	302.6	0.1	38.0	2.86	<0.013	15.17	0.17	0.93	0.008	<0.006	0.267	0.143	0.281	<0,01
PMII-3 (15:00 h)	6.6	49	190	46.8	12.35	54.3	9.36	302.6	0.1	38.7	2.89	0.013	15.86	0.17	1.82	0.009	<0.006	0.361	0.142	0.278	<0,01
PMII-3 (17:00 h)	6.9	49	150	45.3	12.40	55.0	9.37	312.3	0.1	38.7	2.90	<0.013	15.52	0.17	0.85	0.005	<0.006	0.282	0.145	0.287	<0,01
River (11:40 h)	6.5	860	240	5.6	16.90	151.5	6.93	35.4	362.4	2654.4	< 0.05	0.986	<0.017	<0.001	9.80	0.007	<0.006	0.046	0.007	0.114	<0,01
River (16:10 h)	6.5	1100	289	7.4	23.26	192.4	9.76	39.0	470.6	3446.3	< 0.05	1.243	<0.017	<0.001	12.89	0.008	<0.006	0.016	0.009	0.151	<0,01
River (18:15 h)	6.8	1000	256	7.0	22.28	187.2	9.52	39.0	451.0	3333.3	< 0.05	1.113	<0.017	<0.001	11.16	0.006	<0.006	0.014	0.008	0.146	<0,01

Table 7. River and ground water compositions under tidal fluctuations in the study area (08/20/2005)

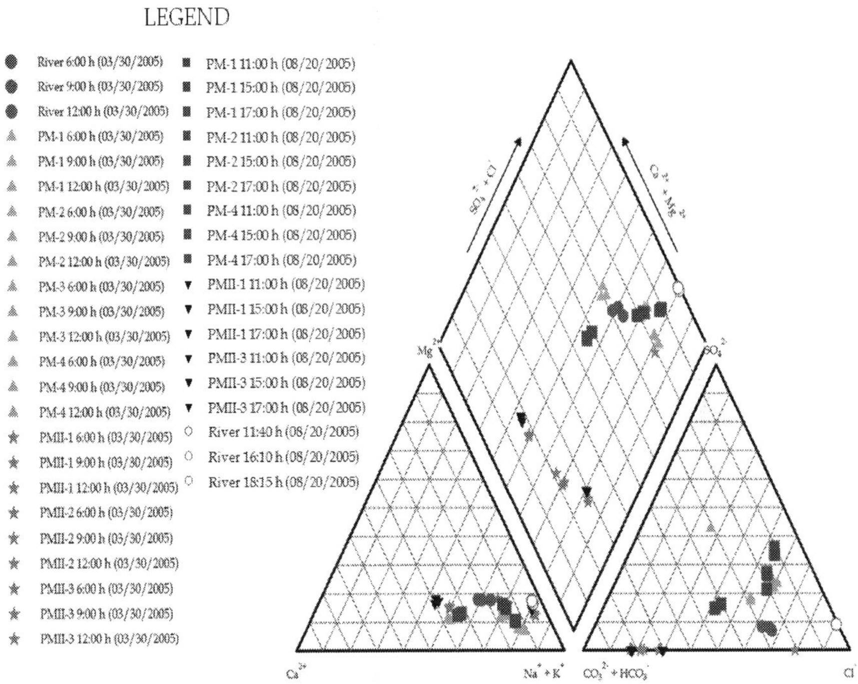

Fig. 13. Piper plot showing compositions of river water and groundwater from shallow and deep aquifers

3.5 Groundwater dynamic and the tide influence

The waters of the shallow and deep aquifers do not suffer direct interference of tides and saline intrusion that enters by the river channel. However, the potentiometric surface of the shallow aquifer has groundwater flow direction from SW to NE, or the river to the inland area of study, both in the rainy season (represented by day 03/30/2005) as during dry season (08/20/2005). During the rainy season, the river flows are high and the daily tide has less influence on the dynamics of groundwater, due to a lower pressure of the tide (the flow direction of the river to the mouth is predominant). However, in the less rainy period, one notices the presence of more saline waters in the river, and an influence of saline intrusion in relation to the fresh waters of the river as the river flows are lower. Thus, the potentiometry of the water in the shallow aquifer can be related to this dynamic. The daily fluctuation of the tide makes the potentiometry of the water in the shallow aquifer, both vertically and laterally.

The waters of the deep aquifer have the flow towards NE-SW, toward the river with lateral variations between one period and another, and vertical (especially during the rainy season).

The waters of the shallow and deep aquifer almost do not suffer the influence by tidal changes and seasonal climate. The waters of the shallow aquifer have low mineralization, are acidic and less reduced, have mixed composition, Na-Cl-SO4 water type, and are similar to the composition of the river water.

The waters of the deep aquifer already show compositional differences, and are characterized as Na-HCO3 water type and Ca-Na-HCO3 water type. These waters have calcium and bicarbonate resulting from the dissolution of calcite from the shells and organic matter in the geological formation. In contact with saline waters, there is ion exchange with sodium and change to Na-HCO3 water type.

The waters of the aquifer do not have significant contribution from the sea water. However, the river waters suffer significant compositional changes between the rainy and less rainy period. During the rainy season the waters are slightly mineralized, although in the dry season, there is an increase of dissolved salts in the waters of the river and its composition becomes typical of sea water (Na-Cl water type).

4. Final considerations

The preservation of plant and zoo debris accumulations is directly related to the sediment and aquatic environments. These characteristics have in turn influenced by the dynamic seasonal climate variations and the tides throughout the Quaternary plain of the Itanhaém River.

Along the estuary there are two clear taphonomy trends differentiated by the joint action of the depositional and aquatic environments. The Preto River has a low oxygen content and a high humic acid content, low salinity and higher stability in the chemical composition of water, due to little influence of the saline intrusion. These factors joined to low energy of the river present an ideal location for the preservation of plant biomass that is deposited in the IHS deposits. However, the preservation of zoo debris is minimal, since the skeletons of carbonate composition are easily dissolved in these conditions.

On the other hand, the Itanhaém and Branco rivers show a greater variability in hydrodynamics and in chemical composition. This variation may be a result of: (1) the Branco river rises in the Serra do Mar in steep areas and its journey through the Quaternary plain is shorter than the Preto River, a fact that favors greater water oxygenation; (2) Branco River presents greater flow than the Preto River, softening the chemical characteristics of the Preto River in the Itanhaém River, and finally (3) the Itanhaém River also receives the direct influence of the saline intrusion.

So the rivers Preto, Branco and Itanhaém have high variability in their dynamics, chemical and sedimentological characteristics that are reflected in the nature of bio debris accumulations.

Unlike the waters of rivers, sub aquatic environment have no significant compositional changes, however, there is influence of tidal oscillations in the groundwater flow dynamics. The potentiometry of the shallow aquifer has a strong influence on the daily fluctuation of the tide; however, it was not observed compositional changes, not with seasonality.

In the shallow aquifer, the waters have the same water type of the river waters in the rainy season. However, in the less rainy period, the composition is maintained and the river becomes more saline. The more acidic and oxidizing environment of groundwater are favorable to the preservation of logs rich in lignin. Yet in the waters of the deep aquifer, carbonate skeletons are preserved, since the water is more basic and saline.

These unique features of the sea water are the conditions that preserve the shells in the coastal assemblies of zoo debris.

The integrated and multidisciplinary studies are important tools in studying environment and paleoenvironmental, also contributing to better understand the processes that can lead or not to the preservation of fossil records. The aquatic environment where the records are being fossilized or suffering the epigenesis is crucial to the preservation of fossils or not.

5. Acknowledgements

The authors would like to acknowledge the financial support of the Foundation for the Support of Research of the State of São Paulo (FAPESP, Process 01/09881-2 and FAPESP, Process 2007/07190-0) and the Brazilian National Council for Scientific and Technological Development (CNPq; post-doctoral fellowship), as well as the collaboration of the Research Center of the Estuary of the Itanhaém river (CPeRio) for making their installations available and assisting with the research in the rivers of the basin, including the provision of a boat and the invaluable assistance of the excellent pilot and guide, José Machado.

6. References

Amaral, P.G.C., Ledru, M.P., Ricardi-Branco, F. & Giannini, P.C.F. (2006). Late Holocene development of a mangrove ecosystem in southeastern Brazil (Itanhaém, state of São Paulo): *Paleogeography, Palaeoclimatology and Palaeoecology*, Vol. 241, No. 3-4, (November 2006) pp. 608-620, ISSN 0031-0182

Angulo, R.J., Souza, M.C. & Pessenda, L.C.R. (2002). O significado das datações ^{14}C na reconstrução de paleoníveis marinhos e na evolução das barreiras quaternárias do litoral paranaense. *Revista Brasileira de Geologia*, Vol. 32, No. 1, pp. 95-106, ISSN 0375-7636

Appelo, C.A.J. & Postma, D. (2007). *Geochemistry, groundwater and pollution*. 2nd Edition. A.A. Balkema Publishers. ISBN 0415364213, Amsterdam, the Netherlands.

Batista Filho, J.J. (2006). A dinâmica das águas subterrâneas no estuário do rio Itanhaém, litoral sul do estado de São Paulo, 06/28/2011, Available from:
http://www.bibliotecadigital.unicamp.br/document/?code=vtls000393402&opt=1

Camargo, A.F.M., Pereira, L.A. & Pereira, A.M.M. (2002). Ecologia da bacia hidrográfica do rio Itanháem. In: *Conceitos de bacias hidrográficas: teorias e aplicações*, A. Schiaretti & A.F.M. Camargo (Eds.), 239-256, Editora da UESC. ISBN 857455099X, Ilhéus, Brazil.

Comitê de Bacias Hidrográficas/Baixada Santista (CBH-BS). (2000). Relatório Zero: Minuta Preliminar do Relatório de Situação dos Recursos Hídricos da UGRHI 7. São Paulo-SP, In:
http://www.sigrh.sp.gov.br/sigrh/ARQS/RELATORIO/CRH/CBH-BS/218/relbsseg.pdf, 05/10/2011, Available from: < http://www.sigrh.sp. gov.br>.

Fetter, C.W. *Applied Hydrogeology*. 4th Edition. Prentice Hall, ISBN 013088239-9, New Jersey, United States of America.

Giannini, P.C.F., Guedes, C.F., Nascimento, D.R., Tanaka, A.P.B., Angulo, R.J., Assine, M. & Souza, M.C. (2009). sedimentology and mophological evolution of the ilha comprida barrier system, southern São Paulo coast. In: *Geology and Geomorphology of Holocene Coastal barriers of Brazil*, S. Dillenburgand & P., Hesp (Eds.), 177-224, Springer-Verlag. ISBN 978-3-540-25008-1 Berlin, Germany.

Holz, M & Simões, M.G. (2002) *Elementos fundamentais de Tafonomia*. Editora da Universidade UFRGS, ISBN 8531406730, Porto Alegre, Brazil

Lopes, M.C. (1984). Água subterrânea no estado de São Paulo - síntese das condições de ocorrência, *Proceedings of ABAS, Third Brazilian Groundwater Congress,* Fortaleza - Ceará, October 1984.

Suguio, K. (1992). *Dicionário de Geologia marinha.* 1 ed, Editora T.A. Queiroz, ISBN 85-7182-001-5, S. Paulo, Brazil

Suguio, K., & Martin, L., (1978) Formações quaternárias marinhas do litoral paulista e sul fluminense: *Boletim IG-USP*, especial number, pp. 1-55, ISSN 0102-6275

Suguio, K. (2004). O papel das variações do nível relativo do mar durante o Quarternário tardio na origem da baixada litorânea de Juréia, SP. In: *Estação ecológica Juréia – Itatins: ambiente físico, flora e fauna.* Marques, O.A.V. & Duleba, W. (eds), pp. 34-41. Holos Editora, ISBN 85-86699-37-3, São Paulo, Brazil

Ricardi-Branco, F., Ianniruberto, M., Silva, M.A. & Branco, F. (2011). Plant debris Accumulations in the Rio Preto sub-basin, Itanhaém, SP, Brazil: insights from geotechnology. *Palaios,Vol.* 26 (May 2011), pp. 264-274. ISSN 0883-1351

Ricardi-Branco, F., Branco, F.C, Garcia, R.F., Faria, F.S, Pereira, S.Y., Portugal, R, Pessenda, L.C., Pereira, P.R.B. (2009) Features of plant accumulations along the Itanhaém river, on the southern coast of the Brazilian state of São Paulo. *Palaios*, Vol. 24, No. 7 (July 2009), pp. 416-424, ISSN 0883-1351

Zincone, S.A, Ricardi-Branco, F., Vidal, A.C. & Pessenda, L.C. (2007). Caracterização de depósitos transgressivos quaternários na bacia do rio Itanhaém, litoral sul do Estado de São Paulo. In: *Paleontologia: Cenários da Vida,* I. S. Carvalho, R.C. T. Cassab, C., Schwanke, M.A, Carvalho, A.C.S., Fernandes, M.A.C.R., Rodrigues, M.S.S., Carvalho, M., Arai & E.Q.O., Oliveira (eds.), 295-307, Editora Interciência. ISBN 978-85-7193-184-8, Rio de Janeiro, Brazil.

25

The Permo-Triassic Tetrapod Faunal Diversity in the Italian Southern Alps

Marco Avanzini[1], Massimo Bernardi[1] and Umberto Nicosia[2]
[1]*Science Museum, Trento,*
[2]*Sapienza University of Rome,*
Italy

1. Introduction

The Permian and Triassic palaeogeography of the Alpine region originated a peculiar geological situation, now well exposed in several sections, in which marine sediments, continental deposits, and volcanites interfinger. The study of the resulting mixed sections allowed to build a framework of biostratigraphic and chronological data in which tetrapod footprints play a key role (e.g., Avanzini & Mietto, 2008).

Global track record is much more abundant than the skeletal record and, although suffering from problems related to a correct attribution to the trackmakers, provides data as reliable as those obtained from skeletal remains (Carrano & Wilson, 2001 and references therein).

During the Permian and the Early-Middle Triassic, tetrapods, and especially reptiles, radiated and entire new land-dwelling groups originated (e.g., the archosaurs). Consequently, in this temporal interval the tetrapod track record shows a huge increase in variability reflecting the morphological diversity spanning from a stem-reptile to a 'mammalian' foot, from basal crocodilomorph to a dinosauromorph foot. How is this pattern documented in the Dolomites region and surrounding areas (Southern Alps, NE Italy)?

In this geographical sector the recent discovery of many new tetrapod footprint-bearing outcrops has yield to a phase of renewed interest for ichnological data. As a result, the stratigraphical, palaeoecological and palaeogeographical importance of tetrapod footprints in this geographical sector is becoming more and more widely acknowledge.

In this contribution we provide an overview of Permian and Triassic tetrapod faunal composition as deduced from the study of several ichnosites located in the Italian Southern Alps with special reference to the pattern exhibited around the PT boundary.

2. Geological setting

In the Permian succession of the Southern Alps tetrapod tracks are found within two tectonosedimentary cycles. The lower cycle is represented by the alluvial-lacustrine continental deposits of the Collio and Tregiovo Formations while the upper cycle is represented by clastic red-beds of the Verrucano Lombardo Formation and Val Gardena Sandstone (Trentino Alto Adige). Towards the top, the latter interfingers with the sulphate evaporites and shallow-marine carbonate sequences of the Bellerophon Fm. The two cycles are put in contact by a regional unconformity documented by erosional surfaces and palaeosoil horizons. Floristic and radiometric data (Cassinis & Doubinger, 1991, 1992; Cassinis et al., 1999, 2002; Schaltegger & Brack, 1999) indicate that the lower cycle is Artinskian-Kungurian in age (Early Permian). The upper cycle has been dated by means of sporomorphs and foraminifers and is considered Wuchiapingian in age (Late Permian, Cassinis et al., 2002).

Triassic units follow. The precise position of the PT boundary in the Southern Alps has been debated for long time (see Broglio Loriga & Cassinis, 1992 for a review). Although some early works put the boundary at or very close to the Bellerophon-Werfen Formation limit (Bosellini, 1964; Assereto et al., 1973; Posenato, 1988), the PT boundary is now thought to lie within (Pasini, 1985; Broglio Loriga et al., 1986; Visscher & Brugman, 1988; Farabegoli & Perri, 1998; Farabegoli et al., 2007) or at the top (Wignall et al., 1996) of the Tesero Member, the lowermost part of the Werfen Fm. The first Triassic tetrapod tracks are found within the terrigenous and terrigenous-carbonate units of the Werfen Fm. and are Olenekian in age (Early Triassic). Several terrigenous and terrigenous-carbonate units of Anisian age (Middle Triassic) follow: Gracilis Fm., Voltago Cgl., Richthofen Cgl., Morbiac dark Limestones. All were deposited in lagoonal-peritidal to continental environments and have yielded well preserved ichnoassociations. No tetrapod tracks are known from the Dolomites' Ladinian (Middle Triassic) which is represented mainly by basinal units (Buchenstein Fm. and Wengen Group) and carbonate platform deposits (Sciliar Dolomite).

3. Permian and Triassic ichnoassociations

Permian and Triassic ichnoassociations in the Southern Alps have been documented since the XIX century (e.g., Curioni, 1870). Recent reviews have been published by Conti et al. (1997), Conti et al. (1999), Conti et al. (2000), Avanzini et al. (2001), Avanzini & Mietto (2008). Here we provide an updated overview on these studies focusing on the contribution that tetrapod ichnology can give to the debates concerning the large scale patterns of extinction and subsequent recovery on lands near the PT boundary (Benton & Twitchett, 2003; Twitchett et al., 2001; Retallack et al., 2006; Sheldon & Chakrabarti, 2010).

In the Alpine area several different types of bioevents were recognized (Conti et al., 2000; Ronchi et al., 2005). According to their relative importance, they had different degrees of utility for correlation. The first-appearance datum (FAD) and last appearance datum (LAD) concern a taxon, at either genus and species level, and are important at the regional level. The first occurrence (FO) and last occurrence (LO) as well as the disappearance event (DE) concern a faunal complex and have very high correlation power.

Here, ichnoassociations have been grouped in four discrete time bins: Artinskian Kungurian (Early Permian) (Fig. 1), Wuchiapingian (Late Permian) (Fig. 2), Olenekian (Early Triassic) and Anisian (Middle Triassic) (Fig. 3). Traditionally, these discrete associations have been considered 'Faunal Units' (FU) (Conti et al., 1997; Conti et al., 1999; Conti et al., 2000; Avanzini et al., 2001; Ronchi et al., 2005).

In the present study: i) no vertical differences within a single bin have been considered even though some are well known; and, ii) only species-level determinations have been considered except where a genus or a higher-level taxon is represented by an undetermined species only (e.g., *Procolophonichnium* isp.), or in the case of *Rhynchosauroides* isp.1 and *Rhynchosauroides* isp.2 of Valentini et al. (2007) which underwent a formal revision and, although not yet renamed, can be considered as distinct ichnospecies.

3.1 Artinskian-Kungurian (Early Permian)

The Permian first cycle ichnoassociation is documented within the Collio and Tregiovo Formations. Data have been recorded in the Collio basin (Geinitz, 1869; Curioni, 1870; Cassinis, 1966; Berruti, 1969; Ceoloni et al., 1987; Conti et al., 1991) and from sediments cropping out in other localities of the Orobic basin (Gümbel, 1880; Dozy, 1935; Casati, 1969; Casati & Forcella, 1988; Nicosia et al., 2000; Santi, 2003, 2005; Santi & Krieger, 2001) and in the Tregiovo and Luco basins (Conti et al., 1997; Avanzini et al., 2008).

The ichnoassociation includes the following ichnospecies (after Conti et al., 1997; Conti et al., 2000, updated) (Fig. 1).

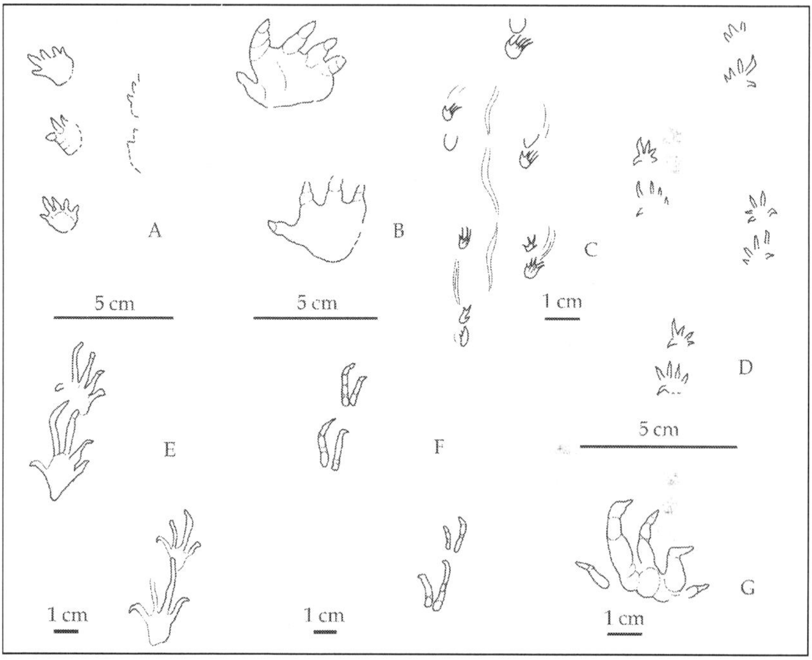

Fig. 1. Artinskian and Kungurian (Early Permian) tetrapod tracks form the Southern Alps. A) *Amphisauropus imminutus* Haubold, 1970; B) *Amphisauropus latus* Haubold, 1970; C) *Batrachichnus salamandroides* (Geinitz, 1861); D) *Camunipes cassinisi* Ceoloni et al., 1987; E) *Dromopus lacertoides* (Geinitz, 1861); *F) Dromopus didactylus* (Moodie, 1896); G) *Varanopus curvidactylus* Moodie, 1929. From Santi & Krieger (1999), Ceoloni et al. (1987), Avanzini et al. (2008).

Amphisauropus imminutus Haubold, 1970
Amphisauropus latus Haubold, 1970
Batrachichnus salamandroides (Geinitz, 1861)
Camunipes cassinisi Ceoloni et al., 1987
Dromopus lacertoides (Geinitz, 1861)
Dromopus didactylus (Moodie, 1896)
Ichniotherium cottae (Pohlig, 1885)

Varanopus curvidactylus Moodie, 1929

Santi (2007) has clearly shown that *Camunipes cassinisi* Ceoloni et al., 1987 is a junior synonym of *Erpetopus* Moodie, 1929.

The Early Permian ichnoassociation is bounded by the *Amphisauropus latus* FO and *Ichniotherium cottae* FO at the base and by the *Dromopus didactylus* LO at the top. This association is very similar to those found in the Permian of various European countries. It is named Collio FU and it can in turn be subdivided into two subunits. The lower subunit (Pulpito subunit) is characterized by the presence of *Amphisauropus latus* Haubold, 1970, *Dromopus lacertoides* (Geinitz, 1861) and *Ichniotherium cottae* (Pohlig, 1885). The upper subunit (Tregiovo subunit) is bounded by the disappearance event (DE) for all but one of the preceding forms, yielding a drastic reduction of faunal diversity; indeed the subunit is monotypic being represented by very frequent specimens of the only relict species *Dromopus didactylus* (Moodie, 1896). This pattern was interpreted as a bioevent, given the relatively unchanged sedimentation conditions.

3.2 Wuchiapingian (Late Permian)

The Permian second cycle ichnoassociation has been recorded in the Bletterbach section (Ceoloni et al., 1988; Conti et al., 1975; 1977; 1980; Leonardi & Nicosia, 1973; Leonardi et al., 1975; Nicosia et al., 1999) and other quasi-coheval outcrops (SS. 48, Col Fratton (Kittl, 1891; Abel, 1926); S. Pellegrino Pass (Conti et al., 1977), Seceda (Valentini et al., 2009), Nova Ponente (Wopfner, 1999), Recoaro (Mietto, 1975; 1981) San Genesio-Meltina Plateau in the Adige basin (Avanzini, unpublished).

The ichnoassociation includes the following ichnospecies (after Avanzini et al., 2001; Conti et al. 2000; Valentini et al., 2007, updated) (Fig. 2).

Chelichnus tazelwurmi Ceoloni et al., 1988
Dicynodontipus geinitzi (Hornstein, 1876)
Ganasauripus ladinus Valentini et al., 2007
Hyloidichnus tirolensis Ceoloni et al., 1988
Ichniotherium accordii Ceoloni et al., 1988
Ichniotherium cottae (Pohlig, 1885)
Janusichnus bifrons Ceoloni et al., 1988
Pachypes dolomiticus Leonardi et al., 1975
Paradoxichnium radeinensis Coeloni et al., 1988
Protochirotherium isp.
Rhynchosauroides pallinii Conti et al., 1977
Rhynchosauroides isp.1 of Valentini et al., 2007
Rhynchosauroides isp.2 of Valentini et al., 2007
Therapsida indet.

Note that the specimens from Val Gardena Fm. ascribed to *Chelichnus tazelwurmi* Ceoloni et al., 1988 and *Ichniotherium cottae* (Pohlig, 1885) are currently under revision and that their nomenclature is expected to change in the near future.

The Upper Permian tetrapod footprint complex is characterized by a diversified fauna almost completely new as compared with other Permian faunas (Ceoloni et al., 1986). Bounded in its lower part by the *Rynchosauroides* and *Protochirotherium* FAD, its records also the *Dycinodontipus* FAD and the *Ichniotherium* LAD. It is named the Bletterbach FU.

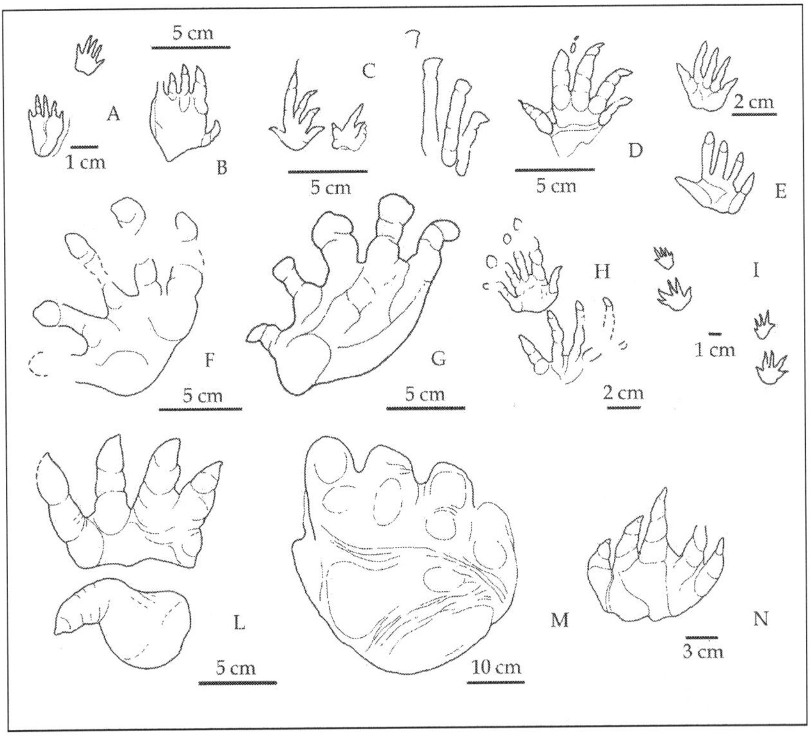

Fig. 2. Wuchiapingian (Late Permian) tetrapod tracks form Southern Alps. A) *Chelichnus tazelwurmi* Coeloni et al., 1988; B) *Dicynodontipus geinitzi* (Hornstein, 1876); C) *Ganasauripus ladinus* Valentini et al., 2007; D) *Rhynchosauroides pallinii* Conti et al., 1977; E) *Hyloidichnus tirolensis* Ceoloni et al., 1988; F) *Ichniotherium accordii* Ceoloni et al., 1988; G) *Ichniotherium cottae* (Pohlig, 1885); I) *Janusichnus bifrons* Ceoloni et al., 1988; M) *Pachypes dolomiticus* Leonardi et al., 1975; H) *Paradoxichnium radeinensis* Ceoloni et al., 1988; L) cf. *Protochirotherium* isp.; M) Therapsida indet. From Conti et al. (1977, 2000), Valentini et al. (2007).

Early Permian and Late Permian associations of the Southern Alps are separated by a stratigraphic break that correspond to a time-gap of 14-27 Ma (Italian IGCP 203 Group, 1986; Cassinis et al., 1988; Cassinis et al., 1999). A similar hiatus, equivalent to part of the Middle Permian, has been recognized in the global Permian tetrapod fossil record and has been termed Olson's gap (Lucas and Heckert, 2001). The gap is thought to be associated with a significant remodelling of the global Permian tetrapod fauna (Lucas and Heckert, 2001; Lucas, 2004). As stated before, the two Southern Alpine FUs have been considered two different 'evolutionary stages' (Conti et al., 2000).

3.3 Olenekian (Early Triassic)

In the studied sector, the Early Triassic is represented by the Werfen Fm.: a complex succession of carbonate, terrigenous and mixed sediments. The formation is divided into nine members and tracks are present in the upper three (Campil Member, Cencenighe Member and San Lucano Member).

Tracks have been found near Recoaro, in Val Gardena (Bulla/Pufels), and Val Travignolo (Conti et al., 2000; Mietto, 1986). Olenekian tracks are generally poorly preserved and some, as those signaled by Leonardi (1967, p. 118, 119), are here discarded. Furthermore (*contra* e.g. Avanzini & Mietto, 2008) levels bearing *Capitosauroides* and *Procolophonichnium* ichnogenera (e.g., cropping out in the Monte Marzola) previously considered to be part of the Werfen Fm. are here reconsidered within the 'Gracilis Formation' which is Anisian (Middle Triassic) in age.

The ichnoassociation includes the following ichnospecies (after Avanzini et al., 2001; Conti et al. 2000, updated) (Fig. 3):

Rhynchosauroides palmatus (Lull, 1942)

Rhynchosauroides schochardti (Rühle von Lilienstern, 1939)

Both *R. palmatus* (Lull, 1942) and *R. schochardti* (Rühle von Lilienstern, 1939) have not been validated by Valentini et al. (2007) in their review of the ichnogenus *Rhynchosauroides* from the Val Gardena Sandstone. No review has been conducted of this ichnogenus in the Werfen Formation. We acknowledge that both ichnospecies could follow in an undetermined *Rhynchosauroides* isp., this is however nearly irrelevant (i.e., Early Triassic ichnodiversity could be 1 instead of 2) to the data analyses presented here.

The Lowermost Triassic of the Southern Alps is characterized by the scarce presence of vertebrates, a factor that is undoubtedly linked to environmental conditions (palaeogeography) which did not favor the permanence in this region of complex and consistent faunal associations. The ichnoassociations are dominated by *Rhynchosauroides* forms that are likely mainly Permian survivor.

3.4 Anisian (Middle Triassic)

Several Anisian ichnoassociations have been found in the Dolomites and surrounding areas. These are found within several formations (Gracilis Fm., Voltago Cgl., Richthofen Cgl., Morbiac dark Limestones) which document transitional continental to marine environments. Tracksites are located in the Braies Dolomites (Northern Dolomites), in the eastern Dolomites, in the upper Val di Non and Val d'Adige, and in the Recoaro-Vallarsa area (Abel, 1926; Brandner, 1973; Mietto, 1987; Avanzini, 1999, 2000, 2002; Avanzini et al., 2001; Avanzini & Leonardi, 2002; Avanzini & Lockley, 2002; Avanzini & Neri, 1998; Avanzini & Mietto, 2008; Todesco et al., 2008; Todesco & Bernardi, 2011).

The ichnoassociation includes the following ichnospecies (after Avanzini & Mietto, 2008, updated) (Fig. 3):

Brachychirotherium paeneparvum Demathieu & Leitz, 1982

Brachychirotherium circaparvum Demathieu, 1967

Capitosauroides cf. *bernburgensis* Haubold, 1971

Chirotherium barthii Kaup, 1835

Chirotherium rex Peabody, 1948

Isochirotherium infernense Avanzini & Leonardi, 2002

Isochirotherium delicatum Courel & Demathieu, 1976

Parasynaptichnium gracilis Mietto, 1987

Procolophonichnium isp.

Rhynchosauroides tirolicus Abel, 1926

Rhynchosauroides peabodyi Faber, 1958

Rotodactylus cf. cursorius Peabody, 1948

Fig. 3. Olenekian (A, B) and Anisian (C-P) (Early and Middle Triassic) tetrapod tracks from Southern Alps. A) *Rhynchosauroides palmatus* (Lull, 1942); B) *Rhynchosauroides schochardti* (Rühle von Lilienstern, 1939); C) *Chirotherium rex* Peabody, 1948; D) *Isochirotherium infernense* Avanzini and Leonardi, 2002; E) cf. *Sphingopus* isp.; F) *Chirotherium barthii* Kaup, 1835; G) tridactyl footprint, possible dinosauriformes; H) *Isochirotherium delicatum* Courel & Demathieu ,1976; I) *Parasynaptichnium gracilis* Mietto 1987; L) *Synaptichnium pseudosuchoides* Nopcsa, 1923; M) *Synaptichnium cameronense* (Peabody, 1948); N) *Brachychirotherium* aff. *Parvum*; O) *Rhynchosauroides tirolicus* Abel, 1926; P) *Rhynchosauroides peabodyi* Faber, 1958. From Avanzini (2003), Avanzini & Mietto (2008).

Rotodactylus lucasi Demathieu & Gand, 1973
Sphingopus isp.
Synaptichnium cameronense (Peabody, 1948)
Synaptichnium pseudosuchoides Nopcsa, 1923
Synaptichnium diabloense Peabody, 1948
Therapsida indet.

In the considered sector, tetrapod tracks are not recorded from Ladinian deposits, in correspondence of a great rise of the sea level. Well documented associations reappear only in the Late Triassic (Tuvalian).

In the Anisian, a progressive increase in the complexity of the ichnoassociations and the size of the taxa from the Bithynian to the Illyrian is documented in the Southern Alps.

The analysis of the stratigraphic distribution of ichnofaunas crossed with sequence stratigraphy and the ammonite biostratigraphy has led to the identification, within the Pelsonian - Illyrian interval of several taxa characterized by a narrow vertical distribution (Avanzini & Mietto, 2008).

While the Aegean-Bithynian interval is characterized by several lizard-like ichnotaxa and small archosauriform (*Parasynaptichnium gracilis* and *Synaptichnium pseudosuchoides* FAD), the middle Pelsonian and the Illyrian are mainly dominated by medium-large chirotheroids (i.e., *Chirotherium barthii, C.* cf. *rex, Isochirotherium herculis*).

4. Ichnodiversity trends trough the Permo-Triassic

The likely producer of an ichnospecies is generally uncertain; tracks and trackmakers can only be associated with certainty when found in association (e.g., Voigt et al., 2007). However, for most studies the identity of the trackmaker does not need to be (and cannot be) determined at a specific or generic level. The range of the possible producers can be constrained within the least inclusive group that bounds all the taxa sharing similar morphological characteristics that fit with the features of the studied tracks (Carrano & Wilson, 2001). In the case of the track record here analysed this implies that different ichnogenera can be attributed to trackmakers belonging to different taxonomic categories. Furthermore, acknowledging that different substrates, behaviours and even different ichnotaxonomic traditions can bias our palaeobiodiversity and abundance estimates, the here suggested taxonomic designations are based on: i) the most conservative found in literature (i.e., broad), and ii) the most widely acknowledge (i.e., some single alternative hypothesis have been discarded) (Tab. 1).

For example the ichnogenus *Procolophonichnium* has been attributed to different biological groups such as procolophonids (Anapsida), therapsids (Synapsida) and basal amniotes (see Klein & Lucas, 2010; Klein et al., in press). The least inclusive group that bounds these taxa is the Amniota and *Procolophonichnium* is here conservatively considered as produced by an indeterminate stem group amniote (i.e., basal amniote). A coarse assignment of trackmaker identity has been shown to be still useful for evolutionary studies (e.g., Wilson and Carrano, 1999; Carrano & Wilson, 2001).

Although possibly unsafe, a link between ichnofaunal and faunal record in the Dolomites area is the only possible source of information about tetrapod life on lands in Permian and Triassic since, as state above, the bony record is nearly absent.

Ichnodiversity and faunal composition through time are shown in Figs. 4 and 5. Taken at face value, available data show that diversity (shear number of ichnospecies) grows through the Permian, then suddenly drops in (or until) the Olenekian and then reaches its maximum in the Anisian. A finer stratigraphic and temporal resolution would show that lowest and highest diversity are separated by circa 6 million years only (Avanzini & Mietto, 2008).

Ichnotaxon	**Attribution**
Amphisauropus	Seymouriamorpha
Batrachichnus	Stem group Amphibia
Brachychirotherium	Archosauria
Capitosauroides	Stem-group Amphibia
Chelichnus	Non-therapsid Eupelycosauria
Chirotherium	Archosauriformes
Dicynodontipus	Non-mammalian Therapsida
Dromopus	Araeoscelida
Erpetopus (*Camunipes*)	Captorhinidae
Ganasauripus	Lepidosauromorpha
Hyloidichnus	Captorhinidae
Ichniotherium	Diadectomorpha
Isochirotherium	Archosauriformes
Janusichnus	Stem-group Amniota
Pachypes	Pareiasauridae
Paradoxichnium	Lepidosauromorpha
Parasynaptichnium	Archosauriformes
Procolophonichnium	Stem-group Amniota
Protochirotherium	Archosauriformes
Rhynchosauroides	Lepidosauromorpha/Eosuchia
Rotodactylus	Dinosauromorpha
Sphingopus	Dinosauromorpha
Synaptichnium	Archosauriformes
Varanopus	Captorhinomorpha

Table 1. Permian-Triassic tetrapod ichnogenera and their inferred trackmakers (mainly after Haubold, 1996, 2000; Klein et al., 2011)

The Artinskian-Kungurian (Early Permian) ichnoassociation (Fig. 1) is comparable with coeval associations around the World (the so-called *Batrachichnus* ichnofacies; Lucas & Hunt, 2006) and document the presence of 5 different groups: amphibians, seymouriamorphs, diadectomorphs, captorhinomorphs and araeoscelids (Fig. 4). These are represented, in similar proportion, by 8 different ichnospecies. The most notable difference with other Lower Permian tracksites is the absence of tracks ascribed to large non-therapsid Eupelycosaur grade trackmakers (e.g. *Dimetropus*).

The richer Wuchiapingian (Late Permian) ichnoassociation (Fig. 2) is constituted of 14 ichnospecies which document the presence of 6 groups: diadectomorphs and captorhinomorphs are still present, together with lepidosauromophs, pareiasaurs, synapsids and archosauriforms (Fig. 4). This ichnoassociations, according to Lucas and Hunt (2006, p. 152), "the most characteristic ichnofauna being that of the Val Gardena Formation" allowed defining a *Pachypes* ichnocoenosis in the *Brontopodus* ichnofacies (Lucas and Hunt, 2006). As widely acknowledge, however, Southern Alps Late Permian ichnoassociation is far richer than elsewhere and, although ichnodiversity-based estimates can be misleading for the problems acknowledge before, it is interesting to note that, on a global scale, Wuchiapingian land vertebrate life experienced a dramatic drop in diversity (the end-Guadalupian crisis; see Benton, 1995; King, 1991; Retallack et al., 2006; Sahney & Benton, 2008).

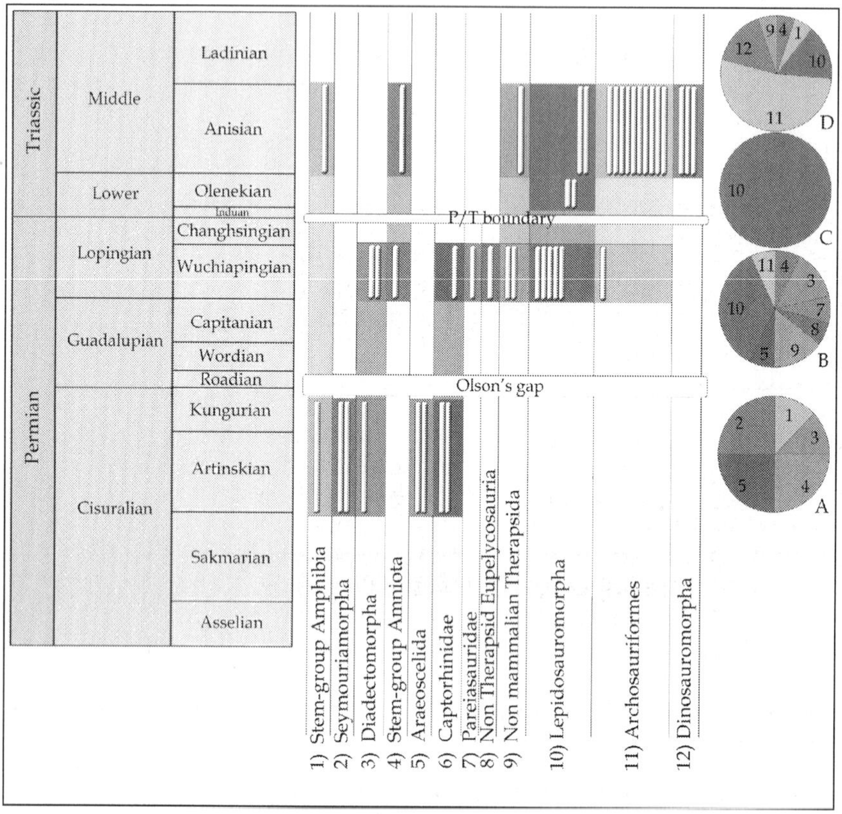

Fig. 1. Tetrapod ichnodiversity and faunal composition trough the Permian and Triassic in the Dolomites region (Southern Alps, NE Italy). White bars represent the ichnospecies listed in the text for each of the four studied ichnoassociation: A) Artinskian-Kungurian (Early Permian), B) Wuchiapingian (Late Permian), C) Olenekian (Early Triassic), and D) Anisian (Middle Triassic). Each color, corresponding to the cardinal numbers 1-12, represent any least inclusive group to which the ichnospecies are referred.

Olenekian (Early Triassic) tracks (Fig. 3) are all found isolated and poorly preserved. This surely has to do with the low preservation potential for tracks in the Southern Alps formations. However the few track-bearing levels within the three upper members of the Werfen Fm. undoubtedly record transitional continental-marine environments with high preservation potential: where tracks could be preserved a monotypic lepidosauromorph fauna is documented (2 ichnospecies). The interpretation of this datum in the considered geographical sector is however poorly supported: both an "ecological bias" (e.g., only the coast dwellers who were able to live on the ephemeral emerged lands left their footprints), or a real "evolutionary pattern" (e.g., recovery in diversity from PT mass extinction was extremely slow) would produce the observed pattern of diversity.

Anisian (Middle Triassic) ichnoassociation is very diverse being constituted by 18 ichnospecies. 5 groups are represented: lepidosauromophs, stem amphibians, "synapsids", archosauriforms and dinosauromorphs (Fig. 4). It is interesting to note that as soon as the Anisian, nearly 70% of the ichnodiversity in the considered geographical sector documents the presence of archosaurs (10 archosaur-produced + 3 dinosauromorph-produced ichnospecies). The pattern is consistent with those observed in Germany, France and Arizona (USA) (see Hunt and Lucas, 2007 and references therein).

5. Discussion and conclusions

Our analysis of the Permo-Triassic faunal composition in the Dolomites region as deduced from the track record well matches the skeletal record with the obvious exclusion of few, very rare forms, dubiously referred to high level taxonomic groups too inclusive to be meaningful (Amphibia, Amniota). In this way, incidentally, the footprint reliability for evolutionary studies was, once again, confirmed. This is in particular true when data are collected from rich and deeply studied ichnoassociations. Once stated the reliability of data, we can analyse the impact of two well known biological crises that have been documented world-wide in the Middle and Late Permian (see Hallam & Wignall, 1997; Retallak et al. 2006; Sheldon, 2006 and references therein) on mid Early Permian to late Early Triassic faunas in the Dolomites.

The effects of the first crisis, the so-called Olson's extinction, are documented in the considered geographical sector by: i) the progressive disappearance of long ranging taxa, typical of the Early Permian (and Carboniferous) as the Seymouriamorpha and the Araeoscelidia; ii) the persistence of groups as the Diadectids and the Captorhinomorphs; iii) the appearance of groups that will have their explosion during the Late Permian (Pareiasaurs and non-mammalian Therapsids). Contemporaneously it is worth of note the appearance of the first Lepidosauromorphs and of rare Archosauromorphs. On the contrary, the events recorded between Late Permian and late Early Triassic, across the end-Permian crisis, suggest a very low survival, and only of the groups that appeared during the Late Permian (Lepidosauromorph, non Mammalian Therapsid, Archosauromorphs). Noticeably, an early appearance and the abundant presence of *Rotodactylus* and *Sphingopus* in the Southern Alps also supports the hypotheses of an early origin, soon after the PT boundary, for the Dinosauromorpha (e.g., Brusatte et al., 2011).

The recognized trends, suggest different controlling factors for the two Permian crises (Fig. 5). The end-Guadalupian "crisis" (Benton, 1995, 1989; King, 1991; Retallack et al., 2006), or "extinction" (Sahney & Benton, 2008) seems to show the typical feature of the depletion of a long-lived association, mainly characterized by ecological replacement of the old taxa with new ones (famous examples, based on bone remains, are the replacement of Caseids with Pareiasaurs in the niche of large plant-eaters and of large Sphenacodontids with the non-mammalian Therapsid among carnivorous), and thus an event controlled by normal environment changes, more or less slow and gradual. The end-Permian crisis, instead, is recorded by the termination of short lived taxa, just appeared during the same Late Permian and can be better interpreted as the result of a mass-extinction, connected to a series of catastrophic events.

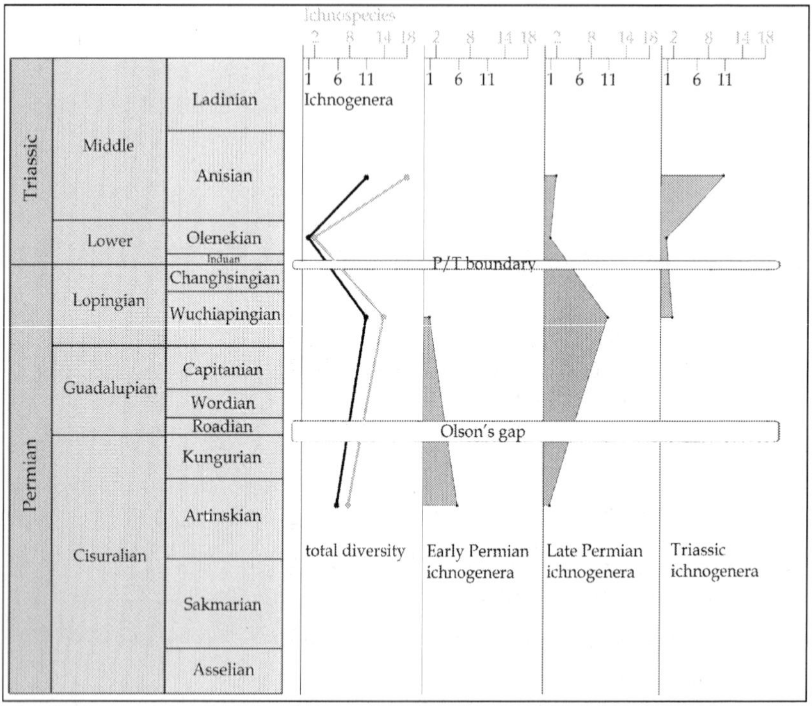

Fig. 5. Tetrapod ichnodiversity trough time in the Southern Alps. Colored polygons represent the duration of each of the studied ichnoassociations.

Thus, paradoxically, it may have been the end-Guadalupian, slow speed, crisis that, anticipating the substitution of obsolescent anatomical and ecological 'models' and forcing the appearance of entirely new taxa, allowed some land-animal groups overcoming the following and much more acute, end-Permian crisis.

6. Acknowledgments

We thank F.M. Petti and R. Todesco (Science Museum, Trento) for constructive discussions and M.C. Deflorian (Science Museum, Trento) for assistance with figures. This study is part of the project "The Permian-Triassic ecological crisis in the Dolomites: extinction and recovery dynamics in Terrestrial Ecosystems" financed by the Promotion of Educational Policies, University and Research Department of the Autonomous Province of Bolzano - South Tyrol. The research was supported also by the Science Museum, Trento and the Sapienza University of Rome.

7. References

Abel, O. (1926). Der erste Fund einer Tetrapodenfährte in den unteren alpinen Trias. *Paläontologische Zeitschrift*, Vol.7, pp. 22-24.

Assereto, R.; Bosellini, A., Fantini Sestini, N. & Sweet, W.C. (1973). The Permian–Triassic boundary in the Southern Alps (Italy). In: T*he Permian and Triassic Systems and their Mutual Boundary*, A., Logan & L.V., Hills (Eds.), *Alberta Society* of *Petroleum* Geologists Memories, Vol.2, pp. 176–199.

Avanzini, M. (1999). New Anisian vertebrate tracks from the Southern Alps. In: *Third International Symposium on lithographic limestones*, S. Renesto, (Ed.), 17-21, suppl. Rivista Museo Civico Scienze Naturali "E.Caffi" Bergamo, Bergamo, Italy.

Avanzini, M. (2000). *Synaptichnium* tracks with skin impressions from the Anisian (Middle Triassic) of the Southern Alps (Val di Non - Italy). *Ichnos*, Vol.7, pp. 243-251.

Avanzini, M. (2002). Dinosauromorph tracks from the middle Anisian (middle Triassic) of the Southern Alps (Valle di Non - Italy). *Bollettino della Società Paleontologica Italiana*, Vol.41, pp. 37-40.

Avanzini, M. (2002). Tetrapod footprints from the Mesozoic carbonate platforms of the Italian Alps. *Zubía*, Vol.21, pp. 175-186.

Avanzini, M.; Ceoloni, P., Conti, M.A., Leonardi, G., Manni, R., Mariotti, N., Mietto, P., Muraro, C., Nicosia, U., Sacchi, E., Santi, G. & Spezzamonte, M. (2001). Permian and Triassic tetrapod ichnofaunal units of Northern Italy: Their potential contribution to continental biochronology. *Natura Bresciana*, Vol.25, pp. 89-107.

Avanzini M.; Ferretti, P., Seppi, R. & Tomasoni, R. (2001). Un grande esemplare di orma 'dinosauroide' *Parachirotherium* Kuhn 1958 sp. dall'Anisico superiore (Illirico) del Sudalpino. *Studi Trentini di Scienze Naturali, Acta Geologica*, Vol.76, pp. 201-204.

Avanzini, M. & Leonardi, G. (2002). *Isochirotherium inferni* ichnosp. n. in the Upper Anisian (Illyrian) of Adige Valley (Bozen, Italy). *Bollettino Società Paleontologica Italiana*, Vol.41, pp. 41-50.

Avanzini, M. & Lockley, M. (2002). Middle Triassic archosaur ontogeny and population structure: interpretation based on *Isochirotherium delicatum* fossil footprints (Southern Alps - Italy). *Palaeogeography, Palaeoclimatology, Palaeoecology*, Vol.185, pp. 391-402.

Avanzini, M. & Mietto P. (2008). Lower and Middle Triassic footprints-based biochronology in the Italian Southern Alps. *Oryctos*, Vol.8, pp. 3-13

Avanzini, M & Neri, C. (1998). Impronte di tetrapodi da sedimenti anisici della Valle di Non (Trentino occidentale - Italia): nota preliminare. *Annali del Museo Civico di Storia Naturale di Ferrara*, Vol.1, pp. 5-19.

Avanzini M.; Neri C., Nicosia, U. & Conti, M.A. (2008). A new Early Permian ichnocoenosis from the "Gruppo vulcanico atesino"(Mt. Luco, Southern Alps, Italy). *Studi Trentini di Scienze Naturali, Acta Geologica*, Vol.83, pp. 231-236.

Benton, M.J. (1989). Mass extinctions among tetrapods and the quality of the fossil record. *Proceedings of the Royal Society, Series B*, Vol.325, pp. 369-386.

Benton, M.J. (1995). Diversification and extinction in the history of life. *Science*, Vol.268, pp. 52–58.

Benton, M.J. & Twitchett, R.J. (2003). How to kill (almost) all life: the end-Permian extinction event. *Trends in Ecology and Evolution*, Vol.18, pp. 358-365.

Berruti G., (1969). Osservazioni biostratigrafiche sulle formazioni continentali pre-quaternarie delle valli Trompia e Sabbia. II. Sulla fauna fossile della Formazione di Collio (alta Val Trompia). *Natura Bresciana, Annali del Museo Civico di Scienze Naturali, Brescia*, Vol.6, pp. 3-32.

Bosellini, A. (1964). Stratigrafia, petrografia e sedimentologia delle facies carbonatiche al limite Permiano-Trias nelle Dolomiti occidentiali. Memorie del Museo di Storia Naturale della Venezia Tridentina, Vol.15, pp. 59-110.

Brandner, R. (1973). Tetrapodenfährten aus der unteren Mitteltrias der Sűdalpen. *Veröffentiichungen der Universiätt Innsbruck*, Vol.86, pp. 57-71.

Broglio-Loriga, C. & Cassinis, G. (1992). The Permo-Triassic boundary in the Southern Alps (Italy) and in adjacent Periadriatic regions. In: *Permo-Triassic events in the eastern Tethys*, W.C. Sweet, Y. Zunyi, J.M. Dickins & Y. Hongfu (Eds.), 78–97, Cambridge University Press, Cambridge, United Kingdom.

Broglio Loriga, C.; Neri, C. & Posenato, R. (1986). The Early macrofaunas of the Werfen Formation and the Permian Triassic boundary in the Dolomites (Southern Alps, Italy). *Studi Trentini di Scienze Naturali, Acta Geologica*, Vol.62, pp. 3-18.

Brusatte, S.L.; Niedźwiedzki, G. & Butler, R.J. (2011). Footprints pull origin and diversification of dinosaur stem-lineage deep into Early Triassic. *Proceedings of the Royal Society of London, Series B*, Vol.278, pp. 1107-1113.

Carrano, M.T. & Wilson, J.A. (2001). Taxon distributions and the tetrapod track record. *Paleobiology*, Vol.27, pp. 563-581.

Casati, P., 1969 - Strutture della Formazione di Collio (Permiano Inferiore) Nelle Alpi Orobie. *Natura*, Vol.60, pp. 301-312.

Casati, P. & Forcella, F. (1988). Alpi Orobie. *'Il Bollettino', CAI-Annuario del Comitato Scientifico centrale*, Vol.87, pp. 81-86.

Cassinis, G. (1966). La Formazione di Collio nell'area-tipo dell'alta Val Trompia (Permiano inferiore bresciano). *Rivista Italiana di Paleontologia e Stratigrafia*, Vol.72, pp. 507-588.

Cassinis, G. & Doubinger, J. (1991). On the geological time of the typical Collio and Tregiovo continental beds in the Southalpine Permian (Italy), and some additional observations. *Atti Ticinensi di Scienze della Terra*, Vol.34, pp. 1-20.

Cassinis, G. & Doubinger, J. (1992). Artinskian to Ufimian palynomorph assemblages from the Central Southern Alps, Italy and their regional stratigraphic implications. In: *Contributions to Eurasian Geology*, A.E.M. Nairn & V. Koroteev (Eds), 9-18, Occasional Publication ESRI New Series 8b, Part I, Columbia University of South Carolina, South Carolina, USA.

Cassinis, G.; Cortesogno, L., Gaggero, L., Massari, F., Neri, C., Nicosia, U. & Pittau, P. (1999). *Stratigraphy and facies of the Permian deposits between Eastern Lombardy and the Western Dolomites*, Field Trip Guidebook, Earth Scences Department, Pavia University, Brescia, Italy.

Cassinis, G.; Massari, F., Neri, C. & Venturini, C. (1988). The continental Permian in the Southern Alps (Italy). A review. *Zeitschrift geologische Wissenshaftliche*, Vol.16, pp. 1117-1126.

Cassinis, G.; Nicosia, U., Pittau, P. & Ronchi, A. (2002). Palaeontological and radiometric data from the Permian continental deposits of the Central-Eastern Southern Alps (Italy), and their stratigraphic implications. Mémoires de l'Association des Géologues du Permien, Vol.2, pp. 53-74.

Ceoloni, P.; Conti, M.A., Mariotti, N. & Nicosia, U. (1986). New Late Permian tetrapod footprints from Southern Alps. *Memorie della Società Geologica Italiana*, Vol.34, pp. 45-65.

Ceoloni, P.; Conti, M.A., Mariotti, N., Mietto, P. & Nicosia, U. (1987). Tetrapod footprints from Collio Formation (Lombardy, northern Italy). *Memorie Scienze Geologiche Università di Padova*, Vol.39, pp. 213-233.

Ceoloni, P.; Conti, M.A., Mariotti, N., Mietto, P., and Nicosia, U. (1988). Tetrapod footprint faunas from Southern and Central Europe. *Zeitschrift Geologische Wissenschaften*, Vol.16, pp. 895-906.

Conti M.A.; Leonardi, G., Mariotti, N. & Nicosia, U. (1975). Tetrapod footprints, fishes and molluscs from the Middle Permian of the Dolomites (N. Italy). *Memorie Geopaleontologiche dell' Università di Ferrara*, Vol.3, pp. 139-150.

Conti M.A.; Leonardi, G., Mariotti, N. & Nicosia, U. (1977). Tetrapod footprints of the "Val Gardena Sandstone" (North Italy). Their paleontological, stratigraphic and paleoenvironmental meaning. *Palaeontographia Italica*, Vol.70, pp. 1-91.

Conti M.A.; Leonardi G., Mariotti, N. & Nicosia, U., (1980). Nuovo contributo alla stratigrafia delle "Arenarie di Val Gardena". *Memorie della Società Geologica Italiana*, Vol.20, pp. 357-363.

Conti, M.A.; Leonardi, G., Mietto, P. & Nicosia, U. (2000). Orme di tetrapodi non dinosauriani del Paleozoico e Mesozoico in Italia. In: *Dinosauri in Italia. Le orme giurassiche dei Lavini di Marco (Trentino) e gli altri resti fossili italiani*, G. Leonardi & P. Mietto (Eds.), 297-320, Accademia Editoriale, Pisa, Italy.

Conti M.A.; Mariotti, N., Manni, R., & Nicosia, U. (1999). Tetrapod footprints in the Southern Alps: an overview. In: *Stratigraphy and facies of the Permian deposits between eastern Lombardy and the western Dolomites*, G. Cassinis, L. Cortesogno, L. Gaggero, F. Massari, C. Neri, U. Nicosia & P. Pittau (Eds.), 137-138, International Field Conference on 'The continental Permian of the Southern Alps and Sardinia (Italy). Regional reports and general correlations', Field Trip Guidebook.

Conti, M.A.; Mariotti, N., Mietto, P. & Nicosia, U. (1991). Nuove ricerche sugli icnofossili della Formazione di Collio in Val Trompia. *Natura Bresciana, Annali del Museo Civico di Storia Naturale Brescia*, Vol.26, pp. 109-119.

Conti, M.A.; Mariotti N., Nicosia, U. & Pittau, P. (1997). Selected bioevents succession in the continental Permian of the Southern Alps (Italy): improvements of intrabasinal and interregional correlations. In: *Late Palaeozoic and Early Mesozoic Circum-Pacific Events and Their Global Correlation*, J.M., Dickins, Y., Zunyi, Y., Yhongfu, S.G., Lucas & S.J., Acharyya (Eds.), 51-65, Cambridge University Press, Cambridge, United Kingdom.

Curioni, G. (1870). Osservazioni geologiche sulla Val Trompia. *Rendiconti Istituto Lombardo Scienze Lettere Arti, Memorie*, Vol.3, pp. 1-60.

Dozy, J.J. (1935). Einige Tierfärhten aus dem Unteren Perm der Bergamasker Alpen. *Paläontologische Zeitschrift*, Vol.17, pp. 45-55.

Farabegoli, E. & Perri, M.C. (1998). Permian/Triassic boundary and early Triassic of the Bulla section (Southern Alps, Italy): Lithostratigraphy, facies and conodont biostratigraphy. *Giornale di geologia*, Vol.60, pp. 292-311.

Farabegoli, E.; Perri, M.C. & Posenato, R. (2007). Environmental and biotic changes across the Permian-Triassic boundary in western Tethys: The Bulla parastratotype, Italy. *Global and Planetary Change*, Vol.55, pp. 109-135.

Geinitz, H.B. (1869). Über Fossile Pflanzenreste aus der Dyas von Val Trompia. *Neue Jährbuch für Mineralogie Geologie Paläontologie*, pp. 456-461.

Gümbel, C.W. (1880). Geognostische Mitteilungen aus den Alpen. VI. Ein geognostischer Streifzug durch die Bergamasker Alpen. *Sitzungsberichte der königl Akademie der Wissenschaften Mathematisch-Naturwissenschaftliche Klasse*, Vol.10, pp. 164-240.

Hallam, A. & Wignall, P.B. (1997). *Mass extinctions and their aftermath.* Oxford University Press, Oxford.

Haubold, H. (1996). Ichnotaxonomie und Klassifikation von Tetrapodenfährten aus dem Perm. *Hallesches Jahrbuch für Geowissenschaften B*, Vol.18, pp. 23-88.

Haubold, H. (2000). The tetrapod tracks of the Permian - state of knowledge and progress 2000. *Hallesches Jahrbuch für Geowissenschaften, B*, Vol.22, pp. 1-16.

Hunt, A.P. & Lucas, S.G. (2007). Late Triassic tetrapod tracks of western North America. *New Mexico Museum of Natural History and Science Bulletin*, Vol.40, pp. 215–230.

Italian IGCP 203 Group (1986). *Permian and Permian-Triassic boundary in the South-Alpine segment of the western Tethys.* Field guide-book, SGI and IGCP Project 203, Pavia.

King, G.M. (1991). Terrestrial tetrapods and the end Permian event: a comparison of analysis. *Historical Biology*, Vol.5, pp. 239–255.

Kittl, E. (1891). Saurier Färthe von Bozen. *Mitteilungen/Österreichischer Touristen-Club Wien*, Vol.3, pp. 7.

Klein, H. & Lucas, S.G. (2010).Tetrapod footprints - their use in biostratigraphy and biochronology of the Triassic. *Geological Society, London, Special Publications*, Vol.334, pp. 419-446.

Klein, H.; Voigt, S., Saber, H., Schneider, J.W., Hminna, A., Fischer, J., Lagnaoui A. & Brosig, A. (In press). First occurrence of a Middle Triassic tetrapod ichnofauna from the Argana Basin (Western High Atlas, Morocco). *Palaeogeography, Palaeoclimatology, Palaeoecology.*

Leonardi, P. (1967). *Le Dolomiti: Geologia dei monti tra Isarco e Piave.* Edizioni Manfrini, Rovereto.

Leonardi P.; Conti, M.A., Leonardi, G., Mariotti, N. & Nicosia, U. (1975). *Pachypes dolomiticus* n. gen. n. sp.; Pareiasaur footprint from the "Val Gardena Sandstone" (Middle Permian) in the Western Dolomites (N. Italy). *Atti dell'Accademia Nazionale dei Lincei, Rendiconti, Classe di Scienze Fisiche Matematiche e Naturali*, Vol.57, pp. 221-232.

Leonardi, G. & Nicosia, U. (1973). Stegocephaloid footprint in the Middle Permian Sandstone ("Grödener Sandsteine") of the western Dolomites. *Annali dell'Università di Ferrara*, Vol.5, pp. 116-124.

Lucas, S.G. (2004). A global hiatus in the Middle Permian tetrapod fossil record. *Stratigraphy*, Vol.1, pp. 47-64.

Lucas, S.G. & Hecker, T.A.B. (2001). Olson's gap: A global hiatus in the record of Middle Permian tetrapods. *Journal of Vertebrate Paleontology*, Vol.21, pp. 75A.

Lucas, S.G. & Hunt, A.P. (2006). Permian tetrapod footprints: biostratigraphy and biochronology. In: *Non-marine Permian biostratigraphy and biochronology*, 179-200, S.G., Lucas, G., Cassinis & J.W., Schneider (Eds.), Geological Society, London, Special Publications, 265.

Mietto P. (1975). Orme di tetrapodi nelle arenarie permiche di Recoaro (Vicenza). *Studi Trentini di Scienze Naturali, Acta Geologica*, Vol.52, pp. 57-67.

Mietto P. (1981). Una grande impronta di pareiasauro nel Permiano di Recoaro (Vicenza). *Rendiconti della Società Geologica Italiana*, Vol.4, pp. 363-364.

Mietto, P. (1986). Orme di tetrapodi nella Formazione di Werfen del Recoarese. *Rivista Italiana di Paleontologia e Stratigrafia*, Vol.92, pp. 321-326.

Mietto, P. (1987). *Parasynaptichnium gracilis* nov. ichnogen., nov. isp. (Reptilia: Archosauria Pseudosuchia) nell'Anisico inferiore di Recoaro (Prealpi vicentine - Italia). *Memorie Scienze Geologiche*, Vol.39, pp. 37-47.

Nicosia, U.; Sacchi, E. & Spezzamonte, M. (1999). New Palaeontological data on the Val Gardena Sandstone. In: *Stratigraphy and facies of the Permian deposits between eastern Lombardy and the western Dolomites,* G. Cassinis, L. Cortesogno, L. Gaggero, F. Massari, C. Neri, U. Nicosia & P. Pittau (Eds.), 33, International Field Conference on 'The continental Permian of the Southern Alps and Sardinia (Italy). Regional reports and general correlations', Abstract book.

Nicosia, U.; Ronchi, A. & Santi, G. (2000). Permian tetrapod footprints from W Orobic Basin (Northern Italy). Biochronological and evolutionary remarks. *Geobios*, Vol.33, pp. 753-768.

Pasini, M. (1985). Biostratigrafia con i foraminiferi del limite Formazione a Bellerophon / Formazione di Werfen fra Recoaro e la Val Badia (Alpi Meridionali). *Rivista Italiana di Paleontologia e Stratigrafia*, Vol.90, pp. 481-510.

Posenato, R. (1988). The Permian/Triassic boundary in the western Dolomites, Italy. Review and proposal. *Annali dell'Università di Ferrara, Scienze della Terra*, Vol.1, pp. 31-45.

Retallack, G.J.; Metzger, C.A., Greaver, T., Jahren, A.H., Smith, R.M.H., Sheldon, N.D. (2006). Middle-Late Permian mass extinction on land. *Geological Society of America Bulletin*, Vol.118, pp. 1398–1411.

Ronchi, A.; Santi, G. & Confortini, F. (2005). Biostratigraphy and facies in the continental deposits of the central Orobic Basin: A key section in the Lower Permian of the southern Alps (Italy). *New Mexico Museum of Natural History & Science Bulletin*, Vol.30, pp. 273-281.

Santi, G. (2003). Early Permian tetrapod ichnology from the Orobic Basin (Southern Alps-Northern Italy). Data, problems, hypotheses. *Bollettino della Società Geologica Italiana, Volume Speciale*, Vol.2, pp. 59-66.

Santi, G. (2005). Lower Permian palaeoichnology from the Orobic Basin (Northern Alps). *GeoAlp*, Vol.2, pp. 77-90.

Santi, G. (2007). A short critique of the ichnotaxonomic dualism *Camunipes-Erpetopus*, Lower Permian ichnogenera from Europe and North America. *Ichnos*, Vol.14, pp.185-191.

Santi, G. & Krieger, C. (2001). Lower Permian tetrapod footprints from Brembana Valley–Orobic Basin (Lombardy, Northern Italy). *Revue de Paléobiologie*, Vol.20, pp. 45-68.

Schaltegger, U. & Brack, P. (1999). Radiometric age constraints on the formation of the Collio Basin (Brescian Prealps). In: Stratigraphy and facies of the Permian deposits between eastern Lombardy and the western Dolomites, G. Cassinis, L. Cortesogno, L. Gaggero, F. Massari, C. Neri, U. Nicosia, P. Pittau (Eds.), 15-25, International Congress on 'The continental Permian of the Southern Alps and Sardinia (Italy). Regional reports and general correlations', Field trip guide-book, 71, Brescia, Italy.

Sahney, S. & Benton, M.J. (2008). Recovery from the most profound mass extinction of all time. *Proceedings of the Royal Society, Series B*, Vol.275, pp. 759-765.

Sheldon, N.D. (2006). Abrupt chemical weathering increase across the Permian–Triassic boundary. *Palaeogeography, Palaeoclimatology, Palaeoecology*, Vol.231, pp. 315-321.

Sheldon, N.D. & Chakrabarti, R. (2010). Mass extinctions, climate change, and enhanced terrestrial weathering?: The End-Permian and End-Guadalupian events compared. *Geochimica et Cosmochimica Acta*, Vol.74S1, p. A944.

Todesco, R. & Bernardi, M. (2011). Una nuova icnoassociazione nel Triassico medio (Anisico) del Trentino meridionale (val Gerlano, Vallarsa). *Studi Trentini di Scienze Naturali*, Vol.88, pp. 203-218.

Todesco, R.; Wachtler, M., Kustatscher, E. & Avanzini, M. (2008). Preliminary report on a new vertebrate track and flora site from Piz da Peres (Anisian-Illyrian): Olang Dolomites, Northern Itlay. *Geo.Alp*, Vol.5, pp. 121-137.

Twitchett, R.J.; Looy, C.V., Morante, R., Visscher, H. & Wignall, P.B. (2001). Rapid and synchronous collapse of marine and terrestrial ecosystems during the end-Permian biotic crisis. *Geology*, Vol.29, pp.351–354.

Valentini, M.; Conti, M.A. & Mariotti, N. (2007). Lacertoid footprints of the Upper Permian Arenaria di Val Gardena formation (Northern Italy). *Ichnos*, Vol.14, pp. 193-218.

Valentini M.; Nicosia, U. & Conti, M.A. (2009). A re-evalution of *Pachypes*, a pareiasaurian track from the Late Permian. *Neues Jahrbuch für Geologie und Paläontologie Abhandlungen*, Vol.251, pp. 71-94.

Visscher, H. & Brugman, W.A. (1988). The Permian-Triassic boundary in the Southern Alps: A palynological approach. *Memorie della Società Geologica Italiana*, Vol.34, pp. 121- 128.

Voigt, S.; Berman, D.S. & Henrici, A.C. (2007). First well established track-trackmaker association of Paleozoic tetrapods based on *Ichniotherium* trackways and diadectid skeletons from the Lower Permian of Germany. *Journal of Vertebrate Paleontology*, Vol.27, pp. 553–570.

Wignall, P.B.; Kozur, H. & Hallam, A. (1996). The timing of palaeoenvironmental changes at the Permo-Triassic boundary using conodont biostratigraphy. *Historical Biology*, Vol.11, pp. 39–62.

Wilson, J.A. & Carrano, M.T. (1999). Titanosaurs and the origin of "wide-gauge" trackways: a biomechanical and systematic perspective on sauropod locomotion. *Paleobiology*, Vol.25, pp. 252–267.

Wopfner, H. (1999). Über Tetrapoden-Fährten, Kohlen und versteinerte Holzer aus dem Grödner Sandstein (Perm) dei Deutschnofen. *Der Schlern*, Vol.73, pp. 23-32.

26

Genomics of Bacteria from an Ancient Marine Origin: Clues to Survival in an Oligotrophic Environment

Luis David Alcaraz[1], Varinia López-Ramírez[2], Alejandra Moreno-Letelier[3], Luis Herrera-Estrella[4], Valeria Souza[5] and Gabriela Olmedo-Alvarez[2]

[1]Department of Genomics and Health, Center for Advanced Research in Public Health, Valencia,
[2]Departamento de Ingeniería Genética, Cinvestav Unidad Irapuato,
[3]Division of Biology, Imperial College London, Silwood Park Campus, Ascot,
[4]Langebio, Cinvestav, Mexico
[5]Departamento de Ecologia Evolutiva, Instituto de Ecología, Universidad Nacional Autónoma de México
[1]Spain
[2,4,5]Mexico
[3]UK

1. Introduction

Genomics has certainly changed the way that biology is studied, and has had a substantial impact on many other scientific disciplines as well. Life and Earth have had an interdependent history since the early establishment of the biogeochemical cycles in the Archean. Genomic sequences provide historical information that can be correlated with the geological record. Thus, it is not surprising that comparative genomics aids in understanding current findings in geological sciences. Genomics allows us to explain the evolutionary history of an organism by analyzing and comparing the set of shared genes with all respective relatives. Additionally, by examining the genes that are unique to some strains or taxonomic groups, we can make inferences about their ecology. Since molecular biology was initially developed in bacterial model organisms and we have extensive knowledge about the enzymes that participate in different biochemical pathways, inferences and functional predictions can be made about numerous sequenced genes. Moreover, due to the energetic and evolutionary costs of preserving a gene in bacteria, where specialists tend to have small size genomes, unique genes may aid in exploiting a given niche. Using comparative genomics, we have undertaken the study of the diversity, evolutionary relationships, and adaptations of bacteria to the oligotrophic (low nutrient content) conditions of the unique ecosystem of Cuatro Cienegas, Coahuila, Mexico. In this chapter, we review how genomics can be used to describe and understand microbial diversity, evolution, and ecology in different environments as well as the relationship with geological data by analyzing current studies, and our own work as a case study (Fig. 1).

2. Bacterial comparative genomics: Clues to evolution, geological history, and ecology

Since the early 1980's, science has had technology to determine the precise nature of DNA, which is the molecule found in all genomes. The era of genomics began with the sequencing of the first bacterial genome, *Haemophilus influenzae,* in 1995. The sequencing of complete genomes has since gained speed and precision, and with the costs of this technology continually decreassing, the sequencing of genomes today that contain millions of base pairs of DNA is possible in just a few days. Over 1,800 bacterial genomes have been sequenced to date, with several hundred more currently in progress. More impressively, the very large genomes of several plants and animals, including the human genome with more than 3 billion bases of DNA, have also been completed. In order to understand how particular genomes give rise to different organisms with very different traits, the ability to compare and contrast features of this wide range of genomes is needed. Moreover, an understanding of the evolutionary history and ecology of the organisms for which we have genomes is required.

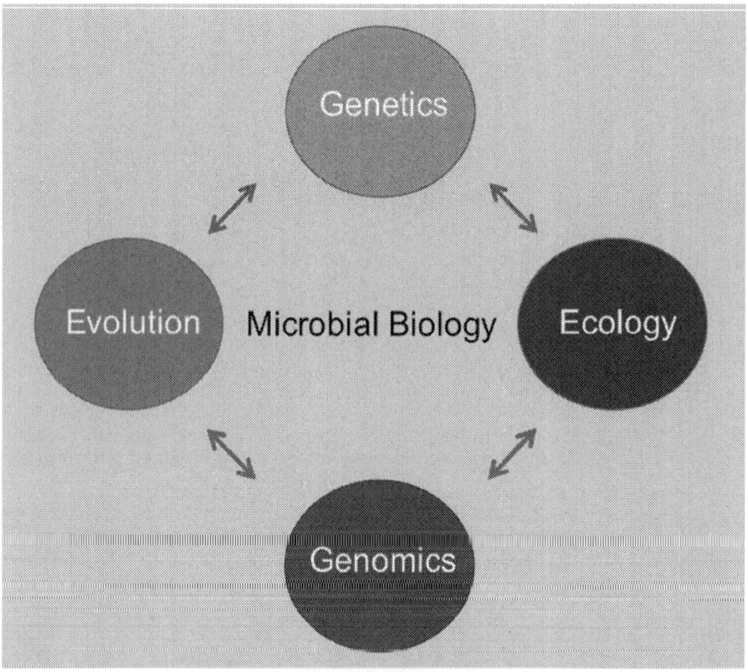

Fig. 1. Integration of Genetics, Genomics, Evolution, and Ecology. Genomics has had a great impact on Biology, particularly in the areas of Genetics, Ecology, and Evolution. Therefore, none of these areas can be studied separately. For microbial Biology, genomic studies have revealed a plethora of microbes that had eluded previous experimental strategies, which has provided the opportunity to understand their influence on the planet's history and present function in the environment.

2.1 How genomic sequences provide clues for understanding the diversity and origin of microorganisms

Classical microbiological classification systems made use of a culturable microbial diversity that only accounted for 1% of the total estimated microbial diversity (Amann et al. 1995; Khan et al. 2010), and due to the historical and clinical importance, it was a classification system that was highly biased towards clinical pathogenic strains. Although the 1% of culturable bacteria is an over-sold idea, it is clear that new culturing media should be developed and tested in order to capture more diversity within a given environment. To date, the most extensively used method for estimating the diversity of bacterial species is sequencing and comparing the 16S rRNA gene, which is an approach that has been used since the early 1970s (Woese & Fox 1977). Moreover, this method has been used in the current culture-free approaches, such as the metagenome surveys, by means of the Next Generation Sequencing (NGS) systems (Metzker 2010) in order to sequence DNA in particular environments (Fig. 2).

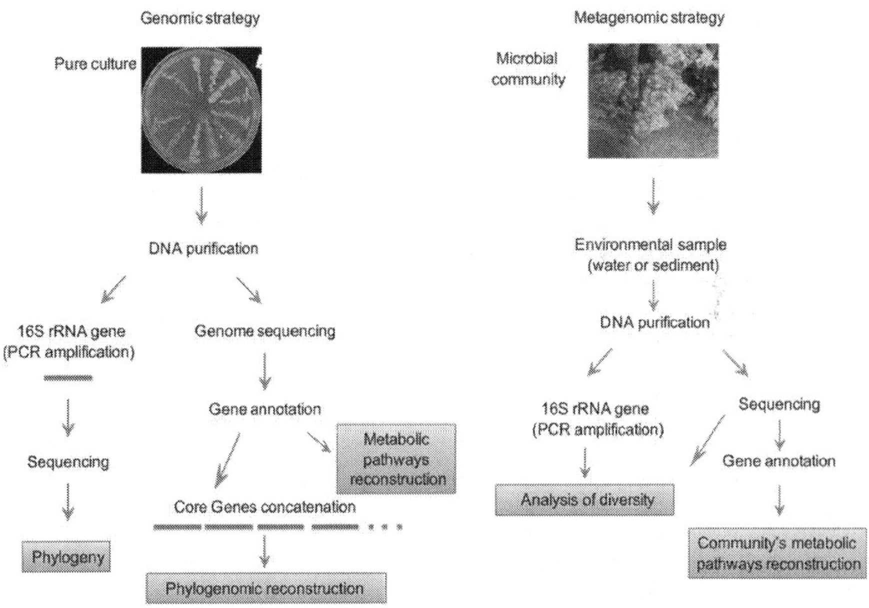

Fig. 2. Sequencing strategies for specific isolates and metagenomes.

The 16S rRNA gene sequencing approach has shed light onto the complex relationships amongst bacteria, and has even predicted the existence of new groups when using culture-free approaches. However, today we can go beyond explaining the evolution of a single gene and try to explore entire genome relationships of bacteria. With the development of NGS technologies, the associated costs are falling, and we can now envision a scenario where a small-sized lab is able to sequence the complete genome of a bacterial strain of interest. With whole genome sequencing, it is possible to perform a deep analysis to explain the relationships of every gene in genomes rather than single gene histories. Moreover, we can also explain the history and presence of bacteria by understanding the set of genes shared by all relatives, which can reveal a common evolutionary lineage as well as the genes that are exclusive to particular strains. The later genes, which are unique to certain strains or taxonomic groups, can also give us insight on the capabilities of an organism to exploit a given niche and will therefore provide information on environmental features as well.

The numerous genome sequencing projects provide clues for elucidating whole genome relatedness and functions that are exclusive to some groups. On another level of complexity, global scale efforts are conducted to attempt to determine the roles of microbes in the global biogeochemical cycles as well as the discovery and inventory of gene diversity, such as that from the Global Ocean Expedition (Rusch et al. 2007). Other projects focus on a deep understanding of the relationships between our body and the microbes with which we live. This project, known as the Human Microbiome Project (Turnbaugh et al. 2007), attempts to shed light on the close relationship between humans and our microbes. Another challenging project has to do with the sequencing of at least one representative strain of all of the major bacteria groups, known as the Genomic Encyclopedia of Bacteria and Archaea (GEBA) (Nelson et al. 2010), and the ambitious Earth Microbiome Project (http://www.earthmicrobiome.org/) will provide a reference framework for conducting comparative genomics and establishing deep relationships amongst microbes.

Some research areas, such as population genetics, are transitioning from an understanding of a very limited number of loci to the analysis of whole genome single nucleotide polymorphisms (SNPs) in order to understand fine differences between the populations of pathogens and free-living bacteria. In contrast, other research areas, such as Molecular Biology, have the power to analyze whole genome regulation by sequencing the entire RNA of an organism that is exposed to a specific condition (transcriptomics) and using the results to predict the function of hypothetical genes. The improvement in the performance of computational resources as well as decreasing costs of these technologies has created tremendous analytical power. It is now possible to rent web services, such as cloud computing, and use highly efficient software online to assemble (Chaisson et al. 2004; Miller et al. 2008; Zerbino & Birney 2008), align (Kurtz et al. 2004), annotate (Aziz et al. 2008; Van Domselaar et al. 2005), and compare the more than one thousand bacterial genomes sequenced to date with in-house datasets. In addition, we are now able to democratically access the experimental and analytical tools to dissect both the evolution and functions of microbial life.

2.2 Core and accessory genomes and the pan genome concept

In recent years, several groups have started to make whole genome comparisons of close relative strains, which has led to new and unexpected results for what was supposed to be members of the same bacterial species, based on 16S rRNA gene analysis, DNA-DNA re-association experiments, and phenotypic traits (Medini et al. 2005; Reno et al. 2009; Tettelin et al. 2005). Comparative genomic analysis has revealed that within species with the exact same 16S rRNA gene sequence, differences in gene content can account for ± 30% of the total coding genes in one strain that are not found in their close relatives. To clearly understand this finding, we can make the comparison that if we look at this in percentages, humans are closer to chimpanzees (1% differences in coding genes) than two bacteria isolates causing the same disease.

From these observations, the genome of a species is now conceptually split into a core genome, which refers to all of the shared genes between analyzed strains, and the accessory genome, which comprises the strain-specific genes. Core genes are those that are found in all genomes from a given taxonomic group and make up the genes that provide essential functions, such as those for DNA conservation and expression (involved in translation, replication, and transcription) as well as central metabolic pathways. These genes are collectively considered housekeeping genes. Core genes are considered likely components of all isolates of a given species or genus. Since selection at these loci exerts a stabilizing rather than a diversifying influence, variations detected in these genes tend to be neutral, or nearly so, and will accumulate in a consecutive, clock-like manner (see section 9). Thus, these genes should be reliable indicators of evolutionary history. For instance, core 16S rRNA gene sequences are frequently used to reconstruct the phylogenies of microbial species, and metabolic or housekeeping genes are commonly used for genotyping.

Accessory genes are those that may or may not be present in a given strain, since they can typically be associated with mobile elements (i. e. phages or transposons), can encode genes for secondary biochemical pathways, or can code for functions that mediate interactions with the environment. However, there are examples of genes that escape detection as core genes because they have been replaced by genes from a different origin (xenologous genes) that serve the same function but cannot be identified by similarity. This is often the case for divergent alleles. In any case, the genes classified as accessory genes are very interesting from an ecological point of view, since the genes in this category would be those that reveal the function of an organism in its environment. In addition, the range of genetic diversity of a species can be discovered in these genes. Pan-genomes, on the other hand, help us to understand the gene repertoire that has yet to be discovered within the group. The pan-genome is defined as the sum of the core genome and the accessory genes. Pan-genomes are also classified as open, where in an accumulative plot there is still no plateau, and closed, where all of the expected genes are present in the group and newly discovered genes are just a product of chance regardless of how many new strains are sequenced. Pan-genomes help us determine when other genome dynamics are shaping interesting phenotypes. For example, when comparing strains of the *Bacillus cereus-anthracis* group (Anderson et al. 2005; Ehling-Schulz et al. 2005; Helgason et al. 2004), it was found that they were very clonal, and that their central pathogenic traits were the result of horizontal gene transfer in mobile elements.

A surprising finding in the many available microbial genomes was the high number of genes that had been acquired by exchange of DNA between microbes. The so-called "horizontal genetic transfer" events, confer a remarkable evolutionary potential to many species. The loss or gain of individual genes or large "genomic islands" accounts for the emergence of several specific metabolic, virulence, or drug resistance phenotypes. In fact, another characteristic of accessory genes is that these are often transferred among bacteria through the aid of phages, plasmids, or by transformation with free DNA. In a population, alleles (variant forms of a gene) may be distributed among the members, and individuals may possess either copy of the gene. These alleles (also called homeoalleles) encode enzymes with identical functions but that may have had a different evolution trajectory, and therefore exhibit differences at the gene level. Individual lineages may exhibit different homeoalleles acquired through horizontal gene transfer, but these homeoalleles may have also been lost from the lineage and replaced by other homeoalleles (reviewed in (Andam &

Gogarten 2011)). Importantly, horizontal gene transfer is responsible for the mosaicism often observed in genomes.

3. Using genes to understand the environment: A case study

3.1 The Cuatro Cienegas Basin: Water in a low phosphorous desert ecosystem

The Cuatro Cienegas Basin (CCB) is located in a valley in the central part of the State of Coahuila, Mexico (26°59'N 102°03'W). The Basin is roughly 84 km^2 in area and an average of 740 m above sea level (Fig. 3). CCB is surrounded by mountains that rise to elevations of approximately 2,500 and 3,000 m. The Geology and Physiography of the region have been thoroughly reviewed in the literature (Minckley 1969). Ancient geologic history indicates that the CCB was at the very nexus of the separation of Pangea, which created what we now know as the Northern hemisphere 220 million years ago. The CCB became isolated from the sea much later, and the subsequent uplifting of the Sierra Madre Oriental occurred approximately 35 million years. The major geological events of the Eocene epoch in northern Mexico corresponded to the genesis and development of the Sierra Madre Oriental, the fold-ranges of the Chihuahua-Coahuila and the Gulf coast plain development (Ferrusquía-Villafranca & González-Guzmán 2005).

Both climatically and geographically, the CCB belongs to the Chihuahuan Desert (Schmidt 1979), which is the second largest desert in North America. The climate is arid, with an average annual precipitation of less than 200 mm, and daytime temperatures in the summer that sometimes exceed 44°C. Despite the dry climate of the CCB, it harbors an extensive system of springs, streams, and pools. Within the valley, spring water flows on the surface and through subsurface channels in karstified alluvium. The main source for the subterranean water in these systems is old water that was deposited there in the late Pleistocene epoch (Wolaver 2008). Water in the CCB is thought to be the relict of a shallow sea that existed 35 million years ago. The oasis water contains low levels of NaCl and carbonates, but is rich in sulfates, magnesium, and calcium.

Vegetation and fauna in the CCB appear typical of an arid zone. Given the combined conditions of habitat diversity and permanence, as well as the isolation of the basin since historic times, elements of the aquatic fauna have undergone adaptive radiation and speciation that has resulted in many endemic organisms (reviewed in Holsinger & Minckley 1971; Minckley 1969).

The spring-fed ecosystems of the CCB are dominated by microbial mats and living stromatolitic features (see Fig. 3) that are supported by an aquatic sulfur cycle and a terrestrial gypsum-based ecology in large parts of the valley.

The most striking feature of the CCB ecosystem is the very low levels of phosphorus in both the water and soil, which presents an extreme elemental stoichiometry with regards to phosphorus (900:150:1- 15820:157:1 C:N:P ratio, respectively) (Elser 2005) when compared to similar environments. Phosphorus is an essential nutrient for multiple cellular processes, including energy and information. However, it is not an abundant element on the planet and can only be obtained from organic detritus or from tectonics and volcanism. Therefore, the availability of phosphorus is a limiting factor for all life forms. Nevertheless, life persevered and the CCB is characterized by a high endemism in all of the domains of life (Minckley 1969; Scanlan et al. 1993) despite the fact that phosphorus levels are below the level of detection (0.3 μM). In addition, the extremely oligotrophic waters are unable to sustain algal growth, which has caused the microbial mats to be the base of the food web (Elser 2005).

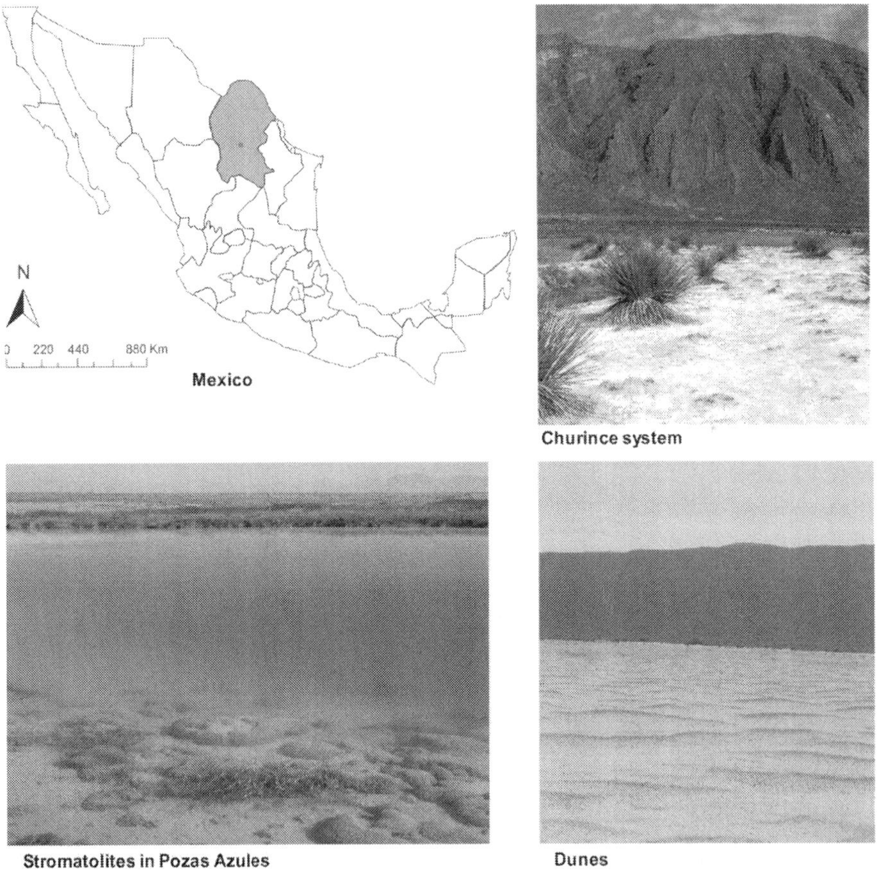

Fig. 3. The Cuatro Cienegas Basin. (a) Location map of Cuatro Cienegas, Coahuila. (b) Photograph of the Churince sytem facing San Marcos, one of the mountains in the valley. (c) One of the ponds at Pozas Azules, where live stromatolites can be observed. (d) White gypsum dunes, which are the result of evaporating water, once covered a vast area in the basin.

3.2 Bacterial genomics of two bacterial isolates from the CCB provide clues for the mechanism of bacterial survival in a highly oligotrophic environment

We have taken different approaches to gain knowledge on the diversity and evolution of microbial communities and to understand how microbes in these communities deal with the oligotrophic conditions in the various water bodies of Cuatro Cienegas. Several microbiological surveys have been carried out, including bacterial isolation and culturing, 16S rRNA gene amplification, genome sequencing, and metagenome sequencing.

Characterization of the microbial taxonomic diversity by sequence analysis of the 16S rRNA genes directly from environmental DNA has revealed that nearly half of the phylotypes from the CCB are closely related to bacteria from marine environments (Souza et al. 2006). This makes sense, given that the geological history of Cuatro Cienegas suggests that a shallow ocean covered the region and some of the water there is most likely a part of the underground water that feeds different ponds. This has led to the hypothesis that bacteria in the CCB water systems are descendants of marine bacteria from Pre-Cambrian (Souza et al. 2006). Systematic studies have also been used to describe new species of bacteria using current criteria (Anderson et al. 2005; Cerritos et al. 2008; Ehling-Schulz et al. 2005; Escalante et al. 2009; Helgason et al. 2004) as well as by attempting to apply multivariate analysis and physical-chemical analyses in order to determine special relationships amongst the bacteria (Cerritos et al. 2011). At the genomic scale, 2 genomes (Alcaraz et al. 2010; Alcaraz et al. 2008) and 3 metagenomes (Breitbart et al. 2009; Desnues et al. 2008; Dinsdale et al. 2008) had been described, in which genes that allow for adaptation to the oligotrophic environment of Cuatro Cienegas have been identified. Two additional metagenomes from microbial mats and several more microbial genomes are currently under analysis. The entire dataset that is currently available from this area has helped us identify shared features across different microbial communities that deal with the same constraints, including phosphorous limitation, and has helped us understand the genetic strategies that are available to cope with these conditions.

However, taxonomy has only provided a single molecular marker approach. In order to gather more evidence on whether bacteria from the CCB could be recent migrants from marine environments or ancient creatures from an old sea, it required more than one gene as a tool. Therefore, we sequenced the whole genome of two isolates: *Bacillus coahuilensis* (Alcaraz et al. 2008; Cerritos et al. 2008) and *Bacillus sp.* isolate m3-13 (Alcaraz et al. 2010). On one hand, an analysis of the gene content and metabolic pathways represented by each genome revealed numerous differences (Table 1), starting with genome size, as *B. coahuilensis* turned out to have the smallest genome of all sequenced *Bacillus* spp. (3.65 Mbp) and an incomplete genome in many functions.

Another gene family overrepresented in the genomic datasets from Cuatro Cienegas includes genes involved in environment sensing mechanisms. For example, all of the two-component (histidine-kinases) as well as the unusual sensitive rhodopsins, such as in the case of *B. coahuilensis,* were found to be overrepresented, indicating that most of the responses of the bacteria that have had their genomes analyzed are environmentally-triggered when compared to generalist and cosmopolitan organisms, such as *B. subtilis.* Moreover, these comparisons also showed an underrepresentation of secondary metabolism genes in the genome of *B. coahuilensis* (Alcaraz et al. 2010; Alcaraz et al. 2008; Desnues et al. 2008)(Fig. 4). Basic capabilities, such as sulfur utilization, were found to be different between the two isolates. *B. coahuilensis* appears not to be able to use inorganic sulfur and depends on organic sources for this element, while *Bacillus* m3-13 is able to use inorganic sulfur. In addition, the two isolates seem to be exposed to different stress, as *B. coahuilensis* has an alkyl hydroperoxide reductase protein coding gene that can aid with oxidative stress. Moreover, this strain also has several sugar transporters and biosynthesis pathways for sugars that help it deal with osmotic stress, such as trehalose, choline, and betaine uptake as well as betaine biosynthesis, all of which are absent in *Bacillus* m3-13.

In contrast, isolate m3-13 of *Bacillus* seemed to have a robust metabolism, starting with a complete urea cycle, capacity for taking up inorganic sulfur, and a wide selection of sugar transporters. Theoretically, the m3-13 strain would be able to use chitin, N-acetylglucosamine, maltose, maltodextrin, and sucrose to ferment lactate. It also has genes

Genomic features (differences)	*B. coahuilensis*	*Bacillus* **m3-13**
Genome size	3,640 Mbp	4,294 Mbp
Nitrogen metabolism	Lacks many genes, incomplete urea cycle	Complete urea cycle (Arginine and Ornithine Degradation)
Bacteriorhodopsin	Genes for **bacteriorhodopsin** and carotene synthesis	Lacks bacteriorhodopsin gene Pigmented, *crtB y crtI* (fitoene synthase and dehydrogenase)
Phosphate utilization genes	**Sulfoquinovose and glycosyl transferase genes**, sulfolipid synthesis	Phosphonate ABC transporter Genes for phosphonate utilization (C-P)
Phosphorous recycling	Alkaline phosphatase	Predicted ATPase related to phosphate starvation-inducible protein PhoH
Membrane Transport	Dipeptide-binding ABC transporter, periplasmic substrate-binding component	
Carbohydrates	Trehalose Uptake and Utilization	Chitin and N-acetylglucosamine Maltose and Maltodextrin Sucrose Fermentations: Lactate D-gluconate and ketogluconates metabolism D-ribose utilization Formaldehyde assimilation: Ribulose monophosphate pathway Beta-galactosidase
Cofactors, Vitamins, and Prosthetic Groups		Cobalamin, heme and siroheme, and thiamin biosynthesis
Sulfur metabolism	Sulfolipid synthesis	Inorganic Sulfur Assimilation: Adenylyl-sulfate reductase Ferredoxin--sulfite reductase; Sulfate permease, Pit-type
Stress Response	Osmotic stress: Choline and Betaine Uptake and Betaine Biosynthesi; Alkyl hydroperoxide reductase	
Virulence, Disease and Defense	Arsenical-resistance protein ACR3, Bacitracin export and resistance	Multidrug Resistance Efflux Pumps

Table 1. Genome differences between *B. coahuilensis* and *Bacillus* m3-13.

involved in D-gluconate and ketogluconates metabolism as well as D-ribose utilization. In addition, most of the amino acid biosynthesis pathways are complete. As expected in a place with extreme phosphorous limitation, some notable genetic features of those genomes are related to their ability to utilize phosphonate, which is discussed in the next section.

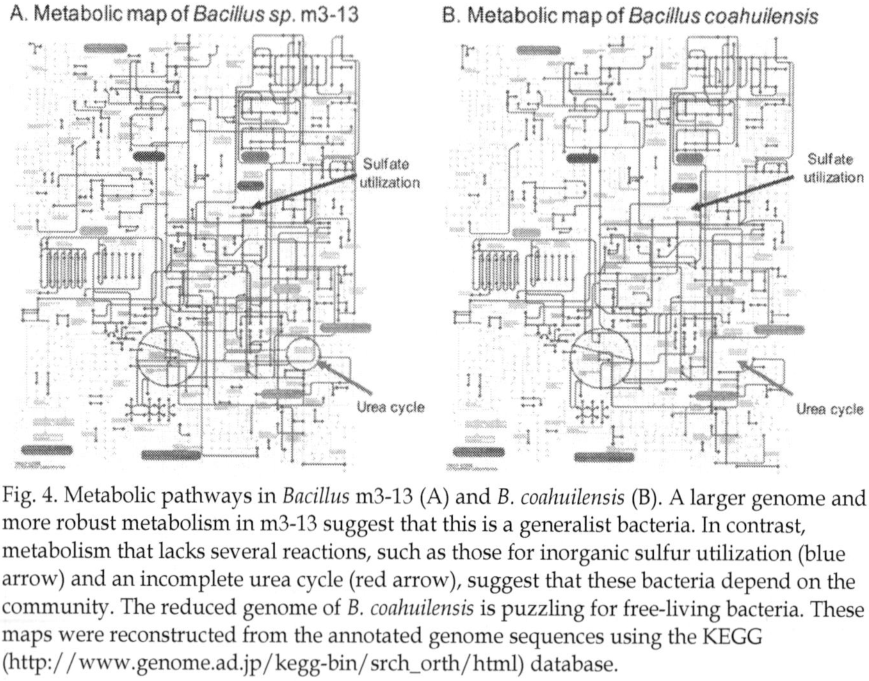

Fig. 4. Metabolic pathways in *Bacillus* m3-13 (A) and *B. coahuilensis* (B). A larger genome and more robust metabolism in m3-13 suggest that this is a generalist bacteria. In contrast, metabolism that lacks several reactions, such as those for inorganic sulfur utilization (blue arrow) and an incomplete urea cycle (red arrow), suggest that these bacteria depend on the community. The reduced genome of *B. coahuilensis* is puzzling for free-living bacteria. These maps were reconstructed from the annotated genome sequences using the KEGG (http://www.genome.ad.jp/kegg-bin/srch_orth/html) database.

4. Strategies of bacteria for dealing with limited phosphorous

Phosphorus is an essential nutrient for multiple processes, such as the synthesis of DNA, RNA, and ATP as well as many other pathways involving phosphorylation (Tetu et al. 2009). However, it is not an abundant element on the planet and can only be obtained from organic detritus or from tectonics and volcanism, and therefore the availability is a limiting factor for all life forms. Since the growth rate and primary productivity are highly dependent on phosphorus (Elser & Hamilton 2007; Elser et al. 2006; Zubkov et al. 2007), bacteria have different mechanisms for the uptake and storage of phosphates in order to cope with this limitation (Adams et al. 2008; Rusch et al. 2007; Tetu et al. 2009). For example, bacteria can use alternative phosphorus sources, use polyphosphates as storage compounds, or employ a highly effective phosphate-recycling mechanism (Fig. 5). Two major phosphate transport systems are involved: the low affinity phosphate inorganic transport (Pit) system, and the high affinity phosphate specific transport (Pst) system (van Veen, 1997). The uptake and assimilation of organic forms of phosphorus, such as phosphonates, require different transporters. Phosphonates are a class of organophosphorus compounds that are characterized by a chemically stable carbon-to-phosphorus (C-P) bond. Although phosphonates are widespread, only microorganisms are able to cleave this bond. The use of alternative phosphorus sources (phosphonates, phosphites, and hypophosphites) was determined by the presence of the high affinity transporters *pnhD* and *ptxB* as well as the C-P lyase genes *phnH* and *htxB* (White & Metcalf 2004).

Some microorganisms have been shown to accumulate relatively large amounts of polyphosphate, which has been hypothesized to have an important role in the response to changes in nutritional status or environmental conditions. Polyphosphate acts as a reservoir of intracellular phosphate, which is a strategy that seems to be particularly important for mobility and the formation of biofilms (Brown & Kornberg 2004). Amongst the genes induced under phosphorus deprivation are *ppA, ppK,* and *ppX* (coding for pyrophosphatase, polyphosphatase kinase, and exopolyphosphatase, respectively), which are involved in polyphosphate metabolism. Finally, extracellular phosphates are recycled by the overexpression of alkaline phosphatases *phoA* and *phoX* (Scanlan et al. 1993). Some bacteria can also utilize phytic acid (inositol hexakisphosphate (IP6)), which is the principal storage form of phosphorus in some plant tissues.

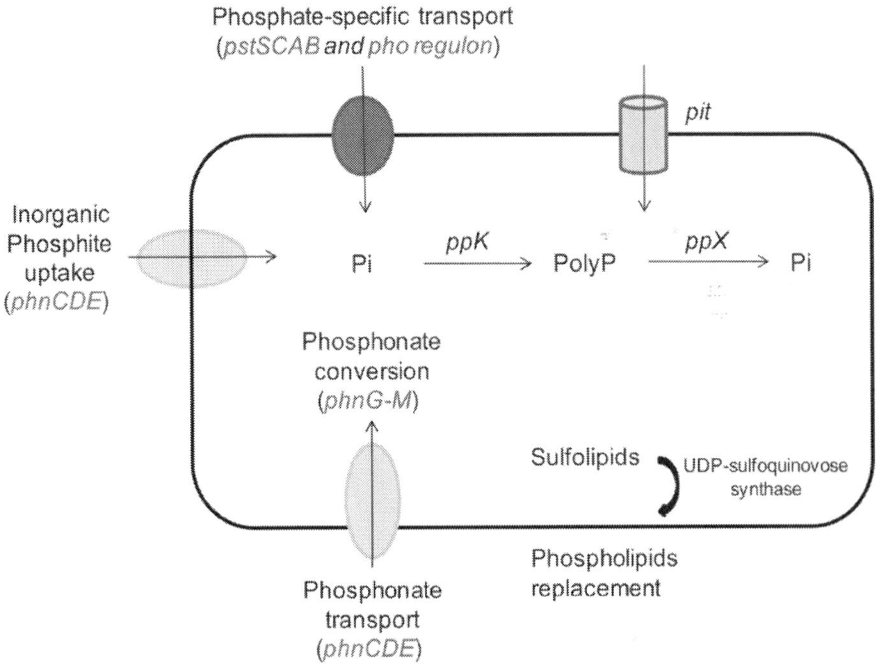

Fig. 5. Strategies for phosphorous utilization by bacteria (adapted from Hirota et al. 2010).

4.1 Phosphourous metabolism in the *Bacillus* genus

Since a low phosphorus concentration is a feature of the CCB basin, we focused on identifying the mechanisms of the *Bacillus* genus as a group, and in particular the isolates from the CCB for dealing with phosphorous. Comparative genomics can reveal the diversity

of mechanisms employed by the *Bacillus* strains where the genomes have been sequenced (Fig. 6). We observed a great diversity in strategy used by different *Bacilli*. For instance, Phytases seem to be restricted to the *B. subtilis-pumilus* group. These genes are a good example of accessory genes and reveal a specific niche for this group of bacteria. However, this same group lacks genes for polyphosphate or phosphonates metabolism. Alkaline phosphatases are widespread in the *Bacillus* genus. The *B. cereus-anthracis-thuringiensis* group is very homogeneous and has quite versatile mechanisms for phosphorous usage.
Among the CCB isolates, *Bacillus* m3-13 appears to be able to use phosphonates, but *B. coahuilensis* cannot, and these isolates have different alkaline phosphatase genes. Most species harbor the high affinity transporter PstS and lack the low affinity transporter Pit, which is present in *B. cereus* and *B. subtilis*. Remarkably, only *B. coahuilensis* has genes for the replacement of cell-membrane phospholipids to sulfolipids (see below). Finally, it has been proposed that genome reduction may be a strategy for reducing phosphorous utilization (Alcaraz et al. 2008; Desnues et al. 2008), and *B. coahuilensis* has the smallest genome among all sequenced *Bacilli* to date.

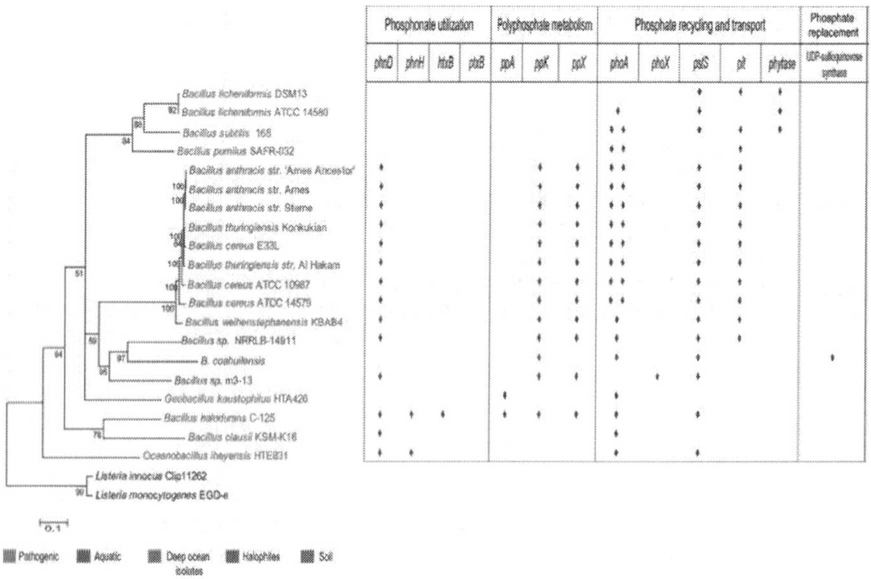

Fig. 6. The *Bacillus* genus phylogeny and the diversity in the strategies for phosphorous usage.

4.2 Sulfolipid synthesis in *B. coahuilensis*

Phospholipids constitute 30% of the total phosphate in most organisms. Interestingly, in plants and cyanobacteria subjected to phosphorous deprivation, phospholipids can be replaced by non-phosphorus lipids (such as sulfo- and galactolipids) to maintain membrane functionality and integrity, and can release phosphorus in order to sustain other cellular processes that require phosphorus (Dormann&Benning 2002). In addition, genes encoding sulfoquinovose synthase (*sqd1*) and glycosyltransferase (*sqdX*), which are the two key enzymes in the synthesis of sulfolipids, are present in *B. coahuilensis*. Thin layer chromatography and mass spectrometry analysis have confirmed the presence of sulfolipids in *B. coahuilensis*.

The remarkable acquisition of constitutively expressed genes that allow *B. coahuilensis* to replace membrane phospholipids with sulfolipids is in agreement with genomic adaptations to extreme phosphate limitation. However, not all *Bacilli* from the CCB ponds use the same strategy to cope with phosphorous limitations. *B. coahuilensis* and its relatives as well as a few other *Bacilli* species have this gene, but we did not find sulfolipids among several other *Bacillus* spp. from the CCB. Therefore, how do other bacteria strains cope with the limiting phosphorous conditions? As explained above, genomic sequencing of another strain, *Bacillus* isolate m3-13, showed that it possessed *phn* genes that code for phosphonate ABC importers, permeases, and a phosphonate lyase. We hypothesize that this strain may use these genes to take up and assimilate phosphonates. Importantly, both strategies seem to be used by bacteria from the Cuatro Cienegas as well as by marine bacteria (Fig. 6).

4.3 Mobile genes involved in phosphorous uptake

As discussed above, growth rate and primary productivity are highly dependent on phosphorus, and bacteria have different mechanisms for the uptake and storage of phosphates in order to cope with this limitation. The sulfoquinovose synthesis operon is absent in all other known *Bacillus* spp. genomes. The *B. coahuilensis* genes are closely related to cyanobacterial *sqd1* and *sqdX*, and the operon arrangement is identical to that in *Synechococcus* sp. PC7942, where these genes participate in the synthesis of sulfolipids (Benning 1998). This finding suggests that the adaptation of *B. coahuilensis* to the extremely low phosphorous concentration of the CCB may have included the acquisition of these genes through horizontal gene transfer (Alcaraz et al. 2008).

Another example of horizontal gene transfer involves genes of the *pho* regulon. The high affinity phosphate transport system (*pst*) is thought to be responsible for phosphate uptake under nutrient stress (Qi et al. 1997; Adams et al. 2008). *Pst* is a typical ABC transport system. Unlike the model bacteria *Escherichia coli* or *B. subtilis*, bacteria from the CCB as well as sequenced marine *Bacillus* lack the low affinity phosphate uptake system and must rely solely on the high affinity transport system. We found two types of operon architectures that harbor the *pst* gene and evaluated their phylogenetic congruity with housekeeping genes. We found high divergence of the two types of *pst*-operons in *Firmicutes* and incongruence with species phylogeny. In contrast to what was expected, the *pst* operon of marine *Bacillus* is not monophyletic, even though marine and the CCB *Bacilli* are resolved as a monophyletic group in the core-gene reconstruction. Therefore, the heterogeneous distribution of the different types of the *pst* operon among closely related species suggests horizontal gene transfer (Moreno-Letelier et al. 2011).

5. Carotenes and gene transfer of bacteriorhodopsin: A good combination in a high-radiation environment

B. coahuilensis has a gene encoding Bacteriorhodopsin, which is a situation similar to the abundance of BR genes in marine environmental samples, and suggests an additional adaptation of marine bacteria in the CCB. The phylogeny of *B. coahuilensis* sensory BR showed that its closest relative is the Anabaena sp. PCC7120 rhodopsin. Evidence for horizontal gene transfer of rhodopsins has recently been obtained from whole genome sequencing and metagenomic projects, and is now thought to be a frequent event in marine bacteria in the photic zone and extreme saline environments. The retinal chromophore of rhodopsin is synthesized as a cleavage product of carotenoids; thus, the combination of carotenoid synthesis and rhodopsin genes has been suggested to be sufficient for rhodopsin function (Frigaard et al. 2006). The genome of *B. coahuilensis* also contains genes encoding *crtB* (phytoene synthase) and *crtICA2* (phytoene dehydrogenases) that could be involved in retinal biosynthesis. The high radiation exposure that is prevalent in shallow waters of the CCB could explain the selection pressure responsible for the maintenance and constitutive expression of the *bsr* gene.

6. Core and pan-genome of the *Bacillus* genus

In order to understand the cohesion of the *Bacillus* genus at the genomic level, we used the core and pan-genomes as the working units and took advantage of the large dataset available. We have analyzed the genes comprising the core genome of the *Bacillus* genus. The core genome of the *Bacillus* spp. that was analyzed (see phylogeny in Fig. 6) contained 814 genes. After annotating each gene in the genome of the two isolates, it was possible to classify and assign the different genes to specific functional categories and reconstruct and compare their metabolic pathways. Figure 7 shows the drawing of a metabolic map representation of the Bacillus pan-genome and Bacillus core genome obtained with the Kyoto Encyclopedia of Genes and Genomes. Using the pan-genome, we are able to understand the variation in functions across a cosmopolitan genus that can survive under harsh conditions, such as the bottom of the sea, hydrothermal vents, hypersaline environments, or even simply the hosts, as is the case for pathogens. The average gene content for the *Bacillus* genus was 4,973 ± 923 and the total pan-genome involved a around 75,000 genes clustered in 19,043 gene families. This is a very large number if we consider that the most recent predictions show that the human genome harbors 20,000 genes (Nelson et al. 2011). From these analyses, it is evident that a vast repertoire of functions is encoded in the *Bacillus* genus, and helps to explain the versatility of these bacteria for living and surviving in harsh conditions.

6.1 The sporulation core and accessory genes

The core genome sequence can help to identify and understand relevant, conserved genes for a trait of an entire group, and we therefore analyzed sporulation (Alcaraz et al. 2010), as the *Bacillus* is a group that is defined as endospore-forming genus. To obtain insight into the biology of this group, we described the relatedness within *Bacillus* using whole genome information to reconstruct their evolutionary history by taking advantage of the dataset available from the complete and draft genomes of 20 Bacilli isolated from a wide range of environments. Nucleotide metabolism, cell motility, and secretion showed little variability. In contrast, the features that were highly represented within the genomes and varied the most among the different genes were related to repair and transcription. Secondary metabolism, as expected, also exhibited variation among the taxonomic groups. Figure 7 shows a comparison between the functions defined by the core genome versus those defined by the pan-genome. The latter, which covers numerous metabolic pathways and reflects the potential of the whole group, explains why this is a cosmopolitan bacterial genus with the capability of colonizing diverse niches.

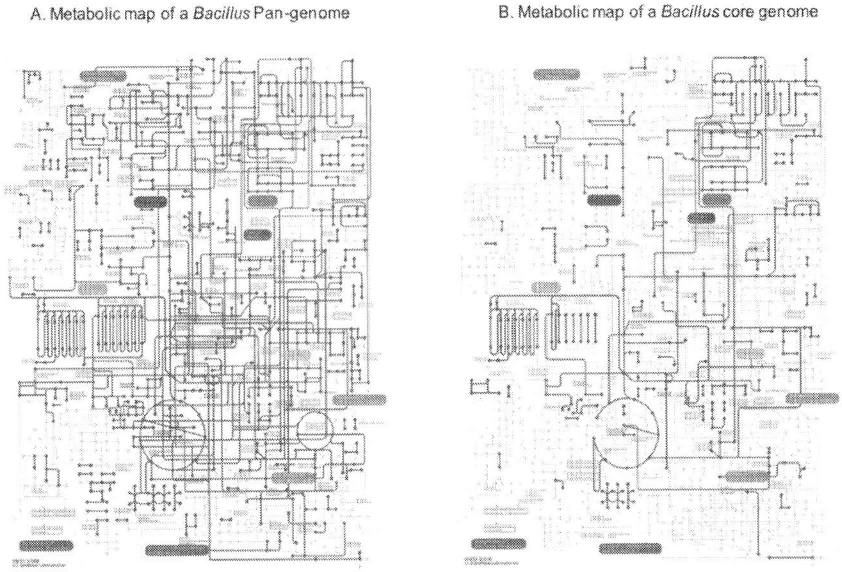

Fig. 7. Comparative metabolic reconstruction for the *Bacillus* pan-genome (A) and the *Bacillus* core genome (B). The combined pool of genes in the sequenced Bacillus genomes reflects the greater metabolic potential of the *Bacillus* genus. (B) The metabolic map of the core genome, which is represented by all of the conserved genes among the sequenced *Bacillus* genomes, reflects the basic housekeeping functions of these genera. The apparent absence of some enzymatic pathways that can be considered essential may be explained by the replacement of genes coding these capacities with xenologous genes. These maps were reconstructed from the annotated genome sequences using the KEGG (http://www.genome.ad.jp/kegg-bin/srch_orth/html) database.

When analyzing the core genome for genes involved in sporulation, we found that less than 52 of the 200 genes were conserved across 20 other *Bacillus* genomes. These genes are known to be essential for completion of the sporulation process in *B. subtilis*, but the fact that only a quarter of them are present in the other strains suggests the adaptability of genes for the same process, and particularly for the signaling circuit that responds to diverse environmental cues. Our study also identified the variable genes involved in general metabolism and allowed for the clustering of genes that made sense of their evolution and ecology.

The "accessory" sporulation genes are strain-specific and reflect the genomic flexibility of the group that allows *Bacillus* to colonize different environments that require different sensors to trigger the developmental response of the bacteria. This comparative strategy allows for the identification of the variable genes involved in the process and allows for the clustering of groups on the basis of their evolution and ecology.

7. Genomics in the aid of phylogeny

The use of the whole core genome to reconstruct the group's phylogeny helped us understand the evolutionary relationships (Alcaraz et al. 2010) (Fig. 6). As stated above, the core genome is thought to be faithful to the evolutionary history of the analyzed strains, and thus many individual genes' phylogenies are dissected to determine which can be used to define a natural group. Some other studies have analyzed what genes are necessary for a strain to be considered part of the same genus. For example, the cosmopolitan groups are either pathogenic, free-living, or extremophiles, such as *Pseudomonas* (Sarkar & Guttman 2004) and *Bacillus* (Alcaraz et al. 2010). Although the 16S rRNA gene has been traditionally used to draw a phylogeny, the resolution is often lost when working within a genus. Many genes can be concatenated and used to build a phylogeny. In fact, all genes shared by different genomes can be used to reconstruct a phylogeny; in the case of *Bacillus*, we used the 814 genes that constitute the core genome to build a *Bacillus* phylogeny. Most of these genes code for the expected housekeeping genes involved in basic cellular functions, as explained above, which fall into the traditional markers for phylogeny. However, we also included hypothetical conserved genes as well as nontraditional categories, such as genes involved in metabolism and transport mechanisms. The use of the whole core genome to reconstruct a group's phylogeny resolves the evolutionary relationships using the most available information. However, there are drawbacks of a low number of representatives of a given genus and the constant release of new genomes, which will eventually help fill in the gaps in the phylogenetic diversity and will help to understand which genes define a group.

8. Molecular clock

Traditionally, it has been difficult to set absolute dates for diversification of lineage events throughout Earth's history. Before DNA sequencing became widely available, the ages of divergence of the major groups of organisms depended solely on fossil information. Since the fossil record is incomplete, the history of lineages that lacked a rich fossil record, such as bacteria, could not be reconstructed (Kuo & Ochman 2009). When the first amino acid sequences of proteins were analyzed, they seemed to change in a rather constant fashion, which led to the hypothesis that a molecular clock existed in protein evolution. Therefore, molecular dating is based on the assumption that if most variation is neutral, then mutations will become fixed in a lineage at a constant rate that is equal to the mutation rate (Bromham & Penny 2003). One direct consequence of the evolution of a sequence at a rate that is relatively constant is that the genetic difference between any two species is proportional to the time that passed since the species last shared a common ancestor.

If the molecular clock hypothesis holds true, it serves as an extremely useful method for estimating evolutionary timescales. This is of particular value when studying organisms such as bacteria, which have left few traces of their biological history in the fossil record. Phylogenetic trees can be reconstructed that shed light onto the Earth's past. A calibration point is needed to obtain an absolute estimate of the age of a clade, and this can be set by using a known date of divergence, such as fossil information and samples from historical sources as well as paleontological, geological, atmospheric, and climatologic records. Based on that calibration point, the absolute date of a divergence can be estimated based on the number of mutations or substitutions (k) per total length of the sequence (n) per unit of time (Fig. 8), and can then be extrapolated to other parts of the phylogeny (Li 1997). This entire methodology assumes that the rates are equal in all branches of a phylogeny, which is not always the case (Ayala 1999). To deal with rate heterogeneity, several methods have been developed in the past few years that make molecular dating more accurate (Drummond & Rambaut 2007).

Geological evidence suggests that the CCB was a shallow marine environment for most of Earth's history, and when CCB became isolated due to the uplift of the Coahuila block, the bacteria that remained became relics of the diversity from the ancient sea. Thus, the high diversity of CCB would be a product of two things: the diversity already present at the time of isolation, and the new community assemblages that arose in this unique aquatic environment (Moreno-Letelier et al. 2011; Souza et al. 2006). *Firmicutes*, an abundant and widespread genera within CCB (Alcaraz et al. 2010; Alcaraz et al. 2008; Cerritos et al. 2011; Moreno-Letelier et al. 2011; Souza et al. 2006) are the focal group of our studies, in part because there are several endemic species within the site, and because the whole genome has been sequenced for both *Exiguobacterium* and *Bacillus* isolates (*Bacillus coahuilensis* m4-4 (Alcaraz et al. 2008); *Bacillus* sp. m3-13 (Alcaraz et al. 2010); and several unpublished data (from drafts *Bacillus* sp. p15.4, *Exiguobacterium* EPVM and 11-28). We have reconstructed the phylogenetic relationships and estimated the divergence times of aerobic *Firmicutes* from CCB and other similar habitats, in order to determine whether the diversity of the basin is a product of a recent adaptive radiation or the bacteria that remained became relics of the diversity we observe today. For our phylogenetic studies, calibration points were obtained from geologic events. The maximum age of the tree can not exceed the estimate of the origin of life on Earth, which is estimated to be approximately 4000 million years ago (Nisbet & Sleep 2001). The node of aerobic *Firmicutes* is set to have an age of 2300 Ma, the date of the Great Oxidation Event (Battistuzzi et al. 2004; Papineau 2010). CCB bacteria and their sister species were also constrained to have a minimum age of 35 Ma in the case of CCB bacteria and marine bacteria. This age corresponds to the final retreat of the Western Interior Seaway and the uplift of the Sierra Madre Oriental that finally isolated Cuatro Cienegas from the ocean (Ferrusquía-Villafranca & González-Guzmán 2005).

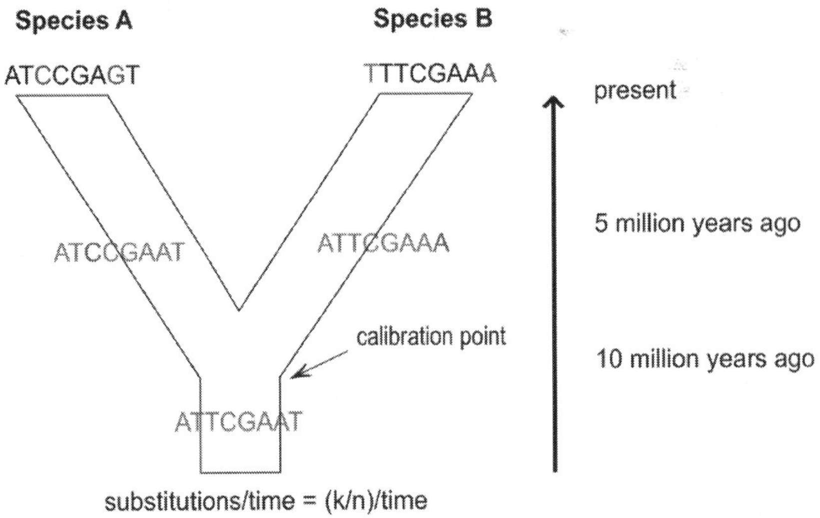

Fig. 8. The concept of the molecular clock based on DNA sequences. A representation of a set of sequences that evolves in a molecular clock fashion, where each branch contains 2 substitutions in 10 million years. For each species k = 2 and n = 8, which equals 0.025 substitutions per million years.

Our results showed that the diversity of *Firmicutes* in Cuatro Cienegas is not the product of a recent adaptive radiation. We suggest instead that ancient lineages became isolated and continued their own evolutionary path. This is especially true for the representatives of *Bacillus*, where the different taxa have diverged form their sister species at different times (Moreno-Letelier et al, under review). Overall, the Cuatro Cienegas Basin reinforces its geological importance since it is a refuge of ancient bacterial lineages, with traits that are relicts of old times, in some cases even from Precambrian times, and others that have arisen in order to survive in this endangered, extreme, and unique environment.

Questions	Experimental strategies
Have we already uncovered all of the microbial diversity in water, sediments, or mats of any pond?	16S rRNA gene libraries, *taxon*-specific libraries, metagenomes
Why is there so much microbial diversity? Does genetic diversity correlate with environmental stability or resource availability, or is it a matter of chance and history?	Explore the reasons for diversity: competition, mesocosm experiments in combination with 16S rRNA gene libraries, *taxon*-specific libraries, metagenomes
In most ponds, microbial mats are in charge of biogeochemical cycles. Therefore, what are the most important functions and are the functions similar in mats from different ponds?	Transcriptome analysis
Organisms from an ancient sea have adapted to the oligotrophic conditions of the ponds. How did they evolve and adapt? How was their genome shaped? What is the role of mobile elements?	Study transposons, plasmids, and phages in genomes and metagenomes
Do communities with extensive genetic diversity also have more functional diversity?	Comparative metagenomics and transriptomics
Do communities respond differently to environmental change?	Mesocosms, transcriptomics

Table 2. Remaining questions on the CCB microbial diversity and evolution, and the genomic approaches to address them.

9. Future directions

The current approaches that combine genomics, transcriptomics, metagenomics, and proteomics together with classical microbiology will continue to contribute to our understanding of microbial activity and strategies for cell survival and growth of bacteria in oligotrophic ecosystems. Additional explorations of the diversity of the bacteria in these ponds using taxon specific oligonucleotides will allow us to determine whether we've actually reached an understanding of diversity with more general approaches, or if some taxons were missed due to the bias of DNA isolation techniques and PCR amplification of 16S rRNA genes with universal primers. Once we confidently identify the members of a community, a future challenge will be to understand the rules of assembly for the organisms within it, including the role of history, competition (for nutrients or antagonic), and migration. In particular, transcriptomics will help us go beyond the identification of the genetic potential of the organisms afforded by the metagenomic approach and will allow us to actually look at the expression of such genes. The presence of specific mRNAs will be the actual reflection of what the microorganisms are sensing from the environment and how they are responding to them. This will be particularly fruitful once we have full genome information for many of the CCB bacterial isolates, since we will be able to assign mRNAs to specific taxons in the communities.

10. Conclusion

The application of genomic approaches to living systems uncovers the genetic bases of functional variation in nature. The revolution in high-throughput DNA sequencing and gene expression technologies has redefined the notion of a 'model' organism. The interrogation of genomes from animals, plants, microbes, or communities of organisms can identify genetic markers of processes at any scale: ecological, physiological, developmental, transcriptional, and others. Challenges lie ahead in the full interpretation of these datasets as well as understanding the connections between the gene information and the metadata of each particular environment. This will require interdisciplinary work between ecologists, microbial geneticists, biogeochemists, and computational biologists.

11. Acknowledgments

Julio Cruz for the Dune's photograph; Eugenio Reynoso for graphic art. G. Olmedo's laboratory team for their comments to the manuscript. This work was supported by a Multidisciplinario project from Cinvestav to GOA and Fondos Mixtos-Conacyt.

12. References

Adams, M. M., Gomez-Garcia, M. R., Grossman, A. R. & Bhaya, D. (2008). Phosphorus deprivation responses and phosphonate utilization in a thermophilic Synechococcus sp. from microbial mats. *J Bacteriol* 190(24): 8171-8184.

Alcaraz, L. D., Moreno-Hagelsieb, G., Eguiarte, L. E., Souza, V., Herrera-Estrella, L., et al. (2010). Understanding the evolutionary relationships and major traits of Bacillus through comparative genomics. *BMC Genomics* 11: 332.

Alcaraz, L. D., Olmedo, G., Bonilla, G., Cerritos, R., Hernandez, G., et al. (2008). The genome of Bacillus coahuilensis reveals adaptations essential for survival in the relic of an ancient marine environment. *Proc Natl Acad Sci U S A* 105(15): 5803-5808.

Amann, R. I., Ludwig, W. & Schleifer, K. H. (1995). Phylogenetic identification and in situ detection of individual microbial cells without cultivation. *Microbiol Rev* 59(1): 143-169.

Andam, C. P.& Gogarten, J. P. (2011). Biased gene transfer in microbial evolution. *Nat Rev Microbiol* 9(7): 543-555.

Anderson, I., Sorokin, A., Kapatral, V., Reznik, G., Bhattacharya, A., et al. (2005). Comparative genome analysis of Bacillus cereus group genomes with Bacillus subtilis. *FEMS Microbiol Lett* 250(2): 175-184.

Ayala, F. J. (1999). Molecular clock mirages. *Bioessays* 21(1): 71-75.

Aziz, R. K., Bartels, D., Best, A. A., DeJongh, M., Disz, T., et al. (2008). The RAST Server: rapid annotations using subsystems technology. *BMC Genomics* 9: 75.

Battistuzzi, F. U., Feijao, A. & Hedges, S. B. (2004). A genomic timescale of prokaryote evolution: insights into the origin of methanogenesis, phototrophy, and the colonization of land. *BMC Evol Biol* 4: 44.

Benning, C. (1998). Biosynthesis and Function of the Sulfolipid Sulfoquinovosyl Diacylglycerol. *Annu Rev Plant Physiol Plant Mol Biol* 49: 53-75.

Breitbart, M., Hoare, A., Nitti, A., Siefert, J., Haynes, M., et al. (2009). Metagenomic and stable isotopic analyses of modern freshwater microbialites in Cuatro Cienegas, Mexico. *Environ Microbiol* 11(1): 16-34.

Bromham, L. & Penny, D. (2003). The modern molecular clock. *Nat Rev Genet* 4(3): 216-224.

Brown, M. R .& Kornberg, A. (2004). Inorganic polyphosphate in the origin and survival of species. *Proc Natl Acad Sci U S A* 101(46): 16085-16087.

Cerritos, R., Eguiarte, L. E., Avitia, M., Siefert, J., Travisano, M., et al. (2011). Diversity of culturable thermo-resistant aquatic bacteria along an environmental gradient in Cuatro Cienegas, Coahuila, Mexico. *Antonie Van Leeuwenhoek* 99(2): 303-318.

Cerritos, R., Vinuesa, P., Eguiarte, L. E., Herrera-Estrella, L., Alcaraz-Peraza, L. D., et al. (2008). Bacillus coahuilensis sp. nov., a moderately halophilic species from a desiccation lagoon in the Cuatro Cienegas Valley in Coahuila, Mexico. *Int J Syst Evol Microbiol* 58(Pt 4): 919-923.

Chaisson, M., Pevzner, P. & Tang, H. (2004). Fragment assembly with short reads. *Bioinformatics* 20(13): 2067-2074.

Desnues, C., Rodriguez-Brito, B., Rayhawk, S., Kelley, S., Tran, T., et al. (2008). Biodiversity and biogeography of phages in modern stromatolites and thrombolites. *Nature* 452(7185): 340-343.

Dinsdale, E. A., Edwards, R. A., Hall, D., Angly, F., Breitbart, M., et al. (2008). Functional metagenomic profiling of nine biomes. *Nature* 452(7187): 629-632.

Dormann, P.& Benning, C. (2002). Galactolipids rule in seed plants. *Trends Plant Sci* 7(3): 112-118.

Drummond, A. J. & Rambaut, A. (2007). BEAST: Bayesian evolutionary analysis by sampling trees. *BMC Evol Biol* 7: 214.

Ehling-Schulz, M., Svensson, B., Guinebretiere, M. H., Lindback, T., Andersson, M., et al. (2005). Emetic toxin formation of Bacillus cereus is restricted to a single evolutionary lineage of closely related strains. *Microbiology* 151(Pt 1): 183-197.

Elser, J. J. (2005). Effects of phosphorus enrichment and grazing snails on modern stromatolitic microbial communities. *Freshwater Biol* 50: 1808–1825.

Elser, J. J. & Hamilton, A. (2007). Stoichiometry and the new biology: the future is now. *PLoS Biol* 5(7): e181.

Elser, J. J., Watts, J., Schampel, J. H. & Farmer, J. (2006). Early Cambrian food webs on a trophic knife-edge? A hypothesis and preliminary data from a modern stromatolite-based ecosystem. *Ecol Lett* 9(3): 295-303.

Escalante, A. E., Caballero-Mellado, J., Martinez-Aguilar, L., Rodriguez-Verdugo, A., Gonzalez-Gonzalez, A., et al. (2009). Pseudomonas cuatrocienegasensis sp. nov., isolated from an evaporating lagoon in the Cuatro Cienegas valley in Coahuila, Mexico. *Int J Syst Evol Microbiol* 59(Pt 6): 1416-1420.

Ferrusquía-Villafranca, I. & González-Guzmán, L. (2005). Northern Mexico's Landscape, Part 2, The biotic setting across space and time. *Biodiversity, Ecosystems and Conservation in Northern Mexico.* J. L. Cartron, G. C.-G. R. F. New York, Oxford University Press. 2: 40-51.

Frigaard, N. U., Martinez, A., Mincer, T. J. & DeLong, E. F. (2006). Proteorhodopsin lateral gene transfer between marine planktonic Bacteria and Archaea. *Nature* 439(7078): 847-850.

Helgason, E., Tourasse, N. J., Meisal, R., Caugant, D. A. & Kolsto, A. B. (2004). Multilocus sequence typing scheme for bacteria of the Bacillus cereus group. *Appl Environ Microbiol* 70(1): 191-201.

Hirota, R., Kuroda, A., Kato, J. & Ohtake, H. (2010). Bacterial phosphate metabolism and its application to phosphorus recovery and industrial bioprocesses. *J Biosci Bioeng* 109(5): 423-432.

Holsinger, J. R.& Minckley, W. L. (1971). A new genus and two new species of subterranean amphipod crustaceans (Gammaridae) from northern Mexico. *Proceedings of the Biological Society of Washington* 83: 25–44.

Khan, M. M., Pyle, B. H. & Camper, A. K. (2010). Specific and rapid enumeration of viable but nonculturable and viable-culturable gram-negative bacteria by using flow cytometry. *Appl Environ Microbiol* 76(15): 5088-5096.

Kuo, C. H. & Ochman, H. (2009). Inferring clocks when lacking rocks: the variable rates of molecular evolution in bacteria. *Biol Direct* 4: 35.

Kurtz, S., Phillippy, A., Delcher, A. L., Smoot, M., Shumway, M., et al. (2004). Versatile and open software for comparing large genomes. *Genome Biol* 5(2): R12.

Li, W. H. (1997). Molecular Evolution Sunderland, Massachusetts, Sinauer Associates.

Medini, D., Donati, C., Tettelin, H., Masignani, V. & Rappuoli, R. (2005). The microbial pan-genome. *Curr Opin Genet Dev* 15(6): 589-594.

Metzker, M. L. (2010). Sequencing technologies - the next generation. *Nat Rev Genet* 11(1): 31-46.

Miller, J. R., Delcher, A. L., Koren, S., Venter, E., Walenz, B. P., et al. (2008). Aggressive assembly of pyrosequencing reads with mates. *Bioinformatics* 24(24): 2818-2824.

Minckley, W. L. (1969). Environments of the Bolson of Cuatro Cienegas, Coahuila, Mexico, with Special Reference to the Aquatic Biota. El Paso, TX, Texas Western Press,.

Moreno-Letelier, A., Olmedo, G., Eguiarte, L. E., Martinez-Castilla, L. & Souza, V. (2011). Parallel Evolution and Horizontal Gene Transfer of the pst Operon in Firmicutes from Oligotrophic Environments. *Int J Evol Biol* 2011: 781642.

Nelson, K. E., Mullany, P., Warburton, P. & Allan, E. (2011). Metagenomics of the Human Body. K. E. Nelson, E. New York, NY: . 165-173-.

Nelson, K. E., Weinstock, G. M., Highlander, S. K., Worley, K. C., Creasy, H. H., et al. (2010). A catalog of reference genomes from the human microbiome. *Science* 328(5981): 994-999.

Nisbet, E. G. & Sleep, N. H. (2001). The habitat and nature of early life. *Nature* 409(6823): 1083-1091.

Papineau, D. (2010). Global biogeochemical changes at both ends of the proterozoic: insights from phosphorites. *Astrobiology* 10(2): 165-181.

Qi, Y., Kobayashi, Y. & Hulett, F. M. (1997). The pst operon of Bacillus subtilis has a phosphate-regulated promoter and is involved in phosphate transport but not in regulation of the pho regulon. *J Bacteriol* 179(8): 2534-2539.

Reno, M. L., Held, N. L., Fields, C. J., Burke, P. V. & Whitaker, R. J. (2009). Biogeography of the Sulfolobus islandicus pan-genome. *Proc Natl Acad Sci U S A* 106(21): 8605-8610.

Rusch, D. B., Halpern, A. L., Sutton, G., Heidelberg, K. B., Williamson, S., et al. (2007). The Sorcerer II Global Ocean Sampling expedition: northwest Atlantic through eastern tropical Pacific. *PLoS Biol* 5(3): e77.

Sarkar, S. F. & Guttman, D. S. (2004). Evolution of the core genome of Pseudomonas syringae, a highly clonal, endemic plant pathogen. *Appl Environ Microbiol* 70(4): 1999-2012.

Scanlan, D. J., Mann, N. H. & Carr, N. G. (1993). The response of the picoplanktonic marine cyanobacterium Synechococcus species WH7803 to phosphate starvation involves a protein homologous to the periplasmic phosphate-binding protein of Escherichia coli. *Mol Microbiol* 10(1): 181-191.

Schmidt, R. H. J. (1979). A climatic delineation of the 'real' Chihuahuan Desert. *J. Arid Environ* 2: 243-250.

Souza, V., Espinosa-Asuar, L., Escalante, A. E., Eguiarte, L. E., Farmer, J., et al. (2006). An endangered oasis of aquatic microbial biodiversity in the Chihuahuan desert. *Proc Natl Acad Sci U S A* 103(17): 6565-6570.

Tettelin, H., Masignani, V., Cieslewicz, M. J., Donati, C., Medini, D., et al. (2005). Genome analysis of multiple pathogenic isolates of Streptococcus agalactiae: implications for the microbial "pan-genome". *Proc Natl Acad Sci U S A* 102(39): 13950-13955.

Tetu, S. G., Brahamsha, B., Johnson, D. A., Tai, V., Phillippy, K., et al. (2009). Microarray analysis of phosphate regulation in the marine cyanobacterium Synechococcus sp. WH8102. *ISME J* 3(7): 835-849.

Turnbaugh, P. J., Ley, R. E., Hamady, M., Fraser-Liggett, C. M., Knight, R., et al. (2007). The human microbiome project. *Nature* 449(7164): 804-810.

Van Domselaar, G. H., Stothard, P., Shrivastava, S., Cruz, J. A., Guo, A., et al. (2005). BASys: a web server for automated bacterial genome annotation. *Nucleic Acids Res* 33(Web Server issue): W455-459.

White, A. K. & Metcalf, W. W. (2004). Two C-P lyase operons in Pseudomonas stutzeri and their roles in the oxidation of phosphonates, phosphite, and hypophosphite. *J Bacteriol* 186(14): 4730-4739.

Woese, C. R. & Fox, G. E. (1977). Phylogenetic structure of the prokaryotic domain: the primary kingdoms. *Proc Natl Acad Sci U S A* 74(11): 5088-5090.

Wolaver, B., Sharp, JM, Rodríguez JM. and Ibarra JC (2008). Delineation of Regional Arid Karstic Aquifers: An Integrative Data Approach. *Ground Water* 46(3): 396–413.

Zerbino, D. R. & Birney, E. (2008). Velvet: algorithms for de novo short read assembly using de Bruijn graphs. *Genome Res* 18(5): 821-829.

Zubkov, M. V., Mary, I., Woodward, E. M. S., Warwick, P. E., Fuchs, B. M., et al. (2007). Microbial control of phosphate in the nutrient depleted North Atlantic subtropical gyre. *Environmental Microbiology* 9(8): 2079-2089.